Computational Acoustics of Noise Propagation in Fluids – Finite and Boundary Element Methods

Steffen Marburg · Bodo Nolte
Editors

Computational Acoustics of Noise Propagation in Fluids – Finite and Boundary Element Methods

With 285 Figures and 29 Tables

Springer

Editors:

Steffen Marburg (Ed.)
Institut für Festkörpermechanik
Technische Universität Dresden
Helmholtzstraße 10
01062 Dresden, Germany
marburg@ifkm.mw.tu-dresden.de

Bodo Nolte (Ed.)
Forschungsanstalt der Bundeswehr
für Wasserschall und Geophysik
Klausdorfer Weg 2–24
24148 Kiel, Germany
BodoNolte@bwb.org

ISBN 978-3-540-77447-1 ISBN 978-3-540-77448-8 (eBook)

DOI 10.1007/978-3-540-77448-8

Library of Congress Control Number: 2008921125

Typesetting and Production: LE-TEX Jelonek, Schmidt & Vöckler GbR, Leipzig, Germany
Cover Design: eStudioCalamar S. L., F. Steinen-Broo, Girona, Spain

Printed on acid-free paper

9 8 7 6 5 4 3 2 1

springer.com

For Mäxi, Jolly & Milan, who was the WDL[1]

[1]das Wunder des Lebens (*The Wonder of life*)

Preface

Low noise constructions receive increasing attention in highly industrialized countries. Consequently, control of noise emission challenges a growing community of engineers. Classically, noise emission is controlled experimentally utilizing the trial and error method and engineering experience. The development of numerical methods such as the finite element and the boundary element method for low frequency acoustic problems and statistic methods for high frequency problems allows simulation of radiation and scattering from arbitrary geometric objects.

For low and medium frequency problems, classical approaches for solution of problems of acoustics favor analytical methods including Fourier series approaches. These approaches are quite powerful and they are still developed further. In particular, if orthogonal eigenfunctions are used as the basis functions of the Fourier series, they converge rapidly. However, if the geometry of the radiator or scatterer becomes more complicated, Fourier series become impractical to use. In these cases, numerical methods can be used more conveniently. The easiest and most straightforward approach consists of the finite difference method. However, finite difference methods suffer from a number of specific problems such as mesh restrictions and dispersion. Alternatively, finite element and boundary element methods use a more complicate mathematical formulation but can be applied in a very general way.

This book deals with finite element and boundary element methods for acoustic problems. Although, the title contains the restriction of the acoustics of fluids, a number of chapters consider solid structures as well. The edition comprises 21 chapters. The first one, i.e. Chapter 0, is a concept chapter. It starts with the derivation of the harmonic wave equation from the fundamental relations of continuum mechanics. It is followed by ten chapters on finite element methods and another ten chapters on boundary element methods. The reader is referred to Chapter 0 and Section 0.6, cf. pages 20–22, to survey the remaining chapters and discuss them related to the formulations which are given in Chapter 0.

This is a book on numerical methods. In the first volume of his series *The Hitchhiker's Guide to the Galaxy*,[2] Douglas Adams formulates "the ultimate answer to life, the universe, and everything." It is a numeric solution: 42, evaluated by the computer Deep Thought. A CPU time of $\approx 2.37 \times 10^{14}$ seconds (7.5 Million years) was required to achieve this interesting result. This book on numerical methods contains contributions written by 42 authors. The number of 42 might indicate that it covers a wide range of topics of computational acoustics. However, the reader should not expect the ultimate answer to the problems of computational acoustics in general. It took the editors ≈ 20 months ($\approx 5.2 \times 10^7$ seconds) of manual work from the idea to the final version of the book. There are many reasons why this book has been completed much faster than the evaluation of Deep Thought. Maybe this was achieved because the overall content is less general than the ultimate answer to life, the universe, and everything. Probably, the major reason for the successful and efficient completion consists in the willingly collaboration of all authors to supply the editors with their contributions. The editors wish to acknowledge that it has been their great pleasure to work together with all authors.

A number of other persons have been relevant for the successful completion of this edition. First of all, we wish to mention Eva Hestermann–Beyerle of the Springer–Verlag in Heidelberg. It is worth mentioning that she encouraged the editors to start with their editorial work. Moreover, Eva Hestermann–Beyerle has continuously supervised the progress of the edition and provided the editors with numerous valuable advice.

The editors wish to thank their close colleagues at the Institute of Solid Mechanics at Technische Universität Dresden and at the Federal Armed Forces Underwater Acoustics and Marine Geophysics Research Institute for their patience and their support. There are many others who contributed to the successful completion of this work. It seems to be impossible to mention all of them. However, the editors are very thankful for every single assistance in during the preparation of this book.

Dresden and Kiel,
October 2007

Steffen Marburg
Bodo Nolte

[2]Douglas Adams' *The Hitchhiker's Guide to the Galaxy* was originally published in 1979 by *Pan Books Ltd., London.*

Contents

List of Contributors

Irfan Ali Friedrich–Alexander–Universität Erlangen–Nürnberg, Lehrstuhl für Strömungsmechanik, Cauerstraße 4, 91058 Erlangen, Germany
irfan@lstm.uni–erlangen.de

R Jeremy Astley Institute of Sound and Vibration Research, University of Southampton, Highfield, Southampton, SO17 1BJ, United Kingdom
rja@isvr.soton.ac.uk

Stefan Becker Friedrich–Alexander–Universität Erlangen–Nürnberg, Lehrstuhl für Strömungsmechanik, Cauerstraße 4, 91058 Erlangen, Germany
sbecker@lstm.uni–erlangen.de

Alfredo Bermúdez Departamento de Matemática Aplicada, Universidade de Santiago de Compostela, 15782 Santiago de Compostela, Spain
mabermud@usc.es

Robert Bernhard Purdue University, Hovde Hall, 610 Purdue Mall, West Lafayette, IN 47907, USA
bernhard@purdue.edu

Haike Brick TFH Berlin – University of Applied Sciences, Department of Mathematics, Physics and Chemistry, Luxemburger Str. 10, 13353 Berlin, Germany
brick@tfh–berlin.de

Dominik Brunner Institute of Applied and Experimental Mechanics, Pfaffenwaldring 9, 70550 Stuttgart, Germany
brunner@iam.uni–stuttgart.de

Zhensheng Chen Acoustics Research Institute, Austrian Academy of Sciences, Wohllebengasse 12–14, A–1040 Vienna, Austria
zhensheng.chen@oeaw.ac.at

Gary Cohen INRIA, Domaine de Voluceau, Rocquencourt – BP 105, 78153 Le Chesnay Cedex, France
gary.cohen@inria.fr

Daniel Dreyer Hanse Wohnbau GmbH, Birkenweg 15, D–22850 Norderstedt, Germany
daniel.dreyer@h–wb.de

Max Escobar Friedrich–Alexander–Universität Erlangen–Nürnberg, Lehrstuhl für Sensorik, Paul–Gordan–Straße 3/5, 91052 Erlangen, Germany
max.escobar@lse.eei.uni–erlangen.de

Otto von Estorff Institute of Modelling and Computation, Hamburg University of Technology, Denickestraße 17, D–21073 Hamburg, Germany
estorff@tu–harburg.de

Pablo Gamallo Departamento de Matemática Aplicada II, Universidade de Vigo, 36310 Vigo, Spain
pgamallo@uvigo.es

Lothar Gaul Institute of Applied and Experimental Mechanics, Pfaffenwaldring 9, 70550 Stuttgart, Germany
gaul@iam.uni–stuttgart.de

Dan Givoli Department of Aerospace Engineering, Technion, Haifa 32000, Israel
givolid@aerodyne.technion.ac.il

Isaac Harari Faculty of Engineering, Tel Aviv University, 69978 Ramat Aviv, Israel
harari@eng.tau.ac.il

Andreas Hauck Friedrich–Alexander–Universität Erlangen–Nürnberg, Lehrstuhl für Sensorik, Paul–Gordan–Str. 3/5, 91052 Erlangen, Germany
andreas.hauck@lse.eei.uni–erlangen.de

Luis Hervella–Nieto Departamento de Matemáticas, Universidade da Coruña, 15707 A Coruña, Spain
luisher@udc.es

Günter Hofstetter Institute for Basic Sciences in Civil Engineering, University of Innsbruck, Technikerstraße 13, A–6020 Innsbruck, Austria
Guenter.Hofstetter@uibk.ac.at

Jeong–Guon Ih Centre for Noise and Vibration Control (NoViC), Department of Mechanical Engineering, Korea Advanced Institute of Science and Technology (KAIST), Science Town, Daejeon 305–701, Korea
J.G.Ih@kaist.ac.kr

Michael Junge Institute of Applied and Experimental Mechanics, Pfaffenwaldring 9, 70550 Stuttgart, Germany
junge@iam.uni–stuttgart.de

Sabine Langer Institute of Applied Mechanics, Technical University Clausthal, Adolph–Roemer-Str. 2a, 38678 Clausthal–Zellerfeld, Germany
s.langer@tu–clausthal.de

Haw–Ling Liew Department of Mechanical Engineering, Faculty of Mechanics and Computation, Stanford University, Stanford, CA, 94305, USA
hlliew@stanford.edu

Manfred Kaltenbacher Friedrich–Alexander–Universität Erlangen–Nürnberg, Lehrstuhl für Sensorik, Paul–Gordan–Straße 3/5, 91052 Erlangen, Germany
manfred@lse.eei.uni–erlangen.de

Herbert Mang Institute for Mechanics of Materials and Structures, Vienna University of Technology, Karlsplatz 13, A–1040 Vienna, Austria
Herbert.Mang@tuwien.ac.at

Steffen Marburg Institut für Festkörpermechanik, Technische Universität Dresden, 01062 Dresden, Germany
marburg@ifkm.mw.tu–dresden.de

Bodo Nolte Forschungsanstalt für Wasserschall und Geophysik, 24148 Kiel, Germany
BodoNolte@BWB.ORG

Martin Ochmann TFH Berlin – University of Applied Sciences, Department of Mathematics, Physics and Chemistry, Luxemburger Str. 10, 13353 Berlin, Germany
ochmann@tfh–berlin.de

Andrew Peplow Marcus Wallenberg Laboratory for Sound and Vibration Research, Department of Aeronautics & Vehicle Engineering, Kungl Tekniska Högskolan, S–100 44, Stockholm, Sweden
atpeplow@kth.se

Steffen Petersen Department of Mechanical Engineering, Stanford University, 488 Escondido Mall, Mail code 3035, Stanford, CA 94305, USA
steffenp@stanford.edu

Peter M. Pinsky Department of Mechanical Engineering, Faculty of Mechanics and Computation, Stanford University, Stanford, CA, 94305, USA
pinsky@stanford.edu

Andrés Prieto Departamento de Matemática Aplicada, Universidade de Santiago de Compostela, 15782 Santiago de Compostela, Spain
maprieto@usc.es

Toru Otsuru Oita University, 700 Dannoharu, Oita 870–1192, Japan
otsuru@cc.oita–u.ac.jp

Rodolfo Rodríguez GI^{2}MA, Departamento de Ingeniería Matemática, Universidad de Concepción, Casilla 160–C, Concepción, Chile
rodolfo@ing–mat.udec.cl

Tetsuya Sakuma Graduate School of Frontier Sciences, University of Tokyo, 5–1–5 Kashiwanoha, Kashiwa, Chiba 277–8563, Japan
sakuma@k.u–tokyo.ac.jp

Duarte Santamarina Departamento de Matemática Aplicada, Universidade de Santiago de Compostela, 15782 Santiago de Compostela, Spain
duartesr@usc.es

Martin Schanz Institute of Applied Mechanics, Graz University of Technology, Technikerstraße 4, 8010 Graz, Austria
m.schanz@tugraz.at

Stefan Schneider CNRS/LMA, 31 chemin Joseph–Aiguier, 13402 Marseille Cedex 20, France
schneider@lma.cnrs–mrs.fr

James P. Tuck–Lee Department of Mechanical Engineering, Faculty of Mechanics and Computation, Stanford University, Stanford, CA, 94305, USA
jtucklee@stanford.edu

Shuo Wang Science Applications International Corporation (SAIC), 2450 NASA Road One, Houston, TX 77058, USA
shuowang_usa@yahoo.com

Ting–Wen Wu Department of Mechanical Engineering, University of Kentucky, Lexington, KY 40506, USA
timwu@engr.uky.edu

Yosuke Yasuda Graduate School of Frontier Sciences, University of Tokyo, 5–1–5 Kashiwanoha, Kashiwa, Chiba 277–8563, Japan
yyasuda@k.u–tokyo.ac.jp

0

A Unified Approach to Finite and Boundary Element Discretization in Linear Time–Harmonic Acoustics

Steffen Marburg[1] and Bodo Nolte[2]

[1] Institut für Festkörpermechanik, Technische Universität, 01062 Dresden, Germany
marburg@ifkm.mw.tu-dresden.de

[2] Forschungsanstalt für Wasserschall und Geophysik, 24148 Kiel, Germany
BodoNolte@BWB.ORG

Summary. This chapter introduces the reader too important physical and mathematical concepts in acoustics. It presents an approach to finite and boundary element techniques for linear time–harmonic acoustics starting from the fundamental axioms of continuum mechanics. Based on these axioms, the wave equation is derived. Using a time–harmonic approximation, the boundary value problem of linear time–harmonic acoustics is formulated in the classic and in the weak form. Subsequently, two types of the weak form are used as the basis for discretization resulting in a Galerkin finite element formulation, in a collocation boundary element formulation and in a Galerkin boundary element formulation. Then, different representations of sources and incident wave–fields in finite and boundary element methods are discussed. In the final part of this chapter, the authors categorize the subsequent twenty chapters of this book. The chapter will be completed by an outlook and some open problems in the development of finite and boundary element techniques from the authors' points of view.

0.1 Introduction

It must be stated in the beginning that due to the wide use of numerical methods, the range of papers about finite and boundary element methods in acoustics is quite difficult to survey. The development of these methods started almost half a century ago and there are a number of monographs and editions, but only a limited number which are solely dedicated to computational acoustics in general and, in particular, to FEM and BEM in acoustics. With respect to finite element methods in acoustics, the authors are aware of the books by Givoli [36] (including a chapter on BEM too) and by Ihlenburg [47]; with respect to boundary element methods, we can mention the editions by Ciskowski and Brebbia [18], von Estorff [29], Wu [110] and the monograph by Kirkup [55]. There are a couple of interesting review papers on FEM and BEM, see for example Harari et al. [42], Harari [41] and Thompson [99], also the cost comparison of traditional FEM and BEM by Harari and Hughes [43]. Often, FEM and BEM for acoustics are discussed in other contexts such as computational methods for unbounded domains, cf. Geers [33] and Givoli [38], together with structural and/or electromagnetic wave propagation [1,44,78] and structural–acoustic optimization and noise control [56,57,76], see also the review paper by Marburg [63]. Numerical methods have been described and reviewed in books on acoustics such as Crighton et al. [23] and Mechel [73]. Furthermore, there are a number of special issues of journals on FEM and BEM (some of them primarily dedicated to treatment of the exterior problem), e.g. [7,37,48,61,62,66,71,80–82,104] and conference proceedings which contain high quality papers of the field, among many others, see for instance [101,108]. Finally, books on fluid–structure interaction problems are often closely related to numerical methods and present specific formulations in detail, see for example [2,19,46,75,83].

It is the purpose of this chapter to present basic formulations of linear time–harmonic acoustics and, in this context, to categorize the remaining twenty chapters of this book. We will start with a short derivation of the linear wave equation. This will be followed by presenting the boundary value problem of time–harmonic acoustics with its partial differential equation (Helmholtz equation) and boundary conditions on one hand, and, on the other hand, a weak form which is the basis for the discretization process. Approximation and discretization will be discussed in Section 0.4. In the following Section 0.5, we will discuss different representations of sources and incident wave fields. In Section 0.6, we categorize the remaining chapters of this book. The chapter will be completed by an outlook and identification of areas for future work in the development of finite and boundary element techniques from the authors' points of view.

0.2 Approach to the Wave Equation

Fundamentals of linear acoustics are based on the basic equations of continuum mechanics. It is assumed that the dimensions of the problem are large with respect to the nanoscale in which the number of molecules is countable. For derivation of the wave equation, we will use the Eulerian representation and, thus, the Eulerian or spatial coordinates.

We consider problems defined in a domain Ω. The complement is denoted by Ω_c. Γ represents the closed boundary of Ω and Ω_c. This configuration includes the direction of the outward normal, pointing into the domain Ω_c as shown in Figure 0.1.

0.2.1 Fundamental Axioms of Continuum Mechanics

For derivation of the wave equation, two fundamental laws of the theory of continuum mechanics are required. These are the principle of conservation of mass and the principle of balance of momentum.

Conservation of Mass

The principle of conservation of mass means that the total mass M of the considered domain Ω

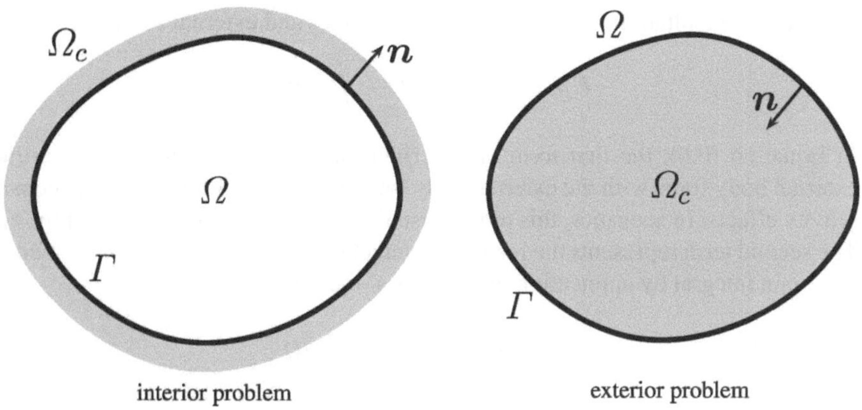

Fig. 0.1 Definition of regions Ω and Ω_c, boundary Γ and outward normal vector $\boldsymbol{n}$.

$$M(t) = \int_\Omega \varrho(\boldsymbol{x},t)d\Omega \tag{0.1}$$

remains constant during the motion, where $\boldsymbol{x}$ and t denote position vector and time. Often, these dependencies will not be shown in this section. The principle of conservation of mass implies that the material derivative (or total time derivative) vanishes, i.e.

$$\dot{M} = \frac{dM}{dt} = \int_\Omega \left(\frac{\partial \varrho}{\partial t} + \varrho \boldsymbol{\nabla} \cdot \boldsymbol{v} \right) d\Omega = 0 \,. \tag{0.2}$$

The material derivative introduces the flow velocity vector $\boldsymbol{v}$ which results from $\partial \boldsymbol{x}/\partial t$. In addition to the global validity of the conservation of mass, we require that it is also valid for an arbitrarily small neighborhood of each material point which implies the local conservation of mass as

$$\frac{\partial \varrho}{\partial t} + \varrho \boldsymbol{\nabla} \cdot \boldsymbol{v} = 0 \,. \tag{0.3}$$

Balance of Momentum

The principle of balance of momentum means that the time rate of change of momentum is equal to the resultant force $\boldsymbol{F}_R$ acting on the body. With momentum vector $\boldsymbol{P}$, also known as the linear momentum vector, this is written as

$$\dot{\boldsymbol{P}} = \frac{d\boldsymbol{P}}{dt} = \boldsymbol{F}_R \,. \tag{0.4}$$

Herein, the momentum vector is given by

$$\boldsymbol{P} = \int_\Omega \varrho \boldsymbol{v} \, d\Omega \tag{0.5}$$

whereas the resultant force combines volume forces and external forces as

$$\boldsymbol{F}_R = \int_\Omega \boldsymbol{b}\varrho \, d\Omega - \int_\Gamma p\boldsymbol{n} \, d\Gamma \, . \tag{0.6}$$

In Equation (0.6), the first term on the right hand side is known as the resultant external body force with the external body force $\boldsymbol{b}$. Using this term, we may consider gravity effects. In acoustics, this term is usually not relevant and, consequently, zero. The second term represents the resultant contact force which can be transformed into a domain integral by application of the Gauss' theorem

$$\int_\Gamma p\boldsymbol{n} d\Gamma = \int_\Omega \boldsymbol{\nabla} p \, d\Omega \, . \tag{0.7}$$

The material derivative of the momentum is given as

$$\begin{aligned} \frac{d\boldsymbol{P}}{dt} &= \frac{d}{dt}\left(\int_\Omega \varrho \boldsymbol{v} \, d\Omega\right) = \int_\Omega \frac{d(\varrho \boldsymbol{v})}{dt} d\Omega = \\ &= \int_\Omega \left[\frac{\partial \varrho}{\partial t}\boldsymbol{v} + \varrho\,(\boldsymbol{\nabla}\cdot\boldsymbol{v})\,\boldsymbol{v} + \varrho\frac{\partial \boldsymbol{v}}{\partial t} + \varrho\,(\boldsymbol{v}\cdot\boldsymbol{\nabla})\,\boldsymbol{v}\right] d\Omega \, . \end{aligned} \tag{0.8}$$

The first two terms of the integrand vanish with respect to the conservation of mass in Equation (0.2) and (0.3), respectively. This yields

$$\frac{d\boldsymbol{P}}{dt} = \int_\Omega \left[\varrho\frac{\partial \boldsymbol{v}}{\partial t} + \varrho\,(\boldsymbol{v}\cdot\boldsymbol{\nabla})\,\boldsymbol{v}\right] d\Omega \, . \tag{0.9}$$

Summarizing these manipulations, we incorporate Equations (0.6), (0.7) and (0.9) into Equation (0.4) to obtain the so–called Euler equation

$$\int_\Omega \left[\varrho\frac{\partial \boldsymbol{v}}{\partial t} + \varrho\,(\boldsymbol{v}\cdot\boldsymbol{\nabla})\,\boldsymbol{v} + \boldsymbol{\nabla} p\right] d\Omega = 0 \tag{0.10}$$

or, in local form,

$$\varrho\frac{\partial \boldsymbol{v}}{\partial t} + \varrho\,(\boldsymbol{v}\cdot\boldsymbol{\nabla})\,\boldsymbol{v} + \boldsymbol{\nabla} p = 0 \, . \tag{0.11}$$

In continuum mechanics, Euler's equations of motion comprise the balance of momentum and the balance of momentum of momentum, also known as the balance of angular momentum. The latter axiom can be neglected since shear effects are not considered herein. Euler's equation (0.11) can be considered as a special local form of Newton's equation of motion $\boldsymbol{F}_R = \partial(m\boldsymbol{v})/\partial t$.

Linearization and Simplification

Commonly, problems of linear acoustics refer to small perturbations of ambient quantities. These ambient quantities are referred to by using the subscript 0. The small fluctuating parts of pressure, density and flow velocity vector are represented as $\tilde{p}$, $\tilde{\varrho}$ and $\tilde{\boldsymbol{v}}$. With this notation, we can substitute for the quantities pressure, density and flow velocity as

$$\begin{aligned} p &= p_0 + \tilde{p}\,, \\ \varrho &= \varrho_0 + \tilde{\varrho}\,, \\ \boldsymbol{v} &= \boldsymbol{v}_0 + \tilde{\boldsymbol{v}}\,. \end{aligned} \tag{0.12}$$

For simplicity for the wave equation approach, we assume that there is no ambient flow, i.e. $\boldsymbol{v}_0 = \mathbf{0}$.

Substituting for the major quantities in Equation (0.3) and considering only first order terms, we write

$$\frac{\partial \tilde{\varrho}}{\partial t} + \varrho_0 \boldsymbol{\nabla} \cdot \tilde{\boldsymbol{v}} = 0\,. \tag{0.13}$$

Similarly, Euler's equation (0.11) is linearized and simplified as

$$\varrho_0 \frac{\partial \tilde{\boldsymbol{v}}}{\partial t} + \boldsymbol{\nabla} \tilde{p} = 0\,, \tag{0.14}$$

where it is assumed that ϱ_0 and p_0 are independent of time and spatial coordinates.

0.2.2 Constitutive Equation

In fluids, sound propagates through pressure waves only. The velocity of the sound pressure wave – better known as the speed of sound – depends on the propagation material. For wave propagation in linear fluid acoustics, the speed of sound is one of two relevant material parameters. It can be understood as the result of mathematical relations of other material parameters which are not solely relevant for our considerations.

The constitutive relations are usually referred to as the equations of state. With respect to thermodynamics, the pressure fluctuation and, thus, sound propagation occurs with negligible heat flow because the changes of the state occur so rapidly that there is no time for the temperature to equalize with the surrounding medium. This is the property of an adiabatic process. If fluctuation amplitudes and frequency remain small enough, the process can be considered as reversible and isentropic.

Derivation of the speed of sound is different for gases, liquids and solids. Since we limit our considerations to fluids herein, we will only discuss derivation of the speed of sound for gases and liquids in what follows.

The speed of sound c may be introduced as a constant to relate the fluctuating parts of pressure and density to each other as

$$\tilde{p} = c^2\, \tilde{\varrho}\,. \tag{0.15}$$

This is equivalent to

$$c = \sqrt{\frac{\partial p}{\partial \varrho}}\,. \tag{0.16}$$

For gases, we will present finding the relation (0.15) whereas for liquids, we will derive the speed of sound based on Equation (0.16).

Gases

We consider ideal gases only. With the specific heat ratio κ, an adiabatic process implies the relation $p\varrho^{-\kappa} = $ constant. Since this relation is valid at any time, it implies first

$$(p_0 + \tilde{p})(\varrho_0 + \tilde{\varrho})^{-\kappa} = p_0 \varrho_0^{-\kappa} \tag{0.17}$$

and is rewritten as

$$1 + \frac{\tilde{p}}{p_0} = \left(1 + \frac{\tilde{\varrho}}{\varrho_0}\right)^{\kappa} . \tag{0.18}$$

The right hand–side is linearized by

$$\left(1 + \frac{\tilde{\varrho}}{\varrho_0}\right)^{\kappa} = 1 + \kappa \frac{\tilde{\varrho}}{\varrho_0} , \tag{0.19}$$

which simplifies Equation (0.18) yielding

$$\tilde{p} = \left(\kappa \frac{p_0}{\varrho_0}\right) \tilde{\varrho} = c^2 \, \tilde{\varrho} \tag{0.20}$$

where the speed of sound is denoted by c. The variable K denoting the adiabatic bulk modulus is introduced as

$$c = \sqrt{\frac{K}{\varrho_0}} = \sqrt{\frac{\kappa \, p_0}{\varrho_0}} . \tag{0.21}$$

Liquids

The adiabatic bulk modulus K for a liquid is defined as

$$K = -V \frac{\partial p}{\partial V} . \tag{0.22}$$

Mass conservation can be formulated as $\varrho V = m = $ constant. Consequently, we can write

$$V = \frac{m}{\varrho} \quad \text{and} \quad \frac{\partial V}{\partial \varrho} = -\frac{m}{\varrho^2} . \tag{0.23}$$

Returning to the adiabatic bulk modulus, using the chain rule and substituting for volume yields

$$K = -\frac{m}{\varrho} \frac{\partial p}{\partial \varrho} \frac{\partial \varrho}{\partial V} = -\frac{m}{\varrho} \frac{\partial p}{\partial \varrho} \left(-\frac{\varrho^2}{m}\right) = \varrho \frac{\partial p}{\partial \varrho} \tag{0.24}$$

which allows the speed of sound to be written as

$$c = \sqrt{\frac{\partial p}{\partial \varrho}} = \sqrt{\frac{K}{\varrho_0}} . \tag{0.25}$$

The adiabatic bulk modulus is sometimes replaced by its reciprocal which is called adiabatic compressibility.

0.2.3 Derivation of the Wave Equation

It is useful for further description to reduce the problem to one variable. Herein, this variable will be the pressure fluctuation which will be referred to as the sound pressure in what follows. The local conservation of mass (0.3) in its linearized form (0.13), the Euler equation as the balance of momentum (0.11) in its linearized form (0.14) and the constitutive relation of Equation (0.15) are all summarized into one partial differential equation, i.e. the wave equation.

For that, we start at the constitutive relation (0.15) which is differentiated twice with respect to time

$$\frac{\partial^2 \tilde{p}}{\partial t^2} = c^2 \frac{\partial^2 \tilde{\varrho}}{\partial t^2} . \tag{0.26}$$

Then, derivatives of the density fluctuations are replaced by the local conservation of mass in linearized form 0.13 which gives

$$\frac{\partial^2 \tilde{p}}{\partial t^2} = -c^2 \varrho_0 \frac{\partial (\boldsymbol{\nabla} \cdot \tilde{\boldsymbol{v}})}{\partial t} = -c^2 \varrho_0 \boldsymbol{\nabla} \cdot \left(\frac{\partial \tilde{\boldsymbol{v}}}{\partial t} \right) . \tag{0.27}$$

Finally, the linearized Euler equation (0.14) is used to substitute for the velocity vector as

$$\frac{\partial^2 \tilde{p}}{\partial t^2} = c^2 \boldsymbol{\nabla} \cdot \boldsymbol{\nabla} p . \tag{0.28}$$

Equation (0.28) is known as the wave equation. Mostly (and in the subsequent sections of this chapter) the scalar product $\boldsymbol{\nabla} \cdot \boldsymbol{\nabla}$ is replaced by the Laplacian Δ. The wave equation is a hyperbolic partial differential equation.

0.3 The Boundary Value Problem

0.3.1 Partial Differential Equation and Boundary Conditions

We consider linear acoustic problems defined in the domain Ω with the complement Ω_c and Γ representing the closed boundary of Ω and Ω_c. The outward normal is pointing into the complementary domain Ω_c as shown in Figure 0.1. The wave equation (0.28)

$$\Delta \tilde{p}(\boldsymbol{x}, t) = \frac{1}{c^2} \frac{\partial^2 \tilde{p}(\boldsymbol{x}, t)}{\partial t^2} \qquad \boldsymbol{x} \in \Omega \subset \mathbb{R}^d \tag{0.29}$$

is valid for the sound pressure $\tilde{p}$. Alternatively, a velocity potential may be used. The space dimension d is three in real applications, but can be two or one in certain cases. To complete a solution, the differential equation requires boundary conditions and initial conditions, which will be specified when used.

For time–harmonic problems, a time dependence is introduced. Herein, we use the time–dependence $\tilde{p}(\boldsymbol{x}, t) = \bar{p}(\boldsymbol{x}) \, \mathrm{e}^{-\,\mathrm{i}\omega t}$.

In further analysis, the time–harmonic sound pressure, or simply pressure, $\bar{p}(\boldsymbol{x})$ will be represented as $\bar{p}(\boldsymbol{x}) = p(\boldsymbol{x})$. Applying the time–harmonic dependence of p

to Equation (0.29) leads to the Helmholtz–equation, or reduced wave equation, for the sound pressure.

$$\text{Helmholtz–equation:} \qquad \Delta p(\boldsymbol{x}) + k^2 p(\boldsymbol{x}) = 0 \qquad \boldsymbol{x} \in \Omega \subset \mathbb{R}^d \,. \tag{0.30}$$

We assume admittance boundary conditions being equivalent to Robin boundary conditions which may degenerate to Neumann boundary conditions if the admittance is zero.

$$\text{Boundary condition:} \qquad v_f(\boldsymbol{x}) - v_s(\boldsymbol{x}) = Y(\boldsymbol{x})\, p(\boldsymbol{x}) \qquad \boldsymbol{x} \in \Gamma \subset \mathbb{R}^{d-1} \,. \tag{0.31}$$

Here, the wave–number $k = \omega/c$ is the quotient of the circular frequency $\omega = 2\pi f$ (f denoting frequency) and the speed of sound c, Y represents the boundary admittance, and the normal fluid particle velocity v_f is related to the normal derivative of the sound pressure p by means of the Euler equation in frequency domain

$$v_f(\boldsymbol{x}) = \frac{1}{a}\frac{\partial p(\boldsymbol{x})}{\partial n(\boldsymbol{x})} = \frac{1}{\mathrm{i}\omega\varrho_0}\frac{\partial p(\boldsymbol{x})}{\partial n(\boldsymbol{x})} \,. \tag{0.32}$$

In Equation (0.32), i is the imaginary unit ($\mathrm{i}^2 = -1$) and ϱ_0 the average density of the fluid. The vector $\boldsymbol{n}(\boldsymbol{x})$ represents the outward normal at the surface point $\boldsymbol{x}$ and $\partial/\partial n(\boldsymbol{x})$ is the normal derivative. Note that for time dependence $\mathrm{e}^{\mathrm{i}\omega t}$ the constant a takes the conjugate value $a = -\mathrm{i}\omega\varrho_0$.

In some cases, it is useful to consider the Dirichlet boundary conditions. The Robin condition as formulated in Equation (0.31) is not suited for this case. Instead, we may use the Robin condition as an impedance boundary condition with the impedance $Z(\boldsymbol{x})$ as

$$Z(\boldsymbol{x}) \Big[v_f(\boldsymbol{x}) - v_s(\boldsymbol{x}) \Big] = p(\boldsymbol{x}) \qquad \text{and} \qquad Z(\boldsymbol{x}) = \frac{1}{Y(\boldsymbol{x})} \,. \tag{0.33}$$

In case of a homogeneous Dirichlet boundary condition, the value of the impedance is zero and thus leading to $p(\boldsymbol{x}) = 0$. Obviously, the inhomogeneous Dirichlet condition results in $p(\boldsymbol{x}) = p_0(\boldsymbol{x})$.

The boundary value problem for the time–harmonic case assumes a locally acting boundary admittance relating particle velocities of the fluid and the underlying structure and the sound pressure. For the several different problems considered in this book, authors consider simplified boundary conditions where either $v_s = 0$ or $Y = 0$.

In addition to fulfilling the Helmholtz equation and the boundary conditions, solutions of external problems require fulfillment of the decay condition at infinity, i.e. the Sommerfeld radiation condition. This is formulated in two steps for the sound pressure as

$$\begin{aligned} p &= O(r^{-\alpha}) && \text{and} \\ \frac{\partial p}{\partial r} - ikp &= o(r^{-\alpha}) && \text{for} \quad r \to \infty \,, \end{aligned} \tag{0.34}$$

with $\alpha = (d-1)/2$ and r denoting the distance between an arbitrary point close to a source. Hence, the first expression of Equation (0.34) formulates the decay rate of the solution of the Helmholtz equation p, whereas the second expression requires the left hand–side to decay faster than $r^{-(d-1)/2}$. A valuable description in a rigorous form is given in Ihlenburg's book [47, page 7–8]. Clearly, the Sommerfeld condition is a decay condition only for $d > 1$. Note that the minus sign on the left hand–side of the second part of the Sommerfeld condition changes into a plus sign if time dependence is chosen to be $\mathrm{e}^{i\omega t}$.

There are many practical problems with speed of sound and fluid density depending on position as $c = c(\boldsymbol{x})$ and $\varrho_0 = \varrho_0(\boldsymbol{x})$, e.g. in underwater acoustics and in atmospheric sound propagation. However, for the sake of simplicity, these cases will not be considered here.

0.3.2 Weak Formulation

A weak formulation is based on introducing the weight function $\chi(\boldsymbol{x})$ and "testing" it with the Helmholtz operator such that

$$\int_\Omega \chi(\boldsymbol{x}) \left[\Delta p(\boldsymbol{x}) + k^2 p(\boldsymbol{x})\right] d\Omega(\boldsymbol{x}) \quad = \quad 0\,. \tag{0.35}$$

Integrating by parts gives

$$\begin{aligned} &\int_\Omega \chi(\boldsymbol{x}) \left[\Delta p(\boldsymbol{x}) + k^2 p(\boldsymbol{x})\right] d\Omega(\boldsymbol{x}) \quad = \\ &\int_\Gamma \chi(\boldsymbol{x}) a v_f(\boldsymbol{x}) d\Gamma(\boldsymbol{x}) - \int_\Omega \left[\nabla\chi(\boldsymbol{x}) \cdot \nabla p(\boldsymbol{x}) - k^2 \chi(\boldsymbol{x}) p(\boldsymbol{x})\right] d\Omega(\boldsymbol{x}) \quad = \quad 0 \end{aligned} \tag{0.36}$$

and then

$$\begin{aligned} \int_\Omega \chi(\boldsymbol{x}) \left[\Delta p(\boldsymbol{x}) + k^2 p(\boldsymbol{x})\right] d\Omega(\boldsymbol{x}) \quad &= \quad \int_\Gamma \chi(\boldsymbol{x}) a v_f(\boldsymbol{x}) d\Gamma(\boldsymbol{x}) + \\ - \int_\Gamma \frac{\partial\chi(\boldsymbol{x})}{\partial n(\boldsymbol{x})} p(\boldsymbol{x}) d\Gamma(\boldsymbol{x}) + &\int_\Omega p(\boldsymbol{x}) \left[\Delta\chi(\boldsymbol{x}) + k^2 \chi(\boldsymbol{x})\right] d\Omega(\boldsymbol{x}) \quad = \quad 0\,. \end{aligned} \tag{0.37}$$

Often, Equation (0.36) represents the starting point for conventional finite element discretizations, e.g. Galerkin method. The second part (lower row) consists of a domain integral and a boundary integral. Similarly, the second part of Equation (0.37) consists of one domain integral and two boundary integrals. This domain integral can be transformed into an integral–free term by using fundamental solutions $G(\boldsymbol{x}, \boldsymbol{y})$ in the sense of distributions. Function G represents the solution of the equation

$$\Delta G(\boldsymbol{x}, \boldsymbol{y}) + k^2 G(\boldsymbol{x}, \boldsymbol{y}) \quad = \quad -\,\delta(\boldsymbol{x}, \boldsymbol{y}) \;. \tag{0.38}$$

It is known as free–space Green's function as well, whereas $\delta(\boldsymbol{x}, \boldsymbol{y})$ is the Dirac or delta function at the origin $\boldsymbol{y}$. In terms of physics, $G(\boldsymbol{x}, \boldsymbol{y})$ can be understood as the sound pressure distribution according to a point source (monopole) in $\boldsymbol{y}$. Together with the harmonic time–dependence of $\mathrm{e}^{-\mathrm{i}\omega t}$, it represents an outgoing wave. We can write G as

$$
\begin{aligned}
G(\boldsymbol{x},\boldsymbol{y}) &= -\frac{1}{2k}\sin\left[kr(\boldsymbol{x},\boldsymbol{y})\right] && \boldsymbol{x},\boldsymbol{y}\in\mathbb{R}^1 \ , \\
G(\boldsymbol{x},\boldsymbol{y}) &= \frac{\mathrm{i}}{4}H_0^1\left(kr(\boldsymbol{x},\boldsymbol{y})\right) && \boldsymbol{x},\boldsymbol{y}\in\mathbb{R}^2 \ \text{and} \qquad (0.39)\\
G(\boldsymbol{x},\boldsymbol{y}) &= \frac{1}{4\pi}\frac{\mathrm{e}^{\mathrm{i}kr(\boldsymbol{x},\boldsymbol{y})}}{r(\boldsymbol{x},\boldsymbol{y})} && \boldsymbol{x},\boldsymbol{y}\in\mathbb{R}^3 \ .
\end{aligned}
$$

with r as the Euclidean distance between field point $\boldsymbol{x}$ and source point $\boldsymbol{y}$ as $r(\boldsymbol{x},\boldsymbol{y}) = |\boldsymbol{x}-\boldsymbol{y}|$. Note that the fundamental solutions are different when the time dependence is chosen to $\mathrm{e}^{\mathrm{i}\omega t}$.

Applying the property of the fundamental solution and the delta function, we find

$$
\begin{aligned}
&\int_\Omega p(\boldsymbol{x})\Big[\Delta G(\boldsymbol{x},\boldsymbol{y}) + k^2 G(\boldsymbol{x},\boldsymbol{y})\Big]\,d\Omega(\boldsymbol{x}) = \\
&\qquad = \int_\Omega p(\boldsymbol{x})\Big[-\delta(\boldsymbol{x},\boldsymbol{y})\Big]\,d\Omega(\boldsymbol{x}) = -c(\boldsymbol{y})p(\boldsymbol{y}) \ . \qquad (0.40)
\end{aligned}
$$

Hence, Equation (0.37) is rewritten as

$$
c(\boldsymbol{y})p(\boldsymbol{y}) + \int_\Gamma \frac{\partial G(\boldsymbol{x},\boldsymbol{y})}{\partial n(\boldsymbol{x})}\,p(\boldsymbol{x})d\Gamma(\boldsymbol{x}) = \int_\Gamma G(\boldsymbol{x},\boldsymbol{y})\,a v_f(\boldsymbol{x})d\Gamma(\boldsymbol{x}) \,. \qquad (0.41)
$$

Equation (0.41) is known as representation formula. For $\boldsymbol{y}\in\Gamma$, it is known as the Kirchhoff–Helmholtz (boundary) integral equation. Note that plus and minus signs of either the first term or the second and the third term may be different if the direction of the normal vector is chosen in opposite direction.

Before entering the discretization process, it will be useful to incorporate the boundary condition (0.31) into the weak formulations (0.36) and (0.41). Furthermore, we substitute for the constant a as

$$
a = s\,k \qquad \text{with} \qquad s = \mathrm{i}\,\varrho_0\,c \,, \qquad (0.42)
$$

which explicitly shows wave–number dependency. Recalling the second part of Equation (0.36), introducing the boundary condition and rearranging only unknown terms on the left hand–side, i.e. terms including the sound pressure p, we find

$$
\begin{aligned}
&\int_\Omega\left[\boldsymbol{\nabla}\chi(\boldsymbol{x})\cdot\boldsymbol{\nabla}p(\boldsymbol{x}) - k^2\chi(\boldsymbol{x})p(\boldsymbol{x})\right]d\Omega(\boldsymbol{x}) - s\,k\int_\Gamma \chi(\boldsymbol{x})Y(\boldsymbol{x})p(\boldsymbol{x})d\Gamma(\boldsymbol{x}) = \\
&\qquad = s\,k\int_\Gamma \chi(\boldsymbol{x})v_s(\boldsymbol{x})d\Gamma(\boldsymbol{x}) \,. \qquad (0.43)
\end{aligned}
$$

We will use Equation (0.43) as the basis for Galerkin discretization using finite elements.

Equation (0.41) modifies to

$$c(\boldsymbol{y})p(\boldsymbol{y}) + \int_\Gamma \frac{\partial G(\boldsymbol{x},\boldsymbol{y})}{\partial n(\boldsymbol{x})} p(\boldsymbol{x}) d\Gamma(\boldsymbol{x}) = = sk \int_\Gamma G(\boldsymbol{x},\boldsymbol{y}) \left[v_s(\boldsymbol{x}) + Y(\boldsymbol{x})p(\boldsymbol{x}) \right] d\Gamma(\boldsymbol{x}) \,. \quad (0.44)$$

In case of $\boldsymbol{y} \in \Gamma$ and, thus, $0 < c(\boldsymbol{y}) < 1$, Equation (0.44) represents a Fredholm integral equation of the second kind. This becomes more obvious if the part of the right hand–side integral which includes the sound pressure p is moved into the integral on the left hand–side

$$c(\boldsymbol{y})p(\boldsymbol{y}) + \int_\Gamma \left[\frac{\partial G(\boldsymbol{x},\boldsymbol{y})}{\partial n(\boldsymbol{x})} - sk\, G(\boldsymbol{x},\boldsymbol{y}) Y(\boldsymbol{x}) \right] p(\boldsymbol{x}) d\Gamma(\boldsymbol{x}) = = sk \int_\Gamma G(\boldsymbol{x},\boldsymbol{y})\, v_s(\boldsymbol{x}) d\Gamma(\boldsymbol{x}) \,. \quad (0.45)$$

While finite element discretization can be simply formulated by choosing appropriate basis and test functions for Equation (0.43) which is a direct outcome of the partial integration of the weak formulation in Equation (0.35), a boundary element formulation requires discretization of the Fredholm integral Equation (0.45). For that, we introduce another test function $\tilde{\chi}(\boldsymbol{y})$ such that

$$\int_\Gamma \tilde{\chi}(\boldsymbol{y}) c(\boldsymbol{y}) p(\boldsymbol{y}) d\Gamma(\boldsymbol{y}) + + \int_\Gamma \tilde{\chi}(\boldsymbol{y}) \left\{ \int_\Gamma \left[\frac{\partial G(\boldsymbol{x},\boldsymbol{y})}{\partial n(\boldsymbol{x})} - sk\, G(\boldsymbol{x},\boldsymbol{y}) Y(\boldsymbol{x}) \right] p(\boldsymbol{x}) d\Gamma(\boldsymbol{x}) \right\} d\Gamma(\boldsymbol{y}) = = sk \int_\Gamma \tilde{\chi}(\boldsymbol{y}) \left\{ \int_\Gamma G(\boldsymbol{x},\boldsymbol{y})\, v_s(\boldsymbol{x}) d\Gamma(\boldsymbol{x}) \right\} d\Gamma(\boldsymbol{y}) \,. \quad (0.46)$$

Equation (0.46) is the basis for collocation and Galerkin discretization using boundary elements.

0.4 Discretization Process

0.4.1 Approximation

Independent of the discretization method, we formulate approximations of our physical quantities. First of all, we approximate the sound pressure $p(\boldsymbol{x})$ as

$$p(\boldsymbol{x}) = \sum_{l=1}^{N} \phi_l(\boldsymbol{x})\, p_l = \boldsymbol{\phi}^T(\boldsymbol{x}) \boldsymbol{p} \,, \quad (0.47)$$

where p_l represents the discrete sound pressure at point $\boldsymbol{x}_l$ and ϕ_l is the $l-$th basis function for our approximation. Further, we assume that similar approximations are formulated for the particle velocity of the structure v_s and the boundary admittance Y

$$v_s(\boldsymbol{x}) = \sum_{j=1}^{\bar{N}} \bar{\phi}_j(\boldsymbol{x})\, v_{sj} \;=\; \bar{\boldsymbol{\phi}}^T(\boldsymbol{x})\boldsymbol{v}_s \qquad \text{and}$$

$$Y(\boldsymbol{x}) = \sum_{k=1}^{\tilde{N}} \tilde{\phi}_k(\boldsymbol{x})\, Y_k \;=\; \tilde{\boldsymbol{\phi}}^T(\boldsymbol{x})\boldsymbol{Y}\,. \tag{0.48}$$

If v_s and Y are explicitly known, these approximations are not necessary for evaluation of the boundary integrals in Equation (0.43). However, there are many practical cases where the structural particle velocity is the result of a finite element simulation and available only as piecewise defined function. Similarly, the boundary admittance may vary locally or results from other evaluations which motivate the piecewise approximation.

The number of basis functions ϕ_l, $\bar{\phi}_j$ and $\tilde{\phi}_k$ is given by N, $\bar{N}$ and $\tilde{N}$, respectively. If the particle velocity of the structure and the boundary admittance are known functions, N accounts for degree of freedoms. Herein, this coincides with the number of nodes of the finite or boundary element mesh. $\bar{N}$ and $\tilde{N}$ may be equal to each other. For BEM with discontinuous boundary elements, it is common that $\bar{N} = \tilde{N} = N$.

0.4.2 Finite Element Method

In many cases, finite element discretization actually refers to a Galerkin discretization of the weak formulation as shown in Equation (0.43). Galerkin discretization means that the basis functions ϕ_l which have been used in Equation (0.47) are substituted for the test function χ. Hence we write

$$\begin{aligned} sk\int_{\Gamma} \phi_l(\boldsymbol{x})\left\{\sum_{j=1}^{\bar{N}} \bar{\phi}_j(\boldsymbol{x})\, v_{sj} + \left[\sum_{k=1}^{\tilde{N}} \tilde{\phi}_k(\boldsymbol{x})\, Y_k\right]\left[\sum_{m=1}^{N} \phi_m(\boldsymbol{x})\, p_m\right]\right\} d\Gamma(\boldsymbol{x}) \,+ \\ -\int_{\Omega}\left\{\nabla\phi_l(\boldsymbol{x})\cdot\nabla\left[\sum_{j=1}^{N}\phi_j(\boldsymbol{x})\,p_j\right] - k^2\phi_l(\boldsymbol{x})\left[\sum_{j=1}^{N}\phi_j(\boldsymbol{x})\,p_j\right]\right\} d\Omega(\boldsymbol{x}) \;=\; 0\,. \end{aligned} \tag{0.49}$$

This can be written in a simpler form by introducing matrices. Herein, we introduce the boundary mass matrix $\boldsymbol{\Theta}$ with entries θ_{lj} as

$$\theta_{lj} \;=\; \int_{\Gamma} \phi_l(\boldsymbol{x})\, \bar{\phi}_j(\boldsymbol{x})\, d\Gamma(\boldsymbol{x})\,, \tag{0.50}$$

the mass matrix $\boldsymbol{M}$ with entries m_{lj} as

$$m_{lj} \;=\; \int_{\Omega} \phi_l(\boldsymbol{x})\, \phi_j(\boldsymbol{x})\, d\Omega(\boldsymbol{x})\,, \tag{0.51}$$

the stiffness matrix $\boldsymbol{K}$ with entries k_{lj} as

$$k_{lj} = \int_{\Omega} \nabla\phi_l(\boldsymbol{x}) \cdot \nabla\phi_j(\boldsymbol{x})\, d\Omega(\boldsymbol{x}) \;, \tag{0.52}$$

and the damping matrix $\boldsymbol{C}$ with entries c_{lj} as

$$\begin{aligned} c_{lj} &= \varrho_0\, c \int_{\Gamma} \phi_l(\boldsymbol{x}) \left[\tilde{\boldsymbol{\phi}}^T(\boldsymbol{x})\, \boldsymbol{Y}\right] \bar{\phi}_j(\boldsymbol{x})\, d\Gamma(\boldsymbol{x}) = \\ &= \int_{\Gamma} \phi_l(\boldsymbol{x}) \left[\tilde{\boldsymbol{\phi}}^T(\boldsymbol{x})\, \tilde{\boldsymbol{Y}}\right] \bar{\phi}_j(\boldsymbol{x})\, d\Gamma(\boldsymbol{x}) \;, \end{aligned} \tag{0.53}$$

where $\tilde{\boldsymbol{Y}}$ assembles the values of the normalized boundary admittances such that $\tilde{Y} = \varrho_0 c Y$ and $\tilde{\boldsymbol{Y}} = \varrho_0 c \boldsymbol{Y}$, respectively. Consequently, we write our system of equations in matrix form as

$$\left(\boldsymbol{K} - \mathrm{i}\, k\, \boldsymbol{C} - k^2 \boldsymbol{M}\right) \boldsymbol{p} = s\, k\, \boldsymbol{\Theta}\, \boldsymbol{v}_s = \boldsymbol{f} \;. \tag{0.54}$$

The left hand–side shows a matrix polynomial in k. Note that, in general, $\boldsymbol{C}$ represents a complex matrix because, in general, the boundary admittance is complex. The system matrix consists of three static matrices, $\boldsymbol{K}$, $\boldsymbol{M}$, and, provided the boundary admittance is independent of the frequency, $\boldsymbol{C}$. $\boldsymbol{K}$ and $\boldsymbol{M}$ are symmetric and positive definite, $\boldsymbol{C}$ is symmetric but not Hermitian.

0.4.3 Boundary Element Method

In engineering literature, the discretization procedure of the integral equation, cf. Equation (0.45), is often omitted or, at least, is abridged such that it can't be identified. Moreover, there are many engineering articles which prefer using the categories of direct and indirect approaches. Often, the direct approach is automatically associated with collocation whereas the indirect approach seems to be virtually linked to the Galerkin discretization. Herein, we limit our consideration to the direct approach. This, however, does not prohibit the use of either collocation or Galerkin discretization methods. Even other discretization methods can be used, e.g. Nyström methods and least squares methods. Similarly, the indirect approach allows different methods of discretization including collocation and Galerkin methods which are the most commonly used techniques for practical applications of the boundary element method.

Collocation

The collocation method requires substituting the Dirac function $\delta(\boldsymbol{y}, \boldsymbol{z})$ for the test function $\tilde{\chi}(\boldsymbol{y})$ in Equation (0.46). It modifies to

$$\int_\Gamma \delta(\boldsymbol{y},\boldsymbol{z})\, c(\boldsymbol{y})\, p(\boldsymbol{y})\, d\Gamma(\boldsymbol{y}) +$$

$$+ \int_\Gamma \delta(\boldsymbol{y},\boldsymbol{z}) \left\{ \int_\Gamma \left[\frac{\partial G(\boldsymbol{x},\boldsymbol{y})}{\partial n(\boldsymbol{x})} - s\,k\, G(\boldsymbol{x},\boldsymbol{y})\, Y(\boldsymbol{x}) \right] p(\boldsymbol{x}) d\Gamma(\boldsymbol{x}) \right\} d\Gamma(\boldsymbol{y}) \;=$$

$$= \; s\,k \int_\Gamma \delta(\boldsymbol{y},\boldsymbol{z}) \left\{ \int_\Gamma G(\boldsymbol{x},\boldsymbol{y})\, v_s(\boldsymbol{x})\, d\Gamma(\boldsymbol{x}) \right\} d\Gamma(\boldsymbol{y})\,. \tag{0.55}$$

The outer integration is known analytically, cf. Equation (0.40). It yields

$$c(\boldsymbol{z})p(\boldsymbol{z}) + \int_\Gamma \left[\frac{\partial G(\boldsymbol{x},\boldsymbol{z})}{\partial n(\boldsymbol{x})} - sk\, G(\boldsymbol{x},\boldsymbol{z}) Y(\boldsymbol{x}) \right] p(\boldsymbol{x}) d\Gamma(\boldsymbol{x}) \;=$$

$$= \; sk \int_\Gamma G(\boldsymbol{x},\boldsymbol{z})\, v_s(\boldsymbol{x}) d\Gamma(\boldsymbol{x})\,, \tag{0.56}$$

which is basically the same expression as shown in Equation (0.45). The major difference between equations (0.45) and (0.56) is that the former is actually a continuous integral equation whereas the latter is valid just for the discrete point $\boldsymbol{z}$. This means that the integral equation is fulfilled at a number of discrete points, i.e. collocation points $\boldsymbol{z}_l$. It is common practice that the collocation points coincide with the nodes of the piecewise formulated approximation of the sound pressure as shown in Equation (0.47). For further considerations we assume that $\phi_l(\boldsymbol{z}_k) = \delta_{lk}$ where δ_{lk} is the Kronecker symbol with $\delta_{lk} = 0$ for $l \neq k$ and $\delta_{lk} = 1$ for $l = k$. Then, applying the approximation of equations (0.47) and (0.48) yields

$$c(\boldsymbol{z}_l)\, p_l +$$

$$+ \int_\Gamma \left\{ \frac{\partial G(\boldsymbol{x},\boldsymbol{z}_l)}{\partial n(\boldsymbol{x})} - skG(\boldsymbol{x},\boldsymbol{z}_l) \left[\sum_{j=1}^{\tilde{N}} \tilde{\phi}_j(\boldsymbol{x}) Y_j \right] \right\} \left[\sum_{k=1}^{N} \phi_k(\boldsymbol{x}) p_k \right] d\Gamma(\boldsymbol{x}) \;=$$

$$= \; s\,k \int_\Gamma G(\boldsymbol{x},\boldsymbol{z}_l) \left[\sum_{m=1}^{\bar{N}} \bar{\phi}_m(\boldsymbol{x})\, v_{sm} \right] d\Gamma(\boldsymbol{x})\,. \tag{0.57}$$

To simplify this equation, we introduce matrices. Matrix $\boldsymbol{G}$ is the system matrix of the single layer potential as

$$g_{lj} \;=\; s\,k \int_\Gamma G(\boldsymbol{x},\boldsymbol{z}_l)\, \bar{\phi}_j(\boldsymbol{x})\, d\Gamma(\boldsymbol{x})\,, \tag{0.58}$$

matrix $\boldsymbol{H}$ contains the integral–free term and the contribution of the double layer potential as

$$h_{lj} \;=\; c(\boldsymbol{z}_l)\, \delta_{lj} + \int_\Gamma \frac{\partial G(\boldsymbol{x},\boldsymbol{z}_l)}{\partial n(\boldsymbol{x})} \phi_j(\boldsymbol{x})\, d\Gamma(\boldsymbol{x}) \tag{0.59}$$

and matrix $\boldsymbol{D}$ which contains the boundary admittance terms as

$$d_{lj} = s\,k \int_\Gamma G(\boldsymbol{x}, \boldsymbol{z}_l) \left[\tilde{\boldsymbol{\phi}}^T(\boldsymbol{x})\,\boldsymbol{Y}\right] \phi_j(\boldsymbol{x})\,d\Gamma(\boldsymbol{x}) =$$

$$= \mathrm{i}\,k \int_\Gamma G(\boldsymbol{x}, \boldsymbol{z}_l) \left[\tilde{\boldsymbol{\phi}}^T(\boldsymbol{x})\,\tilde{\boldsymbol{Y}}\right] \phi_j(\boldsymbol{x})\,d\Gamma(\boldsymbol{x}), \qquad (0.60)$$

where the normalized admittance $\tilde{Y}$ is introduced as in Equation (0.53). Equation (0.57) in matrix form is written as

$$(\boldsymbol{H} - \boldsymbol{D})\,\boldsymbol{p} = \boldsymbol{G}\,\boldsymbol{v}_s = \boldsymbol{f}. \qquad (0.61)$$

There are examples in literature where $\boldsymbol{D} = \boldsymbol{G}\boldsymbol{Y}$ with the diagonal matrix $\boldsymbol{Y}$. This form requires some specified conditions as for example piecewise constant approximation of sound pressure and boundary admittance. The system matrices $\boldsymbol{G}, \boldsymbol{H}$ and $\boldsymbol{D}$ are neither Hermitian nor positive definite in general.

Galerkin Method

The classical Galerkin method requires use of the basis functions ϕ_l for approximation of the sound pressure for the test function $\tilde{\chi}(\boldsymbol{y})$ in Equation (0.46) as

$$\int_\Gamma \phi_l(\boldsymbol{y})\,c(\boldsymbol{y})\,p(\boldsymbol{y})\,d\Gamma(\boldsymbol{y}) +$$

$$+ \int_\Gamma \phi_l(\boldsymbol{y}) \left\{ \int_\Gamma \left[\frac{\partial G(\boldsymbol{x}, \boldsymbol{y})}{\partial n(\boldsymbol{x})} - s\,k\,G(\boldsymbol{x}, \boldsymbol{y})\,Y(\boldsymbol{x}) \right] p(\boldsymbol{x}) d\Gamma(\boldsymbol{x}) \right\} d\Gamma(\boldsymbol{y}) =$$

$$= s\,k \int_\Gamma \phi_l(\boldsymbol{y}) \left\{ \int_\Gamma G(\boldsymbol{x}, \boldsymbol{y})\,v_s(\boldsymbol{x})\,d\Gamma(\boldsymbol{x}) \right\} d\Gamma(\boldsymbol{y}). \qquad (0.62)$$

This time, the double surface integral does not vanish as for collocation. Simplification of the first term becomes possible since the integration over piecewise smooth surface elements allows to set $c(\boldsymbol{y}) = 1/2$. Applying the approximation of equations (0.47) and (0.48) gives (omitting dependencies on $\boldsymbol{x}$ and $\boldsymbol{y}$)

$$\frac{1}{2} \int_\Gamma \phi_l \left[\sum_{j=1}^{N} \phi_j\,p_j \right] d\Gamma +$$

$$+ \int_\Gamma \phi_l \left\langle \int_\Gamma \left\{ \frac{\partial G}{\partial n_x} - s\,k\,G \left[\sum_{k=1}^{\tilde{N}} \tilde{\phi}_k\,Y_k \right] \right\} \left[\sum_{j=1}^{N} \phi_j p_j \right] d\Gamma_x \right\rangle d\Gamma =$$

$$= s\,k \int_\Gamma \phi_l \left\{ \int_\Gamma G \left[\sum_{j=1}^{\bar{N}} \bar{\phi}_j\,v_{sj} \right] d\Gamma \right\} d\Gamma. \qquad (0.63)$$

Similar to the collocation method we introduce matrices $\boldsymbol{G}, \boldsymbol{H}, \boldsymbol{D}$ and the boundary mass matrix $\boldsymbol{\Theta}$ very similar to the one in Equation (0.50). We write the system matrix of the single layer potential $\boldsymbol{G}$ as

$$g_{lj} = s\,k \int_\Gamma \int_\Gamma G(\boldsymbol{x},\boldsymbol{y})\,\phi_l(\boldsymbol{y})\,\bar{\phi}_j(\boldsymbol{x})\,d\Gamma(\boldsymbol{x})d\Gamma(\boldsymbol{y})\;, \tag{0.64}$$

the matrix of the double layer potential $\boldsymbol{H}$ as

$$h_{lj} = \int_\Gamma \int_\Gamma \frac{\partial G(\boldsymbol{x},\boldsymbol{y})}{\partial n(\boldsymbol{x})}\,\phi_l(\boldsymbol{y})\,\phi_j(\boldsymbol{x})\,d\Gamma(\boldsymbol{x})d\Gamma(\boldsymbol{y})\;, \tag{0.65}$$

the matrix which contains the boundary admittance terms $\boldsymbol{D}$ as

$$d_{lj} = s\,k \int_\Gamma \int_\Gamma G(\boldsymbol{x},\boldsymbol{y}) \left[\tilde{\boldsymbol{\phi}}^T(\boldsymbol{x})\,\boldsymbol{Y}\right] \phi_l(\boldsymbol{y})\phi_j(\boldsymbol{x})\,d\Gamma(\boldsymbol{x})d\Gamma(\boldsymbol{y}) =$$

$$= \mathrm{i}\,k \int_\Gamma \int_\Gamma G(\boldsymbol{x},\boldsymbol{y}) \left[\tilde{\boldsymbol{\phi}}^T(\boldsymbol{x})\,\tilde{\boldsymbol{Y}}\right] \phi_l(\boldsymbol{y})\phi_j(\boldsymbol{x})\,d\Gamma(\boldsymbol{x})d\Gamma(\boldsymbol{y})\;, \tag{0.66}$$

where the normalized boundary admittance $\tilde{Y}$ has been used again, and the boundary mass matrix $\boldsymbol{\Theta}$ as

$$\theta_{lj} = \frac{1}{2} \int_\Gamma \phi_l(\boldsymbol{y})\,\phi_j(\boldsymbol{y})d\Gamma(\boldsymbol{y})\;. \tag{0.67}$$

Hence, we write the system of equations in matrix form as

$$(\boldsymbol{\Theta} + \boldsymbol{H} - \boldsymbol{D})\,\boldsymbol{p} = \boldsymbol{G}\,\boldsymbol{v}_s = \boldsymbol{f}\;. \tag{0.68}$$

Similar to collocation BEM, the system matrices $\boldsymbol{G}$, $\boldsymbol{H}$ and $\boldsymbol{D}$ are neither Hermitian nor positive definite in general. It will be shown in Chapters 16 and 19 how to set up a Hermitian system of equations for Galerkin BEM.

0.5 Sources and Incident Wave Fields

0.5.1 A General Approach

The previous considerations have started from the homogeneous Helmholtz equation with inhomogeneous boundary conditions. Now, we consider problems with sources, e.g. monopole sources and, further, incident waves.

Assume the source q located somewhere in Ω, then the Helmholtz equation (0.30) becomes inhomogeneous as

$$\Delta p(\boldsymbol{x}) + k^2 p(\boldsymbol{x}) = -q\;. \tag{0.69}$$

Since Equation (0.69) is a linear inhomogeneous partial differential equation, its solution can be constructed by superimposing the solution of the homogeneous Helmholtz equation (0.30) which is also known as complementary solution p^c and a particular solution p^p which solves the inhomogeneous Equation (0.69). In acoustics, the complementary solution is usually referred to as the scattered pressure p^s while the particular solution is usually referred to as the incident pressure field p^i. Denoting the entire sound pressure by p, it can be represented as

$$p(\boldsymbol{x}) = p^c(\boldsymbol{x}) + p^p(\boldsymbol{x}) = p^s(\boldsymbol{x}) + p^i(\boldsymbol{x}) \tag{0.70}$$

for the sound pressure and

$$v_f(\boldsymbol{x}) = v_f^c(\boldsymbol{x}) + v_f^p(\boldsymbol{x}) = v_f^s(\boldsymbol{x}) + v_f^i(\boldsymbol{x}) \,. \tag{0.71}$$

for the fluid particle velocity. The two most commonly applied sources are the monopole source and the incident wave field.

For the monopole source, we can formulate the source term in Equation (0.69) as

$$q(\boldsymbol{x}) = q_m(\boldsymbol{x}, \boldsymbol{y}) = C\,\delta(\boldsymbol{x}, \boldsymbol{y}) \,, \tag{0.72}$$

thus, looking very similar as the right hand side in Equation (0.38). Actually, the Dirac function in (0.38) accounts for the inhomogeneity which leads to the fundamental solution or Green's function G. Similarly, the source term in Equation (0.72) demands a particular solution as

$$p^i(\boldsymbol{x}, \boldsymbol{y}) = \frac{C}{4\pi} \frac{\mathrm{e}^{\mathrm{i}kr(\boldsymbol{x},\boldsymbol{y})}}{r(\boldsymbol{x}, \boldsymbol{y})} \qquad \boldsymbol{x}, \boldsymbol{y} \in \mathbb{R}^3 \,. \tag{0.73}$$

Assuming the monopole to be a pulsating sphere (radius R) with constant surface sound pressure p_0, the particular solution p^i can be written in terms of the radius $r \geq R$ as

$$p^i(r) = p_0 \frac{R}{r} \mathrm{e}^{\mathrm{i}k(r-R)} \,, \tag{0.74}$$

whereas the monopole solution for a prescribed surface particle velocity v_{f_0} provides the sound pressure distribution as

$$p^i(r) = \varrho_0 c v_{f_0} \frac{R}{r} \frac{\mathrm{i}kR}{1 - \mathrm{i}kR} \mathrm{e}^{\mathrm{i}k(r-R)} \,. \tag{0.75}$$

The particular solution of the particle velocity v_f^i for the monopole with surface pressure prescribed is

$$v_f^i(r) = \frac{\mathrm{i}p_0 R}{\varrho_0 \omega} \frac{1 - \mathrm{i}kr}{r^2} \mathrm{e}^{\mathrm{i}k(r-R)} \frac{\partial r}{\partial n} \,, \tag{0.76}$$

whereas for the monopole with surface particle velocity prescribed we get

$$v_f^i(r) = v_{f_0} \frac{1 - \mathrm{i}kr}{1 - \mathrm{i}kR} \frac{R^2}{r^2} \mathrm{e}^{\mathrm{i}k(r-R)} \frac{\partial r}{\partial n} \,. \tag{0.77}$$

Note that $\partial p^i(r = R)/\partial r = -av_{f_0}$ where the negative sign results from the definition of the normal vector for v_f according to Equation (0.32) and Figure 0.1.

An analogous approach is possible for any other source or even any sink with either known right hand side q of Equation (0.69) or known particular solution p^i. It might even be possible to experimentally determine the particular solution by measuring and analytically reconstructing the sound pressure distribution around an arbitrary radiator under free–field conditions. This can be an option for a loudspeaker in a frequency range where the monopole characteristics cannot be guaranteed.

Although the situation is different for an uniform incident wave field, it is treated in the same way as the monopole source above. The infinite uniform incident wave field can be described by the particular solution

$$p^i(\boldsymbol{x}) \;=\; p_0\, \mathrm{e}^{\mathrm{i}(\boldsymbol{k}\cdot\boldsymbol{x}+\varphi_0)} \,. \tag{0.78}$$

The vector $\boldsymbol{k}$ contains the components which determine the direction of the traveling waves. The magnitude of $\boldsymbol{k}$ is the actual wave–number k, hence $k = |\boldsymbol{k}|$. The reference phase angle φ_0 should be suitably chosen. For the particle velocity this yields

$$v_f^i(\boldsymbol{x}) \;=\; \frac{p_0}{\varrho_0\omega}\,\boldsymbol{k}\cdot\boldsymbol{n}\,\mathrm{e}^{\mathrm{i}(\boldsymbol{k}\cdot\boldsymbol{x}+\varphi_0)} \;=\; \frac{p_0}{\varrho_0 c}\,\frac{\partial k}{\partial n}\,\mathrm{e}^{\mathrm{i}(\boldsymbol{k}\cdot\boldsymbol{x}+\varphi_0)} \,. \tag{0.79}$$

It can be easily seen that the particular solution in Equation (0.78) fulfills the homogeneous Helmholtz equation. Thus, the right hand side q in Equation (0.69) vanishes, $q = 0$. So, the particular solution is just an additional part of the complementary solution which does not necessarily fulfill other conditions such as the Sommerfeld radiation condition (0.34). From the physical point of view, the infinite uniform incident wave field does not make sense since it assumes arbitrarily many (i.e. an infinite number of) energy sources and sinks.

Superposition of complementary solutions p^s, v_f^s and particular solutions p^i, v_f^i provides us with the complete solutions p, v_f. Substitutions can be made at any point of the finite and boundary element approaches in the previous sections which are valid for the complementary solutions only. It is a reasonable approach to start at Equation (0.36) for finite elements and at Equation (0.41) for boundary elements. Performing this, the finite element system of equations (0.54) modifies into

$$\left(\boldsymbol{K} - \mathrm{i}k\boldsymbol{C} - k^2\boldsymbol{M}\right)\boldsymbol{p} \;=\; sk\,\boldsymbol{\Theta}\left(\boldsymbol{v}_s - \boldsymbol{v}_f^i\right) + \left(\boldsymbol{K} - k^2\boldsymbol{M}\right)\boldsymbol{p}^i \;=\; \boldsymbol{f} + \boldsymbol{f}^i \tag{0.80}$$

whereas the boundary element collocation system of equations (0.61) becomes

$$\left(\boldsymbol{H} - \boldsymbol{D}\right)\boldsymbol{p} \;=\; \boldsymbol{G}\left(\boldsymbol{v}_s - \boldsymbol{v}_f^i\right) + \boldsymbol{H}\boldsymbol{p}^i \;=\; \boldsymbol{f} + \boldsymbol{f}^i \,. \tag{0.81}$$

Application to the Galerkin boundary element method is done analogously. Obviously, the particular solution contributes to the right hand side only.

0.5.2 Boundary Integral Equations and Source Terms

The concept of source consideration of the previous subsection is valid for any formulation which is derived from the Helmholtz equation, (0.30) and (0.69). Formally, the inhomogeneous Helmholtz equation can be written in a weak formulation, integrated by parts twice and, thus, changing Equation (0.41) into

$$\begin{aligned} c(\boldsymbol{y})p(\boldsymbol{y}) \;+\; \int_\Gamma \frac{\partial G(\boldsymbol{x},\boldsymbol{y})}{\partial n(\boldsymbol{x})}\,p(\boldsymbol{x})d\Gamma(\boldsymbol{x}) \;=& \\ =\int_\Gamma G(\boldsymbol{x},\boldsymbol{y})\,av_f(\boldsymbol{x})d\Gamma(\boldsymbol{x}) \;+\; \int_Q G(\boldsymbol{x},\boldsymbol{y})\,q(\boldsymbol{x},\boldsymbol{z})dQ(\boldsymbol{x})\,,& \end{aligned} \tag{0.82}$$

where the free field Q is the combination of Ω and Ω_c, hence $\Omega \cup \Omega_c \cup Q$. Often, Equation (0.82) is not suited for practical use.

It is useful to return to Equation (0.41) which has introduced the Kirchhoff–Helmholtz integral equation. It is valid for the source free acoustic field. Therefore, we write

$$c(\boldsymbol{y})p^s(\boldsymbol{y}) + \int_\Gamma \frac{\partial G(\boldsymbol{x},\boldsymbol{y})}{\partial n(\boldsymbol{x})}\, p^s(\boldsymbol{x})d\Gamma(\boldsymbol{x}) \;=\; \int_\Gamma G(\boldsymbol{x},\boldsymbol{y})\, a v_f^s(\boldsymbol{x})d\Gamma(\boldsymbol{x})\,. \quad (0.83)$$

Incorporating (0.70) and (0.71) into (0.83) yields

$$c(\boldsymbol{y})\left[p(\boldsymbol{y}) - p^i(\boldsymbol{y})\right] \;+\; \int_\Gamma \frac{\partial G(\boldsymbol{x},\boldsymbol{y})}{\partial n(\boldsymbol{x})}\left[p(\boldsymbol{x}) - p^i(\boldsymbol{x})\right] d\Gamma(\boldsymbol{x}) \;=$$

$$= \int_\Gamma G(\boldsymbol{x},\boldsymbol{y})\, a\left[v_f(\boldsymbol{x}) - v_f^i(\boldsymbol{x})\right] d\Gamma(\boldsymbol{x}) \quad (0.84)$$

which obviously leads to the system of equations shown in Equation (0.81).

Alternatively, we may assume that the particular solution either fulfills the homogeneous Helmholtz equation, e.g. plane wave solution (0.78) and (0.79), or the source is located in Ω, e.g. the monopole solution in Equations (0.73)–(0.77) with $\boldsymbol{y} \in \Omega$. Then, the particular solution fulfills the homogeneous Helmholtz equation und, thus, the Kirchhoff–Helmholtz integral equation in the complementary domain Ω_c

$$-\tilde{c}(\boldsymbol{y})p^i(\boldsymbol{y}) + \int_\Gamma \frac{\partial G(\boldsymbol{x},\boldsymbol{y})}{\partial n(\boldsymbol{x})}\, p^i(\boldsymbol{x})d\Gamma(\boldsymbol{x}) \;=\; \int_\Gamma G(\boldsymbol{x},\boldsymbol{y})\, a v_f^i(\boldsymbol{x})d\Gamma(\boldsymbol{x}) \quad \text{for } \boldsymbol{y} \in \Omega_c\,. \quad (0.85)$$

Note that the normal vector is pointing into the complementary domain. Further, notice $c(\boldsymbol{y})$ in Equation (0.83) and $\tilde{c}(\boldsymbol{y})$ in Equation (0.85) are related such that $c(\boldsymbol{y}) + \tilde{c}(\boldsymbol{y}) = 1$. This is obvious if either $\boldsymbol{y} \in \Omega$ or $\boldsymbol{y} \in \Omega_c$ but also holds for $\boldsymbol{y} \in \Gamma$, since c is the value for $\boldsymbol{y} \in \Omega$ approaching Γ and $\tilde{c}$ is the value for $\boldsymbol{y} \in \Omega_c$ approaching Γ. Summation of Equations (0.83) and (0.85) together with substituting $p - p^i$ for the complementary solution leads to

$$c(\boldsymbol{y})p(\boldsymbol{y}) + \int_\Gamma \frac{\partial G(\boldsymbol{x},\boldsymbol{y})}{\partial n(\boldsymbol{x})}\, p(\boldsymbol{x})d\Gamma(\boldsymbol{x}) \;=\; \int_\Gamma G(\boldsymbol{x},\boldsymbol{y})\, a v_f(\boldsymbol{x})d\Gamma(\boldsymbol{x}) \;+\; p^i(\boldsymbol{y}) \quad \text{for } \boldsymbol{y} \in \Gamma\,. \quad (0.86)$$

Note that Equation (0.86) can be derived from Equation (0.82) for the case of an arbitrary three–dimensional monopole source, see also Equations (0.72) and (0.73). This yields

$$\int_Q G(\boldsymbol{x},\boldsymbol{y})\, q(\boldsymbol{x},\boldsymbol{z})dQ(\boldsymbol{x}) = \int_Q G(\boldsymbol{x},\boldsymbol{y})\, C\,\delta(\boldsymbol{x},\boldsymbol{z})dQ(\boldsymbol{x}) \;=$$

$$= C\,G(\boldsymbol{y},\boldsymbol{z}) \;=\; \frac{C}{4\pi}\,\frac{\mathrm{e}^{ikr(\boldsymbol{y},\boldsymbol{z})}}{r(\boldsymbol{y},\boldsymbol{z})} \;=\; p^i(\boldsymbol{y},\boldsymbol{z})\,. \quad (0.87)$$

Equation (0.86) is probably the most popular variant to consider sources in boundary element techniques. The final system of equations for the boundary element collocation method becomes

$$(\boldsymbol{H} - \boldsymbol{D})\boldsymbol{p} = \boldsymbol{G}\boldsymbol{v}_s + \boldsymbol{p}^i = \boldsymbol{f} + \boldsymbol{f}^i . \tag{0.88}$$

Again, application to the Galerkin boundary element method is done analogously.

0.5.3 Source Distribution as Boundary Condition

Another very common practice is to apply the particular solution as a boundary condition for the problem. This can be easily explained for the systems of equations (0.80) and (0.81). The complementary solution of the sound pressure $\boldsymbol{p}^s$ is substituted for $\boldsymbol{p} - \boldsymbol{p}^i$. This leads to

$$\left(\boldsymbol{K} - \mathrm{i}k\boldsymbol{C} - k^2\boldsymbol{M}\right)\boldsymbol{p}^s = sk\boldsymbol{\Theta}\left(\boldsymbol{v}_s - \boldsymbol{v}_f^i\right) + \mathrm{i}k\boldsymbol{C}\boldsymbol{p}^i = \boldsymbol{f} + \boldsymbol{f}^i \tag{0.89}$$

for the finite element system and to

$$(\boldsymbol{H} - \boldsymbol{D})\boldsymbol{p}^s = \boldsymbol{G}\left(\boldsymbol{v}_s - \boldsymbol{v}_f^i\right) + \boldsymbol{D}\boldsymbol{p}^i = \boldsymbol{f} + \boldsymbol{f}^i \tag{0.90}$$

for the boundary element system. Note the simplification in case of acoustically rigid surfaces where the matrices $\boldsymbol{C}$ and $\boldsymbol{D}$ vanish in Equations (0.89) and (0.90), respectively. For the single scattering problem, i.e. $\boldsymbol{v}_s = \boldsymbol{0}$, the negative particular solution of the particle velocity remains as the only boundary condition. It can be seen as if the solution $\boldsymbol{p}^s$ must be chosen such that the boundary condition which is induced by the particular solution is balanced.

0.6 Categorization of Subsequent Chapters

In the subsequent twenty chapters of this book, the first ten are basically on finite element methods, the second ten are basically on boundary element methods. These papers cover a wide range of the field of computational acoustics.

Parts I and IV illuminate a number of numerical problems of finite and boundary element methods. Chapters 1, 2 and 11 deal with disretization requirements and different types of elements. This refers to Equation (0.47) for approximation, Equations (0.49) for finite element discretization, and (0.57) for boundary element collocation. It is well–known that classical Galerkin finite element discretization suffers from the pollution effect, which makes the method inefficient for low polynomial degree of element interpolation at high frequencies, see Chapter 1. Chapter 2 presents different types of finite elements whereas different types of (Lagrangian) boundary elements are investigated in Chapter 11. In the context of other problems, specified elements for a displacement formulation of the FEM, i.e. Raviart–Thomas elements, are introduced in Chapter 9. Chapters 3, 12, 13, and 14 deal with different types of efficient numerical techniques in general. This includes efficient solution over a frequency range by Padé approximation while simultaneously performing a model reduction by using a Lanczos method. For a fast boundary element solution, the authors of Chapter 12 discuss the suitability of using iterative solvers. The arising matrix–vector product is efficiently evaluated by applying the multilevel fast multipole technique for collocation. Later on, in Chapter 19, this technique is revisited for the Galerkin boundary element method. Efficiency gains can be reported for the multi-domain BEM which is presented in Chapter 13. The formulation in Chapter 14 refers to waveguide boundary spectral finite elements which can be considered as finite but also as boundary elements. They are based on the assumption that the direction of wave propagation is known and, thus, it can be separated from other directions. Finally, the interface to computational fluid dynamics is touched on by Chapter 4. Therein, the so–called Lighthill acoustic analogy is considered. Lighthill's acoustic analogy rearranges the governing equations of fluid dynamics into a wave–like equation. This wave–like equation contains the wave equation (0.28) on the left hand side and allows the interpretation of the remaining terms as elementary sources, i.e. monopole, dipole and quadrupole.

Parts II and V contain eight chapters devoted to specific problems of external acoustics, i.e. acoustics in unbounded domains. Clearly, the problems of finite element methods have hardly anything in common with those appearing with boundary element methods. All chapters in Part II, i.e. Chapters 5– 8, deal with the problem of mesh truncation for finite elements but, at the same time, fulfilling the Sommerfeld radiation condition (0.34). Chapter 5 provides the reader with a survey of treating this problem and focusses somewhat deeper into the field of absorbing boundary conditions and absorbing layers. The so–called perfectly matched layer is described in more detail in Chapter 6. Therein, an optimal variant of a perfectly matched layer is presented. Chapters 7 and 8 are focussed on infinite elements. The former chapter provides the reader with a survey of infinite elements, whereas, in the latter, a very well–conditioned infinite element is formulated. In Part V, Chapter 15 gives a survey of methods which are designed to waive the so–called irregular frequencies, e.g. different types of the combined Helmholtz integral equation formulation (CHIEF) and the different types of an approach which uses the combination of the Kirchhoff–Helmholtz integral equation (0.45) and its normal derivative. A Galerkin BEM resulting in a symmetric (Hermitian) system matrix is presented in Chapter 16. This method is formulated in very general way since it combines compact radiator shapes as shown in Figure 0.1 and thin structures. This formulation is a combination of the direct and the indirect BEM. Sound propagation above an impedance plane is considered in Chapter 17. Chapter 18 deals with time–domain formulations in the context of BEM. Since this time–domain solution with BEM seems to be quite different from the time–harmonic solution, it is the only chapter dealing with transient problems in the BEM related parts. Time–domain solutions with FEM appear in many chapters of the FEM related parts, in particular, Chapters 4–9 contain passages about time–domain solutions.

Parts III and VI assemble four selected contributions on related problems. More specifically, the fluid–structure interaction chapters 9 and 19 comprise many different subjects of FEM and BEM, respectively. Chapter 9 provides the reader with a survey of many different formulations for fluid structure interaction whereas the authors of Chapter 19 describe the complex formulation of fluid–structure interaction with a structural finite element model and a fluid boundary element model from the theoretical formulation to the industrial application. Chapter 10 gives a totally different view at finite element formulations since, therein, an energy based formulation is presented. Instead of using pressure or displacement as variables, energy terms account for the unknowns in this case. Finally, Chapter 20 surveys inverse boundary element techniques which are used to identify unknown boundary conditions.

The subsequent twenty chapters of this book contain a considerable variety of academic and industrial examples. Among many others, we mention the examples of engines (Chapters 8, 12, 15 and 20), exhaust systems (Chapters 13, 14 and 19), submarine–like structures (Chapter 5 and 19), a vehicle cabin (Chapter 11), hearing simulation (Chapter 3), a vacuum cleaner (Chapter 20), a reverberation room (Chapter 2), an anechoic chamber (Chapter 12) and an amphitheatre (Chapter 18).

0.7 FEM and BEM for Acoustics: Future Work

0.7.1 General Remarks

There is no doubt that finite and boundary element methods will receive further attention, most likely even an ever increasing attention over the next decades. Many of the chapters of this book are completed by an outlook into the future. Open problems are identified at these points. In this section, the authors sketch their personal view of future development and future requirements in the field of FEM and BEM in acoustics. This outlook provides a somewhat detailed discussion of three areas and, then, touches three more areas very briefly. Since it can only be a personal view of the authors, it will not comprise the entire field of finite and boundary element methods in acoustics.

0.7.2 Comparisons and Benchmarks

Clearly, the development of efficient methods will continue. Nowadays, efficient finite element methods and fast boundary element methods are almost optimally performing since they require $\approx O(N)$ operations for solution of the particular problem. (Herein, N represents the degree of freedom of the finite/boundary element model.) Similarly, the storage requirements are estimated as $\approx O(N)$. But even though the methods become more and more efficient and the computers faster and faster, the main problems are stated

- in the details of the methods, e.g. efficient formulation of absorbing boundary conditions for FEM and reliable evaluation of hypersingular integrals on arbitrary element shapes for BEM,
- in the general applicability of methods in commercial codes to make them available to industrial users, i.e. many new and promising methods require substantial

additional work to make them robust and easy–to–use, as it is required for implementation into commercial codes, and

- in the preparation of reliable and reasonable simulation models and, further, in the ability to understand the results. It makes no sense to look at single frequency solutions for real–world cavity problems at wave–numbers $kR > 20$.

It turns out that there are hardly any reliable estimations comparing computational costs between different methods. A nice paper in this context written by Harari and Hughes [43] compares computational expenditure of standard FEM and standard BEM. Their work was based on the assumption that a fixed number of elements per wavelength is sufficient to achieve a certain accuracy. With respect to low–order Lagrangian elements, today we know that this is valid for BEM but does not hold for FEM. Since then – the paper was published 15 years ago – both, FEM and BEM, have made remarkable progress. A new cost comparison must be based on accuracy. One of the best known comparisons between FEM and BEM is the one which was given by Burnett [14]. He compared CPU time in terms of degree of freedom. There is no doubt that FEM performs (much) better than BEM. Usually, for the same degree of freedom and for a reasonable use of the method, standard BEM will give more accurate results then standard FEM for the same degree of freedom. It is the personal opinion of the authors that both methods can be clearly justified for certain applications. Similarly, the authors are not aware of comparisons between different types of absorbing boundary conditions , e.g. between high–order absorbing boundary conditions, infinite elements and perfectly matched layers. Valuable comparisons about the performance of different types of infinite elements were given by Astley [5,6] and by Shirron and Babuška [94]. For BEM, Marburg has published some comparisons. Two papers on investigation of different element types are summarized in Chapter 11 of this book, the results of a paper comparing different methods to waive irregular frequencies are contained in Chapter 15. In the paper by Marburg and Schneider [72], iterative solvers are compared for collocation BEM.

When looking at these activities and at their results, it still seems that further comparisons of methods might illuminate the jungle of methods in the field. It might be possible to evaluate different methods if benchmark problems are created and if more research projects and papers on this problem are accepted. In this context, the paper by Otsuru et al. [85] describes an interesting benchmark project which has been funded in Japan[3].

0.7.3 Mid–Frequency Range and Uncertainty Analysis

A field which is receiving increasing attention in recent years is the so–called mid–frequency range, also known as medium frequency range. The definition refers to the terms of the low and the high frequency ranges. Usually, the low frequency range is understood such that a certain excitation results in a deterministic response. Contrary to the low frequency range, the high frequency range is understood as non–

[3]Further information about this interesting benchmark project is available at its website `http://gacoust.hwe.oita-u.ac.jp/AIJ-BPCA/`

deterministic but essentially statistically performing. A clearly resulting signal due to a specified excitation is impossible to reproduce in practice. The mid–frequency range lies in between. There are different opinions about it, some say that the mid–frequency range does not exist since the deterministic methods can be used up to frequencies where the statistical methods can be used as well. No doubt, the methods supply results whatever these results mean.

It has been mentioned above that real world cavity simulations at a single discrete frequency do not make sense for $kR > 20$. Two major reasons for this are formulated as follows:

1. In practical applications it is hardly possible to control the wave–number that exactly. The sound field is very sensitive with respect to frequency and location of sources and receivers. Thus, it will be completely different if these parameters are slightly changed.
2. Human perception of sound allows to identify the fundamental tone of a musical instrument. But for many room acoustic evaluation quantities, only third octave band results or even just octave band results are relevant.

The first point can even be extended since reflection, diffraction, absorption and transmission of sound waves become extremely sensitive quantities for large wave–numbers. For that reason, numerical simulations for industrial applications in the mid–frequency range should include an averaging over a frequency band and should allow the definition of uncertain physical parameters. A number of research papers in this field can be found in the collection [60].

In uncertainty analysis, we distinguish between possibilistic and probabilistic approaches. The former category comprises different methods of interval mathematics. Among them, fuzzy modeling seems to be the most popular method for analysis. Interesting reviews of fuzzy finite element modeling are found in the monograph by Hanss [40] and in the article by Moens and Vandepitte [74]. With respect to probabilistic modeling, the authors are not aware of a review paper similar to the one mentioned in the context of fuzzy FEM. The most simple approach is known as Monte–Carlo simulation which is computationally very costly. Alternatively, the stochastic process can be approximated using deterministic functions. Apparently, the most popular method for this is known as polynomial chaos approximation. A nice example of using polynomial chaos approximation is in the paper by Elman et al. [27]. Cottereau et al. [20] show how to formulate impedance matrices – which are used to evaluate the sound pressure in terms of particle velocity boundary conditions – in terms of stochastic variables. Historically, the polynomial chaos approximation uses approximation by Hermitian polynomials only [35]. Xiu and Karniadakis [111] introduced the generalized polynomial chaos to consider non–Gaussian distributions too. Among others, they suggest to approximate β–distributions by Jacobi polynomials and, as a special case of the β–distribution, the uniform distribution by Legendre polynomials. A further development consists in the introduction of the multi–element generalized polynomial chaos method [107] which decreases the computational costs significantly if many uncertain variables are considered.

When using the approaches which have been discussed above, either the same boundary value problem is solved again and again, such as in fuzzy models and in Monte Carlo simulations, or the size of the original model is dramatically increased, e.g. for polynomial chaos approximation. Inevitably though, due to computational expenditure, both approaches will require significant advances in efficiency.

Another group of methods in this field has become quite popular in recent years. There have been great efforts to combine deterministic methods such as FEM and BEM with statistical methods such as statistical energy analysis (SEA) [59, 95, 106]. This type of method seems to be most useful if a complex structural–acoustic system contains parts where the modal density is high whereas certain other modes play a dominant role in the overall vibration behavior. This is the case in complex structural assemblies but may even occur in simple structures where bending modes and longitudinal modes are present. Similarly, we may consider a problem of architectural acoustics with rooms which are large in two directions and but small in the third. Even though the room may be too large for a reasonable deterministic model, room acoustic quantities will be severely influenced by the modes – depending on absorption, eventually even by standing waves – which do not appear in statistically based simulation methods.

Soize [96, 97] constructed a reduced model for linear structural systems by describing the structural behavior using an energy operator which is symmetric and positive definite. The dominant eigensubspace of the energy operator adapted to a fixed frequency band is used to construct a reduced model adapted to that frequency band. Apparently, the method can be reasonably applied to problems where the structure exhibits the mid–frequency behavior and the fluid is considered to vibrate in the low–frequency range. A different approach based on stochastic models has been proposed by Ghanem and Sarkarb [34].

There is no doubt that development of efficient methods which deal with parameter uncertainties in finite and boundary element models will be developed further in the future. It is worth mentioning that in some applications of structural acoustics, only the structure involves uncertainties but the fluid is actually well defined. Uncertainty handling which requires many single function evaluations, e.g. fuzzy models, Monte–Carlo simulation, may be limited to multiple evaluations of the structural problem whereas the acoustic problem needs to be evaluated only once. This type of method will be discussed in the context of structural–acoustic optimization.

0.7.4 Acoustic and Structural–Acoustic Optimization

Depending on the specific problem, its formulation and the method of solution, acoustic and structural–acoustic optimization requires a repeated evaluation of an objective function. The number of these evaluations counts between 10 and 10^6 times. The upper limit is an estimate and may be exceeded in some cases. Obviously, this demands efficient analysis techniques for calculation of the objective function. In structural–acoustic optimization, structural models are often analyzed using the finite–element method, whereas acoustic models usually use finite and boundary element models. Additionally, examples based on the Rayleigh integral have been reported. If simulation covers a wide frequency range, the multifield problem has to be solved for a certain number of discrete frequencies. An outer loop of optimization encompasses the entire solution of the multifrequency multifield problem in a single step. This difficulty may explain why comparatively few publications are available in the field of structural–acoustic optimization. There are some review papers available [16, 17, 63].

For the acoustic characteristics to be influenced, we can name the sound power, sound pressure level, directivity patterns or other measures. Approaches to formulate an objective function to judge the desired properties can be categorized into three different groups, the first one being the sound pressure level at one or more specified points, primarily utilized for closed domains [28, 64, 86]. For general–purpose noise control in open domains, the emitted sound power accounted for the objective function in a number of papers [10, 16, 50, 57, 58]. The remaining third category comprehend other proposals for objective functions [17, 39, 100].

Another aspect that may be utilized for efficient evaluation of acoustic data is based on the circumstance that many structural modifications do not affect the fluid system properties. It is clear that thickness modifications, material data adjustment or even small modifications of the radiating surface – small with respect to the fluid's wavelength [63, 64] – will hardly affect system data of the fluid. Unfortunately, until now, this feature has only been utilized reasonably for internal problems. Simplified evaluation of emitted sound power would require too much data storage. Many contributions to structural–acoustic optimization may have been resolved without repeatedly solving the acoustic problem.

To omit repeated fluid analysis, it is required to efficiently reconstruct or, in other words, factorize and store, the impedance matrix $\boldsymbol{Z}$ with $\boldsymbol{p} = \boldsymbol{Z}\boldsymbol{v}_s$, cf. Equations (0.54), (0.61) and (0.68). Alternatively, the surface sound pressure can be substituted by introducing the transfer impedance matrix $\boldsymbol{Z}_T$ as $\boldsymbol{p}_i = \boldsymbol{Z}_T\boldsymbol{v}_s$ where $\boldsymbol{p}_i$ contains the sound pressure values at discrete points within the field Ω. The authors are aware of three concepts which have been followed to omit repeated fluid analysis:

- *Usage of modal data for interior problems:* Pal and Hagiwara [86] reported calculation of fluid modes first and, then, used them repeatedly in each optimization step for fully coupled structural–acoustic analysis. Owing to modal reduction, reconstruction of the impedance matrix $\boldsymbol{Z}$ performs efficiently once the relevant (frequency–independent) modes have been determined. Modal data are only evaluated for the non–damped case. Often, acoustic damping, i.e. absorption, is frequency dependent. In such a case, the admittance condition can be approximated with respect to frequency. Then, the modes are evaluated in certain frequency intervals, cf. [15]. In general, this concept is well–known for interior acoustics.
- *Usage of modal data for exterior problems:* Using modal data in exterior acoustics is not common. In 1990/91, four papers were published on eigenvalues and modes of the real part of the implicit frequency dependent impedance matrix $\boldsymbol{Z}_R$ [13, 25, 88, 91]. This real symmetric matrix is used for sound power evaluation as $P = 1/2\boldsymbol{v}_s^H\boldsymbol{Z}_R\boldsymbol{v}_s$ with superscript ^{H} denoting the conjugate complex transposed. It was found that, in particular for low frequencies, the eigenvalues

of $\boldsymbol{Z}_R$ decay rapidly and the sound power value is dominated by only a few eigenvectors. Therefore, storage of even frequency dependent modes can be efficient for optimization purposes as shown by Kessels [53]. It is common to call these eigenvectors "radiation modes", a term which was most likely coined by Cunefare [24]. Frequency independent modes have been used for sound power reconstruction by Marburg, cf. [65]. This method aims on the reconstruction of the impedance matrix $\boldsymbol{Z}$ using finite and conjugated Astley–Leis infinite elements, cf. Chapters 7 and 8 of this edition. It it still an open question how to find a reduced modal basis and, then, how to evaluate only modal data which are required for the reduced basis. Minimization of the radiated sound power using these modes has been reported in the paper [68].

- *Storage of transfer impedance matrices:* Originally referred to as acoustic sensitivities [21, 26, 49], the transfer impedance vector, also known as the acoustic transfer vector and the vector of acoustic influence coefficients, is an efficient tool for repeated fluid analysis if the objective function considers only one or a few data recovery points. Marburg has established this concept in numerous article starting with [70] and, most of them, reviewed in [63]. Finally, a comprehensive package on evaluation and use of influence coefficients as acoustic transfer functions and their further development in acoustic transfer matrices as well as efficient interpolation in a large frequency range has been the subject of a patent application by Cremers et al. [22].

It can be concluded here that, with respect to repeated fluid analysis over frequency ranges in structural–acoustic optimization, there are very efficiently performing methods for interior problems as long as local quantities account for the objective function. However, global quantities and large frequency range solutions require much more computation and future work should focus at their development. It can be one idea to continue with model reduction techniques and frequency interpolation as discussed in Chapter 3.

Another field which belongs to acoustic and structural–acoustic optimization is known as sensitivity analysis. For further discussion of this topic in the context of structural–acoustic optimization, we refer to Marburg [63] and the more recent paper [30].

In the book, Koopmann and Fahnline [57] emphasized that efficient analysis techniques account for the basis of optimization. Though regarding the entire analysis, fast solutions of the fluid problem or even of the coupled problem involving radiation into open space and considering frequency ranges are strongly desired. There are a couple of ideas to perform efficient analysis over a large frequency range. Insofar, this book can be understood as a contribution to structural–acoustic optimization.

0.7.5 Other Problems

There are many other areas in which further developments of numerical methods are relevant and which future developments will cover. We will touch on three more areas in what follows but, of course, it is impossible to mention all requirements for future development of finite and boundary element methods in acoustics.

Among these areas, we refer to more general multifield analyses. In addition to structural–acoustic analysis, this includes acoustics coupled with electromagnetic fields, thermal fields, flow and porous media. Recent work on sound propagation in and radiation from porous media was published by Bermúdez et al. [11, 12], by Atalla et al. [8, 87] and by Tanneau et al. [98]. The book by Schanz describes wave propagation in poroelastic and viscoelastic materials [92]. Batifol et al. [9] considers combined passive and active noise control and, thus, combines poroelastic and electromagnetic–acoustic problems, for the latter see [31, 51, 52, 93]. Sound propagation in flow by solution of Galbrun's equation is receiving increasing attention, cf. [102]. Numerical simulation of sound propagation in flow together with computational fluid dynamics is an attractive field for simulation of musical wind instruments. However, it seems as if this has only been discovered recently, cf. [89].

Another hot topic is modeling. The authors are aware of only a few papers on structural–acoustic modeling [67]. It seems to be more popular to combine modeling techniques with flow or with other multifield analysis [103] and with applications in bio–mechanics/–acoustics [32, 84], also [90]. Yet, the authors are not aware how to categorize modeling in a general way. Usually this subject appears very variable and ad hoc. Hopefully, the future will provide us with more general modeling strategies.

Identification is closely related to modeling. This field is often referred to as inverse acoustics. Chapter 20 is reviewing this area in more detail. Most activities in this field are related to identification of boundary source terms, i.e. the distribution of the particle velocity, either v_f or v_s, see for example [105, 109] and Nolte [79]. A methodology which is not so common but, apparently, receives substantial attention by practitioners, determines admittance boundary conditions by using methods of inverse acoustics. The authors are aware of only a few, approximately ten contributions, in this field. Essentially, the papers stem from four sources, i.e. Kim and Kim [54], Nava et al. [77], Hepberger et al. [45] and Marburg et al. [3, 4, 69]. Basically, these papers show the potential of admittance identification. Further development might establish these methods as a valuable tool for identification and modeling in structural acoustics.

0.8 Conclusions

In this chapter, we have presented a unified approach to basic finite and boundary element formulations of linear time–harmonic acoustics. Starting with a brief review of literature on finite and boundary elements, i.e. limited to books and review papers, the harmonic wave equation was derived from fundamental axioms of continuum mechanics. After formulation of the boundary value problem of linear time–harmonic acoustics in the classical form, the weak form has been used as the starting point for discretization. Discretization methods in this chapter comprised a Galerkin method for finite elements and for boundary elements and, also, collocation for boundary element discretization. In another section, we have presented and discussed ways to consider sources and incident wave fields. The remaining part of this chapter has dealt with categorization of the subsequent twenty chapters of this book and with a personal view of the authors identifying future requirements and areas of future work in finite and boundary element methods for acoustics.

The remaining chapters of this edition confirm that FEM and BEM can be applied to a large variety of problems in computational acoustics. The authors of these chapters point out advantages of methods and limitations. It is one matter of concern of this book to show that there is no general best method. There maybe methods which are best suited for certain applications. However, it is very likely that these fail for other cases. The subsequent chapters combine a variety of the state of the art of the finite and boundary element research in one edition.

References

1. Ainsworth M, Davies P, Duncan D, Martin P, Rynne B (eds) (2003) Topics in computational wave propagation and inverse problems. Springer–Verlag, Berlin–Heidelberg–New York
2. Amini S, Harris PJ, Wilton DT (1992) Coupled boundary and finite element methods for the solution of the dynamic fluid–structure interaction problem. Springer–Verlag, Berlin
3. Anderssohn R, Marburg S (2007) Non–linear approach to approximate acoustic boundary admittance in cavities. Journal of Computational Acoustics 15:63–79
4. Anderssohn R, Marburg S, Großmann C (2006) FEM–based reconstruction of sound pressure field damped by partially absorbing boundary conditions. Mechanics Research Communication 33:851–859
5. Astley RJ (2000) Infinite elements for wave problems: A review of current formulations and an assessment of accuracy. International Journal for Numerical Methods in Engineering 49:951–976
6. Astley RJ, Coyette J–P (2001) The performance of spheroidal infinite elements. International Journal of Numerical Methods in Engineering 52:1379–1396
7. Astley J, Gerdes K, Givoli D, Harari I (eds) (2000) Special issue: Finite elements for wave propagation. Journal of Computational Acoustics 8(1)
8. Atalla N, Sgard F, Amedin CK (2006) On the modeling of sound radiation from poroelastic materials. Journal of the Acoustical Society of America 120:1990-1995
9. Batifol C, Zielinski TG, Ichchou MN, Galland M–A (2007) A finite–element study of a piezoelectric/poroelastic sound package concept. Smart Materials & Structures 16:168–177
10. Belegundu AD, Salagame RR, Koopmann GH (1994) A general optimization strategy for sound power minimization. Structural Optimization 8:113–119
11. Bermúdez A, Ferrín JL, Prieto A (2006) Finite element solution of new displacement/pressure poroelastic models in acoustics. Computer Methods in Applied Mechanics and Engineering 195:1914–1932
12. Bermúdez A, Hervella–Nieto L, Prieto A, Rodríguez R (2007) Validation of acoustic models for time harmonic dissipative scattering problems. Journal of Computational Acoustics 15:95–121
13. Borgiotti GV (1990) The power radiated by a vibrating body in an acoustic fluid and its determination from acoustic measurements. Journal of the Acoustical Society of America 88:1884–1893

14. Burnett D (1994) A three–dimensional acoustic infinite element based on a prolate spheroidal multipole expansion. Journal of the Acoustical Society of America 96:2798–2816
15. Chen ZS, Hofstetter G, Mang HA (1993) A 3d boundary element method for determination of acoustic eigenfrequencies considering admittance boundary conditions. Journal of Computational Acoustics 1:455–468
16. Christensen ST, Sorokin SV, Olhoff N (1998) On analysis and optimization in structural acoustics. Part I: Problem formulation and solution techniques. Structural Optimization 16:83–95
17. Christensen ST, Sorokin SV, Olhoff N (1998) On analysis and optimization in structural acoustics. Part II: Exemplifications for axisymmetric structures. Structural Optimization 16:96–107
18. Ciskowski RD, Brebbia CA (eds) (1991) Boundary elements in acoustics. Computational Mechanics Publications and Elsevier Applied Science, Southampton–Boston
19. Conca C, Planchard J, Vanninathan M (1997) Fluids and periodic structures. John Wiley & Sons, Chichester
20. Cottereau R, Clouteau D, Soize C (2007) Construction of a probabilistic model for impedance matrices. Computer Methods in Applied Mechanics and Engineering 196:2252–2268
21. Coyette J–P, Wynendaele H, Chargin MK (1993) A global acoustic sensitivity tool for improving structural design. Proceedings – SPIE The International Society for Optical Engineering 1923:1389–1394
22. Cremers L, Guisset P, Meulewaeter L, Tournour M (2000) A computer–aided engineering method for predicting the acoustic signature of vibrating structures using discrete models. Great Britain Patent No. GB 2000–16259
23. Crighton D, Dowling A, Ffowcs–Williams J, Heckl M, Leppington F (1992) Modern methods in analytical acoustics (Lecture Notes). Springer–Verlag, London
24. Cunefare KA (1990) The design sensitivity and control of acoustic power radiated by three-dimensional structures. PhD Thesis, The University of Pennsylvania
25. Cunefare KA (1991) The minimum multi–modal radiation efficiency of baffled finite beams. Journal of the Acoustical Society of America 90:2521–2529
26. Dong J, Choi KK, Kim N–H (2004) Design optimization of structural–acoustic problems using FEA–BEA with adjoint variable method. ASME Journal of Mechanical Design 126:527–533
27. Elman HC, Ernst OG, O'Leary DP, Stewart M (2005) Efficient iterative algorithms for the stochastic finite element method with application to acoustic scattering. Computer Methods in Applied Mechanics and Engineering 194:1037–1055
28. Engelstad SP, Cunefare KA, Powell EA, Biesel V (2000) Stiffener shape design to minimize interior noise. Journal of Aircraft 37:165–171
29. Estorff O von (ed) (2000) Boundary elements in acoustics: Advances and applications. WIT Press, Southampton
30. Fritze D, Marburg S, Hardtke H–J (2005) FEM–BEM–coupling and structural–acoustic sensitivity analysis for shell geometries. Computers & Structures 83:143–154
31. Füldner M, Dehe A, Aigner R, Lerch R (2005) Analytical analysis and finite element simulation of advanced membranes for silicon microphones. IEEE Sensors Journal, 5(5):857-863, 2005
32. Gan RZ, Sun Q, Feng B, Wood MW (2006) Acoustic–structural coupled finite element analysis for sound transmission in human ear – Pressure distributions. Medical Engineering & Physics 28:395–404

33. Geers TL (ed) (1998) Iutam Symposium on computational methods for unbounded domains. Proceedings of the IUTAM Symposium held in Boulder, Colorado. Kluwer Academic Publishers, Dortrecht
34. Ghanem R, Sarkarb A (2003) Reduced models for the medium–frequency dynamics of stochastic systems. Journal of the Acoustical Society of America 113:834–846
35. Ghanem R, Spanos P (1991) Stochastic finite elements: a spectral approach. Springer–Verlag, Berlin
36. Givoli D (1992) Numerical methods for problems in infinite domains. Elsevier, Amsterdam
37. Givoli D, Harari I (eds) (1998) Special issue: Exterior problems of wave propagation Computer Methods in Applied Mechanics and Engineering 164(1)&(2)
38. Givoli D (1999) Recent advances in the DtN finite element method for unbounded domains. Archives of Computational Methods in Engineering. State of the art reviews 6:71–116
39. Hambric SA (1995) Approximation techniques for broad–band acoustic radiated noise design optimization problems. Journal of Vibration and Acoustics 117:136–144
40. Hanss M (2005) Applied fuzzy arithmetic. An introduction with engineering applications. Springer–Verlag, Berlin
41. Harari I (2006) A survey of finite element methods for time–harmonic acoustics. Computer Methods in Applied Mechanics and Engineering 195:1594–1607
42. Harari I, Grosh K, Hughes TJR, Malhotra M, Pinsky PM, Stewart JR, Thompson LL (1996) Recent development in finite element methods for structural acoustics. Archives of Computational Methods in Engineering 3:131–309
43. Harari I, Hughes TJR (1992) A cost comparison of boundary element and finite element methods for problems of time–harmonic acoustics. Computer Methods in Applied Mechanics and Engineering 97:77–102
44. Helmig R, Mielke A, Wohlmuth BI (2006) Multifield problems in solid and sluid mechanics (Lecture Notes in Applied and Computational Mechanics.) Springer–Verlag, Berlin–Heidelberg–New York
45. Hepberger A, Volkwein S, Diwoky F, Priebsch H–H (2006) Impedance identification out of pressure data's with a hybrid measurement–simulation methodology up to 1kHz. Proceedings of ISMA 4513–4523
46. Howe MS (2004) Acoustics of fluid–structure interactions. Cambridge Monographs on Mechanics
47. Ihlenburg F (1998) Finite element analysis of acoustic scattering. Springer–Verlag, New York
48. Ihlenburg F (ed) (2003) Special issue: Computational methods for vibroacoustic simulation in the medium-frequency range. Journal of Computational Acoustics 11(2)
49. Ishiyama S–I, Imai M, Maruyama S–I, Ido H, Sugiura N, Suzuki S (1988) The application of ACOUST/BOOM – A noise level prediction and reduction code. SAE–paper 880910:195–205
50. Jog CS (2002) Topology design of structures subjected to periodic load. Journal of Sound and Vibration 253:687–709
51. Kaltenbacher M (2007) Numerical simulation of mechatronic sensors and actuators. Springer–Verlag, Berlin–Heidelberg–New York
52. Kaltenbacher M, Ettinger K, Lerch R, Tittmann B (1998) Finite element analysis of coupled electromagnetic acoustic systems. IEEE Transactions on Magnetics 35:1610–1613
53. Kessels PHL (2001) Engineering toolbox for structural–acoustic design. Applied to MRI–scanners. PhD Thesis, Technische Universiteit Eindhoven

54. Kim Y–K, Kim Y–H (1999) Holographic reconstruction of active sources and surface admittance in an enclosure. Journal of the Acoustical Society of America 105:2377–2383
55. Kirkup SM (1998) The boundary element method in acoustics. Integrated Sound Software, Heptonstall
56. Kollmann FG (1999) Maschinenakustik. Grundlagen, Meßtechnik, Berechnung, Beeinflussung. Springer–Verlag, Berlin–Heidelberg
57. Koopmann GH, Fahnline JB (1997) Designing quiet structures: A sound power minimization approach. Academic Press, San Diego–London
58. Lamancusa JS, Eschenauer HA (1994) Design optimization methods for rectangular panels with minimal sound radiation. AIAA Journal 32:472–479
59. Langley RS, Bremner P (1999) A hybrid method for the vibration analysis of complex structural–acoustic systems. Journal of the Acoustical Society of America 105:1657–1671
60. Mace BR, Worden K, Manson G (eds) (2005) Special Issue: Uncertainty in structural dynamics. Journal of Sound and Vibration 288(3)
61. Magoul'es F (ed) (2005) Special issue: Innovative computational methods for wave propagation. Journal of Computational Acoustics 13(3)
62. Magoul'es F (ed) (2006) Special issue: High performance computing for wave propagation. Journal of Computational Acoustics 14(1)
63. Marburg S (2002) Developments in structural–acoustic optimization for passive noise control. Archives of Computational Methods in Engineering. State of the art reviews, 9:291–370
64. Marburg S (2002) Efficient optimization of a noise transfer function by modification of a shell structure geometry. Part I: Theory. Structural and Multidisciplinary Optimization 24:51–59
65. Marburg S (2006) Normal modes in external acoustics. Part III: Sound power evaluation based on frequency–independent superposition of modes. Acta Acustica united with Acustica 92:296–311
66. Marburg S (ed) (2003) Special Issue: Numerical methods. Journal of Computational Acoustics 11(3)
67. Marburg S, Beer H–J, Gier J, Hardtke H–J, Rennert R, Perret F (2002) Experimental verification of structural–acoustic modeling and design optimization. Journal of Sound and Vibration 252:591–615
68. Marburg S, Dienerowitz F, Fritze D, Hardtke H–J (2006) Case studies on structural–acoustic optimization of a finite beam. Acta Acustica united with Acustica 92:427–439
69. Marburg S, Hardtke H–J (1999) A study on the acoustic boundary admittance. Determination, results and consequences. Engineering Analysis with Boundary Elements 23:737–744
70. Marburg S, Hardtke H–J, Schmidt R, Pawandenat D (1997) An application of the concept of acoustic influence coefficients for the optimization of a vehicle roof. Engineering Analysis with Boundary Elements 20:305–310
71. Marburg S, Nolte B (eds) (2005) Special Issue: Boundary element methods. Journal of Computational Acoustics 13(1)
72. Marburg S, Schneider S (2003) Performance of iterative solvers for acoustic problems. Part I. Solvers and effect of diagonal preconditioning. Engineering Analysis with Boundary Element 27:727–750
73. Mechel FP (ed) (2002) Formulas of acoustics. Springer–Verlag, Berlin–Heidelberg–New York

74. Moens D, Vandepitte D (2006) Recent advances in non–probabilistic approaches for non–deterministic dynamic finite element analysis. Archives of Computational Methods in Engineering 13:389–464
75. Morand HJP, Ohayon R (1995) Fluid structure interaction. John Wiley & Sons, Chichester
76. Munjal ML (ed) (2002) IUTAM Symposium on designing for quietness. Kluwer Academic Publishers, Dortrecht–Boston–London
77. Nava GP, Sato Y, Yasuda Y, Sakamoto S (2006) An inverse method for in–situ estimation of acoustic surface impedance targeting inverse sound rendering in rooms. In: Proceedings of the Inter–Noise 2006, Honolulu (CD)
78. Nédélec JC (2000) Acoustic and electromagnetic equations. Integral representations for harmonic problems. Springer–Verlag, New York
79. Nolte B (2005) Reconstruction of sound sources by means of a proper boundary element formulation. Journal of Computational Acoustics 13:187–201
80. Nolte B, Marburg S (eds) (2007) Special Issue: Efficient numerical methods. Journal of Computational Acoustics 15(1)
81. Nolte B, Marburg S (eds) (2007) Special Issue: Wave–based numerical methods. Journal of Computational Acoustics 15(4)
82. Ochmann M (ed) (2003) Special issue: Computational acoustics: FEM, BEM, structural optimization, and additional numerical methods. Acta Acustica united with Acustica 89(1)
83. Ohayon R, Soize C (1998) Structural acoustics and vibration. Academic Press, New York
84. Oliveira Rosa M, Pereira JC, Grellet M, Alwan A (2003) A contribution to simulating a three-dimensional larynx model using the finite element method. Journal of the Acoustical Society of America 114:2893–2905
85. Otsuru T, Sakuma T, Sakamoto S (2005) Constructing a database of computational methods for environmental acoustics. Acoustical Science and Technology 26:221–224
86. Pal C, Hagiwara I (1993) Dynamic analysis of a coupled structural–acoustic problem. Simultaneous multi–modal reduction of vehicle interior noise level by combined optimization. Finite Elements in Analysis and Design 14:225–234
87. Panneton R, Atalla N (1997) An efficient finite element scheme for solving the three–dimensional poroelasticity problem in acoustics. Journal of the Acoustical Society of America 101:3287–3298
88. Photiadis DM (1990) The relationship of singular value decomposition to wave–vector filtering in sound radiation problems. Journal of the Acoustical Society of America, 88:1152–1159
89. Richter A, Stiller J, Grundmann R (2007) Stabilized discontinuous Galerkin methods for flow–sound interaction. Journal of Computational Acoustics 15:123–143
90. Rusch PA, Dhingra AK (2002) Numerical and experimental investigation of the acoustic and flow performance of intake systems. Journal of Vibration and Acoustics 124:334–339
91. Sarkissian A (1991) Acoustic radiation from finite structures. Journal of the Acoustical Society of America 90:574–578
92. Schanz M (2001) Wave propagation in viscoelastic and poroelastic continua: a boundary element approach. Lecture Notes in Applied Mechanics. Springer–Verlag, Berlin–Heidelberg–New York
93. Schinnerl M, Kaltenbacher M, Langer U, Lerch R (2007) An efficient method for the numerical simulation of magneto–mechanical sensors and actuators. European Journal for Applied Mathematics 18:233–271

94. Shirron JJ, Babuška I (1998) A comparison of approximate boundary conditions and infinite element methods for exterior Helmholtz problems. Computer Methods in Applied Mechanics and Engineering 164:121–139
95. Shorter PJ, Langley RS (2005) Vibro–acoustic analysis of complex systems. Journal of Sound and Vibration 288:669–699
96. Soize C (1998) Reduced models in the medium–frequency range for general external structural–acoustic systems. Journal of the Acoustical Society of America 103:3393–3406.
97. Soize C (1999) Reduced models for structures in the medium–frequency range coupled with internal acoustic cavities. Journal of the Acoustical Society of America 106:3362–3374.
98. Tanneau O, Lamary P, Chevalier Y (2006) A boundary element method for porous media. Journal of the Acoustical Society of America 120:1239–1251
99. Thompson LL, (2006) A review of finite–element methods for time–harmonic acoustics. Journal of the Acoustical Society of America 119:1315–1330
100. Tinnsten M, Esping B, Jonsson M (1999) Optimization of acoustic response. Structural Optimization 18:36–47
101. Tolstoy A, Teng Y–C, Shang EC (eds) (2004) Theoretical and computational acoustics 2003. World Scientific Publishing, Singapore
102. Treyssède F, Gabard G, Ben Tahar M (2003) A mixed finite element method for acoustic wave propagation in moving fluids based on an Eulerian–Lagrangian description. Journal of the Acoustical Society of America 113:705–716
103. Tsay H–S, Yeh F–H (2006) Finite element frequency–domain acoustic analysis of open–cell plastic foams. Finite Elements in Analysis and Design 42:314–339
104. Turkel E (ed) (1998) Special issue: Absorbing boundary conditions. Applied Numerical Mathematics 27(4)
105. Visser R (2004) A boundary element approach to acoustic radiation and source identification. PhD thesis. University of Twente, Enschede
106. Vlahopoulos N, Zhao X (1999) Basic development of hybrid finite element method for midfrequency structural vibrations. AIAA Journal 37:1495–1505
107. Wan X, Karniadakis GE (2006) Multi–element generalized polynomial chaos for arbitrary probability measures. SIAM Journal of Scientic Computing 28:901–928
108. Whiteman JR (ed) (2000) The mathematics of finite elements and applications X. MAFELAP 1999. Elsevier, Oxford
109. Williams EG (1999) Fourier acoustics. Sound radiation and nearfield acoustical holography. Academic Press, San Diego–London
110. Wu TW (ed) (2000) Boundary element acoustics: Fundamentals and computer codes. WIT Press, Southampton
111. Xiu D, Karniadakis GE (2002) The Wiener–Askey polynomial chaos for stochastic differential equations. SIAM Journal of Scientic Computing 24:619–644

Part I

FEM: Numerical Aspects

1

Dispersion, Pollution, and Resolution

Isaac Harari

Faculty of Engineering, Tel Aviv University, 69978 Ramat Aviv, Israel
harari@eng.tau.ac.il

Summary. Standard computational methods are unable to cope with wave phenomena at short wave lengths due to resolutions required to control dispersion and pollution errors, leading to prohibitive computational demands. Dispersion analysis provides a useful tool in both characterizing the performance of conventional methods and designing improved techniques. Since computation naturally separates the scales of a problem according to the mesh size, multiscale considerations provide a useful framework for viewing these difficulties and developing methods to counter them.

1.1 Introduction

Computational acoustics has been an area of active research for almost half a century, also related to other fields of application, such as geophysics, meteorology, electromagnetics, etc. The challenge of efficient computation, at high wave numbers in particular, has been designated as one of the problems still unsolved by current numerical techniques [61]. Standard computational methods are unable to cope with wave phenomena at short wave lengths because they require a prohibitive computational effort in order to resolve the waves and control numerical dispersion errors. The failure to adequately represent subgrid scales misses not only the fine–scale part of the solution, but often causes severe pollution of the solution on the resolved scale as well. Many current discretization techniques are being developed in response to the challenge of controlling such errors effectively.

The Helmholtz equation describes time–harmonic acoustic and electromagnetic waves. The indefinite Helmholtz operator may lose ellipticity with increasing wave number, since in that case its weak form no longer induces a norm. This is related to the pollution effect, in which Galerkin finite element solutions with continuous low–order piecewise polynomials differ significantly from the best approximation [2], due to spurious dispersion in the computation, unless the mesh is sufficiently refined. In practical terms, pollution leads to a substantial increase in the cost of the finite element solution of the Helmholtz equation at higher wave numbers.

1.2 Conventional Galerkin Approximation

Domain–based methods such as finite elements are suitable for solving interior problems as well as exterior radiation and scattering problems in bounded domains that have been truncated by absorbing boundary conditions, infinite elements, or absorbing layers, see, e.g., the book [39]. Historically, boundary element schemes based on integral equations [12, 45, 55], which do not require special treatment of the unbounded domain, were the preferred computational method in acoustics due to the reduced dimensionality of the domain leading to fewer degrees of freedom. Over a decade ago it became apparent that finite elements can be more efficient on large–scale problems because of the structure of their matrices in comparison to the global nature for boundary element discretization [8, 31]. While this conclusion becomes less obvious with the recent incorporation of fast multipole methods [11, 26], finite element methods retain the advantages of robustness and ease of integration with other discrete models in coupled problems.

1.2.1 Formulation

Let $\Omega \subset \mathbb{R}^d$ be a d–dimensional region with smooth boundary Γ. The outward unit vector, normal to Γ, is denoted $\mathbf{n}$. For simplicity, consider the following Neumann boundary–value problem: find $p\colon \overline{\Omega} \to \mathbb{R}$ such that

$$\mathcal{L}p = f \qquad \text{in } \Omega \tag{1.1}$$

$$\frac{1}{a}\frac{\partial p}{\partial n} = h \qquad \text{on } \Gamma\,. \tag{1.2}$$

Here, $f\colon \Omega \to \mathbb{R}$ and $h\colon \Gamma \to \mathbb{R}$ are given functions; $\mathcal{L}p = -\Delta p - k^2 p$ is the self–adjoint, indefinite, Helmholtz operator with given wave number k, and a is a given coefficient. Generalization of the following results to problems with other types of boundary conditions, including radiation conditions representing unbounded domains, is straightforward.

The standard variational form is stated in terms of the set of functions $\mathcal{V} = H^1(\Omega)$: find $p \in \mathcal{V}$ such that

$$\int_\Omega (\nabla v \cdot \nabla p - v\, k^2 p)\, d\Omega = \int_\Omega v\, f\, d\Omega + \int_\Gamma v\, h\, d\Gamma, \quad \forall v \in \mathcal{V}\,. \tag{1.3}$$

The left–hand side is related to the Helmholtz operator via integration by parts, for sufficiently smooth $p, v \in \mathcal{V}$.

Remark 1. The Hilbert space $H^1(\Omega)$, the class of admissible functions for this problem, consists of all functions which have square integrable derivatives. Typical, continuous, piecewise polynomial, finite element functions are admissible in this sense.

Remark 2. The weak operators that appear in (1.3) can be related to inner products that induce norms, by conjugating one of the complex–valued functions v or p. In typical finite element implementations, in which such functions are represented by linear combinations of complex–valued coefficients and real–valued basis functions, this operation is of no consequence. There is a difference in specialized methods that use complex–valued bases.

The Galerkin finite element approximation is constructed by replacing the functions p and v with finite dimensional counterparts, typically continuous, piecewise polynomials. The support of these functions is defined by a mesh partition of the domain into non–overlapping regions (element domains). This set of functions is denoted $\mathcal{V}^h \subset \mathcal{V}$. The discrete problem is: find $p^h \in \mathcal{V}^h$ such that

$$\int_{\Omega} \left(\nabla v^h \cdot \nabla p^h - v^h\, k^2 p^h\right) d\Omega = \int_{\Omega} v^h\, f\, d\Omega + \int_{\Gamma} v^h\, h\, d\Gamma, \quad \forall v^h \in \mathcal{V}^h . \tag{1.4}$$

This approach is optimal for the Laplace operator in the sense that it minimizes the error in the energy norm — the H^1 seminorm in this case. This property assures good performance of the computation at any mesh resolution, i.e., high coarse–mesh accuracy. However, good numerical performance at any mesh resolution is not guaranteed by the standard finite element method for the Helmholtz equation. In this case, finite element computation can become prohibitively expensive in the presence of rapid oscillations.

Remark 3. In bounded domains, the indefinite Helmholtz equation may admit resonance, namely non–unique solutions at discrete wave numbers. Due to spurious dispersion in computation, this is manifested in the form of ill–conditioned discrete equations at *adjacent* wave numbers, in which the exact solution is unique. With improved approximation, these two wave numbers draw closer. Solutions in unbounded domains are unique at all wave numbers. In order to employ domain–based discretization methods such as those being considered, a problem is formulated in a bounded domain. Suitable techniques for defining this problem must guarantee the uniqueness of its solution.

1.2.2 Spurious Dispersion

Dispersion analysis of numerical methods for the Helmholtz equation (1.1) examines the dependence of the numerical error on mesh resolution as well as mesh orientation, by comparison to exact, free–space solutions of the constant–coefficient, homogeneous equation, typically in the form of plane waves. These ideas can be extended to cylindrical and spherical waves as well [28].

The analysis measures the performance of the standard Galerkin method, but also provides a tool for the design of improved methods. Dispersion analysis is used to determine the parameters of least–squares stabilized methods [30], preferred series representations for auxiliary functions and guidelines for the number of terms to be retained in bubble–enriched methods [29], and preferred element configurations in Lagrange–multiplier discontinuous methods [35]. Dispersion analysis is also used to improve the performance of the perfectly matched layer [34].

A homogeneous, isotropic continuum is nondispersive. This is usually no longer the case for discrete representations. Each numerical method is characterized by an approximate wave number which accounts for numerical dispersion.

One–dimensional Analysis

Plane wave dispersion analysis of the standard Galerkin method in one dimension represents the case of a uniform *d*–dimensional Cartesian mesh, aligned with the direction of propagation.

Orient the positive x–axis in the direction of propagation. An exact, free–space solution to the Helmholtz equation (1.1) in one dimension, with constant wave number and in the absence of sources, is a plane wave

$$p = \mathrm{e}^{\mathrm{i}kx} \,. \tag{1.5}$$

Here, $\mathrm{i} = \sqrt{-1}$ is the imaginary unit. Consider a uniform mesh of two–noded linear elements of size h, with nodes at $x_n = nh$, $n \in \mathbb{Z}$. Nodal values of the exact solution (1.5) are

$$p(x_n) = \left(\mathrm{e}^{\mathrm{i}kh}\right)^n \,. \tag{1.6}$$

Dispersion analysis considers corresponding nodal values of finite element solutions in the form

$$p_n = \left(\mathrm{e}^{\mathrm{i}k^h h}\right)^n \,. \tag{1.7}$$

Here, $p_n = p^h(x_n)$. The following analysis determines the dependence of the approximate wave number k^h on the mesh resolution

$$G = \frac{2\pi}{kh} \tag{1.8}$$

indicating the number of nodal points per wavelength, in terms of the non–dimensional product kh.

Due to the local support of finite element shape functions, the global coefficient matrix is banded (all the non–zero entries are concentrated in a relatively narrow band along the main diagonal). Each internal node in a mesh of linear elements is shared by two elements. The banded matrix is tridiagonal in one dimension, i.e., the equation for each internal node depends only on the values of the nodes belonging to those two elements — three nodes altogether forming a typical three–node stencil.

The Galerkin method (1.4) yields the following algebraic equation at each interior node

$$-p_{n_1} + 2p_n - p_{n+1} - \frac{(kh)^2}{6}\left(p_{n_1} + 4p_n + p_{n+1}\right) = 0 \,. \tag{1.9}$$

Substituting the plane wave form (1.7) leads to

$$(kh)^2 = 6\,\frac{1-\cos(k^h h)}{2+\cos(k^h h)} \,. \tag{1.10}$$

Thus, the Galerkin dispersion relation for a Cartesian mesh aligned with a plane wave is

$$k^h h = \arccos\left[\frac{1-(kh)^2/3}{1+(kh)^2/6}\right] . \tag{1.11}$$

This relation represents propagation ($k^h \in \mathbb{R}$) in the range $kh < \sqrt{12}$, a resolution of approximately two nodal points per wavelength and over, see (1.8). This is a special case of the results presented in the next section.

Anisotropy

Exact, free–space solutions to the multidimensional Helmholtz equation (1.1) with constant wave number and in the absence of sources, are plane waves

$$p = \mathrm{e}^{\mathrm{i}\mathbf{k}\cdot\mathbf{x}} . \tag{1.12}$$

Here, $|\mathbf{k}| = k$. For a plane wave propagating in the θ direction in two dimensions, $\mathbf{k}^T = k\langle c, s\rangle$, $c = \cos\theta$, and $s = \sin\theta$.

In contrast to exact solutions in isotropic continua, numerical solutions are anisotropic in the sense that they depend on the orientation of the mesh with respect to the direction of propagation, in addition to mesh resolution. This phenomenon is known for both Cartesian [13–16, 27, 46, 51–53, 59] and triangular [16, 51] mesh topologies.

Bilinear Quadrilaterals

The finite element representation of a plane wave parallel to element faces on a uniform three–dimensional mesh of hexahedra is identical to propagation at an arbitrary direction in two dimensions. Consider a uniform two–dimensional mesh of four–noded bilinear quadrilateral elements of size h, aligned with the global axes, with nodal points located at (mh, nh), $m, n \in \mathbb{Z}$. Values of a plane wave in two dimensions (1.12), oriented at an angle θ to the mesh, at the nodal points are

$$p(mh, nh) = \left(\mathrm{e}^{\mathrm{i}khc}\right)^m \left(\mathrm{e}^{\mathrm{i}khs}\right)^n . \tag{1.13}$$

Dispersion analysis considers corresponding nodal values of finite element solutions in the form

$$p^h(mh, nh) = \left(\mathrm{e}^{\mathrm{i}k^h hc}\right)^m \left(\mathrm{e}^{\mathrm{i}k^h hs}\right)^n . \tag{1.14}$$

The following analysis determines the dependence of the approximate wave number k^h on the mesh resolution and orientation. Each internal node in a structured planar mesh of bilinear quadrilaterals is shared by four elements. In the banded global coefficient matrix, the equation for each internal node depends only on the values of the nodes belonging to those four elements — nine nodes altogether. Consequently, the dispersion analysis considers such a typical nine–point patch, Figure 1.1.

Substituting the plane wave form (1.14) into the nine–point stencil that arises at any interior node yields the following Galerkin dispersion relation for a Cartesian mesh aligned with element faces parallel to a plane wave

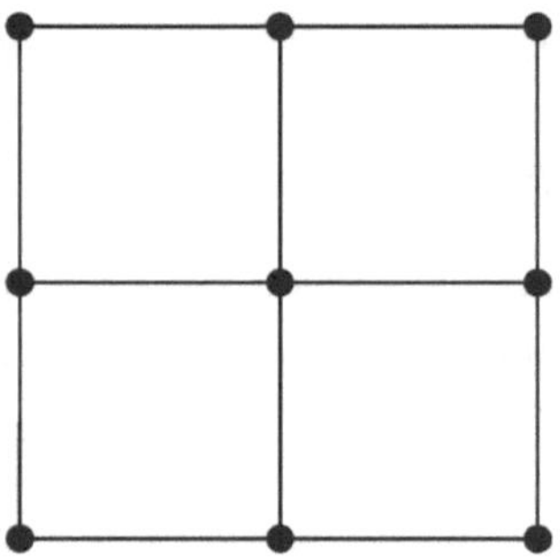

Fig. 1.1 Nine–node Cartesian patch

$$(kh)^2 = 6\,\frac{4 - \cos(k^h hc) - \cos(k^h hs) - 2\cos(k^h hc)\cos(k^h hs)}{[2 + \cos(k^h hc)]\,[2 + \cos(k^h hs)]} \tag{1.15}$$

The variation with respect to the direction of propagation θ is a manifestation of anisotropy.

This is an implicit relation for k^h. The response is a symmetric function of orientation, with a periodicity of $\pi/2$. Consequently, it is sufficient to examine the response between 0 and $\pi/4$. When the mesh is aligned with the wave (e.g., $\theta = 0$) this leads to the one–dimensional plane wave dispersion relation (1.11). The other extreme case occurs when the face diagonals are aligned with the wave (e.g., $\theta = \pi/4$)

$$k^h h = \sqrt{2}\,\arccos\left[\frac{1 - (kh)2/6}{1 + (kh)2/12}\right] \tag{1.16}$$

Values of $k^h h$ satisfying the implicit relation (1.15) for intermediate orientations are obtained numerically, and shown in Figure 1.2, for different levels of the mesh resolution, see (1.8). Note that the bilinear element solution is *more* dispersive when the mesh is aligned *with* waves. The dispersion varies approximately 50% with the orientation.

Linear Triangles

A mesh of square elements provides a canonical form for the analysis of quadrilateral mesh topologies. There is no such canonical form for triangles.

Hexagonal Mesh. Consider a uniform, two–dimensional mesh of linear triangles obtained by periodic application of a hexagonal patch, aligned with the global axes in the manner implied in Figure 1.3. Nodes are located at (mh, nh) as before, but m and n need not be integers. Expressions for a plane wave at the nodal points (1.13) and the form assumed of the finite element solution (1.14) are retained.

Due to the local nature of the finite element equations, the dispersion analysis considers a typical seven–point patch, Figure 1.3. Substituting the plane wave form (1.14) into the seven–point stencil that arises at any interior node yields the following dispersion relation for a hexagonal mesh

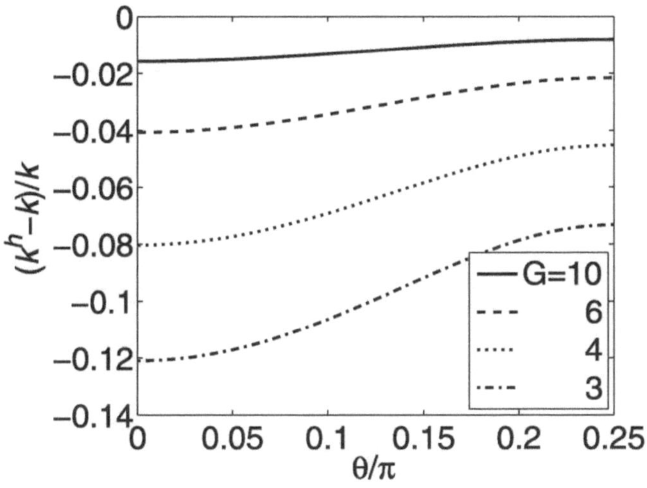

Fig. 1.2 Dispersion of the bilinear element at various resolutions, $G = 2\pi/(kh)$

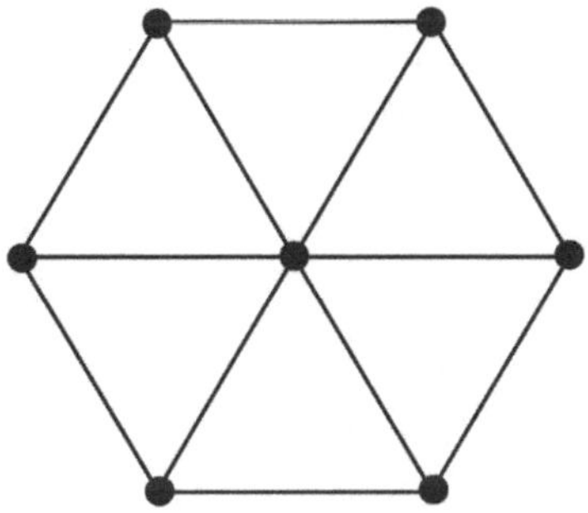

Fig. 1.3 Seven–node hexagonal patch

$$(kh)^2 = 8\,\frac{3 - \left[\cos(k^h hc) + 2\cos(k^h hc/2)\cos(\sqrt{3}k^h hs/2)\right]}{3 + \cos(k^h hc) + 2\cos(k^h hc/2)\cos(\sqrt{3}k^h hs/2)} \tag{1.17}$$

This implicit relation for k^h is a symmetric function of orientation, with a periodicity of $\pi/3$. Consequently, it is sufficient to examine the response between 0 and $\pi/6$. Values of $k^h h$ satisfying the implicit relation (1.17) are obtained numerically, and shown in Figure 1.4, for different levels of the mesh resolution, see (1.8). The dispersion error is similar to the corresponding mean values of the bilinear elements, but the hexagonal mesh exhibits very little anisotropy.

Union Jack. The Cartesian mesh of bilinear elements and the hexagonal mesh of linear triangles are each made up of a single repeating stencil. However, this isn't always so. Consider, for example a mesh of linear triangles in the so–called Union Jack pattern. In this case, the mesh is made up of two repeating patterns: a nine–node patch and a five–node patch, Figure 1.5. In keeping with common practice, e.g. [16],

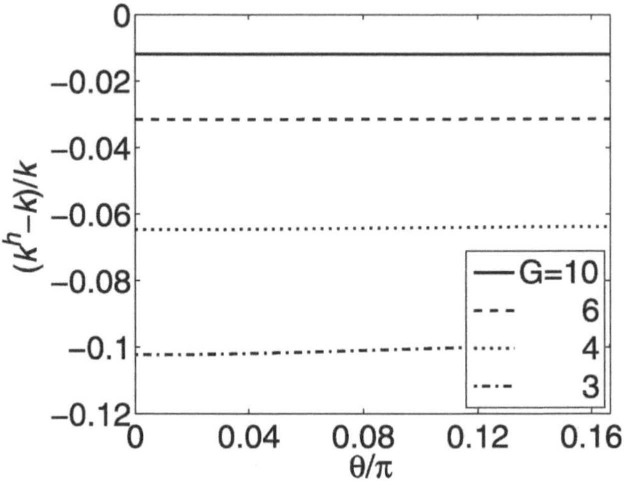

Fig. 1.4 Dispersion of the hexagonal mesh of linear triangles at various resolutions, $G = 2\pi/(kh)$

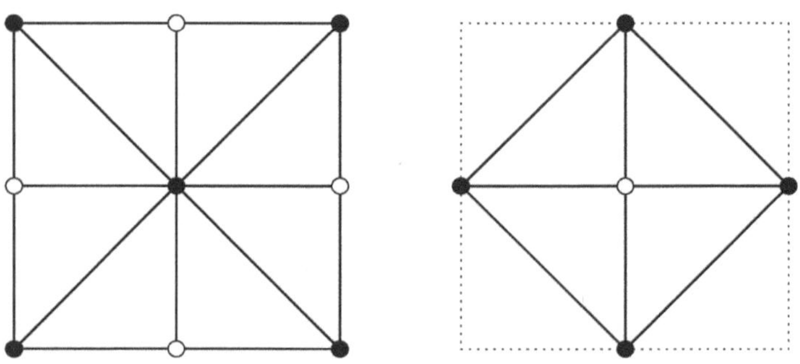

Fig. 1.5 A nine–node patch (left) and a five–node in the Union Jack pattern of triangles.

the element size h to taken as smallest side of the triangle for the purpose of the following analysis.

Two types of nodes are identified in order to account for the two stencils, those at the center of nine–node and five–node patches (full and empty circles in Figure 1.5, respectively). The form assumed of the finite element solution (1.14) is generalized to include different amplitudes for the two types of nodes

$$p^h(mh, nh) = A_1 \left(e^{ik^h hc}\right)^m \left(e^{ik^h hs}\right)^n \tag{1.18}$$

$$p^h(mh, nh) = A_2 \left(e^{ik^h hc}\right)^m \left(e^{ik^h hs}\right)^n \tag{1.19}$$

Each of the two stencil equations is considered a homogeneous linear algebraic equation for the unknown amplitudes, which can be expressed in matrix form

$$\begin{bmatrix} C_{11} & C_{12} \\ \text{sym.} & C_{22} \end{bmatrix} \begin{Bmatrix} A_1 \\ A_2 \end{Bmatrix} = \mathbf{0} \tag{1.20}$$

Here,

$$C_{11} = \left[2 - (kh)^2/3\right] - (kh)^2 \cos(k^h hc)\cos(k^h hs)/6 \tag{1.21}$$

$$C_{12} = -\left[1 + (kh)^2/12\right]\left[\cos(k^h hc) + \cos(k^h hs)\right] \tag{1.22}$$

$$C_{22} = \left[2 - (kh)^2/6\right] \tag{1.23}$$

Nontrivial solutions exist when the coefficient matrix is singular, i.e., when the determinant vanishes. Thus, the dispersion relation is the characteristic equation

$$\begin{aligned} &\left[2 + 2(kh)^2/3 - (kh)^4/72\right]\cos(k^h hc)\cos(k^h hs) \\ &+ \left[1 + (kh)^2/6 + (kh)^4/144\right]\left[\cos^2(k^h hc) + \cos^2(k^h hs)\right] \\ &\qquad - \left[4 - (kh)^2 + (kh)^4/18\right] = 0 \end{aligned} \tag{1.24}$$

This implicit relation for k^h is a symmetric function of orientation, with a periodicity of $\pi/2$. Consequently, it is sufficient to examine the response between 0 and $\pi/4$. Values of $k^h h$ satisfying the implicit relation (1.24) are obtained numerically, and shown in Figure 1.6, for different levels of the mesh resolution, see (1.8). The dispersion error is slightly larger than the corresponding values of the hexagonal mesh and the bilinear elements. The Union Jack mesh exhibits dispersion variation of approximately 35% with the orientation, less than the Cartesian mesh and more than the hexagonal mesh. Note that the Union Jack pattern is less dispersive when the triangle legs are aligned with waves.

1.2.3 Pollution and Resolution

Standard Galerkin finite element solutions with low–order piecewise polynomials differ significantly from the best approximation, due to spurious dispersion in the computation, unless the mesh is sufficiently refined. This phenomenon, related to the indefiniteness of the Helmholtz operator and originally derived by Garding's inequality [6], is known as the pollution effect [2, 3, 24]. In practical terms, this leads to an increase in the cost of the finite element solution of the Helmholtz equation at higher wave numbers.

This section presents a priori error estimates and meshing guidelines for linear elements. Error estimates for *hp* finite elements, as well as a posteriori error estimates, can be found elsewhere [39, 54].

Error Estimates

The following estimates from the literature are stated without proof. These results are rigorous, but provide mostly qualitative guidelines for meshing.

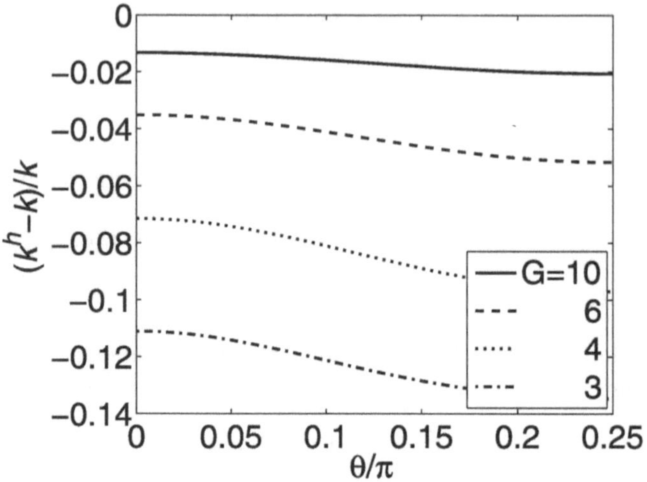

Fig. 1.6 Dispersion of the Union Jack mesh of linear triangles at various resolutions, $G = 2\pi/(kh)$

Dispersion analysis indicates that the finite element error depends on mesh resolution via the non–dimensional product kh, see (1.8). This is a *local* parameter. However, due to the accumulation of dispersion, the resolution is not sufficient to determine accuracy. This phenomenon is called pollution to indicate dependence on a global parameter. This result, as was first stated in [6], can be expressed (under some restrictions to the data) in the L_2 norm as

$$\frac{\|p^h - p\|}{\|p\|} \leq C_1(kh)^2 + C_2 kL(kh)^2 \tag{1.25}$$

Here, L is a length scale characterizing the geometry of the computational domain. (At times this scale is only implied in the mathematical texts that deal with this topic.) The *global* non–dimensional parameter kL is a measure of the number of wavelengths in the computational domain, which clearly effect accuracy. The degree of resolution needed to attain a given level of accuracy increases with the number of wavelengths.

Later work [40] reports results related to a one–dimensional model problem on the unit interval. For a uniform mesh of linear elements of size h, the error of the finite element solution is estimated [41] as

$$\|p^h - p\| \leq C_1(1 + C_2 kL)\left(\frac{h}{\pi}\right)^2 \|p''\| \tag{1.26}$$

This estimate also shows that the error may increase with the number of wavelengths.

A fundamental result provides for a *sharp* bound [42] on the error in the H^1–seminorm

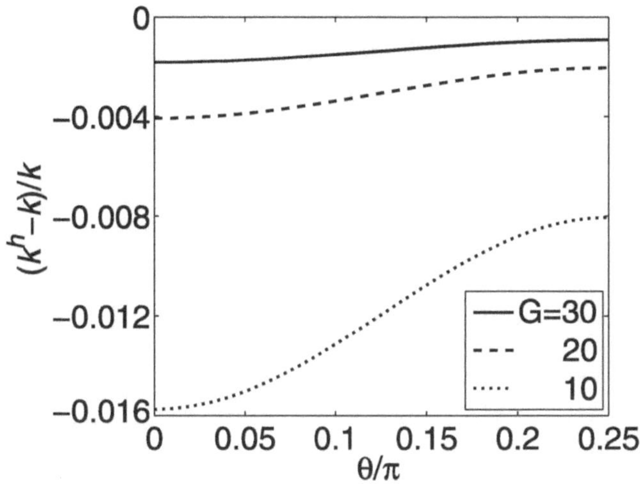

Fig. 1.7 Dispersion of the bilinear element at higher resolutions, $G = 2\pi/(kh)$

$$\frac{|p^h - p|_1}{|p|_1} \leq C_1 kh + C_2 kL(kh)^2 \tag{1.27}$$

The first term arises from the interpolation error. However, the presence of the second term points to potential pollution, i.e., the solution is not quasioptimal, equivalent to the best approximation, unless the mesh is sufficiently refined. For the solution to be in the asymptotic range, where pollution is absent and the error depends on resolution only, the product $(kL)(kh)$ must be small.

Computational experiments, see, e.g. [42], show that this estimate is sharp. Similar a priori estimates have been derived for the *hp* finite element method [43,54].

Meshing Guidelines

The error estimates provide general insight into the performance of computation, and often serve as the basis for meshing guidelines, see, e.g., [42]. In the following, intuitive reasoning based on the understanding of pollution as the accumulation of dispersion error leads to practical guidelines.

Values of $k^h h$ satisfying the dispersion relation for bilinear elements (1.15) at higher levels of mesh refinement than those considered in the dispersion analysis of Section 1.2.2 are shown in Figure 1.7. The mean dispersion errors (with respect to mesh orientation θ) are listed in Table 1.1.

The phase lag associated with the dispersion error at these resolutions is virtually unnoticeable, even over three wavelengths, Figure 1.8. However the pollution error, the accumulation of dispersion, is evident, Figure 1.9. Thus, a resolution of 10 points per wavelength keeps the error at approximately 5%, for propagation of one wavelength, but a resolution of 20 points per wavelength is required for the same accuracy after three wavelengths, and higher resolution is required for longer distances

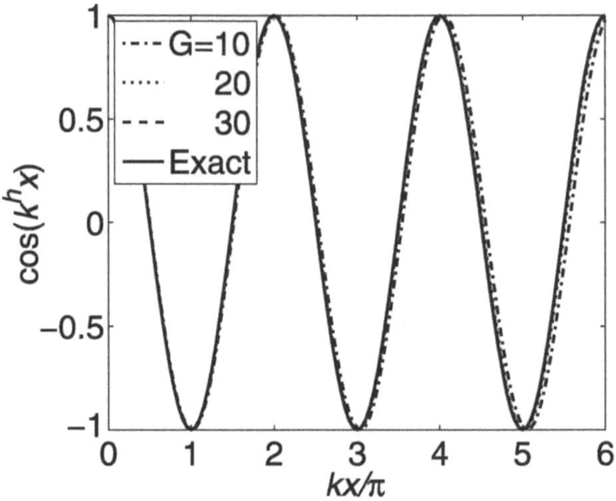

Fig. 1.8 Phase lag associated with the dispersion error of the bilinear element.

Table 1.1 Mean dispersion errors of the bilinear element.

G	$\left\lvert\dfrac{k^h - k}{k}\right\rvert$ (%)
10	1.2
20	0.3
30	0.1

of propagation. This illustration provides a useful, albeit intuitive, tool for deciding on mesh resolutions for practical computation.

1.3 Recent Developments

Standard Galerkin finite element solutions with low–order piecewise polynomials were seen to differ significantly from the best approximation, due to spurious dispersion in the computation, unless the mesh is sufficiently refined. This phenomenon, known as the pollution effect, is related to the indefiniteness of the Helmholtz operator. In practical terms, this leads to an increase in the cost of the finite element solution of the Helmholtz equation at higher wave numbers.

1.3.1 Variational Multiscale Framework

Numerous approaches to alleviating the above deficiency have been proposed. Several such related methods can be derived by the Variational Multiscale (VMS) approach [36, 37]. Higher–order Galerkin [17] and wavelet methods [4] are also used.

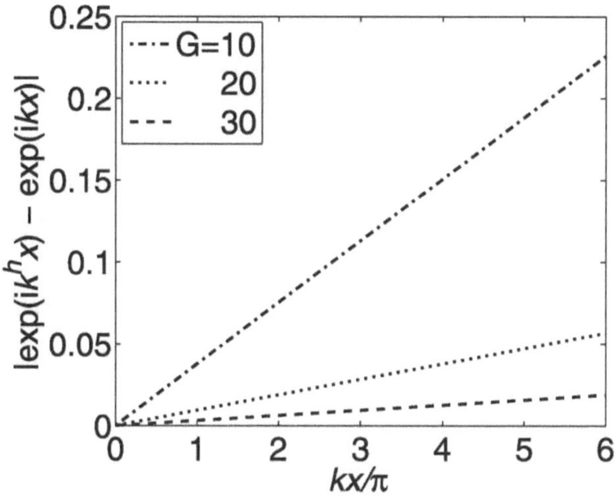

Fig. 1.9 Illustration of the pollution error of the bilinear element.

For simplicity of presentation, consider a homogeneous Neumann problem ($h = 0$). The Galerkin approximation (1.4) may be stated in abstract terms: find $p^h \in \mathcal{V}^h$ such that

$$a(v^h, p^h) = (v^h, f), \quad \forall v^h \in \mathcal{V}^h \tag{1.28}$$

The weak operator is $a(v, p) = (\nabla v, \nabla p) - (v, k^2 p)$, where $(\cdot, \cdot)$ is the $L_2(\Omega)$ inner product.

By the VMS method we consider an overlapping sum decomposition of the solution. The separation of scales is within the continuum description, in reference to the numerical mesh employed. In finite element computation we have

$$u^h = p^{\mathrm{P}} + p^{\mathrm{E}} \tag{1.29}$$

Here, $p^{\mathrm{P}} \in \mathcal{V}^{\mathrm{P}}$ is based on standard, finite element polynomials, representing coarse scales that are resolved by a given mesh, and $p^{\mathrm{E}} \in \mathcal{V}^{\mathrm{E}}$ is an enhancement or enrichment, representing fine or subgrid scales, satisfying the direct sum relationship

$$\mathcal{V}^h = \mathcal{V}^{\mathrm{P}} \oplus \mathcal{V}^{\mathrm{E}} \tag{1.30}$$

Such a decomposition of the solution into a linear part and a bubble was already considered [20]. The determination of the fine scales is key to the multiscale representation. Various implementations approach this issue in different ways, see Section 1.3.2.

Substituting the overlapping sum (1.29) (and its weighting function counterpart) into the variational formulation (1.28), leads to a decomposed form [37]

$$a(v^{\mathrm{P}}, p^{\mathrm{P}}) + (\mathcal{L}^* v^{\mathrm{P}}, p^{\mathrm{E}}) = (v^{\mathrm{P}}, f) \tag{1.31}$$

$$a(v^{\mathrm{E}}, p^{\mathrm{P}}) + a(v^{\mathrm{E}}, p^{\mathrm{E}}) = (v^{\mathrm{E}}, f) \tag{1.32}$$

The second term in the left–hand side of Equation (1.31) is integrated by parts, leading to an interpretation of $\mathcal{L}^* v^{\mathrm{P}}$ as a Dirac distribution on the entire domain, with integrals over element interiors and jump terms integrated across element boundaries [37].

Remark 4. The notation of the adjoint operator $\mathcal{L}^*$ is intended to retain the general context of the equations, although it has no bearing on the self–adjoint Helmholtz operator in practice.

Equation (1.32) provides a formula for the unresolved, fine scales

$$p^{\mathrm{E}} = M^{\mathrm{E}}(\mathcal{L}p^{\mathrm{P}} - f) \tag{1.33}$$

in terms of the integral, generally nonlocal, operator M^{E} which depends on the space of fine scales [37]. This formula is substituted into (1.31) to eliminate the fine scales

$$a(v^{\mathrm{P}}, p^{\mathrm{P}}) + (\mathcal{L}^* v^{\mathrm{P}}, M^{\mathrm{E}} \mathcal{L} p^{\mathrm{P}}) = (v^{\mathrm{P}}, f) + (\mathcal{L}^* v^{\mathrm{P}}, M^{\mathrm{E}} f) \tag{1.34}$$

This equation for the coarse scales includes the nonlocal effect of the fine scales. Various approximations arise from different treatments of the fine scales.

1.3.2 Fine Scales

A simple approach is to employ a bubble representation of the fine scales (1.33), thereby localizing the effect of the fine scales. Solving a homogeneous Dirichlet, element–level, problem for the fine scales is the approach of residual–free bubbles (RFB) [21], with the variational equation

$$a(v^{\mathrm{P}}, p^{\mathrm{P}}) + (\mathcal{L}^* v^{\mathrm{P}}, p^{\mathrm{E}})_{\widetilde{\Omega}} = (v^{\mathrm{P}}, f) \tag{1.35}$$

(Here p^{E} is the bubble–based enrichment and $\widetilde{\Omega}$ denotes the union of element interiors such that $\overline{\Omega} = \overline{\widetilde{\Omega}}$.) A related bubble–based method is nearly optimal Petrov–Galerkin [5]. The explicit integration over element interiors supersedes the distributional interpretation in this case. Employing an element Green's function leads to a similar result [36], related to RFB [7]. The deficiency of the loss of global effects inherent in local approaches may be overcome by employing nonconforming methods [22]. The relationship of VMS methods based on *fine–scale* Green's functions to optimal stabilized methods with global and local character is described in [38].

Stabilized methods of adjoint type, also called "unusual stabilized finite element methods" [23], may be derived in the VMS framework as well, and are related to RFB

$$a(v^{\mathrm{P}}, p^{\mathrm{P}}) - (\mathcal{L}^* v^{\mathrm{P}}, \tau \mathcal{L} p^{\mathrm{P}})_{\widetilde{\Omega}} = (v^{\mathrm{P}}, f) - (\mathcal{L}^* v^{\mathrm{P}}, \tau f)_{\widetilde{\Omega}} \tag{1.36}$$

The structure of the second term on the left–hand side of (1.34) indicates that the mesh–dependent stability parameter τ provides an algebraic approximation of the integral operator M^{E}.

In practice, for the self–adjoint Helmholtz operator, this method is form–identical to the Galerkin/least–squares (GLS) method [32]

$$a(v^{\mathrm{P}},p^{\mathrm{P}}) + (\mathcal{L}v^{\mathrm{P}},\tau\mathcal{L}p^{\mathrm{P}})_{\widetilde{\Omega}} = (v^{\mathrm{P}},f) + (\mathcal{L}v^{\mathrm{P}},\tau f)_{\widetilde{\Omega}} \tag{1.37}$$

(the only difference is in the sign of the stability parameter). Stabilized methods stand out among the numerous improved approaches, by combining substantial improvement in performance with extremely simple implementation. The stability parameter is usually defined by dispersion considerations [32, 33, 59], which don't account for unstructured meshes, although improved performance in computation is not limited to structured meshes [33,56]. There is recent progress in the definition of the stability parameter for distorted elements [44]. The VMS distributional interpretation motivated the development of a stabilized method that includes the inter–element jump terms [53], that are usually omitted in the local approach.

The related method of Galerkin–gradient/least–squares (GGLS)

$$a(v^{\mathrm{P}},p^{\mathrm{P}}) + (\nabla\mathcal{L}v^{\mathrm{P}},\tau^{\mathrm{G}}\nabla\mathcal{L}p^{\mathrm{P}})_{\widetilde{\Omega}} = (v^{\mathrm{P}},f) + (\nabla\mathcal{L}v^{\mathrm{P}},\tau^{\mathrm{G}}\nabla f)_{\widetilde{\Omega}} \tag{1.38}$$

was originally developed in order to stabilize problems governed by the modified Helmholtz equation [19], and was later shown to be effective on the Helmholtz equation as well [30]. The GLS and GGLS methods are quite similar for linear finite elements. In fact, both produce *identical* solutions on structured meshes of linear elements (for constant–coefficient Dirichlet problems with uniform source distributions) [30]. Numerical comparisons of the two methods in more elaborate configurations show that their performance is similar [33].

1.3.3 Plane Wave Methods

An alternative approach that has appeared predominantly in time–harmonic acoustic applications is to base the fine scales on free–space solutions of the homogeneous differential equation (for example, plane waves in the case of the Helmholtz equation). These functions are often readily available, but typically global and hence require specialized treatment in practice. The Generalized finite element method (GFEM) [58] is a recent extension of the partition of unity method (PUM) [49], applied to acoustics [48], in which the free–space homogeneous solutions are multiplied by conventional finite element shape functions. The piecewise polynomial shape functions localize the free–space homogeneous solutions and provide inter–element continuity. In PUM, the product of free–space homogeneous solutions and finite element shape functions constitutes the entire approximation, whereas in GFEM only the fine scales are based on this product, together with conventional finite element functions for the coarse scales, thus alleviating the severe ill–conditioning to which PUM is susceptible. The efficient integration of oscillatory functions is also a crucial issue in these methods.

1.3.4 Discontinuous Methods

Similar ideas for incorporating features of the differential equation in the approximation, but in discontinuous frameworks with specialized treatment for inter–element continuity, go back to the weak element method [25], as well as the recent ultra weak variational formulation [9] and least–squares method [50]. Similarly, the variational theory of complex rays is based on a formulation in which admissible solution satisfy the differential equation, and inter–element continuity is enforced weakly by average flux–type quantities. This method was developed for structural vibrations [47], and its extension to acoustics should be straightforward. Such formulations closely resemble Trefftz approaches, see, e.g., [57]. As in PUM, the special basis functions in these methods *replace* the standard finite element polynomials. Various implementations of hybrid approaches combining finite elements with wave–based methods have been suggested [10, 60]. Some discontinuous formulations employ finite element polynomials instead of oscillatory basis functions [1].

In the discontinuous enrichment method (DEM), standard finite element polynomials are retained for the coarse scales, and enriched within each element by nonconforming free–space homogeneous solutions representing fine scales, with continuity enforced in the variational formulation [18]. The strategy that underlies DEM is based on the assumption that particular solutions are usually well resolved, and thus may be considered coarse scales. The fine scales should therefore contain solutions of the homogeneous partial differential equation. This interpretation of the fine scales differs somewhat from that of conventional multiscale numerical representations. Weak enforcement of inter–element continuity permits the use of free–space solutions, i.e., $\mathcal{V}^{\mathrm{E}}$ is spanned by solutions of

$$\mathcal{L}p^{\mathrm{E}} = 0 \qquad \text{in } \mathbb{R}^d \tag{1.39}$$

that are not already represented in the polynomial basis, leading to relative ease of implementation, yet retaining global, fine–scale effects.

The discontinuous Galerkin approximation is stated in terms of the set of functions $\mathcal{V}^h \subset L_2(\Omega) \cap H^1\left(\widetilde{\Omega}\right)$, with Lagrange multiplier approximations $\lambda^h \in \mathcal{W}^h \subset H^{-1/2}\left(\widetilde{\Gamma}\right)$ defined on the union of element interiors $\widetilde{\Gamma}$ (and corresponding weights μ^h). The hybrid variational formulation that underlies DEM, may be decomposed as

$$a(v^{\mathrm{P}}, p^{\mathrm{P}}) + a(v^{\mathrm{P}}, p^{\mathrm{E}}) - \langle \lambda^h, v^{\mathrm{P}} \rangle \;\; = (v^{\mathrm{P}}, f) \tag{1.40}$$

$$a(v^{\mathrm{E}}, p^{\mathrm{P}}) + a(v^{\mathrm{E}}, p^{\mathrm{E}}) - \langle \lambda^h, v^{\mathrm{E}} \rangle_{\widetilde{\Gamma}} = (v^{\mathrm{E}}, f) \tag{1.41}$$

$$-\langle \mu^h, p^{\mathrm{P}} \rangle - \langle \mu^h, p^{\mathrm{E}} \rangle_{\widetilde{\Gamma}} \qquad\qquad = 0 \tag{1.42}$$

Here, $\langle \cdot, \cdot \rangle$ is the duality pairing between $H^{-1/2}(\Gamma)$ and $H^{1/2}(\Gamma)$. Allowing for discontinuities, the weak operator in this case is $a(v, p) = (\nabla v, \nabla p)_{\widetilde{\Omega}} - (v, k^2 p)$. Element–level basis functions for p^{E} that satisfy (1.39) for constant k are plane waves of the form (1.12). Early numerical testing of DEM exhibits little pollution [18] (the error depends primarily on resolution), perhaps due to the discontinuous nature of the approximation. This striking result could have significant consequences regarding the performance of the method.

1.4 Conclusions

The development of efficient discretization schemes for acoustics is a challenge due to the numerical difficulties that arise in the solution of wave problems, particularly at high wave numbers. Since computation naturally separates the scales of a problem according to the mesh size, multiscale considerations provide a useful framework for viewing these difficulties and developing methods to counter them. Tremendous progress has been made in recent years. The diversity of these contributions demonstrates both the breadth of the numerical methodology which is now applied to acoustic problems, and the many possibilities that exist for future research in this area.

References

1. Alvarez GB, Loula AFD, Dutra do Carmo EG, Rochinha FA (2006) A discontinuous finite element formulation for Helmholtz equation. Computer Methods in Applied Mechanics Engineering 195:4018–4035
2. Babuška I, Ihlenburg F, Paik ET, Sauter SA (1995) A generalized finite element method for solving the Helmholtz equation in two dimensions with minimal pollution. Computer Methods Applied Mechanics Engineering 128:325–359
3. Babuška I, Sauter SA (1997) Is the pollution effect of the FEM avoidable for the Helmholtz equation considering high wave numbers? SIAM Journal on Numerical Analysis 34:2392–2423, (2000) Reprinted in SIAM Review 42:451–484
4. Bao G, Wei GW, Zhao S (2003) Numerical solution of the Helmholtz equation with high wavenumbers. International Journal for Numerical Methods in Engineering 59:389–408
5. Barbone PE, Harari I (2001) Nearly H^1–optimal finite element methods. Computer Methods in Applied Mechanics and Engineering 190:5679–5690
6. Bayliss A, Goldstein CI, Turkel E (1985) On accuracy conditions for the numerical computation of waves. Journal of Computational Physics 59:396–404
7. Brezzi F, Franca, LP, Hughes TJR, Russo A (1997) $b = \int g$. Computer Methods in Applied Mechanics and Engineering 145:329–339
8. Burnett DS (1994) A three–dimensional acoustic infinite element based on a prolate spheroidal multipole expansion. Journal of the Acoustical Society of America 96:2798–2816, (1995) Erratum Ibid 97:2607
9. Cessenat O, Després B (1998) Application of an ultra weak variational formulation of elliptic PDEs to the two–dimensional Helmholtz problem. SIAM Journal on Numerical Analysis 35:255–299
10. Cessenat O, Després B (2003) Using plane waves as base functions for solving time harmonic equations with the ultra weak variational formulation. Journal of Computational Acoustics 11:227–238
11. Chen J–T, Chen K–H (2004) Applications of the dual integral formulation in conjunction with fast multipole method in large–scale problems for 2D exterior acoustics. Engineering Analysis with Boundary Elements 28:685–709, (2004) Erratum Ibid 28:995
12. Chertock G (1971) Integral equation methods in sound radiation and scattering from arbitrary structures. NSRDC Technical Report 3538, David W Taylor Naval Ship Research and Development Center, Bethesda
13. Cherukuri HP (2000) Dispersion analysis of numerical approximations to plane wave motions in an isotropic elastic solid. Computational Mechanics 25:317–328

14. Christon MA (1999) The influence of the mass matrix on the dispersive nature of the semi–discrete second–order wave equation. Computer Methods in Applied Mechanics an Engineering 173:147–166
15. Christon MA, Voth TE (2000) Results of von Neumann analyses for reproducing kernel semi–discretizations. International Journal for Numerical Methods in Engineering 47:1285–1301
16. Deraemaeker A, Babuška I, Bouillard P (1999) Dispersion and pollution of the FEM solution for the Helmholtz equation in one, two and three dimensions. International Journal for Numerical Methods in Engineering 46:471–499
17. Dey S, Shirron JJ, Couchman LS (2001) Mid–frequency structural acoustic and vibration analysis in arbitrary, curved three–dimensional domains. Computers & Structures 79:617–629
18. Farhat C, Harari I, Franca LP (2001) The discontinuous enrichment method. Computer Methods in Applied Mechanics and Engineering 190:6455–6479
19. Franca LP, Dutra do Carmo EG (1989) The Galerkin gradient least–squares method. Computer Methods in Applied Mechanics and Engineering 74:41–54
20. Franca LP, Farhat C (1995) Bubble functions prompt unusual stabilized finite element methods. Computer Methods in Applied Mechanics and Engineering 123:299–308
21. Franca LP, Farhat C, Macedo AP, Lesoinne M (1997) Residual–free bubbles for the Helmholtz equation. International Journal for Numerical Methods in Engineering 40:4003–4009
22. Franca LP, Madureira AL, Valentin F (2002) Modeling multiscale phenomena via finite element methods. In: Mang HA, Rammerstorfer FG, Eberhardsteiner J (eds) Proceedings of the Fifth World Congress (WCCM V), Vienna University of Technology
23. Franca LP, Valentin F (2000) On an improved unusual stabilized finite element method for the advective–reactive–diffusive equation. Computer Methods in Applied Mechanics and Engineering 189:1785–1800
24. Gerdes K, Ihlenburg F (1999) On the pollution effect in FE solutions of the 3D–Helmholtz equation. Computer Methods in Applied Mechanics and Engineering 170:155–172
25. Goldstein CI (1986) The weak element method applied to Helmholtz type equations. Applied Numerical Mathematics 2:409–426
26. Greengard L, Huang J, Rokhlin V, Stephen W (1998) Accelerating fast multipole methods for the Helmholtz equation at low frequencies. IEEE Computational Science and Engineering 5:32–38
27. Harari I (1997) Reducing spurious dispersion, anisotropy and reflection in finite element analysis of time–harmonic acoustics. Computer Methods in Applied Mechanics and Engineering 140:39–58
28. Harari I (2001) Finite element dispersion of cylindrical and spherical acoustic waves. Computer Methods in Applied Mechanics and Engineering 190:2533–2542
29. Harari I Gosteev K (2007) Bubble–based stabilization for the Helmholtz equation. International Journal for Numerical Methods in Engineering 70:1241–1260
30. Harari I, Hughes TJR (1991) Finite element methods for the Helmholtz equation in an exterior domain: Model problems. Computer Methods in Applied Mechanics and Engineering 87:59–96
31. Harari I, Hughes TJR (1992) A cost comparison of boundary element and finite element methods for problems of time–harmonic acoustics. Computer Methods in Applied Mechanics and Engineering 97:77–102
32. Harari I, Hughes TJR (1992) Galerkin/least–squares finite element methods for the reduced wave equation with nonreflecting boundary conditions in unbounded domains. Computer Methods in Applied Mechanics and Engineering 98:411–454

33. Harari I, Magoulès F (2004) Numerical investigations of stabilized finite element computations for acoustics. Wave Motion 39:339–349
34. Harari I, Slavutin M, Turkel E (2000) Analytical and numerical studies of a finite element PML for the Helmholtz equation. Journal of Computational Acoustics 8:121–137
35. Harari I, Tezaur R, Farhat C (2006) A study of higher–order discontinuous Galerkin and quadratic least–squares stabilized finite element computations for acoustics. Journal of Computational Acoustics 14:1–19
36. Hughes TJR (1995) Multiscale phenomena: Green's functions, the Dirichlet–to–Neumann formulation, subgrid scale models, bubbles and the origins of stabilized methods. Computer Methods in Applied Mechanics and Engineering 127:387–401
37. Hughes TJR, Feijóo GR, Mazzei L, Quincy JB (1998) The variational multiscale method – a paradigm for computational mechanics. Computer Methods in Applied Mechanics and Engineering 166:3–24
38. Hughes TJR, Sangalli G (2007) Variational multiscale analysis: the fine–scale Green's function, projection, optimization, localization, and stabilized methods. SIAM Journal on Numerical Analysis 45:539–557
39. Ihlenburg F (1998) Finite element analysis of acoustic scattering. Springer–Verlag, New York
40. Ihlenburg F (2003) The medium–frequency range in computational acoustics: Practical and numerical aspects. Journal of Computational Acoustics 11:175–193
41. Ihlenburg F, Babuška I (1995) Dispersion analysis and error estimation of Galerkin finite element methods for the Helmholtz equation. International Journal for Numerical Methods in Engineering 38:3745–3774
42. Ihlenburg F, Babuška I (1995) Finite element solution of the Helmholtz equation with high wave number. Part I: The h version of the FEM. Computer Mathematics Applied 30:9–37
43. Ihlenburg F, Babuška I (1997) Finite element solution of the Helmholtz equation with high wave number. Part II: The h–p version of the FEM. SIAM Journal on Numerical Analysis 34:315–358
44. Kechroud R, Soulaimani A, Saad Y, Gowda S (2004) Preconditioning techniques for the solution of the Helmholtz equation by the finite element method. Mathematical Computer Simulation 65:303–321
45. Kleinman RE, Roach GF (1974) Boundary integral equations for the three–dimensional Helmholtz equation. SIAM Review 16:214–236
46. Krenk S (1985) Optimal formulation of simple finite elements. Variational Methods in Engineering In: Proceedings of the 2nd International Conference, University of Southampton 9.3–9.16
47. Ladevèze P, Arnaud L, Rouch P, Blanzé C (2001) The variational theory of complex rays for the calculation of medium–frequency vibrations. Engineering Computations 18:193–214
48. Laghrouche O, Bettess P (2000) Short wave modelling using special finite elements. Journal of Computational Acoustics 8:189–210
49. Melenk JM, Babuška I (1996) The partition of unity method finite element method: Basic theory and applications. Computer Methods in Applied Mechanics and Engineering 139:289–314
50. Monk P, Wang D–Q (1999) A least–squares method for the Helmholtz equation. Computer Methods in Applied Mechanics and Engineering 175:121–136
51. Mullen R, Belytschko T (1982) Dispersion analysis of finite element semidiscretizations of the two–dimensional wave equation. International Journal for Numerical Methods in Engineering 18:11–29

52. Oberai AA, Pinsky PM (2000) A numerical comparison of finite element methods for the Helmholtz equation. Journal of Computational Acoustics 8:211–221
53. Oberai AA, Pinsky PM (2000) A residual–based finite element method for the Helmholtz equation. International Journal for Numerical Methods in Engineering 49:399–419
54. Oden JT, Prudhomme S, Demkowicz L (2005) A posteriori error estimation for acoustic wave propagation problems. Archives of Computational Methods in Engineering 12:343–389
55. Shaw RP (1988) Integral equation methods in acoustics. In: Brebbia CA (ed) Boundary Elements X 4:221–244, Computational Mechanics Publications, Southampton
56. Stewart JR, Hughes TJR (1997) h–adaptive finite element computation of time–harmonic exterior acoustics problems in two dimensions. Computer Methods in Applied Mechanics and Engineering 146:65–89
57. Stojek M (1998) Least–squares Trefftz–type elements for the Helmholtz equation. International Journal for Numerical Methods in Engineering 41:831–849
58. Strouboulis T, Babuška I, Copps K (2000) The design and analysis of the generalized finite element method. Computer Methods in Applied Mechanics and Engineering 181:43–69
59. Thompson LL, Pinsky PM (1995) A Galerkin least–squares finite element method for the two–dimensional Helmholtz equation. International Journal for Numerical Methods in Engineering 38:371–397
60. van Hal B, Desmet W, Vandepitte D, Sas P (2003) A coupled finite element–wave based approach for the steady–state dynamic analysis of acoustic systems. Journal of Computational Acoustics 11:285–303
61. Zienkiewicz OC (2000) Achievements and some unsolved problems of the finite element method. International Journal for Numerical Methods in Engineering 47:9–28

2

Different Types of Finite Elements

Gary Cohen[1], Andreas Hauck[2], Manfred Kaltenbacher[3], and Toru Otsuru[4]

[1] INRIA, Domaine de Voluceau, Rocquencourt – BP 105,
78153 Le Chesnay Cedex, France
gary.cohen@inria.fr

[2] Friedrich–Alexander–Universität Erlangen–Nürnberg, Lehrstuhl für Sensorik,
Paul–Gordan–Str. 3/5, 91052 Erlangen, Germany
andreas.hauck@lse.eei.uni-erlangen.de

[3] Friedrich–Alexander–Universität Erlangen–Nürnberg, Lehrstuhl für Sensorik,
Paul–Gordan–Str. 3/5, 91052 Erlangen, Germany
manfred@lse.eei.uni-erlangen.de

[4] Oita University, 700 Dannoharu, Oita 870–1192, Japan
otsuru@cc.oita-u.ac.jp

Summary. Our contribution concentrates on the numerical solution of acoustic wave problems applying the Finite Element Method (FEM). After a short introduction, we provide a detailed discussion about the concepts of FEM and apply it to the acoustic wave equation. We introduce the standard (Lagrangian) FE basis functions for approximating the continuous acoustic pressure as well as an alternative approach utilizing spline functions. In a second part we focus on two methods to improve the accuracy of the FE solution: (i) increasing the order of the FE basis functions using higher order elements (p–FEM) and (ii) applying spectral FEM to a mixed formulation of the wave equation.

2.1 Introduction

Among numerical methods of many kinds, FEM is advantageous in its broad range of adaptability, e.g. its applicability to problems with heterogeneous materials, exhibiting complex geometric structures and different kinds of nonlinearities. In the 1950s, FEM was firstly developed as "matrix method of structural analysis", and after that, FEM has been applied to solve various problems of acoustics, see e.g., [2, 16, 25, 26, 32, 42, 49].

Generally speaking, FEM needs a volume discretization of the computational domain of the system under consideration, which results in a remarkable increase of number of unknowns, especially when a three dimensional domain is considered. However, the arising matrices in a FE formulation can be evaluated rather simple and have a sparse structure, which makes their computation efficient and easy to optimize using further parallel/vector processing. In many cases, iterative algebraic solvers help to reduce the memory requirements and to enlarge the computational efficiency when solving the arising algebraic system of equations, see, e.g., [24, 44, 54].

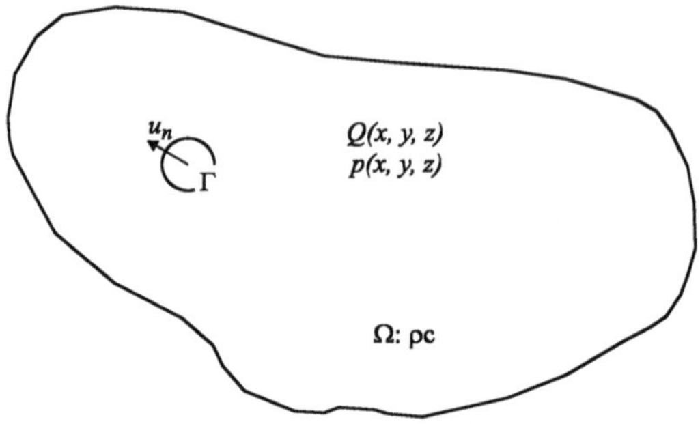

Fig. 2.1 Sound field Ω.

The accuracy of FE formulations is characterized by the numerical interpolation and pollution errors [33], see also Chapter 1 of this book. Enhanced FE formulations concentrate on reducing these errors while keeping the efficiency of standard FEM. Within this context, we will discuss the application of p–FEM and spectral FEM to computational acoustics. For an overview of different enhanced FE formulations we refer, e.g., to [28, 43].

The rest of our contribution is organized as follows: First, we perform a discussion of the basics of the FEM applied to non-dissipative and dissipative sound fields. The derivations will be based on the energy functionals of the acoustic field and we will introduce Lagrange polynomial as well as spline polynomial basis functions to approximate the acoustic pressure. An eigenfrequency analysis and sound field computation of a reverberation chamber will demonstrate the requirements for the spatial discretization in order to obtain a given accuracy. In the following section, we generalize the concept of basis functions to hierarchical functions of higher order (p–FEM). Therewith, we demonstrate the superiority of p–FEM compared to standard FEM concerning the accuracy. In the last section, we present an enhanced mixed spectral method for the efficient computation of the acoustic wave equation in the time domain. Therewith, we will demonstrate, how the special construction of this method results in a computational scheme, which is even faster and less memory consuming than traditional finite difference methods.

2.2 Formulation of Finite Element Method for Acoustical Problems

2.2.1 Sound Field without Dissipation

In a three dimensional domain Ω, see Figure 2.1, with a given sound pressure distribution $p(x, y, z)$, the kinetic and potential energies of sound at an angular frequency ω are of the following form

$$T = \frac{1}{2}\frac{1}{\rho\omega^2}\int_{\Omega} (\nabla p \cdot \nabla p)\ \mathrm{d}V \tag{2.1}$$

and

$$U = \frac{1}{2}\frac{1}{\rho c^2}\int\limits_{\Omega} p^2 \,\mathrm{d}V\,. \tag{2.2}$$

Here, T and U denote the kinetic and potential energy with ρ and c being the density and speed of sound, respectively. For simplification and a better understanding, we restrict our derivation to a computational domain, where c and ρ are no functions of the space (homogeneous media). The work W performed by an external force with sound pressure $p(x,y,z)$ and displacement $u_n(x,y,z)$, that is normal to the surface area Γ, can be written as

$$W = \int\limits_{\Gamma} u_n \, p \,\mathrm{d}S\,. \tag{2.3}$$

Then, the total energy Π is

$$\Pi = U - T - W, \tag{2.4}$$

and, based on the principle of minimum potential energy, i.e. $\delta\Pi = 0$, one can derive the governing equations of the sound field. It is also possible to derive the same result as a weak form solution by applying Galerkin's method to the wave equation, see Section 2.4.

As already mentioned, the use of the FE method requires the discretization of the whole computational domain Ω. For the discretization triangular as well as quadrilateral finite elements are used in 2D and tetrahedral as well as hexahedral finite elements in 3D. Hereafter, by help of the FEM, the sound pressure p at an arbitrary point Q within an element e can be approximated using both nodal sound pressure vector $\mathbf{p}_e$ and interpolation function (also called shape function) vector $\boldsymbol{\phi}$ as

$$p(x,y,z) \simeq \sum_{a=1}^{n_{\mathrm{en}}} \phi_a(x,y,z) p_a^e = \boldsymbol{\phi}^T \mathbf{p}_e\,. \tag{2.5}$$

In (2.5) n_{en} denotes the number of element nodes, ϕ_a the shape function for node a and p_a^e the acoustic pressure corresponding to finite element node a. With (2.5), the energy terms T and U are rewritten to

$$\begin{aligned} T = &\sum_{e=1}^{n_{\mathrm{e}}} \left\langle \frac{1}{2}\frac{1}{\rho\omega^2} \int\limits_{\Omega_e} \left\{ \left[\frac{\partial}{\partial x}\left(\boldsymbol{\phi}^T\mathbf{p}_e\right)\right]^2 + \left[\frac{\partial}{\partial y}\left(\boldsymbol{\phi}^T\mathbf{p}_e\right)\right]^2 \right.\right. \\ &\left.\left. + \left[\frac{\partial}{\partial z}\left(\boldsymbol{\phi}^T\mathbf{p}_e\right)\right]^2 \right\} \mathrm{d}V \right\rangle \\ = &\sum_{e=1}^{n_{\mathrm{e}}} \mathbf{p}_e^T \left[\frac{1}{2}\frac{1}{\rho\omega^2} \int\limits_{\Omega_e} \left(\frac{\partial\boldsymbol{\phi}}{\partial x}\frac{\partial\boldsymbol{\phi}^T}{\partial x} + \frac{\partial\boldsymbol{\phi}}{\partial y}\frac{\partial\boldsymbol{\phi}^T}{\partial y} + \frac{\partial\boldsymbol{\phi}}{\partial z}\frac{\partial\boldsymbol{\phi}^T}{\partial z} \right) \mathrm{d}V \right] \mathbf{p}_e \end{aligned} \tag{2.6}$$

$$U = \sum_{e=1}^{n_{\mathrm{e}}} \left(\mathbf{p}_e^T \frac{1}{2}\frac{1}{\rho c^2} \int\limits_{\Omega_e} \boldsymbol{\phi}\boldsymbol{\phi}^T \,\mathrm{d}V \,\mathbf{p}_e \right) \tag{2.7}$$

and the work W gives

$$W = \sum_{e=1}^{n_e} \left(\mathbf{p}_e^T u_n \int_{\Gamma_e} \phi \, \mathrm{d}S \right) . \tag{2.8}$$

Therewith, the following equation holds for each element e

$$\Pi_e = \frac{1}{2\rho\omega^2} \mathbf{p}_e^T \mathbf{k}^e \mathbf{p}_e - \frac{1}{2\rho} \mathbf{p}_e^T \mathbf{m}^e \mathbf{p}_e - \mathbf{p}_e^T u_n \, \mathbf{w}_e \, . \tag{2.9}$$

In (2.9) $\mathbf{k}^e$ denotes the element stiffness matrix, $\mathbf{m}^e$ the element mass matrix and $\mathbf{w}^e$ the element right hand side, which computes as follows

$$\mathbf{k}^e = \int_{\Omega_e} \left(\frac{\partial \phi}{\partial x} \frac{\partial \phi^T}{\partial x} + \frac{\partial \phi}{\partial y} \frac{\partial \phi^T}{\partial y} + \frac{\partial \phi}{\partial z} \frac{\partial \phi^T}{\partial z} \right) \mathrm{d}V \tag{2.10}$$

$$\mathbf{m}^e = \frac{1}{c^2} \int_{\Omega_e} \phi \phi^T \, \mathrm{d}V \tag{2.11}$$

$$\mathbf{w}^e = \int_{\Gamma_e} \phi \, \mathrm{d}S \, . \tag{2.12}$$

Finally, by incorporating $\delta \Pi = 0$, the following discretized matrix equation can be obtained

$$\left(\mathbf{K} - \omega^2 \mathbf{M} \right) \mathbf{p} = \rho \omega^2 u_n \mathbf{W}, \tag{2.13}$$

where $\mathbf{W}$ is called the distribution vector. The global matrices, $\mathbf{K}$ and $\mathbf{M}$, are obtained by the assembly procedure of the element matrices, $\mathbf{k}^e$ and $\mathbf{m}^e$. For details we refer, e.g., to [9, 31, 35].

2.2.2 Sound Field with Dissipation

When there exists a certain dissipation in the system Ω, we need to take both the dissipation energy J and the adjoint system into account [34]. Since the adjoint system works to represent the increasing energy, which corresponds to the dissipating energy in the original system, the principle of minimum potential energy comes to be applicable. Therewith, the energies are in the form of

$$T = \frac{1}{4\rho\omega^2} \int_{\Omega} \left(\boldsymbol{\nabla} p \cdot \boldsymbol{\nabla} \bar{p}^* + \boldsymbol{\nabla} \bar{p} \cdot \boldsymbol{\nabla} p^* \right) \mathrm{d}V \tag{2.14}$$

$$U = \frac{1}{4\rho c^2} \int_{\Omega} \left(p \bar{p}^* + \bar{p} p^* \right) \mathrm{d}V \tag{2.15}$$

$$W = \frac{1}{4} \int_{\Gamma} \left(u_n \bar{p}^* + \bar{u}_n p^* + u_n^* \bar{p} + \bar{u}_n^* p \right) \mathrm{d}S, \tag{2.16}$$

where $(^{-})$ and $(*)$ denote the adjoint system and complex conjugate, respectively. To model the dissipation of a wall surface, let us consider that an area A_z is an impedance boundary with normal impedance ratio z_{n} relative to air. Then, J can be expressed by

$$J = \frac{\mathrm{i}}{4\omega\rho c} \int\limits_{A_z} \left(\frac{1}{z_{\mathrm{n}}} p\bar{p}^* - \frac{1}{\bar{z}_{\mathrm{n}}} \bar{p}p^* \right) \mathrm{d}A \qquad \text{with } \mathrm{i} = \sqrt{-1}\,, \tag{2.17}$$

and the total energy becomes

$$\Pi = U - T - W + J\,. \tag{2.18}$$

Let us apply the FE method and incorporate $\delta\Pi = 0$. Then the element dissipation matrix $\mathbf{c}^e$ for an element e with impedance z_{n} becomes

$$\mathbf{c}^e = \frac{1}{c} \int\limits_{\Omega_e} \frac{1}{z_{\mathrm{n}}} \boldsymbol{\phi}\boldsymbol{\phi}^T \,\mathrm{d}S\,, \tag{2.19}$$

where the domain of integration is the surface area that has the impedance value of z_{n}. Finally, the discretized matrix equation for the sound field with dissipation can be written as follows

$$(\mathbf{K} + \mathrm{i}\omega\mathbf{C} - \omega^2\mathbf{M})\,\mathbf{p} = \rho\omega^2 u_n \mathbf{W}\,. \tag{2.20}$$

On the other hand, it is also easy for FEM to model the sound field in absorbing media or materials with varying material properties. Here, based on the one dimensional Rayleigh model for porous materials [37], the element matrices, $\tilde{\mathbf{m}}^e$, $\tilde{\mathbf{c}}^e$ and $\tilde{\mathbf{k}}^e$ are defined using the above mentioned element matrices for air ($\mathbf{m}^e$ and $\mathbf{k}^e$) as follows [17, 39]

$$\begin{aligned} \tilde{\mathbf{m}}^e &= \epsilon\, K_{\mathrm{s}} \tilde{\Omega} \mathbf{m}^e \\ \tilde{\mathbf{c}}^e &= \epsilon \frac{R\tilde{\Omega}}{\rho} \mathbf{m}^e \qquad \epsilon = \frac{\mathrm{i}\omega\rho}{(R + \mathrm{i}\omega\tilde{\rho})} \\ \tilde{\mathbf{k}}^e &= \epsilon\, \mathbf{k}^e\,. \end{aligned} \tag{2.21}$$

In (2.21) K_{s}, $\tilde{\Omega}$, R and $\tilde{\rho}$ are respectively structural constant, porosity, flow resistivity and density of the porous material.

2.2.3 Finite Elements for Three–dimensional Acoustics

In order to model a three–dimensional sound field by FEM, the hexahedral element is straightforward to implement and easy applicable in practical problems. Therefore, various elements with 8, 20, 27 or 32 nodes have been proposed, e.g., [16, 18, 34]. Hereafter, hexahedral 8–node, cf. Figure 2.2), and 27–node, cf. Figure 2.3), elements are described and their accuracy within two different applications is compared.

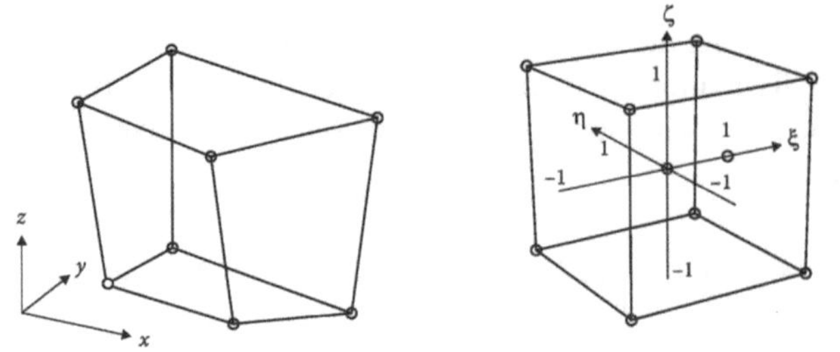

Fig. 2.2 Hexahedral 8–node element; (left): In the global coordinate system. Eight nodes are located at $\mathrm{P}_i(x_i, y_i, z_i)$ on corners; (right): In the local coordinate system. Nodes are located at $\mathrm{P}_i(\xi_i, \eta_i, \zeta_i)$ on corners, where $\xi_i, \eta_i, \zeta_i = 1$ or -1.

Shape Functions

In most cases, the shape of an element is interpolated using the same functions as for the approximation of the physical quantity (in our case the acoustic pressure). In general, Lagrange polynomials given by (2.22) are often used for constructing a finite element[5]

$$\hat{\varphi}_{a_k}(\hat{x}_k) = \frac{\prod\limits_{j=1,(j\neq a_k)}^{r+1} (\hat{x}_k - \hat{x}_{j_k})}{\prod\limits_{j=1,(j\neq a_k)}^{r+1} (\hat{x}_{a_k} - \hat{x}_{j_k})} \qquad a_k = 1, ..., r+1 \,. \tag{2.22}$$

In (2.22) r denotes the order of the Lagrange polynomial, $\hat{x}_k$ the local coordinate in k–direction and $\hat{x}_{j_k}$ the interpolation points, classically regular spaced. The functions φ_{a_k} satisfy the relation

$$\hat{\varphi}_{a_k}(\hat{x}_{j_k}) = \delta_{a_k j_k} \qquad \forall j_k = 1, ..., r+1 \,.$$

The basis functions in 2D (quadrilateral) or 3D (hexahedron) can be constructed by a tensor product of the one–dimensional functions as follows

$$\hat{\phi}_a(\hat{\mathbf{x}}) = \prod_{k=1}^{d} \hat{\varphi}_{a_k}(\hat{x}_k) \qquad \hat{\mathbf{x}} = (\hat{x}_1\,, \hat{x}_2\,, \hat{x}_3)^T = (\xi\,, \eta\,, \zeta)^T = \boldsymbol{\xi}\,. \tag{2.23}$$

In (2.23), d is the space dimension, a the corresponding FE node, in which the basis function is defined, and $\hat{\mathbf{x}}$ the vector of the local coordinates. Therewith, we arrive at he following explicit expression for the basis functions of the hexahedral element

$$\hat{\phi}_a(\xi, \eta, \zeta) = \hat{\varphi}_{a_1}(\xi)\hat{\varphi}_{a_2}(\eta)\hat{\varphi}_{a_3}(\zeta)\,. \tag{2.24}$$

[5]In general, all symbols with ^ are defined on the reference element (local coordinate system)

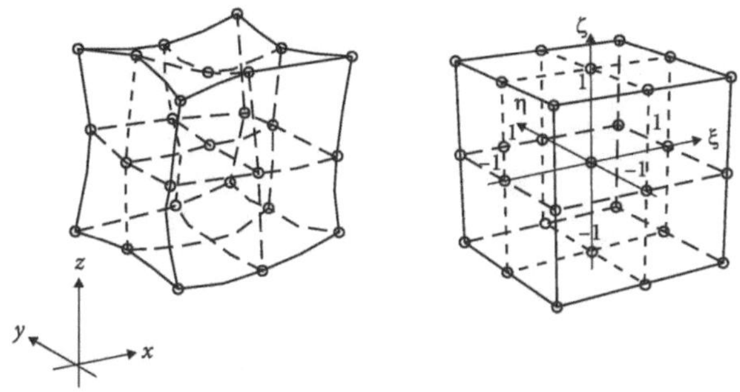

Fig. 2.3 Hexahedral 27–node element; (left): In the global coordinate system, (right): In the local coordinate system. Twenty-seven nodes are located at $P_i(\xi_i, \eta_i, \zeta_i)$, where $\xi_i, \eta_i, \zeta_i = 1$ or -1 (edge or corner), = 0 (otherwise).

By using (2.24), we obtain the basis functions for an isoparametric hexahedral 8–node element by setting $r = 1$, or for a 27–node element by setting $r = 2$. An explicit expression for basis functions of the 8–node element, e.g. looks like

$$\hat{\phi}_a(\hat{\mathbf{x}}) = \frac{1}{8}(1 + \xi_a \xi)(1 + \eta_a \eta)(1 + \zeta_a \zeta), \; a = 1, \ldots, 8 \,, \tag{2.25}$$

where ξ_a, η_a, ζ_a denote the local corner coordinates of the hexahedron.

For the computation of a three–dimensional sound field, as in our case, we can collect all basis functions $\hat{\phi}_a$ with $a = 1, 2, \cdots, (r+1)^3$ of the finite element in a vector $\boldsymbol{\phi}$. Here, $\boldsymbol{\phi}$ corresponds to the vector of weight functions which are used to approximate the element's shape defined by the coordinates (x, y, z), and the acoustic pressure p at any location $\boldsymbol{\xi}$ within this element as follows

$$x(\boldsymbol{\xi}) = \sum_{a=1}^{(r+1)^3} \hat{\phi}_a(\boldsymbol{\xi}) x_a \,, \qquad y(\boldsymbol{\xi}) = \sum_{a=1}^{(r+1)^3} \hat{\phi}_a(\boldsymbol{\xi}) y_a \,, \qquad z(\boldsymbol{\xi}) = \sum_{a=1}^{(r+1)^3} \hat{\phi}_a(\boldsymbol{\xi}) z_a \tag{2.26}$$

$$p(\boldsymbol{\xi}) \simeq p^h(\boldsymbol{\xi}) = \sum_{a=1}^{(r+1)^3} \hat{\phi}_a(\boldsymbol{\xi}) p_a^e = \boldsymbol{\phi}^T \, \mathbf{p}^e \,. \tag{2.27}$$

In (2.26) and (2.27) $\hat{x}_a$, $\hat{y}_a$ and $\hat{z}_a$ denote the local coordinates of the reference element, whereas p and p^h are the analytical and the discretized pressure distribution, respectively.

Especially in practical problem settings, arbitrary shaped elements can occur. Therewith it is important to transform the coordinates within the calculation of the element matrices using (2.10)–(2.12), as well as within the numerical integration (e.g. Gauss–Legendre quadrature) from the local to a global coordinate system using the Jacobian matrix.

In general, finite elements can be constructed based on different kinds of polynomial interpolation functions. Here, a set of finite elements utilizing natural cubic

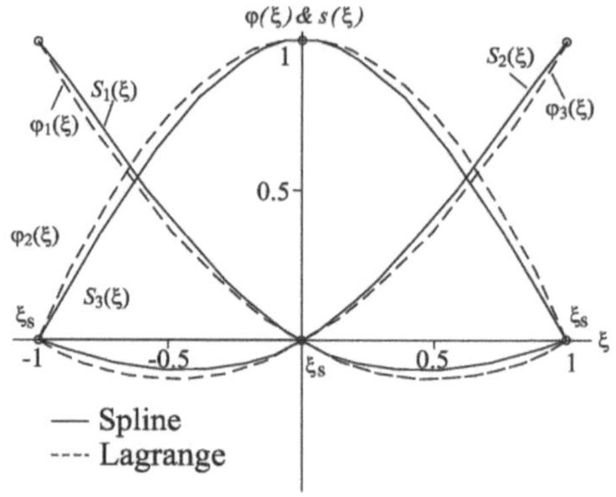

Fig. 2.4 Comparison of shape functions $\varphi(\xi)$ and $S(\xi)$.

splines of degree 3 with continuity C^2 is introduced [40]. Using the spline function S_{a_k}, that is defined by

$$\text{if } \hat{\mathbf{x}}_{a_k} = \pm 1 : \quad S_{a_k}(\hat{x}_k) = \begin{cases} 0.25\hat{x}_k^3 + 0.75\hat{x}_k^2 + 0.5\hat{x}_{a_k}\hat{x}_k & : \hat{x}_k \in [-1,0] \\ -0.25\hat{x}_k^3 + 0.75\hat{x}_k^2 + 0.5\hat{x}_{a_k}\hat{x}_k & : \hat{x}_k \in [\ 0,1] \end{cases} \tag{2.28}$$

$$\text{if } \hat{\mathbf{x}}_{a_k} = \ 0 : \quad S_{a_k}(\hat{x}_k) = \begin{cases} -0.5\hat{x}_k^3 - 1.5\hat{x}_k^2 + 1 & : \hat{x}_k \in [-1,0] \\ 0.5\hat{x}_k^3 - 1.5\hat{x}_k^2 + 1 & : \hat{x}_k \in [\ 0,1] \end{cases}$$

the shape function ϕ_a can be rewritten as

$$\phi_a(\boldsymbol{\xi}) \;=\; S_{a_1}(\xi)\, S_{a_2}(\eta)\, S_{a_3}(\zeta). \tag{2.29}$$

The related values and matrices are obtained following the procedure mentioned above. Figure 2.4 visualizes the two types of shape functions of third order, and in the next section we will discuss the different results when applying these interpolation functions to two acoustic problem settings.

2.2.4 Applications

Eigenfrequency Analysis

Firstly, a simple sound field in a square tube with hard walls, the diameter of which is small enough comparing to sound's wavelength, see Figure 2.5), is investigated by the above discussed FEM. Eigenfrequencies of the sound field can be obtained by solving

$$(\mathbf{K} - \omega^2 \mathbf{M})\, \mathbf{p} = 0\,. \tag{2.30}$$

In (2.30), ω^2 and $\mathbf{p}$ denote eigenvalue and eigenmode of the eigenproblem, respectively. The relative error is defined as

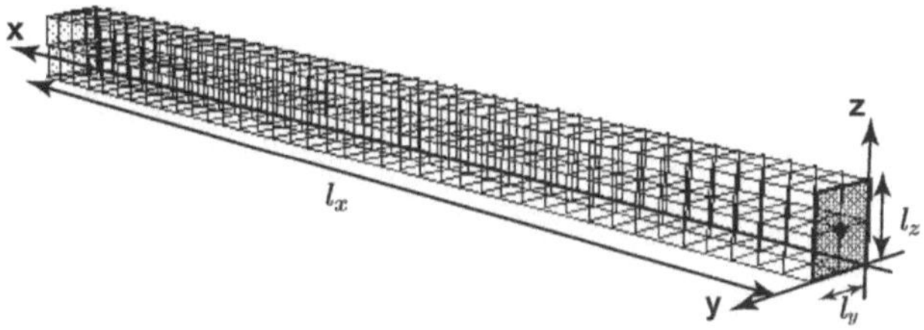

Fig. 2.5 A tube to be analyzed by FEM.

$$\varepsilon = \frac{\omega_{\mathrm{FEM}} - \omega_{\mathrm{analytic}}}{\omega_{\mathrm{analytic}}} \times 100\,\%\,, \tag{2.31}$$

where $\omega_{\mathrm{analytic}}$ and ω_{FEM} denote the eigenfrequencies obtained by analytic solution and by FEM, respectively.

In Figure 2.6 the relative errors are compared that are obtained from eigenfrequency computations conducted by applying the following three element types: hexahedral 8–node (Lin8, for short) and 27–node elements that are constructed based on the Lagrange polynomial function (Lag27), and hexahedral 27–node element (Spl27) constructed based on the spline polynomial functions. As can bee seen, the two kinds of 27–node elements show far better approximation accuracy than that of the 8-node element because of their higher order interpolation functions. Moreover, Spl27 shows a better and more stable nature than the other two elements.

We want to emphasize, that at $\lambda/d = 4$ all the elements resulted in almost the same error (about $10\,\%$). The reason can easily be understood by taking a glance at Figure 2.7. At the frequency, i.e. 5th eigen mode, all three nodes are located on identical lines repeatedly, which results the interpolated waveform to be an identical zig–zag shape. Therefore, there arises no difference between the types of interpolation functions in this case.

Thus, it is important to know the basic behavior of finite elements in relation to frequency–spatial resolution, or λ/d. In general, it can be said that the stable nature of Spl27 shows its excellent applicability for a broad range of problems in acoustics.

Sound Field Analysis of a Reverberation Chamber

To show how accurate the FEM is when three–dimensional sound fields are analyzed, we investigate in the sound field analysis of a reverberation chamber [52].

The setup and FE mesh of the reverberation chamber to be analyzed is illustrated in Figure 2.8. For the computation of the FE matrices in (2.20), we apply the introduced 27–node hexahedron with spline polynomial functions (Spl27). A sound source, radiating 1/3 octave band white noise centered at 200 Hz, is assumed. We solved (2.20) frequency for frequency with a 1 Hz step. For comparison to measured data, we summed up the 1/3 octave band sound pressure level distribution on a plane at 1.2 m height. The degrees of freedom of the FE computation were set to 16767, which satisfies the requirement mentioned above, $\lambda/d > 4.4$, to ensure that the accuracy of the eigenfrequency approximation error is within $1\,\%$. As for the impedance

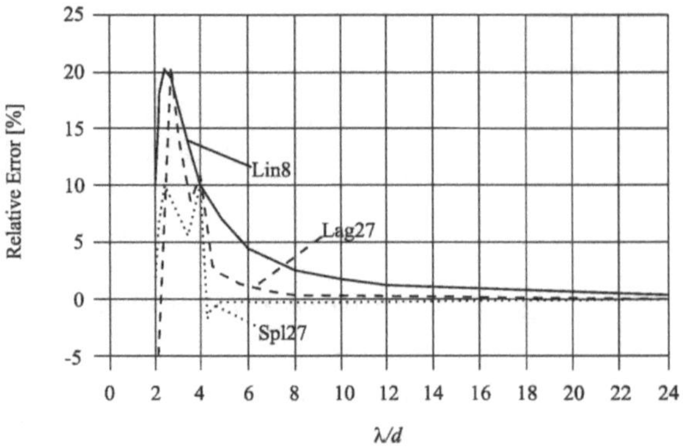

Fig. 2.6 Comparison of eigenvalue approximation error in eigenfrequency evaluation among element types and spatial resolutions. Hexahedral 8– and 27–node elements with Lagrange polynomial functions(Lin8, Lag27) and hexahedral 27–node elements with spline polynomial function(Spl27) are compared. Spatial resolution is defined by wavelength(λ)/nodal–distance(d). The symbol denotes the number of elements in x–direction ($l_x = 1 \sim 20$). [40]

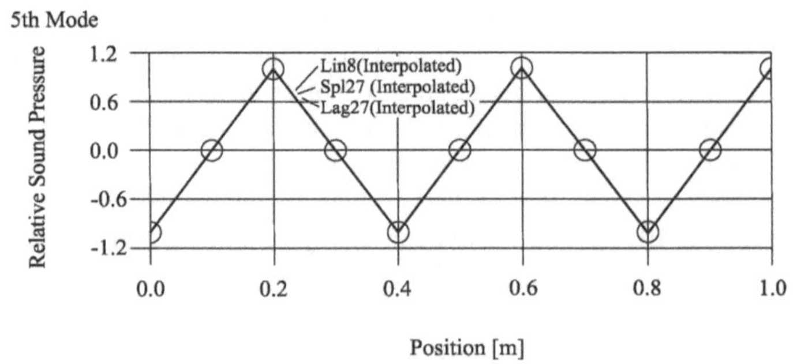

Fig. 2.7 Waveforms of 5th mode interpolated by Lin8, Lag27 and Spl27.

value, the real part of the impedance was calculated from sound absorption coefficient for concrete wall by assuming the impedance's imaginary part to be zero. The result is compared with measured data in Figure 2.9, and as can be seen good agreement was achieved.

2.3 Hierarchical Finite Elements for Acoustics

The so called p–version of the FEM (p–FEM) allows to increase the polynomial order of the basis functions, while keeping the computational mesh fixed, see, e.g., [51]. One can show, that in the case of an analytic solution, there is an exponential convergence rate and we can achieve an optimal control of the dispersion error, which has been extensively studied in [1]. The dispersion error is characterized by the difference between the physical wave number k and the wave number k^h of the FE solution (for details see also the contribution *Dispersion Pollution, and Resolution*,

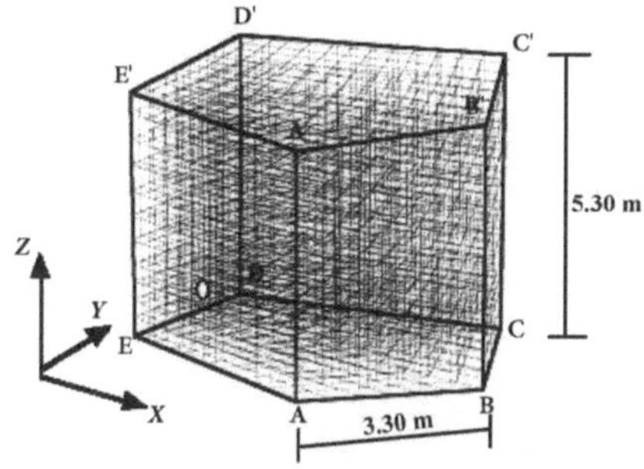

Fig. 2.8 Schematic drawing of a reverberation chamber (at Oita University) and its FE spatial discretization mesh. o: sound source.

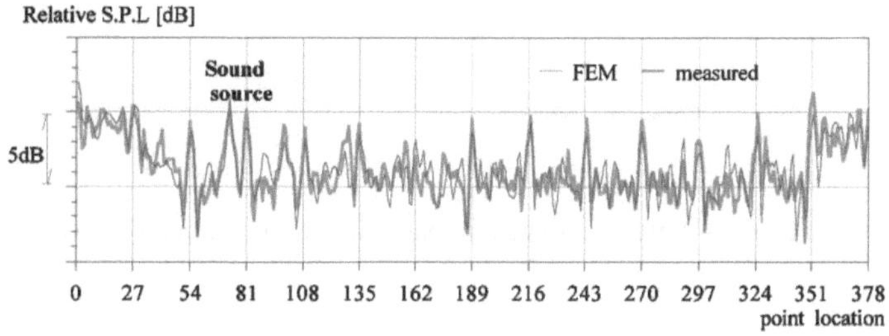

Fig. 2.9 Comparison of relative sound pressure distribution level maps in the reverberation chamber obtained by the finite element method (FEM) and by measurement (measured).

Chapter 1, in this book). The proposed analysis in [1,9] provides a concrete guideline for choosing the order r of the FE basis functions and the mesh size h in order to virtually eliminate the dispersion error. This relation as a function of the polynomial order r of the FE basis is given by

$$r + \frac{1}{2} > \frac{kh}{2} + C(kh)^{1/3} \,. \tag{2.32}$$

In (2.32), C denotes a constant, which can be set to one for practical purpose. In addition, we can precisely define the dispersion error in the small wave number limit $kh << 1$ and the high wave number limit $kh >> 1$ as follows

$$\cos(k^h h) - \cos(kh) = \frac{1}{2}\left[\frac{r!}{(2r)!}\right]^2 \frac{(kh)^{2r+2}}{2r+1} + O(kh)^{2r+4} \qquad kh << 1$$

$$\cos(k^h h) - \cos(kh) = \frac{\sin(kh)}{2}\left[\frac{khe}{2(r+1)}\right]^{2r+1} \qquad kh >> 1 \,.$$

2.3.1 Hierarchical FE Basis Functions

In the FE context, the continuous acoustic pressure p gets approximated by a *finite* dimensional subset of interpolation functions, defined on a FE mesh. In element local

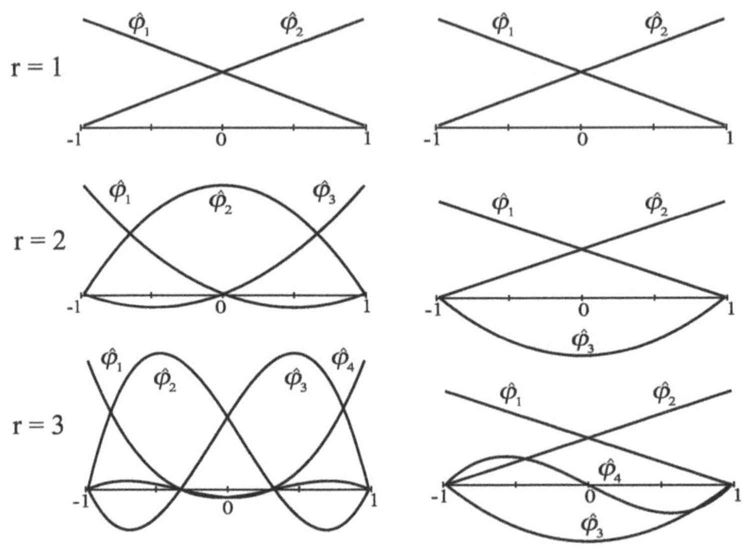

Fig. 2.10 Lagrange– (left) and Legendre based ansatz functions (right) up to order 3.

coordinates, this reads as

$$p(\boldsymbol{\xi}) \simeq p^h(\boldsymbol{\xi}) = \sum_{a=1}^{n_{\rm eq}} \hat{\phi}_a(\boldsymbol{\xi})\, p_a \,, \tag{2.33}$$

where $p^h(\boldsymbol{\xi})$ is the approximated acoustic pressure, $\hat{\phi}_a(\boldsymbol{\xi})$ the FE basis (ansatz) function related to the unknown a, p_a its coefficient and $n_{\rm eq}$ the number of unknowns.

In the case of standard Lagrangian elements, $\hat{\phi}_a$ are defined by the corner coordinates and p_a are the related values of the function in these nodes, see Section 2.2.3. Let us illustrate the idea by a one–dimensional finite element. Therewith, the ansatz functions of first order on the unit domain Ω $[-1, 1]$ are defined as follows

$$\hat{\phi}_1(\xi) = \frac{1-\xi}{2} \qquad \hat{\phi}_2(\xi) = \frac{1+\xi}{2} \,. \tag{2.34}$$

However, one disadvantage of the Lagrangian basis is that for each polynomial degree, one needs a new set of shape functions, see left subfigure of Figure 2.10, which prevents the use of mixed approximation orders within one FE–mesh.

In contrast, hierarchical basis functions are defined in such a way that every basis of order r is contained in the set of functions of order $r+1$, see right subfigure of Figure 2.10. In this work, we make use of the Legendre based interpolation functions as given in [51], which are defined as follows

$$\begin{aligned} \hat{\varphi}_1(\xi) &= \frac{1}{2}(1-\xi) \qquad \hat{\varphi}_2(\xi) = \frac{1}{2}(1+\xi) \\ \hat{\varphi}_i(\xi) &= \hat{\mu}_{i-1}(\xi), \qquad i = 3, \ldots, r+1 \,. \end{aligned} \tag{2.35}$$

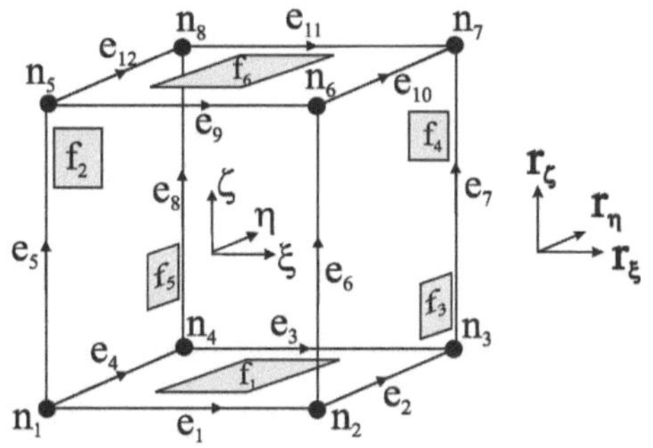

Fig. 2.11 Reference hexahedron with directional polynomial degrees r_ξ, r_η, r_ζ.

In (2.35), $\hat{\mu}_j$ denotes the integrated Legendre polynomial $\hat{L}_j$, which computes as follows

$$\hat{\mu}_j(\xi) = \sqrt{\frac{2j-1}{2}} \int\limits_{-1}^{\xi} \hat{L}_{j-1}(\hat{x}) \, \mathrm{d}\hat{x} \,, \tag{2.36}$$

$$\hat{L}_j(\hat{x}) = \frac{1}{2^j!} \frac{d^j}{d\hat{x}^j} (\hat{x}^2 - 1)^j \,. \tag{2.37}$$

Explicit expressions for $\hat{\varphi}_j$ and $\hat{\mu}_j$ can be found e.g. in [51]. Note that only the first two functions $\hat{\varphi}_1$ and $\hat{\varphi}_2$ contribute to the value at the ends of the unit interval $[-1, 1]$, whereas all other functions $\hat{\varphi}_j$ of higher order $j > 2$ give only a non-zero value within the interval. Therefore, they are also called *internal modes* or *bubble modes*.

Now, let us transfer the introduced basis functions to the three–dimensional case. Therewith, we choose the hexahedron element as displayed in Figure 2.11. To build the ansatz functions in 3D, we simply take the Cartesian product of the one–dimensional basis functions. As we have three element local directions ξ, η, ζ, we can accordingly choose three different approximation orders r_ξ, r_η, r_ζ. For the hexahedron the resulting basis functions can be distinguished into the following types:

1. *Nodal modes*: The 8 nodal modes are the standard trilinear functions known from the first order Lagrange element

$$\hat{\phi}^{1,1,1}_{n_a} = \frac{1}{8}(1 + \xi_a \xi)(1 + \eta_a \eta)(1 + \zeta_a \zeta), \; a = 1, \ldots, 8 \,, \tag{2.38}$$

   where ξ_a, η_a, ζ_a denote the local coordinates of the reference element vertices in $[-1, 1]$.
2. *Edge modes*: The edge modes are defined individually for each edge and their number depends on the approximation order in the direction of the edge. The modes corresponding to edge 1, e.g. are

$$\hat{\phi}^{r_\xi,1,1}_{e1} = \frac{1}{4} \hat{\mu}_{r_\xi}(\xi)(1 - \eta)(1 - \zeta) \,, \tag{2.39}$$

where $\hat{\mu}_r$ denote the related one–dimensional integrated Legendre basis function as given in (2.36).

3. *Face modes*: Also the face modes are defined for each face separately, e.g. for face 1

$$\hat{\phi}_{f_1}^{r_\xi, r_\eta, 1} = \frac{1}{2}(1-\zeta)\hat{\mu}_{r_\xi}(\xi)\hat{\mu}_{r_\eta}(\eta)\ . \tag{2.40}$$

4. *Internal modes*: The internal modes are only defined within the elements interior and vanish at its boundaries. They are defined as

$$\hat{\phi}_{\text{int}}^{r_\xi, r_\eta, r_\zeta} = \hat{\mu}_{r_\xi}(\xi)\hat{\mu}_{r_\eta}(\eta)\hat{\mu}_{r_\zeta}(\zeta)\ . \tag{2.41}$$

The number of nodal functions is always 8, whereas the number of edge, face and internal modes depends on the choice of r_ξ, r_η and r_ζ. In order to guarantee a C^0–continuous approximation, the inter–element orientation of edge and surface functions has to be considered. Also, in order to achieve a convergent solution, the exact geometry of the continuous domain has to be known for the computation of global derivatives $\partial\hat{\phi}_a/\partial\mathbf{x}$. For further details we refer, e.g., to [21,50]. In addition, we want to stress, that an anisotropic choice of approximation order, e.g. different polynomial order in x–, y– and z–direction may allow to minimize the number of unknowns. This technique is especially suitable, e.g., for thin mechanical structures, see [22,29].

At this point we want to note, that in the presence of singularities the exponential convergence of the p–FEM reduces to an algebraic one. In such cases, a proper combination of h–refinement (mesh refinement) and a local increase of the polynomial order (p–FEM), which is called the hp–method, will regain the exponential convergence. For the pioneering works of this method we refer to [3, 4] and for a recent textbook including practical aspects of implementation we refer to [19].

2.3.2 Numerical Investigations

Accuracy Study of p–FEM

Let us consider the simple case of an acoustic wave in a channel with given Dirichlet boundary conditions on the left and right end, and homogeneous Neumann boundary conditions elsewhere, see Figure 2.12. We keep the discretization fixed by 20 finite elements and change the wave number k, see Table 2.1, as well as the order r of the FE basis functions. For comparison we compute the accumulated error E_a

$$E_\text{a} = \sqrt{\frac{\sum\limits_{a=1}^{n_\text{n}} \left(p_a^h - p_a\right)^2}{\sum\limits_{a=1}^{n_\text{n}} p_a^2}} \qquad \begin{aligned} &p_a \text{ ... analytic solution at node } a \\ &p_a^h \text{ ... FE solution in node } a \end{aligned}$$

with n_n the number of finite element nodes. The analytic solution p computes as

$$p(x) = \cos(kx) - \frac{\cos(kL)}{\sin(kL)}\sin(kx)\,.$$

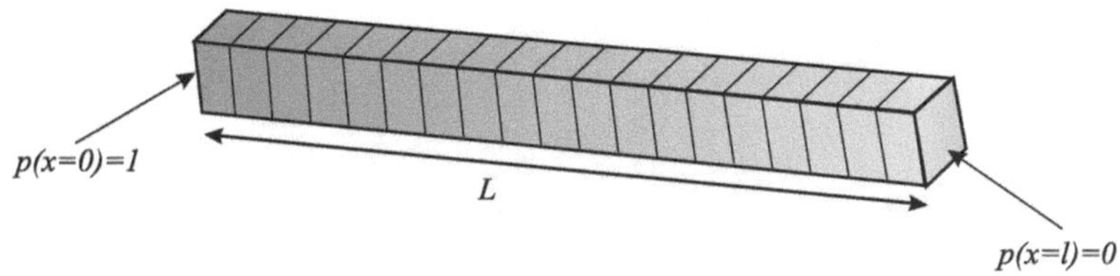

Fig. 2.12 Acoustic wave in a channel with given Dirichlet boundary conditions.

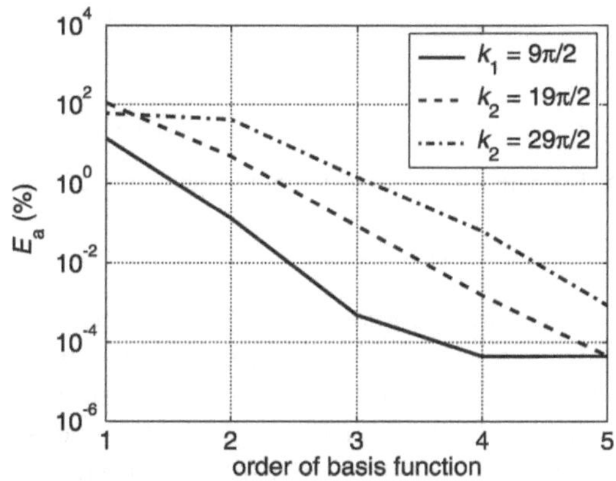

Fig. 2.13 Accumulated error as a function of the order of the FE basis functions.

Table 2.1 Choice of wave number k and resulting number of finite elements per wavelength.

k	$9\pi/2$	$19\pi/2$	$29\pi/2$
λ/h	8.9	4.2	2.7

Figure 2.13 clearly demonstrates that we effectively reduce the error by increasing the order q of the finite element basis functions. However, as shown by the case $k = 29/2\,\pi$, we do not fulfill (2.32) on the mesh with $r = 1$ and so no reduction of the error is achieved by increasing the order of the FE basis functions from $r = 1$ to $r = 2$. This clearly demonstrates, that (2.32) is a very important and practical guideline for acoustic computations.

p–FEM versus h–FEM

Within the following practical example, we perform a comparison between h–FEM and p–FEM concerning the accuracy as well as the elapsed CPU time. For this purpose, we investigate the acoustic noise generated due to a strong current within a blowout coil. If a short circuit to ground occurs, the current is flowing through the

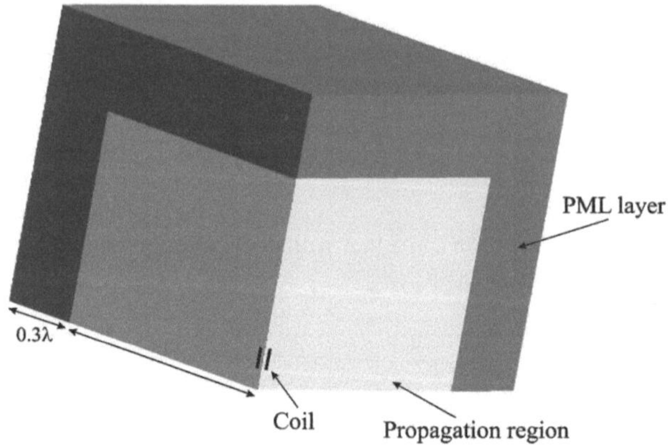

Fig. 2.14 Domain for the acoustic field computation of a blowout coil.

blowout coil, and the resulting Lorentz–forces have twice the frequency of the current, which results due to the interaction of the current with the magnetic field. These forces will lead to vibrations of the coil. The setup of the simulation domain for computing the acoustic field is displayed in Figure 2.14. Our goal is to compute the overall radiated acoustic power. To guarantee a free radiation, we have attached to the propagation region a Perfectly Matched Layer (PML) region (thickness about one third of the acoustic wavelength λ) with an inverse distance damping function [35]. The starting mesh consists of 6 finite elements per wavelength, both within the propagation as well as PML region. The acoustic power is directly computed at the interface between the propagation and PML region. Therewith, in the overall error the performance of the PML is included.

Now in a first step, we keep the spatial discretization fixed and increase for each new computation the order of the FE basis functions. Afterwards we refine the mesh from computation to computation, so that we follow the classical h–FEM approach. For both cases we compute the total radiated acoustic power, which is displayed in Figure 2.15 as a function of the number of unknowns. It can be clearly seen, that both methods overestimate the sound power level for a too coarse discretization and converge to the correct sound power level as the number of unknowns is increased. However, the p–FEM is quite superior to h–FEM concerning the number of unknowns.

In addition, we have measured the overall CPU time of each simulation run including reading the input files, setting up the FE matrices, solving the algebraic system of equations and writing the result file. For all cases, we solve the algebraic systems of equations by a sparse direct solver [45]. Table 2.2 informs about these elapsed CPU times for both methods as a function of the unknowns. The results clearly demonstrate the current drawback of p–FEM, that due to the increase of the order of the FE basis functions, the arising FE matrices get denser and both the computation of these matrices and the solution of the arising algebraic system of equa-

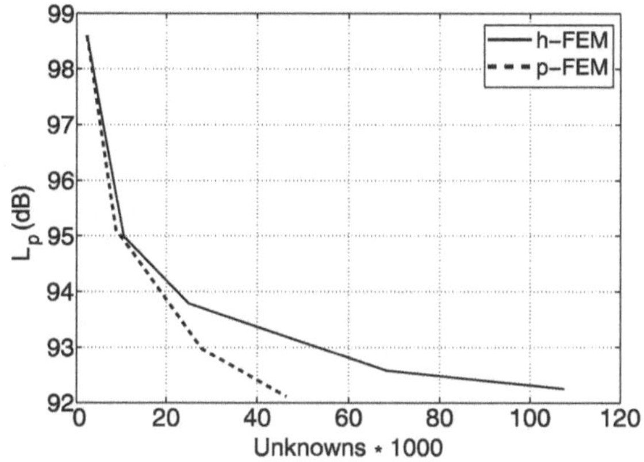

Fig. 2.15 Acoustic power L_P(db) re. 1 pW as a function of the number of unknowns for h–FEM and p–FEM

Table 2.2 Computational time versus unknowns for p–FEM and h–FEM; computations have been performed on a Leneovo T60, 2.167GHz

	p–FEM		h–FEM	
r	unknowns	CPU time	unknowns	CPU time
1	2.352	0.9 s	2.352	0.9 s
2	8.848	5.2 s	10.580	4.2 s
3	15.344	21.6 s	24.986	10.5 s
4	27.856	89.9 s	68.413	47.9 s
5	46.384	319.1 s	107.500	99.4 s

tions is quite time consuming. Current research concentrates especially on fast methods for the computation of the FE matrices and special adapted solvers for the algebraic system of equations. Especially iterative solvers with pre–conditioners based on domain decomposition methods can reduce the solution time tremendously [46], and thus make p–FEM a very attractive method for practical problem settings.

2.4 Mixed Spectral Finite Elements

For a long time, engineers were reluctant to solve wave equations in the time domain using FE methods. The presence of a non–diagonal mass matrix which had to be inverted at each time–step did not fit to the explicit schemes in time generally used for hyperbolic equations. However, although easier to implement and providing faster algorithms, finite difference methods (FDM) meet their limit in solving realistic problems which involve complex geometries.

A first step to overcome the problem of the inversion of a mass matrix was the use of Gauss–Lobatto quadrature rules which lead to mass–lumping, i.e. a diagonal mass matrix [30,55]. This new technique, later defined as a "spectral element method" [41], was then applied to the wave equation [14,53]. A second step was realized by introducing a mixed formulation of this method which increased the gain in storage and computational time for higher–order approximations [9,10]. This mixed formulation was successfully applied to many wave equations such as elastodynamics [11] and Reissner–Mindlin model for plates [13] and is a challenging alternative to finite difference methods (FDM) in terms of computational time and storage. A more detailed description of the method can be found in [9].

Our purpose is to solve the inhomogeneous acoustic wave equation in time domain [6]

$$\frac{1}{\rho(\mathbf{x})c(\mathbf{x})}\frac{\partial^2 p}{\partial t^2}(\mathbf{x},t) - \boldsymbol{\nabla}\cdot\frac{1}{\rho(\mathbf{x})}\boldsymbol{\nabla} p(\mathbf{x},t) = g(\mathbf{x},t)\,, \tag{2.42}$$

where $\mathbf{x} \in \Omega \subset I\!R^d$ ($d = 2$ or 3), $\boldsymbol{\nabla} = (\partial/\partial x_1, ..., \partial/\partial x_d)^T$, ρ the acoustic density, c the speed of sound and g a given source term. As an additional simplification, we suppose that

$$\frac{\partial p}{\partial n} = 0 \text{ on } \partial\Omega, \tag{2.43}$$

where $\partial\Omega$ is the boundary of Ω and $\mathbf{n}$ is the exterior normal to $\partial\Omega$. The variational formulation of (2.42)–(2.43) reads

$$\frac{\mathrm{d}^2}{\mathrm{d}t^2}\int\limits_{\Omega}\frac{1}{\rho c}pu\,\mathrm{d}V + \int\limits_{\Omega}\frac{1}{\rho}\boldsymbol{\nabla} p\cdot\boldsymbol{\nabla} u\,\mathrm{d}V = \int\limits_{\Omega} gu\,\mathrm{d}V, \quad \forall u \in H^1(\Omega). \tag{2.44}$$

In (2.44) the functional space $H^1(\Omega)$ is defined as

$$H^1(\Omega) = \left\{u \in L^2(\Omega) \text{ such that } \forall j = 1..d, \partial u/\partial x_j \in L^2(\Omega)\right\}. \tag{2.45}$$

Now, let the domain Ω be composed of finite elements K_e (quadrilaterals in 2D and hexahedra in 3D)

$$\mathcal{M}_h = \bigcup_{e=1}^{n_e} K_e\,. \tag{2.46}$$

We introduce the mapping $\mathbf{F}_e$ of a reference element $\hat{K}$ to its representation in the global coordinate system such that $\mathbf{F}_e(\hat{K}) = K_e$. On this mesh, we define the following finite–dimensional subspace of $H^1(\Omega)$ (which is also a subspace of C^0)

$$U_h^r = \left\{u_h \in H^1(\Omega) \text{ such that } u_{h|K_e^\circ}\mathbf{F}_e \in Q_r\right\}, \tag{2.47}$$

where Q_r is the Lagrange polynomial space of order r, see Section 2.2.3.

Now, after taking the test functions in U_h^r and integrating over Ω for all the basis functions of U_h^r, (2.44) results in the following ordinary differential system

[6] This equation can be easily extended to an inhomogeneous wave equation including anisotropy, which is often used in geophysics or electromagnetism as an approximate model of the real phenomena.

$$\mathbf{M}\frac{\mathrm{d}^2\mathbf{p}}{\mathrm{d}t^2}(t) + \mathbf{K}\,\mathbf{p}(t) = \mathbf{G}(t)\,. \tag{2.48}$$

In practice, the coefficients of $\mathbf{M}$ and $\mathbf{K}$ are derived from the assembly procedure of their element matrices $\mathbf{m}^e$ and $\mathbf{k}^e$, which compute as follows

$$\mathbf{m}^e = [m^e_{ab}]\,, \qquad m^e_{ab} = \int\limits_{K_e} \frac{1}{\hat{\rho}\hat{c}^2}\,\phi_{e,a}\,\phi_{e,b}\,\mathrm{d}V = \int\limits_{\widehat{K}} \frac{1}{\hat{\rho}\hat{c}^2}\,|J_e|\hat{\phi}_a\,\hat{\phi}_b\,\mathrm{d}\hat{V} \tag{2.49}$$

$$\begin{aligned}\mathbf{k}^e = [k^e_{ab}]\,, \qquad k^e_{ab} &= \int\limits_{K_e} \frac{1}{\hat{\rho}}\boldsymbol{\nabla}\phi_{e,a}\cdot\boldsymbol{\nabla}\phi_{e,b}\,\mathrm{d}V \\ &= \int\limits_{\widehat{K}} \frac{1}{\hat{\rho}}\,|J_e|\,\mathbf{J}_e^{-1}\left(\mathbf{J}_e^T\right)^{-1}\hat{\boldsymbol{\nabla}}\hat{\phi}_a\cdot\hat{\boldsymbol{\nabla}}\hat{\phi}_b\,\mathrm{d}\hat{V}\,,\end{aligned} \tag{2.50}$$

where $\mathbf{J}$ is the Jacobian matrix of $\mathbf{F}_e$, $|J_e| = \det \mathbf{J}_e$ and $\hat{\boldsymbol{\nabla}}$ the $\boldsymbol{\nabla}$ operator in $\hat{\mathbf{x}}$ coordinates (local coordinate system). Since ρ and c depend on the space coordinates, we also have to apply the mapping, e.g., $\hat{\rho} = \rho \circ \mathbf{F}_e$ and $\hat{c} = c \circ \mathbf{F}_e$.

The general framework of the FEM for the acoustic wave equation in time domain is now being set, and we are now going to enumerate the main drawbacks of such a formulation and show how to overcome them.

2.4.1 Spectral Elements

Mass–Lumping

A classical approximation of (2.48) is obtained by a centered second–order finite difference (leapfrog) scheme[7] as follows

$$\mathbf{M}\,\frac{\mathbf{p}^{n+1} - 2\mathbf{p}^n + \mathbf{p}^{n-1}}{\Delta t^2} + \mathbf{K}\,\mathbf{p}^n = \mathbf{G}^n\,. \tag{2.51}$$

The solution of (2.51) reads

$$\mathbf{p}^{n+1} = 2\mathbf{p}^n - \mathbf{p}^{n-1} - \Delta t^2\mathbf{M}^{-1}\left(\mathbf{K}\,\mathbf{p}^n - \mathbf{G}^n\right)\,. \tag{2.52}$$

According to (2.52) it is clear, that even with an explicit scheme in time, (2.48) requires a matrix inversion at each time–step to be solved. This is an important drawback of FEM versus FDM for transient wave equations.

In order to overcome this problem, we use a mass–lumping technique, i.e. we replace the matrix $\mathbf{M}$ by a diagonal one. This technique requires

[7] We shall not deal here with the time discretization. In a few words, although only being of second–order, the leapfrog scheme is the most frequently one used for high–order finite elements. This is due to the unconditional instability of higher–order finite difference schemes in time as shown in [27]. Sophisticated higher–order schemes, such as the modified equation method or symmetrical schemes, can be used but rise some technical problems. More details can be found in [9].

1. to compute the integrals involved in the mass matrix by a quadrature rule,
2. to make the interpolation points coincide with the quadrature points.

The two points above lead to a diagonal matrix as follows: Let us consider the elementary mass integral (neglecting the material parameters) defined in (2.49). Then, we have

$$m_{ab} = \int_{\widehat{K}} |J_e| \hat{\phi}_a \, \hat{\phi}_b \, \mathrm{d}\hat{V} \simeq \sum_{i \in \{1..r+1\}^d} \hat{\omega}_i |J_e(\hat{\boldsymbol{\xi}}_i)| \prod_{k=1}^{d} \hat{\varphi}_{a_k}(\hat{\boldsymbol{x}}_{i_k}) \prod_{k=1}^{d} \hat{\varphi}_{b_k}(\hat{\boldsymbol{x}}_{i_k})$$

$$\simeq \sum_{i \in \{1..r+1\}^d} \hat{\omega}_i |J_e(\hat{\boldsymbol{\xi}}_i)| \prod_{k=1}^{d} \delta_{a_k i_k} \prod_{k=1}^{d} \delta_{b_k i_k} \qquad (2.53)$$

where $\hat{\omega}_i = \prod_{k=1}^{d} \hat{\omega}_{i_k}$, $\hat{\omega}_{i_k}$ and $\hat{\xi}_{i_k}$ are the weights and the points of the 1D quadrature rule on the interval $[-1, 1]$. According to (2.53), we have $m_{ab} \neq 0$ if and only if $a_k = b_k$, $\forall k = 1, \ldots, d$, so that only the diagonal contributions of the mass matrix are not equal to zero (in this case, $i_k = a_k = b_k$).

Now, from a numerical point of view, one must be sure that the error committed by the numerical integration is not larger than the error of approximation. This last point is fixed by Ciarlet in a lemma in [8, cf. exercises 4.1.4 ff.] which states that in 1D, when we use a polynomial interpolation of order r, the quadrature rule must be exact for polynomials of order less or equal to $2r - 1$. This property in 1D holds for $\widehat{K}$ which is obtained by a product of 1D elements. For non–regular elements, a numerical study provides an L^2–error in $O(h^{r+1})$ in 2D and in $O(h^r)$ in 3D [23].

Newton–Cotes quadrature rules are not accurate enough to satisfy Ciarlet's lemma. Gauss rules are accurate enough but their points do not fit with conforming elements, since they do not contain the ends of the interval. The only quadrature rule which fulfills all the requirements above is the Gauss–Lobatto quadrature rule whose points are the roots of the derivatives of Legendre polynomials plus the ends of the interval and which are exact for polynomials of order less or equal to $2r - 1$. The weights and points of this quadrature rule up to order 6 are given in Table 2.3. Approximate values up to order 10 are given in [9]. Of course, interpolation points are no longer regularly spaced and must be located at the quadrature points.

From the stiffness matrix point of view, the use of Gauss–Lobatto quadrature rules reduces the number of computations. It actually provides a complexity in $O(r^4)$ instead of $O(r^6)$ in 2D and in $O(r^5)$ instead of $O(r^9)$ in 3D. Moreover, only coplanar degrees of freedom provide non–zero interactions, which induces a substantial gain of storage in 3D [23].

Finally, we replaced problem (2.48) by

$$\mathbf{M}_{\mathrm{diag}} \frac{\mathrm{d}^2\mathbf{p}}{\mathrm{d}t^2}(t) \; + \; \mathbf{K}' \, \mathbf{p}(t) \; = \mathbf{G}(t), \qquad (2.54)$$

where $\mathbf{M}_{\mathrm{diag}}$ is a diagonal mass matrix and $\mathbf{K}'$ a new stiffness matrix, easier to compute than $\mathbf{K}$.

This technique of mass–lumping was constructed by Hennart [30] and Young [55] independently.

Table 2.3 Exact values of the abscissae of the points and of the weights of the Gauss–Lobatto quadrature rules for the semi–interval $[-1, 0]$ (weights and abscissae are symmetrical w.r.t. 0).

	$i = 1$	$i = 2$	$i = 3$	$i = 4$
$r = 1\ (\hat{\xi})$	-1			
$r = 1\ (\hat{\omega})$	1			
$r = 2\ (\hat{\xi})$	-1	0		
$r = 2\ (\hat{\omega})$	$1/3$	$4/3$		
$r = 3\ (\hat{\xi})$	-1	$-1/\sqrt{5}$		
$r = 3\ (\hat{\omega})$	$1/6$	$5/6$		
$r = 4\ (\hat{\xi})$	-1	$-\sqrt{5/7}$	0	
$r = 4\ (\hat{\omega})$	$1/10$	$49/90$	$32/45$	
$r = 5\ (\hat{\xi})$	-1	$-\sqrt{147 + 42\sqrt{7}}/21$	$-\sqrt{147 - 42\sqrt{7}}/21$	
$r = 5\ (\hat{\omega})$	$1/15$	$(14 - \sqrt{7})/30$	$(14 + \sqrt{7})/30$	
$r = 6\ (\hat{\xi})$	-1	$-\sqrt{495 + 66\sqrt{15}}/33$	$-\sqrt{495 - 66\sqrt{15}}/33$	0
$r = 6\ (\hat{\omega})$	$1/21$	$62/175 - \sqrt{15}/50$	$62/175 + \sqrt{15}/50$	$256/525$

Avoiding Runge's Phenomenon

One of the advantages of quadrilateral or hexahedral elements is the straightforward derivation of higher–order finite elements from 1D Lagrange interpolation polynomials. Unfortunately, these polynomials present a major drawback known as Runge's phenomenon. This phenomenon can be observed on the interpolation of the function $1/(1 + x^2)$ over an interval containing $[-5, 5]$. Under these conditions, the interpolation polynomials produce oscillations between the interpolation points which grow with their order. This phenomenon seems to disqualify Lagrange polynomials for higher–order approximation.

A first palliative to this problem was the use of Chebyshev points which were considered as the optimal points to get rid of this phenomenon for a long time. FEM approximations were called spectral elements because of their ability of having an exponential decay versus the order of approximation. These elements were in particular used to model seismic waves in geophysics [47].

Actually, other non–regularly spaced points, such as Gauss or Gauss–Lobatto quadrature points, can avoid oscillatory interpolations. In other words, our mass–lumped elements are spectral elements too[8]. In Figure 2.16, we compare the L^2–error committed on the Lagrange interpolation of function $1/(1 + x^2)$ over the interval $[-5, 5]$.

[8] Gauss rules are used for discontinuous Galerkin methods, as described in [12].

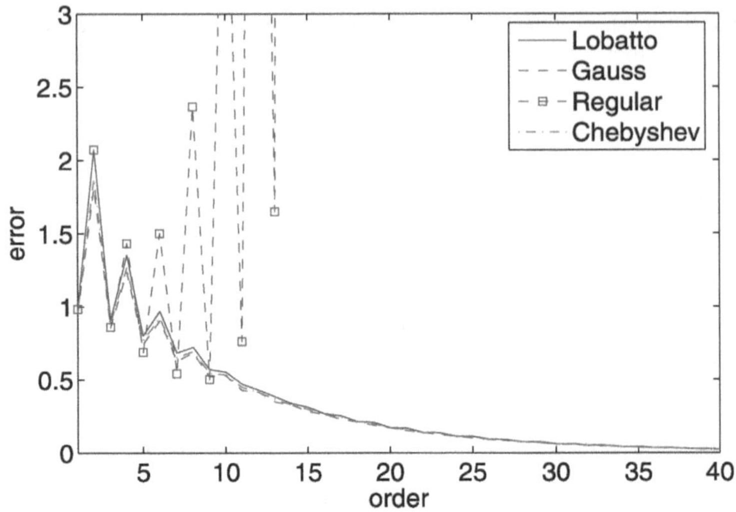

Fig. 2.16 L^2–error interpolation with different points.

Spectral Triangular or Tetrahedral Elements

An interesting issue would be to construct spectral elements for triangular or tetrahedral meshes. Actually this can be done in the same way as for quadrilateral or hexahedral elements. The only change is that the quadrature rule must be exact for P_{2r-2} polynomials for a P_r interpolation. Such rules can be easily constructed. Unfortunately, they generally present null or negative weights which lead to unconditionally unstable schemes [14]. A palliative to this problem lies in the introduction of additional degrees of freedom and in the extension of the polynomial space of approximation. In 2D, the new space $\widetilde{P}_r$ is such that $P_r \subset \widetilde{P}_r \subset P_{r+1}$ which leads to the addition of a few number of degrees of freedom (1 for P_2 and 2 for P_3). In 3D, we have $P_r \subset \widetilde{P}_r \subset P_{r+2}$. Such an inclusion produces more than twice more degrees of freedom to get the proper quadrature rule (23 instead of 10 for P_2). This leads to a very heavy class of elements which are not very efficient. More details can be found in [14, 38]. Another way to get spectral triangular or tetrahedral elements by degeneration of quadrilateral or hexahedral elements is described in [20, 48].

2.4.2 Mixed Spectral Elements

A New Formulation

Using Gauss–Lobatto rules provides a substantial gain of time by avoiding mass matrix inversion. However, the new formulation still requires the storage of the stiffness matrix which is very important. In order to overcome this problem, we introduce the mixed formulation of spectral elements. For this purpose, we first write the wave equation defined in (2.42) as a first order system[9]

$$\frac{1}{c^2(\mathbf{x})}\frac{\partial p}{\partial t}(\mathbf{x},t) = -\boldsymbol{\nabla}\cdot\mathbf{v}(\mathbf{x},t) + G(\mathbf{x},t), \tag{2.55}$$

$$\rho(\mathbf{x})\frac{\partial \mathbf{v}}{\partial t}(\mathbf{x},t) = -\boldsymbol{\nabla} p(\mathbf{x},t), \tag{2.56}$$

where p denotes the acoustic pressure, $\mathbf{v}$ the acoustic particle velocity and $\partial G/\partial t = g$. We write the variational formulation of the system above as follows: Find p and $\mathbf{v}$ such that $p(.,t) \in H^1(\Omega)$ and $\mathbf{v}(.,t) \in [L^2(\Omega)]^d$ ($d = 2, 3$) and

$$\frac{\mathrm{d}}{\mathrm{d}t}\int_\Omega \frac{1}{c^2}\, p\, q\, \mathrm{d}V = \int_\Omega \mathbf{v}\cdot\boldsymbol{\nabla} q\,\mathrm{d}V + \int_\Omega G\, q\,\mathrm{d}V \quad \forall q \in H^1(\Omega) \tag{2.57}$$

$$\frac{\mathrm{d}}{\mathrm{d}t}\int_\Omega \rho\mathbf{v}\cdot\boldsymbol{\psi}\,\mathrm{d}V = -\int_\Omega \boldsymbol{\nabla} p\cdot\boldsymbol{\psi}\,\mathrm{d}V \quad \forall \boldsymbol{\psi} \in [L^2(\Omega)]^d. \tag{2.58}$$

Now, the FE formulation of (2.57–2.58) reads as: Find p_h and $\mathbf{v}_h$ such that $p_h(.,t) \in U_h^r$ and $\mathbf{v}_h(.,t) \in \mathbf{V}_h^r$ and

$$\frac{\mathrm{d}}{\mathrm{d}t}\int_\Omega \frac{1}{c^2}\, p_h\, q_h\, \mathrm{d}V = \int_\Omega \mathbf{v}_h\cdot\boldsymbol{\nabla} q_h\,\mathrm{d}V + \int_\Omega G\, q_h\,\mathrm{d}V \quad \forall q_h \in U_h^r \tag{2.59}$$

$$\frac{\mathrm{d}}{\mathrm{d}t}\int_\Omega \rho\mathbf{v}_h\cdot\boldsymbol{\psi}_h\,\mathrm{d}V = -\int_\Omega \boldsymbol{\nabla} p_h\cdot\boldsymbol{\psi}_h\,\mathrm{d}V \quad \forall \boldsymbol{\psi}_h \in \mathbf{V}_h^r, \tag{2.60}$$

where U_h^r is defined as in (2.47) and[10]

$$\mathbf{V}_h^r = \{\mathbf{v}_h \in \left[L^2(\Omega)\right]^2 \text{ such that } \forall K_e \in \mathcal{M}_h,\ |J_e|\mathbf{J}_e^{-1}\mathbf{v}_{h|K_e}\circ\mathbf{F}_e \in \left[Q_r\right]^d\}. \tag{2.61}$$

The Stiffness Matrix

Actually, the transform defined by $\mathbf{v}_{h|K_e}\circ\mathbf{F}_e = 1/|J_e|\,\mathbf{J}_e\,\hat{\mathbf{v}}_h$ is well–known as Piola's transform and is used to derive $H(div)$–conforming elements (elements with continuity of the normal component) from the reference element on a non–regular mesh. However, this transform seems to have no use in our formulation, in which $\hat{\mathbf{v}}_h$ belongs to $\left[L^2(\Omega)\right]^2$, i.e. has no continuity requirement. In fact, the advantage of this transform will appear in the computation of the stiffness matrices of (2.59–2.60).

Let us first define the basis functions of $\mathbf{V}_h^r$ $[= span(\mathcal{B})]$ as follows

[9] Actually, this system is the original form of the wave equation which is a 0th order linearization of Euler equations.

[10] Classically, a mixed formulation is such that $p(.,t) \in L^2(\Omega)$ and $\mathbf{v}(.,t) \in H(div,\Omega)$. We could define a spectral method in these spaces, but it would not be more efficient than standard spectral elements.

$$\forall \boldsymbol{\psi}_{e,a}^{\ell} \in \mathcal{B}, \quad \boldsymbol{\psi}_{e,a}^{\ell} = \varphi_{e,a}\,\mathbf{e}_{\ell}, \quad e = 1..n_e, \; a \in \{1..r+1\}^d, \; \ell = 1..d, \tag{2.62}$$

where $\varphi_{e,a}$ is the restriction of a basis function of U_h^r to K_e and $\mathbf{e}_\ell$ is a basis vector of $I\!R^d$. After decomposing the basis functions of U_h^r and $\mathbf{V}_h^r$, the element stiffness integrals can be derived from the following elementary integrals

$$\begin{aligned} k_{ab} &= \int_{K_e} \boldsymbol{\psi}_{e,a}^{\ell} \cdot \boldsymbol{\nabla}\phi_{e,b}\,\mathrm{d}V = \int_{\widehat{K}} |J_e|\boldsymbol{\psi}_{e,a}^{\ell} \circ \mathbf{F}_e \cdot \boldsymbol{\nabla}\phi_{e,b} \circ \mathbf{F}_e\,\mathrm{d}\hat{V} \\ &= \int_{\widehat{K}} \frac{|J_e|}{|J_e|}\mathbf{J}_e\,\hat{\boldsymbol{\psi}}_a^{\ell} \cdot \left(\mathbf{J}_e^T\right)^{-1} \hat{\boldsymbol{\nabla}}\hat{\phi}_b\,\mathrm{d}\hat{V} \\ &= \int_{\widehat{K}} \hat{\boldsymbol{\psi}}_a^{\ell} \cdot \hat{\boldsymbol{\nabla}}\hat{\phi}_b\,\mathrm{d}\hat{V}\,, \end{aligned} \tag{2.63}$$

where $\hat{\boldsymbol{\psi}}_a^{\ell} = \hat{\varphi}_a \mathbf{e}_\ell$. In (2.63), we can see that, thanks to Piola's transform, the knowledge of these elementary integrals on $\widehat{K}$ enables one to know them on any element K_e of the mesh. In other words, because of the discontinuity of the basis functions of $\mathbf{V}_h^r$, (2.63) provides the entire stiffness matrix of (2.60) and the matrix of (2.59) providing its assembly whose cost is negligible for high–order elements. So, the above property enables us to avoid the storage of stiffness matrices except on the reference element $\widehat{K}$, which implies a huge gain of storage.

Besides, the gain of storage induced by Piola's transform, the tensor character of basis functions on $\widehat{K}$ and the use of Gauss–Lobatto quadrature rules lead to very sparse stiffness matrices, as one can see in the following 3D example, in which we set $\ell = 1$.

$$\begin{aligned} K_{ab} &= \int_{\widehat{K}} \hat{\varphi}_{a_1}\hat{\varphi}_{a_2}\hat{\varphi}_{a_3}\hat{\varphi}'_{b_1}\hat{\varphi}_{b_2}\hat{\varphi}_{b_3}\,\mathrm{d}\hat{V} \\ &\simeq \sum_{i\in\{1..r+1\}^d} \hat{\omega}_i\,\hat{\varphi}_{a_1}(\hat{\xi}_{i_1})\hat{\varphi}_{a_2}(\hat{\xi}_{i_2})\hat{\varphi}_{a_3}(\hat{\xi}_{i_3})\hat{\varphi}'_{b_1}(\hat{\xi}_{i_1})\hat{\varphi}_{b_2}(\hat{\xi}_{i_2})\hat{\varphi}_{b_3}(\hat{\xi}_{i_3}) \\ &\simeq \sum_{i\in\{1..r+1\}^d} \hat{\omega}_i\,\delta_{a_1 i_1}\delta_{a_2 i_2}\delta_{a_3 i_3}\hat{\varphi}'_{b_1}(\hat{\xi}_{i_1})\delta_{b_2 i_2}\delta_{b_3 i_3}, \quad \text{with } \hat{\omega}_i = \prod_{k=1}^{d} \hat{\omega}_{i_k} \end{aligned} \tag{2.64}$$

which implies that

$$k_{ab} = \begin{cases} \hat{\omega}_a \hat{\varphi}'_{b_1}(\hat{\xi}_{a_1}), & \forall a_1 = 1..r+1, \text{ if } a_2 = b_2 \text{ and } a_3 = b_3 \\ 0 \quad \text{otherwise.} \end{cases} \tag{2.65}$$

The result in (2.65) shows that, for a given basis function $\hat{\varphi}_b$ and for a direction $\mathbf{e}_\ell$, the only non–zero interactions with this basis function are provided by functions located on the line defined by the point $\hat{\mathbf{x}}_a$ and vector $\mathbf{e}_\ell$, which produces, of course, a very sparse matrix and a substantial gain of computational time.

The Mass Matrix

Obviously, the mass matrix involved in (2.59) is the $\mathbf{M}_{\text{diag}}$ matrix defined in (2.54). Let us now look at the effect of Piola's transform on the mass matrix in (2.60). As for the element stiffness matrix, the element mass matrix (neglecting the material parameters) is derived from the following elementary integrals

$$\begin{aligned} m_{ab} = \int\limits_{K_e} \boldsymbol{\psi}^{\ell}_{e,a} \cdot \boldsymbol{\psi}^{m}_{e,b} \, \mathrm{d}V &= \int\limits_{\widehat{K}} |J_e| \, \boldsymbol{\psi}^{\ell}_{e,a} \circ \mathbf{F}_e \cdot \boldsymbol{\psi}^{m}_{e,b} \circ \mathbf{F}_e \, \mathrm{d}\hat{V} \\ &= \int\limits_{\widehat{K}} \frac{|J_e|}{|J_e|} \mathbf{J}_e \, \hat{\boldsymbol{\psi}}^{\ell}_a \cdot \frac{1}{|J_e|} \mathbf{J}_e \, \hat{\boldsymbol{\psi}}^{m}_b \, \mathrm{d}\hat{V} \\ &= \int\limits_{\widehat{K}} \frac{1}{|J_e|} \mathbf{J}^T_e \, \mathbf{J}_e \, \hat{\boldsymbol{\psi}}^{\ell}_a \cdot \hat{\boldsymbol{\psi}}^{m}_b \, \mathrm{d}\hat{V} \, . \end{aligned} \tag{2.66}$$

By setting $\mathbf{A}_e = 1/|J_e|\mathbf{J}^T_e\mathbf{J}_e$, computing the integral by using a Gauss–Lobatto quadrature rule and taking into account (2.62), we get, by using the notation of (2.53)

$$\begin{aligned} m_{ab} &\simeq \sum_{i \in \{1..r+1\}^d} \hat{\omega}_i \prod_{k=1}^{d} \hat{\varphi}_{a_k}(\hat{\xi}_{i_k}) \prod_{k=1}^{d} \hat{\varphi}_{b_k}(\hat{\xi}_{i_k}) \mathbf{A}_e(\hat{\boldsymbol{\xi}}_i) \, \mathbf{e}_\ell \cdot \mathbf{e}_m \\ &\simeq \sum_{i \in \{1..r+1\}^d} \hat{\omega}_i \prod_{k=1}^{d} \delta_{a_k i_k} \prod_{k=1}^{d} \delta_{b_k i_k} \mathbf{A}_e(\hat{\boldsymbol{\xi}}_i) \, \mathbf{e}_\ell \cdot \mathbf{e}_m \, . \end{aligned} \tag{2.67}$$

So, we have

$$m_{ab} = \begin{cases} \hat{\omega}_a \mathbf{A}_e(\hat{\boldsymbol{\xi}}_a) \, \mathbf{e}_\ell \cdot \mathbf{e}_m, & \text{if } a_k = b_k \\ 0 \quad \text{otherwise.} \end{cases} \tag{2.68}$$

Due to to the discontinuity of functions of $\mathbf{V}^r_h$, (2.68) provides a symmetric block–diagonal mass matrix for (2.60). This block–diagonal structure is very convenient since the matrix can be inverted blockwise. In practice, the upper (or lower) triangular part of the inverse of each block will be stored in the algorithm for solving the system.

We want to note at this point, that the proposed method is equivalent to spectral elements, which has been proven by a theorem of equivalence [9]. In summary, mixed spectral elements are a powerful way to solve the acoustics equation. Even by using split tetrahedra, which provide very distorted elements, they remain much more efficient than methods based on pure tetrahedra. Moreover, it is easy to parallelize. This technique was extended to edge elements [15] and discontinuous Galerkin methods for Maxwell's equations [12]. This last method is very flexible and enables to solve a large class of linear hyperbolic equations. In particular, it has been used to solve linearized Euler equations modeling acoustic waves in flow [6].

Table 2.4 Ratios for the storage for different methods

	$r=1$	$r=2$	$r=3$	$r=4$	$r=5$	$r=6$
$s2sp(s2cl)/s2m$	0.35	1.08	2.09	3.38	4.96	6.82
$s3cl/s3m$	0.27	1.48	4.09	8.55	15.31	24.85
$s3sp/s3m$	0.15	0.72	1.57	2.82	4.59	6.97
$s3t/s3m$	0.19	0.87	2.01	3.62	5.70	8.24

Table 2.5 Ratios for the CPU time for standard and mixed spectral elements

	$r=1$	$r=2$	$r=3$	$r=4$	$r=5$	$r=6$
$c2sp/c2m$	0.20	0.46	0.71	0.97	1.23	1.48
$c3sp/c3m$	0.12	0.42	0.81	1.22	1.67	2.12

2.4.3 Computational Issues

Efficiency of the Method

As shown above, the stiffness matrix of mixed spectral elements only needs the storage of a $d \times d$ block–diagonal matrix, which seems to induce a substantial gain compared to other methods. In Table 2.4, we evaluate this gain versus different other methods. $sdcl$, $sdsp$ and sdm represent the storage required by the stiffness matrix for classical finite elements (without mass–lumping), standard spectral elements and mixed spectral elements in d dimensions. $s3t$ is the storage required by classical tetrahedral finite elements. In 2D, classical and spectral elements need the same storage. The advantage of low–order approximations comes from the few number of diagonal interactions for low–order stiffness matrices. Anyhow, the gain realized by mixed spectral elements in high order is obvious, particularly versus classical finite elements. For these elements, as well for tetrahedral elements, the gain of storage must be multiplied by almost 2 when one takes into account the storage required by the mass matrix.

Another feature of the mixed formulation is the sparsity of the stiffness matrices induced by the orthogonality properties. In Table 2.5, we compare the computational time for standard and mixed spectral methods. $cdsp$ and cdm are the number of operations required by standard and mixed spectral elements in d dimensions. Here, the gain is much less dramatic, due to the simplifications of the stiffness matrix of standard spectral elements induced by the use of a quadrature rule [23]. The bad behaviour of low–order approximations is due to the time of assembly which is very important, since most of the degrees of freedom are located on the boundary of the elements. One can globally say that the use of mixed elements is efficient from the 4th order of approximation[11]. Of course, comparison with other methods is of no need because of the huge time required by the inversion of the mass matrix.

[11] In practice, the number of operations seems to be underevaluated compared to the actual CPU time. So, the mixed formulation can be useful from the third order of approximation.

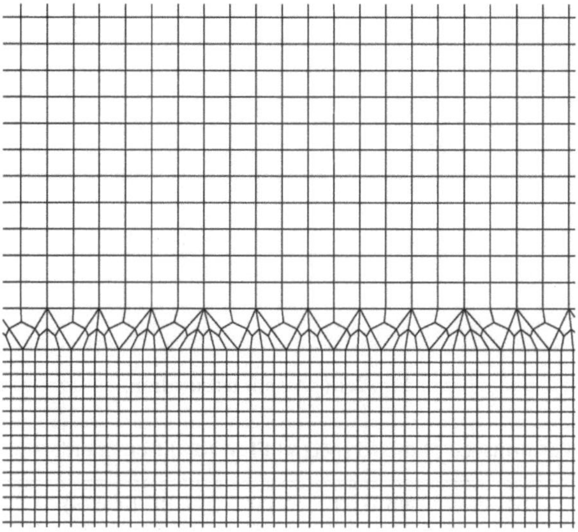

Fig. 2.17 Split 2D elements.

Meshing Problems

Producing a quadrilateral or a hexahedral mesh for complex geometries is not always obvious. In 2D, most of the mesh generators are able to mesh any domain by almost only quadrilaterals plus a few triangles. From such a mesh, one can derive a mesh composed of only quadrilaterals by splitting triangles into quadrilaterals as shown in Figure 2.17 (of course, one must start from a mesh with twice larger elements).

In 3D, such a technique is not possible. Most of mesh generators mix hexahedra with other elements such as pyramids and prisms[12]. Unfortunately, pyramids can not be split into hexahedra in a direct way. So, for reasonably complex domains, one can construct meshes by extrusion or, for more complex domains, one can derive hexahedral meshes from tetrahedral meshes by splitting tetrahedra into four hexahedra as shown in Figure 2.18. Of course, we get distorted hexahedra which can require a smaller mesh size than regular ones to get a good accuracy.

Performance of the Method

In order to illustrate the gain of storage and CPU time realized by high–order elements, we provide in Table 2.6 the results for the resolution of the wave equation on a regular mesh for different orders of approximation. The relative L^2–error permitted is around 5%. This table is an obvious illustration of the advantage of higher–order methods.

[12]CUBIT mesh generator seems to be able to construct hexahedral meshes for any geometry, but this construction can be quite difficult and requires a lot of manipulations from the user.

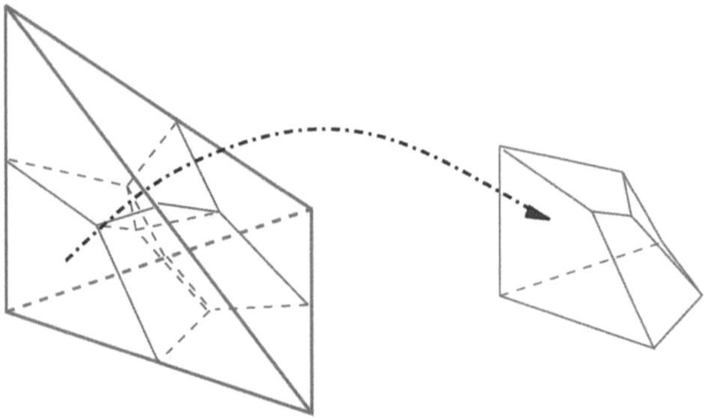

Fig. 2.18 A split tetrahedron.

Table 2.6 Performance of mixed spectral FEM for different orders of approximation.

Order	Nb. ele.	Nb. d.o.f	Δt	CPU time (s)	L^2–Error (%)
2	10000	40401	0.604e-5	556	5.2
3	1296	11881	0.882e-5	84	5.4
4	484	7921	0.908e-5	49	4.8
5	196	5041	0.961e-5	26	4.9
6	100	3721	0.971e-5	18	5.1

A Numerical Illustration

In this section, we give a numerical illustration of the capacity of the method. The case treated is an acoustic approximate model of a 2D geophysical experiment called the "foothills model". This model is a cross section of a layered medium with a complex external boundary (composed of hills). The length of the domain is 4200 m and its maximum height is a little less than 3000 m. The velocities vary from 1700 m/s (in the top layers) to 5500 m/s (in the bottom layers). The source (described by a Gaussian function in space of radius 42 m), located at the middle of the upper boundary, is a second–order Ricker function (second derivative of a Gaussian function) with a frequency equal to 20 Hz.

We mesh this domain by adapting the mesh to the velocity zones such that the edges of the elements are located at the interfaces. We have about two elements per wavelength in each direction. This leads to a mesh with 3141 elements plus 552 elements for the PML surrounding the domain except for the top boundary, Figure 2.20. Unbounded domain is modeled by using PML [9]. We solve the acoustics equation for $0 \leq t \leq 2$ s in its mixed formulation by using Q_5 (92 986 degrees of freedom in p) then Q_8 elements (237 409 degrees of freedom in p). Both approximations provide exactly the same solution, which should mean that we have obtained the converged solution. Snapshots of the solution are given in Figure 2.21.

Fig. 2.19 The different layers in the Foothills model.

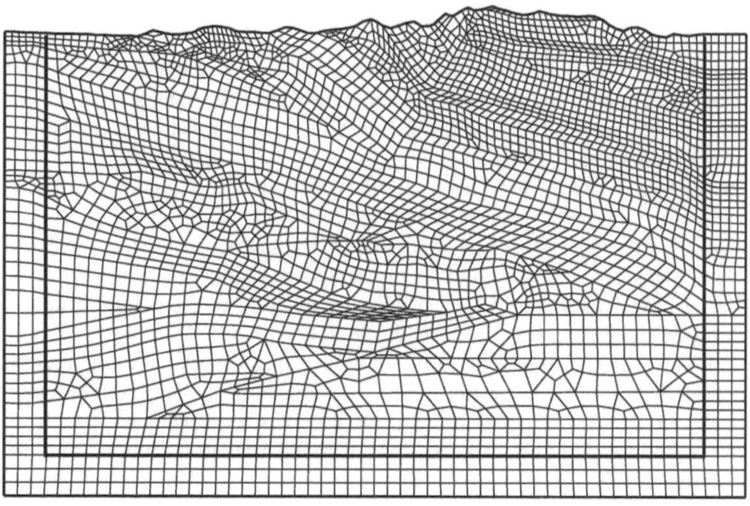

Fig. 2.20 The adapted mesh with its PML zone around it.

2.5 Outlook

In the recent years tremendous progress in the numerical computation of wave problems has been achieved, which is confirmed by the number of published papers as well as organized workshops and conferences in this area. Computational acoustics applying the FEM will remain an intensive topic of research. In addition to the pre-

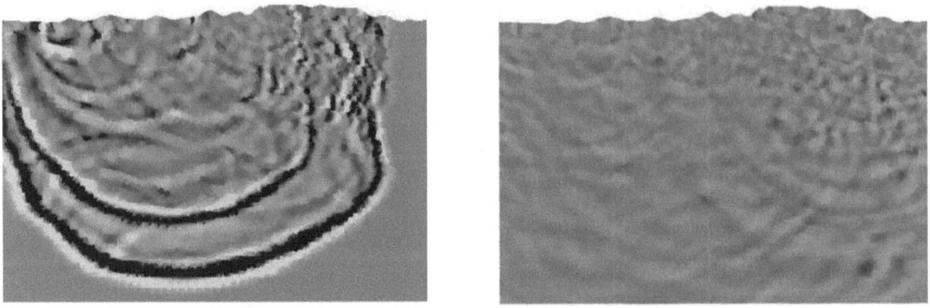

Fig. 2.21 Snapshots of the solution at $t = 0.8$ s (*left*) and $t = 1.6$ s (*right*)

sented enhanced FE formulations an increasing effort can be noticed for methods, which use instead of polynomial basis functions so called wave–like functions. The extension of FEM, called the partition of unity (PUM) [5] utilizes plane wave basis functions. This approach has been also utilized in least squares FEM [36] and the ultra weak variational formulation UWVF [7].

References

1. Ainsworth M (2004) Discrete dispersion relation for hp–version finite element approximation at high wave number. SIAM Journal of Numerical Analysis 42:553–575
2. Aziz KA, Kellogg RB (1981) Finite element analysis of a scattering problem. Mathematics of Computation 37:261–271
3. Babuška I, Guo BQ (1994) The p and h–p versions of the finite element method, basic principles and properties. SIAM Review 36:578–632
4. Babuška I, Guo BQ (1996) The approximation properties of the h-p version of the finite element method. Computer Methods in Applied Mechanics and Engineering 36:319–346
5. Babuska I, Melenk JM (1997) The partition of unity method. International Journal of Numerical Methods in Engineering 40:727–758
6. Castel N, Cohen G, Duruflé M (2007) Discontinuous Galerkin spectral method on hexahedral elements for aeroacoustic. Journal of Computational Acoustics (submitted)
7. Cessenat O, Després B (2003) Using plane waves as base functions for solving time harmonic equations with the ultra weak variational formulation. Journal of Computational Acoustics 11:227–238
8. Ciarlet PG (1987) The finite element method for elliptic problems. North–Holland, Amsterdam
9. Cohen G (2001) Higher order numerical methods for transient wave equations. Springer–Verlag, Berlin–Heidelberg
10. Cohen G, Fauqueux S (2000) Mixed finite elements with mass–lumping for the transient wave equation. Journal of Computational Acoustics 8:171–188
11. Cohen G, Fauqueux S (2005) Mixed spectral finite elements for the linear elasticity system in unbounded domains. SIAM Journal on Scientific Computing 26:864–884
12. Cohen G, Ferrieres X, Pernet S (2006) A spatial high–order hexahedral discontinuous Galerkin methods to solve Maxwell's equations in time domain. Journal of Computational Physics 217:340–363

13. Cohen G, Grob P (2007) Mixed higher–order spectral finite elements for Reissner–Mindlin equations. SIAM Journal on Scientific Computing 29:986–1005
14. Cohen G, Joly P, Roberts JE, Tordjman N (2001) Higher order triangular finite elements with mass lumping for the wave equation, SIAM Journal on Numerical Analysis 38:2047–2078
15. Cohen G, Monk P (1999) Mur–Nédélec finite element schemes for Maxwell's equations. Computer Methods in Applied Mechanics and Engineering 169:197–217
16. Craggs A (1972) The use of simple three dimensional acoustic finite elements for determining the natural modes and frequencies of complex shaped enclosures. Journal of Sound and Vibration 23:331–339
17. Craggs A (1978) A finite element model for rigid porous absorbing materials. Journal of Sound and Vibration 61:101–111
18. Craggs A (1997) Acoustic modeling: finite element method. In: Crocker MJ (ed) Encyclopedia of acoustics. John Wiley & Sons, New York 165–172
19. Demkowicz L (2006) Computing with hp finite elements. I. One– and two–dimensional elliptic and Maxwell problems. Chapman & Hall/CRC Press, Taylor and Francis, Boca Raton–London–New York
20. Dubiner M (1991) Spectral methods on triangles and other domains. SIAM Journal on Scientific Computing 6:345–390
21. Düster A (2003) High order finite elements for three–dimensional, thin–walled nonlinear continua. PhD Thesis, Technische Universität München, Lehrstuhl für Bauinformatik, Munich
22. Düster A, Broeker H, Rank E (2001) The p–version of the finite element method for three–dimensional curved thin walled structures. International Journal for Numerical Methods in Engineering 52:673–703
23. Duruflé M (2006) Integration numérique et éléments finis d'ordre élevé appliqués aux équations de Maxwell en régime harmonique. PhD Thesis, Université de Paris–Dauphine
24. Freund RW, Nachtigal NM (1991) QMR: a quasi–minimal residual method for non–Hermitian linear systems. Numerische Mathematik 60:315–339
25. Gladwell GML (1965) A finite element method for acoustics. In: Proceedings of the Fifth International Congress of Acoustics, Liege L33:1–4
26. Goldstein CI (1981) The finite element method with non–uniform mesh size applied to exterior Helmholtz problem. Numerische Mathematik 38:61–82
27. Hairer E, Nørsett SP, Wanner G (1993) Solving ordinary differential equations I: nonstiff problems. Springer–Verlag, New York, 2nd edition
28. Harari I (2006) A survey of finite element methods for time–harmonic acoustics. Computer Methods in Applied Mechanics and Engineering 195:1594–1607
29. Hauck A, Kaltenbacher M, Lerch R (2006) Simulation of thin piezoelectric structures using anisotropic hierarchic finite elements. In: Proceedings of the IEEE Ultrasonics Symposium, Vancouver 476–479.
30. Hennart J–P, Sainz E, Villegas M (1979) On the efficient use of the finite element method in static neutron diffusion calculations. Computational Methods in Nuclear Engineering 1:3–87
31. Hughes TJR (1987) The finite element method. Prentice–Hall, Englewood Cliffs
32. Ihlenburg F (1998) Finite element analysis of acoustic scattering. Springer–Verlag, New York
33. Ihlenburg F, Babuška I (1995) Finite element solution of the Helmholtz equation with high wave number. Part I: The h version of the FEM. Computer Mathematics Applied 30:9–37

34. Kagawa Y (1981) Vibration · acoustical engineering by finite element method. Baifukan, Tokyo (in Japanese)
35. Kaltenbacher M (2007) Numerical simulation of mechatronic sensors and actuators. Springer, Berlin–Heidelberg–New York
36. Monk P, Wang D (1999) A least squares method for the Helmholtz equation. Computer Methods in Applied Mechanics and Engineering 175:121–136
37. Morse PM, Ingard KU (1968) Theoretical acoustics. McGraw–Hill, New York
38. Mulder WA (2001) Construction and application of higher–order mass–lumped finite elements for the wave equation. Journal of Computational Acoustics 9:671–680
39. Otsuru T (1996) Finite elemental analysis of rooms. In: Summaries of the Annual Meeting, Architectural Institute of Japan 157–158 (in Japanese)
40. Otsuru T, Tomiku R (2000) Basic characteristics and accuracy of acoustic element using spline function in finite element sound field analysis. Journal of the Acoustical Society of Japan, Series E 21:87–95
41. Patera AT (1984) A spectral element method for fluid dynamics – laminar flow in a channel expansion. Journal of Computational Physics 54:468–488
42. Petyt M, Leas J, Koopmann GH (1976) A finite element method for determining the acoustic modes of irregular shaped cavities. Journal of Sound and Vibration 45:495–502
43. Pluymers B (2006) Wave based modelling methods for steady–state vibro–acoustics. PhD Thesis, Catholic University of Leuven
44. Saad Y (2003) Iterative methods for sparse linear systems. SIAM, Philadelphia, 2nd edition
45. Schenk O, Gaertner K (2004) Solving unsymmetric sparse systems of linear equations with pardiso. Journal of Future Generation Computer Systems 20:475–487
46. Schöberl J, Melenk JM, Pechstein CG, Zaglmayr SC (2007) Additive Schwarz preconditioning for p–version triangular and tetrahedral finite elements. IMA Journal of Numerical Analysis, doi:10.1093/imanum/drl046
47. Seriani G, Priolo E (1994) Spectral element method for acoustic wave simulation in heterogeneous media. Finite Elements in Analysis and Design 16:337–348
48. Sherwin SJ, Karniadakis GE (1995) A triangular spectral element method; applications to the incompressible Navier–Stokes equations. Computer Methods in Applied Mechanics and Engineering 123:189–229
49. Shuku T, Ishihara K (1973) The analysis of the acoustic field in irregularly shaped rooms by the finite element method. Journal of Sound and Vibration 29:67–75
50. Šolín P, Segeth K, Doležel I (2004) Higher–order finite element methods. Chapman & Hall/CRC Press, Boca Raton–London–New York
51. Szabó B, Babuška I (1991) Finite element analysis. John Wiley & Sons, New York
52. Tomiku R, Otsuru T, Takahashi Y (2002) Finite element sound field analysis of diffuseness in reverberation rooms. Journal of Asian Architecture and Building Engineering 1:33–39
53. Tordjman N (1995) Éléments finis d'ordre élevé avec condensation de masse pour l'équation des ondes. Phd Thesis, Université de Paris IX–Dauphine
54. Vorst HA van der (1992) Bi–CGSTAB: a fast and smoothly converging variant of Bi–CG for the solution of nonsymmetric linear systems. SIAM Journal on Scientific Statistical Computing 13:631–644
55. Young LC (1978) An efficient finite element method for reservoir simulation. In: Proceeding of the 53rd Annual Fall Technical Conference and Exhibition of the Society of Petroleum Engineers of AIME, Houston

3

Multifrequency Analysis using Matrix Padé–via–Lanczos

James P. Tuck–Lee, Peter M. Pinsky, and Haw–Ling Liew

Department of Mechanical Engineering, Faculty of Mechanics and Computation, Stanford University, Stanford, CA, 94305, USA
pinsky@stanford.edu

Summary. In many problems in acoustics, the frequency response is required over a large range of frequencies in order to characterize the behavior of the system. The simplest approach based on solving the matrix system of equations arising from a finite element discretization of the problem across a frequency range using a direct or iterative solver at each frequency can be prohibitively expensive. However, in modeling a physical problem, the analyst may be interested only in the solution on a subset of the computational domain, called the partial field. This smaller field of interest may be used to form a Padé approximation for the matrix equations in the frequency domain, constructing a reduced–order model that can be solved efficiently over multiple frequencies to compute the response of the partial field. In this chapter, the matrix Padé–via–Lanczos algorithm is applied to structural acoustics and for problems with non–reflecting boundary conditions. The approach is extended by the introduction of an adaptive scheme which can automatically span a frequency range of interest, producing an efficient and robust algorithm for multifrequency analysis.

3.1 Introduction

The dynamic behavior for many linear systems, including problems in acoustics and elasticity, can be characterized by the frequency response of the system. The frequency response assumes a time–harmonic input to the system, and computes the corresponding time–harmonic response as a function of the input frequency, characterized by circular frequency ω. Key characteristics of the system such as its natural frequencies can be determined from the frequency analysis. The dynamic response to any input condition may also be approximated using Fourier analysis of the loads and the application of the previously computed frequency response. Unfortunately, the computation of the frequency response can be computationally prohibitive. The standard approach involves solving a matrix system separately at every sampling frequency within the range desired, which can become very expensive with large systems of equations and poor conditioning of the system matrices.

In this chapter, a multifrequency approach is presented which provides an alternative to solving the full system at every frequency. Instead, a reduced order model is constructed around a reference frequency using the matrix Padé–via–Lanczos algorithm, which can be used to compute efficiently the response on a subset of the original domain. We will review first the original application of the Padé approximation to single degree–of–freedom systems and its extension to matrix systems. This will be applied to problems in both structural acoustics and exterior acoustics with radiating boundary conditions. Finally, we will present a new adaptive algorithm to compute accurately the response over a given frequency range of interest, and its practical application to complex problems in structural acoustics.

3.2 Padé Approximation

The use of the Padé approximation to construct reduced order models first appeared in the context of circuit modeling, but has been applied widely to other frequency–dependent problems including acoustics [3, 8–10], electromagnetics [11], and structural dynamics [3]. For Helmholtz–like problems, one can formally write the solution as:

$$\mathbf{d} = \left(\mathbf{K} - \omega^2\mathbf{M}\right)^{-1}\mathbf{F}\,, \tag{3.1}$$

where $\mathbf{K}$ and $\mathbf{M}$ are the $N \times N$ stiffness and mass matrices, and $\mathbf{F}$ is the force vector. For m different load cases, both $\mathbf{d}$ and $\mathbf{F}$ will be of size $N \times m$. In the present work, a partial field solution is defined to be a selected partition of the full solution vector, $\mathbf{d}$. For example, in many problems the solution is needed only at a restricted subset of the finite element nodes. In this case, we can write

$$\mathbf{H} = \mathbf{E}^T\left(\mathbf{K} - \omega^2\mathbf{M}\right)^{-1}\mathbf{F} \tag{3.2}$$

for a boolean selection matrix $\mathbf{E}$, of size $N \times p$, for p degrees of freedom in the partial field. In general, $\mathbf{H}$ could also be a linear combination of components of $\mathbf{d}$. We will show that insisting on a partial field solution such that $p << N$ is an important condition for the efficiency of the MPVL algorithm.

The partial field solution can be rewritten with respect to some reference frequency ω_0 as:

$$\mathbf{H} = \mathbf{E}^T\left(\mathbf{K} - \omega_0^2\mathbf{M} - (\omega^2 - \omega_0^2)\mathbf{M}\right)^{-1}\mathbf{F} \tag{3.3}$$

$$= \mathbf{E}^T\left(\mathbf{A}_0 - \sigma\mathbf{M}\right)^{-1}\mathbf{F} \tag{3.4}$$

$$= \mathbf{E}^T\left[\mathbf{A}_0\left(\mathbf{I} - \sigma\mathbf{A}_0^{-1}\mathbf{M}\right)\right]^{-1}\mathbf{F}\,, \tag{3.5}$$

$$\mathbf{H}(\sigma) = \mathbf{L}^T\left(\mathbf{I} - \sigma\mathbf{A}\right)^{-1}\mathbf{R}\,, \tag{3.6}$$

where

$$\sigma = \omega^2 - \omega_0^2\,, \tag{3.7}$$

$$\mathbf{A}_0 = \left(\mathbf{K} - \omega_0^2\mathbf{M}\right)\,, \tag{3.8}$$

$$\mathbf{A} = \mathbf{A}_0^{-1}\mathbf{M}\,, \tag{3.9}$$

$$\mathbf{L} = \mathbf{E}\,, \tag{3.10}$$

$$\mathbf{R} = \mathbf{A}_0^{-1}\mathbf{F}\,. \tag{3.11}$$

This will be referred to as the standard shifted form. For now, we will treat both $\mathbf{L}$ and $\mathbf{R}$ as vectors $\mathbf{l}$ and $\mathbf{r}$ respectively ($p = m = 1$), leaving $\mathbf{H}$ a scalar equation.

The Padé approximant of order n [5] is a rational function of the form

$$H_n(\sigma) = \frac{b_{n-1}\sigma^{n-1} + \ldots + b_1\sigma + b_0}{a_n\sigma^n + \cdots + a_1\sigma + 1}, \tag{3.12}$$

that matches the first $2n$ coefficients (in the scalar case) of the Taylor expansion of $H(\sigma)$:

$$H(\sigma) = \mathbf{l}^T(\mathbf{I} + \sigma\mathbf{A} + \sigma^2\mathbf{A}^2 + \ldots)\mathbf{r} = \sum_{i=0}^{\infty} \sigma^i \mathbf{M}_i\,. \tag{3.13}$$

The coefficients $\mathbf{M}_i$ are called the moments of $H(\sigma)$. The advantage is that as a rational series, this approximation can directly represent the significant poles and zeros of the system, leading to a better approximation than the Taylor expansion over a frequency range.

In asymptotic waveform expansion (AWE), these moments are computed directly [7]. Accordingly, one uses a partial fraction decomposition to write the Padé approximant as

$$H_n(\sigma) = \sum_{i=1}^{n} \frac{\zeta_i\lambda_i}{1 - \sigma\lambda_i} \tag{3.14}$$

and solves for the poles $1/\lambda_i$ via the characteristic polynomial and then for the residuals ζ_i. The problem with this approach is that finding the characteristic polynomial directly involves an extremely ill–conditioned matrix system, leading rapidly to numerical instability and stagnation.

However, the Lanczos process offers an alternative formulation, as shown by Feldmann and Freund [12], and by Gallivan, Grimme, and Van Dooren [15]. Usage of this algorithm allows for the computation of a much higher order Padé approximant before stagnation occurs. This algorithm has also been extended to multiple starting vectors (block matrices $\mathbf{L}$ and $\mathbf{R}$) by Aliaga et al. as an approach to compute the matrix–Padé approximation, which has been widely applied to acoustics [25, 38], electromagnetics [35, 39], and fluid–structure dynamics [23]. Other variations for generating the Padé approximation have also been made using multiple reference frequencies (see also, rational Krylov methods), but will not be considered herein.

3.2.1 Padé–via–Lanczos

As used by Feldmann and Freund [12] for single starting vectors, the nth right and left Krylov subspaces induced by the matrix $\mathbf{A}$ and vectors $\mathbf{r}$ and $\mathbf{l}$ can be written as

$$\mathcal{K}_n(\mathbf{A}, \mathbf{r}) = \text{span}\{\mathbf{r}, \mathbf{A}\mathbf{r}, \ldots, \mathbf{A}^{n-1}\mathbf{r}\}\,, \tag{3.15}$$

$$\mathcal{K}_n(\mathbf{A}^T, \mathbf{l}) = \text{span}\{\mathbf{l}, \mathbf{A}^T\mathbf{l}, \ldots, (\mathbf{A}^T)^{n-1}\mathbf{l}\}\,. \tag{3.16}$$

The Krylov subspaces are computed using Lanczos basis vectors $\mathbf{v}_i$ and $\mathbf{w}_i$ such that

$$\text{span}\,\{\mathbf{v}_i\}_1^n = \mathcal{K}_n(\mathbf{A}, \mathbf{r}) , \tag{3.17}$$

$$\text{span}\,\{\mathbf{w}_i\}_1^n = \mathcal{K}_n(\mathbf{A}^T, \mathbf{l}) . \tag{3.18}$$

These basis vectors are computed using an iterative unsymmetric Lanczos algorithm, as detailed in Feldmann and Freund:

The Lanczos Algorithm:

0\. Set $\mathbf{v}_0 = \mathbf{w}_0 = 0, \quad \delta_0 = 1, \quad \hat{\mathbf{v}}_1 = \mathbf{r}, \quad \hat{\mathbf{w}}_1 = \mathbf{l}.$

For $i = 1, 2, \ldots, n$:

1. $\rho_i = \|\hat{\mathbf{v}}_i\|_2, \quad \eta_i = \|\hat{\mathbf{w}}_i\|_2$
2. $\mathbf{v}_i = \hat{\mathbf{v}}_i/\rho_i, \quad \mathbf{w}_i = \hat{\mathbf{w}}_i/\eta_i, \quad \delta_i = \mathbf{w}_i^T\mathbf{v}_i$
3. $\alpha_i = \mathbf{w}_i^T\mathbf{A}\mathbf{v}_i/\delta_i, \quad \beta_i = \eta_i\delta_i/\delta_{i-1}, \quad \gamma_i = \rho_i\delta_i/\delta_{i-1}$
4. $\hat{\mathbf{v}}_{i+1} = \mathbf{A}\mathbf{v}_i - \alpha_i\mathbf{v}_i - \beta_i\mathbf{v}_{i-1}$
   $\hat{\mathbf{w}}_{i+1} = \mathbf{A}^T\mathbf{w}_i - \alpha_i\mathbf{w}_i - \gamma_i\mathbf{w}_{i-1}$

It should be noted that on rare occasions, breakdown is encountered where δ_i equals zero. A look–ahead procedure can be used to avoid these situations, where the biorthogonality condition between vectors $\mathbf{v}_i$ and $\mathbf{w}_i$ is relaxed in favor of biorthogonality of blocks thereof (i.e. groups of vectors $i \le n$). The reader is referred to Feldmann and Freund [12] for a discussion of this technique.

If we denote the matrices of these basis vectors $\mathbf{V}_n$ and $\mathbf{W}_n$, these satisfy the biorthogonality condition

$$\mathbf{W}_n^T\mathbf{V}_n = \mathbf{\Delta}_n = \text{diag}\{\delta_1, \delta_2, \ldots, \delta_n\} . \tag{3.19}$$

Also, using the Lanczos algorithm, we construct two tridiagonal matrices, $\mathbf{T}_n$ and $\tilde{\mathbf{T}}_n$

$$\begin{aligned} \mathbf{T}_n(i,i) = \tilde{\mathbf{T}}_n(i,i) &= \alpha_i , \\ \mathbf{T}_n(i,i+1) &= \beta_{i+1} , \\ \mathbf{T}_n(i+1,i) &= \rho_{i+1} , \\ \tilde{\mathbf{T}}_n(i,i+1) &= \gamma_{i+1} , \\ \tilde{\mathbf{T}}_n(i+1,i) &= \eta_{i+1} , \end{aligned} \tag{3.20}$$

such that

$$\mathbf{W}_n^T\mathbf{A}\mathbf{V}_n = \mathbf{\Delta}_n\mathbf{T}_n = \tilde{\mathbf{T}}_n^T\mathbf{\Delta}_n . \tag{3.21}$$

In this way, $\mathbf{T}_n$ is the projection of $\mathbf{A}$ onto $\mathcal{K}_n(\mathbf{A}, \mathbf{r})$ and $\tilde{\mathbf{T}}_n$ is the projection onto $\mathcal{K}_n(\mathbf{A}^T, \mathbf{l})$. We also observe that $\mathbf{v}_1\rho_1 = \mathbf{r}$ and $\mathbf{w}_1\eta_1 = \mathbf{l}$.

We can use these definitions to summarize the recurrences in the Lanczos algorithm as:

$$\mathbf{V}_n\mathbf{T}_{n-1} = \begin{cases} \mathbf{r} & \text{if } n = 1\,, \\ \mathbf{A}\mathbf{V}_{n-1} & \text{if } n > 1\,, \end{cases} \tag{3.22}$$

$$\mathbf{W}_n\tilde{\mathbf{T}}_{n-1} = \begin{cases} \mathbf{l} & \text{if } n = 1, \\ \mathbf{A}^T\mathbf{W}_{n-1} & \text{if } n > 1\,. \end{cases} \tag{3.23}$$

Finally, we construct the reduced–order equations (in the absence of look–ahead). First, we approximate the inversion by projection onto the Lanczos basis vectors

$$(\mathbf{I} - \sigma\mathbf{A})^{-1} \approx \mathbf{V}_n \left(\mathbf{W}_n^T\mathbf{I}\mathbf{V}_n - \sigma\mathbf{W}_n^T\mathbf{A}\mathbf{V}_n\right)^{-1}\mathbf{W}_n^T\,. \tag{3.24}$$

Thus, we can approximate Equation 3.6 as

$$\mathbf{H}_n(\sigma) = \mathbf{l}^T\mathbf{V}_n \left(\mathbf{W}_n^T\mathbf{I}\mathbf{V}_n - \sigma\mathbf{W}_n^T\mathbf{A}\mathbf{V}_n\right)^{-1}\mathbf{W}_n^T\mathbf{r} \tag{3.25}$$

$$= \mathbf{l}^T\mathbf{V}_n \left(\mathbf{I} - \sigma\mathbf{T}_n\right)^{-1}\boldsymbol{\Delta}_n^{-1}\mathbf{W}_n^T\mathbf{r} \tag{3.26}$$

$$= \eta_1\mathbf{w}_1^T\mathbf{V}_n \left(\mathbf{I} - \sigma\mathbf{T}_n\right)^{-1}\boldsymbol{\Delta}_n^{-1}\mathbf{W}_n^T\mathbf{v}_1\rho_1 \tag{3.27}$$

$$= \eta_1\rho_1\delta_1\mathbf{e}_1 \cdot \left(\mathbf{I} - \sigma\mathbf{T}_n\right)^{-1}\mathbf{e}_1 \tag{3.28}$$

for $\mathbf{e}_1$ the first unit vector in $\mathbb{R}^n$. Noting that $\eta_1\rho_1\delta_1 = \mathbf{l}^T\mathbf{r}$, we can write this as

$$\mathbf{H}_n(\sigma) = \mathbf{l}^T\mathbf{r} \cdot \mathbf{e}_1^T\left(\mathbf{I} - \sigma\mathbf{T}_n\right)^{-1}\mathbf{e}_1\,. \tag{3.29}$$

The equivalence of this equation to the Padé approximation can be shown by taking the eigendecomposition of $\mathbf{T}_n = \mathbf{U}_n\boldsymbol{\Sigma}_n\mathbf{U}_n^{-1}$, such that $\boldsymbol{\Sigma}_n = \text{diag}(\lambda_i)$. Now, we can write this as

$$\mathbf{H}_n = \mathbf{l}^T\mathbf{r} \cdot \mathbf{e}_1^T\left(\mathbf{I} - \sigma\mathbf{U}_n\boldsymbol{\Sigma}_n\mathbf{U}_n^{-1}\right)^{-1}\mathbf{e}_1 \tag{3.30}$$

$$= \mathbf{l}^T\mathbf{r} \cdot \mathbf{e}_1^T\mathbf{U}_n\left[\mathbf{I} - \sigma\text{diag}(\lambda_i)\right]^{-1}\mathbf{U}_n^{-1}\mathbf{e}_1 \tag{3.31}$$

$$= \mathbf{l}^T\mathbf{r} \cdot \mathbf{e}_1^T\mathbf{U}_n\left[\frac{1}{1 - \sigma\lambda_i}\right]\mathbf{U}_n^{-1}\mathbf{e}_1 \tag{3.32}$$

$$= \sum_{i=1}^{n}\frac{\zeta_i\lambda_i}{1 - \sigma\lambda_i} \tag{3.33}$$

as in Equation 3.14, for

$$\mathbf{l}^T\mathbf{r}^T\mu_i\nu_i = \zeta_i\lambda_i\,, \tag{3.34}$$

where μ_i and ν_i are the components of

$$\mu^T = \mathbf{e}_1^T\mathbf{U}_n\,, \tag{3.35}$$

$$\nu = \mathbf{U}_n^{-1}\mathbf{e}_1\,. \tag{3.36}$$

3.3 Matrix Padé–via–Lanczos

For multiple starting vectors, the Lanczos–type process developed by Aliaga et al. [1] creates the block Krylov matrices

$$\mathbf{K}(\mathbf{A}, \mathbf{R}) = \text{colspan}\{\mathbf{R}, \mathbf{A}\mathbf{R}, \ldots, \mathbf{A}^{N-1}\mathbf{R}\}\,, \tag{3.37}$$

$$\mathbf{K}(\mathbf{A}^T, \mathbf{L}) = \text{colspan}\{\mathbf{L}, \mathbf{A}^T\mathbf{L}, \ldots, (\mathbf{A}^T)^{N-1}\mathbf{L}\} \tag{3.38}$$

for starting block vectors $\mathbf{R} \in \mathbb{R}^{m \times N}$ and $\mathbf{L} \in \mathbb{R}^{p \times N}$.

In order to remove linearly dependent vectors from the subspace definition due to the presence of block starting vectors, we delete the linear dependent columns. This is called "exact deflation", and modifies the form of the Krylov matrices

$$\mathbf{K}^{dl}(\mathbf{A}, \mathbf{R}) = \text{colspan}\{\mathbf{R}_1, \mathbf{A}\mathbf{R}_2, \ldots, \mathbf{A}^{j-1}\mathbf{R}_j\}\,, \tag{3.39}$$

$$\mathbf{K}^{dl}(\mathbf{A}^T, \mathbf{L}) = \text{colspan}\{\mathbf{L}_1, \mathbf{A}^T\mathbf{L}_2, \ldots, (\mathbf{A}^T)^{k-1}\mathbf{L}_k\} \tag{3.40}$$

for $j, k \leq N$. Here, $\mathbf{R}_i$ and $\mathbf{L}_i$ are submatrices of $\mathbf{R}_{i-1}$ and $\mathbf{L}_{i-1}$ respectively, removing columns if deflation occurs. The n–th Krylov subspace, $\mathcal{K}_n$, is defined as spanning the first n columns of $\mathbf{K}^{dl}$. If no deflation occurs, then this corresponds to the column space for $j = \lfloor n/p \rfloor$ and $k = \lfloor n/m \rfloor$.

The Krylov subspaces are computed using Lanczos–type basis vectors $\mathbf{v}_i$ and $\mathbf{w}_i$ similar to the single vector case, such that:

$$\text{span}\,\{\mathbf{v}_i\}_1^n = \mathcal{K}_n(\mathbf{A}, \mathbf{R})\,, \tag{3.41}$$

$$\text{span}\,\{\mathbf{w}_i\}_1^n = \mathcal{K}_n(\mathbf{A}, \mathbf{L})\,. \tag{3.42}$$

If we denote the matrices of these basis vectors $\mathbf{V}_n$ and $\mathbf{W}_n$, these satisfy the biorthogonality condition

$$\mathbf{W}_n^T\mathbf{V}_n = \boldsymbol{\Delta}_n\,, \tag{3.43}$$

where $\boldsymbol{\Delta}_n \in \mathbb{R}^{n \times n}$ is now block diagonal. The Lanczos–type algorithm used to construct these bases can be summarized using a recurrence relationship (ignoring deflation and look–ahead)

$$\mathbf{V}_n\mathbf{T}_{n-m} = \begin{cases} [\mathbf{r}_1 \quad \mathbf{r}_2 \ldots \mathbf{r}_n] & \text{if } 1 \leq n \leq m\,, \\ \mathbf{A}\mathbf{V}_{n-m} & \text{if } n > m\,, \end{cases} \tag{3.44}$$

$$\mathbf{W}_n\tilde{\mathbf{T}}_{n-p} = \begin{cases} [\mathbf{l}_1 \quad \mathbf{l}_2 \ldots \mathbf{l}_n] & \text{if } 1 \leq n \leq p\,, \\ \mathbf{A}^T\mathbf{W}_{n-p} & \text{if } n > p\,, \end{cases} \tag{3.45}$$

where $\mathbf{T}_n$ and $\tilde{\mathbf{T}}_n \in \mathbb{R}^{n \times n}$ are banded with half–bandwidth n, and $\mathbf{r}_i$ and $\mathbf{l}_i$ are the columns of $\mathbf{R}$ and $\mathbf{L}$, respectively. We define Δ_n such that

$$\mathbf{W}_n^T\mathbf{A}\mathbf{V}_n = \boldsymbol{\Delta}_n\mathbf{T}_n = \tilde{\mathbf{T}}_n^T\boldsymbol{\Delta}_n \tag{3.46}$$

and introduce $\boldsymbol{\eta}_n \in \mathbb{R}^{n \times m}$ and $\boldsymbol{\rho}_n \in \mathbb{R}^{n \times p}$ that satisfy

$$\mathbf{W}_n \boldsymbol{\eta}_n = \mathbf{L}\,, \tag{3.47}$$

$$\mathbf{V}_n \boldsymbol{\rho}_n = \mathbf{R}\,. \tag{3.48}$$

Using these identities on the standard shifted form leads to a reduced–order system of equations, as was the case for single starting vectors. We can again approximate Equation 3.6 as

$$\mathbf{H}_n(\sigma) = \mathbf{L}^T \mathbf{V}_n \left(\mathbf{W}_n^T \mathbf{I} \mathbf{V}_n - \sigma \mathbf{W}_n^T \mathbf{A} \mathbf{V}_n\right)^{-1} \mathbf{W}_n^T \mathbf{R} \tag{3.49}$$

$$= \mathbf{L}^T \mathbf{V}_n \left(\mathbf{I} - \sigma \mathbf{T}_n\right)^{-1} \boldsymbol{\Delta}_n^{-1} \mathbf{W}_n^T \mathbf{R} \tag{3.50}$$

leading to the reduced–order matrix equation

$$\mathbf{H}_n(\sigma) = \boldsymbol{\eta}_n^T \boldsymbol{\Delta}_n \left(\mathbf{I} - \sigma \mathbf{T}_n\right)^{-1} \boldsymbol{\rho}_n\,, \tag{3.51}$$

which is the block Krylov subspace projection of the standard shifted form. Additionally, Freund [13] has proved that this is also the matrix Padé approximant of the system, in that

$$\mathbf{H}_n(\sigma) = \mathbf{H}(\sigma) + \mathcal{O}(\sigma^{t(n)}) \tag{3.52}$$

$$t(n) \leq \lfloor n/p \rfloor + \lfloor n/m \rfloor\,, \tag{3.53}$$

where integer $t(n)$ is maximal. Thus, this process is called the matrix Padé–via–Lanczos (MPVL) algorithm.

The reader is referred to Aliaga et al. [1] for a detailed description of this algorithm, and the application of deflation and look–ahead. The resulting matrix equation from this algorithm constitutes an oblique projection of the original matrix system onto a Krylov subspace, and forms an efficient reduced–order model, as $\mathbf{T}_n$ is a much smaller, banded matrix compared to the original system matrices of size N. It should also be observed that the quality of the resulting matrix Padé approximant, as represented by $t(n)$, is a function of the number of Krylov iterations (n), the size of the partial field (p) and the number of load cases (m). Thus, increasing the size of the starting block vectors will generally require an increase in the size of the Krylov subspace to maintain similar accuracy.

3.3.1 Complexity Comparison

The main cost for the MPVL algorithm is the inversion of the shifted matrix around the reference frequency, $\mathbf{A}_0^{-1}$, an $\mathcal{O}(N\log N)$ operation using modern sparse direct solvers and reordering algorithms. The construction of the Lanczos vectors involve matrix–vector products and orthogonalization of $\mathcal{O}(N)$ per iteration. In contrast, each frequency step using the reduced matrix system only requires an inversion of a banded diagonal matrix, $\mathcal{O}(n)$, and matrix–vector operations of the same order. Thus, for $n << N$, the frequency step computations are much more computationally efficient than the initial cost of factorization. In summary, the complexity of this algorithm may be viewed as $\mathcal{O}(N\log N + nN + kn)$ for k frequency steps, compared to $\mathcal{O}(kN\log N)$ for a direct solution.

3.4 Application to Structural Acoustics

Problems in structural acoustics present several challenges to traditional solution methods when solving for the response at multiple frequencies. The size, structure, and conditioning of the system matrices make classical solution techniques impractical when solving over the wide range of frequencies desired. 3D coupled problems lead to matrices which are not easily optimized for bandwidth, which hampers the effectiveness of direct profile solvers. The system matrices often have large condition numbers that slow the convergence of iterative solutions, arising from the unsymmetric matrix structure, physical dimensions, and different wave speeds in the elastic and acoustic media. In problems involving thin shells, this ill–conditioning is even more pronounced, further degrading iterative solver performance. Most robust preconditioners that would typically be used in static analysis have to be recomputed at every frequency step, as the condition of the system matrix changes. This also applies to recently developed preconditioners for Helmholtz problems, such as multigrid methods [27]. Although these preconditioners may be reused across multiple adjacent frequencies (or the cost reduced using a previous preconditioner), a large number of preconditioners would need to be constructed for a large frequency range of interest, which can be very costly for such extremely ill–conditioned problems.

Block solution and domain decomposition methods, e.g. [26], face similar challenges. For example, using the Schur complement, one would perform a direct solve (or a large block iterative solve) of the acoustic block (which may be easily preconditioned, such as by SSOR), followed by another inversion of the Schur complement (which would remain difficult to precondition). This gives the advantage of solving two smaller systems, but still leaves the necessity of doing so at every frequency. Alternatively, iteration between the acoustic and structural domains via loose coupling can lead to slow convergence behavior when the two fields are highly dependent on each other as they are in structural acoustics.

Modal decomposition of unsymmetric matrices requires the solution of two generalized eigenvalue problems, which can be costly for large problems. Further, depending on the loading conditions, accurate results between resonances may require a large range of eigenvalues and corresponding eigenvectors, which also lead to impractical memory requirements for large sparse systems. For these reasons, modal decomposition techniques on the tightly coupled problem also proves to be impractical. Diagonalizing the individual systems independently leads to large dense coupling matrices which again leads to increased computation.

In contrast, the reduced–order matrix Padé–via–Lanczos (MPVL) algorithm has proved to be a viable alternative. As the Padé approximation for the system around an expansion frequency, it provides an inexpensive partial field solution across a wide frequency range that is less affected by matrix conditioning than traditional solution methods. Furthermore, as a Krylov subspace projection, it requires a relatively small basis, improving on both computation and storage compared to modal reduction.

3.4.1 Formulation

We consider a thin plate coupled to an acoustic domain. The Helmholtz equation for the acoustic domain may be written as a function of pressure, P

$$\nabla^2 P + k^2 P = 0 \quad \text{in } \Omega\,, \tag{3.54}$$
$$P = g \quad \text{on } \delta\Omega_g\,, \tag{3.55}$$
$$\nabla P \cdot \mathbf{n}_a = h \quad \text{on } \delta\Omega_h \tag{3.56}$$

for wave number k, density ρ_a, prescribed boundary conditions g and h, and outward normal vector $\mathbf{n}_a$. Similarly, the linear elasticity equations may be written in terms of displacements, $\mathbf{u}$

$$\nabla \cdot \boldsymbol{\sigma} + \omega^2 \rho_e \mathbf{u} = 0 \quad \text{in } \mathbb{B}\,, \tag{3.57}$$
$$\mathbf{u} = \mathbf{g} \quad \text{on } \delta\mathbb{B}_{\mathbf{g}}\,, \tag{3.58}$$
$$\boldsymbol{\sigma} \cdot \mathbf{n}_e = \mathbf{h} \quad \text{on } \delta\mathbb{B}_{\mathbf{h}}\,, \tag{3.59}$$
$$\boldsymbol{\sigma} = \mathbb{C} : \nabla^S \mathbf{u} \tag{3.60}$$

for linearized strain $\nabla^S \mathbf{u}$, corresponding stress $\boldsymbol{\sigma}$, elasticity tensor $\mathbb{C}$, density ρ_e, prescribed boundary conditions $\mathbf{g}$ and $\mathbf{h}$, and outward normal vector $\mathbf{n}_e$. Coupling on the structural acoustic interface is governed by the relationship between pressures, stresses, and velocity

$$P\mathbf{n}_e = -\boldsymbol{\sigma} \cdot \mathbf{n}_e \quad \text{on } \delta\Omega \cup \delta\mathbb{B}\,, \tag{3.61}$$
$$\nabla P = -\mathrm{i}\omega\rho_a \mathbf{v} = \omega^2 \rho_a \mathbf{u} \quad \text{on } \delta\Omega \cup \delta\mathbb{B}\,. \tag{3.62}$$

We define the normal on the interface as $\mathbf{n} = \mathbf{n}_e = -\mathbf{n}_a$. This corresponds to the variational equation

$$\begin{aligned} &\int_\Omega \left(\nabla Q \cdot \nabla P - k^2 QP\right) d\Omega + \int_{\delta\Omega_h \cup \delta\mathbb{B}} Q\left(\omega^2 \rho_a \mathbf{u} \cdot \mathbf{n}\right) d\Gamma + \\ &\int_{\mathbb{B}} \left(\nabla^S \mathbf{w} : \mathbb{C} : \nabla^S \mathbf{u} - \mathbf{w}\omega^2 \rho_e \mathbf{u}\right) d\mathbb{B} + \int_{\delta\Omega_h \cup \delta\mathbb{B}} \mathbf{w} \cdot (P\mathbf{n}) d\Gamma = \\ &\int_{\mathbb{B}} \mathbf{w} \cdot \mathbf{f} d\mathbb{B} + \int_{\delta\Omega_h} Qh d\Gamma + \int_{\delta\mathbb{B}_{\mathbf{h}}} \mathbf{w} \cdot \mathbf{h} d\Gamma \end{aligned} \tag{3.63}$$

for Q and $\mathbf{w}$ the variations of P and $\mathbf{u}$, respectively.

To model the thin plate, we have used an 8–node solid shell element developed by Vu–Quoc and Tan [37]. In order to prevent locking behavior, a combination of mixed interpolation (assumed natural strains) and enhancing strains is added to the stiffness formulation ($\mathbf{K}_e$). Two shell elements through the thickness were used to resolve the simply supported boundary conditions on the mid–surface of the plate. The acoustic medium is modeled by linear tetrahedra, due to the potential complexity of the acoustic volume geometries in general application that would prohibit an ordered hexahedral mesh. The two domains are coupled using node–on–node coupling at the

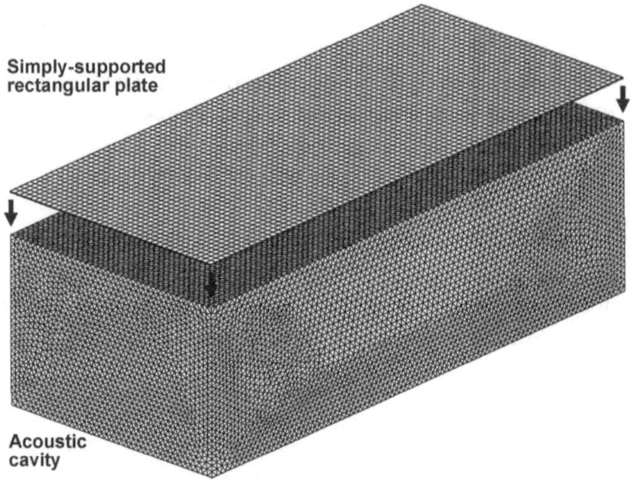

Fig. 3.1 Illustrative example of a simply–supported plate coupled to an acoustic cavity with reflecting boundary conditions.

interface, integrating over quadrilaterals. It should be noted that this results in an inconsistent interpolation of the pressure distribution on the interface, which introduces an $\mathcal{O}(h)$ error, for mesh size h. We have used a typical $(\mathbf{u}, p)$ formulation for the coupled equations (for ω strictly greater than zero), resulting in the unsymmetric matrix equations

$$\left\{ \begin{bmatrix} \mathbf{K}_a & \mathbf{0} \\ \mathbf{K}_c^T & \mathbf{K}_e \end{bmatrix} - \omega^2 \begin{bmatrix} \frac{1}{c^2}\mathbf{M}_a & -\rho\mathbf{K}_c \\ \mathbf{0} & \mathbf{M}_e \end{bmatrix} \right\} \begin{bmatrix} \mathbf{p} \\ \mathbf{d} \end{bmatrix} = \begin{bmatrix} \mathbf{F}_P \\ \mathbf{F}_u \end{bmatrix}, \tag{3.64}$$

where

$$\mathbf{K}_c = \int_{\delta\Omega_h \cup \delta\mathbb{B}} N^T \mathbf{n}^T [N] d\Gamma \tag{3.65}$$

for N the shape function array for the pressures on the coupling quadrilaterals and $[N]$ the matrix equivalent for the displacements. $\mathbf{K}_e$ and $\mathbf{M}_e$ are the stiffness and mass matrices associated with the uncoupled structural equations [37], and $\mathbf{K}_a$ and $\mathbf{M}_a$ are the stiffness and mass matrices arising from the Helmholtz equation for linear acoustics. Finally, for a partial field solution using selector matrix $\mathbf{E}$, this can be written in the familiar form:

$$\mathbf{d}_r = \mathbf{E}^T \left(\mathbf{K} - \omega^2\mathbf{M}\right)^{-1} \mathbf{F}. \tag{3.66}$$

3.4.2 Illustrative Example

For the numerical example, we consider a simply supported thin aluminium plate with dimensions 0.5461 m x 0.2413 m x 0.9144 mm, coupled to an acoustic box

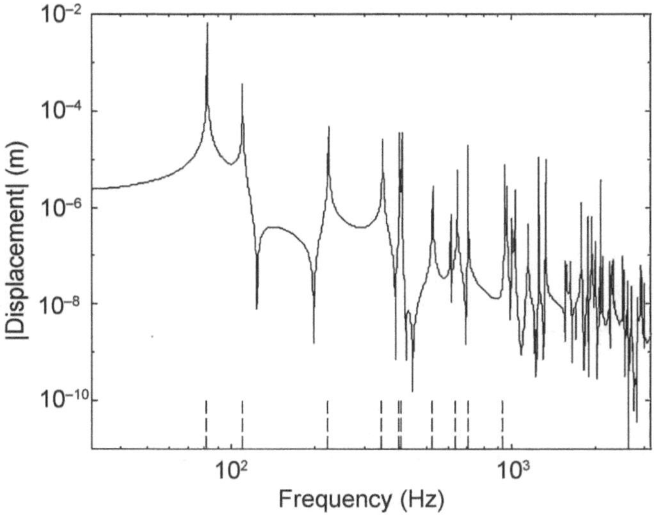

Fig. 3.2 Magnitude of displacement for center of plate computed using the finite element model, compared with ten computed natural frequencies (– – –) using the algorithm by Pretlove [32]. 2000 frequency steps computed, distributed evenly on a log scale.

of the same length and width, and a depth of 0.16764 m filled with air. Figure 3.1 depicts this geometry, with the plate (hexahedra) on the upper surface of the acoustic cavity (tetrahedra). This example problem corresponds (using the same dimensions and material properties) to one analyzed by A.J. Pretlove in 1965 [32] using sine and cosine expansions to compute the natural frequencies of the coupled system. The resulting finite element mesh consisted of 140,000 nodes corresponding to 160,000 degrees of freedom, and a total of 750,000 elements in order to span a large frequency range. The finite element model was computed using a custom in–house code and the MUMPS solver [2] in parallel on 6 processors on a distributed memory cluster. In our implementation using centralized assembled matrix input, the only calculations done in parallel were the full matrix inversion and matrix–vector operations involving the matrix factors. In all, 4 million nonzero entries were required for the original matrix, and 180 million nonzero entries for its factors using PORD reordering [34] to reduce fill–in. The system is evaluated for a uniform pressure load on the upper surface of the plate at 2000 frequency steps sampled logarithmically from 30 Hz to 3 kHz (at $f = 10^k$ Hz, for $k = 1.5$ to 3.5 in increments of 0.001).

Figure 3.2 shows the agreement between several natural frequencies computed using the algorithm by Pretlove [32], and direct solutions of the full matrix equations under the distributed load, plotting the displacement of the center of the plate versus frequency. This serves to validate and benchmark our finite element model using MUMPS [2] as a direct solver. The solution times, given by the wall clock on the cluster, resulted in 3 minutes for the initial symbolic factorization, and 2.2 minutes for each frequency step solution. For 2000 frequency steps shown in the figure,

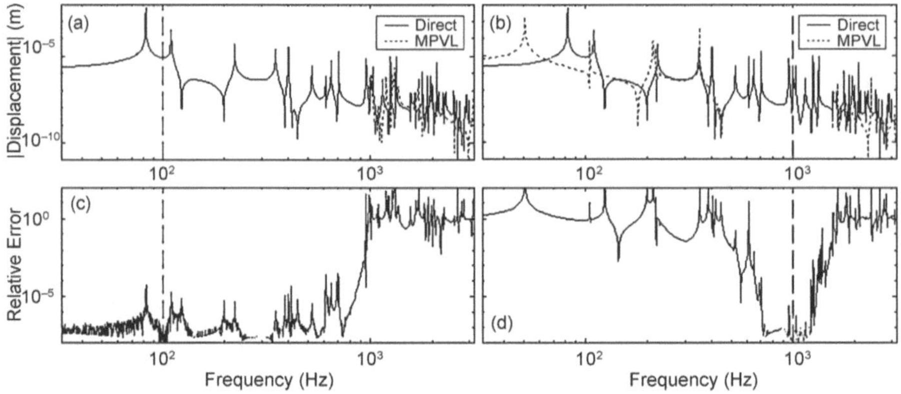

Fig. 3.3 Comparison between direct and MPVL solution at center of plate, about (a) 100 Hz and (b) 1000 Hz. Relative error between direct and MPVL solution about (c) 100 Hz and (d) 1000 Hz, $n = 100$ iterations.

distributed uniformly on the log scale shown, this took approximately 80 hours of computational time.

3.4.3 Performance of MPVL

We will use the structural acoustics example to examine the behavior of MPVL. The three degrees of freedom at the center node of the plate were chosen for the partial field of interest for this problem. Choosing to use 100 Lanczos iterations, the MPVL solution was computed in only 20 minutes for the same 2000 frequency steps as the direct solution above, using the same MUMPS solver and parallel implementation. As expected, as a matrix Padé approximant, the reduced–order system matches the behavior of the original system matrices very well around the selected reference frequency. Figure 3.3 shows the comparison between the MPVL solution and the direct solution for two different reference frequencies, at 100 Hz and 1 kHz. The relative error is computed as the norm of the difference between the MPVL and direct partial field solution, divided by the magnitude of the response. A sharply defined window of accuracy is observed in both cases around the reference frequencies. It is clear (even on a logarithmic scale) that the window sizes are not equal. This is partly due to the number of resonances within the frequency range. This can be seen directly from the order of the Padé approximant, which determines how many poles of the system are represented.

Additionally, the accuracy of the Lanczos–type algorithm is affected by the ability to construct new Lanczos basis vectors, which in turn is limited by finite precision and round–off error [30]. In exact arithmetic, the Lanczos algorithm should converge to the original matrix system. However, the size of the original system, condition number, and machine precision all impact the maximum number of iterations before we can no longer calculate new orthogonal Lanczos basis vectors [29]. This can

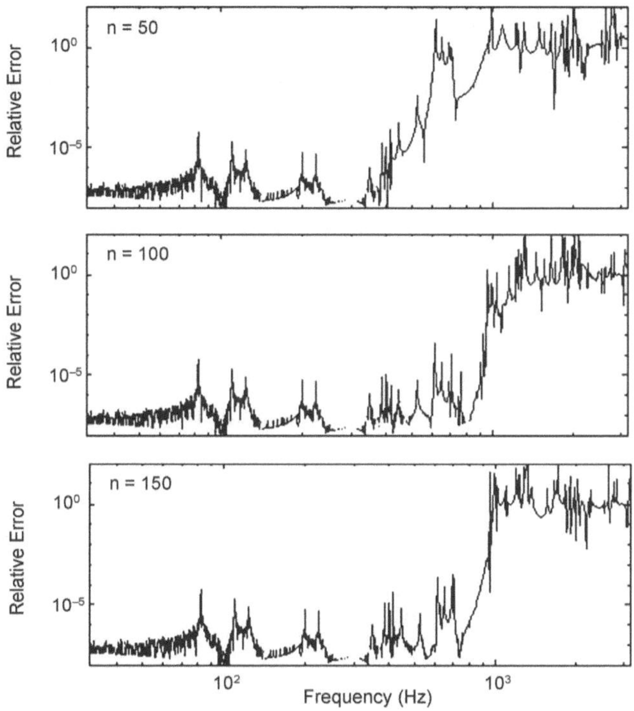

Fig. 3.4 Relative error between direct and MPVL solution about 100 Hz for $n = 50, 100$, and 150 iterations.

be seen in Figure 3.4. Increasing the number of Lanczos iterations from 50 to 100 clearly shows an increase in the window size. However, increasing to 150 iterations shows no such gains. Thus, we cannot rely on a higher–order reconstruction giving a more accurate result. Furthermore, the reduced order model produced from the 150 iterations is half again the size of the one produced for 100 iterations, requiring more computation without any gain in the accuracy of the solution. Unfortunately, the number of iterations before this behavior occurs varies by problem and reference frequency.

3.5 Extension to Acoustics in an Exterior Domain

An example of an extension of MPVL is one involving an acoustic domain with a non–reflecting boundary condition as described in Malhotra and Pinsky [25] and Wagner et al. [38]. This is an example that does not appear to conform to the requirement that it can be expressed in the standard shifted form necessary for the MPVL algorithm. However, we will show that under certain conditions the algorithm may indeed be applied, extending the algorithm to the case of exterior acoustics.

3.5.1 Formulation

We use the Dirichlet–to–Neumann (DtN) map described by Keller and Givoli [20] to represent the non–reflecting boundary condition on a finite acoustic domain boundary. The governing equations can then be written as a function of pressure, P and wave number k

$$\nabla^2 P + k^2 P = 0 \quad \text{in } \Omega\,, \tag{3.67}$$

$$P = g \quad \text{on } \delta\Omega_g\,, \tag{3.68}$$

$$\nabla P \cdot \mathbf{n} = h \quad \text{on } \delta\Omega_h\,, \tag{3.69}$$

$$\nabla P \cdot \mathbf{n} = -\mathcal{B}_{DtN}(P) \quad \text{on } \delta\Omega_{DtN}\,. \tag{3.70}$$

For a 2D problem with a circular truncation boundary of radius R, the DtN condition can be written as

$$\mathcal{B}_{DtN}(P)(\mathbf{x}) = \sum_{n=0}^{\infty} z_n(kR) \int_{\delta\Omega_{DtN}} s_n(\mathbf{x},\xi)P(\xi)d\Gamma_\xi, \quad \mathbf{x} \in \delta\Omega_{DtN}\,, \tag{3.71}$$

$$z_n(kR) = \frac{\alpha_n k H_n'(kR)}{2\pi R H_n(kR)}, \quad n > 0\,, \tag{3.72}$$

$$s_n = \cos n(\theta - \theta')\,, \tag{3.73}$$

$$\alpha_n = \begin{cases} 1, & n = 0 \\ 2, & n > 0\,, \end{cases} \tag{3.74}$$

where θ and θ' denote the angle of a point on the circle (such that $d\Gamma_\xi = Rd\theta'$), H_n are spherical Hankel functions of the first kind, z_n are the radial–impedance coefficients, and s_n are the surface harmonics. In order to ensure uniqueness of the solution, we modify the DtN formulation for a finite expansion N_{DtN}, using the approach by Grote and Keller [18] by adding and subtracting the first–order Bayliss–Gunzburger–Turkel local boundary operator

$$\mathcal{B}_1 = \left(\mathrm{i}k - \frac{1}{2R}\right) P(\mathbf{x}) \tag{3.75}$$

such that

$$\mathcal{B}^h_{DtN} = \tilde{\mathcal{B}}^h + \mathcal{B}_1 \tag{3.76}$$

and

$$\tilde{\mathcal{B}}^h = \sum_{n=0}^{N_{DtN}} \tilde{z}_n \int_{\delta\Omega_{DtN}} s_n(\mathbf{x},\xi)P(\xi)d\Gamma_\xi, \quad \mathbf{x} \in \delta\Omega_{DtN}\,, \tag{3.77}$$

$$\tilde{z}_n = z_n - \frac{\alpha_n}{2\pi R}\left(\mathrm{i}k - \frac{1}{2R}\right). \tag{3.78}$$

This produces the variational equation

$$\int_{\Omega} (\nabla Q \cdot \nabla P - k^2 QP)\, d\Omega + \int_{\delta\Omega_{DtN}} Q\mathcal{B}_1(P) d\Gamma \ldots$$
$$+ \int_{\delta\Omega_{DtN}} Q\tilde{\mathcal{B}}(P) d\Gamma = \int_{\delta\Omega_h} Qh d\Gamma\,. \tag{3.79}$$

In discretized form with N degrees of freedom, this becomes:

$$[\mathbf{K} - k^2\mathbf{M} + \mathbf{B}_1(k) + \tilde{\mathbf{K}}_{DtN}(k)]\mathbf{p} = \mathbf{f}\,, \tag{3.80}$$

where $\mathbf{K}$ and $\mathbf{M}$ are the typical stiffness and mass matrices for the Helmholtz equation, and

$$\mathbf{B}_1 = \int_{\delta\Omega_{DtN}} \left(\mathrm{i}k - \frac{1}{2R}\right) \mathbf{N}\mathbf{N}^T d\Gamma\,, \tag{3.81}$$

$$\tilde{\mathbf{K}}_{DtN} = \int_{\delta\Omega_{DtN}} \mathbf{N}\tilde{\mathcal{B}}^h(\mathbf{N}) d\Gamma\,. \tag{3.82}$$

It can be shown [24] that $\tilde{\mathbf{K}}_{DtN}$ is a dense matrix, but can be written as a low–rank update of order $N_{mod} = (2N_{DtN} + 1)$ to the matrix system, such that

$$\left[\mathbf{K} - k^2\mathbf{M} + \mathbf{B}_1(k) + \mathbf{V}\Lambda(k)\mathbf{V}^T\right]\mathbf{p} = \mathbf{f} \tag{3.83}$$

for $\Lambda(k)$ a diagonal matrix of impedance coefficients (of size $N_{mod} \times N_{mod}$), and $\mathbf{V}$ is independent of frequency, representing the discrete surface harmonics (of size $N_{mod} \times N$).

In standard shifted form, we make a further assumption that $\mathbf{B}_1(k)$ does not vary substantially around the reference frequency, such that we can write

$$\left[\mathbf{A}_0 - \sigma_i\mathbf{M} + \mathbf{V}\Lambda(\sigma_i)\mathbf{V}^T\right]\mathbf{p} = \mathbf{f} \tag{3.84}$$

for

$$\mathbf{A}_0 = \mathbf{K} - k_0^2\mathbf{M} + \mathbf{B}_1(k_0)\,. \tag{3.85}$$

If we desire the partial field solution, this can be rewritten as

$$\mathbf{p}_{red} = \mathbf{E}^T[\mathbf{A}_0 - \sigma_i\mathbf{M} + \mathbf{V}\Lambda(\sigma_i)\mathbf{V}^T]^{-1}\mathbf{f}\,. \tag{3.86}$$

Using the Sherman–Morrison–Woodbury formula, the partial field solution can be computed as

$$\mathbf{p}_{red} = \mathbf{E}^T(\mathbf{A}_0 - \sigma_i\mathbf{M})^{-1}\mathbf{f} - \mathbf{E}^T[\mathbf{A}_0 - \sigma_i\mathbf{M}]^{-1}\mathbf{V} \ldots$$
$$\cdot [\Lambda(\sigma_i)^{-1} + \mathbf{V}^T(\mathbf{A}_0 - \sigma_i\mathbf{M})^{-1}\mathbf{V}]^{-1}\mathbf{V}^T(\mathbf{A}_0 - \sigma_i\mathbf{M})^{-1}\mathbf{f} \tag{3.87}$$

or, using a block matrix representation

$$\mathbf{H} = \begin{bmatrix} \mathbf{H}_{Ef}\ \mathbf{H}_{EV} \\ \mathbf{H}_{Vf}\ \mathbf{H}_{VV} \end{bmatrix} = \left[\mathbf{E}\ \mathbf{V}\right]^T (\mathbf{A}_0 - \sigma_i\mathbf{M})^{-1} \left[\mathbf{f}\ \mathbf{V}\right]\,, \tag{3.88}$$

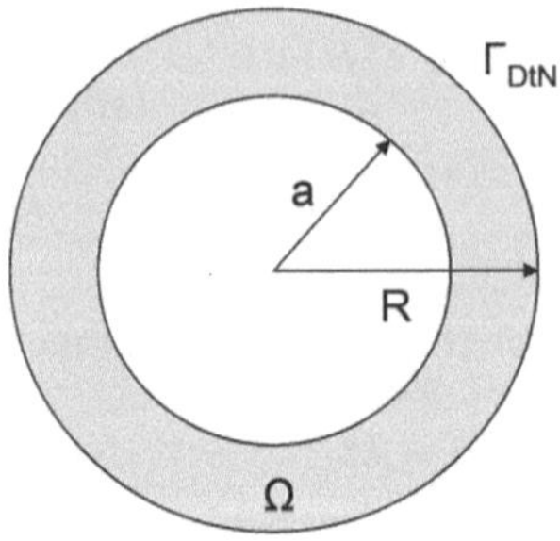

Fig. 3.5 Radiating circle with truncating boundary.

where $\mathbf{H}_{EF} \in \mathbb{R}^p$, $\mathbf{H}_{EV} \in \mathbb{R}^{p \times N_{mod}}$, $\mathbf{H}_{VF} \in \mathbb{R}^{N_{mod}}$ and $\mathbf{H}_{VV} \in \mathbb{R}^{N_{mod} \times N_{mod}}$, such that

$$\mathbf{p}_{red} = \mathbf{H}_{Ef} - \mathbf{H}_{EV}\left(\Lambda^{-1} + \mathbf{H}_{VV}\right)^{-1}\mathbf{H}_{Vf}\,. \tag{3.89}$$

The block matrix can then be computed using the MPVL algorithm, for

$$\mathbf{H} = \mathbf{L}^T\left(\mathbf{I} - \sigma\mathbf{A}\right)^{-1}\mathbf{R}\,, \tag{3.90}$$
$$\mathbf{L} = \left[\mathbf{E}\ \mathbf{V}\right]\,, \tag{3.91}$$
$$\mathbf{R} = \mathbf{A}_0^{-1}\left[\mathbf{f}\ \mathbf{V}\right]\,. \tag{3.92}$$

It should be noted that if the local support vector ($\mathbf{V}$) consists of a linear combination of the degrees of freedom of interest ($\mathbf{E}$), then one should not add $\mathbf{V}$ to the left starting vector. Rather, one can reconstruct $\mathbf{H}_{VF}$ and $\mathbf{H}_{VV}$ using linear combinations from $\mathbf{H}_{EF}$ and $\mathbf{H}_{EV}$, respectively. Finally, it should be observed that we can also compute the pressure values on the DtN boundary via modal parameters that can be found from a related computation

$$\mathbf{H}_{mod} = \mathbf{D}\mathbf{V}^T[\mathbf{A}_0 - \sigma_i\mathbf{M} + \mathbf{V}\Lambda(\sigma_i)\mathbf{V}^T]^{-1}\mathbf{f} \tag{3.93}$$

using the Sherman–Morrison–Woodbury formula and the same block matrices as above, for $\mathbf{D}$ a diagonal matrix of orthonormalization constants related to the discrete surface harmonics $\mathbf{V}$.

3.5.2 Example

For a numerical example, we take a sound radiating circle of radius $a = 0.01$m with a prescribed flux condition, and a truncating circular boundary of radius $R = 0.015$m, as reported in Wagner et al. [38], see Figure 3.5 The prescribed flux condition is

$$\nabla p \cdot \mathbf{n}|_{r=a,\theta} = \sum_{n=1}^{N_{DtN}} q_n \cos(n\theta)\,, \tag{3.94}$$

where q_n is chosen randomly. We choose four points for the partial field solution: one on the radiating circle, one on the DtN boundary, and two in the acoustic medium.

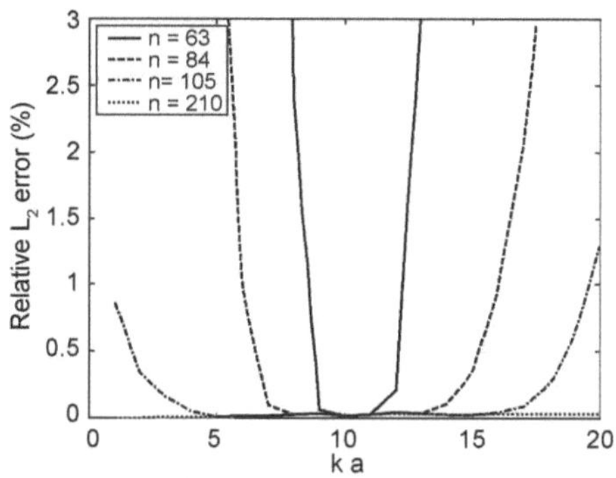

Fig. 3.6 Relative error QMR and MPVL solution on DtN boundary.

In this 2D example, we choose a mesh with 30,160 quadrilateral elements in the domain, and 753 1D elements on the DtN boundary. This gives a minimum of 25 elements per wavelength at $ka = 20$, and a total of 30914 unknowns. We truncate the DtN series at $N_{DtN} = 10$. We compare the solution with different numbers of Lanczos iterations and a reference frequency at $ka = 10$ to one found using a QMR iterative solver with SSOR preconditioning [28], and compute a L_2 relative error on the DtN boundary. As shown in Figure 3.6, the series is convergent over the frequency range studied with increasing numbers of Lanczos iterations. It should be noted, however, that we do not observe the considerable reduction in computational costs in this problem, as the matrix equations are well–conditioned using the SSOR preconditioner. Thus, the cost of the direct solve at the reference frequency $\mathcal{O}(N \log N + kn)$ must compete against the cheaper iterative solves for every frequency step $\mathcal{O}(kN)$. Nevertheless, this example shows the extensibility of the algorithm beyond simple mass–stiffness systems.

3.6 Error Estimate and Adaptation

A means of estimating the accuracy of the algorithm around a given reference frequency, at a given iteration n, may be found by comparing the results with those generated by a previous iteration $m < n$. As the series is convergent until new basis vectors can no longer be prescribed, one can view convergence between iterations as indications of regions of accuracy. In other words, we assume that if the the matrix Padé approximant converges, it will converge to the original system. This can be shown from the convergence behavior of the formal Stieltjes expansion and the Padé approximant. As this comparison uses the results from a previous iteration to construct the projected matrices, the only additional computations needed are the actual response of the reduced system at each frequency step. Figure 3.7 shows the comparison for the numerical example in structural acoustics for 150 iterations, using the relative error against the direct solution, compared with the solution for 100 iterations. Again, as the partial–field solution is small (3 degrees of freedom in this

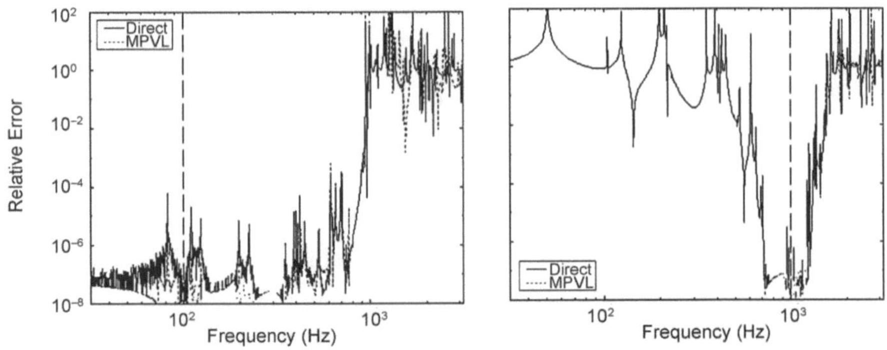

Fig. 3.7 Comparison of relative error of MPVL solution at center of plate, about (a) 100 Hz and (b) 1000 Hz for $n = 150$ iterations, against both the direct solution and MPVL with $n = 100$ iterations.

example), this error computation is inexpensive to make at every frequency step. The results show a very close correlation, at both reference frequencies, between the two relative error computations.

Similar error estimates have been made for block–Arnoldi Krylov methods by Heres [19]. In that thesis, a comparison between sequential iterations of the approximations is used to quantify the approximation error. In addition, Grimme [17] describes both error detection and a general adaptive approach in his thesis, in the context of selecting interpolation points for multi–point Krylov projection methods. Another paper by Zhou et al. used a similar comparison to evaluate convergence of MPVL for fixed frequency windows [39].

Alternatively, Bai and Ye [4] introduce a mathematical error bound for scalar–valued PVL ($m = p = 1$) for small shift parameters, $|\sigma| < \|\mathbf{A}\|^{-1}$

$$|H(\sigma) - H_n(\sigma)| \leq \left|\mathbf{l}^T\mathbf{r}\right| \left|\sigma^2\tau_{1n}(\sigma)\tau_{n1}(\sigma)\right| \frac{\|\hat{\mathbf{w}}_{n+1}\| \, \|\hat{\mathbf{v}}_{n+1}\|}{1 - |\sigma| \, \|\mathbf{A}\|}, \tag{3.95}$$

where

$$\tau_{1n}(\sigma) = \mathbf{e}_1^T(\mathbf{I} - \sigma\mathbf{T}_n)^{-1}\mathbf{e}_n\,, \tag{3.96}$$

$$\tau_{n1}(\sigma) = \mathbf{e}_n^T(\mathbf{I} - \sigma\mathbf{T}_n)^{-1}\mathbf{e}_1\,. \tag{3.97}$$

However, this does not resolve significant finite arithmetic errors that are well known in the Lanczos process. Also, this is unfortunately limited to a small shift parameter, as the largest singular value of the matrix equation is related to the mesh size, and may become very large in comparison to the frequency range desired.

3.6.1 Adaptive Windowing Algorithm

We use the error estimate from the approximation convergence to create a simple adaptive algorithm for computing the response across a given frequency range [36].

The adaptive windowing scheme is a sequential bisection algorithm to generate sufficient reference frequencies to span a given frequency range, $[\omega_{min}, \omega_{max}]$. Given a frequency resolution governed by a constant frequency step size $\Delta\omega$, a starting number of Lanczos iterations n_{min}, and an iteration step size Δn:

1. Given $[\omega_{min}, \omega_{max}]$, set the reference frequency to the midpoint of this domain, either in a linear or logarithmic sense

$$\omega_0 = (\omega_{min} + \omega_{max})/2 \,.$$

2. Set $m = n_{min}$ and $n = n_{min} + \Delta n$.
3. Use the block Lanczos–type algorithm [1] to construct the reduced–order matrices at ω_0, for both m and n Lanczos iterations

$$\begin{aligned} &\eta_m, \rho_m, \mathbf{\Delta}_m, \mathbf{T}_m \,, \\ &\eta_n, \rho_n, \mathbf{\Delta}_n, \mathbf{T}_n \,. \end{aligned}$$

4. Increment frequency step i away from the reference frequency (in either direction), and compare the response of the two reduced–order systems

$$\begin{aligned} \sigma_i &= (\omega_0 \pm i\Delta\omega)^2 - \omega_0^2 \,, \\ \mathbf{H}_m(\sigma_i) &= \boldsymbol{\eta}_m^T \mathbf{\Delta}_m \left(\mathbf{I} - \sigma \mathbf{T}_m\right)^{-1} \boldsymbol{\rho}_m \,, \\ \mathbf{H}_n(\sigma_i) &= \boldsymbol{\eta}_n^T \mathbf{\Delta}_n \left(\mathbf{I} - \sigma \mathbf{T}_n\right)^{-1} \boldsymbol{\rho}_n \,, \\ e &= \|\mathbf{H}_n - \mathbf{H}_m\|_2 \,/\, \|\mathbf{H}_n\|_2 \,. \end{aligned}$$

5. If the relative error exceeds some given error tolerance, either
   a) Increment m and n by Δn and return to (3), adding to existing Lanczos iterations (observe that the new m is in fact the previous n). Continue (4) from last frequency step.
   b) If (a) does not result in more valid frequency step increments, or n exceeds some predefined limit, denote current step as the edge of the current window: ω_1 or ω_2 (depending on direction).

   Otherwise, save $\mathbf{H}_n(\sigma_i)$ and repeat from (4) until an edge is detected or the range of the current frequency window is exhausted.
6. If either edge of the current window lies within the desired frequency range, generate a new frequency range using the remainder. Repeat from (1) using the new frequency ranges: $[\omega_{min}, \omega_1]$ and $[\omega_2, \omega_{max}]$.

3.6.2 Effect of Finite Arithmetic

The negative influence of finite precision computation in the Lanczos algorithm is well known. Paige [29] conducted extensive error analysis of the symmetric vector Lanczos algorithm in 1976, showing the growth of orthogonality loss as a function of the machine precision. Although not directly applicable to this particular application (unsymmetric block Lanczos), the results are nevertheless instructive. Error analysis is also presented recently in Liew and Pinsky [23]. One result of this analysis is that reducing the largest singular value of the shifted matrix **A** will reduce the round–off error and improve the frequency range of validity for the MPVL expansion. This motivates using the diagonal to normalize the original matrix equations, as an inexpensive preconditioner. However, for the poorly conditioned matrices of the coupled problem, this still leaves a significant factor for the error bound on the Lanczos process.

Running numerical examples of MPVL on a parallel machine has shown specific examples of the effects of rounding error in the reduced–order model. This effect is visible due to randomness in communication times between processors, which changes the order of execution of the computation, leading to different round–off errors between subsequent executions of the Lanczos algorithm. Unfortunately, the adaptive algorithm described above does not always detect the gradual accumulation of round–off error when comparing successive Lanczos iterations. This problem was not evident in the illustrative problem, but a more complex example described below illustrates this failure in the adaptive algorithm, leading to larger errors than were observed by comparisons between iterations. However, while these errors were not detectable by sequential iterations in the adaptive windowing algorithm, the errors were observable by comparing two different executions of the same MPVL expansion where round–off was the only differentiating factor.

One possible solution for this problem is to increase the detection of round–off error accumulation using the results observed from successive MPVL expansions in the parallel code. By restarting the Lanczos iterations when comparing between m and n iterations, one accumulates a new set of round–off errors for each MPVL approximation when executing on a parallel machine due to randomness in the execution order caused by communication time between processors. If these round–off errors cause deviations in the desired results, the independent Lanczos iterations will show greater differences when comparing between approximations $\mathbf{H}_m$ and $\mathbf{H}_n$. Thus, using a restart of the Lanczos algorithm, instead of continuing the series when adding additional iterations, adds an additional check on the accuracy of the solution by magnifying the sensitivity of the method to detect round–off accumulation.

3.6.3 Complexity Estimate

This method will generate a number of reference frequencies within the given frequency range, until an adequate solution is found for the entire range. The main computational cost involves the factorization of the shifted matrix at each reference frequency, $\mathcal{O}(jN\log N)$ for j total reference frequencies needed over k total frequency steps, compared to $\mathcal{O}(kN\log N)$ for the direct solution. The MPVL expansions are bounded by $\mathcal{O}(jn_{max}i_{max}N)$, for n_{max} the maximum number of Lanczos iterations and i_{max} the maximum number of increments to the number of iterations allowed in step (5a). Finally, each frequency requires two minor computations using the reduced matrices (m and n) and a comparison between the two. Thus, the frequency sweep complexity is bounded by $\mathcal{O}(2kn_{max})$. The algorithm attempts to minimize

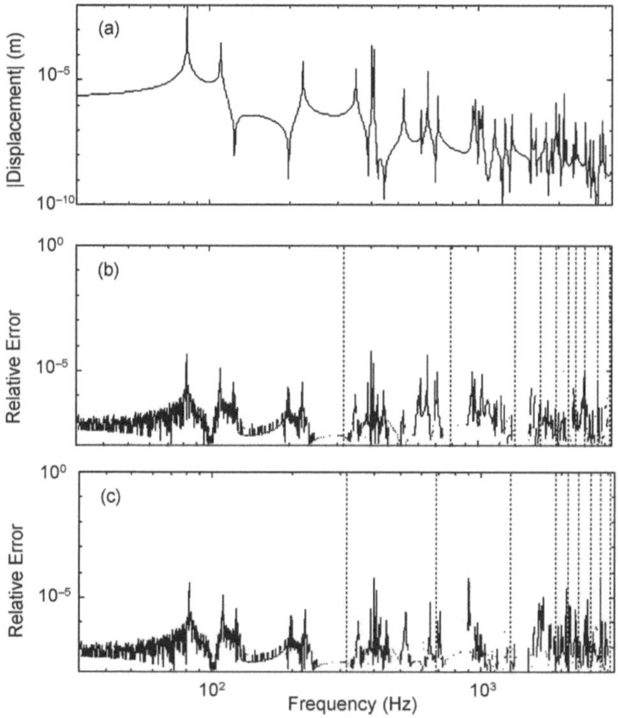

Fig. 3.8 Relative error between direct and adaptive MPVL solution, with computed reference frequencies shown, for (b) $n_{min} = 50$, $\Delta n = 50$, and (c) $n_{min} = 85$, $\Delta n = 20$. $e_{tol} = 0.001\%$.

the number of reference frequencies (j) by maximizing n_{max} for each reference frequency, with i_{max} the cost of attempting to produce the minimum such n_{max} due to restarting the Lanczos process. This is by necessity more expensive than the basic MPVL algorithm, although both n_{max} and i_{max} are very small compared to k and N. However, this method is more robust compared to the unknown accuracy of the original MPVL algorithm over the given frequency range.

3.7 Numerical Examples

3.7.1 Illustrative Example

Figure 3.8 shows the results of the adaptive windowing algorithm, also indicating the reference frequencies used, for the example problem in Section 3.4.2. Here, we have chosen two sets of algorithmic parameters: $n_{min} = 50$ and $\Delta n = 50$, and $n_{min} = 85$ and $\Delta n = 20$, both with an error tolerance of 0.001%. As shown, both choices give an accurate solution across the entire frequency range used, from 30

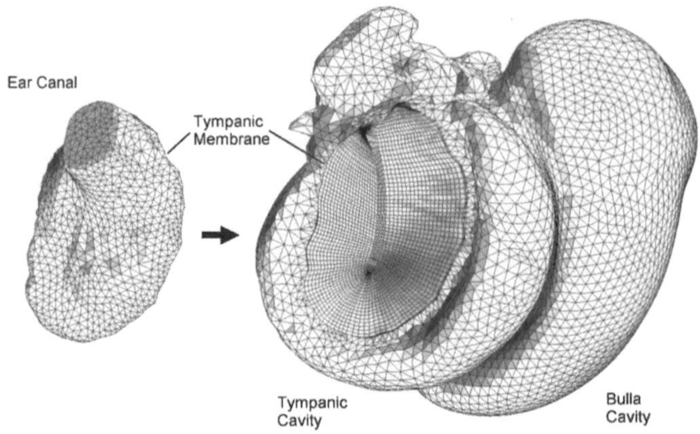

Fig. 3.9 Coupled finite element model of ear canal, middle ear cavities, and tympanic membrane from μCT imaging of domestic cat cadaver.

Hz to 3 kHz, which would not be possible with only a single reference frequency, while maximizing the window size for each reference frequency needed. The only results showing greater error than the desired tolerance are at the resonance and anti–resonance frequencies, which is to be expected. Due to the sensitivity of the Lanczos algorithm to round–off error accumulation and the nature of parallel computing, we found that even using the same parameters generally does not achieve repeatability in the size of the accuracy windows, the corresponding reference frequencies, and the time of execution.

The first set of parameters took 5 hours and used 10 reference frequencies. The second only took 3.7 hours using 9 reference frequencies (using fewer iterations per frequency). However, both show similar results in the relative error, demonstrating the robustness of the algorithm with respect to the chosen parameters. It should be noted that for the first case, there were two isolated frequency steps, corresponding to abrupt resonances. Because we compute the exact solution at the reference frequency before the MPVL algorithm, one should recognize that the MPVL approximation is not needed for single frequency calculations.

3.7.2 Hearing Simulation

An example with complex geometry is also presented from the application of hearing simulation. In this example, the middle ear of a domestic cat, including the acoustic spaces of the ear canal and middle ear cavities and the thin shell structure of the eardrum were determined using μCT imaging. Figure 3.9 shows the resulting geometry used in the model, based on manual segmentation and manipulation using Raindrop Geomagic and MSC.Patran. For the sake of illustration, we have assumed isotropic material properties for the eardrum.

An extension to the established MPVL algorithm has been added to model the ossicles, the bones of the middle ear. Here, we use a scalar transfer function to model the contribution of the ossicles and cochlea [33]. This appears as a local mixed boundary condition with complicated frequency dependence at the center of the eardrum. Essentially, a local Robin boundary condition applied at a point can be expressed as

$$\mathbf{F}_{\mathbf{d}} = \lambda(\omega)\mathbf{d}\,, \tag{3.98}$$

where λ is a function of frequency, and $\mathbf{F}_{\mathbf{d}}$ is the corresponding force for a displacement $\mathbf{d}$ at the point of application. One can write this as an addition to the matrix system of equations

$$\mathbf{K}_{Robin} = \lambda(\omega)\mathbf{V}\mathbf{V}^T\,, \tag{3.99}$$

where $\mathbf{V} \in \mathbb{R}^N$ is the corresponding global support vector for the point. This can be used with the Sherman–Morrison–Woodbury formula discussed for the treatment of DtN boundary conditions, with the advantage that this is only a rank–one update.

The finite element model, employing the same shell element formulation and acoustic elements as the illustrative example, consisted of 81,000 degrees of freedom from 230,000 acoustic elements and 10,000 solid shell elements. Figure 3.10 shows a comparison of the direct solution with the adaptive MPVL solution. The degrees of freedom of interest for this analysis were chosen to be the normal displacement of the eardrum at its center (the umbo, represented by four nodes in the vicinity) and the pressure in the ear canal (represented by three nodes near the center of the ear canal). For this problem, the direct solution for 1500 frequency steps took 130 hours to run on 6 processors, whereas an adequate adaptive MPVL solution spent 105 minutes, needing 6 reference frequencies across the domain (using $n_{min} = 85$, $\Delta n = 20$, and $e_{tol} = 0.001\%$). Figure 3.10 shows the computed middle ear impedance (ear canal pressure resulting from a uniform volume velocity in the ear canal) and comparisons between the direct solution and adaptive MPVL results (the relative error in the 2–norm of the reduced degrees of freedom). One can readily see the effect of restarting the Lanczos algorithm in the adaptive scheme. Without restart, the adaptive algorithm does not detect the gradual accumulation of error that results in differences of up to 0.3%, much higher than our specified tolerance. With restart, the round–off accumulation is detected, and the frequency window is appropriately terminated to keep the error within the desired tolerance.

3.8 Conclusion

The application of the matrix Padé–via–Lanczos algorithm to linear multifrequency analysis of structural acoustics for interior and exterior domains has been presented. The method provides the multifrequency solution for a subset of the complete solution vector; the method is optimal for a few selected "interesting" degrees of freedom of interest and becomes less efficient as the number of selected variables is increased. Thus the utility of the method is contingent on the analyst being able to identify a reduced number of solution variables of interest, for example those that exist in a computational subdomain, which we have called a "partial field." If a partial field

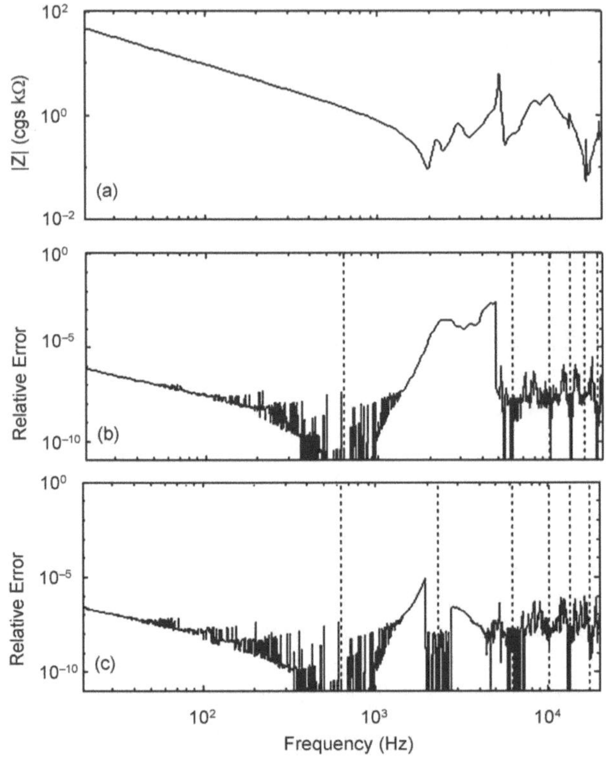

Fig. 3.10 Comparison between direct and adaptive MPVL solution. (a) Computed middle ear impedance (pressure in ear canal/volume velocity input). (b) Relative error without Lanczos restart. (c) Relative error with Lanczos restart. Algorithmic parameters used: $n_{min} = 85$, $\Delta n = 20$, and $e_{tol} = 0.001\%$.

of interest can be identified, then the method provides a very efficient approach for solving for this field.

Care must be taken for problems with more complicated frequency dependence, such as the example given for Dirichlet–to–Neumann boundary conditions. Another drawback is that the accuracy of the method is limited to a window of frequencies surrounding a reference frequency. The size of the window for a given accuracy depends on the number of Lanczos iterations but this fact cannot be exploited practically because of roundoff and other errors. To address this issue, we have introduced an adaptive algorithm to choose reference frequencies in order to generate an accurate response over a desired frequency range. This addition corrects the primary limitations of the MPVL method as previously applied, making a more robust method for practical usage.

The examples presented demonstrate the computational efficiency of this method compared to modern sparse solvers for poorly conditioned problems which would prove problematic for more conventional alternatives.

References

1. Aliaga JI, Boley DL, Freund RW, Hernández VA (2000) Lanczos–type method for multiple starting vectors. Mathematics of Computation 69:1577–1601
2. Amestoy PR, Duff IS, L'Excellent JY (2000) Multifrontal parallel distributed symmetric and unsymmetric solvers. Computer Methods in Applied Mechanics and Engineering 184:501–520
3. Avery P, Farhat C, Reese G (2007) Fast frequency sweep computations using a multi-point Padé–based reconstruction method and an efficient iterative solver. International Journal for Numerical Methods in Engineering 69:2848–2875
4. Bai Z, Ye Q (1999) Error bound for reduced system model by Padé approximation via the Lanczos process. IEEE Transaction on Computer–Aided Design 18:133–141
5. Baker G Jr, Graves–Morris P (1996) Padé approximants. 2nd edition Cambridge University Press, New York
6. Bayliss A, Gunzburger M, Turkel E (1982) Boundary conditions for the numerical solution of elliptic equations in exterior domains. SIAM Journal on Applied Mathematics 42:430–451
7. Chiprout E, Nakhla MS (1994) Asymptotic waveform evaluation and moment matching for interconnect analysis. Kluwer Academic, Norwell
8. Coyette J, Lecomte C, Migeot J, Blanche J, Rochette M, Mirkovic G (1999) Calculation of vibro–acoustic frequency response functions using a single frequency boundary element solution and a Padé expansion. Acustica united with Acta Acustica 85:371–377
9. Djellouli R, Farhat C, Tezaur R (2001) A fast method for solving acoustic scattering problems in frequency bands. Journal of Computational Physics 168:412–432
10. Ekinci A, Atalar A (1998) A circuit theoretical method for efficient finite element analysis of acoustical problems. In: Proceedings IEEE Ultrasonics Symposium, Sendai, Japan, 1251–1254
11. Erdemli YE, Reddy CJ, Volakis JL (1999) AWE technique in frequency domain electromagnetics. Journal of Electromagnetic Wave Applications. 13:359–378
12. Feldmann P, Freund RW (1995) Efficient linear circuit analysis by Padé approximation via the Lanczos process. IEEE Transaction on Computer–Aided Design 14:639–649
13. Freund RW (1995) Computation of matrix Padé approximations of transfer functions via a Lanczos–type process. In: Chiu CK, Schumaker LL (eds) Approximation Theory VIII, Vol. 1: Approximation and Interpolation. World Scientific Publishing Co., Inc., Singapore, 215–222
14. Freund RW (1999) Krylov–subspace methods for reduced–order modeling in circuit simulation. Technical Report, Numerical Analysis Manuscript No. 99–3–17. Bell Laboratories: Murray Hill, N.J.
15. Gallivan K, Grimme E, Van Dooren P (1994) Asymptotic wave–form evaluation via a Lanczos method. Applied Mathematics Letters 7:75–80
16. Gong J, Volakis J (1996) AWE implementation for electromagnetic FEM analysis. Electronics Letters 32:2216–2217
17. Grimme EJ (1997) Krylov projection methods for model reduction. PhD Thesis, Department of Electrical and Computer Engineering, University of Illinois at Urbana–Champaign
18. Grote MJ, Keller JB (1995) On nonreflecting boundary conditions. Journal of Computational Physics 122:231–243
19. Heres PJ (2005) Robust and efficient Krylov subspace methods for model order reduction. PhD Thesis, Department of Mathematics and Computer Science, Eindhoven University of Technology

20. Keller J, Givoli D (1989) Exact non–reflecting boundary conditions. Journal of Computational Physics 82:172–192
21. Kuzuoglu M, Mittra R (1999) Finite element solution of electromagnetic problems over a wide frequency range via the Padé approximation. Computer Methods in Applied Mechanics and Engineering 169:263–277
22. Lanczos C (1950) An iteration method for the solution of the eigenvalue problem of linear differential and integral operators. Journal of Research of the National Bureau of Standards 45:255–282
23. Liew HL, Pinsky PM (2007) Matrix Padé via Lanczos solutions for vibrations of fluid–structure interaction. (in preparation)
24. Mahlotra, M (1996) Iterative solution of large–scale acoustics problems. PhD Thesis, Department of Mechanical Engineering, Stanford University
25. Malhotra M, Pinsky PM (2000) Efficient computation of multi–frequency far–field solutions of the Helmholtz equation using Padé approximation. Journal of Computational Acoustics 8:223–240
26. Mandel J (2002) An iterative substructuring method for coupled fluid–solid acoustic problems. Journal of Computational Physics 177:95–116
27. Mandel J, Popa M (2005) Iterative solvers for coupled fluid–solid scattering. Applied Numerical Mathematics 54:194–207
28. Oberai AA, Malhotra M, Pinsky PM (1998) On the implementation of the Dirichlet–to–Neumann radiation condition for iterative solution of the Helmholtz equation. Applied Numerical Mathematics 27:443–464
29. Paige CC (1976) Error analysis of the Lanczos algorithm for tridiagonalizing a symmetric matrix. Journal of Institute of Mathematics and Its Applications 18:341–349
30. Parlett BN, (1980) A new look at the Lanczos algorithm for solving symmetric systems of linear equations. Linear Algebra and Its Applications 29:323–346
31. Parlett BN, Taylor DR, Liu ZA (1985) A look–ahead Lanczos algorithm for unsymmetric matrices. Mathematics of Computation 44:105–124
32. Pretlove AJ (1965) Free vibrations of a rectangular panel backed by a closed rectangular cavity. Journal of Sound and Vibration 2:197–209
33. Puria S, Allen JB (1998) Measurement and model of the cat middle ear: Evidence of tympanic membrane acoustic delay. Journal of the Acoustical Society of America 104:3463–3481
34. Schulze J (2001) Towards a tighter coupling of bottom–up and top–down sparse matrix ordering methods. BIT Numerical Mathematics 41:800–841
35. Slone R, Lee R (2000) Applying Padé via Lanczos to the finite element method for electromagnetic radiation problems. Radio Science 35:331–340
36. Tuck–Lee JP, Pinsky PM (2007) Adaptive frequency windowing for multifrequency solutions in structural acoustics based on the matrix Padé–via–Lanczos algorithm. International Journal for Numerical Methods in Engineering, accepted for publication
37. Vu–Quoc L, Tan XG (2003) Optimal solid shells for non–linear analyses of multilayer composites. Computer Methods in Applied Mechanics and Engineering 192:975–1059
38. Wagner M, Pinsky PM, Oberai AA, Malhotra M (2003) A Krylov subspace projection method for simultaneous solution of Helmholtz problems at multiple frequencies. Computer Methods in Applied Mechanics and Engineering 192:4609–4640
39. Zhou TD, Dvorak SL, Prince JL (2003) Application of subspace projection approaches for reduced–order modeling of electromagnetic systems. IEEE Transactions on Advanced Packaging 26:353–360

4

Computational Aeroacoustics based on Lighthill's Acoustic Analogy

Manfred Kaltenbacher[1], Max Escobar[2], Stefan Becker[3], and Irfan Ali[4]

[1] Friedrich–Alexander–Universität Erlangen–Nürnberg, Lehrstuhl für Sensorik, Paul–Gordan–Straße 3/5, 91052 Erlangen, Germany
manfred@lse.eei.uni-erlangen.de

[2] Friedrich–Alexander–Universität Erlangen–Nürnberg, Lehrstuhl für Sensorik, Paul–Gordan–Straße 3/5, 91052 Erlangen, Germany
max.escobar@lse.eei.uni-erlangen.de

[3] Friedrich–Alexander–Universität Erlangen–Nürnberg, Lehrstuhl für Strömungsmechanik, Cauerstraße 4, 91058 Erlangen, Germany
sbecker@lstm.uni-erlangen.de

[4] Friedrich–Alexander–Universität Erlangen–Nürnberg, Lehrstuhl für Strömungsmechanik, Cauerstraße 4, 91058 Erlangen, Germany
irfan@lstm.uni-erlangen.de

Summary. Aeroacoustics is concerned with sound, generated by turbulent fluid motion or aerodynamic forces interacting with surfaces. Although no complete scientific theory of the generation of noise by aerodynamic flows has been established, most practical aeroacoustic analysis relies upon the so–called *Acoustic Analogy*, whereby the governing equations of fluid dynamics are rewritten into a wave like equation. The most common and a widely–used formulation is Lighthill's aeroacoustic analogy, which was proposed by James Lighthill in the 1950s, when studying noise, generated by jet engines. Computational Aeroacoustics (CAA) is the application of numerical methods to find approximate solutions of the governing equations for specific aeroacoustic problems. Most of the proposed methods numerically solved on high performance computers (HPC), are employed in a hybrid two–step approach. The first step consists in computing the turbulent flow field from which the acoustic source terms are evaluated. Subsequently the acoustic radiation is computed. Therewith, no feedback from the acoustic field to the turbulent flow is considered.

4.1 Introduction

In the last years, manufacturers have started to consider the aerodynamic noise level in many industrial applications as a relevant design parameter (e.g. airplanes, wind turbines, cars, air conditioning systems, etc.). Because of this growing demand for reducing noise levels and for fulfilling noise regulations, there is a great motivation to investigate in basic aeroacoustic phenomena and mechanisms of sound generation and propagation. Besides experimental methods, a great need exists to develop numerical tools to simulate aeroacoustic phenomena.

Since the beginning of computational aeroacoustics several numerical methodologies have been proposed, each of these trying to overcome the challenges that the specific problems under investigation pose for an effective and accurate computation of the radiated sound. The main difficulties which have to be considered for the simulation of flow noise problems include [9, 26]:

- *Energy disparity and acoustic inefficiency:* There is a large disparity between the energy of the flow in the non–linear field and the acoustic energy in the far field (the hydrodynamic pressure p is in the range of $10^4 - 10^6$ Pa, whereas the acoustic pressure p' is in the range of $10^{-5} - 10$ Pa). Therefore, the radiated acoustic energy of an unsteady flow is a very small fraction of the total energy in the flow. In general, the total radiated power of a turbulent jet scales with $O(U^8/c_0^5)$, and for a dipole source arising from pressure fluctuations on surfaces inside the flow scales with $O(U^6/c_0^3)$, where U denotes the characteristic flow velocity and c_0 the speed of sound.
- *Length scale disparity:* Large disparity also occurs between the size of an eddy in the turbulent flow and the wavelength of the generated acoustic noise. Low Mach number eddies have a characteristic length scale l, velocity U, an integral time scale l/U and a frequency ω. This eddy will then radiate acoustic waves of the same characteristic frequency, but with a much larger length scale, expressed by the acoustic wavelength λ

$$\lambda \propto c_0 \frac{l}{U} = \frac{l}{M}.$$

- *Preservation of multipole character:* The numerical analysis must preserve the multipole structure of the acoustic source, which applies to quadrupoles representing turbulent eddies or forces acting on a surface inside the flow. In order to estimate the source strength, it is necessary to resolve the whole structure of the source.
- *Dispersion and dissipation:* Any discrete form of the acoustic wave equation cannot precisely represent the dispersion relation of the acoustic sound. Numerical discretization in space and time converts the original non–dispersive system into a dispersive discretized one, which exhibits wave phenomena of two kinds:
  1. Long wavelength components approaching the solution of the original partial differential equation as the grid is refined.
  2. Short wavelength components (spurious waves) without counterpart in the original equation evolving in the numerical scheme disturbing the solution.

  The wave equation shows a non–dissipative behavior; as such, dissipative errors must be avoided by a numerical scheme, in which both the amplitude and phase of the wave are of crucial importance [2, 8].
- *Flows with high Mach and Reynolds number:* Aeroacoustic problems often involve both high Mach and high Reynolds numbers. Flows at a high Mach number may induce new non–linear sources and convective effects while flows at a high Reynolds number introduce multiple scale difficulties due to the disparity between the acoustic wavelength λ and the size of the energy dissipating eddies.

- *Simulation of unbounded domains:* As a main issue for the simulation of unbounded domains using volume discretization methods remains the boundary treatment which needs to be applied to avoid the reflection of the outgoing waves on the truncating boundary of the computational domain. This remains to be a very active field of research and several numerical formulations have been developed for both transient and harmonic analyses.

Currently available aeroacoustic methodologies overcome only some of these broad range of physical and numerical issues, which restricts their applicability, making them, in many cases, problem dependent methodologies. The application of Direct Numerical Simulation (DNS) is becoming more feasible with the permanent advancement in computational resources. However, due to the large disparities of length and energy scales between fluid and acoustic fields, DNS remains restricted to low Reynolds number flows. In a DNS, all relevant scales of turbulence are resolved and no turbulence modeling is employed. Some relevant work has been done in this direction by Freund [21], however, the simulation of practical problems involving high Reynolds numbers requires very high resolutions which are still far beyond the capabilities of current supercomputers [52]. In this respect, more recently a promising heterogeneous domain decomposition technique [51] has been proposed, as a way to overcome the scales disparities and still allow a direct simulation. In this manner, the equations, numerical methods, grids, and time steps within each subdomain could be adapted to meet the local physical requirements. Meanwhile, hybrid methodologies still remain as the approach most commonly used for aeroacoustic computations, due to the practical advantages provided by the separate treatment of fluid and acoustic computations. In these schemes, the computational domain is split into a non–linear source region and a wave propagation region, and different physical models are used for the flow and acoustic computations. Herewith, first a turbulence model is used to compute the unsteady flow in the source region. Secondly, from the fluid field, acoustic sources are evaluated, which are used as input for the computation of the acoustic propagation. Therewith, a common aspect to hybrid methodologies lies in the assumption that no significant physical influence occurs from the acoustic propagation to the flow field.

Figure 4.1 shows the typical numerical methods which are employed when using any of these hybrid methodologies. For a precise prediction of the flow induced noise, Large Eddy Simulation (LES) is mainly used for solving the flow and providing the acoustic sources. In the LES method the large scales are directly resolved and the effect of the small scales on the large scales are modeled [7]. Although still with a relatively high computational cost, with the LES method it is possible to simulate turbulent flows with high Reynolds numbers and complex geometries. Therefore, this method has been adopted by researchers as one of the standard methods for computing the turbulent near field in CAA problems. Additionally we find the group of combined RANS (Reynolds Averaged Navier Stokes)/LES methods [24]. This type of methods blend statistical approaches with LES, in order to yield enhanced predictions of both turbulence statistics and unsteady flow dynamics at a fraction of the cost of traditional LES. Under these RANS/LES methods we find Detached Eddy Simu-

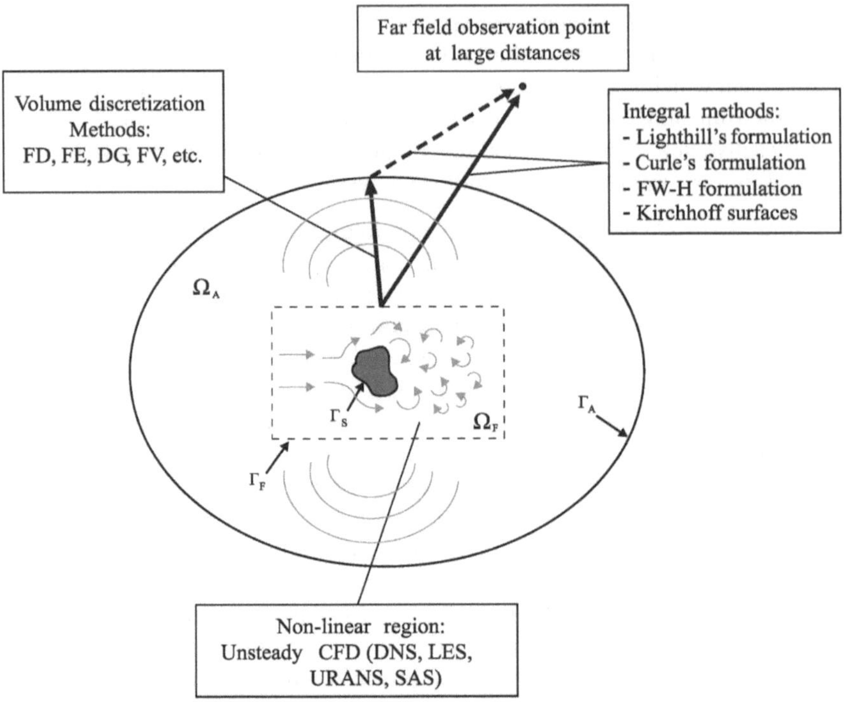

Fig. 4.1 Schematic depicting some of the possible strategies when using an aeroacoustic hybrid approach.

lation (DES) [44] which allows the turbulence model to transition from an uRANS (unsteady Reynolds Averaged Navier Stokes) method for attached boundary layers to a LES in separated regions. Another recently developed approach is the Scale–Adaptive Simulation method (SAS) [38] which instead of producing the large–scale unsteadiness, typically observed in uRANS simulations, it adjusts to the already resolved scales in a dynamic way and allows the development of a turbulent spectrum in the detached regions. It therefore behaves in a way much similar to a DES model, but without the explicit grid dependence in the RANS regime [38]. A thorough review of the current ongoing research in CFD methods used for CAA simulations can be found in [53].

Among the group of hybrid approaches, integral methods remain widely used in CAA for solving open domain problems in large acoustic domains like airframe noise, landing gear noise simulation, fan (turbines) noise, rotor noise, etc. One reason which motivates the use of integral formulations in such applications is that, in general, their acoustic sources can be considered to be compact and only an extension of the acoustic solution at a few points in the far field is desired. Therefore, in such cases, integral methods as Lighthill's acoustic analogy, see, e.g., [23, 31], Curle's formulation, see, e.g., [10], Ffwocs Williams and Hawkings formulation, see, e.g., [6, 10, 19], Kirchhoff method, see, e.g., [6, 17] or extension thereof, see, e.g., [11, 18, 35, 36] are computationally less expensive than volume discretization methods where a whole discretization of the acoustic domain is required. Integral approaches are derived by using free field Green's functions, being the Ffwocs Williams and Hawkings formulation the most widely used among these. The reason is that it is generalized to account for noise generation effects due to moving surfaces immersed in a turbulent flow.

On the other hand, for confined aeroacoustic problems, where non–compact solid boundaries are present, or if structural/acoustic effects are considered, it is more appropriate to use a volume discretization method to account for the interactions between the solid surfaces and the flow–induced noise directly in the acoustic simulation. In such cases, integral formulations would require a priori knowledge of a hard–wall Green's function that is not known for complex geometries [42]. Furthermore, volume discretization methods are capable to include the effects of wave propagation in non–uniform background flows.

Among the volume discretization methods used in CAA we find finite differences (FD), discontinuous Galerkin (DG) or finite volume (FV) schemes, generally employed to solve aeroacoustic formulations based on Linearized Euler Equations, see, e.g., [5, 27, 37, 50], or systems of equations alike: Acoustic Perturbation Equation, see [16], Linearized Perturbed Compressible Equation, see [49]. Additionally, we find the Finite Element (FE) method used to solve the variational formulation of Lighthill's acoustic analogy [3, 43]. Figure 4.1 also depicts the general configuration when using these methods. Herewith, Ω_F denotes the region where the flow field is firstly computed and where the acoustic sources are interpolated from the fluid simulation to the acoustic computation. The acoustic propagation region is given by $\Omega_F \cup \Omega_A$ where the acoustic field is computed in a second step, by solving the inhomogeneous wave equation or a corresponding set of equations depending on the CAA methodology followed. Since interior methods require the whole discretization of the propagation domain, usually they are used to compute the radiated sound until an intermediate region in the far field (i.e. until Γ_A in Figure 4.1), before moving to an integral formulation in which the acoustic solution from the volume discretization method at the interface is used as input for computing pressure levels at further points. Such a combined scheme has been presented, e.g., by Manoha et al. [37] solving the Euler equations for the intermediate inhomogeneous flow region and a 3D Kirchhoff method for the far field noise. For a more detailed discussion on the different CAA methodologies we refer to [15].

4.2 Hybrid Approach Using Lighthill's Acoustic Analogy

4.2.1 Lighthill's Acoustic Analogy

Many hybrid approaches employed in CAA, including the one presented here, are based on the inhomogeneous wave equation as derived by Lighthill in [31]. This formulation allows the calculation of the acoustic radiation from relatively small regions of turbulent flow embedded in a homogeneous fluid. For the derivation of the inhomogeneous wave equation Lighthill starts from the continuity and momentum equations which, by using the summation convention, can be written as [23, 31]

$$\frac{\partial \rho}{\partial t} + \frac{\partial \rho u_j}{\partial x_j} = 0 \tag{4.1}$$

$$\rho \frac{\partial u_i}{\partial t} + u_j \frac{\partial \rho u_i}{\partial x_j} = - \frac{\partial p}{\partial x_i} - \frac{\partial \tau_{ij}}{\partial x_j} . \tag{4.2}$$

In (4.1) and (4.2) ρ denotes the density of the fluid, v_i the i–th component of the flow velocity vector $\mathbf{v}$, p the overall pressure and τ_{ij} the (i, j)th component of the viscous stress tensor. For a Stokesian gas it can be expressed in terms of the velocity gradients by

$$\tau_{ij} = -\mu \left(\frac{\partial u_i}{\partial x_j} + \frac{\partial u_j}{\partial x_i} - \frac{2}{3} \mu \delta_{ij} \frac{\partial u_k}{\partial x_k} \right) , \tag{4.3}$$

where μ is the viscosity of the fluid and δ_{ij} the Kronecker delta.

Multiplying the continuity Equation (4.1) by u_i, adding the result to the momentum equation, and combining terms yields

$$\frac{\partial \rho u_i}{\partial t} = - \frac{\partial}{\partial x_j} (\rho u_i u_j + \delta_{ij} p + \tau_{ij}) , \tag{4.4}$$

which after adding and subtracting the term $c_0^2 \partial \rho / \partial x_i$ (c_0 denotes the speed of sound), becomes

$$\frac{\partial \rho u_i}{\partial t} + c_0^2 \frac{\partial \rho}{\partial x_i} = - \frac{\partial T_{ij}}{\partial x_i} . \tag{4.5}$$

In (4.5), T_{ij} denotes Lighthill's stress tensor given as

$$T_{ij} = \rho u_i u_j + \delta_{ij} [(p - p_0) - c_0^2 (\rho - \rho_0)] + \tau_{ij} \tag{4.6}$$

with ρ_0 and p_0 the atmospheric values of density and pressure respectively. Now it is possible to differentiate (4.1) with respect to t, take the divergence of (4.5), and subtract the results to obtain Lighthill's inhomogeneous wave equation solving for the acoustic density ρ' [23, 31]

$$\frac{\partial^2 \rho'}{\partial t^2} - c_0^2 \frac{\partial^2 \rho'}{\partial x_i^2} = \frac{\partial^2 T_{ij}}{\partial x_i \partial x_j} . \tag{4.7}$$

An important aspect in Lighthill's acoustic analogy, in order to be able to compute the noise radiation, is the assumption that Lighthill's stress tensor is a known source term or at least can be evaluated to a certain degree of approximation. Additionally, this source term is assumed to vanish outside the turbulent region. Indeed, for a turbulent flow embedded in a uniform atmosphere at rest, Lighthill's stress tensor $T_{ij} = \rho u_i u_j + \delta_{ij} [(p - p_0) - c_0^2 (\rho - \rho_0)] + \tau_{ij}$ can be neglected outside the turbulent region itself. In this outer region the velocity u_i consists only of the small motions characteristics of sound. Furthermore, this velocity appears quadratically in the calculation of the tensor as $\rho u_i u_j$. Moreover, the effects of viscosity and heat conduction are expected to cause only a slow damping due to the conversion of acoustic energy into heat and to have a significant effect only over very large distances. Therefore, it should be possible to neglect τ_{ij} entirely [23].

The term $(p - p_0) - c_0^2(\rho - \rho_0)$, becomes of importance only for anisotropic media, when the Mach number in the acoustic domain is significantly different from the one in the fluid domain. Otherwise, for isentropic flows in which $(p - p_0)/p_0$ and $(\rho - \rho_0)/\rho_0$ are very small, the isentropic relation

$$(p - p_0) = c_0^2(\rho - \rho_0) \tag{4.8}$$

can be assumed. Therefore, the resulting approximate of Lighthill's tensor is given by

$$T_{ij} \approx \rho u_i u_j \, . \tag{4.9}$$

Lighthill pointed out in [31], that the resulting approximate tensor could also be found from an approach which makes approximations in the equations of motion right from the beginning of the derivation of Lighthill's inhomogeneous equation, but that approximations at that early stage could introduce dipole or monopole sources whose contribution to the radiated sound might be relatively large.

The explicit separation of propagation and generation as derived by Lighthill has arised many discussions ever since, which have motivated the derivation of improved acoustic analogy formulations. A difficulty of Lighthill's equation is the interpretation of the source term where mean flow effects on the wave propagation are included [46]. In such cases, a compressible velocity field is required for the evaluation of the source term. In order to obtain a formulation to describe the noise propagation in a transversally sheared mean flow, Lilley [32,33] proposed a third–order wave operator. The mean flow and any refraction it causes are explicitely considered in this wave operator. However, this formulation does not provide any obvious simplification of the sources, still leaving complex combinations of terms which must be modeled in making predictions. Since there is no clear simplification that comes about by including more propagation physics into the wave operator, it can therefore be argued that the Lighthill approach is no worse off than such a more sophisticated approach. Since acoustic analogies are exact, an accurate representation of the source in the original Lighthill's analogy provides the correct sound despite the multiple physical effects grouped into it. Since different acoustic analogies require different source models, it is difficult to make clear direct comparisons between them and thus, difficult to judge if one is preferable to others [22].

4.2.2 Weak Formulation

The weak formulation of the initial/boundary–value problem can be derived applying the method of weighted residuals and by making use of a usual space of functions

$$\vartheta = \{u(\cdot,t) \mid u(\mathbf{x}),t) \in H^1 \, , \, \mathbf{x} \in \Omega\} \, , \tag{4.10}$$

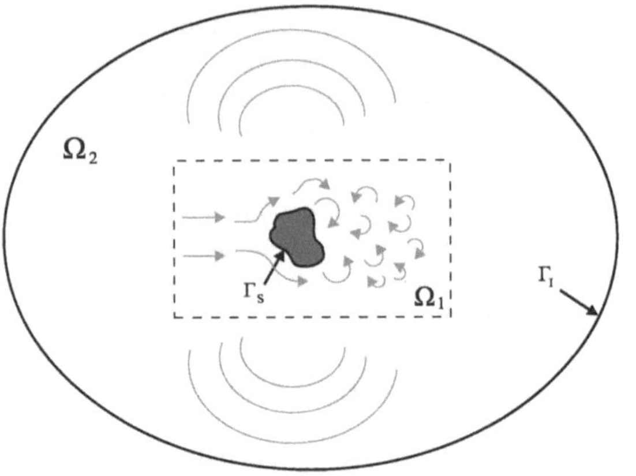

Fig. 4.2 General setup for a flow–induced noise problem

where H^1 denotes the Sobolev space defined as, see e.g. [1],

$$H^1 = \{u \in L^2 | \partial u / \partial x_i \in L^2\} \tag{4.11}$$

and L^2 the space of square integrable functions. After multiplication by an appropriate test function w from the space of functions given in (4.10) and integration over the whole domain Ω, see Figure 4.2, the problem can be formulated in the integral form as (using now the acoustic pressure p' as the unknown physical quantity)

$$\int_\Omega \frac{1}{c_0^2} \frac{\partial^2 p'}{\partial t^2} w \, d\Omega - \int_\Omega \frac{\partial^2 p'}{\partial x_i^2} w \, d\Omega = \int_\Omega \frac{\partial^2 T_{ij}}{\partial x_i \partial x_j} w \, d\Omega$$

$$\int_\Omega p'(\mathbf{x}, 0) \, w \, d\Omega = \int_\Omega p'_0 \, w \, d\Omega$$

$$\int_\Omega \frac{\partial p'}{\partial t}(\mathbf{x}, 0) \, w \, d\Omega = \int_\Omega \dot{p'}_0 \, w \, d\Omega \qquad \forall \; w \in H^1 \, . \tag{4.12}$$

Now, we apply Green's integral theorem in the first equation in (4.12) to the second spatial derivative of p' as well as to the acoustic source term containing T_{ij}. This operation results in the following relations

$$\int_\Omega w \frac{\partial^2 p'}{\partial x_i^2} \, \mathrm{d}\Omega = \int_{\Gamma_\mathrm{S} \cup \Gamma_\mathrm{I}} w \frac{\partial p'}{\partial \mathbf{n}} \, \mathrm{d}\Gamma - \int_\Omega \frac{\partial w}{\partial x_i} \frac{\partial p'}{\partial x_i} \, \mathrm{d}\Omega \tag{4.13}$$

$$\int_\Omega w \frac{\partial^2 T_{ij}}{\partial x_i \partial x_j} \, \mathrm{d}\Omega = \int_{\Gamma_\mathrm{S}} w \frac{\partial T_{ij}}{\partial x_j} n_i \, \mathrm{d}\Gamma - \int_\Omega \frac{\partial w}{\partial x_i} \frac{\partial T_{ij}}{\partial x_j} \, \mathrm{d}\Omega \, . \tag{4.14}$$

Herewith, it is important to emphasize, that the boundary integral in (4.14) is just over the surface Γ_S of any solid/elastic body since we assume T_{ij} to be zero at the limits the computational domain Γ_{I}, whereas in (4.13) we have to integrate over Γ_{S} as well as over Γ_{I}, see Figure 4.2. By using the momentum conservation law and assuming a solid body, e.g. $u_i n_i = 0$ where $\mathbf{u}$ denotes the velocity of the body, we can express the surface integral in (4.14) by

$$\int_{\Gamma_{\mathrm{S}}} w \frac{\partial T_{ij}}{\partial x_j} n_i \, \mathrm{d}\Gamma = - \int_{\Gamma_{\mathrm{S}}} w \frac{\partial p'}{\partial \mathbf{n}} \, \mathrm{d}\Gamma \,. \tag{4.15}$$

Therewith, we can combine the two surface integrals to a single one just performed over the outer boundary Γ_{I}. This remaining surface integral corresponds to the natural boundary condition associated with the weak formulation and will be employed later for applying absorbing boundary conditions. Finally, we arrive at the weak form of (4.7): Find $p' \in H^1$ such that

$$\begin{aligned} \int_{\Omega} \frac{1}{c_0^2} w \frac{\partial^2 p'}{\partial t^2} \, \mathrm{d}\Omega + \int_{\Omega} \frac{\partial w}{\partial x_i} \frac{\partial p'}{\partial x_i} \, \mathrm{d}\Omega - \int_{\Gamma_{\mathrm{I}}} w \frac{\partial p'}{\partial \mathbf{n}} \, \mathrm{d}\Gamma \\ = - \int_{\Omega} \frac{\partial w}{\partial x_i} \frac{\partial T_{ij}}{\partial x_j} \, \mathrm{d}\Omega \end{aligned} \tag{4.16}$$

is fulfilled for all $w \in H^1$. For the flow induced noise computation, we will consider two different methods for treating the outer boundary Γ_{I}. In the transient case, we apply first order absorbing boundary conditions [14], which results in the following simple substitution

$$\frac{\partial p'}{\partial \mathbf{n}} = \frac{1}{c_0} \frac{\partial p'}{\partial t} \,. \tag{4.17}$$

However, this boundary condition is just optimal for acoustic waves impinging orthogonal onto the boundary Γ_{I}, whereas for all other waves numerical reflections occur. Therewith, a limitation of the computational domain within the acoustic near field will result in worse results. For the harmonic case, we will apply a recently developed Perfectly Matched Layer (PML) technique [29]. For this method, we enclose our computational domain by an additional damping layer, see Figure 4.3, in which the acoustic waves are damped and where at the interface between the propagation and damping region no reflections occur. Therewith, with such a PML technique, we can limit the computational domain even in the acoustic near field.

4.2.3 Spatial Discretization

The semidiscrete Galerkin formulation is obtained from the weak formulation of the equation after discretization of the domain with finite elements. The entire domain, $\Omega = \Omega_1 \cup \Omega_2$, is discretized by an approximate decomposition $\sum_h$ of Ω into $\mathbb{R}^2$ or $\mathbb{R}^3$ finite elements. From our previously defined space ϑ, we choose a finite dimensional space $\vartheta_h \subset \vartheta$. We can now write a variational equation of the form of (4.16) in terms of $p'_h \in \vartheta_h$, using the nabla differential operator for better readability and neglecting the term on the boundary Γ_{I}

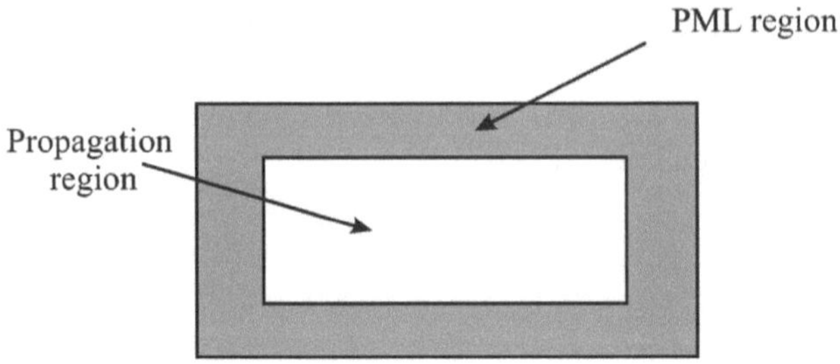

Fig. 4.3 Computational setup: propagation region surrounded by a PML–region.

$$\int_{\Omega_h} \frac{1}{c_0^2} \ddot{p'}_h \, w_h \, d\Omega + \int_{\Omega_h} \nabla p'_h \cdot \nabla w_h \, d\Omega = - \int_{\Omega_h} (\nabla \cdot T_{ij}) \cdot \nabla w_h \, d\Omega \,. \quad (4.18)$$

Using standard nodal finite elements, we can approximate the continuous pressure perturbation p' as well as the test function w in terms of the interpolation functions as[5]

$$p' \approx p'_h(t) = \sum_{i=1}^{n_{\rm eq}} N_i(\mathbf{x}) p'_i(t) \quad (4.19)$$

$$w \approx w_h = \sum_{i=1}^{n_{\rm eq}} N_i(\mathbf{x}) \, c_i \,, \quad (4.20)$$

where $N_i(\mathbf{x})$ denote appropriate interpolation functions and c_i the corresponding weights. From (4.18), we can now write the semidiscrete Galerkin formulation in matrix form as

$$\mathbf{M}\ddot{\mathbf{p}}'(t) + \mathbf{K}\mathbf{p}'(t) = \mathbf{f} \,, \quad (4.21)$$

where the matrices **M** and **K** are computed as follows

$$\mathbf{M} = [M_{ij}] \quad M_{ij} = \int_{\Omega} \frac{1}{c_0^2} N_i N_j \, d\Omega \quad (4.22)$$

$$\mathbf{K} = [K_{ij}] \quad K_{ij} = \int_{\Omega} \left(\frac{\partial N_i}{\partial x_1}\frac{\partial N_j}{\partial x_1} + \frac{\partial N_i}{\partial x_2}\frac{\partial N_j}{\partial x_2} + \frac{\partial N_i}{\partial x_3}\frac{\partial N_j}{\partial x_3} \right) d\Omega \,. \quad (4.23)$$

Finally, the right–hand–side vector including the acoustic sources reads

$$\mathbf{f} = \{f_i\}$$

$$f_i = - \int_{\Omega} \left(\frac{\partial N_i}{\partial x_1}, \frac{\partial N_i}{\partial x_2}, \frac{\partial N_i}{\partial x_3} \right) \cdot \frac{\partial T_{ij}}{\partial x_j} \, d\Omega \,. \quad (4.24)$$

[5] The evaluation of the source terms using the same kind of interpolation functions is considered in Section 4.2.6, while at this point they are treated as given continuous values sampled at the finite element nodes.

4.2.4 Time Discretization

In order to discretize our problem in the time domain, the Newmark scheme is typically applied, see e.g. [28]. Let us start from the semidiscrete Galerkin formulation in matrix form

$$\mathbf{M}\ddot{\mathbf{p}}'_{n+1} + \mathbf{K}\mathbf{p}'_{n+1} = \mathbf{f}_{n+1} \tag{4.25}$$

with $\mathbf{M}$ the mass matrix, $\mathbf{K}$ the stiffness matrix, $\mathbf{f}$ the right hand side, $\mathbf{p}'$ the vector of unknowns at the finite element nodes and $\ddot{\mathbf{p}}'$ its second derivative with respect to time. For hyperbolic partial differential equations the *Newmark* scheme is generally used. Therewith, we have, see e.g. [28],

$$\mathbf{p}'_{n+1} = \mathbf{p}'{}_n + \Delta t\, \dot{\mathbf{p}}'_n + \frac{\Delta t^2}{2}\left[(1-2\beta_{\mathrm{H}})\ddot{\mathbf{p}}'_n + 2\beta_{\mathrm{H}}\ddot{\mathbf{p}}'_{n+1}\right] \tag{4.26}$$

$$\dot{\mathbf{p}}'_{n+1} = \dot{\mathbf{p}}'_n + \Delta t\left[(1-\gamma_{\mathrm{H}})\ddot{\mathbf{p}}'_n + \gamma_{\mathrm{H}}\ddot{\mathbf{p}}'_{n+1}\right] . \tag{4.27}$$

In (4.26)–(4.27) n denotes the time step counter, Δt the time step value and β_{H}, γ_{H} the integration parameters. Substituting $\mathbf{p}'_{n+1}$ and $\dot{\mathbf{p}}'_{n+1}$ according to (4.26) and (4.27) in (4.25) leads to the following algebraic system of equations

$$\mathbf{M}^*\ddot{\mathbf{p}}'_{n+1} = \mathbf{f}_{n+1} - \mathbf{K}\left[\mathbf{p}'_n + \gamma_{\mathrm{H}}\Delta t\, \dot{\mathbf{p}}'_n + \frac{\Delta t^2}{2}(1-2\beta_{\mathrm{H}})\ddot{\mathbf{p}}'_n\right] \tag{4.28}$$

$$\mathbf{M}^* = \mathbf{M} + \beta_{\mathrm{H}}\Delta t^2\,\mathbf{K} . \tag{4.29}$$

Writing the solution process for one time step as a *predictor–corrector* algorithm we arrive at the following formulations, also known as the *effective mass formulation.*

- Perform predictor step:

$$\tilde{\mathbf{p}}' = \mathbf{p}'{}_n + \Delta t\, \dot{\mathbf{p}}'_n + (1-2\beta_{\mathrm{H}})\frac{\Delta t^2}{2}\ddot{\mathbf{p}}'_n \tag{4.30}$$

$$\tilde{\dot{\mathbf{p}}}' = \dot{\mathbf{p}}'_n + \Delta t\,(1-\gamma_{\mathrm{H}})\ddot{\mathbf{p}}'_n \tag{4.31}$$

- Solve algebraic system of equations:

$$\mathbf{M}^*\ddot{\mathbf{p}}'{}_{n+1} = \mathbf{f}_{n+1} - \mathbf{K}\tilde{\mathbf{p}}' \tag{4.32}$$

$$\mathbf{M}^* = \mathbf{M} + \beta\Delta t^2\,\mathbf{K} \tag{4.33}$$

- Perform corrector step:

$$\mathbf{p}'_{n+1} = \tilde{\mathbf{p}}' + \beta_{\mathrm{H}}\Delta t^2\,\ddot{\mathbf{p}}'_{n+1} \tag{4.34}$$

$$\dot{\mathbf{p}}'_{n+1} = \tilde{\dot{\mathbf{p}}}' + \gamma_{\mathrm{H}}\mathbf{\Delta t}\,\ddot{\mathbf{p}}'_{\mathbf{n+1}} \tag{4.35}$$

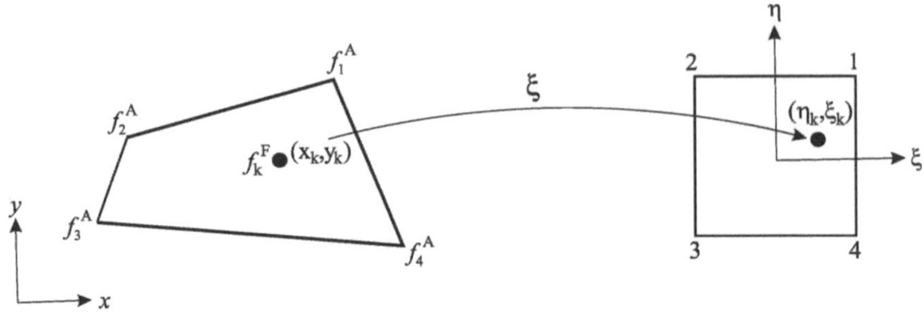

Fig. 4.4 Global to local mapping for conservative interpolation

4.2.5 Harmonic Formulation

By applying a harmonic analysis, it is possible to compute the sound radiation for specific frequency components present in the acoustic sources. In this way, we obtain the complex acoustic pressure at each node in the numerical domain. For deriving the harmonic counterpart we apply a Fourier–transformation to the semidiscrete Galerkin formulation (4.25), obtaining the following complex algebraic system of equations

$$\left(-\omega^2\mathbf{M} + \mathbf{K}\right)\hat{\mathbf{p}} = \hat{\mathbf{f}}\,, \tag{4.36}$$

where the source term $\hat{\mathbf{f}}$ represents the complex nodal acoustic sources, which are obtained by applying a Fourier transformation to the dataset of transient nodal sources interpolated from the fluid computation.

4.2.6 Coupling Scheme

A crucial point is the transformation of the acoustic sources from the computed flow data to the acoustic grid. In order to preserve the acoustic energy, we perform an integration over the source volume (corresponds to the computational flow region) within the FE formulation and project the results to the nodes of the fine flow grid, which has to be interpolated to the coarser acoustic grid. Therewith, our interpolation has to be conservative in order to preserve the total acoustic energy. As illustrated in Figure 4.4, we have to find for each nodal source f_k^{F} in which finite element of the acoustic grid it is located. Then, we compute from the global position (x_k, y_k) its position (ξ_k, η_k) in the reference element. This is in the general case a non–linear mapping and is solved by a Newton scheme. Now, with this data we can perform a bilinear interpolation and add the contribution of f_k^{F} to the nodes of the acoustic grid by using the standard finite element basis functions N_i

$$f_i^{\mathrm{A}} = N_i(\xi_k, \eta_k) f_k^{\mathrm{F}}\,.$$

Therewith, by this procedure the interpolation preserves the overall sum of the acoustic sources.

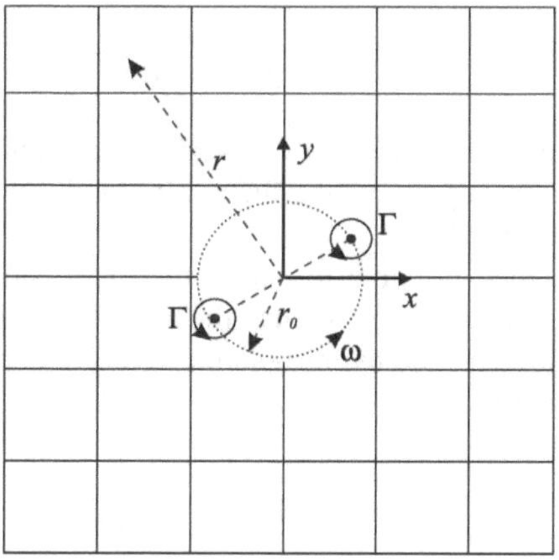

Fig. 4.5 Schematic diagram of corotating vortices.

4.3 Validation of the Computational Scheme

For validation, we consider the acoustic far field caused by an incompressible, purely unsteady vortical flow. The inhomogeneous wave equation is forced with the acoustic sources obtained from the analytical hydrodynamic field induced by a co–rotating vortex pair. CAA investigations using these types of flow fields are a practical way to validate the simulation of flow–induced noise problems [12,13,16,30,34]. The resulting acoustic field represents the basic acoustic field generated by turbulent shear flows, jet flows, edge tones, etc. [30].

4.3.1 Theoretical Approach

The acoustic sources in the flow region are computed from the hydrodynamic quantities of the flow field induced by a spinning vortex pair. This corotating vortex pair consists of two point vortices which are separated by a fixed distance of $2r_0$ with circulation intensity Γ. The schematic of the corotating vortices is presented in Figure 4.5. These vortices rotate around each other with a period $T = 8\pi^2 r_0^2/\Gamma$. Each vortex induces on the other a velocity $v_\theta = \Gamma/(4\pi r_0)$. The configuration results in a rotating speed $\omega = \Gamma/(4\pi r_0^2)$, and rotating Mach number $Ma_r = v_\theta/c_0 = \Gamma/(4\pi r_0 c_0) = 2\pi r_0/T c_0$. The rotating noncircular streamlines are directly associated with the hydrodynamic field of the rotating quadrupole [30]. The incompressible, inviscid flow can be determined numerically by the evaluation of a complex potential function $\Phi(z,t)$ [13, 16, 20]

$$\Phi(z,t) = \frac{\Gamma}{2\pi \mathrm{i}} \ln(z-b) + \frac{\Gamma}{2\pi \mathrm{i}} \ln(z+b) = \frac{\Gamma}{2\pi \mathrm{i}} \ln z^2 (1 - \frac{b^2}{z^2}) , \qquad (4.37)$$

where $z = r\mathrm{e}^{\mathrm{i}\theta}$ and $b = r_0 \mathrm{e}^{\mathrm{i}\omega t}$. For $|z/b| \gg 1$, (4.37) can be approximated by

$$\Phi(z,t) \approx \frac{\Gamma}{\pi \mathrm{i}} \ln z - \frac{\Gamma}{2\pi \mathrm{i}} \left(\frac{b}{z}\right)^2 = \Phi_0 + \Phi_1 \,. \tag{4.38}$$

The first term on the right hand side of (4.38) represents a steady vortical flow, whereas the second term represents the fluctuation with the fundamental frequency due to the vortex motion [30].

From (4.37) it is then possible to derive the expressions representing the hydrodynamic quantities required to compute the acoustic sources when following the acoustic analogy approach. The hydrodynamic velocity is obtained by differentiating (4.37) with respect to z

$$u_x - \mathrm{i} u_y = \frac{\partial \Phi(z,t)}{\partial z} = \frac{\Gamma}{\mathrm{i}\pi} \frac{z}{z^2 - b^2} \,. \tag{4.39}$$

In the acoustic computation a linear propagation is assumed outside the fluid region, governed by the homogeneous acoustic wave equation.

For comparison with the numerical results, the analytical solution of the acoustic pressure fluctuations from the corotating vortex pair, obtained with the matched asymptotic expansion (MAE) method, is given by [41]

$$p' = \frac{\rho_0 \Gamma^4}{64\pi^3 r_0^4 c_0^2} \left[J_2(kr)\cos(\Psi) - Y_2(kr)\sin(\Psi) \right] , \tag{4.40}$$

where $k = 2\omega/c_0$, $J_2(kr), Y_2(kr)$ are the second–order Bessel functions of the first and second kind and $\Psi = 2(\omega t - \theta)$. An equivalent expression for this quadrupole–like solution has also been derived by Mitchell et al. [39] starting from the vortex sound theory proposed by Möhring [40].

4.3.2 Numerical Investigation

In the computations, we have evaluated the flow field induced by the spinning vortex pair in a numerical region with dimensions 200 m × 200 m. This region corresponds to the region, where the acoustic nodal sources for the inhomogeneous wave equation are computed. The acoustic propagation is computed in a larger numerical domain with dimensions 400 m × 400 m, as the one depicted in Figure 4.6 where the inner square domain corresponds to the source region. For evaluating the complex potential function the spinning radius is chosen to be $r_0 = 1$ m, the circulation intensity $\Gamma = 1.00531\,\mathrm{m}^2/\mathrm{s}$ and the speed of sound $c_0 = 1$ m/s. This results in a wave length $\lambda \approx 39$ m and a rotating Mach number $Ma_r = 0.0796$.

In Lighthill's acoustic analogy we require the velocity vector field to compute the velocity gradients contained in the acoustic source term. This aspect makes it necessary to evaluate the fluid field using a much finer resolution than that of the acoustic field, so that the velocity gradients can be computed accurately. Therefore, we employ our coupling approach as follows. The first step consists in evaluating analytically the velocity field and then numerically the right hand side of our FE formulation, see (4.24), on a fine fluid grid. Secondly, after interpolation of the acoustic

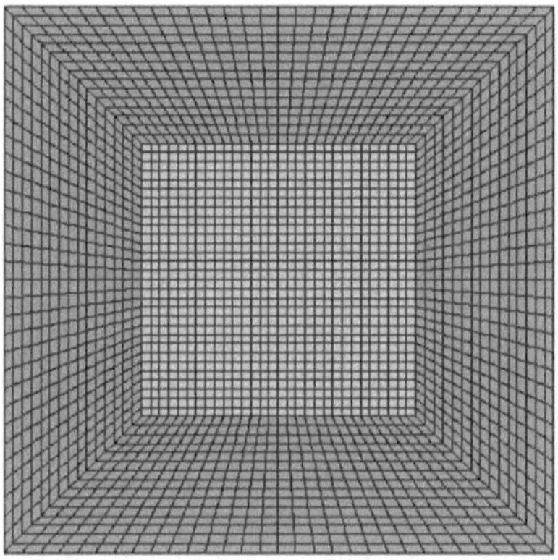

Fig. 4.6 Configuration of numerical domain as used for the computation of the acoustic propagation induced by the corotating vortex pair.

nodal loads from the fine fluid grid to the coarser acoustic grid, we solve the inhomogeneous wave equation to obtain the acoustic solution.

An additional issue for the computation is the singularity of the velocity field at the point vortices. Since the acoustic contribution of the region near the point vortices is important for the final acoustic field, a cut–off of the sources at these region, as proposed in [13], does not produce good results. Therefore, the application of a vortex core model, as proposed in [16,30,45,47], is required to obtain the desingularized tangential velocity field around these points. A desingularized kernel following the Scully model [4,48] and presented in [45,47] is applied at each point vortex for radii $r_{\mathrm{vortex}} < 0.15\,\mathrm{m}$. Within these regions the expressions for the velocity components are given by

$$u_x = -\frac{\Gamma}{2\pi}\frac{y}{r_{\mathrm{core}} + x^2 + y^2}\,;\, u_y = \frac{\Gamma}{2\pi}\frac{x}{r_{\mathrm{core}} + x^2 + y^2}\,, \tag{4.41}$$

where the coordinates (x, y) are taken with respect to each vortex and r_{core} corresponds to the vortex core radius, which is the distance from the point vortex where the maximum tangential velocity values occur. For the computations, a value $r_{\mathrm{core}} = 0.10\,\mathrm{m}$ has been chosen.

For the computations presented here we have used a grid size h_{f}, which corresponds to $r_{\mathrm{core}}/h_{\mathrm{f}} = 5$ in a region around the corotating vortices with dimensions $6\,\mathrm{m} \times 6\,\mathrm{m}$. Outside this area the grid is coarsened in radial directions. For the transient acoustic computation, the grid is correspondingly coarsened in the coupled region and it extends 500 m in all directions to avoid the effects of boundary reflections on the solution. In this case the element size for the acoustic grid in the region next to the spinning vortices is chosen to be $h_{\mathrm{a}} = 0.1\,\mathrm{m}$ which corresponds to $h_{\mathrm{a}}/h_{\mathrm{f}} = 5$. Since the applied vortex core model is an approximation of the velocity field near the point vortices, the evaluation of the velocity gradients at these regions still remains very sensible and numerical oscillations are still present even with a very fine

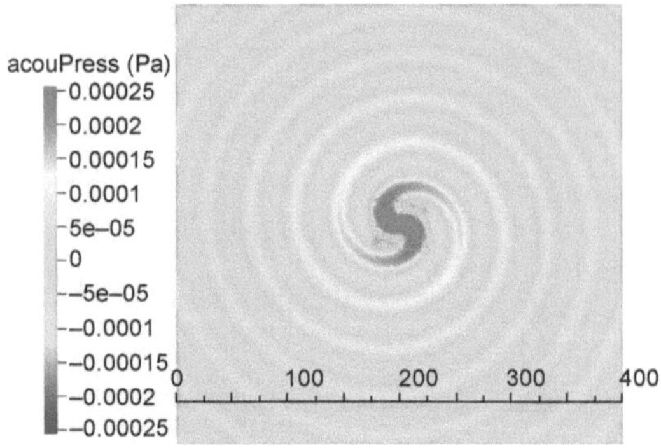

Fig. 4.7 Contour plot of acoustic pressure field at time $t = 413$ s computed using a temporal ramping of the initial acoustic sources. Distance scale in meters.

discretization. Therefore, transient results of the acoustic pressure field show some numerical noise as can be observed from the contour plot of the acoustic field at time $t = 413$ s presented in Figure 4.7, where a time step size $\Delta t = 0.5$ s was used.

To completely suppress transient numerical noise in the acoustic pressure field, the interpolated transient acoustic sources are transformed to the frequency domain and a harmonic analysis is carried out for the main frequency component of the problem, $f = 1/T \approx 0.026$ Hz. A comparison of the acoustic field from the numerical results with the analytical solution is presented in Figure 4.8. Good agreement in both the spiral pattern as well as in amplitudes is found in the far field acoustic pressure except at the center of the computational domain, where the evaluation of the analytical solution resulted in a non–physical behavior. Figure 4.9 compares the decay of the acoustic pressure along the positive x–axis between the transient result, the harmonic result and the analytical solution. Apart from the spurious noise in the transient simulation at distances $x < 40$ m within the source region, numerical decays present good correspondence with the far–field analytical values.

4.4 Flow–induced Noise from 3D Wall Mounted Cylinders

We investigate the flow–induced noise from wall–mounted cylinders using two different geometry profiles, a standard square cylinder geometry and a cylinder with elliptic shape in the downstream direction. The general set–up for the coupled simulation is depicted in Figure 4.10. Ω_1 denotes the region where the flow field is computed and where the acoustic sources are interpolated from the fluid grid to the acoustic grid. The region encompassing Ω_1 and Ω_2 corresponds to the region where the acoustic propagation is computed. Two different geometry profiles evaluated in the numerical investigations are depicted in Figures 4.11 and 4.12. Based on the crossflow side length of D $=$ 20 mm, the Reynolds number *Re* for both fluid calculations results in about 13.000.

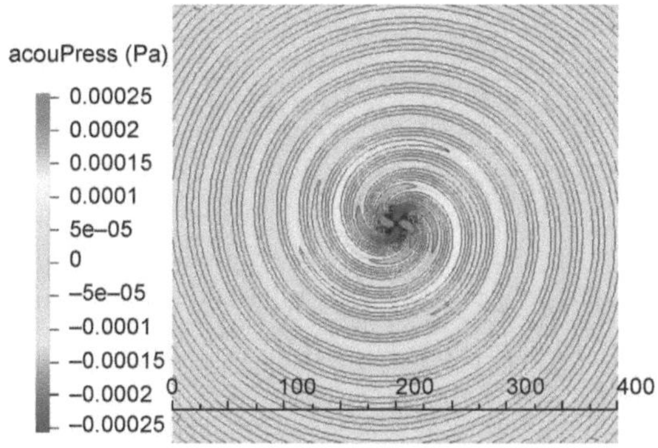

(a) Acoustic pressure field at frequency $f = 0.026$ Hz, from hybrid computation following Lighthill's acoustic analogy

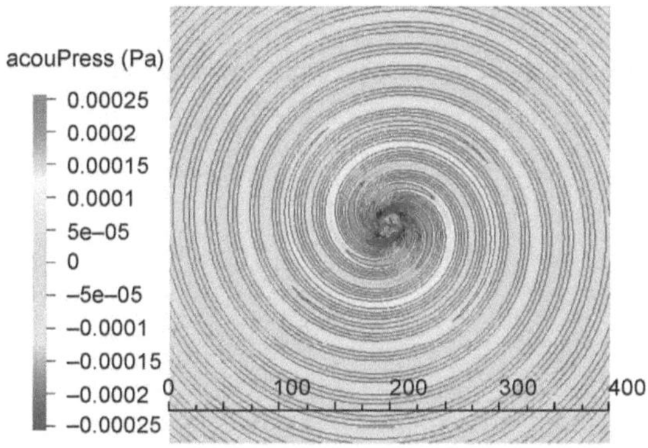

(b) Analytical acoustic pressure field obtained using MAE

Fig. 4.8 Comparison of sound pressure field obtained numerically using Lighthill's acoustic analogy with analytical solution obtained using MAE method. Distance scale in meters.

4.4.1 Flow Computations

The flow–induced noise investigations for the square cylinder and the square cylinder with elliptical afterbody are based on fluid results computed with `ANSYS--CFX`.[6] The numerical domain is described in Figure 4.13, where D $=$ 20 mm. For both configuration a block profile inflow velocity of $u_x = 10$ m/s have been applied. The boundary conditions used in the fluid computation with respect to the configuration from Figure 4.13 are described in Table 4.1. For the simulation of the flow a turbulence modeling approach based on SAS (Scale Adaptive Simulation) was employed. The SAS approach allows to use coarser grids than those used in standard LES (Large Eddy Simulation) computations. For both configurations, the compu-

[6] See `www.ansys.com/cfx`, ANSYS–CFX software

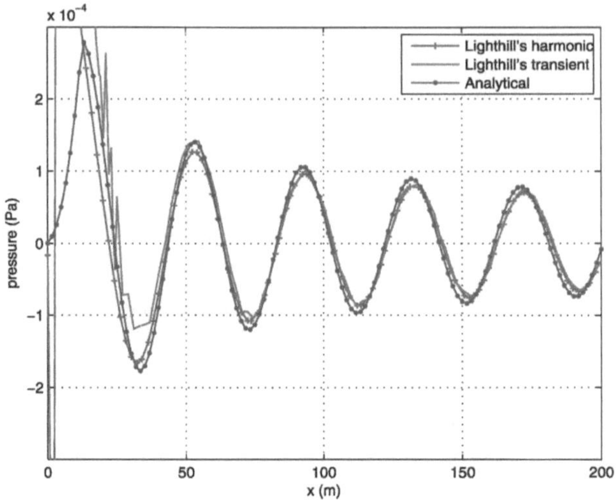

Fig. 4.9 Decay of the acoustic pressure values along the x–axis

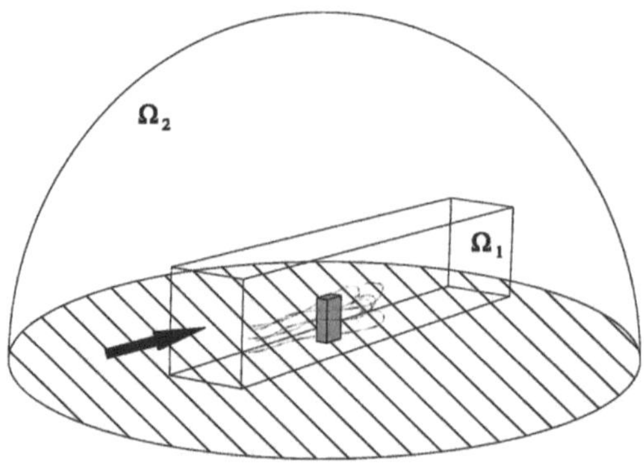

Fig. 4.10 Schematic representation of the hybrid domain as used for the 3D computations.

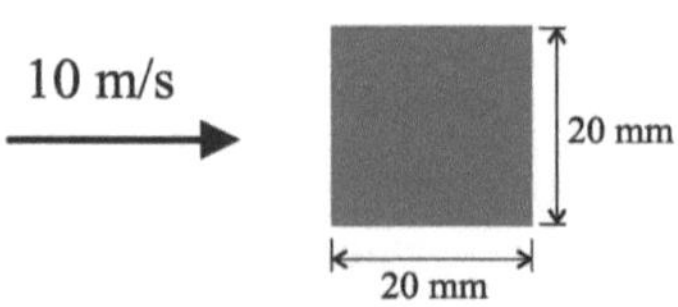

Fig. 4.11 Square cylinder profile

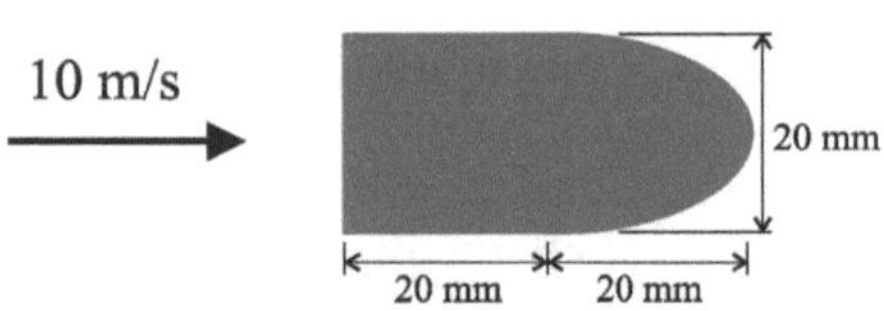

Fig. 4.12 Square cylinder with elliptical afterbody

tational grid consists of about 1.1 million cells, and for time step size a value of $\Delta t_\mathrm{f} = 20\,\mu\mathrm{s}$ has been chosen.

Figure 4.14 shows the obtained transient flow fields at one instantaneous time step. The visualization displays isosurfaces of $\Omega^2 - S^2 = 100000\,\mathrm{s}^{-2}$ colored with the eddy viscosity, where Ω representing here the vorticity and S the strain rate.

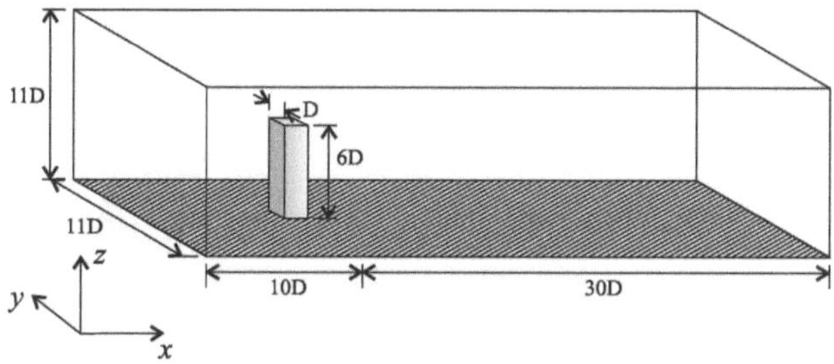

Fig. 4.13 Numerical domain used for fluid computations (D = 20 mm)

Table 4.1 Boundary conditions used for fluid computations

Position	Boundary Condition
X = 0	block inlet profile, 10 m/s
X = 40D	convective exit boundary
Z = 11D	symmetry boundary condition
Y = 0, Y = 11D	symmetry boundary condition
wall	no slip boundary condition

4.4.2 Acoustic Computations

Square Wall–Mounted Cylinder

For the square cylinder profile, acoustic simulations have been performed in time and frequency domain, using the results from the fluid computations presented in the previous section. Directly coupled transient computations were performed, using an acoustic grid with radius $r = 3\,\mathrm{m}$, consisting of 3.683.670 tetrahedral elements with quadratic basis functions which resulted in 4.986.115 finite element nodes. The coupled region has been discretized with 135.173 nodes. Since the conservative interpolation scheme using `MpCCI`[7] is restricted to volume interpolation of linear 3D elements, the acoustic sources have been interpolated on the 17.925 corner (linear) nodes of the quadratic tetrahedral elements.

Figure 4.15 presents the numerical grid in which first order absorbing boundary conditions have been applied on the hemisphere boundary using 51.896 triangular surface elements. A closer view of the coupled region depicting the edges of the tetrahedral elements on the bottom plane and on the cylinder surface is shown in Figure 4.16. The isosurfaces of the near–field acoustic pressure obtained from the flow computations are displayed in Figure 4.17. The time step size for the 3D acoustic computation was chosen to be $\Delta t_a = 10 \cdot \Delta t_\mathrm{f} = 200\,\mu\mathrm{s}$. For these low frequency tonal noise problems, further quantitative analysis of 3D transient results is restricted due to the large wavelength expected for these specific problems ($\lambda \approx 6.2\,\mathrm{m}$ for

[7]See: `http://www.mpcci.org`, Mesh–based parallel Code Coupling Interface – MpCCI 3.0 Specifications, Fraunhofer Institute, Germany

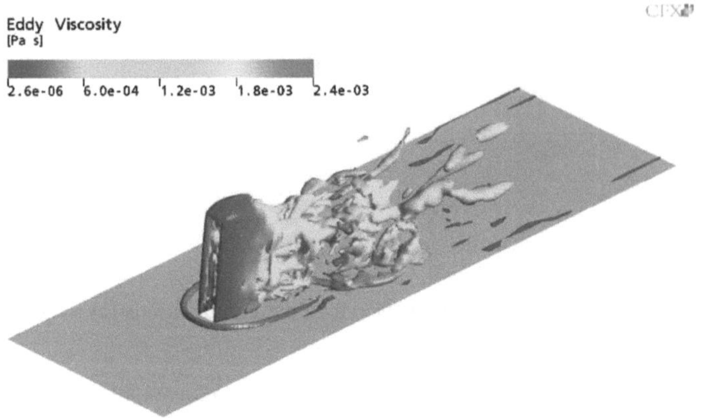

(a) Square cylinder

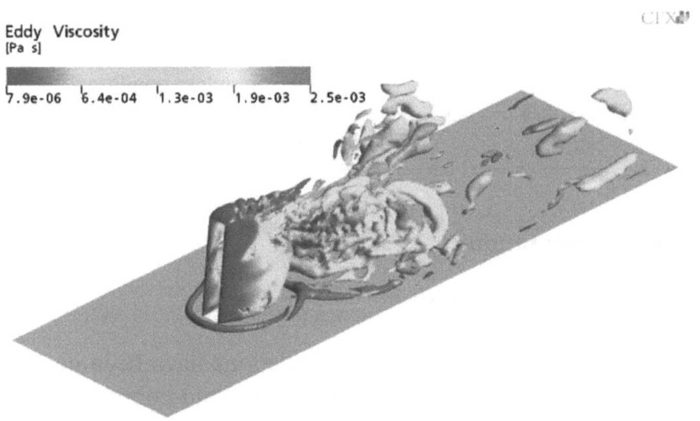

(b) Square cylinder with elliptical afterbody

Fig. 4.14 Instantaneous visualization of transient flow field.

$f = 55\,\mathrm{Hz}$) and the high computational cost due to the very large number of elements required to cover higher frequency signals present in the acoustic sources. Therefore, for more practical analysis of this type of aeolian tone problems, we perform harmonic analyses for the main frequency components. The main advantage of computing in frequency domain is that the acoustic field is solved only for relevant frequency components, avoiding in this way high frequency transient numerical noise. Furthermore, in the 3D computations we can also use our enhanced PML method [29], which allows smaller acoustic domains producing almost no spurious reflections. This aspect significantly reduces the computational time.

In Figure 4.18 the configuration of the simulation domain for the harmonic computation showing the monitoring points used for directivity analysis is presented. Due to the rectangular configuration of the acoustic domain, the discretization could be performed using hexahedron elements with quadratic basis functions, which in-

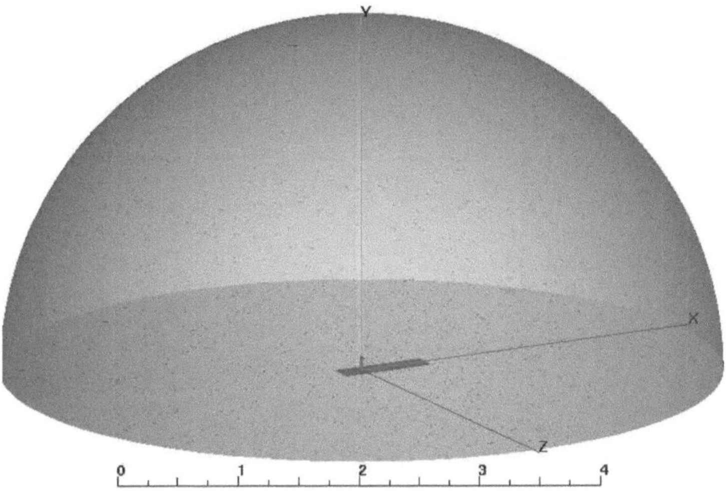

Fig. 4.15 Visualization of complete acoustic grid used for transient computations with radius $r = 3\,\mathrm{m}$ showing coupled region at the center of the domain. Distance scale in meter.

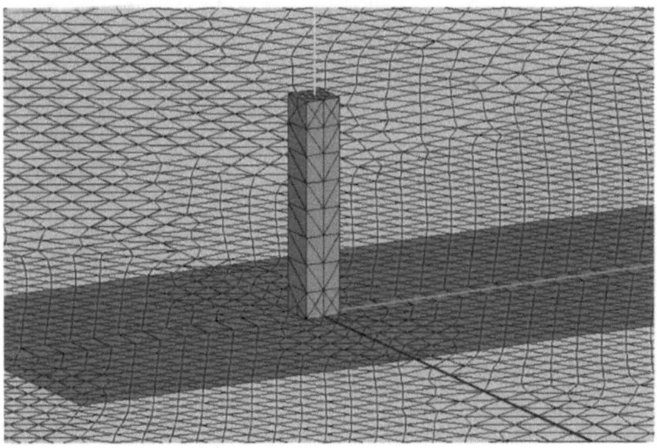

Fig. 4.16 Close up of acoustic grid used for transient computations depicting edges of tetrahedra on the bottom plane and on the cylinder surface.

creases the accuracy of the computation. A spatial discretization has been chosen, in which a ratio of $h_\mathrm{a}/h_\mathrm{f} = 10$ in the region directly around the cylinder sufficed to produce mesh independent results. Following this discretization criteria three different mesh resolutions have been tested with the number of linear nodes in the coupled region ranging from 8.092 to 24.177. Figure 4.19 presents results for the amplitudes at a radius $r = 1\,\mathrm{m}$ using the three different spatial discretizations. While the finest mesh produces a smoother directivity pattern, for the other grids only minimal differences are observed in the amplitudes, which demonstrates the robustness of our interpolation scheme.

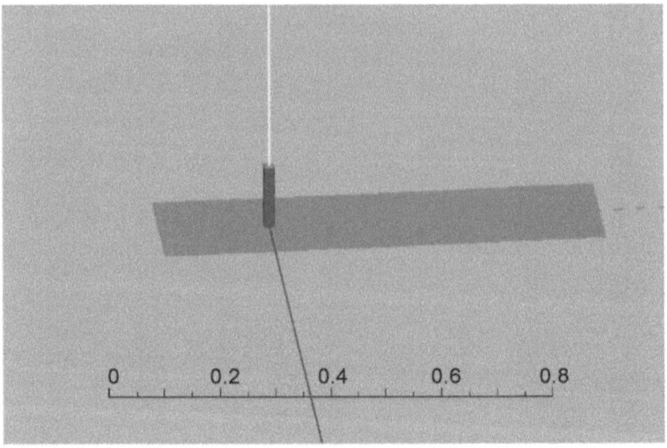

(a) View point

(b) Acoustic pressure isosurfaces

Fig. 4.17 Isosurfaces of acoustic field for values ± 2 Pa at time t=10 ms obtained using fluid results from CFX-ANSYS simulations.

Figure 4.20 presents several isosurfaces of the acoustic pressure clipped at yz–plane at the main frequency component present in the computation, $f = 55$ Hz. Values of the isosurfaces range from 5 mPa to 54 mPa, the outermost corresponding to a sound pressure level (SPL) of about 48 dB.

Wall–mounted Cylinder with Elliptic Profile

Similarly as for the square cylinder profile, harmonic acoustic computations have been performed for the case using the profile from Figure 4.12 and its SPL values and directivity pattern have been analyzed. The acoustic sources have been evaluated using the velocity components from the CFD results computed with CFX–ANSYS using SAS turbulence modeling presented in Figure 4.14. The acoustic computation has been performed using a similar configuration for the computational domain as the one depicted in Figure 4.18, except for the cylinder geometry. Figure 4.21 presents a

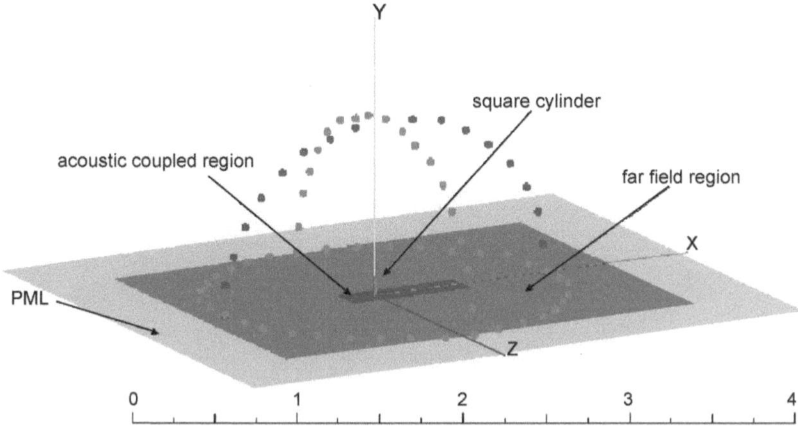

Fig. 4.18 Schematic drawing of the acoustic domain used for the harmonic computation showing points used for directivity analysis. Distance scale in m.

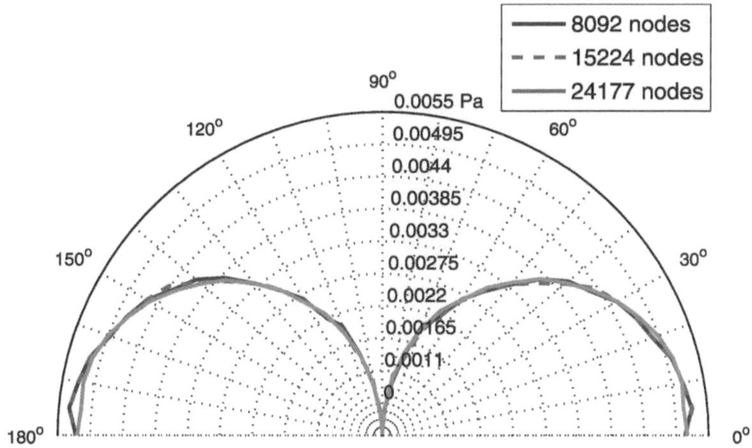

Fig. 4.19 Comparison of amplitudes at radius $r = 1$ m on the crossflow yz–plane using three different spatial discretizations in the acoustic coupled region

close–up of the acoustic grid showing the corresponding elliptic profile employed in this case. Regarding the acoustic coupled region, the fine acoustic sources from the fluid resolution have been interpolated on 21160 hexahedron corner nodes.

In the acoustic computations, the main frequency component found for this problem was $f = 39$ Hz. For this frequency value, isosurfaces of the acoustic pressure clipped at the yz–plane are presented in Figure 4.22. Values of the isosurfaces range from 22 mPa to 100 mPa and the outermost isosurface corresponds in this case to 61 dB and, similarly as in the case using the square cylinder profile, the directivity pattern for this main frequency value results in a dipole–like acoustic field.

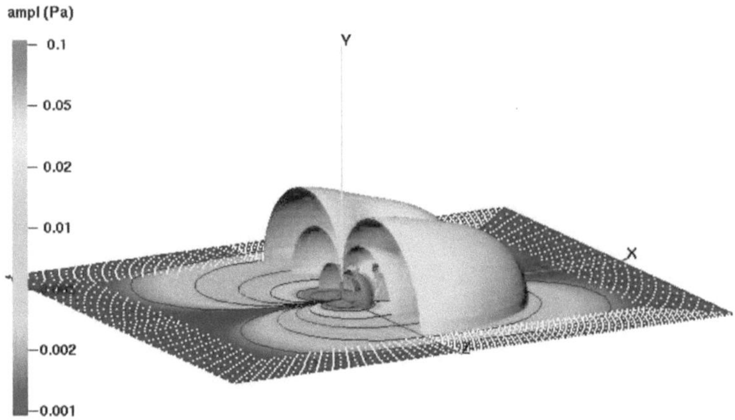

Fig. 4.20 Isosurface of acoustic pressure for wall–mounted square cylinder at $f = 55\,\mathrm{Hz}$, clipped at yz–plane. Dotted region represents PML.

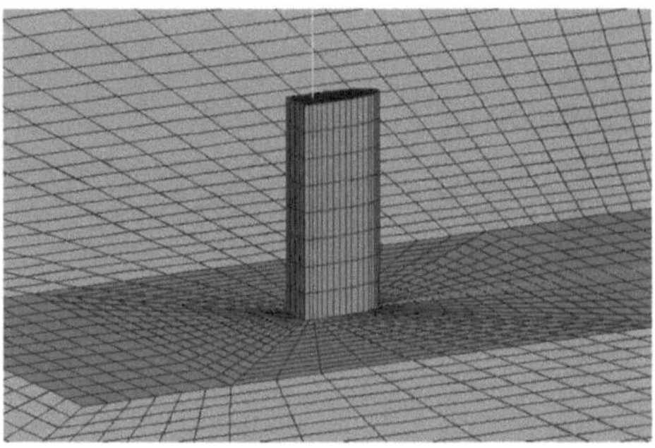

Fig. 4.21 Close up of acoustic grid for the case of the elliptic profile used for harmonic computations

Table 4.2 Comparison of main frequency values found in simulations and measurements

f_{main}	Square Profile	Elliptic Profile
Simulation	55 Hz	39 Hz
Measurements	53 Hz	36 Hz

Evaluation of 3D Results

In the following, acoustic results from the harmonic computations for the two wall–mounted cylinder profiles are evaluated and compared with experimental results from [25]. Table 4.2 presents the main frequency components found in the computations and the frequency values found in the measurements carried out in an anechoic wind tunnel [25]. In Table 4.3 a comparison of the SPL values between the numerical and the experimental results is presented for both cylinder profiles. It

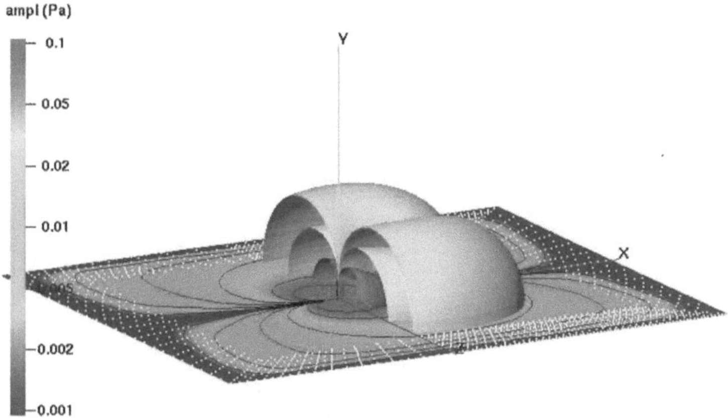

Fig. 4.22 Isosurface of acoustic pressure for wall–mounted square cylinder with elliptic profile at $f = 39\,\text{Hz}$, clipped at yz–plane. Dotted region represents PML

can be noticed, that for both geometries slightly higher SPL values are obtained in the numerical simulations. At this point it is important to note that, due to practical reasons, in the experimental case the cylinders were mounted at the center of a wall with a crossflow length of $L_{\text{exp}} = 0.66\,\text{m}$, whereas in the simulation the wall covers the complete domain width $L_{\text{sim}} = 2.22\,\text{m}$. Reflection of the acoustic waves on the larger wall from the simulation domain is one of the reasons for the higher SPL values at the monitoring points compared. Furthermore, the relative difference of the SPL values from both profiles compare well to the measured value. In Table 4.3 this difference is observed to be 16 dB in the experiments and 14 dB in the numerical computations. Finally, directivity plots of the SPL levels at a radius $r = 1\,\text{m}$ away of the flow–induced noise for the two wall–mounted cylinders are presented in Figure 4.23. In this plot it can be observed the significantly higher amplitudes at all monitoring points obtained in the flow–induced noise computation for the cylinder with elliptic profile.

4.5 Conclusion

The numerical computation of flow induced noise is quite challenging due to the energy and length scale disparity of the flow field compared to the generated acoustic noise. The proposed method applicable to low Mach number problems is based on SAS flow computations and the FE formulation of Lighthill's acoustic analogy. The computation of the acoustic noise due to a flow around wall mounted cylinders with different geometries demonstrate the applicability of the numerical scheme.

Due to the complexity, computational aeroacoustics will remain a challenging topic for research both concerning the mathematical modeling as well as the dedicated numerical methods.

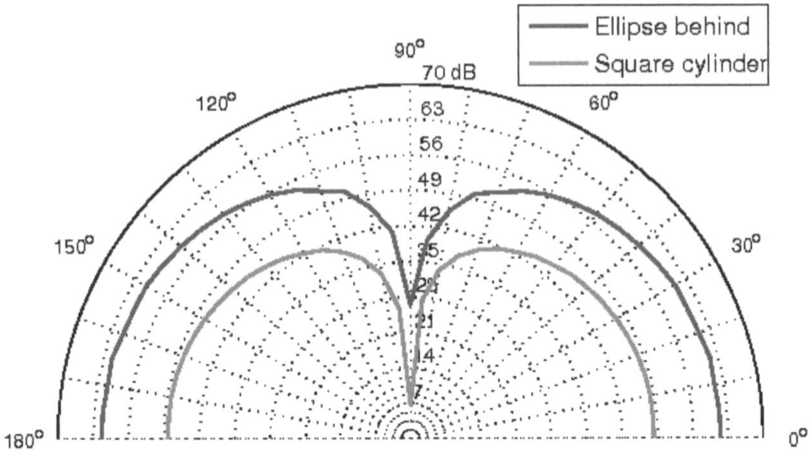

Fig. 4.23 Directivity patterns for two cylinder profiles at radius $r = 1.0$ m on the crossflow yz–plane

Table 4.3 Numerical and experimental SPL values of the flow–induced noise for the two cylinder profiles evaluated at 0° on the yz–plane

SPL value	Square Profile	Elliptic Profile	Relative SPL difference
Simulation	47 dB	61 dB	14 dB
Measurements	44 dB	60 dB	16 dB

References

1. Adams RA (1975) Sobolev spaces. Academic Press, New York
2. Ainsworth M (2004) Discrete dispersion relation for hp–version finite element approximation at high wave number. SIAM Journal of Numerical Analysis 42:553–575
3. Ali I, Escobar M, Hahn C, Kaltenbacher M, Becker S (2004) Numerical and experimental investigations of flow induced noise around a square cylinder. In: 10th AIAA/CEAS Aeroacoustics Conference, AIAA–2004–3004, Manchester
4. Bhagwat M, Gordon J, (2002) Generalized viscous vortex model for application to free–vortex wake and aeroacoustic calculations. In: 58th Annual Forum of the American Helicopter Society International, Montreal
5. Bogey C, Bailly C, Juvé D (2002) Computation of flow noise using source terms in linearized Euler's equations. AIAA Journal 40:235–243
6. Brentner K, Farassat F (1998) An analytical comparison of the acoustic analogy and Kirchhoff formulation for moving surfaces. AIAA Journal 36:1379–1386
7. Breuer M (2001) Direkte numerische Simulation und large–eddy simulation turbulenter Strömungen auf Hochleistungsrechnern. Shaker Verlag, Herzogenrath
8. Cohen G (2001) Higher order numerical methods for transient wave equations. Springer–Verlag, Berlin–Heidelberg
9. Crighton DG (1993) Computational aeroacoustics for low Mach number flows. In: Hardin JC, Hussaini MY (eds) Computational aeroacoustics. Springer–Verlag, New York 50–68

10. Curle N (1995) The influence of solid boundaries upon aerodynamic sound. Proceedings of the Royal Society of London A 231:505–514
11. Di Francescantonio P (1997) A new boundary integral formulation for the prediction of sound radiation. Journal of Sound and Vibration 202:491–509
12. Dumbser M (2005) Arbitrary high order schemes for the solution of hyperbolic conservation laws in complex domains. PhD Thesis, Universität Stuttgart
13. Ekaterinaris J (1999) New formulation of Hardin–Pope equations for aeroacoustics. AIAA Journal 37:1033–1039
14. Engquist B, Majda A (1977) Absorbing boundary conditions for the numerical simulation of waves. Mathematics of Computation 31:629–651
15. Escobar M (2007) Finite element simulation of flow–induced noise applying Lighthill's acoustic analogy. PhD Thesis, Universität Erlangen–Nürnberg
16. Ewert R, Schröder W (2003) Acoustic perturbation equations based on flow decomposition via source filtering. Journal of Computational Physics 188:365–398
17. Farassat F (2001) Acoustic radiation from rotating blades – the Kirchhoff method in aeroacoustics. Journal of Sound and Vibration 239:785–800
18. Farassat F, Myers MK (1988) Extension of Kirchhoff's formula to radiation from moving surfaces. Journal of Sound and Vibration 123:451–560
19. Ffowcs–Williams JE, Hawkings DL (1969) Sound radiation from turbulence and surfaces in arbitrary motion. Philosophical Transactions A of the Royal Society 321–342
20. Fortenbach R (2006) Mehrskalenmodellierung von aeroakustischen Quellen in schwach kompressiblen Strömungen. PhD Thesis, Universität Stuttgart
21. Freund JB, Lele SK, Moin P (2000) Direct numerical simulation of a Mach 1.92 turbulent jet and its sound field. AIAA Journal 38:2023–2031
22. Freund JB, Lele SK, Wei M (2004) The robustness of acoustic analogies. In: Proceedings of the Summer Program, Center for Turbulence Research, Stanford University 241–252
23. Goldstein ME (1976) Aeroacoustics. McGraw–Hill, New York
24. Goldstein ME (2006) A hybrid RANS/LES approach for predicting jet noise. In: 12th AIAA/CEAS Aeroacoustics Conference AIAA–2006–2631, Cambridge
25. Hahn C, Kaltenbacher M, Lerch R, Becker S, Durst F (2006) Influence of geometry on the generation of aeroacoustic noise. In: Proceedings of the ISMA International Conference on Noise & Vibration, CD–ROM–Proceedings, Leuven 619–631
26. Hardin JC (1993) Regarding numerical considerations for computational aeroacoustics. In: Hardin JC, Hussaini MY (eds) Computational aeroacoustics. Springer–Verlag, New York 216–228
27. Hardin JC, Pope DS, (1994) An acoustic/viscous splitting technique for computational aeroacoustics. Theoretical and Computational Fluid Dynamics 6:323–340
28. Hughes TJR (2000) The finite element method. Prentice–Hall, Englewood Cliffs
29. Kaltenbacher M (2007) Numerical simulation of mechatronic sensors and actuators. Springer–Verlag, Berlin–Heidelberg–New York
30. Joo Lee D, Ok Koo S (1995) Numerical study of sound generation due to a spinning vortex pair. AIAA Journal 33:20–26
31. Lighthill MJ (1952) On sound generated aerodynamically. I: General theory. Proceedings of the Royal Society of London, Series A 211:564–587
32. Lilley GM (1972) The generation and radiation of supersonic jet noise. Vol IV: Theory of turbulence generated jet noise, noise radiation from upstream sources, and combustion noise. Part II: Generation of sound in a mixing region. Technical Report 4, U.S. Air Force Aeropropulsion Laboratory (AFAPL–TR–72–53)
33. Lilley GM (1974) On the noise from jets. Technical Report, AGARD, CP–131

34. Liow K, Thompson MC, Hourigan K (2001) Computation of acoustic waves generated by a co–rotating vortex pair. In: Proceedings of the 14th Australasian Fluid Mechanics Conference, Adelaide
35. Lockard D (2002) A comparison of Ffowcs Williams–Hawkings solvers for airframe noise applications. In: Proceedings of 8th AIAA/CEAS Aeroacoustics Conference, AIAA–2002–2580, Breckenridge
36. Lyrintzis AS (2003) Surface integral methods in computational aeroacoustics – From the (CFD) near field to the (Acoustic) far–field, International Journal of Aeroacoustics 2 2:95–128
37. Manoha E, Herrero C, Ben Khelil S, Guillen P, Sagaut P, Mary I (2002) Numerical prediction of airfoil aerodynamic noise. In: Proceedings of 8th AIAA/CEAS Aeroacoustics Conference, AIAA–2002–2573, Breckenridge
38. Menter F, Egorov Y (2005) A scale adaptive simulation model using two–equation models. In: 43rd AIAA Aerospace Sciences Meeting and Exhibit, AIAA–2005–1095, Reno
39. Mitchell B, Lele SK, Moin P (1995) Direct computation of the sound from a compressible co–rotating vortex pair. Journal of Fluid Mechanics 285:181–202
40. Möhring W (1978) On vortex sound at low Mach number. Journal of Fluid Mechanics 85:685–691
41. Müller EA, Obermeier F (1967) The spinning vortices as a source of sound. In: Proceedings of the Conference on Fluid Dynamics of Rotor and Fan Supported Aircraft at Subsonic Speeds (AGARD CP–22) 22.1–22.8
42. Oberai A, Ronaldkin F, Hughes TJR (2000) Computational procedures for determining structural–acoustic response due to hydrodynamic sources. Computer Methods in Applied Mechanics and Engineering 190:345–361
43. Oberai A, Ronaldkin F, Hughes TJR (2002) Computation of trailing–edge noise due to turbulent flow over an airfoil. AIAA Journal 40:2206–2216
44. Paliath U, Morris PJ (2005) Prediction of jet noise from circular beveled nozzles. In: Proceedings of the 11th AIAA Aeroacoustics Conference AIAA–2005–3096, Monterey
45. Peters MCAM (1993) Aeroacoustic sources in internal flows. PhD Thesis, Eindhoven University of Technology, Eindhoven
46. Ribner HS (1996) Effect of jet flow on jet noise via an extension to the Lighthill model. Journal of Fluid Mechanics 321:1–24
47. Schram C, Anthoine J, Hirschberg A (2005) Calculation of sound scattering using Curle's analogy for non–compact bodies. In: Proceedings of the 11th AIAA Aeroacoustics Conference, AIAA–2005–2836, Monterey
48. Scully MP, (1975) Computation of helicopter rotor wake geometry and its influence on rotor harmonic loads. MIT Publications ARSL TR 178–1, Cambridge
49. Seo J, Moon Y (2006) Linearized perturbed compressible equations for low Mach number aeroacoustics. Journal of Computational Physics 218:702–719
50. Tam CKW, Webb JC (1993) Dispersion–relation–preserving finite difference schemes for computational acoustics. Journal of Computational Physics 107:262–281
51. Utzmann JT, Schwartzkopff T, Dumbser M, Munz CD (2006) Heterogeneous domain decomposition for CAA. AIAA Journal 44:2234–2250
52. Uzun A, Lyrintzis AS, Blaisdell GA (2004) Coupling of integral acoustics methods with LES for jet noise prediction. In: Proceedings of AIAA Aerospace Sciences Meeting and Exhibit, AIAA–2004–0517 Reno
53. Wagner C, Hüttl T, Sagaut P (eds) (2007) Large–eddy simulation for acoustics. Cambridge University Press, New York

Part II

FEM: External Problems

5

Computational Absorbing Boundaries

Dan Givoli

Department of Aerospace Engineering, Technion, Haifa 32000, Israel
givolid@aerodyne.technion.ac.il

Summary. The subject of this chapter is the treatment of artificial boundaries in wave problems. Artificial boundaries are introduced when the problem under study is associated with an unbounded medium, yet one is interested (or is forced) to solve the problem in a finite computational domain. In this context the artificial boundaries are often called absorbing boundaries, for reasons that will be explained. After discussing the difficulties involved, the major milestones that have been set in the development of absorbing boundaries are surveyed. These include the classical absorbing boundary conditions, exact nonlocal conditions, absorbing layers, perfectly matched layers and high–order local boundary conditions. Infinite elements and boundary element methods are also mentioned. Examples from previous publications are given.

5.1 Computational Absorbing Boundaries — What are they?

Many problems in acoustics, as well as in other fields of application like geophysics, oceanography and meteorology, involve waves in an *unbounded medium.* The solution of such problems using the Finite Element (FE) method or other domain–type methods usually requires the use of a *finite* computational domain in which the entire calculation is to be done. Thus, one has to introduce an *artificial boundary* that encloses the region of interest. An example is described in Figure 5.1, which is taken from a paper of Farhat et al. [13], showing a three–dimensional FE mesh for underwater acoustics simulation in the vicinity of a submarine. The artificial boundary, shaped like an ellipsoid in this case, is clearly seen.

To describe a well–posed mathematical problem in the finite computational domain, some boundary conditions must be imposed on the artificial boundary. These boundary conditions are the subject of the present chapter. We illustrate their significance using a simple model problem.

5.1.1 The Sommerfeld Radiation Condition

Consider the setup described in the left part of Figure 5.2. Waves in the unbounded domain $\mathcal{R}$ are scattered from a given obstacle with boundary Γ. First we consider

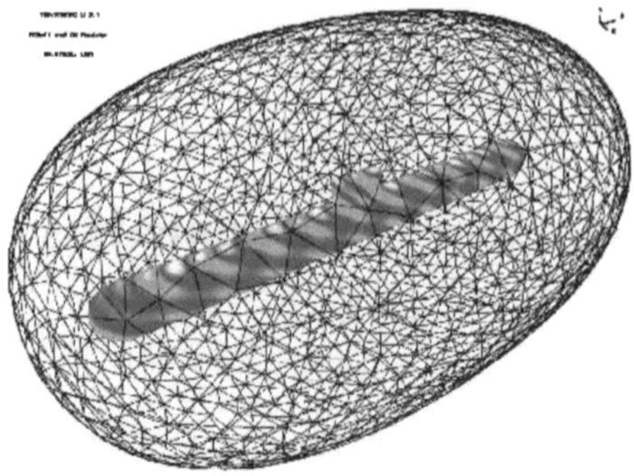

Fig. 5.1 FE mesh for underwater acoustics simulation outside a submarine. The outer surface of the mesh is an artificial boundary. Reprinted by permission from "FETI–DPH: A dual–primal domain decomposition method for acoustic scattering," Farhat C, Avery P, Tezaur R, Li J (2005), Journal of Computational Acoustics 13:499–524, World Scientific Press.

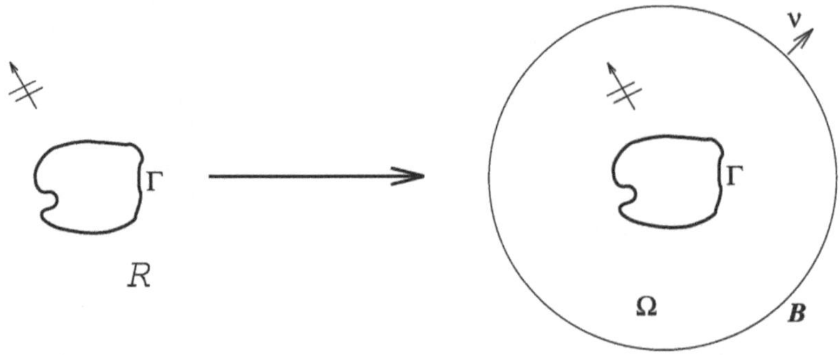

Fig. 5.2 Replacing an unbounded domain problem by a problem in a bounded domain.

time–harmonic (i.e., single frequency) waves, which are solutions of the Helmholtz equation

$$\nabla^2 u + k^2 u = 0 \quad \text{in} \quad \mathcal{R} \, . \tag{5.1}$$

Some boundary condition, which is derived from the incident wave, is given on Γ, say,

$$u = g \quad \text{on} \quad \Gamma \, . \tag{5.2}$$

In addition, the so–called *Sommerfeld radiation condition* is imposed. This condition states that waves at infinity are *outgoing*. The Sommerfeld radiation condition can be written in several forms; a common form which is adopted here is

$$\lim_{r \to \infty} r^{(d-1)/2} \left(\frac{\partial u}{\partial r} - iku \right) = 0 \, . \tag{5.3}$$

Here r is the radial coordinate, and d is the spatial dimension (d =1, 2 or 3). A proof that the Sommerfeld condition (5.3) is indeed an appropriate radiation condition is given in Courant and Hilbert [11, p 315–318]. In [16] this is shown directly for three special types of symmetrical waves: normal plane waves ($d = 1$), cylindrical waves ($d = 2$) and spherical waves ($d = 3$), depending only on the single coordinate r.

The one–dimensional case ($d = 1$) is exceptional. The Sommerfeld condition (5.3) becomes

$$\frac{\partial u}{\partial r} - iku = 0 \tag{5.4}$$

as $r \to \infty$. (Actually the partial derivative is an ordinary derivative when $d = 1$.) It is easy to show that in this case (5.4) holds not only at infinity but also *at any finite* r. More precisely, if a number r_0 exists such that the wave number k is constant in the semi–infinite interval $r_0 \leq r \leq \infty$, then (5.4) holds at any r in that interval. In two and three dimensions the same is not true. The condition (5.4) (or more precisely (5.3)) holds at infinity but not at any finite point.

Now we consider the time–dependent case, using the scalar wave equation as a simple model. Again with the setup in the left part of Figure 5.2, the scattering problem can be stated as follows:

$$\frac{\partial^2 u}{\partial t^2} = c^2 \nabla^2 u \quad \text{in} \quad \mathcal{R}\,, \tag{5.5}$$

$$u(\boldsymbol{x}, t) = g(\boldsymbol{x}, t) \qquad , \qquad \boldsymbol{x} \in \Gamma\,, \tag{5.6}$$

$$u(\boldsymbol{x}, 0) = u_0(\boldsymbol{x}) \qquad ; \qquad \frac{\partial u}{\partial t}(\boldsymbol{x}, 0) = v_0(\boldsymbol{x}) \qquad , \qquad \boldsymbol{x} \in \mathcal{R}\,. \tag{5.7}$$

Here g, u_0 and v_0 are given functions, and u_0 and v_0 are assumed to vanish outside a finite region.

As opposed to the time–harmonic case, a radiation condition at infinity does not appear in the statement of the time–dependent problem. While an elliptic equation such as (5.1) must be accompanied by a boundary condition at infinity in order to be well posed, no such condition needs to be prescribed in the hyperbolic problem. However, it can be shown that the solution u of (5.5)–(5.7) *does* satisfy a "Sommerfeld condition" at infinity which is the analogue of (5.3), namely

$$\lim_{\substack{r \to \infty \\ r + ct = \text{const.}}} r^{(d-1)/2} \left(\frac{\partial u}{\partial r} + \frac{1}{c} \frac{\partial u}{\partial t} \right) = 0\,. \tag{5.8}$$

See, e.g., [11]. As before, the Sommerfeld condition (5.8) asserts that waves at infinity are outgoing. Note that $(1/c)\partial u / \partial t$ in (5.8) is replaced by $-iku$ in (5.3).

Analogously to the time–harmonic case, it is easy to show that *in one dimension* ($d = 1$) the condition

$$\frac{\partial u}{\partial r} + \frac{1}{c} \frac{\partial u}{\partial t} = 0 \tag{5.9}$$

holds not only at infinity, as implied by (5.8), but at any finite r as well. More precisely, if a number r_0 exists such that c is constant and $u_0 \equiv 0$, $v_0 \equiv 0$ hold in the semi–infinite interval $r_0 \leq r \leq \infty$, then (5.9) holds at any r in that interval. To see this, we note that the general solution of the wave equation in one dimension is

$$u(r,t) = F(r+ct) + G(r-ct) \, , \tag{5.10}$$

where F and G are arbitrary functions. The F–term represents an *incoming wave* (i.e. a wave moving to the left), whereas the G–term represents an *outgoing wave* (i.e. a wave moving to the right). Since no incoming waves from infinity are allowed, the solution must have the form

$$u(r,t) = G(r-ct) \, . \tag{5.11}$$

From this we easily get,

$$\frac{\partial u}{\partial r} + \frac{1}{c}\frac{\partial u}{\partial t} = 1 \cdot G'(r-ct) + \frac{1}{c}(-c)G'(r-ct) = 0 \, , \tag{5.12}$$

which is the same as (5.9).

5.1.2 A Fundamental Difficulty

Now we consider the numerical solution of the problem (5.1)–(5.3) in the general two– and three–dimensional case. An artificial boundary $\mathcal{B}$ is introduced to make the computational domain finite. Thus the computational domain, denoted Ω, is bounded internally by Γ and externally by $\mathcal{B}$: see the right part of Figure 5.2. In order that the statement of the problem in Ω be complete, we need to impose a boundary condition on $\mathcal{B}$. This boundary condition must have the property that outgoing waves impinging on $\mathcal{B}$ from inside the computational domain Ω are transmitted through $\mathcal{B}$ without any reflection. At first sight the construction of such a condition seems to be a simple matter, but this apparent simplicity is deceptive. If $\mathcal{B}$ is a circle or a sphere, the obvious naive choice is a *Sommerfeld–like boundary condition* of the form (5.4) on $\mathcal{B}$. More generally, it is tempting to try the boundary condition

$$\frac{\partial u}{\partial \nu} - iku = 0 \quad \text{on} \quad \mathcal{B} \, , \tag{5.13}$$

where $\frac{\partial u}{\partial \nu}$ is the normal derivative of u on $\mathcal{B}$. However, it is a well known fact that this boundary condition typically produces large *spurious reflection* of waves from $\mathcal{B}$. In other words, it usually leads to large errors in the computed solution. This fact has been demonstrated in many publications; see, e.g., the numerical experiments in [16].

A similar situation occurs in the time–dependent case. On $\mathcal{B}$ one has to prescribe a boundary condition which should mimic the effect of the Sommerfeld radiation condition (5.8) at infinity. Again, the simplest condition is the Sommerfeld–like condition

$$\frac{\partial u}{\partial \nu} + \frac{1}{c}\frac{\partial u}{\partial t} = 0 \quad \text{on} \quad \mathcal{B} \, , \tag{5.14}$$

which reduces to (5.9) if $\mathcal{B}$ is a circle or a sphere. However, as in the time–harmonic case, it turns out that this boundary condition leads in many cases to large spurious reflections.

Thus, the apparently simple statement that the boundary $\mathcal{B}$ is "transparent" and that outgoing waves should "pass through it" without giving rise to reflections, is in practice quite difficult to enforce.

This gap between the simplicity of stating the goal and the difficulty in achieving it has been a source of frustration, and has led to a lot of research. During the last 35 years or so, there has been a considerable amount of work to devise boundary conditions that reduce the amount of spurious reflection. Results of this effort can be found in the literature related to various fields, such as acoustics, geophysics, oceanography, meteorology, gasdynamics, hydrodynamics and electromagnetics. The geometry and governing equations considered in these fields are sometimes different, but the goals and techniques are similar. These boundary conditions have been called by various names, including non–reflecting, absorbing, radiating, silent, transmitting, transparent, open, free–space, pulled–back, and one–way boundary conditions. (We remark that "open boundary conditions" are slightly more general, allowing for incoming waves as well. We shall not discuss this matter here.) In acoustics and geophysics the most common name for such conditions is *Absorbing Boundary Conditions* (ABC) , and this is how we shall refer to them in this chapter. This name expresses the fact that the artificial boundary $\mathcal{B}$ effectively serves as an "absorber" of the energy carried by the waves impinging on $\mathcal{B}$. However, one must be careful using this interpretation, since in general this artificial absorber is not equivalent to a "physical" wave absorber, certainly not to a simple absorber with constant acoustical properties.

We have seen the analogy between the frequency–dependent and the time–dependent cases. The term $(1/c)\partial u/\partial t$ in the time–dependent Sommerfeld condition (5.9) is replaced by $-iku$ in the frequency–dependent Sommerfeld condition (5.4). The same analogy holds with respect to ABCs. Thus, any ABC which was devised for the time–dependent case, can be adapted to the frequency–dependent case by simply replacing every occurrence of the operator $(1/c)\partial/\partial t$ by $-ik$. The opposite — adapting a frequency–dependent ABC to the time–dependent case — is not straight forward, unless the frequency–dependent ABC depends on the wave number k in a sufficiently simple way.

The use of an artificial boundary $\mathcal{B}$ in a numerical calculation enables one to eliminate the far–field domain (namely the domain exterior to $\mathcal{B}$) prior to computation. Therefore, it is clear that the ABC on $\mathcal{B}$ is solely responsible for the correct representation of this eliminated domain and the physical phenomena that occur therein. In fact, some methods of constructing ABCs are based on the properties of the far–field solution, and others assume (implicitly or explicitly) that the far–field domain possesses some "regularity." In acoustical and geophysical problems it is often assumed that the medium outside $\mathcal{B}$ is linear, isotropic and homogeneous, and that it contains no sources. Relaxing these assumptions makes the problem of designing good ABCs significantly more difficult. On the other hand, inside Ω the only limitations are those dictated by the capabilities of the interior numerical scheme employed (e.g., the FE method).

5.1.3 Why is Constructing a Good ABC Difficult?

In devising a new ABC one should have in mind at least some of the following goals:

1. *Well–posedness:* The problem in Ω including the ABC on $\mathcal{B}$ is mathematically well posed.
2. *Accuracy on the continuous level:* The problem in Ω including the ABC on $\mathcal{B}$ is a good approximation of the original problem in the infinite domain. In other words, the amount of spurious reflection generated by the ABC is small.
3. *Scheme compatibility:* The ABC on $\mathcal{B}$ is highly compatible with the numerical scheme used in Ω.
4. *Stability:* The numerical method employed including the ABC on $\mathcal{B}$ must result in a stable numerical scheme.
5. *Accuracy on the discrete level:* The error generated by the numerical scheme due to the use of the discrete ABC on $\mathcal{B}$ is small.
6. *Efficiency:* The use of the ABC on $\mathcal{B}$ does not involve a large computational effort.
7. *Ease of implementation:* The ABC is easy to implement and to incorporate in an existing code.
8. *Generality:* The ABC performs well in a variety of situations, i.e., with different angles of incidence, different geometries, different types of media — dispersive, layered, etc., different types of waves (P, S, evanescent, ...), and so on.

In the list above, properties 1 and 2 have to do with the continuous problem, prior to discretization, whereas properties 3–7 deal with the discrete problem. Property 8 is associated with both the continuous and discrete levels. Usually, the combination of properties 2 and 3 implies property 5, although it is easier to check property 5 directly. Also, roughly speaking, the satisfaction to a high degree of properties 1–4 usually implies the convergence of the numerical scheme. Of course one has to prove that the method ensures convergence, but this is more readily done after the ABC has been introduced and not in the process of designing it.

The combination of properties 5 and 6, namely the reduction of spurious reflections to a minimum in an efficient way, has been the main object of most researchers. Most ABCs perform well if the artificial boundary $\mathcal{B}$ is set far away from all sources or scatterers. In fact, in solving time–dependent problems one can set $\mathcal{B}$ sufficiently far away so that no waves would reach it in the time interval in which the solution is sought. However, this would result in a large computational domain Ω, and is therefore inefficient. Hence, the ABC has to perform well even when set quite close to the source or scatterer. In addition, it must not be in itself so complicated as to require a large computational effort.

Constructing an ABC which possesses all 8 properties to a satisfactory level is extremely difficult. This explains why the quest for an excellent ABC has not ended after more than three decades of research.

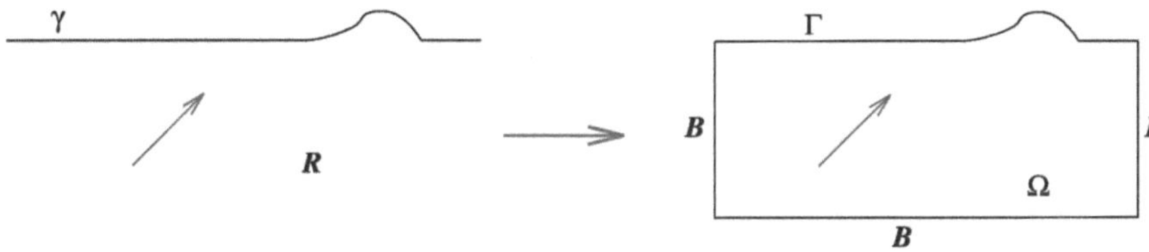

Fig. 5.3 Replacing a problem in a half–space by one in a bounded domain.

5.1.4 Absorbing Boundaries in the Broad Sense

The term "absorbing boundaries" can be generalized to include additional types of computational treatment for the numerical solution of problems in unbounded domains. For example, one may employ an *absorbing layer* rather than an absorbing boundary. In this case, the artificial part of the computational domain which is responsible for absorbing the outgoing waves is a layer of finite thickness. We shall refer to both absorbing boundaries and layers in the sequel. We shall also refer to *infinite elements* and to BEM schemes since they are also closely related to the subject of this chapter.

Review papers on the broader subject of absorbing schemes include those of Givoli [15], Tsynkov [33] and Hagstrom [22]. Chapter 3 in Ihlenburg's book [25] is also dedicated to it. Special journal issues which include many papers on absorbing boundaries include [2]– [35].

5.2 Milestones in the Development of Absorbing Boundaries and Layers

5.2.1 Low Order Absorbing Boundary Conditions (ABCs)

Early attempts to use ABCs based on the Sommerfeld operator led to a mixed success. In geophysics, where the equations of elastodynamics govern, the Lysmer and Kuhlemeyer ABC [28] was devised in 1969 and was immediately accepted for practical use. The geophysical setup is shown in Figure 5.3; the problem in the earth halfspace is replaced by a problem in a finite domain. The Lysmer–Kuhlemeyer ABC is equivalent to a sequence of dashpots at the absorbing boundaries, as illustrated in Figure 5.4. Similar ABCs have been quite popular in weather prediction applications.

However, the big step forward has occurred in the late 70's and early 80's, when a number of improved ABCs — now regarded as classical — have been proposed. The most well–known ones are the ABC sequences of Engquist and Majda [12] and of Bayliss and Turkel [3]. The 2nd–order ABCs in these two sequences have become especially popular, and are still commonly used today. We shall not list here all the other low order ABCs that have been proposed during the years; for a full survey and many related references see the review papers by Givoli [15], Tsynkov [33] and Hagstrom [22].

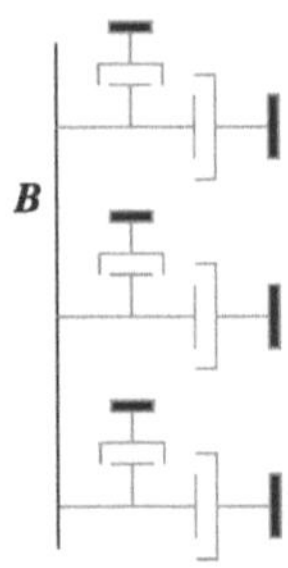

Fig. 5.4 The boundary dashpot model of Lysmer and Kuhlemeyer [28].

These ABCs referred to here are of *low order*. This does not necessarily mean that they cannot be written in theory up to high order; however, they cannot be used in practice beyond a certain low order, typically two or three. The reason for this is that in a "classical–type" ABC sequence, a high order means the presence of *high derivatives* in the ABC, which are difficult to handle numerically. For example, consider the Jth–order Bayliss–Turkel ABC [3] for the scalar wave equation in three dimensions with a spherical artificial boundary $\mathcal{B}$ of radius R:

$$\left[\prod_{j=1}^{J}\left(\frac{1}{c}\frac{\partial}{\partial t}+\frac{\partial}{\partial r}+\frac{2j-1}{R}\right)\right]u=0 \qquad \text{on} \quad \mathcal{B}\,. \tag{5.15}$$

Although J in (5.15) may be theoretically large, this ABC, at least in its original form (5.15), cannot be implemented up to an arbitrarily high order due to the high derivatives appearing in it. As is apparent from (5.15), the Jth–order condition involves a product of J first–order normal and time differential operators and thus leads to Jth–order differential operators whose discrete representation for an arbitrarily high J is not practical. Later we shall see how this difficulty can be alleviated. However, it must be emphasized the low order ABCs, such as the 2nd–order Bayliss–Turkel ABC (i.e., (5.15) with J=2) or other well designed ABCs, are very useful in practical applications and in many cases lead to satisfactory numerical results.

For the sake of demonstration, we mention here three additional ABCs that were proposed in the context of geophysics, for applications in oil exploration and seismology. The first ABC is that of Randall [29]. Figure 5.5, taken from [29], shows, for a problem of elastic waves, a comparison between the results obtained with the Randall ABC (upper plot) and the 2nd–order Engquist–Majda ABC (lower plot). In both cases, the ABC is applied along the upper and lower boundaries. A source in the form of a space–time pulse is activated at $t = 0$ near the left boundary. The two snapshots are shown at $t = 72$. The left wave fronts correspond to shear waves, whereas the right wave fronts correspond to faster pressure waves. The wave fronts obtained with the Engquist–Majda ABC differ significantly from those obtained with the Randall ABC; the former include strong reflections from the upper and lower boundaries, which are totally spurious. The reason that the 2nd–order Engquist–Majda ABC gives rise to strong reflections in this example is that the waves impinge on

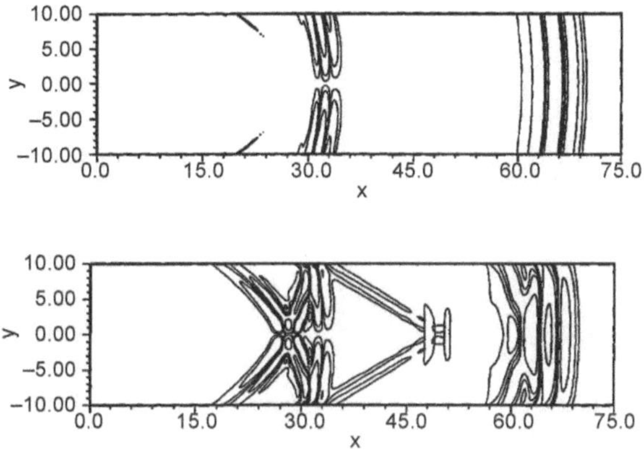

Fig. 5.5 Comparison between the results obtained with the Randall ABC (upper plot) and the 2nd–order Engquist–Majda ABC (lower plot). Reprinted by permission after "Absorbing Boundary Condition for the Elastic Wave Equation," Randall CJ (1988), Geophysics 53:611–624, Society of Exploration Geophysicists.

the boundaries very obliquely, while the Engquist–Majda ABC is known to perform well when the incidence is reasonably close to normal.

Renaut and Petersen, in a 1989 paper [30], also considered the second–order Engquist–Majda ABC, which is based on a rational approximation of the square–root function. They discussed the choice of parameters in this ABC, namely the best way to perform the rational approximation. In particular they compared two choices: the Padé approximation, which is regarded as the standard one in the Engquist–Majda construction, and the Chebychev approximation. Figure 5.6 shows the results of this comparison for a problem of acoustic waves. The source is located slightly under the ground surface. Each plot shows a "seismic map," namely the results from 96 "numerical hydrophones" arranged in a straight horizontal line slightly below the source. The differences between the Padé map (left) and the Chebychev map (right) are clear; the authors argue that the latter is the accurate one, whereas the standard Padé approximation yields significant spurious reflections in this case.

Zhu, in a 1999 paper [36], devised a new ABC for elastic waves. One of the numerical examples in [36] is illustrated in Figure 5.7; it involves a three–layer medium (having different elastic properties in each layer) with a point source located in the middle layer. A vertical line of "numerical geophones" is located left to the source. Figure 5.8 (taken from [36]) shows a comparison between the results obtained with the Zhu ABC (upper pair of plots) and the 2nd–order Engquist–Majda ABC (lower pair). In each pair, the left plot shows the vertical displacement while the right plot shows the horizontal displacement. The wave fronts seen in the upper plots are the physical reflections from the two interfaces between the layers and from the upper surface. The lower plots include additional reflections, which are spurious. This establishes the superiority of the Zhu ABC over the classical Engquist–Majda ABC in this case.

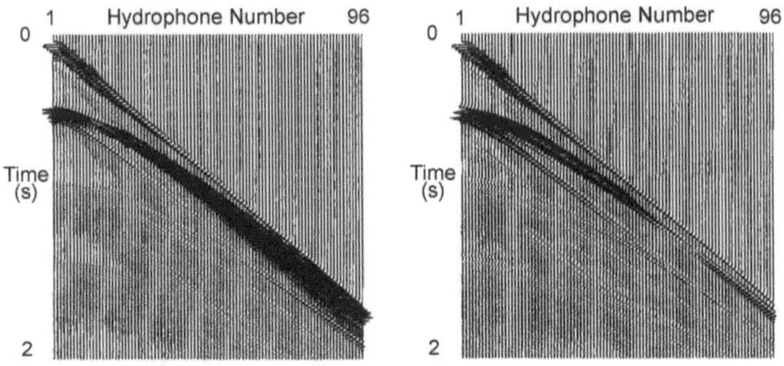

Fig. 5.6 Results obtained by Renaut and Petersen using the 2nd–order Engquist–Majda ABC, for parameters corresponding to a standard Padé approximation (left) and to a Chebychev approximation (right). Reprinted by permission after "Stability of Wide–Angle Absorbing Boundary Conditions for the Wave Equation," Renaut RA, Petersen J (1989), Geophysics 54:1153–1163, Society of Exploration Geophysicists.

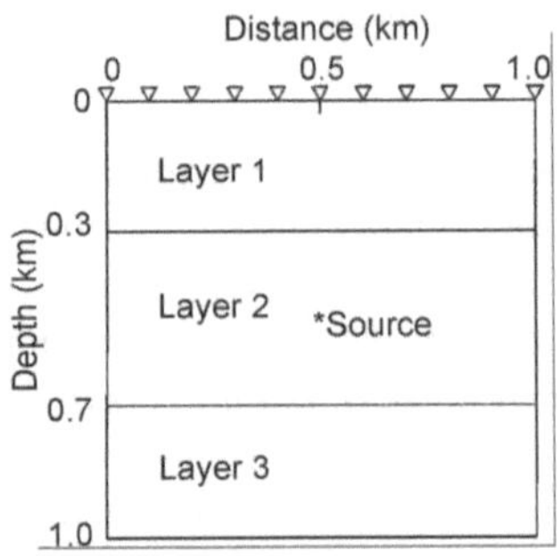

Fig. 5.7 Setup for the Zhu example of a three–layer medium. Reprinted by permission after "A Transparent Boundary Technique for Numerical Modeling of Elastic Waves," Zhu J (1999), Geophysics 64:963–966, Society of Exploration Geophysicists.

These examples, and a lot of other evidence, show that one must be very careful when employing a standard ABC for the solution of a specific problem. In many cases the classical ABCs perform well, but caution must be taken when using them in demanding situations such as very oblique incidence angles, layered media, etc.

5.2.2 Boundary Integral and Boundary Element Methods (BEM)

Although strictly speaking Boundary Integral Methods and Boundary Element Methods (BEM) are not "absorbing boundary" methods, we mention them for completeness. These methods, which have many variations, became popular in the solution of wave problems in the late 70's and mid 80's. They are especially suitable to problems in unbounded domains, since they reduce the problem to one on the physical boundary, and thus no artificial boundary has to be introduced. However, there are a few serious difficulties associated with them, a discussion of which is outside the scope

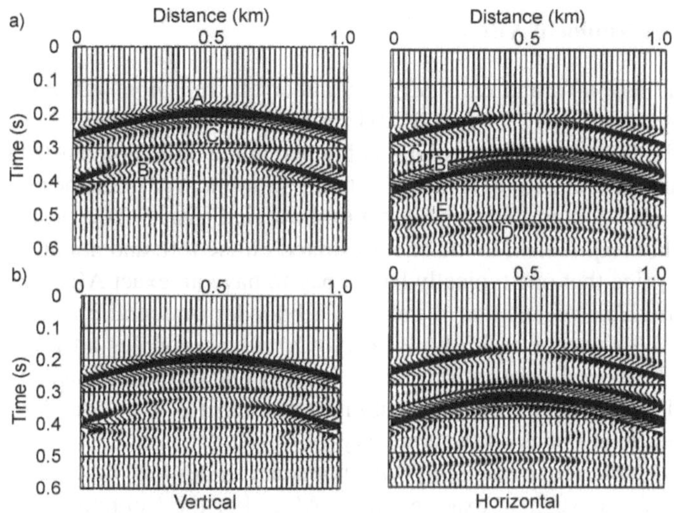

Fig. 5.8 The Zhu example: comparison between the results obtained with the Zhu ABC (upper pair of plots) and the 2nd–order Engquist–Majda ABC (lower pair). Reprinted by permission after same source as in Figure 5.7.

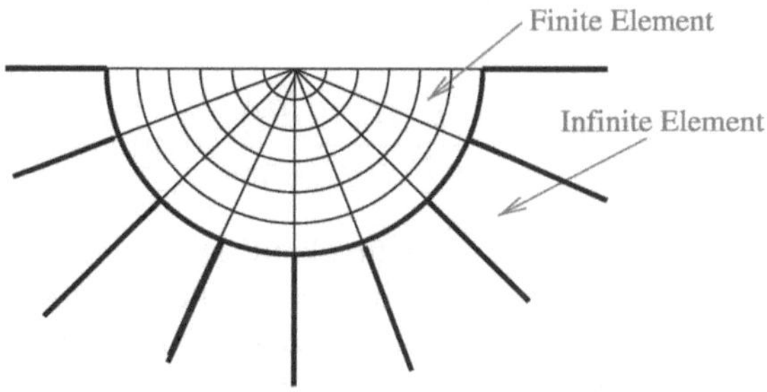

Fig. 5.9 Setup for the use of infinite elements.

of this chapter. We point the reader to the delightful historical account by Rizzo [31] on the subject.

5.2.3 Classical Infinite Elements

In about the same period (late 70's – mid 80's) Infinite Elements were invented, by Bettess and Zienkiewicz [5, 6]. An infinite element is a semi–infinite prism (or strip in two dimensions) associated with shape functions which attempt to represent the far–field behavior of the solution. See the setup illustrated in Figure 5.9. The basic idea underlying infinite elements is extremely attractive and simple. However, the classical infinite elements of the 70's and 80's performed well only for "static" problems (e.g., Laplace's equation and elastostatics) and the right construction for wave problems was yet to be found.

5.2.4 Exact Nonlocal ABCs

During the late 80's and early 90's a number of *exact nonlocal* ABCs have been devised. An exact ABC is an ABC which has the property that when it is used on the artificial boundary $\mathcal{B}$, the solution obtained in the finite domain Ω enclosed by $\mathcal{B}$ (prior to any discretization) is *identical* to the solution in Ω of the original problem in the unbounded domain. Thus, in the setup shown in Figure 5.2 the problem illustrated on the right is *equivalent* to the one illustrated on the left, and not an approximation of it. The price that one typically has to pay to have an exact ABC is that the ABC is *nonlocal*. This means that it involves a boundary integral operator (rather than a differential operator as in the local ABCs) which couples all the points on the boundary $\mathcal{B}$. After discretization, the nonlocality manifests itself in that each node on the boundary "interacts" not only with its neighbors (as in the standard FE method) but with all other nodes on the boundary.

The *Dirichlet–to–Neumann (DtN)* boundary condition, devised by Keller and Givoli [20, 27] is such an exact nonlocal ABC. For the Helmholtz equation in two dimensions, when the artificial boundary $\mathcal{B}$ is a circle of radius R, the DtN condition is

$$\frac{\partial u}{\partial r}(R,\theta) = -\sum_{n=0}^{\infty}{}' \int_0^{2\pi} m_n(\theta-\theta')u(R,\theta')\, R\, d\theta' \,, \tag{5.16}$$

where

$$m_n(\theta-\theta') = -\frac{k}{\pi R}\frac{H_n^{(1)\prime}(kR)}{H_n^{(1)}(kR)} \cos n(\theta-\theta') \,. \tag{5.17}$$

Here $H_n^{(1)}$ is the Hankel function of the first kind. The prime after the sum indicates that a factor of $1/2$ multiplies the term with $n = 0$. The operator involved in (5.16) is called the DtN map. In the context of acoustics, the DtN condition can be perceived as a nonlocal *impedance boundary condition*, relating the normal velocity ($\partial u/\partial r$) and the pressure (u) on the boundary.

The DtN ABC has been proved to be very powerful for frequency–dependent problems, when applicable; see, e.g., the review paper [17]. However, the method is limited in that in many important situations it is very difficult, or even impossible, to express the DtN map in closed analytic form. These situations include the cases of inhomogeneous and anisotropic media, special geometries, and generally the time–dependent case.

The ABC of Tsynkov et al. [32,34] is also an exact nonlocal ABC. See additional discussion on the pros and cons of exact nonlocal vs. approximate local ABCs in [15]– [22].

5.2.5 Absorbing Layers

The 80's saw the first development of *absorbing layers*, that were also called damping layers or filters. As in the case of ABCs, an artificial boundary $\mathcal{B}$ is first introduced to make the computational domain Ω finite. Then, enclosing Ω and adjacent

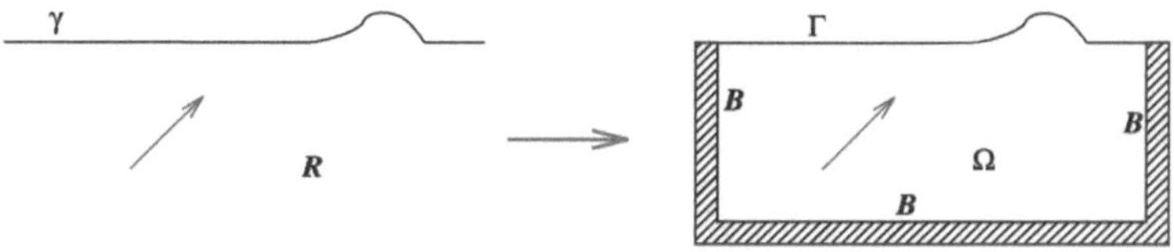

Fig. 5.10 Setup for the use of an absorbing layer.

to $\mathcal{B}$, a narrow region serves as a filter; in it the governing equations undergo some modifications intended to absorb the outgoing waves. Thus, an absorbing scheme has a similar role to that of an ABC, but instead of operating on the boundary $\mathcal{B}$ it operates in a layer near $\mathcal{B}$. See Figure 5.10.

Israeli and Orszag apparently devised the first absorbing layer [26]. They analyzed and discussed several variants of the scheme. For example, consider the one–dimensional wave equation

$$\frac{\partial^2 u}{\partial t^2} = c^2 \frac{\partial^2 u}{\partial x^2} \,. \tag{5.18}$$

Inside the Israeli–Orszag damping layer, this equation is replaced by

$$\frac{\partial^2 u}{\partial t^2} = c^2 \frac{\partial^2 u}{\partial x^2} + \mu(x) \frac{\partial^3 u}{\partial x^2 t} - \nu(x) \frac{\partial u}{\partial t} \,. \tag{5.19}$$

Here $\mu(x) \geq 0$ and $\nu(x) \geq 0$ are chosen functions associated with viscous damping and with friction–type damping (or "Newtonian cooling"), respectively. Additional absorbing layers have been proposed later by other authors; see, e.g., the survey in [15]. These constructions laid the foundation to the next milestone.

5.2.6 Perfectly Matched Layer (PML)

In the mid 90's the Perfectly Matched Layer (PML) [4] was invented by Bérenger. Later many authors made additional contributions, by analyzing, improving and extending the PML in various ways. We mention the work by Chew and Liu [8] in acoustics, showing that the PML is obtained by an analytic continuation in the complex plane. We shall not list all the other related contributions; see a partial survey (covering the first five years of PML development) in [22]. An updated review on PML is a major work still to be done.

A PML is an absorbing layer equipped with two basic properties:

- It is designed to have *zero reflection* at the interface $\mathcal{B}$ for *any* plane wave;
- It is designed to make the solution *decay exponentially* insider the layer.

These two properties ensure excellent wave absorbance, at least on the continuous level. A wave outgoing from Ω enters the layer without any reflection, and then decays exponentially. By the time it arrives at the outer boundary of the PML it is very weak. Then it may reflect back into the PML, it decays exponentially again, and by the time it reaches the interface $\mathcal{B}$ on its way back into Ω it is too weak to cause any damage.

We briefly present the mathematical basis of the PML in the simplest context. Consider the wave equation (5.5) in the three dimensional space (x_1, x_2, x_3). We write it as a system of two first–order equations:

$$\frac{\partial \boldsymbol{v}}{\partial t} = -\boldsymbol{\nabla} u \quad , \quad \frac{\partial u}{\partial t} = -c^2 \boldsymbol{\nabla} \cdot \boldsymbol{v} \, . \tag{5.20}$$

Now we apply the Laplace transform in time to obtain:

$$-\mathrm{i}\omega \hat{\boldsymbol{v}} = -\boldsymbol{\nabla} \hat{u} \quad , \quad -\mathrm{i}\omega \hat{u} = -c^2 \boldsymbol{\nabla} \cdot \hat{\boldsymbol{v}} \, . \tag{5.21}$$

These are the equations which hold in Ω. Inside the PML, we modify these equations by applying *coordinate stretching*. Namely, we replace $\boldsymbol{\nabla}$ by $\boldsymbol{\nabla}_s$, where

$$\boldsymbol{\nabla}_s = \sum_{j=1}^{3} \boldsymbol{e}_j \frac{1}{S_j} \frac{\partial}{\partial x_j} \quad , \quad S_j = 1 + \mathrm{i}\frac{\omega_j}{\omega} \, . \tag{5.22}$$

Here the $\boldsymbol{e}_j$ are the Cartesian unit vectors, and the ω_j are the PML parameters. If x_1 is the PML thickness direction, then typically $\omega_2 = \omega_3 = 0$ while $\omega_1 = \sigma(x_1)$ is called the PML damping function, and is a smooth increasing function of x_1 (say the parabola $\sigma(x_1) = Ax_1^2$).

Thus, in the PML we have

$$-\mathrm{i}\omega \hat{\boldsymbol{v}} = -\sum_{j=1}^{3} \boldsymbol{e}_j \frac{1}{S_j} \frac{\partial \hat{u}}{\partial x_j} \quad , \quad -\mathrm{i}\omega \hat{u} = -c^2 \sum_{j=1}^{3} \frac{1}{S_j} \frac{\partial \hat{v}_j}{\partial x_j} \, . \tag{5.23}$$

Now we *split* the equations, by introducing the new variables $\hat{u}^{(j)}$:

$$-\mathrm{i}\omega \hat{v}_j = -\frac{1}{S_j} \frac{\partial \hat{u}}{\partial x_j} \quad , \quad -\mathrm{i}\omega \hat{u}^{(j)} = -c^2 \frac{1}{S_j} \frac{\partial \hat{v}_j}{\partial x_j} \quad , \quad \hat{u} = \sum_{j=1}^{3} \hat{u}^{(j)} \, . \tag{5.24}$$

Substituting the expression for S_j from (5.22) we get

$$(-\mathrm{i}\omega + \omega_j)\hat{v}_j = -\frac{\partial \hat{u}}{\partial x_j} \quad , \quad (-\mathrm{i}\omega + \omega_j)\hat{u}^{(j)} = -c^2 \frac{\partial \hat{v}_j}{\partial x_j} \, . \tag{5.25}$$

These constitute six equations ($j = 1, 2, 3$) for the six unknowns $\hat{v}_j$ and $\hat{u}^{(j)}$. The $\hat{u}$ is not an additional unknown, owing to the third relation in (5.24).

Now we apply the inverse Laplace transform to these equations in order to return to the time domain:

$$\frac{\partial v_j}{\partial t} + \omega_j v_j = -\frac{\partial u}{\partial x_j} \quad , \quad \frac{\partial u^{(j)}}{\partial t} + \omega_j u^{(j)} = -c^2 \frac{\partial v_j}{\partial x_j} \quad , \quad u = \sum_{j=1}^{3} u^{(j)} \, . \tag{5.26}$$

These are the basic PML equations in the time–dependent case: six equations ($j = 1, 2, 3$) for the six unknowns v_j and $u^{(j)}$.

In the *frequency–dependent case* (i.e., for time–harmonic waves), these equations can be reduced to the single equation

$$\sum_{j=1}^{3} \mu_j \frac{\partial^2 u}{\partial x_j^2} + k^2 u = 0 \, , \tag{5.27}$$

where

$$\mu_j = -\frac{k^2}{(\omega_j - \mathrm{i}\omega)^2} \, . \tag{5.28}$$

This has the form of an *anisotropic* Helmholtz equation. Thus, the implementation of the PML in the frequency–dependent case is especially simple.

The PML has the distinct advantage that on the continuous level it is "perfect" by construction. Indeed it perform extremely well in many circumstances, especially for high–frequency waves. However, there are still a number of PML–related issues that remain open and are subject to current research. These include:

- While the PML is perfect on the continuous level, it is not perfect on the discrete level. In fact, it has been an unpleasant surprise to realize that in some cases the PML performs poorly when incorporated in a discrete model, especially in low frequencies. The PML seems to be more sensitive to discretization than the classical ABCs. A good design of an ABC on the continuous level usually guarantees good performance on the discrete level; this does not seem to be the general case for PML.
- The performance of the PML is sensitive to the choice of the PML parameters, i.e., the PML thickness and the PML damping function $\sigma(x_1)$. For example, there is a clear tradeoff in choosing the rate in which $\sigma(x_1)$ increases; on one hand $\sigma(x_1)$ should increase rapidly to generate a sufficient amount of damping, but on the other hand a rapid variation of $\sigma(x_1)$ requires a fine discretization (many finite elements) inside the PML, which is inefficient.
- In some cases the basic PML may give rise to weak numerical instabilities. To a large extent these have been resolved, by modifying the basic formulation.
- The PML is not associated with a clear notion of convergence. Whereas in the case of high–order ABCs (to be discussed later), with a fixed location of $\mathcal{B}$, one can approach the exact solution arbitrarily close (up to the discretization error) by increasing the order of the ABC, in the case of a PML with a fixed thickness this is not possible.
- The basic PML is designed to perfectly absorb propagating plane waves but it has no control over other types of waves, like evanescent waves. This issue has been addressed in later versions of PML formulations.
- Analysis of the PML is difficult. In particular, in the elastodynamic case (relevant to problems in geophysics), the behavior of the PML in the time domain is not yet well understood.

These open issues are the reason that research on the PML is very active even more than a decade after its invention.

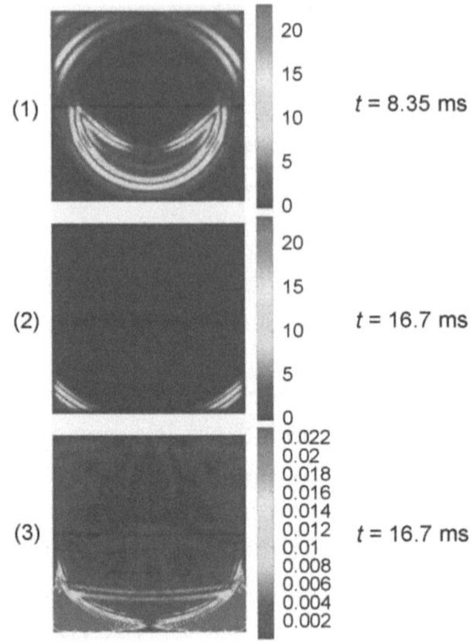

Fig. 5.11 The Collino–Tsogka PML applied to a problem of a two–material medium. Reprinted by permission after "Application of the Perfectly Matched Absorbing Layer Model to the Linear Elastodynamic Problem in Anisotropic Heterogeneous Media," Collino F, Tsogka C (2001), Geophysics 66:294–307, Society of Exploration Geophysicists.

We demonstrate the power of PML by presenting some results from the 2001 paper of Collino and Tsogka [10], who extended the PML for elastic wave problems in inhomogeneous and anisotropic media. Figure 5.11 shows the results obtained for a two–material medium. The material interface is a horizontal line passing in the middle of the computational domain. A point source is located just below the interface. The two upper snapshots of the velocity norm (at times $t = 8.35$ms and $t = 16.7$ms) show that the waves are well absorbed by the PML. The lowest snapshot at a very late time is magnified by a factor of 1000 and shows that the remaining waves have a very small magnitude.

Collino and Tsogka [10] also applied their PML to a highly heterogeneous medium, whose wave speed distribution is described in Figure 5.12. The result, which involves a highly complicated wave pattern, is shown in Figure 5.13. The three plots shown correspond to three PML thicknesses. It is seen that beyond a certain PML thickness the numerical results "converge."

5.2.7 High–performance Infinite Elements

The work of Burnett [7] and of Astley [1] in the 90's was another important milestone in the treatment of waves in unbounded domains. These authors devised high–performance infinite elements especially suitable for wave problems. These basic constructions were later extended and further improved by the same authors as well as by others. The Astley and the Burnett infinite elements are constructed in sepa-

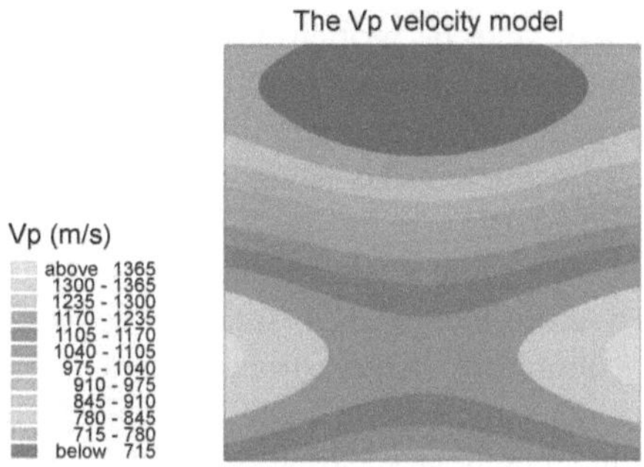

Fig. 5.12 The heterogeneous medium taken in the Collino–Tsogka example. Reprinted by permission after the same source as in Figure 5.11.

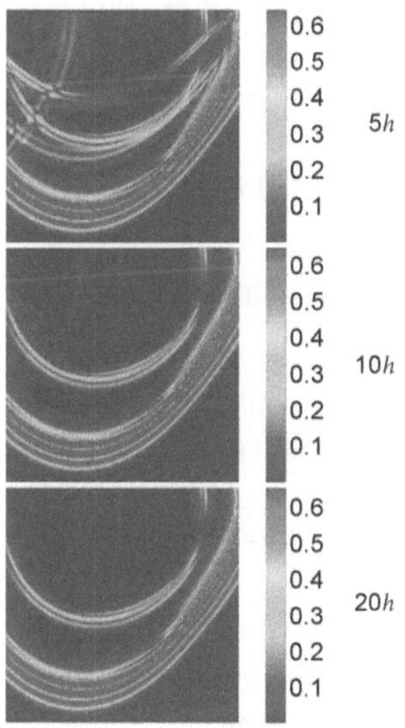

Fig. 5.13 The Collino–Tsogka PML applied to a problem of the heterogeneous medium shown in Figure 5.12. Reprinted by permission after the same source as in Figure 5.11.

rable coordinate systems, e.g., a spheroidal system, and have been shown to be extremely efficient in solving acoustic scattering problems in full space. See, e.g., [25] and relevant papers in [2, 19].

5.2.8 High–order Absorbing Boundary Conditions (ABCs)

We come to the last milestone discussed in this chapter: local high–order ABCs. According to our definition, a *high–order ABC* is an ABC which not only can be shown in theory to be of high–order accuracy, but can also be *implemented in practice up to any desired order*. As was discussed in Section 5.2.1, the second requirement does not hold with the "classical" ABCs. For example, recall that while the the Jth–

order Bayliss–Turkel ABC can be written down up to any order J, as is done in (5.15), it has been implemented in practice only up to second order due to the high derivatives that appear when J is large. Genuinely high–order ABCs involve no high derivatives. The price of achieving this is the need to introduce *auxiliary variables* into the formulation of the ABC.

The subject of high–order ABCs is reviewed in [14]. Collino [9] can be regarded as the pioneer of high–order local ABCs, since he devised the first such ABC in 1993. We shall briefly summarize his construction here.

Consider the two–dimensional wave equation (5.5). We assume that the domain includes a straight artificial boundary $\mathcal{B}$ (one side of a wave guide or of a rectangular computational "box"), and we let x be the direction normal to $\mathcal{B}$ and y the direction tangent to it. If we Fourier–transform (5.5) in x, y and t (or formally replace the partial derivatives $\frac{\partial}{\partial x}$, $\frac{\partial}{\partial y}$ and $\frac{\partial}{\partial t}$ by $-\mathrm{i}k_x$, $-\mathrm{i}k_y$ and $-\mathrm{i}\omega$, respectively) we get the standard dispersion relation

$$-\omega^2/c^2 + k_x^2 + k_y^2 = 0\,, \tag{5.29}$$

where ω is the angular frequency and k_x and k_y are the x and y wave numbers. From (5.29) we have

$$k_x \bar{\bar{u}} = (\omega/c)\sqrt{1 - k_y^2 c^2/\omega^2}\; \bar{\bar{u}}\,, \tag{5.30}$$

where $\bar{\bar{u}}$ is the Fourier transform of u in x, y and t. In (5.30) we have taken the positive root which corresponds to outgoing waves. Now we take the inverse Fourier transform of (5.30) in the x direction only. This yields

$$\frac{\partial}{\partial x}\bar{u} + \mathrm{i}(\omega/c)\sqrt{1 - k_y^2 c^2/\omega^2}\; \bar{u} = 0 \qquad \text{on} \quad \mathcal{B}\,. \tag{5.31}$$

Here $\bar{u}$ is the inverse Fourier transform of $\bar{\bar{u}}$ in x. We denote $z = k_y^2 c^2/\omega^2$. Then we approximate the irrational function $\sqrt{1 - z^2}$ appearing in (5.31) by a rational function, using the Padé approximation:

$$\sqrt{1 - z} \simeq 1 - \sum_{j=1}^{J} \frac{\beta_j z}{1 - \alpha_j z} \qquad \text{for} \quad z \in [0, 1]\,, \tag{5.32}$$

$$\alpha_j = \cos^2\left(\frac{j\pi}{2J+1}\right) \quad , \qquad \beta_j = \frac{2\sin^2\left(\frac{j\pi}{2J+1}\right)}{2J+1}\,. \tag{5.33}$$

We substitute this approximation in (5.31) to obtain

$$\frac{\partial}{\partial x}\bar{u} + \mathrm{i}(\omega/c)\bar{u} - \mathrm{i}(\omega/c)\sum_{j=1}^{J} \frac{\beta_j k_y^2}{\omega^2/c^2 - \alpha_j k_y^2}\, \bar{u} = 0 \qquad \text{on} \quad \mathcal{B}\,. \tag{5.34}$$

Up to this point the development is similar to that used by Engquist and Majda [12]. However, at this point Collino introduces the *auxiliary variables* (in the transformed plane) $\bar{\phi}_j$ defined by

$$\bar{\phi}_j = \frac{k_y^2}{\omega^2/c^2 - \alpha_j k_y^2}\, \bar{u}\,. \tag{5.35}$$

With this definition (5.34) gives

$$\frac{\partial}{\partial x}\bar{u} + \mathrm{i}(\omega/c)\bar{u} - \mathrm{i}(\omega/c)\sum_{j=1}^{J}\beta_j\bar{\phi}_j = 0 \qquad \text{on} \quad \mathcal{B}\,. \tag{5.36}$$

From (5.35) we also have

$$(\omega^2/c^2)\bar{\phi}_j - \alpha_j k_y^2\bar{\phi}_j = k_y^2\bar{u} \qquad \text{on} \quad \mathcal{B}\,. \tag{5.37}$$

Finally we take the inverse Fourier transform of (5.36) and (5.37) in y and t and obtain

$$\frac{\partial}{\partial x}u + \frac{1}{c}\frac{\partial}{\partial t}u - \frac{1}{c}\sum_{j=1}^{J}\beta_j\frac{\partial}{\partial t}\phi_j = 0 \qquad \text{on} \ \mathcal{B}\,, \tag{5.38}$$

$$\frac{1}{c^2}\frac{\partial^2}{\partial t^2}\phi_j - \alpha_j\frac{\partial^2}{\partial y^2}\phi_j - \frac{\partial^2}{\partial y^2}u = 0 \quad , \qquad j = 1,\ldots,J \ \text{on} \ \mathcal{B}\,. \tag{5.39}$$

Equations (5.38) and (5.39) constitute the Collino ABC of order J. In total these are $J+1$ coupled equations on $\mathcal{B}$ for the $J+1$ functions u, $\phi_1, \ldots \phi_J$.

The Collino ABC has the general structure shared by all the high–order local ABCs. It involves no high derivatives beyond second order, and it involves no normal ($x-$) derivatives of any auxiliary variable ϕ_j. The latter fact is essential since the auxiliary variables "live" only along the boundary $\mathcal{B}$.

An additional discussion on high–order ABCs can be found in [14]. Very recent high–order ABCs include those of Givoli and Neta [21] and Hagstrom and Warburton [24]. The latter ABC has proved to be especially powerful, and has been extended in various directions in [18,23] by Hagstrom, Givoli and co–workers.

We demonstrate the use of a high order ABC by presenting a numerical example taken from the 2003 paper of Givoli and Neta [21]. Figure 5.14 shows five sets of snapshots for a problem of acoustic wave propagation in a wave guide. In each set, the upper plot represents the reference solution in a long domain, which can be regarded as the "exact solution." The middle plot corresponds to the truncated–domain solution, where the 5th–order Givoli–Neta ABC is applied on the east boundary. The lower plot corresponds to a solution obtained with a crude 1st–order ABC. Clearly, the 5th–order ABC produces solutions that agree excellently with the reference solution, whereas the 1st–order ABC gives rise to significant spurious reflections.

5.3 Concluding Remarks

With the improvement of computational methods for the solution of wave problems in unbounded media, as encountered in acoustics, geophysics and other fields of application, the need for and interest in accurate schemes for treating artificial boundaries is ever increasing. Absorbing boundary schemes for basic problems, such as

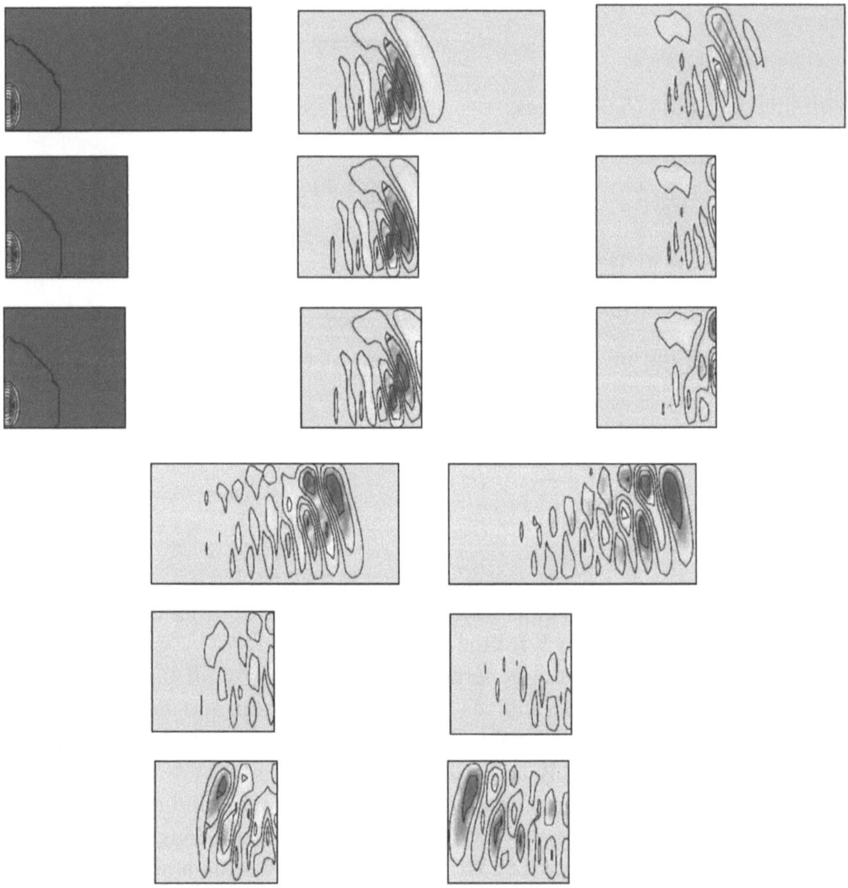

Fig. 5.14 A wave–guide problem: five sets of snapshots comparing the reference solution (upper plot), the solution obtained with the 5th–order Givoli–Neta ABC (middle plot) and the solution obtained with a 1st–order ABC (lower plot). Reprinted from "High–Order Non–Reflecting Boundary Scheme for Time–Dependent Waves," Givoli D, Neta B (2003), Journal of Computational Physics 186:24–46, with permission from Elsevier.

exterior frequency–dependent problems in linear, homogeneous isotropic media are well developed. Method for more complicated scenarios are emerging.

Among the recently developed methods, worth mentioning as especially powerful methods are the *Perfectly Matched Layer* (PML) method, and the method of *high–order Absorbing Boundary Conditions* (ABCs). A lot of research is still ahead of us before these two types of methods will become routine tools in industrial use for complicated problems in acoustics and geophysics. Especially difficult is the time–dependent case. The list of desired properties outlined in Section 5.1.3 is demanding. It should also be noted that nonlinear wave problems have hardly been considered in this context due to the immense difficulty involved in designing a good absorbing boundary scheme for such problems. However, the current state of affairs is extremely promising, and it is reasonable to expect that exciting advances are still ahead of us.

Acknowledgment

The author acknowledges the support of the Fund for the Promotion of Research at the Technion and the fund provided through the Lawrence and Marie Feldman Chair in Engineering. The author is grateful to the following publishers who permitted the use of figures from previous publications: Society of Exploration Geophysicists (Figures 5.5, 5.6, 5.7, 5.8, 5.11, 5.12 and 5.13); World Scientific Press (Figure 5.1); and Elsevier (Figure 5.14). The author would also like to thank the editors of this book for their help in acquiring these permissions.

References

1. Astley RJ (1996) Transient wave envelope elements for wave problems. Journal of Sound and Vibration 192:245–261
2. Astley J, Gerdes K, Givoli D, Harari I (eds) (2000) Special issue: Finite elements for wave propagation. Journal of Computational Acoustics 8(1)
3. Bayliss A, Turkel E (1980) Radiation boundary conditions for wave–like equations. Communications on Pure and Applied Mathematics 33:707–725
4. Bérenger JP (1994) A perfectly matched layer for the absorption of electromagnetic waves. Journal of Computational Physics 114:185–200
5. Bettess P (1977) Infinite elements. International Journal for Numerical Methods in Engineering 11:53–64
6. Bettess P, Zienkiewicz OC (1977) Defraction and refraction of surface waves using finite and infinite elements. International Journal for Numerical Methods in Engineering 11:1271–1290
7. Burnett DS (1994) A three dimensional acoustic infinite element based on a prolate spheroidal multipole expansion. Journal of the Acoustical Society of America 96:2798–2816
8. Chew WC, Liu QH (1996) Perfectly matched Layers for elastodynamics: a new absorbing boundary condition. Journal of Computational Acoustics 4:341–359
9. Collino F (1993) High order absorbing boundary conditions for wave propagation models. In: Kleinman (ed) Proceedings of the 2nd International Conference on Mathematical and Numerical Aspects of Wave Propagation. SIAM, Philadelphia
10. Collino F, Tsogka C (2001) Application of the perfectly matched absorbing layer model to the linear elastodynamic problem in anisotropic heterogeneous media. Geophysics 66:294–307
11. Courant R, Hilbert D (1962) Methods of mathematical physics. Wiley, New York
12. Engquist B, Majda A (1979) Radiation boundary conditions for acoustic and elastic calculations. Communications on Pure and Applied Mathematics 32:313–357
13. Farhat C, Avery P, Tezaur R, Li J (2005) FETI–DPH: A dual–primal domain decomposition method for acoustic scattering. Journal of Computational Acoustics 13:499–524
14. Givoli D (2004) High–order local non–reflecting boundary conditions: a review. Wave Motion 39:319–326

15. Givoli D (1991) Non–reflecting boundary conditions: A review. Journal of Computational Physics 94:1–29
16. Givoli D (1992) Numerical methods for problems in infinite domains. Elsevier, Amsterdam
17. Givoli D (1999) Recent advances in the DtN finite element method for unbounded domains. Archives of Computational Methods in Engineering. State of the art reviews 6:71–116
18. Givoli D, Hagstrom T, Patlashenko I (2006) Finite element formulation with high–order absorbing boundary conditions for time–dependent waves. Computer Methods in Applied Mechanics and Engineering 195:3666–3690
19. Givoli D, Harari I (eds) (1998) Special issue: Exterior problems of wave propagation. Computer Methods in Applied Mechanics and Engineering 164(1)&(2)
20. Givoli D, Keller JB (1990) Non–reflecting boundary conditions for elastic waves. Wave Motion 12:261–279
21. Givoli D, Neta B (2003) High–order non–reflecting boundary scheme for time–dependent waves. Journal of Computational Physics 186:24–46
22. Hagstrom T (1999) Radiation boundary conditions for the numerical simulation of waves. Acta Numerica 8:47–106
23. Hagstrom T, de Castro M., Givoli D, Tsemach D (2007) Local high order absorbing boundary conditions for time–dependent waves in guides. Journal of Computational Acoustics 15:1–22
24. Hagstrom T, Warburton T (2004) A new auxiliary variable formulation of high–order local radiation boundary conditions: corner compatibility conditions and extensions to first–order systems. Wave Motion 39:327–338
25. Ihlenburg F (1998) Finite element analysis of acoustic scattering. Springer–Verlag, New York
26. Israeli M, Orszag SA (1981) Approximation of radiation boundary conditions. Journal of Computational Physics 41:115–135
27. Keller JB, Givoli D (1989) Exact non–reflecting boundary conditions. Journal of Computational Physics 82:172–192
28. Lysmer J, Kuhlemeyer RL (1969) Finite dynamic model for infinite media. Journal of the Engineering Mechanics Division ASCE 95:859–877
29. Randall CJ (1988) Absorbing boundary condition for the elastic wave equation. Geophysics 53:611–624
30. Renaut RA, Petersen J (1989) Stability of wide–angle absorbing boundary conditions for the wave equation. Geophysics 54:1153–1163
31. Rizzo FJ (1989) The boundary element method — some early history: a personal view. In: Beskos DE (ed) Boundary Elements in Structural Analysis. ASCE, New York 1–16
32. Ryaben'kii VS, Tsynkov SV (1995) Artificial boundary conditions for the numerical solution of external viscous flow problems. SIAM Journal on Numerical Analysis 32:1355–1389
33. Tsynkov SV (1998) Numerical solution of problems on unbounded domains, A Review. Applied Numerical Mathematics 27: 465–532
34. Tsynkov SV, Turkel E, Abarbanel S (1996) External flow computations using global boundary conditions. AIAA Journal 34:700–706
35. Turkel E (ed) (1998) Special issue: Absorbing boundary conditions. Applied Numerical Mathematics 27(4)
36. Zhu J (1999) A transparent boundary technique for numerical modeling of elastic waves. Geophysics 64:963–966

6

Perfectly Matched Layers

Alfredo Bermúdez[1], Luis Hervella–Nieto[2], Andrés Prieto[3], and Rodolfo Rodríguez[4]

[1] Departamento de Matemática Aplicada, Universidade de Santiago de Compostela, 15782 Santiago de Compostela, Spain
mabermud@usc.es

[2] Departamento de Matemáticas, Universidade da Coruña, 15707 A Coruña, Spain
luisher@udc.es

[3] Departamento de Matemática Aplicada, Universidade de Santiago de Compostela, 15782 Santiago de Compostela, Spain
maprieto@usc.es

[4] GI^2MA, Departamento de Ingeniería Matemática, Universidad de Concepción, Casilla 160–C, Concepción, Chile
rodolfo@ing-mat.udec.cl

Summary. This is a survey of some recent developments on the so called "Perfectly Matched Layer" (PML) method. We take as model the scattering problems in linear acoustics. First, the Cartesian PML equations are described in the time domain for the split Berenger and the unsplit Zhao–Cangellaris formulations. The energy estimates existing in the literature are revised, and the coupled fluid/PML problem is introduced.

Next, the pressure formulation of the Cartesian PML model is derived in the frequency domain. We show that a PML method based on a non–integrable absorbing function allows recovering the exact solution in the physical domain of interest, in the framework of plane waves with oblique incidence. We revise the theoretical results that state the well–posedness of the continuous model for the acoustic scattering problem. Finally, we illustrate with some numerical results the efficiency and accuracy of the Cartesian PML approach and compare different absorbing profiles.

Finally, we introduce the pressure formulation of the radial PML model in the frequency domain and revise the theoretical results that assess the accuracy of this technique in the continuous model. Under convenient assumptions, we show its convergence when the thickness of the PML becomes larger and its exactness when a non–integrable absorbing function is used. The numerical accuracy of this approach is also illustrated.

6.1 Introduction

One problem to be tackled for the numerical solution of any scattering problem in an unbounded domain is truncating the computational domain without perturbing too much the solution of the original problem. In an ideal framework, this truncation should satisfy, at least, three properties: efficiency, easiness of implementation, and robustness. In fact, the typical first step for the numerical solution by either finite elements or finite differences is *to choose boundary conditions to replace the Sommerfeld radiation condition at infinity*, see, for instance, [26].

Several numerical techniques have been developed with this purpose: boundary element methods, infinite element methods, Dirichlet–to–Neumann operators based on truncating Fourier expansions, absorbing boundary conditions, etc. The potential advantages of each of them have been widely studied in the literature, see, for instance, [3,21,31,40], and [26] for a classical review on this subject.

The last mentioned technique, absorbing boundary conditions (ABCs), can be used to preserve the computational efficiency of the numerical method. Those of Bayliss and Turkel [6], Givoli [25], Engquist and Majda [23], and Feng [24] are among the most widely used. However, in spite of the simple implementation of lowest order ABCs, good accuracy is only achieved for higher order ones [43], because these conditions are not fully non–reflecting on the truncated boundary of the computational domain. As a consequence, high accuracy using ABCs leads to a substantial computational cost and increases the difficulty of implementation. Recently, a promising way has been open: high order ABCs not involving high derivatives, see [27,47].

Let us remark that, if the domain of the original problem is truncated with a sphere, then the Dirichlet–to–Neumann (DtN) boundary condition is exactly known, see [26,38]. However, this boundary condition involves an infinite series which must be truncated for its numerical use. Moreover, the exact DtN condition is non local, leading to dense blocks in the linear system to be solved when a finite element method is used.

An alternative approach to deal with the truncation of unbounded domains is the so called *Perfectly Matched Layer* (PML) method, which was introduced by Berenger [9–11]. It is based on simulating an absorbing layer of anisotropic damping material surrounding the domain of interest, like a thin sponge which absorbs the scattered field radiated to the exterior of this domain. This method is known as "perfectly matched" because the interface between the physical domain and the absorbing layer does not produce spurious reflections inside the domain of interest, as it is the case with ABCs.

This method has been applied to different problems. It was initially settled for Maxwell's equations in electromagnetism [9] and subsequently used for the scalar Helmholtz equation [28,41,46], advective acoustics [2,7,30], shallow water waves [39], elasticity [5,20], poroelastic media [48], and other hyperbolic problems, see for instance [37] among many other papers. We focus our attention on time–harmonic scattering problems in linear acoustics, i.e., on the scalar Helmholtz equation.

In the deduction of the PML [9], Berenger used an artificial splitting of the physical unknowns to force the tangential components of both velocities (in the acoustic medium and in the PML) to coincide on the interface for any frequency and any angle of incidence, thus guaranteeing the absence of spurious reflections, see also [31]. However, from a mathematical point of view, this non physical splitting has been shown unnecessary to state the PML equations. Throughout this chapter we present not only the classical split Berenger formulation but also the widely used unsplit Zhao–Cangellaris formulation [49] which does not require the splitting trick.

Let us recall that some authors, see for instance [1], based on the fact that the PML model is a zero–order perturbation of a weakly hyperbolic first order system, see [33], assume that the Berenger PML model could be ill–posed, despite of its good numerical accuracy. However, nowadays it is well known that this is not the case. Moreover, it is shown in [8] that the well posedness of the time–domain PML formulations does not depend either on the non–negative character of the absorbing functions in the model.

On the other hand, in the framework of time–harmonic wave propagation, Chew and Weedon [17] and Rappaport [42] showed that the PML equations can be obtained by means of a complex–valued coordinate stretching. Related to this, Lassas et al. [34, 36] showed that the PML, and in general a family of absorbing conditions, can be obtained by using complex Riemannian metric tensors.

Furthermore, in spite of the fact that the PML method has been originally settled in Cartesian coordinates, Collino and Monk [19] proposed a similar complex–valued change of coordinates to build a PML on polar coordinates, which leads to the so–called radial PML method, see also [12].

In practice, since the PML has to be truncated at a finite distance of the domain of interest, its external boundary produces artificial reflections. Theoretically, these reflections are of minor importance because of the exponential decay of the acoustic waves inside the PML. In fact, for Helmholtz–type scattering problems, Lassas and Somersalo [35] proved, using boundary integral equation techniques, that the approximate solution obtained by the radial PML method converges exponentially to the exact solution in the computational domain as the thickness of the layer tends to infinity. This result was generalized by Hohage et al. [29] using techniques based on the pole condition. Similarly, Bécache et al. [7] proved an analogous result for the convected Helmholtz equation in a duct.

When the problem is discretized to be numerically solved, the approximation error typically becomes larger. Increasing the thickness of the PML may be a remedy, although not always available because of computational cost. An alternative is to take larger values for the absorbing function involved in the complex–valued coordinate stretching. However, Collino and Monk [18] showed that this methodology may produce an error growth in the solution of the discrete problem. Consequently, an optimization problem arises: given a data set and a mesh, to choose the optimal absorbing function to minimize the error.

In this framework, Asvadurov et al. [4] proposed a pure imaginary stretching to optimize the PML error. They recovered exponential error estimates using finite–difference grid optimization. However, to the best of our knowledge, the optimization problem is still open in that there is no optimal criterion to choose a bounded absorbing function independently of data and meshes. As an alternative procedure to avoid this drawback, we proposed to use absorbing functions with unbounded integral on the PML, see [12–14].

The outline of this chapter is as follows. In Section 6.2, the Cartesian PML equations are described in the time domain for the split Berenger and the unsplit Zhao–

Cangellaris formulations, in the framework of a two–dimensional initial boundary value problem for the linear acoustic model in an unbounded domain. After a revision of the energy estimates found in the literature for the PML formulations, we introduce the coupled fluid/PML problem as an initial boundary value problem in a bounded domain.

In Section 6.3, we derive the pressure formulation of the Cartesian PML model in the frequency domain. As a preliminary study, we include the analysis of the propagation of plane waves with oblique incidence in a two–dimensional unbounded domain. In this framework, we show that a PML method based on a non–integrable absorbing function allows recovering the exact solution in the physical domain of interest. Moreover we revise the theoretical results that state the well–posedness of the continuous model. Finally, we illustrate with some numerical results the efficiency and accuracy of the Cartesian PML approach. A comparison using different absorbing profiles is also done.

In Section 6.4, we introduce the pressure formulation of the radial PML model in the frequency domain and revise the theoretical results that assess the accuracy of this technique in the continuous model. Under convenient assumptions, we show its convergence when the thickness of the PML becomes larger and its exactness when a non–integrable absorbing function is used. The numerical accuracy of this approach is illustrated with a problem with known exact solution.

Finally, some further research topics and open problems are briefly discussed in Section 6.5.

6.2 Time–domain Cartesian PML Equations

We illustrate the construction of the Cartesian PML equations and show some energy estimates for two different formulations. With this purpose, consider the two–dimensional wave equation written as a first–order pressure–velocity system:

$$\frac{\partial P}{\partial t} + \rho_0 c^2 \operatorname{div} \boldsymbol{V} = 0, \tag{6.1}$$

$$\rho_0 \frac{\partial \boldsymbol{V}}{\partial t} + \mathbf{grad}\, P = \mathbf{0}, \tag{6.2}$$

where P is the pressure, $\boldsymbol{V} = (V_1, V_2)$ is the velocity, and c and ρ_0 are the acoustic speed and the mass density of the fluid at rest, respectively.

Although we will focus on the Cartesian PML equations in two dimensions, similar considerations can be done on other systems of coordinates, see for instance the radial two–dimensional PML equations in the time domain derived by Collino and Monk [19]. Moreover, the extension to three dimensions is straightforward, see for instance [11] or [45].

We have chosen the two most widely used formulations: the original one introduced by Berenger [9] and the Zhao–Cangellaris formulation [49]. Both are derived from the first–order system (6.1)–(6.2). Let us remark that other kind of formulations involving second order derivatives in time are described only in terms of the pressure or the displacement field, see for instance [46] or [32].

6.2.1 Split Formulation. Berenger Equations

We state the equations governing the pressure and velocity fields in the PML, which we denote by $\widehat{P}$ and $\widehat{\boldsymbol{V}} = (\widehat{V}_1, \widehat{V}_2)$, respectively. With this aim, we use the splitting technique developed originally by Berenger [9–11].

We write the pressure as the addition of two terms without any physical meaning, $\widehat{P}_j$, $j = 1, 2$, such that the mass conservation law

$$\frac{\partial \widehat{P}}{\partial t} + \rho_0 c^2 \frac{\partial \widehat{V}_1}{\partial x_1} + \rho_0 c^2 \frac{\partial \widehat{V}_2}{\partial x_2} = 0,$$

can be rewritten in terms of the split pressure and the velocity fields as follows:

$$\frac{\partial \widehat{P}_j}{\partial t} + \rho_0 c^2 \frac{\partial \widehat{V}_j}{\partial x_j} = 0, \qquad j = 1, 2,$$

$$\widehat{P} = \widehat{P}_1 + \widehat{P}_2 .$$

Analogously, from (6.2), the components of the velocity field satisfy

$$\rho_0 \frac{\partial \widehat{V}_j}{\partial t} + \frac{\partial \widehat{P}}{\partial x_j} = 0, \qquad j = 1, 2 .$$

The PML method consists of adding dissipative or damping terms in the equations above, as follows:

$$\frac{\partial \widehat{P}_j}{\partial t} + \sigma_j \widehat{P}_j + \rho_0 c^2 \frac{\partial \widehat{V}_j}{\partial x_j} = 0, \qquad j = 1, 2, \tag{6.3}$$

$$\rho_0 \left(\frac{\partial \widehat{V}_j}{\partial t} + \sigma_j \widehat{V}_j \right) + \frac{\partial \widehat{P}}{\partial x_j} = 0, \qquad j = 1, 2, \tag{6.4}$$

$$\widehat{P} = \widehat{P}_1 + \widehat{P}_2, \tag{6.5}$$

where each σ_j is a non–negative absorption coefficient, which only depends on the space variable x_j, $j = 1, 2$. The resulting Equations (6.3)–(6.5) are the so–called *split formulation* of the PML equations.

To study the well posedness of the hyperbolic system (6.3)–(6.5), we show some results derived by Bécache and Joly [8] via Fourier analysis, in two different cases, both with constant absorbing functions.

Theorem 1. *Let $\sigma_1 = 0$ and σ_2 be a positive constant. If the initial conditions are such that $\widehat{P}(\cdot, 0) \in L^2(\mathbb{R}^2)$ and $\widehat{\boldsymbol{V}}(\cdot, 0) \in L^2(\mathbb{R}^2)^2$, then the solution of (6.3)–(6.5) satisfies*

$$\begin{aligned} &\frac{1}{\rho_0 c^2} \left\| \widehat{P}(\cdot, t) \right\|_{L^2(\mathbb{R}^2)} + \rho_0 \left\| \widehat{\boldsymbol{V}}(\cdot, t) \right\|_{L^2(\mathbb{R}^2)^2} \\ &\qquad \leq C \left[\frac{1}{\rho_0 c^2} \left\| \widehat{P}(\cdot, 0) \right\|_{L^2(\mathbb{R}^2)} + \rho_0 \left\| \widehat{\boldsymbol{V}}(\cdot, 0) \right\|_{L^2(\mathbb{R}^2)^2} \right], \end{aligned}$$

and

$$\frac{1}{\rho_0 c^2}\left[\left\|\widehat{P}_1(\cdot,t)\right\|_{\mathrm{H}^{-1}(\mathbb{R}^2)}+\left\|\widehat{P}_2(\cdot,t)\right\|_{\mathrm{H}^{-1}(\mathbb{R}^2)}\right] \\ \leq Ct\left[\frac{1}{\rho_0 c^2}\left\|\widehat{P}(\cdot,0)\right\|_{\mathrm{L}^2(\mathbb{R}^2)}+\rho_0\left\|\widehat{\boldsymbol{V}}(\cdot,0)\right\|_{\mathrm{L}^2(\mathbb{R}^2)^2}\right],$$

where C is a positive constant independent of σ_2.

This theorem shows that the physical quantities satisfy the same estimates as in the acoustic equations and the loss of regularity only affects the artificial unknowns $\widehat{P}_j$, $j=1,2$.

Analogous results clearly hold for the PML equations with $\sigma_2=0$ and σ_1 being a positive constant. On the other hand, if both absorbing functions are not null, the following energy estimate holds true [8]:

Theorem 2. *Let $\sigma_1=\sigma_2=:\sigma$ be a positive constant. If the initial conditions are such that $\widehat{\mathrm{P}}(\cdot,0)\in\mathrm{L}^2(\mathbb{R}^2)$ and $\widehat{\boldsymbol{V}}(\cdot,0)\in\mathrm{L}^2(\mathbb{R}^2)^2$, then the solution of (6.3)–(6.5) satisfies*

$$\frac{1}{\rho_0 c^2}\left\|\widehat{P}(\cdot,t)\right\|_{\mathrm{L}^2(\mathbb{R}^2)}+\rho_0\left\|\widehat{\boldsymbol{V}}(\cdot,t)\right\|_{\mathrm{L}^2(\mathbb{R}^2)^2} \\ \leq \mathrm{e}^{-2\sigma t}\left[\frac{1}{\rho_0 c^2}\left\|\widehat{P}(\cdot,0)\right\|_{\mathrm{L}^2(\mathbb{R}^2)}+\rho_0\left\|\widehat{\boldsymbol{V}}(\cdot,0)\right\|_{\mathrm{L}^2(\mathbb{R}^2)^2}\right].$$

This result shows the dissipative behavior of the PML model with respect to time, in this case. The proof of analogous exponentially decay estimates when the absorbing functions σ_j are not constant remains open.

6.2.2 Unsplit Formulation. Zhao–Cangellaris Equations

Next, we introduce the Cartesian PML following the Zhao–Cangellaris formulation [49]. Let us remark that although the solutions of this and the split Berenger formulation are different and even have different regularity, the solution of each of them can be deduced from the other under some compatibility assumptions on the initial conditions, see [8].

First, we restrict our attention to the case where both absorbing functions are not null. We derive the Zhao–Cangellaris formulation from the Berenger equations (6.3)–(6.5). With this purpose, we introduce new fields $\boldsymbol{V}^\star=(V_1^\star,V_2^\star)$ and $P^\star$ related to the velocity and the pressure, respectively, by means of the following ordinary differential equations:

$$\frac{\partial V_1^\star}{\partial t}=\left(\frac{\partial}{\partial t}+\sigma_2\right)\widehat{V}_1,$$

$$\frac{\partial V_2^\star}{\partial t}=\left(\frac{\partial}{\partial t}+\sigma_1\right)\widehat{V}_2,$$

$$\frac{\partial^2 P^\star}{\partial t^2}=\left(\frac{\partial}{\partial t}+\sigma_1\right)\left(\frac{\partial}{\partial t}+\sigma_2\right)\widehat{P}.$$

From (6.3), applying $\partial/\partial t+\sigma_2$ to the equation satisfied by $\widehat{P}_1$ and $\widehat{V}_1$, and $\partial/\partial t+\sigma_1$ to the equation satisfied by $\widehat{P}_2$ and $\widehat{V}_2$, we have

$$\left(\frac{\partial}{\partial t}+\sigma_1\right)\left(\frac{\partial}{\partial t}+\sigma_2\right)\widehat{P}_1+\rho_0 c^2\left(\frac{\partial}{\partial t}+\sigma_2\right)\frac{\partial \widehat{V}_1}{\partial x_1}=0\,,$$

$$\left(\frac{\partial}{\partial t}+\sigma_1\right)\left(\frac{\partial}{\partial t}+\sigma_2\right)\widehat{P}_2+\rho_0 c^2\left(\frac{\partial}{\partial t}+\sigma_1\right)\frac{\partial \widehat{V}_2}{\partial x_2}=0\,.$$

Now, if we add both equations, take into account (6.5), the differential equations for $\boldsymbol{V}^\star$ and $P^\star$ above, and the fact that each absorbing function σ_j only depends on x_j, then integrating in time we obtain

$$\frac{\partial P^\star}{\partial t}+\rho_0 c^2 \operatorname{div} \boldsymbol{V}^\star=0\,.$$

In summary, the equations of the Zhao–Cangellaris formulation are the following:

$$\frac{\partial P^\star}{\partial t}+\rho_0 c^2 \operatorname{div} \boldsymbol{V}^\star=0\,, \tag{6.6}$$

$$\rho_0\left(\frac{\partial}{\partial t}+\sigma_1\right)\widehat{V}_1+\frac{\partial \widehat{P}}{\partial x_1}=0\,, \tag{6.7}$$

$$\rho_0\left(\frac{\partial}{\partial t}+\sigma_2\right)\widehat{V}_2+\frac{\partial \widehat{P}}{\partial x_2}=0\,, \tag{6.8}$$

$$\frac{\partial V_1^\star}{\partial t}=\left(\frac{\partial}{\partial t}+\sigma_2\right)\widehat{V}_1\,, \tag{6.9}$$

$$\frac{\partial V_2^\star}{\partial t}=\left(\frac{\partial}{\partial t}+\sigma_1\right)\widehat{V}_2\,, \tag{6.10}$$

$$\frac{\partial^2 P^\star}{\partial t^2}=\left(\frac{\partial}{\partial t}+\sigma_1\right)\left(\frac{\partial}{\partial t}+\sigma_2\right)\widehat{P}\,. \tag{6.11}$$

From a computational point of view this unsplit formulation has one advantage: the spatial differential operators to be solved are the same as in the original hyperbolic system for the acoustic fluid. Therefore, to implement it in a computer code, we only have to solve three additional ordinary differential equations which do not require further spatial discretization.

From a mathematical point of view, the Zhao–Cangellaris formulation allows proving the well–posedness of the Cartesian PML in the time domain for non–constant absorbing functions, under the only assumption of being bounded.

For the sake of simplicity, let us assume that the domain of Equations (6.6)–(6.11) is the whole $\mathbb{R}^2$. Consider the following first–order energy–like quantity:

$$\mathcal{E}_1(t) := \frac{1}{2}\left\{ \frac{1}{\rho_0 c^2} \left\| \widehat{P}(\cdot,t) \right\|^2_{\mathrm{L}^2(\mathbb{R}^2)} + \rho_0 \left\| \widehat{\boldsymbol{V}}(\cdot,t) \right\|^2_{\mathrm{L}^2(\mathbb{R}^2)^2} \right.$$
$$\left. + \left\| \boldsymbol{V}^\star(\cdot,t) - \boldsymbol{V}(\cdot,t) \right\|^2_{\mathrm{L}^2(\mathbb{R}^2)^2} + \left[\sigma_1\sigma_2 \widetilde{P}(\cdot,t), \widetilde{P}(\cdot,t) \right]_{\mathrm{L}^2(\mathbb{R}^2)} \right\},$$

where

$$\widetilde{P}(\cdot,t) := \int_0^t \widehat{P}(\cdot,s)\,ds\,.$$

The following result has been proved by Bécache and Joly [8]:

Theorem 3. *Let σ_1 and σ_2 be positive bounded functions. If the initial conditions are such that $\widehat{P}(\cdot,0), P^\star(\cdot,0) \in \mathrm{L}^2(\mathbb{R}^2)$ and $\widehat{\boldsymbol{V}}(\cdot,0), \boldsymbol{V}^\star(\cdot,0) \in \mathrm{L}^2(\mathbb{R}^2)^2$, then the solution of (6.6)–(6.11) satisfies*

$$\mathcal{E}_1(t) \le \mathcal{E}_1(0)\mathrm{e}^{3t\left(\|\sigma_1\|_\infty + \|\sigma_2\|_\infty\right)}\,. \tag{6.12}$$

An analogous derivation can be done for the case where only one of the absorbing functions is not null. For instance, if the absorbing terms are assumed only in the x_1 direction (i.e., $\sigma_2 = 0$), then (6.9) can be simplified to $V_1^\star = \widehat{V}_1$ and (6.11) to

$$\frac{\partial P^\star}{\partial t} = \left(\frac{\partial}{\partial t} + \sigma_1\right)\widehat{P}\,.$$

An estimate similar to that of the previous theorem also holds true in this case, see [8]. Moreover, since now $\sigma_2 = 0$, the term involving $\widetilde{P}$ does not appear in the energy–like quantity $\mathcal{E}_1(t)$.

Let us remark that Theorem 3 does not provide a very precise information about the solution of the Zhao–Cangellaris formulation, since these results allow even an exponential growth of the solution with respect to time. However, this is not what happens in the numerical simulations. As in the case of the Berenger model, a deeper characterization of the time behavior of this PML formulation remains open.

6.2.3 Coupled Fluid/PML Problem

Now, we are in a position to state the coupled fluid/PML problem, which will allow us to write explicitly the coupling conditions on the interface between the fluid and the PML domains. With this purpose, we use the equations governing the fluid motion (6.1)–(6.2) and the Zhao–Cangellaris formulation (6.6)–(6.11).

Let Ω be a bounded domain of $\mathbb{R}^2$ occupied by an obstacle to the propagation of acoustic waves with a totally reflecting boundary. We use the geometrical notation introduced in Figure 6.1: Ω_{F} is the physical domain (i.e., the domain where the fluid motion has to be computed) and the rectangular box Ω_{A} is the PML.

We use subscripts $_{\mathrm{F}}$ and $_{\mathrm{A}}$ (as in P_{F}, $\boldsymbol{V}_{\mathrm{A}}^\star$, etc.) to distinguish the restrictions of different variables to the domains Ω_{F} and Ω_{A}, respectively. In particular, P_{A} and $\boldsymbol{V}_{\mathrm{A}}$ denote the pressure and the velocity fields in the PML domain Ω_{A} (instead of $\widehat{P}$ and $\widehat{\boldsymbol{V}}$, as above).

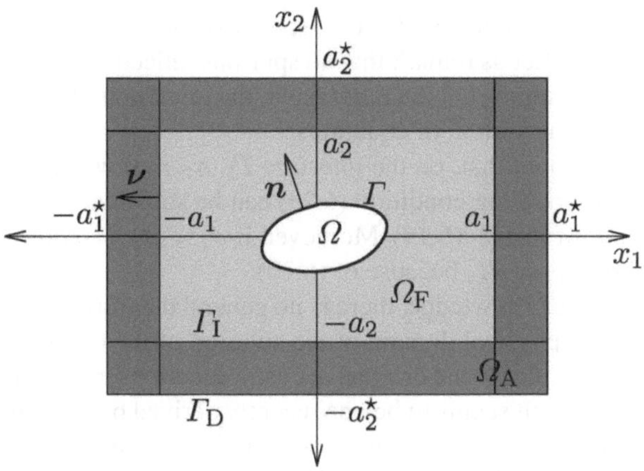

Fig. 6.1 Cartesian PML on a two–dimensional domain.

The equations governing the coupled fluid/PML problem in the time domain are the following:

$$\frac{\partial P_F}{\partial t} + \rho_0 c^2 \operatorname{div} \boldsymbol{V}_F = 0 \qquad \text{in } \Omega_F, \tag{6.13}$$

$$\rho_0 \frac{\partial \boldsymbol{V}_F}{\partial t} + \mathbf{grad}\, P_F = \mathbf{0} \qquad \text{in } \Omega_F, \tag{6.14}$$

$$\frac{\partial P_A^\star}{\partial t} + \rho_0 c^2 \operatorname{div} \boldsymbol{V}_A^\star = 0 \qquad \text{in } \Omega_A, \tag{6.15}$$

$$\rho_0 \left(\frac{\partial}{\partial t} + \sigma_1 \right) V_{A,1} + \frac{\partial P_A}{\partial x_1} = 0 \qquad \text{in } \Omega_A, \tag{6.16}$$

$$\rho_0 \left(\frac{\partial}{\partial t} + \sigma_2 \right) V_{A,2} + \frac{\partial P_A}{\partial x_2} = 0 \qquad \text{in } \Omega_A, \tag{6.17}$$

$$\frac{\partial V_{A,1}^\star}{\partial t} = \left(\frac{\partial}{\partial t} + \sigma_2 \right) V_{A,1} \qquad \text{in } \Omega_A, \tag{6.18}$$

$$\frac{\partial V_{A,2}^\star}{\partial t} = \left(\frac{\partial}{\partial t} + \sigma_1 \right) V_{A,2} \qquad \text{in } \Omega_A, \tag{6.19}$$

$$\frac{\partial^2 P_A^\star}{\partial t^2} = \left(\frac{\partial}{\partial t} + \sigma_1 \right) \left(\frac{\partial}{\partial t} + \sigma_2 \right) P_A \qquad \text{in } \Omega_A, \tag{6.20}$$

$$\boldsymbol{V}_F \cdot \boldsymbol{n} = 0 \qquad \text{on } \Gamma, \tag{6.21}$$

$$P_F = P_A \qquad \text{on } \Gamma_I, \tag{6.22}$$

$$\boldsymbol{V}_F \cdot \boldsymbol{\nu} = \boldsymbol{V}_A \cdot \boldsymbol{\nu} \qquad \text{on } \Gamma_I, \tag{6.23}$$

$$P_A^\star = 0 \qquad \text{on } \Gamma_D, \tag{6.24}$$

combined with adequate initial conditions. The absorbing functions vanish in the physical domain; more precisely, we assume $\sigma_j = 0$ for $|x_j| < a_j$, $j = 1, 2$.

The coupling conditions (6.22)–(6.23) preserve the continuity of the pressure and the normal velocity. Let us remark that no spurious reflections arise on $\Gamma_{\rm I}$ because of these coupling conditions. On the other hand, the outer boundary condition (6.24) is necessary to close the system of equations.

Taking into account that, on the interface $\Gamma_{\rm I}$, $\sigma_1 = 0$ when $\nu_2 \neq 0$ and $\sigma_2 = 0$ when $\nu_1 \neq 0$, the coupling condition (6.23) can be substituted by $\boldsymbol{V}_{\rm F} \cdot \boldsymbol{\nu} = \boldsymbol{V}_{\rm A}^{\star} \cdot \boldsymbol{\nu}$, as a consequence of (6.18)–(6.19). Moreover, if $\sigma_1 = \sigma_2 = 0$ on $\Gamma_{\rm I}$, then (6.22) can be substituted by $P_{\rm F} = P_{\rm A}^{\star}$, because of (6.20).

To the best of our knowledge, there is no general theoretical error estimate of the convergence in the physical domain of the solution of the coupled fluid/PML problem (6.13)–(6.24) to that of the original acoustic equations in the unbounded domain. The only existing result seems to be one recently derived by Diaz and Joly [22], who prove exponential convergence to the exact solution of the scattering problem in the L^∞–norm, in the particular case of a point source acting in the fluid domain.

6.3 Time–harmonic Cartesian PML Equations

To write the PML equations in the frequency domain, we can use either the split Berenger equations or the Zhao–Cangellaris formulation. Here, we use the former.

If we look for time–harmonic solutions of (6.3)–(6.5), then we arrive at the corresponding equations in the frequency domain:

$$\widehat{p}_j = -\frac{\rho_0 c^2}{\sigma_j - \mathrm{i}\omega} \frac{\partial \widehat{v}_j}{\partial x_j}, \qquad j = 1, 2, \tag{6.25}$$

$$\widehat{v}_j = -\frac{1}{\rho_0(\sigma_j - \mathrm{i}\omega)} \frac{\partial \widehat{p}}{\partial x_j}, \qquad j = 1, 2, \tag{6.26}$$

$$\widehat{p} = \widehat{p}_1 + \widehat{p}_2. \tag{6.27}$$

Here and thereafter, we use lower case letters ($\widehat{p}$, $\widehat{v}_j$, etc.) to denote the amplitudes of the corresponding magnitudes ($\widehat{P}$, $\widehat{V}_j$, etc.).

By substituting (6.26) in (6.25), we have

$$\widehat{p}_j = c^2 \frac{1}{\sigma_j - \mathrm{i}\omega} \frac{\partial}{\partial x_j} \left(\frac{1}{\sigma_j - \mathrm{i}\omega} \frac{\partial \widehat{p}}{\partial x_j} \right), \qquad j = 1, 2,$$

and using this expression in (6.27), we obtain

$$-\omega^2 \rho_0 \widehat{p} - \rho_0 c^2 \sum_{j=1}^{2} \frac{\mathrm{i}\omega}{\sigma_j - \mathrm{i}\omega} \frac{\partial}{\partial x_j} \left(\frac{\mathrm{i}\omega}{\sigma_j - \mathrm{i}\omega} \frac{\partial \widehat{p}}{\partial x_j} \right) = 0. \tag{6.28}$$

Let us remark that the complex change of variable introduced by Chew and Weedon [17],

$$\widehat{x}_j(x_j) := x_j + \frac{\mathrm{i}}{\omega} \int_0^{x_j} \sigma(s)\, ds, \qquad j = 1, 2, \tag{6.29}$$

allows recovering formally the weights in (6.28) from

$$\frac{\partial x_j}{\partial \widehat{x}_j} = -\frac{\mathrm{i}\omega}{\sigma_j(x_j) - \mathrm{i}\omega}. \tag{6.30}$$

Hence, the PML equation (6.28) can be seen as the Helmholtz equation in the new complex coordinate system $(\widehat{x}_1, \widehat{x}_2)$:

$$-\rho_0\omega^2\widehat{p} - \rho_0 c^2 \left(\frac{\partial^2 \widehat{p}}{\partial \widehat{x}_1^2} + \frac{\partial^2 \widehat{p}}{\partial \widehat{x}_2^2}\right) = 0.$$

Thus, the construction of the PML equation can be understood as a complex stretching of coordinates.

We restrict our attention to symmetric absorbing functions, i.e., σ_j such that $\sigma_j(-x_j) = \sigma_j(x_j)$. Moreover, since $\sigma_j = 0$ for $|x_j| < a_j$, from now on we consider that σ_j is only defined in $[a_j, a_j^\star)$. Finally, we introduce the following additional coefficients, that will be used for the weights of the PML equation:

$$\gamma_j(x_j) := \begin{cases} 1, & \text{if } |x_j| \le a_j, \\ \dfrac{\sigma_j(|x_j|) - \mathrm{i}\omega}{-\mathrm{i}\omega} = 1 + \dfrac{\mathrm{i}}{\omega}\sigma_j(|x_j|), & \text{if } a_j \le |x_j| < a_j^\star. \end{cases}$$

Therefore, after multiplying (6.28) by $\gamma_1\gamma_2$, the PML equation reads as follows:

$$-\rho_0\omega^2\gamma_1\gamma_2\widehat{p} - \rho_0 c^2 \left[\frac{\partial}{\partial x_1}\left(\frac{\gamma_2}{\gamma_1}\frac{\partial \widehat{p}}{\partial x_1}\right) + \frac{\partial}{\partial x_2}\left(\frac{\gamma_1}{\gamma_2}\frac{\partial \widehat{p}}{\partial x_2}\right)\right] = 0.$$

6.3.1 Cartesian PML for a Scattering Problem

We deal with a time–harmonic acoustic scattering problem in an unbounded exterior 2D domain. In particular, we consider the following exterior Helmholtz problem with Neumann boundary data:

$$-\Delta p - k^2 p = 0 \qquad \text{in } \mathbb{R}^2 \setminus \bar{\Omega}, \tag{6.31}$$

$$\frac{\partial p}{\partial \boldsymbol{n}} = g \qquad \text{on } \Gamma, \tag{6.32}$$

$$\lim_{r\to\infty} \sqrt{r}\left(\frac{\partial p}{\partial r} - ikp\right) = 0, \tag{6.33}$$

where p is the unknown amplitude of the pressure wave and $k := \omega/c$ is the wave number, ω being the angular frequency of the waves and c the sound speed of the fluid.

We introduce the PML on the x_1 and x_2 directions to truncate the unbounded domain, as shown in Figure 6.1. If we denote by p_{F} and p_{A} the pressure fields in Ω_{F} and Ω_{A}, respectively, then the equations governing the motion of the coupled fluid/Cartesian PML problem reads:

$$-\Delta p_{\mathrm{F}} - k^2 p_{\mathrm{F}} = 0 \quad \text{in } \Omega_{\mathrm{F}}, \tag{6.34}$$

$$-\left[\frac{\partial}{\partial x_1}\left(\frac{\gamma_2}{\gamma_1}\frac{\partial p_{\mathrm{A}}}{\partial x_1}\right) + \frac{\partial}{\partial x_2}\left(\frac{\gamma_1}{\gamma_2}\frac{\partial p_{\mathrm{A}}}{\partial x_2}\right)\right] - k^2\gamma_1\gamma_2 p_{\mathrm{A}} = 0 \quad \text{in } \Omega_{\mathrm{A}}, \tag{6.35}$$

$$\frac{\partial p_{\mathrm{F}}}{\partial \boldsymbol{n}} = g \quad \text{on } \Gamma, \tag{6.36}$$

$$p_{\mathrm{F}} = p_{\mathrm{A}} \quad \text{on } \Gamma_{\mathrm{I}}, \tag{6.37}$$

$$\frac{\partial p_{\mathrm{F}}}{\partial \boldsymbol{\nu}} = \frac{\nu_1}{\gamma_1}\frac{\partial p_{\mathrm{A}}}{\partial x_1} + \frac{\nu_2}{\gamma_2}\frac{\partial p_{\mathrm{A}}}{\partial x_2} \quad \text{on } \Gamma_{\mathrm{I}}, \tag{6.38}$$

$$p_{\mathrm{A}} = 0 \quad \text{on } \Gamma_{\mathrm{D}}. \tag{6.39}$$

The absorbing functions play a key role in the behavior of the PML equations, as well as in the accuracy and robustness of the discrete method to numerically solve the problem. As usual, we assume that σ_j are non negative, continuous, and monotonically increasing for $x_j \in [a_j, a_j^\star)$. However, we do not impose that $\sigma_j(a_j) = 0$, so that it does not necessarily vanish on the interface between the fluid and the PML domains.

According to Collino and Monk [19], the existence and uniqueness of solution to the coupled fluid/Cartesian PML problem (6.34)–(6.39), except for a countable number of frequencies, can be proved by adapting to this case the proof of Theorem 2 from that reference:

Theorem 4. *Let $g \in H^{-\frac{1}{2}}(\Gamma)$ and σ_j, $j = 1,2$, be bounded, continuous, non–negative and monotonically increasing functions. Then, problem (6.34)–(6.39) has a unique solution $(p_{\mathrm{F}}, p_{\mathrm{A}})$ in $\mathrm{H}^1(\Omega_{\mathrm{F}}) \times \mathrm{H}^1(\Omega_{\mathrm{A}})$ for every real k, except at most for a discrete set of values of k.*

Let us remark that there is no general theoretical result on the convergence of the coupled fluid/PML solution to the original solution of the scattering problem, when the thickness of the Cartesian PML is enlarged.

Anyway, classical PML techniques do not rely on larger layers, but on taking bounded absorbing functions such that their integrals are large enough. The most widely used are the following quadratic functions, which take the value $\sigma^\star$ at $x_j = a_j^\star$:

$$\sigma_j^{\mathrm{Q}}(x_j) := \sigma^\star \frac{(x_j - a_j)^2}{(a_j^\star - a_j)^2}, \qquad x_j \in [a_j, a_j^\star]. \tag{6.40}$$

Obviously, other kind of bounded absorbing functions, as linear or constant functions, can be considered as well, see, for instance, [7, 9, 19].

The PML with quadratic absorbing functions turns out efficient in practice, provided the value $\sigma^\star$ is correctly tuned. However, there is no theoretical result that confirms the numerical evidence. Instead, we have proposed in [14] to use unbounded absorbing functions such that

$$\int_{a_j}^{a_j^\star} \sigma_j(s)\, ds = +\infty.$$

For instance,

$$\sigma_j^{\mathrm{U}}(x_j) := \frac{c}{a_j^\star - x_j}, \qquad x_j \in [a_j, a_j^\star). \tag{6.41}$$

Also other kind of non–integrable functions can be used. For a more detailed comparison of different singular absorbing functions, see [14].

There are not theoretical results guaranteeing the use of this kind of singular absorbing functions for the Cartesian PML, either. However, we will show below some numerical evidence of the good accuracy of this approach. Moreover, we will show in Section 6.4 below that a similar technique for radial PML leads to an exact recovery of the pressure in the physical domain. The theoretical exactness of Cartesian PML with non–integral absorbing functions is a current research topic.

6.3.2 Plane Wave Analysis

To illustrate the mathematical behavior of the PML equations, we consider a problem with known exact solution: the propagation of two–dimensional acoustic plane waves with oblique incidence. This simple test will provide valuable information for the design of an efficient PML method.

Consider the following time–harmonic problem posed in the right half–space:

$$-\Delta p - k^2 p = 0, \qquad x_1 > 0,\ x_2 \in \mathbb{R}, \tag{6.42}$$

$$p(0, x_2) = \mathrm{e}^{\mathrm{i}k_2 x_2}, \qquad x_2 \in \mathbb{R}, \tag{6.43}$$

$$\lim_{x_1 \to +\infty} \left(\frac{\partial p}{\partial x_1} - \mathrm{i}k_1 p \right) = 0, \tag{6.44}$$

where $k = \omega / c$ is the wave number as above, $k_1 := k \cos\theta$ and $k_2 := k \sin\theta$, θ being the incidence angle. It is well–known that the solution of this problem is the plane wave

$$p(x_1, x_2) = \mathrm{e}^{\mathrm{i}(k_1 x_1 + k_2 x_2)}. \tag{6.45}$$

We introduce a PML in the vertical strip $a_1 < x_1 < a_1^\star$, to truncate the unbounded domain in the x_1–direction, see Figure 6.2. The strip $0 < x_1 < a_1$ plays the role of the physical domain.

Analogously to the scattering problem above, the pressure fields p_{F} and p_{A} in Ω_{F} and Ω_{A}, respectively, satisfy

$$-\Delta p_{\mathrm{F}} - k^2 p_{\mathrm{F}} = 0, \qquad 0 < x_1 < a_1, \tag{6.46}$$

$$-\frac{1}{\gamma_1}\frac{\partial}{\partial x_1}\left(\frac{1}{\gamma_1}\frac{\partial p_{\mathrm{A}}}{\partial x_1}\right) - \frac{\partial^2 p_{\mathrm{A}}}{\partial x_2^2} - k^2 p_{\mathrm{A}} = 0, \qquad a_1 < x_1 < a_1^\star, \tag{6.47}$$

$$p_{\mathrm{F}}(0, x_2) = \mathrm{e}^{\mathrm{i}k_2 x_2}, \tag{6.48}$$

$$p_{\mathrm{F}}(a_1, x_2) = p_{\mathrm{A}}(a_1, x_2), \tag{6.49}$$

$$\frac{\partial p_{\mathrm{F}}}{\partial x_1}(a_1, x_2) = \frac{1}{\gamma_1(a_1)}\frac{\partial p_{\mathrm{A}}}{\partial x_1}(a_1, x_2), \tag{6.50}$$

$$p_{\mathrm{A}}(a_1^\star, x_2) = 0, \tag{6.51}$$

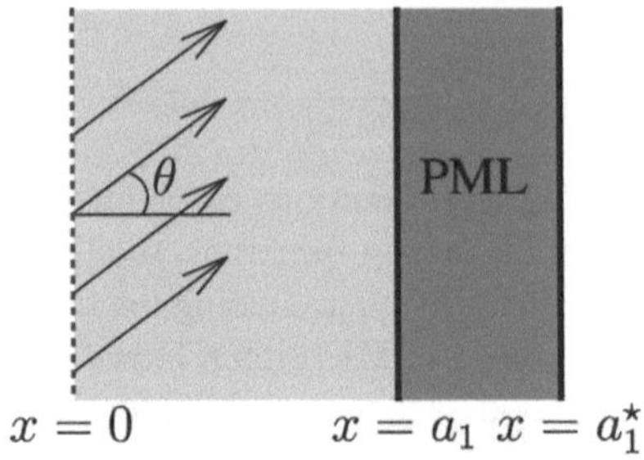

Fig. 6.2 PML in the x_1–direction for plane waves with oblique incidence.

The solution to this problem can be written as a superposition of plane waves:

$$p_{\mathrm{F}}(x_1,x_2) = \left(I\mathrm{e}^{\mathrm{i}k_1x_1} + R_{\mathrm{F}}\mathrm{e}^{-ik_1x_1}\right)\mathrm{e}^{\mathrm{i}k_2x_2}, \qquad x_1 \in [0,a_1),$$
$$p_{\mathrm{A}}(x_1,x_2) = \left[T\mathrm{e}^{\mathrm{i}k_1\widehat{x}_1(x_1)} + R_{\mathrm{A}}\mathrm{e}^{-ik_1\widehat{x}_1(x_1)}\right]\mathrm{e}^{\mathrm{i}k_2x_2}, \qquad x_1 \in [a_1,a_1^\star),$$

where $\widehat{x}_1(x_1) := x_1 + \frac{\mathrm{i}}{\omega}\int_{a_1}^{x_1}\sigma(s)\,ds$. In the expressions above, I is the amplitude of the incident wave, T that of the wave transmitted to the PML, and R_{F} and R_{A} are the amplitudes of the reflected waves in the physical domain and in the absorbing layer, respectively.

Hence, from (6.48)–(6.51), we obtain

$$I = T = 1 - R_{\mathrm{F}}, \tag{6.52}$$

$$R_{\mathrm{F}} = R_{\mathrm{A}} = \frac{\mathrm{e}^{2\mathrm{i}k_1a_1^\star}}{\mathrm{e}^{2\mathrm{i}k_1a_1^\star} - \mathrm{e}^{\frac{2\cos\theta}{c}\int_{a_1}^{a_1^\star}\sigma_1(s)\,ds}}. \tag{6.53}$$

Notice that the latter implies that no spurious reflection arises at $x_1 = a_1$ (which is the main feature of the PML technique); the terms involving R_{F} and R_{A} arise as a consequence of the waves reflected at $x_1 = a_1^\star$.

Summarizing, we have obtained the following analytical expression for the solution to the PML problem above:

$$p_{\mathrm{F}}(x_1,x_2) = \left[(1-R_{\mathrm{A}})\mathrm{e}^{\mathrm{i}k_1x_1} + R_{\mathrm{A}}\mathrm{e}^{-ik_1x_1}\right]\mathrm{e}^{\mathrm{i}k_2x_2},$$

$$p_{\mathrm{A}}(x_1,x_2) = \Big[(1-R_{\mathrm{A}})\mathrm{e}^{\mathrm{i}k_1x_1}\mathrm{e}^{-\frac{\cos\theta}{c}\int_{a_1}^{x_1}\sigma_1(s)\,ds} + R_{\mathrm{A}}\mathrm{e}^{-ik_1x_1}\mathrm{e}^{\frac{\cos\theta}{c}\int_{a_1}^{x_1}\sigma_1(s)\,ds}\Big]\mathrm{e}^{\mathrm{i}k_2x_2}.$$

Expression (6.53) for the reflection coefficient R_{A} shows that the larger the integral $\int_{a_1}^{a_1^\star}\sigma_1(s)\,ds$, the closer R_{A} to zero and, consequently, the closer p_{F} to the solution (6.45) of problem (6.42)–(6.44) in the physical domain. Indeed, straightforward computations lead to

$$\int_0^{a_1}|p(x_1,x_2) - p_{\mathrm{F}}(x_1,x_2)|^2\,dx_1 = |R_{\mathrm{A}}|^2\,\frac{2k_1a_1 - \sin(2k_1a_1)}{k_1}. \tag{6.54}$$

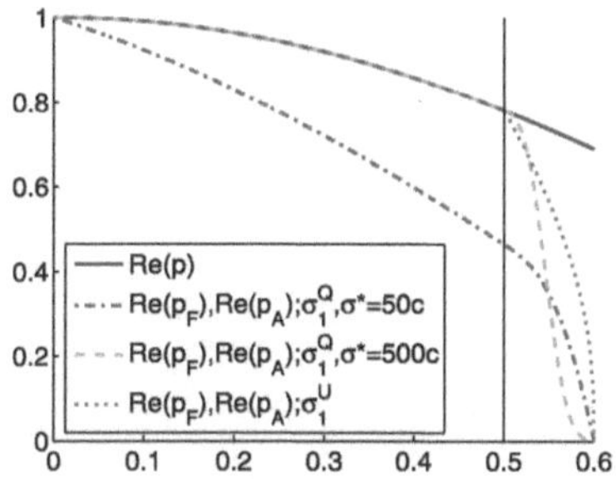

Fig. 6.3 Real part of exact and approximate pressures for $\theta = 3\pi/8$.

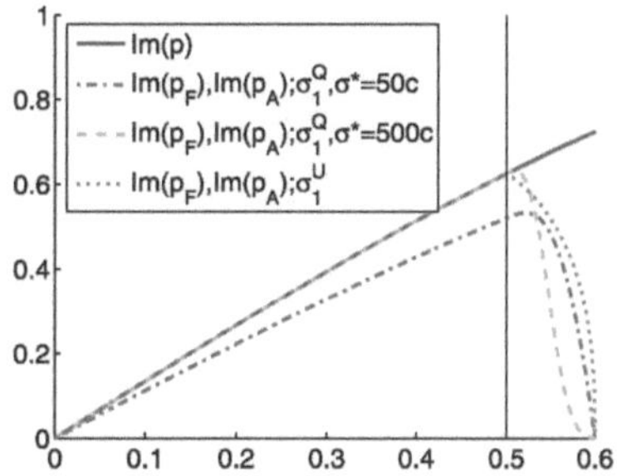

Fig. 6.4 Imaginary part of exact and approximate pressures for $\theta = 3\pi/8$.

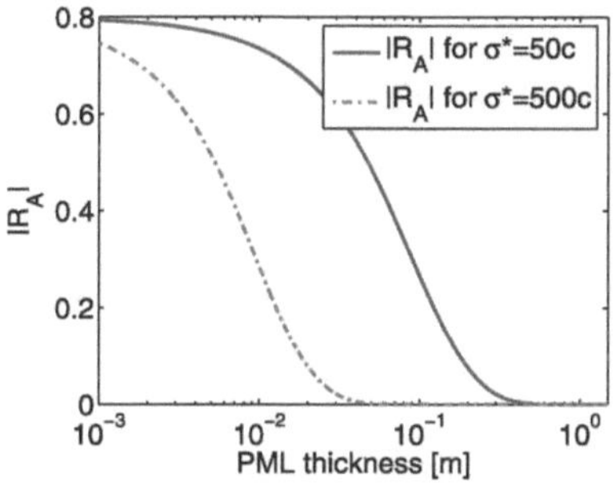

Fig. 6.5 Absolute value of the reflection coefficient for σ_1^{Q} versus PML thickness.

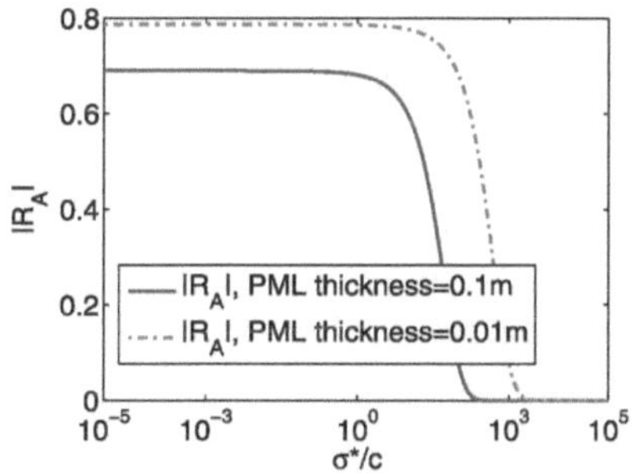

Fig. 6.6 Absolute value of the reflection coefficient for σ_1^{Q} versus $\sigma^\star/c$.

It is clear that if we use a non–integrable function, for instance σ_1^{U} as defined in (6.41), then $R_{\mathrm{A}} = 0$ and, consequently, the resulting p_{F} will coincide exactly with (6.45) in the physical domain.

To illustrate this behavior we take the following parameters: $a_1 = 0.5\,\mathrm{m}$, $a_1^\star = 0.6\,\mathrm{m}$, $\omega = 1200\ \mathrm{rad/s}$, $c = 340\,\mathrm{m/s}$, and $\theta = \frac{3\pi}{8}$ rad. We compare the absorbing functions σ_1^{Q} and σ_1^{U} defined in (6.40) and (6.41), respectively.

In Figures 6.3 and 6.4 we show that, when using σ_1^{Q}, p_{F} approximates the exact solution p as $\sigma^\star$ becomes larger. The reflection coefficient is in this case $R_{\mathrm{A}} = 0.26$ for $\sigma^\star = 50c$, and $R_{\mathrm{A}} = 2.88 \times 10^{-6}$ for $\sigma^\star = 500c$. In the same figure, we can see that the error vanishes when using σ_1^{U}.

In Figures 6.5 and 6.6 we show the dependence of R_{A} with respect to the thickness of the PML and to the value of $\sigma^\star$, respectively, when σ_1^{Q} is used. As expected, R_{A} decreases when even the thickness or $\sigma^\star$ are larger.

In Figure 6.7 we show the dependence of R_{A} with respect to the angle of incidence of the plane wave, when σ_1^{Q} is used. It is important to emphasize that, in this case, the reflection coefficient (and hence the error) increases as the angle of

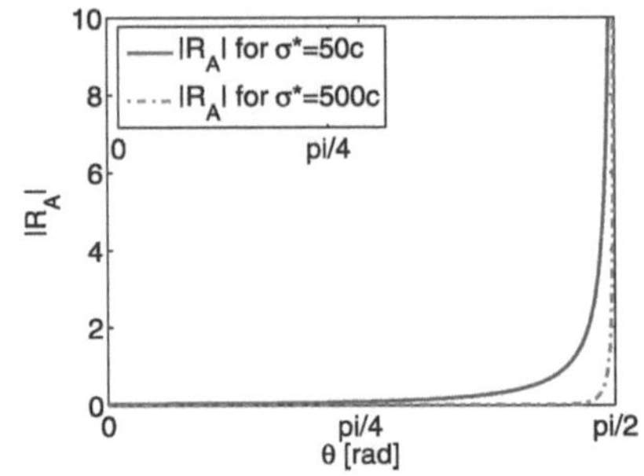

Fig. 6.7 Absolute value of the reflection coefficient for σ_1^{Q} versus incidence angle.

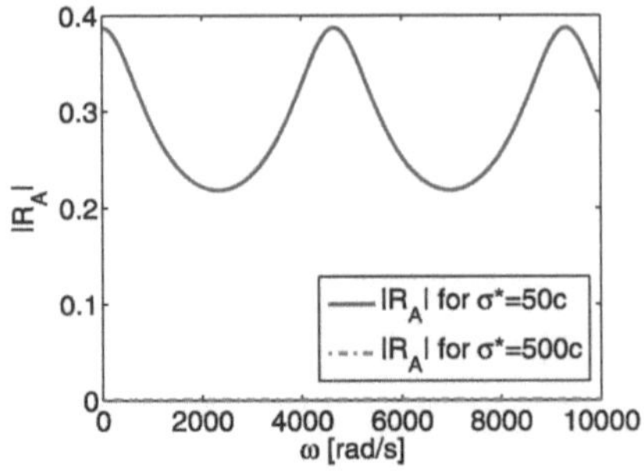

Fig. 6.8 Absolute value of the reflection coefficient for σ_1^{Q} versus frequency.

incidence is larger, whereas the error is null for any angle of incidence when σ_1^{U} is used.

Finally, in Figure 6.8 we show the dependence of R_{A} with respect to the frequency ω when σ_1^{Q} is used. For this test we have taken again $\theta = \frac{3\pi}{8}$ as the angle of incidence. We observe that R_{A} achieves periodically maximum values for certain frequencies. Again, we want to remark that using σ_1^{U} in the PML we recover $R_{\mathrm{A}} = 0$ for any value of the frequency.

6.3.3 Numerical Results

Now we return to the scattering problem with Cartesian PML (6.34)–(6.39). We describe a finite element method for its numerical solution and show that the resulting discrete problem is well posed for certain unbounded absorbing functions.

We consider a partition in triangles of the physical domain Ω_{F} and a partition in rectangles of the absorbing layer Ω_{A}, both meshes matching on the common interface Γ_{I}, as shown in Figure 6.9. As usual, h denotes the mesh–size.

The reason why we use such hybrid meshes is that triangles are more adequate to fit the boundary of the obstacle, whereas rectangles allow us to compute explicitly the integrals involving the absorbing functions which appear in the elements in the layer. This is not strictly necessary, since these integrals can be efficiently computed by standard quadrature rules, see [14]. However, in what follows, we will consider exact integration, to be able to assess the accuracy of the proposed PML independently of quadrature errors.

We compute approximations p_{F}^h and p_{A}^h of p_{F} and p_{A}, respectively, by using linear triangular finite elements for the former and bilinear rectangular finite elements for the latter. The degrees of freedom defining the finite element solution are the values of p_{F}^h and p_{A}^h at the vertices of the elements. Notice that, because of the transmission condition (6.37), the values of p_{F}^h and p_{A}^h must coincide at the vertices on the interface.

Moreover, we impose the Dirichlet boundary condition (6.39) to the finite element solution. Hence, p_{A}^h does not have degrees of freedom on the outer boundary.

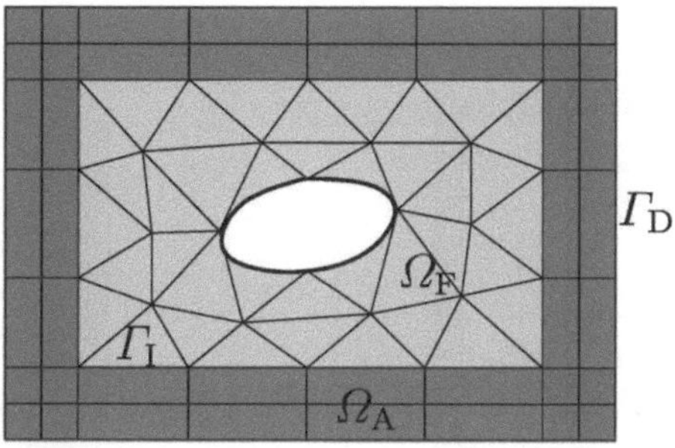

Fig. 6.9 Hybrid mesh on PML and physical domain.

This fact will be essential for the resulting discrete problem to be well posed when non–integrable absorbing functions are used.

Standard arguments in this finite element framework lead to the following discretization of the weak form of problem (6.34)–(6.39):

$$\int_{\Omega_{\mathrm{F}}} \mathbf{grad}\, p_{\mathrm{F}}^h \cdot \mathbf{grad}\, \bar{q}_{\mathrm{F}}^h \, dx_1\, dx_2 - \int_{\Omega_{\mathrm{F}}} k^2 p_{\mathrm{F}}^h \bar{q}_{\mathrm{F}}^h \, dx_1\, dx_2 + \int_{\Omega_{\mathrm{A}}} \frac{\gamma_2}{\gamma_1} \frac{\partial p_{\mathrm{A}}^h}{\partial x_1} \frac{\partial \bar{q}_{\mathrm{A}}^h}{\partial x_1} \, dx_1\, dx_2$$
$$+ \int_{\Omega_{\mathrm{A}}} \frac{\gamma_1}{\gamma_2} \frac{\partial p_{\mathrm{A}}^h}{\partial x_2} \frac{\partial \bar{q}_{\mathrm{A}}^h}{\partial x_2} \, dx_1\, dx_2 - \int_{\Omega_{\mathrm{A}}} k^2 \gamma_1 \gamma_2 p_{\mathrm{A}}^h \bar{q}_{\mathrm{A}}^h \, dx_1\, dx_2 = \int_{\Gamma} g \bar{q}_{\mathrm{F}}^h \, ds,$$

for all functions q^h, continuous in $\Omega_{\mathrm{F}} \cup \Omega_{\mathrm{A}}$, piecewise linear in Ω_{F}, piecewise bilinear in Ω_{A}, and vanishing on Γ_{D}.

Straightforward computations show that, when using the absorbing function σ_j^{U} defined in (6.41), the coefficients of the finite element matrices are bounded, in spite of the singularity of σ_j^{U} on Γ_{D}. For a detailed proof, see [14].

On the other hand, when using the absorbing function σ_j^{Q} defined in (6.40), the standard procedure to minimize the spurious reflections produced at the outer boundary of the PML consists of taking large values for $\sigma^\star$. However, larger values of $\sigma^\star$ lead to larger discretization errors. Therefore, $\sigma^\star$ cannot be chosen arbitrarily large because, otherwise, the discretization errors would be dominant, deteriorating the overall accuracy of the method.

As shown in [18], for a given problem and a given mesh there is an optimal value of $\sigma^\star$ leading to minimal errors. Unfortunately, such optimal value depends strongly on the problem data as well as on the particular mesh. Thus, in practice, it is necessary to find a reasonable value of $\sigma^\star$ in advance. No theoretical procedure to tune this parameter is known to date. Some efforts have been done by Sjögreen and Petersson [44], but the dependence of σ^* with respect to the mesh has not been avoided.

We have applied the PML method with quadratic and non–integrable absorbing functions to a scattering problem with known analytical solution to compare the accuracy of the numerical results.

Consider problem (6.31)–(6.33) where the obstacle Ω is the unit circle centered at the origin. Given any inner point $\boldsymbol{x}^0 := (x_1^0, x_2^0)$ of this circle, it is well known

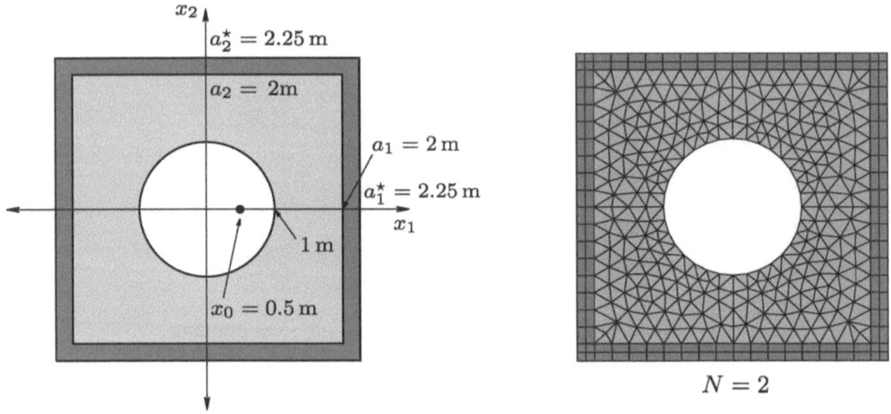

Fig. 6.10 Domains and mesh in the scattering problem.

that the function

$$p(\boldsymbol{x}) = \frac{\mathrm{i}}{4} \mathrm{H}_0^{(1)} \left(k \left| \boldsymbol{x} - \boldsymbol{x}^0 \right| \right)$$

satisfies (6.31) and (6.33). Therefore, we take $g = \partial p / \partial \boldsymbol{n}$, so that p is the unique solution of (6.31)–(6.33).

In our experiments we have taken $x_1^0 = 0.5\,\mathrm{m}$, $x_2^0 = 0$, and $k = \omega / c$, with $c = 340\,\mathrm{m/s}$ and different values of the frequency ω. For our computational domain we have taken $a_1 = a_2 = 2.0\,\mathrm{m}$ and $a_1^\star = a_2^\star = 2.25\,\mathrm{m}$, see Figure 6.10.

We have used uniform refinements of the mesh shown in Figure 6.10; the number N of elements through the thickness of the PML is used to label each mesh.

To measure the accuracy we have estimated the relative error in the L^2–norm in Ω_{F} as follows:

$$\text{Error} := \frac{\left\| p_{\mathrm{F}}^h - \Pi_h p \right\|_{\mathrm{L}^2(\Omega_{\mathrm{F}})}}{\left\| \Pi_h p \right\|_{\mathrm{L}^2(\Omega_{\mathrm{F}})}}, \tag{6.55}$$

where p_{F}^h is the numerical solution in Ω_{F} and $\Pi_h p$ is the Lagrange interpolant of the exact solution p.

In Table 6.1 we compare the errors of the PML method with absorbing functions σ_j^{U} and σ_j^{Q}. For the latter, we have used the optimal value of $\sigma^\star$, which is also reported in the table. We also include in the table the condition number κ of the system matrix for each discrete problem.

Let us remark that, in both cases, we recover an order $\mathcal{O}(h^2)$ of convergence in the L^2–norm, which is optimal for the finite elements used for the discretization.

A significant advantage of the proposed unbounded absorbing functions can be clearly appreciated from this table. This is particularly remarkable for the lowest frequencies, but the errors with the quadratic absorbing functions are larger in all cases, even though the optimal value of $\sigma^\star$ has been always used. Moreover, in spite

Table 6.1 Comparison of PML methods with unbounded and quadratic absorbing functions.

		Non–integrable σ_j^{U} (6.41)		Quadratic σ_j^{Q} (6.40)		
ω(rad/s)	Mesh	Error(%)	κ	$\sigma^\star$	Error(%)	κ
	$N=2$	0.763	6.7e+02	22.28 c	11.644	4.7e+02
250	$N=4$	0.131	5.1e+03	29.57 c	3.675	5.0e+03
	$N=8$	0.029	4.1e+04	38.37 c	1.134	4.6e+04
	$N=2$	1.700	1.1e+02	27.67 c	7.602	1.1e+02
750	$N=4$	0.447	7.0e+02	35.52 c	2.291	9.4e+02
	$N=8$	0.109	5.6e+03	43.49 c	0.698	8.2e+03
	$N=2$	6.958	2.7e+02	27.89 c	11.620	2.9e+02
1250	$N=4$	1.946	1.1e+03	36.94 c	3.336	1.7e+03
	$N=8$	0.430	9.7e+03	45.70 c	0.919	1.5e+03

of the singular character of σ_j^{U}, the condition numbers of the resulting matrices are essentially of the same order of magnitude as those of σ_j^{Q}.

On the other hand, Table 6.1 also shows that the optimal value of $\sigma^\star$ strongly depends on the problem data (the frequency ω in this case) and the mesh. The errors and the condition numbers κ would be significantly larger if values larger than the optimal $\sigma^\star$ were used.

As a conclusion, the proposed PML method with unbounded absorbing functions σ_j^{U} clearly beats the classical choice of bounded absorbing functions. Moreover, it overcomes the problem of determining optimal parameters.

Next, we report the results obtained by applying the proposed PML strategy to solve two "real life" Helmholtz problems, see [15] for other numerical experiments.

The first one consists of computing the acoustic propagation in a horizontal section of an open elliptic amphitheatre. In architectural acoustics, it is well known that in partially elliptic rooms, an acoustic source acting at one of the focus of the ellipse leads to maximum values for the pressure at the other focus.

We have modeled this by means of the domain shown in Figure 6.11. We have used a refinement of the mesh plotted in the same figure with 36214 triangles and 4352 rectangles. We have taken as wave number $k = 4\pi$ and acoustic speed $c = 340\,\mathrm{m/s}$.

Figures 6.12 and 6.13 show the wave field. We observe that there are two local maximums of the modulus of the pressure, one at each focus of the ellipse.

The last example resembles the well–known double–slit interference experiment developed by Thomas Young in 1801 to infer the wave–like nature of light. This model mimics the plane–wave excitation with two waveguides leading to slits in a screen and computes the diffraction pattern in a domain surrounding the apertures.

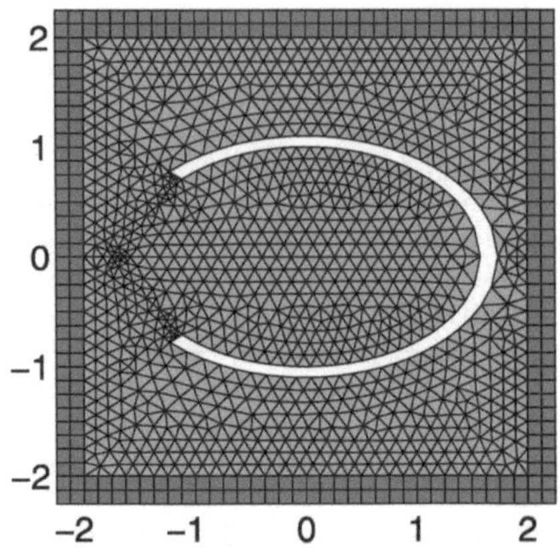

Fig. 6.11 Domain and coarse mesh for the elliptic section.

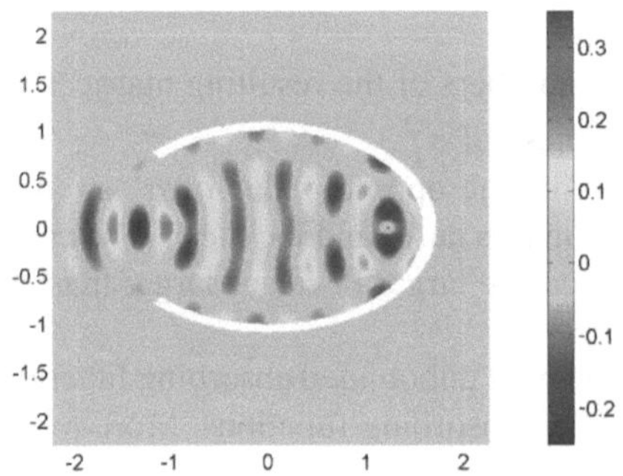

Fig. 6.12 Real part of the pressure field generated by an excitation at one focus of the ellipse.

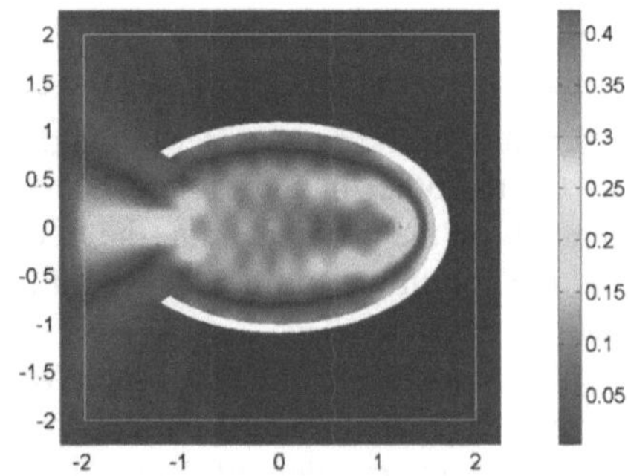

Fig. 6.13 Modulus of the pressure field generated by an excitation at one focus of the ellipse.

We have used the domain shown in Figure 6.14 and a refinement of the mesh plotted in the same figure with 51424 triangles and 6656 rectangles. We have taken as wave number $k = 8\pi$ and acoustic speed $c = 340\,\mathrm{m/s}$.

Figures 6.15 and 6.16 show the wave field in a neighborhood of the double–slit.

6.4 Time–harmonic Radial PML Equations

Radial PML methods are based on simulating dissipation in an annular domain, $\Omega_{\mathrm{A}} := \{\boldsymbol{x} \in \mathbb{R}^2 : R < |\boldsymbol{x}| < R^\star\}$, surrounding the physical domain of interest, Ω_{F}, as shown in Figure 6.17. Let us remark that, now, the unit normal vector $\boldsymbol{\nu}$ on Γ_{I} coincides with the radial canonical basis vector.

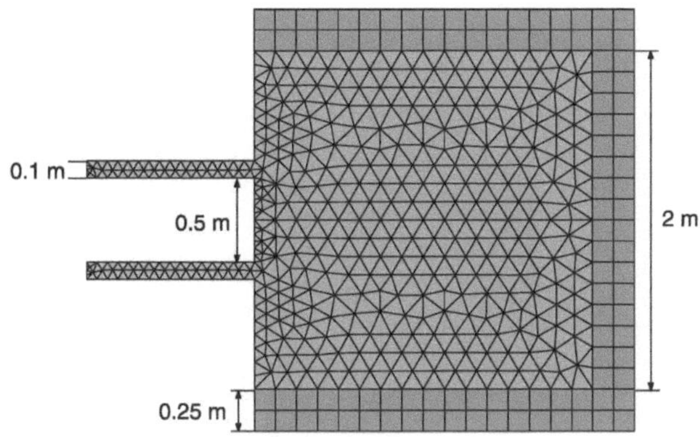

Fig. 6.14 Domain and coarse mesh for the double–slit interference test.

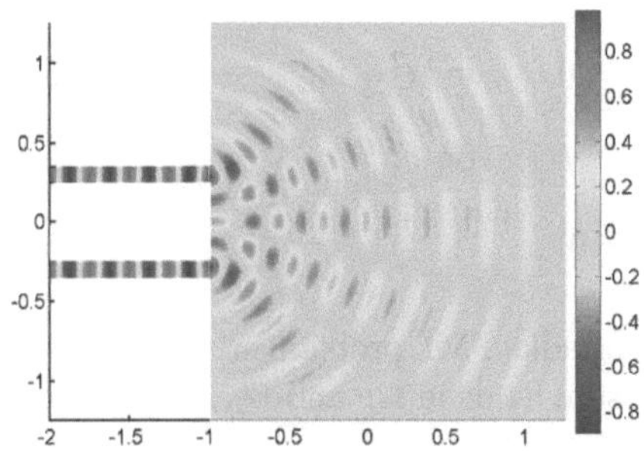

Fig. 6.15 Real part of the pressure field generated by an excitation of the wave-guides.

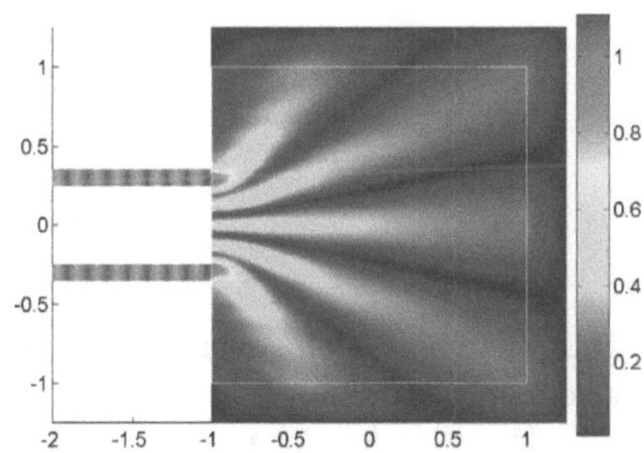

Fig. 6.16 Modulus of the pressure field generated by an excitation of the wave-guides.

The construction of the radial PML equation is done by means of a complex–valued radial stretching proposed by Collino and Monk [19], which leads to the following partial differential equation written in polar coordinates for the pressure $\widehat{p}$ in the PML domain:

$$-\frac{1}{r}\left\{\frac{\partial}{\partial r}\left[\frac{\widehat{\gamma}(r)r}{\gamma(r)}\frac{\partial \widehat{p}}{\partial r}\right]+\frac{\gamma(r)}{\widehat{\gamma}(r)r}\frac{\partial^2 \widehat{p}}{\partial \theta^2}\right\}-\gamma(r)\widehat{\gamma}(r)k^2\widehat{p}=0 \qquad \text{in } \Omega_{\mathrm{A}},$$

where

$$\gamma(r) := 1+\frac{\mathrm{i}}{\omega}\sigma(r) \qquad \text{and} \qquad \widehat{\gamma}(r) := 1+\frac{\mathrm{i}}{\omega r}\int_R^r \sigma(s)\,ds,$$

with the absorbing function σ being non negative, monotonically increasing and smooth in $[R, R^\star)$.

We consider now a scattering problem with Dirichlet data on the boundary of the obstacle:

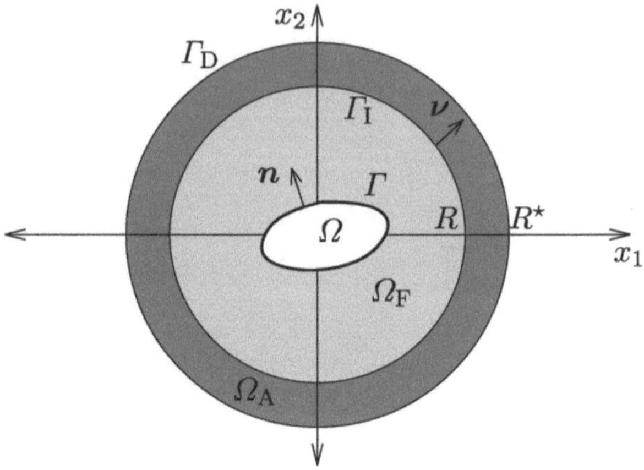

Fig. 6.17 Radial PML on a two–dimensional domain.

$$-\Delta p - k^2 p = 0 \qquad \text{in } \mathbb{R}^2 \setminus \bar{\Omega}, \tag{6.56}$$

$$p = f \qquad \text{on } \Gamma, \tag{6.57}$$

$$\lim_{r\to\infty} \sqrt{r}\left(\frac{\partial p}{\partial r} - \mathrm{i}kp\right) = 0. \tag{6.58}$$

The equations which govern the coupled fluid/radial PML problem are the following, where once more p_F and p_A denote the pressure fields in Ω_F and Ω_A, respectively:

$$-\Delta p_F - k^2 p_F = 0 \qquad \text{in } \Omega_F, \tag{6.59}$$

$$-\frac{1}{r}\left[\frac{\partial}{\partial r}\left(\frac{\widehat{\gamma} r}{\gamma}\frac{\partial p_A}{\partial r}\right) + \frac{\gamma}{\widehat{\gamma} r}\frac{\partial^2 p_A}{\partial \theta^2}\right] - \gamma\widehat{\gamma}k^2 p_A = 0 \qquad \text{in } \Omega_A, \tag{6.60}$$

$$p_F = f \qquad \text{on } \Gamma, \tag{6.61}$$

$$p_F = p_A \qquad \text{on } \Gamma_I, \tag{6.62}$$

$$\frac{\partial p_F}{\partial r} = \frac{1}{\gamma}\frac{\partial p_A}{\partial r} \qquad \text{on } \Gamma_I, \tag{6.63}$$

$$p_A = 0 \qquad \text{on } \Gamma_D. \tag{6.64}$$

Contrary to the case of Cartesian PML, for the radial PML there exist theoretical results about existence and uniqueness as well as convergence, when the absorbing function is bounded.

In the following theorem we collect the results derived by Bramble and Pasciak [16] for three–dimensional problems, under some specific assumptions on the absorbing function. In particular, $\sigma(r)$ is assumed to be constant for r large enough, see [16] for details.

Theorem 5. *Let σ be sufficiently smooth and such that $\sigma(R) = \sigma'(R) = 0$ and $\sigma(r) = \sigma_0$ (constant) for r large enough. For $f \in \mathrm{H}^{\frac{1}{2}}(\Gamma)$ and $R^\star$ sufficiently large,* *there exists a unique solution* $(p_{\mathrm{F}}, p_{\mathrm{A}}) \in \mathrm{H}^1(\Omega_{\mathrm{F}}) \times \mathrm{H}^1(\Omega_{\mathrm{A}})$ *of the coupled problem (6.59)–(6.64). Moreover,*

$$\|p - p_{\mathrm{F}}\|_{\mathrm{L}^2(\Omega_{\mathrm{F}})} \leq C\mathrm{e}^{-2\sigma_0 k R^\star} \|f\|_{\mathrm{H}^{\frac{1}{2}}(\Gamma)},$$

where p *is the solution of the scattering problem (6.56)–(6.58).*

On the other hand, we have recently analyzed in [12] the existence and uniqueness of solution for the coupled fluid/radial PML problem when a non–integrable function is used. With this purpose we have introduced the non–standard weighted Sobolev space

$$\mathcal{V} := \left\{ v \in \mathrm{H}^1_{\mathrm{loc}}(\Omega_{\mathrm{A}}) : \|v\|_{\mathcal{V}} < \infty \right\},$$

where

$$\|v\|^2_{\mathcal{V}} := \int_R^{R^\star} \int_{-\pi}^{\pi} \left| \frac{\widehat{\gamma}(r) r}{\gamma(r)} \right| \left| \frac{\partial v}{\partial r} \right|^2 d\theta\, dr + \int_R^{R^\star} \int_{-\pi}^{\pi} \left| \frac{\gamma(r)}{\widehat{\gamma}(r) r} \right| \left| \frac{\partial v}{\partial \theta} \right|^2 d\theta\, dr$$
$$+ \int_R^{R^\star} \int_{-\pi}^{\pi} |\widehat{\gamma}(r)\gamma(r) r| \, |v|^2 \, d\theta\, dr.$$

We have proved in [12] the exactness of the coupled problem when the absorbing function is not integrable, with the unique restriction of this absorbing function being smooth enough in the PML domain. We state this result in the following theorem, see [12] for details:

Theorem 6. *Let* σ *be sufficiently smooth and such that* $\int_R^{R^\star} \sigma(s)\, ds = \infty$. *For* $f \in \mathrm{H}^{\frac{1}{2}}(\Gamma)$, *there exists a unique solution* $(p_{\mathrm{F}}, p_{\mathrm{A}}) \in \mathrm{H}^1(\Omega_{\mathrm{F}}) \times \mathcal{V}$ *of the coupled problem (6.59)–(6.63). Moreover,* p_{F} *coincides in* Ω_{F} *with the solution* p *of the scattering problem (6.56)–(6.58).*

Let us remark that although the previous theorem does not claim that (6.64) is satisfied, a kind of homogeneous Dirichlet boundary condition on the outer boundary of the PML domain Ω_{A} is implicitly contained in the definition of the space $\mathcal{V}$. This boundary condition will be enforced in the discretization of the coupled problem.

6.4.1 Numerical Results

In what follows, we introduce a standard finite element discretization of the weak form of problem (6.59)–(6.64). For this purpose, we use meshes $\mathcal{T}_h$ of curved elements which correspond to standard quadrilaterals in polar coordinates. As usual, h denotes the mesh–size. Each element must be completely contained either in $\bar{\Omega}_{\mathrm{F}}$ or in $\bar{\Omega}_{\mathrm{A}}$. We use isoparametric bilinear elements in polar coordinates. Moreover, we take advantage of the fact that Ω_{A} is an annular domain by using curved rectangles in Ω_{A}. Thus, the finite–element space is

$$\mathcal{H}_h := \big\{(q_{\mathrm{F}}^h, q_{\mathrm{A}}^h) \in \mathcal{C}(\Omega_{\mathrm{F}}) \times \mathcal{C}(\Omega_{\mathrm{A}}) : \; q_{\mathrm{A}}^h = 0 \text{ on } \Gamma_{\mathrm{D}}, \; q_{\mathrm{F}}^h = q_{\mathrm{A}}^h \text{ on } \Gamma_{\mathrm{I}},$$
$$q_{\mathrm{F}}^h|_K, \, q_{\mathrm{A}}^h|_K \text{ isoparametric bilinear in } K \;\; \forall K \in \mathcal{T}_h \big\}.$$

The boundary condition $q_{\mathrm{A}}^h = 0$ on Γ_{D} in the definition of $\mathcal{H}_h$ turns out necessary for the inclusion $\mathcal{H}_h \subset \mathrm{H}^1(\Omega_{\mathrm{F}}) \times \mathcal{V}$ to hold. For a non–integrable absorbing function σ, as for instance that in (6.65) below, this boundary condition is also sufficient, see [14] for other feasible choices of σ.

Let f_h be a convenient approximation of f. The discrete weak problem associated with the coupled fluid/PML consists in finding $(p_{\mathrm{F}}^h, p_{\mathrm{A}}^h) \in \mathcal{H}_h$ such that $p_{\mathrm{F}}^h = f_h$ on Γ and

$$\int_{\Omega_{\mathrm{F}}} \mathbf{grad}\, p_{\mathrm{F}}^h \cdot \mathbf{grad}\, \bar{q}_{\mathrm{F}}^h + \int_{\Omega_{\mathrm{A}}} \left(\frac{\widehat{\gamma}}{\gamma} \frac{\partial p_{\mathrm{A}}^h}{\partial r} \frac{\partial \bar{q}_{\mathrm{A}}^h}{\partial r} + \frac{\gamma}{\widehat{\gamma} r^2} \frac{\partial p_{\mathrm{A}}^h}{\partial \theta} \frac{\partial \bar{q}_{\mathrm{A}}^h}{\partial \theta} \right)$$
$$- k^2 \left(\int_{\Omega_{\mathrm{F}}} p_{\mathrm{F}}^h \bar{q}_{\mathrm{F}}^h + \int_{\Omega_{\mathrm{A}}} \gamma \widehat{\gamma} p_{\mathrm{A}}^h \bar{q}_{\mathrm{A}}^h \right) = 0,$$

for all $(q_{\mathrm{F}}^h, q_{\mathrm{A}}^h) \in \mathcal{H}_h$ with $q_{\mathrm{F}}^h = 0$ on Γ.

In what follows we report some numerical results obtained with a computer code implementing the perfectly matched layer method with the non–integrable absorbing function

$$\sigma^{\mathrm{U}}(s) := \frac{c}{R^\star - s}, \qquad s \in [R, R^\star). \tag{6.65}$$

To illustrate the performance of this approach, we use again the fundamental solution of the Helmholtz problem

$$p(\boldsymbol{x}) := \frac{\mathrm{i}}{4} \mathrm{H}_0^{(1)}(k\,|\boldsymbol{x}|),$$

to obtain a scattering problem with known analytical solution p, by taking $f := p|_\Gamma$ as the Dirichlet data of problem (6.56)–(6.58).

In this numerical experiment we have taken $k = \omega / c$ with $c = 340\,\mathrm{m/s}$ and $\omega = 750\,\mathrm{rad/s}$. We have considered a circular obstacle of radius $R_\star = 1\,\mathrm{m}$ and the following data for the PML layer: $R = 2\,\mathrm{m}$ and $R^\star = 2.1\,\mathrm{m}$, see Figure 6.18.

Because of the boundary condition $p_{\mathrm{A}}^h = 0$ on Γ_{D}, all the integrals involved in the finite element method are finite, see [12, 14] for details. To evaluate these integrals, we have used a Simpson adaptive rule to reduce the effect of the numerical errors arising from quadrature as much as possible. However, it is shown in [14] that standard quadrature rules lead to numerical results essentially of the same accuracy.

We have used uniform refinements of the mesh shown in Figure 6.18; the number N of elements through the thickness of the PML is used to label each mesh. Specifically, meshes corresponding to $N = 2$, 4 and 8, have 264, 1008 and 3936 degrees of freedom, respectively.

In Figure 6.19 we show the real and imaginary parts of the solution computed for the fluid/PML coupled problem with the mesh corresponding to $N = 8$. The solution is plotted in the fluid domain and in the PML, as well.

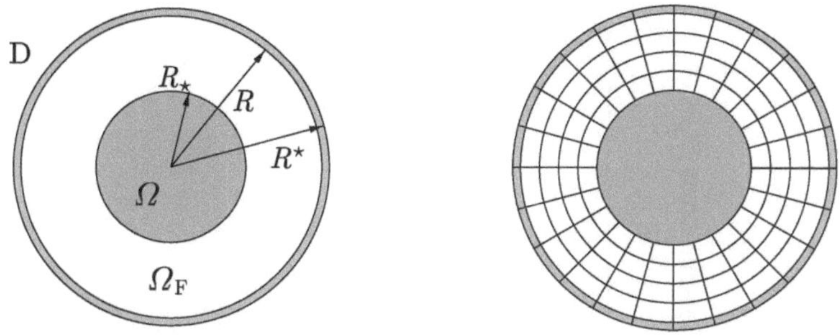

Fig. 6.18 Domains and mesh ($N = 1$) in the scattering problem.

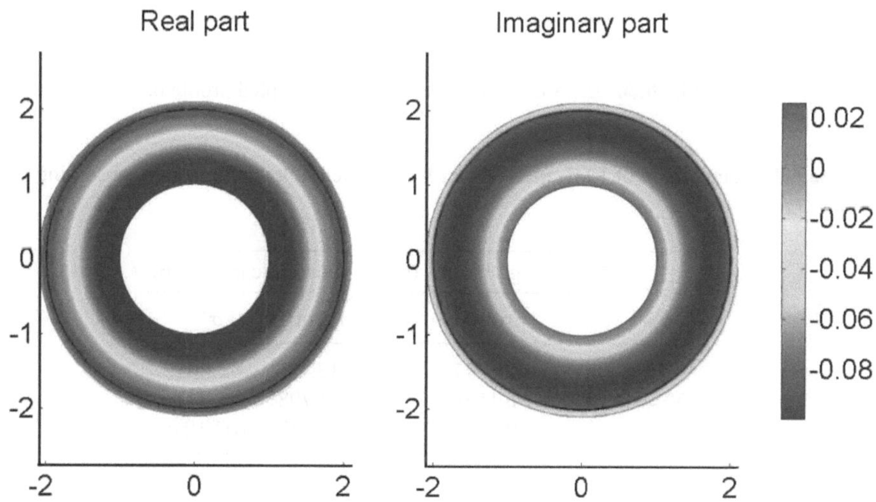

Fig. 6.19 Solution of the fluid/PML coupled problem. Mesh $N = 8$

To assess the order of convergence of the proposed numerical method, we show in Figure 6.20 the error curve computed in the fluid domain (log–log plot of error (6.55) versus mesh–size). It can be seen from this figure that an order of convergence $\mathcal{O}(h^2)$ is again achieved. Let us recall that this is the optimal order in L^2–norm for the finite elements used for the discretization.

To end this section, we compare the numerical performance of this technique with that of a PML based on the quadratic function

$$\sigma^{Q}(s) := \sigma^\star \frac{(s-R)^2}{(R^\star - R)^2}, \qquad s \in [R, R^\star]. \tag{6.66}$$

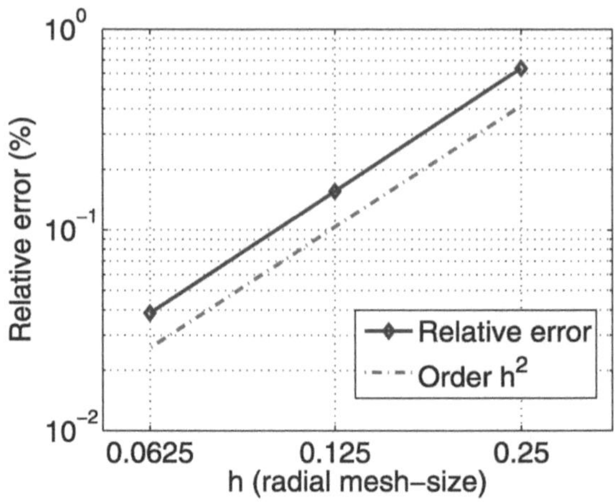

Fig. 6.20 Error curve for the fluid/PML coupled problem.

Table 6.2 Comparison of PML methods with non–integrable and quadratic absorbing functions.

	Non–integrable σ^{U} (6.65)	Quadratic σ^{Q} (6.66)	
Mesh	Error(%)	Error(%)	$\sigma^\star$
$N = 2$	0.6377	13.7867	61.56 c
$N = 4$	0.1558	3.7986	84.84 c
$N = 8$	0.0386	1.1373	107.12 c

Recall that, for a given problem and a given mesh, there is an optimal value of $\sigma^\star$ leading to minimal errors, which depends strongly on the problem data and the particular mesh. This optimal $\sigma^\star$ has to be tuned in advance.

In Table 6.2, we compare the errors of both PML methods applied to the same test problem as above. For the quadratic absorbing function, we have used the optimally tuned value of $\sigma^\star$, which is also reported in the table. It can be seen again that this optimal value changes significantly from one mesh to another.

Table 6.2 shows that the errors of the PML method with the non–integrable absorbing function are once more significantly smaller than those of the classical PML technique. On the other hand, another benefit of our proposed PML method is that there is no need of fitting any non–physical parameter.

6.5 Further Research

One of the most challenging topics in the context of time–harmonic Cartesian PML is the proof of error estimates for the discrete finite element solution of the coupled fluid/PML problem. These kind of estimates have an inherent difficulty in the case of non–integrable absorbing functions, since they involve non–standard weighted Sobolev spaces similar to that presented in Section 6.4.

Another computational challenge is to solve fluid/structure problems in unbounded domains, where the elastic solid and the coupling interfaces between fluid and solid are unbounded and, hence, they have to be truncated by the PML.

On the other hand, there are several open problems in the context of time–domain PML, for instance, to obtain energy estimates for non–constant absorbing functions, even in the case of classical bounded ones. In this context, some numerical aspects of the stability of finite difference schemes started to be well understood only recently.

The mathematical analysis of the PML technique in the time domain using non–integrable absorbing functions is also an open research line. Some preliminary results show that the exact solution of the time–domain scattering problem is recovered. Existence and uniqueness of solution have been also recently proved for a pressure/velocity formulation of this problem by using weighted energy norms and the Fourier–Laplace transform. Moreover, the time–domain implementation does not require any extra computational cost and leads to better numerical results than classical PML techniques with bounded absorbing functions.

Acknowledgments

A. Bermúdez, L. Hervella–Nieto and A. Prieto were partially supported by MEC (Spain) research project DPI2004–05504–C02–02 and Xunta de Galicia project PGIDIT05PXIC20705PN (Spain). L. Hervella–Nieto was also partially funded by MEC (Spain) research project MTM2004–05796–C02–01. R. Rodríguez was partially funded by FONDAP in Applied Mathematics (Chile).

References

1. Abarbanel S, Gottlieb D (1997) A mathematical analysis of the PML method. Journal of Computational Physics 134:357–363
2. Abarbanel S, Gottlieb D, Hesthaven JS (1999) Well–posed perfectly matched layers for advective acoustics. Journal of Computational Physics 154:266–283
3. Astley RJ (2000) Infinite elements for wave problems: A review of current formulations and an assessment of accuracy. International Journal for Numerical Methods in Engineering 49:951–976
4. Asvadurov S, Druskin V, Guddati MN, Knizhnerman L (2003) On optimal finite–difference approximation of PML. SIAM Journal on Numerical Analysis 41:287–305
5. Basu U, Chopra AK (2003) Perfectly matched layers for time–harmonic elastodynamics of unbounded domains: theory and finite–element implementation. Computer Methods in Applied Mechanics and Engineering 192:1337–1375

6. Bayliss A, Turkel E (1980) Radiation boundary conditions for wave–like equations. Communications on Pure and Applied Mathematics 33:707–725
7. Bécache E, Bonnet–Benn Dhia AS, Legendre G (2004) Perfectly matched layers for the convected Helmholtz equation. SIAM Journal on Numerical Analysis 42:409–433
8. Bécache E, Joly P (2002) On the analysis of Berenger's perfectly matched layers for Maxwell's equations. ESAIM: Mathematical Modelling and Numerical Analysis – Modélisation Mathématique et Analyse Numérique 36:87–119
9. Berenger JP (1994) A perfectly matched layer for the absorption of electromagnetic waves. Journal of Computational Physics 114:185–200
10. Berenger JP (1996) Perfectly matched layer for the FDTD solution of wave–structure interaction problems. IEEE Transactions Antennas Propagation 44:110–117
11. Berenger JP (1996) Three–dimensional perfectly matched layer for the absorption of electromagnetic waves. Journal of Computational Physics 127:363–379
12. Bermúdez A, Hervella–Nieto L, Prieto A, Rodríguez R (2007) An exact bounded perfectly matched layer for time–harmonic scattering problems. SIAM Journal on Scientific Computing, accepted for publication.
13. Bermúdez A, Hervella–Nieto L, Prieto A, Rodríguez R (2004) An exact bounded PML for the Helmholtz equation. Comptes Rendus de l'Académie des Sciences – Série I – Mathématiques 339:803–808
14. Bermúdez A, Hervella–Nieto L, Prieto A, Rodríguez R (2007) An optimal perfectly matched layer with unbounded absorbing function for time–harmonic acoustic scattering problems. Journal of Computational Physics 223:469–488
15. Bermúdez A, Hervella–Nieto L, Prieto A, Rodríguez R (2007) Validation of acoustic models for time harmonic dissipative scattering problems. Journal of Computational Acoustics 15:95–121
16. Bramble JH, Pasciak JE (2007) Analysis of a finite PML approximation for the three dimensional time–harmonic Maxwell and acoustic scattering problems. Mathematics of Computation 76:597–614
17. Chew WC, Weedon WH (1994) A 3D perfectly matched medium from modified Maxwell's equations with stretched coordinates. IEEE Microwave and Optical Technology Letters 7:599–604
18. Collino F, Monk P (1998) Optimizing the perfectly matched layer. Computer Methods in Applied Mechanics and Engineering 164:157–171
19. Collino F, Monk P (1998) The perfectly matched layer in curvilinear coordinates. SIAM Journal on Scientific Computing 19:2061–2090
20. Collino F, Tsogka C (2001) Application of the PML absorbing layer model to the linear elastodynamic problem in anisotropic heterogeneous media. Geophysics 66:294–307
21. Demkowicz L, Ihlenburg F (2001) Analysis of a coupled finite–infinite element method for exterior Helmholtz problems. Numerische Mathematik 88:43–73
22. Diaz J, Joly P (2006) A time domain analysis of PML models in acoustics. Computer Methods in Applied Mechanics and Engineering 195:3820–3853
23. Engquist B, Majda A (1977) Absorbing boundary conditions for numerical simulation of waves. Mathematics of Computation 31:629–651
24. Feng K (1983) Finite element method and natural boundary reduction. In: Proceedings of the International Congress of Mathematicians, Vol 1, 2 (Warsaw, 1983), 1439–1453, PWN, Warsaw, 1984.
25. Givoli D (1991) Non–reflecting boundary conditions. Journal of Computational Physics 94:1–29
26. Givoli D (1992) Numerical methods for problems in infinite domains. Elsevier, Amsterdam

27. Givoli D, Neta B (2004) High–order non–reflecting boundary scheme for time–dependent waves. Journal of Computational Physics 186:24–46
28. Harari I, Slavutin M, Turkel E (2000) Analytical and numerical studies of a finite element PML for the Helmholtz equation. Journal of Computational Acoustics 8:121–137
29. Hohage T, Schmidt F, Zschiedrich L (2004) Solving time–harmonic scattering problems based on the pole condition. II: Convergence of the PML Method. SIAM Journal on Mathematical Analysis 35:547–560
30. Hu FQ (2001) A stable, perfectly matched layer for linearized Euler equations in unsplit physical variables. Journal of Computational Physics 173:455–480
31. Ihlenburg F (1998) Finite element analysis of acoustic scattering. Springer–Verlag, New York
32. Komatitsch D, Tromp J (2003) A perfectly matched layer absorbing boundary condition for the second–order seismic wave equation. Geophysical Journal International 154:146–153
33. Kreiss HO, Lorenz J (1989) Initial–boundary problems and the Navier–Stokes equation. Academic Press, New York
34. Lassas M, Liukkonen J, Somersalo E (2001) Complex Riemannian metric and absorbing boundary conditions. Journal de Mathématiques Pures et Appliquées 80:739–768
35. Lassas M, Somersalo E (1998) On the existence and convergence of the solution of PML equations. Computing 60:229–242
36. Lassas M, Somersalo E (2001) Analysis of the PML equations in general convex geometry. Proceedings of the Royal Society of Edinburgh. Section A. Mathematics 131:1183–1207
37. Lions JL, Métral J, Vacus O (2002) Well–posed absorbing layer for hyperbolic problems. Numerical Mathematics 92:535–562
38. Masmoudi M (1987) Numerical solution for exterior problems. Numerical Mathematics 51:87–101
39. Navon IM, Neta B, Hussaini MY (2004) A perfectly matched layer approach to the linearized shallow water equations models. Monthly Weather Review 132:1369–1378
40. Nédélec JC (2000) Acoustic and electromagnetic equations. Integral representations for harmonic problems. Springer–Verlag, New York
41. Qi Q, Geers TL (1998) Evaluation of the perfectly matched layer for computational acoustics. Journal of Computational Physics 139:166–183
42. Rappaport CM (1996) Interpreting and improving the PML absorbing boundary condition using anisotropic lossy mapping of space. IEEE Transactions on Magnetics 32:968–974
43. Shirron JJ, Babuska I (1998) A comparison of approximate boundary conditions and infinite element methods for exterior Helmholtz problems. Computer Methods in Applied Mechanics and Engineering 164:121–139
44. Sjögreen B, Petersson NA (2005) Perfectly matched layers for Maxwell's equations in second order formulation. Journal of Computational Physics 209:19–46
45. Teixeira FL, Chew WC (1998) Extension of the PML absorbing boundary condition to 3D spherical coordinates: scalar case. IEEE Transactions on Magnetics 34:2680–2683
46. Turkel E, Yefet A (1998) Absorbing PML boundary layers for wave–like equations. Applied Numerical Mathematics 27:533–557
47. van Joolen VJ, Neta B, Givoli D (2005) High–order Higdon–like boundary conditions for exterior transient wave problems. International Journal for Numerical Methods in Engineering 63:1041–1068
48. Zeng YQ, HeJQ, Liu, QH (2006) The application of the perfectly matched layer in numerical modeling of wave propagation in poroelastic media. Geophysics 66:1258–1266

49. Zhao L, Cangellaris AC (1996) A general approach for the development of unsplit–field time–domain implementations of perfectly matched layers for FDTD grid truncation. IEEE Microware and Guided Wave Letters 6:209–211

7

Infinite Elements

R Jeremy Astley

Institute of Sound and Vibration Research, University of Southampton, Highfield, Southampton, SO17 1BJ, United Kingdom
rja@isvr.soton.ac.uk

Summary. Infinite elements are used to represent the effect of far field radiation on unbounded finite element acoustic models. They have several advantages over alternative boundary treatments for such problems. For example, they provide a direct numerical estimate of the solution at all points in the outer domain and by using elements of arbitrary radial order they give an anechoic boundary condition which is accurate to an arbitrary order of approximation. In this chapter infinite elements are introduced with reference to a simple one–dimensional formulation and then extended to two and three dimensions. The application of the method to transient problems is also touched upon.

7.1 Preliminaries

The solution of the acoustic wave equation on an unbounded domain presents particular problems for computation. When conventional grid–based or element–based methods are used, they must of necessity be applied to a *bounded* computational region which contains a discrete number of grid points or nodes. This implies the existence of a truncation surface through which acoustic energy must propagate without being reflected back into the computational domain.

A generic exterior problem of this type is illustrated in Figure 7.1(a). Here the sound field in an unbounded domain Ω which lies outside a radiating or scattering surface Γ, is truncated at a computational boundary Γ_c. The computational domain which lies between the surfaces Γ and Γ_c is denoted by Ω_c. The development of "anechoic" boundary conditions which can be applied on Γ_c and which are effective for arbitrary radiative fields poses a major challenge for domain–based numerical methods.

In this chapter the Infinite Element (IE) concept is introduced as a practical technique for treating such problems. Conventional Finite Element (FE) models will be used to represent the solution within the inner domain Ω_c. A variety of other methods currently exist for terminating such models. Local Non–Reflecting Boundary Conditions (NRBCs) are perhaps the most common. These are applied on the boundary itself and generally involve the normal and/or tangential derivative(s) of the solution [17, 18]. Non–local NRBCs are also well developed for such problems. Galerkin

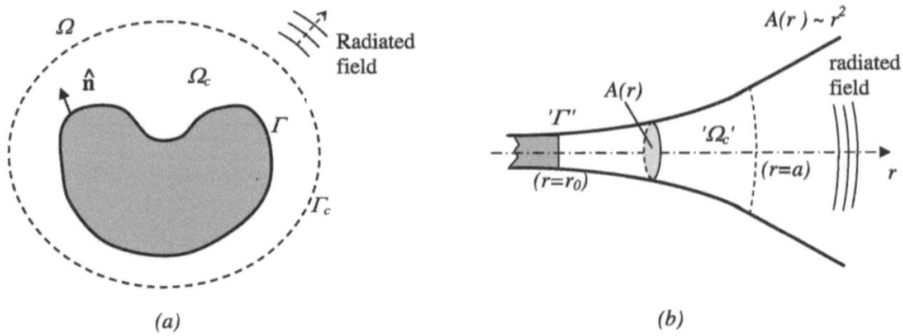

Fig. 7.1 (a) The exterior Helmholtz problem. (b) The acoustic horn.

mode–matching and the closely related FE–Dirichlet–to–Neumann (FE–DtN) approach [2] are well known examples. Such methods generally involve integration over the truncation boundary and are less appealing for large 3D models where it is important to exploit fully the sparsity of the FE matrices.

Alternatively, the computational domain can be extended beyond the truncation boundary to include an "absorbing" or "perfectly matched" layer within which a non–physical operator is used to progressively remove the outgoing disturbance [19] so that it passes through the truncation boundary without significant reflection.

In all of these approaches, the location of the truncation surface determines the size of the numerical problem in the inner domain, and hence the computational effort that is required to generate a solution. It also influences strongly the effectiveness of the truncation boundary. As a general rule, the more distant the truncation boundary from the source, the more accurate is the non–reflecting boundary condition or absorbing layer. On the other hand, the demands of mesh resolution and pollution error [15] in the inner domain make it desirable to *reduce* rather than *increase* the size of this region. An ideal treatment at the truncation boundary is therefore one that can be applied with great accuracy as close as possible to the radiating or scattering surface.

The Infinite Element (IE) method tackles the problem of providing an accurate anechoic termination in a way which differs radically from the methods reviewed above. It resembles a local NRBC to the extent that it provides a local high order anechoic termination to an inner FE model, but is similar also to an absorbing layer in that it relies on an extension of the domain of the numerical solution beyond the truncation boundary of the FE mesh. Unlike an absorbing layer however, the IE solution is not dissipative but propagates the solution accurately to the far field. In effect, the IE method avoids the issue of truncation altogether by extending the numerical discretization to an *infinite* boundary. In most IE formulations, moreover, the acoustic pressure at all exterior points however distant from the source can be recovered directly from the discrete FE/IE solution. This is not the case in the other methods mentioned above, for which far field data must be obtained from a secondary calculation, usually by integrating the computed solution over a near field Kirchhoff surface.

To understand the logic behind the various IE formulations which will be presented in the remainder of this chapter, it is necessary first to define the equations and boundary conditions which govern an exterior problem in acoustics.

7.2 The Unbounded Helmholtz Problem

7.2.1 The Governing Equations

The unknown, time dependent acoustic pressure in an exterior domain Ω, as shown in Figure 7.1(a), will be denoted by $\tilde{p}(\boldsymbol{x}, t)$. In the case of a radiation problem $\tilde{p}$ is the total field. In the case of a scattering problem it is the scattered component. In either event $\tilde{p}$ is a solution of the inviscid, linearised Euler equations for a compressible fluid. In the absence of mean flow, $\tilde{p}$ is a solution of the homogeneous wave equation,

$$\nabla^2 \tilde{p} = \frac{1}{c^2} \frac{\partial^2 \tilde{p}}{\partial t^2} \quad \text{in} \quad \Omega, \tag{7.1}$$

where c is the local sound speed. For steady time–harmonic excitation at radian frequency ω, $\tilde{p}$ can be written

$$\tilde{p}(\boldsymbol{x}, t) \;=\; \Re\left\{ p(\boldsymbol{x}, \omega)\, \mathrm{e}^{\mathrm{i}\omega t} \right\} \tag{7.2}$$

and the complex amplitude $p(\boldsymbol{x}, \omega)$ then satisfies the Helmholtz equation

$$\nabla^2 p + k^2 p = 0 \quad \text{in} \quad \Omega, \tag{7.3}$$

where $k = \omega / c$. If the case of a radiation problem, for which the acoustic field is generated by a prescribed normal surface acceleration $a_n(\boldsymbol{x}, \omega)$ on Γ, the appropriate boundary condition which must be satisfied on Γ is

$$\nabla p \cdot \boldsymbol{n} \;=\; -\rho\, a_n \quad \text{on} \quad \Gamma, \tag{7.4}$$

where $\boldsymbol{n}$ is an outward unit normal, as shown in Figure 7.1(a), and ρ is the ambient fluid density. A *scattering* problem can be specified without loss of generality by posing an equivalent radiation problem for the scattered component.

7.2.2 The Sommerfeld Radiation Condition

The requirement that only outwardly propagating components are present at large distances from the radiating surface must be imposed on unbounded time harmonic solutions to satisfy causality. This is assured provided that the Sommerfeld radiation condition is satisfied [41], *viz*

$$r^{\alpha} \left\{ \frac{\partial p}{\partial r} + \mathrm{i}kp \right\} \to 0 \quad \text{as} \quad r \to \infty, \tag{7.5}$$

where $\alpha = 1$ and $\alpha = \frac{1}{2}$ for three–dimensional and two–dimensional problems respectively. In all that follows, we consider only three–dimensional (or axisymmetric) problems, for which $\alpha = 1$.

The Sommerfeld condition be can approximated by removing the limit from Equation (7.5) and equating the left hand side to zero at large distances from the source. This gives an impedance–type condition of the form

$$\frac{\partial p}{\partial r} = -ikp \quad \text{at} \quad r = R, \tag{7.6}$$

where R is large. In physical terms equation (7.6) represents a "plane damper" or "ρc impedance" in acoustic terminology. It tells us that at sufficiently large distances from the radiator, the unbounded solution behaves locally like a plane wave propagating radially outwards with impedance ρc. This is a primitive form of NRBC but is of limited practical use since it holds only at large distances from the radiating surface and requires a large computational domain.

7.3 Infinite Elements. A One–dimensional Tutorial

The infinite element concept for acoustics is introduced first in application to the acoustic horn equation, a simple one–dimensional analogy for two and three dimensional acoustic radiation.

7.3.1 The Acoustic Horn

At low frequencies, plane wave propagation can be assumed in a variable area duct. The acoustic pressure amplitude then depends only on the axial distance r and satisfies the one–dimensional horn equation

$$\frac{1}{A(r)}\left\{\frac{d}{dr}\left[A(r)\frac{dp}{dr}\right]\right\} + k^2 p = 0 \tag{7.7}$$

where $A(r)$ is the cross–sectional area of the horn. A physical representation of the horn problem is shown in Figure 7.1(b). If the sound field is driven by a piston at $r = r_0$, then

$$\frac{dp}{dr} = -\rho\, a_r \quad at \quad r = r_0, \tag{7.8}$$

where a_r is the complex axial acceleration of the piston. In the case of a conical horn, $A(r) \sim r^2$ as $r \to \infty$, and the asymptotic solution of Equation (7.7) consists of outward and inward propagating "spherical" waves

$$p(r) = \frac{A^+}{r}\mathrm{e}^{-ikr} + \frac{A^-}{r}\mathrm{e}^{+ikr}. \tag{7.9}$$

The outward propagating component, $\frac{A^+}{r}\mathrm{e}^{-ikr}$, satisfies the three–dimensional Sommerfeld condition, Equation (7.5),

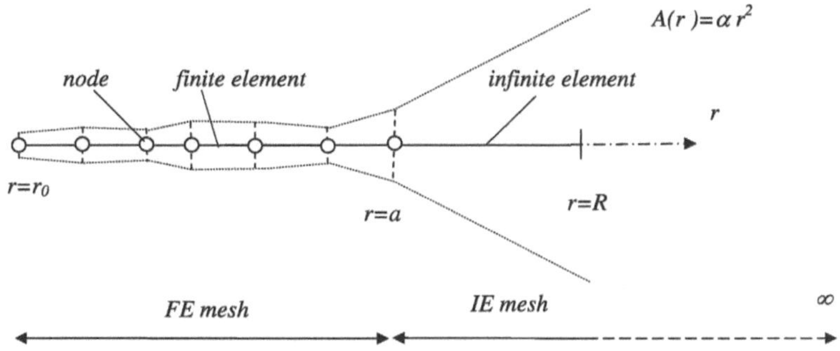

Fig. 7.2 Computational mesh for the horn problem.

$$r\left\{\frac{dp}{dr} + \mathrm{i}kp\right\} \to 0 \qquad as \qquad r \to \infty. \tag{7.10}$$

The horn problem posed by Equation (7.7) and satisfying conditions (7.8) and (7.10), is therefore analogous to the multi–dimensional Helmholtz problem posed by Equation (7.3) and satisfying conditions (7.4) and (7.5).

7.3.2 The Mesh

A discrete model for the horn is formed by subdividing the unbounded interval $[r_0, \infty]$ at $r = a$ into a finite inner domain, $[r_0, a)$, and an unbounded outer domain $[a, \infty]$ as indicated in Figure 7.2. The inner domain is modelled by an arbitrary mesh of two–noded finite elements lying along the r axis. The outer domain is represented by a single "infinite" element. An analogy can clearly be drawn between the boundary Γ_c which defines the extent of an inner finite element computational domain in the multidimensional problem, and the section $r = a$ which performs the same function in the horn problem.

In the first instance, the "infinite" element which occupies the region $r \geq a$ will be truncated at $r = R$, and an approximate "ρc" impedance applied in place of the Sommerfeld condition

$$\frac{dp}{dr} = -\mathrm{i}kp \qquad at \qquad r = R. \tag{7.11}$$

This replaces the Sommerfeld condition in the initial stages of the numerical formulation. The limit as $R \to \infty$ is taken only as a last step in the formulation *after* the discrete equations have been formed. This ensures that contributions to these equations from the far field boundary are correctly accounted for as "$R \to \infty$" [6].

7.3.3 The Variational Statement

A weak variational statement of the horn problem is obtained by multiplying equation (7.7) by a test function, $w(r, \omega)$, and integrating from $r = r_0$ to $r = R$. Integration by parts and the substitution of boundary conditions (7.8) and (7.11) then gives

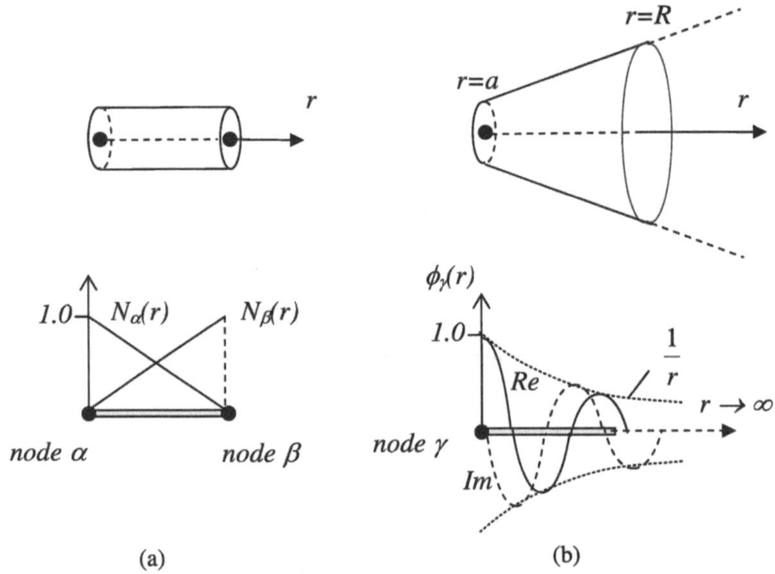

Fig. 7.3 Element geometry and shape functions for finite and infinite, acoustic line elements. (a) A two–noded finite line element. (b) A single–node infinite element.

$$\int_{r_0}^{R} \left\{ \frac{dp}{dr}\frac{dw}{dr} - k^2 wp \right\} A(r)dr + \mathrm{i}k \left[A(r)wp \right]_{r=R} - \rho \left[A(r)wa_r \right]_{r=r_0} = 0. \tag{7.12}$$

This must be satisfied by a trial solution $p(r,\omega)$ for all piecewise continuous test functions w.

7.3.4 The Trial Solution

A trial solution for the acoustic pressure in the horn is written in terms of global "shape" functions, $\phi_\alpha(r,\omega), \alpha = 1, 2, \ldots, n$, which are defined for the combined finite/infinite element mesh of Figure 7.2. Each shape function ϕ_α is associated with a particular node, α, and satisfies

$$\phi_\alpha(r_\beta, \omega) = \delta_{\alpha\beta} \tag{7.13}$$

where r_β is the coordinate of node β and $\delta_{\alpha\beta}$ is the Kronecker delta in the usual notation. The shape functions exist globally, but are defined locally (within each element) in the usual way.

The shape functions for a typical finite element are shown in Figure 7.3(a). They are linear in r and take values of 1.0 and 0.0 at each node of the element. The shape function $\phi_\gamma(r,\omega)$ for an infinite element is shown in Figure 7.3(b), and will be discussed later. It is sufficient at this stage to know that it is continuous and satisfies condition (7.13). A continuous trial solution can therefore be written for the acoustic pressure

$$p(r,\omega) = \sum_{\alpha=1}^{n} q_\alpha(\omega)\phi_\alpha(r,\omega) \tag{7.14}$$

where n is the total number of nodes. The parameters $q_\alpha(\omega), \alpha = 1, 2..n$, are unknown coefficients. As a consequence of equation (7.13), however, they have direct physical meaning as nodal values of the pressure amplitude, i.e. $q_\alpha(\omega) \equiv p(r_\alpha, \omega)$.

7.3.5 The Discrete Equations

The trial solution (7.14) is substituted into the variational statement (7.12) and the resulting expression is evaluated for set of n test functions $w_\alpha(r,\omega), (\alpha = 1, 2..n)$. In the limit as $R \to \infty$, this gives a set of algebraic equations

$$\boldsymbol{A}(\omega)\boldsymbol{q}(\omega) \ = \ \boldsymbol{f}(\omega), \tag{7.15}$$

where the solution vector $\boldsymbol{q}$ contains the unknown trial coefficients, $q_\alpha(\omega)$, and the components of $\boldsymbol{A}$ and $\boldsymbol{f}$ are given by;

$$A_{\alpha\beta} \ = \ \lim_{R\to\infty}\left\{ \int_{r_0}^{R} \left[\frac{dw_\alpha}{dr}\frac{d\phi_\beta}{dr} - k^2 w_\alpha \phi_\beta \right] A(r)dr - \mathrm{i}k \left[w_\alpha \phi_\beta A(r) \right]_{r=R} \right\}, \tag{7.16}$$

and

$$f_\alpha = \left[\rho w_\alpha a_r A(r) \right]_{r=r_0} . \tag{7.17}$$

In practice, the trial and test functions, ϕ_α and w_α, are defined explicitly within elements. The above expressions are therefore evaluated element by element and "assembled" in the usual way.

7.3.6 The Finite Element Contribution

In the FE region, the trial and test functions are chosen so that $\phi_\alpha(r,\omega) = w_\alpha(r,\omega) = N_\alpha(r)$, where $N_\alpha(r)$ is the standard shape function for a two–noded (linear) element, see Figure 7.3(a). The resulting contribution $A^{fe}_{\alpha\beta}$ to the coefficient matrix $\boldsymbol{A}$ can be written in terms of frequency–independent, acoustic stiffness and mass matrices, i.e.

$$A^{fe}_{\alpha\beta} = K^{fe}_{\alpha\beta} - \omega^2 M^{fe}_{\alpha\beta}, \tag{7.18}$$

where

$$K^{fe}_{\alpha\beta} = \int_{r_0}^{a} \frac{dN_\alpha}{dr}\frac{dN_\alpha}{dr} A(r)dr, \quad \text{and} \quad M^{fe}_{\alpha\beta} = \frac{1}{c^2}\int_{r_0}^{a} N_\alpha N_\beta A(r)dr. \tag{7.19}$$

These are not affected by the limiting process as $R \to \infty$ and are the "standard", symmetric FE expressions for the mass and stiffness of a conventional acoustic mesh.

7.3.7 The Infinite Element Contribution

The contribution, $A^{ie}_{\alpha\beta}$, to the coefficient matrix $\boldsymbol{A}$ from the infinite element mesh is given by

$$A^{ie}_{\alpha\beta} = \lim_{R\to\infty} \left\{ \int_a^R \left[\frac{dw_\alpha}{dr}\frac{d\phi_\beta}{dr} - k^2 w_\alpha \phi_\beta \right] A(r)dr + \mathrm{i}k \left[[w_\alpha \phi_\beta A(r)] \right]_{r=R} \right\}. \tag{7.20}$$

This will be zero in the current instance unless $\alpha = \gamma$ and $\beta = \gamma$, since (γ) is the only node in the infinite element domain. To proceed further we must define the trial and test functions, ϕ_γ and w_γ, for such an element.

The Infinite Element Shape Functions

Two methods are used to construct the shape functions for acoustic infinite elements. In the case of the one–dimensional infinite element presented here, both methods yield identical matrices, but this is not generally the case in two and three dimensions.

In the first of these approaches, which we will term the *separable* method, the trial and test functions are derived a separable functions of the physical coordinates which define the element. In the second, the *mapped* approach, the trial and test functions are derived in a mapped space to which the physical element is transformed.

The separable formulation relies on the edges or faces of the element being aligned with iso–surfaces in an orthogonal curvilinear coordinate system. It was formalised by Burnett [22] specifically for a spheroidal coordinate system but can equally be applied in spherical coordinates [30, 40].

The mapped method relies on a mapping which transforms the infinite element to a unit square or block. In particular, a radial dimension r is transformed to a mapped coordinate s where

$$s = 1 - \frac{2a}{r} \quad \text{or} \quad r = \frac{2a}{1-s}. \tag{7.21}$$

In the case of the one–dimensional element of Figure 7.3(b) this transformation is used as it stands to map the unbounded element on the interval $r \in [a, \infty]$ onto the interval $s \in [-1, +1]$ in the mapped space. The trial basis is then defined in terms of the mapped coordinate s. This procedure was developed by Bettess [20, 43] for two–dimensional surface gravity waves, and applied first to acoustics by Gorranson [31]. The mapped and separable approaches give the same shape and test functions for the one–dimensional element of Figure 7.3(b). Without loss of generality, all expressions will be presented here in terms of the physical coordinate r.

The infinite element is defined by a single node γ at its inner edge. The associated shape function ϕ_γ is defined as

$$\phi_\gamma(r, \omega) = \left\{ \frac{a}{r} \right\} \mathrm{e}^{-\mathrm{i}k(r-a)}. \tag{7.22}$$

The oscillatory nature of the shape function is illustrated in Figure 7.3(b). Note also that ϕ_γ takes the required value of unity at node γ. The *rationale* behind choosing this particular function for ϕ_γ derives from expression (7.9), in that expression (7.22) is able to represent *exactly* the outwardly propagating component of the solution in the far field. This is the essential feature of the infinite element concept, that the trial functions in each element are chosen so that they include asymptotic, far field solutions of the continuous problem.

The Test Functions

Three types of test function w_γ will be considered. Each gives a different set of equations. They are defined here in terms of the physical coordinate, r, but can be rewritten in terms of mapped coordinate, s, if required, by using expression (7.21).

1. *The "unconjugated" formulation.* The test basis function w_γ is chosen to be the shape function, $\phi_\gamma(r,\omega)$, i.e.

$$w_\gamma(r,\omega) = \phi_\gamma(r,\omega) = \left\{\frac{a}{r}\right\} \mathrm{e}^{-ik(r-a)}. \tag{7.23}$$

   This is a traditional *Bubnov–Galerkin* process giving a symmetric coefficient matrix. The Burnett element [22] is formulated in this way, as is the Bettess element [43], except that in the latter case ϕ_γ and w_γ are expressed in terms of s.

2. *The "conjugated" formulation.* The test basis function is defined to be the complex conjugate of the shape function. i.e

$$w_\gamma(r,\omega) = \phi_\gamma(r,\omega)^* = \left\{\frac{a}{r}\right\} \mathrm{e}^{+ik(r-a)}. \tag{7.24}$$

   This formulation was proposed by Gerdes [30] and investigated by Shirron [40]. It is the simplest of the conjugated schemes but is less accurate for multi–dimensional problems than the closely related Astley–Leis conjugated element [30].

3. *The Astley–Leis formulation.* This also uses the complex conjugate of the shape function but incorporates a geometric weighting proportional to $1/r^2$. The test function w_γ is given by;

$$w_\gamma(r,\omega) = \left\{\frac{a}{r}\right\}^2 \phi_\gamma(r,\omega)^* = \left\{\frac{a}{r}\right\}^3 \mathrm{e}^{+ik(r-a)}. \tag{7.25}$$

   This was first applied by Astley and Eversman in the 1980s to large but finite "wave envelope" elements [1,10]. Infinite elements based on the same idea were formulated later for spherical [13, 14] and spheroidal [3] element geometries. The same elements were investigated by Gerdes, Demkowicz and Shirron [30, 40] and found to fall within a general variational framework proposed by Leis [36].

The Element Matrix

The infinite element of Figure 7.3(b) is defined by a single degree of freedom at node γ. The element matrix is therefore of order 1, contributing a single entry $A^{ie}_{\gamma\gamma}$ to the coefficient matrix $\boldsymbol{A}$. This is given by

$$A^{ie}_{\gamma\gamma} = \lim_{R\to\infty}\left\{ \underbrace{\int_a^R \left[\frac{dw_\gamma}{dr}\frac{d\phi_\gamma}{dr} - k^2 w_\gamma \phi_\gamma\right] A(r)dr}_{a} + \mathrm{i}k \underbrace{\left[w_\gamma \phi_\gamma A(r)\right]_{r=R}}_{b} \right\}. \tag{7.26}$$

Terms "a" in the above expression arise from the integral over the element volume; terms "b" arise from contributions on the "infinite" boundary. The above expression can be evaluated explicitly for each of the three formulations listed above. By setting $A(r) = A_0(\frac{r}{a})^2$ for a conical expansion, this gives the following expressions:

The Unconjugated Element:

The trial and test functions are

$$w_\gamma = \phi_\gamma = \left\{\frac{a}{r}\right\} \mathrm{e}^{-\mathrm{i}k(r-a)} \tag{7.27}$$

These give after some manipulation

$$\Delta A_{\gamma\gamma} = \lim_{R\to\infty} A_0 \left\{ \underbrace{\left[\frac{1}{a} + \mathrm{i}k - \left(\frac{1}{R} + \mathrm{i}k\right)\mathrm{e}^{-2\mathrm{i}k(R-a)}\right]}_{a} + \mathrm{i}k \underbrace{\left[\mathrm{e}^{-2\mathrm{i}k(R-a)}\right]}_{b} \right\} \tag{7.28}$$

The boundary term "b" in this case is not well–defined as $R \to \infty$. However an equal but opposite term occurs in integral "a" which exactly cancels with the boundary term so that the coefficient itself is well defined. The same behaviour occurs in two–dimensional and three–dimensional elements of this type; while the individual volume and boundary contributions are undefined, they combine to give a well defined coefficient [6]. This behaviour must be treated carefully in any numerical treatment of these integrals.

The Conjugated Element – A:

The trial and test functions are

$$\phi_\gamma = \left\{\frac{a}{r}\right\} \mathrm{e}^{-\mathrm{i}k(r-a)}, \quad \text{and} \quad w_\gamma = \left\{\frac{a}{r}\right\} \mathrm{e}^{+\mathrm{i}k(r-a)}, \tag{7.29}$$

which give

$$A^{ie}_{\gamma\gamma} = \lim_{R\to\infty} A_0 \left\{ \underbrace{\left[\frac{1}{a} - \frac{1}{R}\right]}_{a} + \mathrm{i}k \underbrace{\left[1\right]}_{b} \right\}. \tag{7.30}$$

This expression is somewhat simpler than for the unconjugated case, since the integrand contains no spatially oscillatory terms, products of the factors $\mathrm{e}^{-\mathrm{i}k(r-a)}$ and $\mathrm{e}^{-\mathrm{i}k(r-a)}$ occurring in all terms and cancelling identically. In this case also the volume term ("a") and the boundary term ("b") are finite and well defined in the limit as $R \to \infty$.

The Conjugated Element – B (Astley–Leis):

The trial and test functions are

$$\phi_\gamma = \left\{\frac{a}{r}\right\} e^{-ik(r-a)}, \qquad \text{and} \qquad w_\gamma = \left\{\frac{a}{r}\right\}^3 e^{+ik(r-a)} \tag{7.31}$$

These give

$$A^{ie}_{\gamma\gamma} = \lim_{R\to\infty} A_0 \left\{ \underbrace{\left[\frac{1}{a} + ik - \frac{a^2}{R^3} - \frac{ika^2}{R^2}\right]}_{a} + ik \underbrace{\left[\frac{a^2}{R^2}\right]}_{b} \right\}. \tag{7.32}$$

Once again the products of the spatially harmonic factor and its conjugate cancel leaving a relatively simple expression for the integrand. In this case however the contribution at the infinite boundary vanishes in the limit as $R \to \infty$.

The Limiting Case as $R \to \infty$***:***

All three of the expressions for $A^{ie}_{\gamma\gamma}$ — (7.28), (7.30) and (7.32) — converge to the same limiting value as $R \to \infty$, i.e.

$$A^{ie}_{\gamma\gamma} = \left[\frac{1}{a} + ik\right]. \tag{7.33}$$

When substituted into equation (7.15) this gives an exact solution for the pressure field in the region $r > a$. If the finite element mesh is removed, for example, and the piston placed at node γ, the resulting model has one degree of freedom $q_\gamma (= p_\gamma)$, and equation (7.15) becomes

$$A_{\gamma\gamma}(\omega)p_\gamma(\omega) = f_\gamma(\omega), \tag{7.34}$$

where $f_\gamma = -\rho_0 A_0 a_r$ and $A_{\gamma\gamma}$ is given by expression (7.33). If a_r is written in terms of a piston velocity u_r $(= a_r/i\omega)$, Equation (7.34) defines a computed impedance relationship for the pressure and velocity on the piston face, i.e.

$$\frac{p_\gamma}{u_r} = \rho_0 c_0 \frac{ika}{1 + ika}. \tag{7.35}$$

This is the *exact* value of radial impedance for an outwardly propagating spherical wave. The pressure computed at the piston is therefore obtained without approximation for a given piston acceleration. Moreover, the computed solution sampled at any point within the unbounded domain will also be "exact" by virtue of trial solution (7.22) which can exactly model an outwardly propagating wave. If the solution is correct at $r = a$ it will therefore be correct at all points in the outer region. This would *not* necessarily hold if a finite element mesh were present for $r < a$. The accuracy of the computed solution at the inner edge of the infinite element domain, and at all points beyond, would then be governed by the resolution of the inner mesh.

7.4 Infinite Elements in Three Dimensions

The infinite element formulations which have been described in the preceding section were based on a trial solution which mimicked an outwardly propagating wave for this one–dimensional case. In two–dimensional and three–dimensional problems, an infinite number of outwardly propagating "multipole" solutions exist. These are used to construct trial functions for multi–dimensional infinite elements by using a broadly similar approach to that described above. Before attempting to define suitable trial or test functions for such elements, it is helpful to review the known far field behaviour of multi–dimensional, unbounded solutions to the Helmholtz equation. Three–dimensional and axisymmetric problems will be considered in the remainder of this chapter. Similar arguments can be applied to plane acoustics but these will not be presented here.

7.4.1 The Asymptotic Form of a Radiating Solution

Solutions of the Helmholtz equation in three dimensions, bounded or unbounded, can be expanded in terms of separable solutions in spherical coordinates r, θ and ϕ. The general expression is,

$$\begin{aligned} p(\boldsymbol{x},\omega) = &\sum_{n=0}^{\infty}\sum_{m=0}^{n} h_n^{(2)}(kr)P_n^m(\cos\theta)\left\{A_{nm}\sin(m\phi) + B_{nm}\cos(m\phi)\right\} \\ &+ \sum_{n=0}^{\infty}\sum_{m=0}^{n} h_n^{(1)}(kr)P_n^m(\cos\theta)\left\{C_{nm}\sin(m\phi) + D_{nm}\cos(m\phi)\right\} \end{aligned} \tag{7.36}$$

where $h_n^{(1)}()$ and $h_n^{(2)}()$ are spherical Hankel functions of the first and second kind, $P_n^m()$ is a Legendre function, and n and m are integers > 0. The Hankel functions can be written as finite expansions

$$h_n^{(1)}(kr) = \frac{1}{kr}\left\{\alpha_0 + \frac{\alpha_1}{kr} + \ldots + \frac{\alpha_n}{(kr)^n}\right\}\mathrm{e}^{+ikr}, \quad (\alpha_j, j = 0\ldots n \text{ constants}), \tag{7.37}$$

$$h_n^{(2)}(kr) = \frac{1}{kr}\left\{\beta_0 + \frac{\beta_1}{kr} + \ldots + \frac{\beta_n}{(kr)^n}\right\}\mathrm{e}^{-ikr}, \quad (\beta_j, j = 0\ldots n \text{ constants}). \tag{7.38}$$

and correspond to inwardly and outwardly propagating modes respectively. If it is assumed that only outwardly propagating solutions are present, the second summation in expression (7.36) can be omitted and the remaining terms involving Hankel functions of the second kind rearranged to give a general expression of the form

$$p(\boldsymbol{x},\omega) = \mathrm{e}^{-ikr}\sum_{n=1}^{\infty}\frac{G_n(\theta,\phi,\omega)}{r^n}, \tag{7.39}$$

where $G_n(\theta,\phi,\omega)$ is a directivity function associated with the nth inverse power of r. This will be termed a "multipole expansion". In the far field the first term dominates

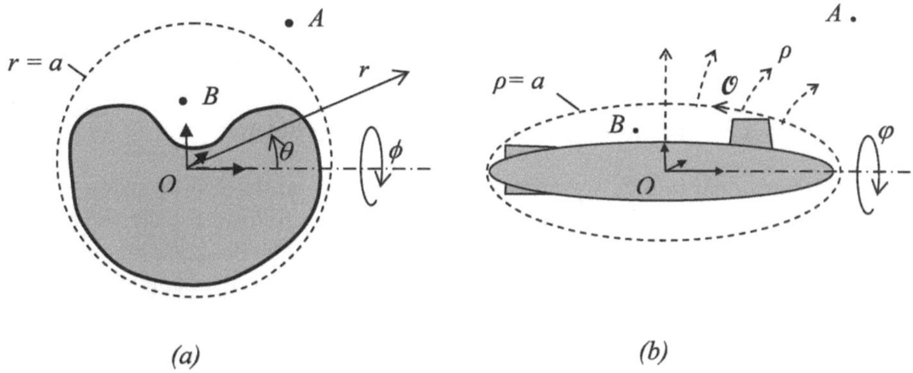

Fig. 7.4 The Atkinson–Wilcox criterion: (a) Spherical coordinates. (b) Spheroidal coordinates.

giving the well known relationship between acoustic pressure, radial distance, and far field directivity,

$$p(\boldsymbol{x},\omega) \sim \frac{\mathrm{e}^{-ikr}}{r} D(\theta,\phi,\omega). \tag{7.40}$$

where $D(\theta,\phi,\omega) = G_1(\theta,\phi,\omega)$. Clearly one expects that the multipole expansion (7.39) will hold at large distances from a radiating or scattering object, but the extent to which it is valid also in the near field is not obvious. The **Atkinson–Wilcox** theorem answers this question [42]. It states that the sound field at any point which lies entirely outside a circumscribing sphere, which itself encloses all radiating and scattering sources, can be written as a multipole expansion of type (7.39), and that this expansion is absolutely and uniformly convergent. The principle is illustrated in Figure 7.4(a) in which point "A" lies outside a circumscribing sphere of radius $r = a$ and therefore satisfies the Atkinson–Wilcox criterion, whereas point "B" does not.

The Atkinson–Wilcox criterion can also be interpreted in terms of near and far field contributions. That is to say, the leading term determines the far field directivity as defined in expression (7.40), and the remaining terms then contribute to the near and intermediate field. If expression (7.39) is truncated, as will be the case when trial solutions are defined for multi–dimensional infinite elements, the number of terms which are retained determines the extent to which the truncated expansion it is able to resolve near field effects.

The Atkinson–Wilcox expansion was derived for a spherical coordinate system. A similar result holds for spheroidal or ellipsoidal coordinate systems [22]. That is to say expression (7.39) holds without modification when spheroidal coordinates ρ, ϑ and φ are substituted for spherical coordinates r, θ and ϕ. The implications are illustrated in Figure 7.4(b) where a spheroidal surface $\rho = a$ encloses a slender radiating body. As in Figure 7.4(a), the point A satisfies the spheroidal version of the Atkinson–Wilcox criterion whereas the point B does not. Clearly the inner region $\rho \leq a$ in which the multipole expansion cannot be used can be greatly reduced by using a spheroidal rather than a spherical surface when dealing with "long thin" objects.

7.4.2 A Weak Variational Statement for Exterior Problems

Multi–dimensional infinite element formulations will be derived by using a weak variational statement for the unbounded problem. This follows closely the process already described in Section 7.3 for the one–dimensional model equation.

A weak statement of the unbounded Helmholtz problem is obtained by considering a large but finite domain Ω_R which is bounded by a spherical surface Γ_R at $r = R$ as shown in Figure 7.5(a). The Sommerfeld radiation condition is approximated initially by a "ρc" termination on Γ_R. The acoustic pressure amplitude $p(\boldsymbol{x}, \omega)$ must then satisfy

$$\nabla^2 p + k^2 p = 0 \quad \text{in} \quad \Omega_R, \tag{7.41}$$

$$\nabla p \cdot \hat{\boldsymbol{n}} = \rho a_n \quad \text{on} \quad \Gamma, \tag{7.42}$$

$$\nabla p \cdot \hat{\boldsymbol{n}}_R = -ikp \quad \text{on} \quad \Gamma_R, \tag{7.43}$$

where $\hat{\boldsymbol{n}}$ and $\hat{\boldsymbol{n}}_R$ are outward unit normals on Γ and Γ_R, see Figure 7.5(a). As $R \to \infty$, the original unbounded problem is recovered.

A weak variational statement is obtained by multiplying Equation (7.41) by a test function $w(\boldsymbol{x}, \omega)$ and integrating over Ω_R. Boundary conditions (7.42) and (7.43) are imposed in the usual way by applying the divergence theorem to give an integral relationship

$$\int_{\Omega_R} \{\nabla w \cdot \nabla p - k^2 wp\} \, dR + \mathrm{i}k \int_{\Gamma_R} wpdS \; - \int_{\Gamma} \rho w a_n dS \; = \; 0\,, \tag{7.44}$$

which must be satisfied by $p(\boldsymbol{x}, \omega)$ for any test function $w(\boldsymbol{x})$ which is continuous on Ω_R [34].

7.4.3 The Subdivision into Elements

The region Ω_R is subdivided into a mesh consisting of finite and infinite elements. The finite elements lie adjacent to the radiating surface. The infinite elements form a single layer adjacent to the far field boundary Γ_R. The two subregions are separated by an interface Γ_c. This arrangement is illustrated in Figure 7.5(b). The infinite elements are aligned in the radial direction and match compatibly to the finite element mesh, element for element and node for node on the interface Γ_c.

7.4.4 The Discrete Equations

A trial solution is constructed based on global "shape" functions $\{\phi_\alpha(\boldsymbol{x}, \omega), \alpha = 1, 2, \ldots, n\}$ where n is the total number of nodes in the finite and infinite element meshes. These functions are defined explicitly over each element, and implicitly over

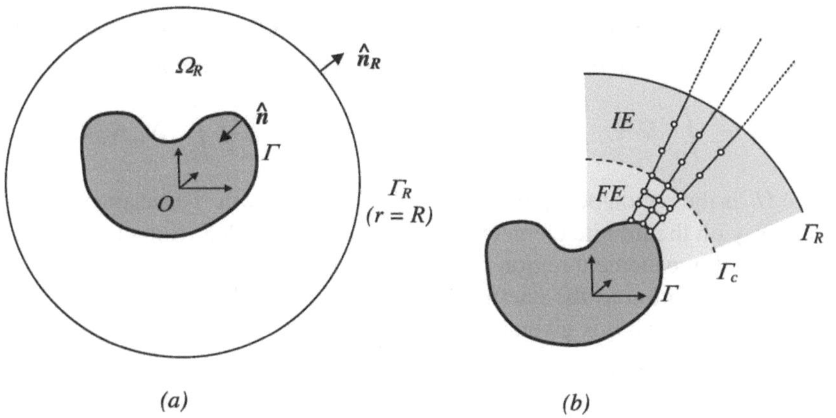

Fig. 7.5 The FE/IE model:(a) Modified domain with a far field boundary. (b) Finite element (FE) and infinite element (IE) meshes.

the entire mesh. Each function $\phi_\alpha(\boldsymbol{x}, \omega)$ is associated with a particular node, α, and is continuous across element boundaries. It takes the value of unity and its "own" node and zero at all other nodes. These properties are sufficient to ensure that the unknown coefficients $q_\alpha(\omega)$ in a continuous trial solution of the form

$$p(\boldsymbol{x}, \omega) = \sum_{\alpha=1}^{n} q_\alpha(\omega)\phi_\alpha(\boldsymbol{x}, \omega) \tag{7.45}$$

correspond to nodal values of the acoustic pressure amplitude. A system of linear equations for these coefficients is then obtained by substituting expression (7.45) into the variational statement (7.44), and evaluating the resulting expression for a complete set of test functions $\{w_\alpha(\boldsymbol{x}, \omega), \alpha = 1, 2, ...n\}$. In the limit as $R \to \infty$, this gives

$$\boldsymbol{A}(\omega)\boldsymbol{q}(\omega) \;=\; \boldsymbol{f}(\omega) \;, \tag{7.46}$$

where the components of $\boldsymbol{A}$ and $\boldsymbol{f}$ are given by;

$$A_{\alpha\beta} = \lim_{R\to\infty} \left\{ \int_{\Omega_R} \left(\nabla w_\alpha \cdot \nabla \phi_\beta \;-\; k^2 w_\alpha \phi_\beta \right) dR \;+\; \mathrm{i}k \int_{\Gamma_R} w_\alpha \phi_\beta dS \right\}, \tag{7.47}$$

$$f_\alpha = \int_\Gamma \rho w_\alpha a_n dS, \qquad \text{where} \qquad \alpha, \beta = 1, \ldots, n \,. \tag{7.48}$$

These expressions are entirely analogous to their one dimensional counterparts; (7.15),(7.16) and (7.17). They are evaluated and "assembled" element by element in the usual way.

As in the one–dimensional case, the trial and test functions, ϕ_α and w_α in the finite element region are defined to be conventional finite element shape functions. That is to say, $\phi_\alpha(\boldsymbol{x}, \omega) \;=\; w_\alpha(\boldsymbol{x}, \omega) \;=\; N_\alpha(\boldsymbol{x})$, where $N_\alpha(\boldsymbol{x})$ is the frequency–independent shape function for node α. The contribution to the coefficient matrix,

$A^{fe}_{\alpha\beta}$ say, can be written once more, *c.f.* expression (7.18), in terms of symmetric, acoustic stiffness and mass matrices $\boldsymbol{K}^{fe}$ and $\boldsymbol{M}^{fe}$. These are given by

$$K^{fe}_{\alpha\beta} = \int_{\Omega_c} \nabla N_\alpha \cdot \nabla N_\beta dR, \quad \text{and} \quad M^{fe}_{\alpha\beta} = \frac{1}{c^2}\int_{\Omega_c} N_\alpha N_\beta dR, \tag{7.49}$$

where Ω_c is the region occupied by the finite element mesh. These are not dependent in any way on the infinite element model and are assembled from element contributions in the finite element region in the usual way.

The contribution to the coefficient matrix $\boldsymbol{A}$ from elements in the IE region is less straightforward. It is given by

$$A_{\alpha\beta} = \lim_{R\to\infty}\left\{\int_{\Omega_R-\Omega_c} \left(\nabla w_\alpha \cdot \nabla\phi_\beta - k^2 w_\alpha \phi_\beta\right) dR + ik\int_{\Gamma_R} w_\alpha\phi_\beta dS\right\} \tag{7.50}$$

Such contributions are evaluated element by element and will depend upon the trial and test functions — ϕ_β and w_α — which are chosen within each infinite element.

In Sections 7.4.5 and 7.4.6 to follow, "mapped" and "separable" approaches will be described for constructing suitable shape functions. "Conjugated" and "unconjugated" strategies will be presented for selecting test functions.

7.4.5 A Separable Infinite Element

A separable formulation will now be presented for an axisymmetric infinite element. Conjugated and unconjugated formulations will be described.

Element Geometry and Shape Functions

The topology of the axisymmetric infinite element is shown in Figure 7.6, where r, θ and ϕ (ϕ is not used for axisymmetric problems) are spherical polar coordinates in the usual notation. The lateral boundaries of the element are formed by the surfaces $\theta = \theta_i$ and $\theta = \theta_j$. Inner and outer boundaries are located at $r = a$ and $r = R$. The nodes of the element are labelled $i_1, i_2, ..i_q$ and $j_1, j_2, ..j_q$. They are located at radial distances $r = r_1, r_2, ..r_q$ along each lateral face, where $r_1 = a$. It is assumed here that the element is defined by two nodes in the θ direction. Higher order elements with more nodes in the θ direction can readily be formulated but will not be discussed here.

Trial basis functions $\phi_\alpha(\boldsymbol{x},\omega)$ and $\phi_\beta(\boldsymbol{x},\omega)$ which correspond to nodes $\alpha = i_\mu$ or $\beta = j_\nu$ where μ and ν lie between 1 and q, are defined as the product of interpolation functions in the radial and transverse directions,

$$\phi_\alpha(\mathbf{x},\omega) = g_i(\theta)f_\mu(r,\omega), \text{ and } \phi_\beta(\mathbf{x},\omega) = g_j(\theta)f_\nu(r,\omega), \qquad \mu,\nu = 1,2,\ldots,q. \tag{7.51}$$

The θ functions are defined by linear interpolation as

$$g_i(\theta) = \frac{\theta-\theta_j}{\theta_i-\theta_j} \quad \text{and} \quad g_j(\theta) = \frac{\theta-\theta_i}{\theta_j-\theta_i}. \tag{7.52}$$

The radial functions are based on the terms in the Atkinson–Wilcox expansion truncated after q terms. They are chosen to satisfy the compatibility requirement that $f_\mu(r_\nu,\omega)=\delta_{\mu\nu}$. It is not difficult to show that both requirements are satisfied by the family of functions

$$f_\mu(r,\omega) \;=\; l_\mu(r)\,\mathrm{e}^{-\mathrm{i}k\psi_\mu(r)}, \quad \mu=1,2,\ldots,q, \tag{7.53}$$

where $l_\mu(r)$ is a Lagrange interpolation function in $1/r$, and where $\mathrm{i}k\psi_\mu(r)$ is a phase function corresponding to outwardly propagating spherical wave, i.e. $l_\mu(r)$ and $\psi_\mu(r)$ are given by

$$l_\mu(r) \;=\; \left(\frac{r_\mu}{r}\right)^q \prod_{\nu=1(\neq\mu)}^{q} \frac{(r-r_\nu)}{(r_\mu-r_\nu)} \quad \text{and} \quad \psi_\mu(r)=r-r_\mu. \tag{7.54}$$

Expressions (7.52), (7.53) and (7.54) ensure that the original trial solution (7.45) is a truncated version of the Atkinson–Wilcox expansion in which the number of radial nodes q corresponds to the number of multipole terms. This tells us that the trial solution is able to represent the true solution as q increases, provided that the spherical surface Γ_c fully encloses the radiating body.

The Unconjugated Formulation

The test functions for the unconjugated formulation are equated to the trial functions, $w_\alpha(\boldsymbol{x},\omega)=\phi_\alpha(\boldsymbol{x},\omega)$ to give a symmetric Galerkin scheme. When the appropriate expressions for $g_i(\theta)$ and $f_\mu(r,\omega)$, Equations (7.52), (7.53) and (7.54), are substituted into the general expression for the $A^{ie}_{\alpha\beta}$, Equation (7.50), the resulting integrals over $\Omega_R-\Omega_c$ can be written as the product of line integrals over r and θ which involve the radial and transverse shape functions only. The contribution ΔA from an individual element is given by

$$\Delta A^{ie}_{\alpha\beta} \;=\; B^{(1)}_{\mu\nu}\,C^{(1)}_{ij} \;+\; B^{(2)}_{\mu\nu}\,C^{(2)}_{ij}\,, \tag{7.55}$$

where

$$C^{(1)}_{ij} = \int_{\theta_i}^{\theta_j} g_i g_j \sin\theta d\theta\,, \qquad C^{(2)}_{ij} = \int_{\theta_i}^{\theta_j} \frac{dg_i}{d\theta}\frac{dg_j}{d\theta}\sin\theta d\theta\,, \tag{7.56}$$

$$B^{(1)}_{\mu\nu} = 2\pi\mathrm{e}^{\mathrm{i}k(r_\mu+r_\nu)}\lim_{R\to\infty}\left\{\int_a^R\left[\left(\frac{dl_\mu}{dr}-\mathrm{i}kl_\mu\right)\left(\frac{dl_\nu}{dr}-\mathrm{i}kl_\nu\right)-k^2l_\mu l_\nu\right]\mathrm{e}^{-2\mathrm{i}kr}r^2dr \right.$$
$$\left. +\,\mathrm{i}kR^2l_\mu(R)l_\nu(R)\mathrm{e}^{-2\mathrm{i}kR}\right\}, \tag{7.57}$$

and

$$B^{(2)}_{\mu\nu} = 2\pi\mathrm{e}^{\mathrm{i}k(r_\mu+r_\nu)}\lim_{R\to\infty}\left\{\int_a^R l_\mu l_\nu\mathrm{e}^{-2\mathrm{i}kr}dr\right\}. \tag{7.58}$$

Numerical integration of the above radial integrals is not straightforward since the

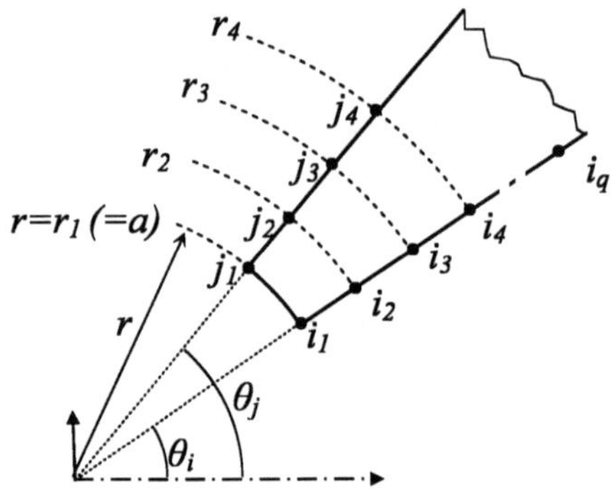

Fig. 7.6 Geometry and nodal topology of an axisymmetric separable infinite element.

integrands include the highly oscillatory factor $\mathrm{e}^{-2\mathrm{i}kr}$. Each can be reduced, however to a linear combination of integrals of the form

$$I_m = \lim_{R\to\infty} \left\{ \int_a^R r^{-m} \mathrm{e}^{-2\mathrm{i}kr}\, dr \right\}, \qquad m = 0, 1, \ldots, 2q, \tag{7.59}$$

and these can be evaluated by using a recursion formula for $m \geq 1$. In the case $m = 0$, however the integral is undefined as $R \to \infty$. The line integral in expression (7.57) contains such terms and is therefore indeterminate. A careful inspection of the full expression for ΔA reveals however that the undefined terms in the integral cancel exactly with boundary terms at $r = R$, and that expression (7.57) as a whole *does* therefore converge to a well defined limit as $R \to \infty$ (see also comments on the equivalent one dimensional element in Section 7.3.7). The integration of these terms is discussed in some detail by Burnett [22] and by Astley and Bettess [6] in the context of the numerical integration of mapped elements where the same issue arises. With this adjustment the evaluation of the infinite element contributions proceeds without difficulty.

The Conjugated Astley–Leis Formulation

Conjugated versions of the separable element start from the same trial solution as the unconjugated element — expressions (7.51), (7.52) and (7.53) — but the test function $w_\alpha(\boldsymbol{x}, \omega)$ is then chosen to be the complex conjugate of the corresponding shape function $\phi_\alpha(\boldsymbol{x}, \omega)$, with or without a geometric factor. Only the formulation *with* a geometric factor $(r_\alpha/r)^2$ — the "Astley–Leis" formulation[1] — will be described here. A typical test function $w_\alpha(\boldsymbol{x}, \omega)$ for node $\alpha = i_\mu$ is given by

$$w_\alpha(\boldsymbol{x}, \omega) = g_i(\theta) \left(\frac{r_\mu}{r}\right)^2 f_\mu^*(r, \omega) = g_i(\theta) \left(\frac{r_\mu}{r}\right)^2 l_\mu(r) \mathrm{e}^{+\mathrm{i}k\psi_\mu(r)}, \tag{7.60}$$

where * denotes the complex conjugate. Expression (7.50) then becomes

[1] see previous comments in Section 7.3.7

$$\Delta A^{ie}_{\alpha\beta} = D^{(1)}_{\mu\nu} C^{(1)}_{ij} + D^{(2)}_{\mu\nu} C^{(2)}_{ij}, \tag{7.61}$$

where the matrices $C^{(1)}$ and $C^{(2)}$ which arise from integration in the θ direction are as previously defined, but where the "radial" matrices $D^{(1)}$ and $D^{(2)}$ have components

$$\begin{aligned} D^{(1)}_{\mu\nu} &= 2\pi r_\mu^2 \mathrm{e}^{-\mathrm{i}k(r_\mu - r_\nu)} \lim_{R\to\infty} \left\{ \int_a^R \left[\left(\frac{dl_\mu}{dr}\frac{dl_\nu}{dr} - 2\frac{l_\mu}{r}\frac{dl_\nu}{dr} \right) + \right.\right. \\ &\left.\left. + \mathrm{i}k \left(l_\mu \frac{dl_\nu}{dr} - l_\nu \frac{dl_\mu}{dr} + 2\frac{l_\mu l_\nu}{r} \right) \right] dr + \mathrm{i}k l_\mu(R) l_\nu(R) \right\}, \end{aligned} \tag{7.62}$$

and

$$D^{(2)}_{\mu\nu} = 2\pi r_\mu^2 \mathrm{e}^{-\mathrm{i}k(r_\mu - r_\nu)} \lim_{R\to\infty} \left\{ \int_a^R \frac{1}{r^2} l_\mu l_\nu dr \right\}. \tag{7.63}$$

Several important points distinguish these expressions from their unconjugated counterparts. First the oscillatory factors e^{+ikr} and e^{-ikr} which occur in the test and trial functions, cancel when multiplied together within each integral. As a consequence, the resulting integrand is largely independent of frequency and varies smoothly with r. Second, no indefinite terms occur in the limit as $R \to \infty$ and indeed the contribution on Γ_R disappears in the limit. The integrals themselves are particularly simple to evaluate, involving integrands of the form $(1/r^m)$. The resulting coefficient matrix is however no longer symmetric.

7.4.6 A Mapped Infinite Element

While the philosophy which underpins the mapped formulation is somewhat different from that of the separable element, the resulting shape functions and element matrices are very similar. Indeed, when the geometry of the mapped element is restricted to that of a separable element, the two elements are for all practical purposes identical. The main strength of the mapped formulation is that the shape of the mapped element is more general than that of its separable counterpart, a distinct advantage from the meshing point of view. The mapped approach will be presented here for an axisymmetric element whose topology is similar to that used in the previous section. Once again, unconjugated and conjugated formulations will be presented.

The Element Mapping

An axisymmetric mapped infinite element is shown in Figure 7.7(a). The dependent variable $p(\boldsymbol{x}, \omega)$ is defined at nodes $i_1, i_2, ..i_q$ and $j_1, j_2, ..j_q$ along each radial edge of the element. The main difference between the current element and the separable element of Figure 7.6 is that the inner boundary of the element need no longer be aligned with the "theta" direction. Two types of nodes are used in the element definition. These are, *"variable"* nodes at which nodal degrees of freedom are defined (nodes $i_1, i_2, \ldots, j_1, j_2, \ldots$ et cetera), and *"mapping"* nodes which are used to

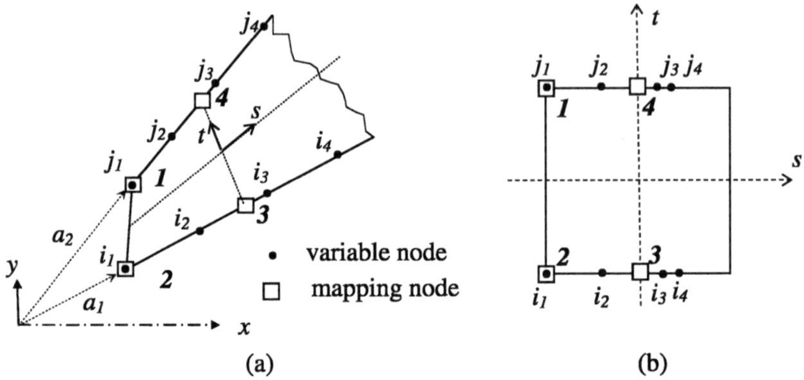

Fig. 7.7 Geometry and nodal topology of an axisymmetric, mapped infinite element. (a) Physical element. (b) Parent element.

transform the physical element to a parent domain. Both types of node are shown in Figure 7.7(a). This element has four mapping nodes labelled 1, 2, 3 and 4. These are located at the inner corners of the element (at $r = a_1$ and $r = a_2$) and on each radial side at $r = 2a_1$ and $r = 2a_2$. It is not necessary for mapping nodes, such as 3 and 4, to coincide with a variable node although this is often the case. If the coordinates of the mapping nodes are denoted by (x_i, y_i), $i = 1, 2, \ldots, 4$ in the plane of revolution (y is interpreted here as a polar radius about the x axis) a transformation from a parent (s, t) space to the physical (x, y) plane is given by

$$x = \sum_{i=1}^{4} M(s,t)\, x_i \ , \qquad y = \sum_{i=1}^{4} M(s,t)\, y_i \tag{7.64}$$

where

$$\begin{aligned} M_1(s,t) &= -\frac{(1-t)s}{(s-1)} \,, & M_2(s,t) &= -\frac{(1+t)s}{(s-1)} \,, \\ M_3(s,t) &= -\frac{(1-t)(1+s)}{2(s-1)} \,, & M_4(s,t) &= -\frac{(1+t)(1+s)}{2(s-1)} \end{aligned} \tag{7.65}$$

This mapping has the following properties:

- The infinite element is transformed to a square "parent" element of side 2 in the (s, t) space, as shown in figure 7.7(b).
- The s coordinate and the radial distance r along each side of the element are related by

$$s = 1 - 2a_i/r \ , \qquad i = 1 \ \text{or} \ 2 \,. \tag{7.66}$$

- Points at "∞" in the physical element map to points on the line $s = 1$ in the parent element.

Note that expression (7.66) is identical to expression (7.21) used in section 7.3.7.

The Trial Solution

The trial functions $\phi_\alpha(\boldsymbol{x}, \omega)$ for the mapped element are defined in terms of the mapped coordinates. Conventional polynomial Lagrangian shape functions are used in s and t multiplied by an outwardly propagating term $\mathrm{e}^{-\mathrm{i}k(r-r_\alpha)}$. From expression (7.66) it is simple to show that the phase factor "$r - r_\alpha$" in the above term can be written in terms of s and t as

$$\psi_\alpha(s,t) \;=\; r \,-\, r_\alpha \;=\; a(t)\,\frac{2}{1-s} \,-\, r_\alpha, \tag{7.67}$$

where $a(t)$ is the value of a interpolated along the inner edge of the element from $= a_1$ at mapping node 1 to a_2 at mapping node 2. The trial function for node $\alpha = i_\mu$ is then defined to be

$$\phi_\alpha(\boldsymbol{x}, \omega) \;=\; N_\alpha(s,t)\,\mathrm{e}^{-\mathrm{i}k\psi_\alpha(s,t)}, \tag{7.68}$$

where N_α is the standard polynomial shape function for node α in the mapped element defined as if the original corner nodes at "∞" (located at $(1, \pm 1)$ in the mapped element) were part of the element topology. This ensures that a factor "$(1 - s)$" occurs in each shape function, corresponding to a factor $2a/r$ in the physical space. This means that the trial solution has the correct asymptotic behaviour, as $\frac{1}{r}\mathrm{e}^{-\mathrm{i}kr}$, for large values of r. Moreover since N_α is a polynomial of order q in s, and since s is proportional to $1/r$, N_α can be written as an expansion in inverse powers of r. Expression (7.68) therefore conceals a truncated multipole expansion as defined by expression (7.39), which includes terms up to and including those of order $(\frac{1}{r})^q$. While the derivation of this result is quite different from that used for the separable element, the trial solution itself is almost identical. Indeed if the geometry of the mapped and separable elements are made the same by setting $a_1 = a_2 = a$, the shape functions given by expressions (7.51)–(7.54) and those given by expression (7.68) are the same, except that the transverse interpolation in the separable element is written in terms of a true circular arc distance, whereas the transverse interpolation in the mapped element is measured along a straight line between the two nodes. This difference becomes indistinguishable when a fine mesh is used, or when higher order quadratic elements with curved sides are employed.

The Unconjugated "Bettess" Formulation

The unconjugated formulation proceeds in the same way as for the separable case with the definition of test functions which are identical to the trial functions, i.e

$$w_\alpha(\boldsymbol{x}, \omega) \;=\; \phi_\alpha(\boldsymbol{x}, \omega) \;=\; N_\alpha(s,t)\,\mathrm{e}^{-\mathrm{i}k\psi_\alpha(s,t)}, \tag{7.69}$$

where N_α and ψ_α are defined in the parent rather than the physical space. The substitution of these expressions into Equation (7.50) gives the element contribution to the coefficient matrix. Since all the integrands are written in terms of s and t, it is convenient to perform the integrals over the parent element. As in the separable case, highly oscillatory factors occur in the integrands, the product of wave–like factors $e^{-ik\psi_\alpha(\boldsymbol{x})}$ and $e^{-ik\psi_\beta(\boldsymbol{x})}$, in the $\alpha\beta$ term. While Gauss–Legendre rules can be used in the t direction special integration rules must be developed to integrate these terms in the s direction. These are detailed by Bettess [43] and will not be discussed here except to note that ill–defined terms which occur in the volume integral and on the boundary Γ_R once again cancel exactly as $R \to \infty$ and can therefore be omitted from the numerical integration without approximation [6].

The Conjugated "Astley–Leis" Mapped Formulation

The test functions for conjugated mapped elements are selected in the same way as for the separable element. That is to say, in the case of the Astley–Leis element, the test function $w_\alpha(\boldsymbol{x}, \omega)$ is defined to be the complex conjugate of the shape function, multiplied by a geometric factor proportional to $(1/r^2)$, *i.e*

$$w_\alpha(\boldsymbol{x}, \omega) \;=\; D_\alpha(\boldsymbol{x})\, \phi^*_\alpha(\boldsymbol{x}, \omega) \;=\; D_\alpha(\boldsymbol{x})\, N_\alpha(\boldsymbol{x})\, e^{+ik\psi_\alpha(s,t)} \,, \tag{7.70}$$

where $D_\alpha(\boldsymbol{x}) = (r_\alpha/r)^2$. This factor can be written in terms of the mapped coordinates by using relationship (7.66),

$$D_\alpha(\mathbf{x}) = \left\{ \frac{r_\alpha(1-s)}{2a(t)} \right\}^2 . \tag{7.71}$$

Substitution of (7.68) and (7.70) into (7.50) then gives, after some careful re–arrangement of terms

$$\Delta A^{ie}_{\alpha\beta} \;=\; e^{-ik(r_\alpha - r_\beta)} \left[\Delta K^{ie}_{\alpha\beta} \;+\; i i\omega \Delta C^{ie}_{\alpha\beta} \;-\; \omega^2 \Delta M^{ie}_{\alpha\beta} \right] \tag{7.72}$$

where $\Delta\mathbf{K}^{ie}$,$\Delta\mathbf{C}^{ie}$ and $\Delta\mathbf{M}^{ie}$ are frequency independent stiffness damping and mass matrices for a specific element whose components are given by

$$\Delta K^{ie}_{\alpha\beta} = \lim_{R\to\infty} \left\{ \int_{(\Omega_R - \Omega_c)^e} \Big(N_\alpha \nabla D_\alpha + N_\alpha \nabla D_\alpha \Big) \cdot \nabla N_\beta d\Omega \right\}, \tag{7.73}$$

$$\Delta C^{ie}_{\alpha\beta} = \frac{1}{c} \lim_{R\to\infty} \left\{ \int_{(\Omega_R - \Omega_c)^e} D_\alpha \Big(N_\alpha \nabla \psi_\alpha \cdot \nabla N_\beta - N_\beta \nabla \psi_\beta \cdot \nabla N_\alpha + - N_\alpha N_\beta \nabla \psi_\alpha \Big) d\Omega + \int_{\Gamma^e_R} D_\alpha N_\alpha N_\beta d\Gamma \right\}, \tag{7.74}$$

$$\Delta M^{ie}_{\alpha\beta} = \frac{1}{c^2} \lim_{R\to\infty} \left\{ \int_{(\Omega_R - \Omega_c)^e} D_\alpha N_\alpha N_\beta \Big[1 - \nabla \psi_\alpha \cdot \nabla \psi_\beta \Big] d\Omega \right\}. \tag{7.75}$$

The oscillatory factors $e^{\pm ik\psi_{\alpha,\beta}}$ in the trial and test functions cancel within each integral, leaving integrands that are well behaved and smooth in the (s,t) space. Gauss–Legendre quadrature can therefore be used in both directions. Also, the individual mass, stiffness and damping matrices are independent of frequency, and can be assembled once only for multi–frequency calculations.

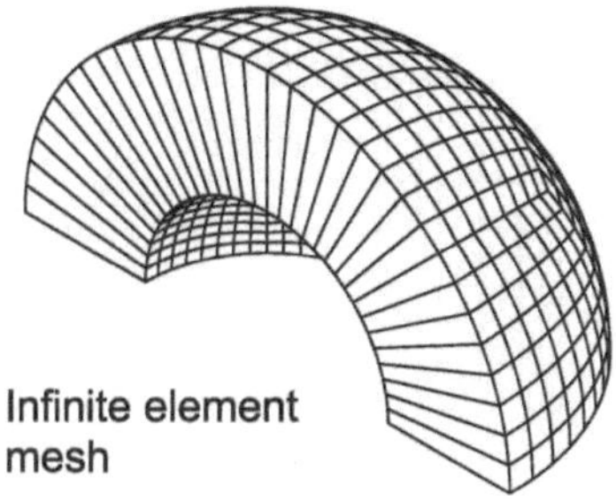

(a) Mesh.

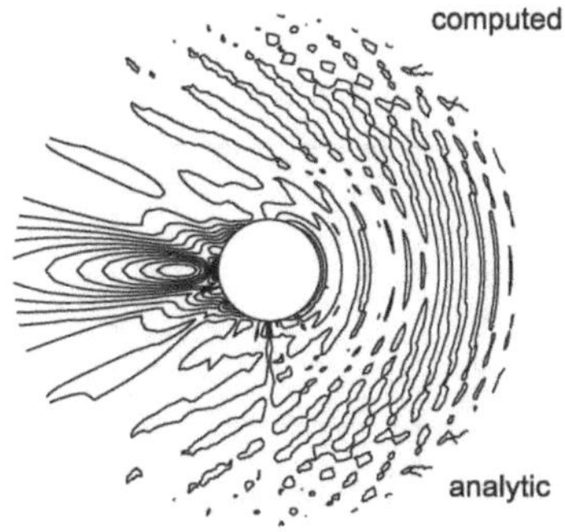

(b) Solution contours, $kR = 10.0$.

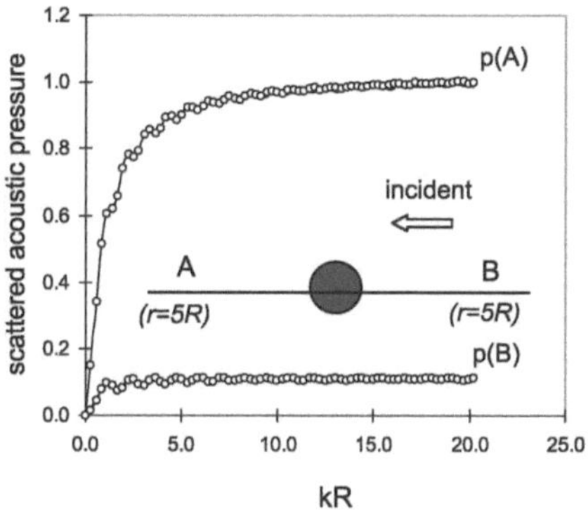

(c) Computed versus exact response at points A and B.

Fig. 7.8 Performance of conjugated elements. Scattering by a hard sphere. (Reused with permission from R Astley, G Macaulay, J–P Coyette, and L Cremers , *The Journal of the Acoustical Society of America* **103**, 49–63(1998). Copyright 1998, Acoustical Society of America).

7.5 Accuracy and Performance

The formulations of sections 7.4.5 and 7.4.6 give a choice of four types of element; separable–unconjugated, separable–conjugated, mapped–unconjugated, and mapped–conjugated. As already noted, the mapped and separable formulations give virtually identical elements when the element faces are aligned with the coordinate directions. The main distinction is therefore between the conjugated and unconjugated schemes.

The accuracy and computational performance of conjugated and unconjugated schemes have been assessed in various studies which compare computed and analytic results for known benchmark problems [13, 14, 22]. Fewer comparisons have been made which match the performance of the conjugated and unconjugated formulations directly [8, 30, 40]. All of the formulations involve the solution of a set of linear equations of order n, where n is the total number of degrees of freedom in the model. This dominates the solution time, and n can therefore be used as a com-

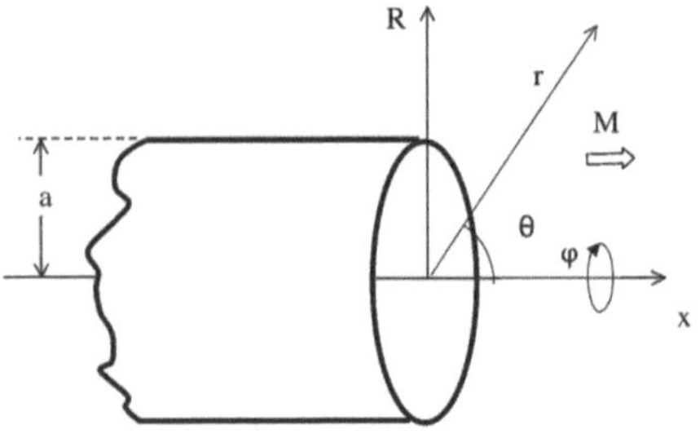

(a) Problem geometry.

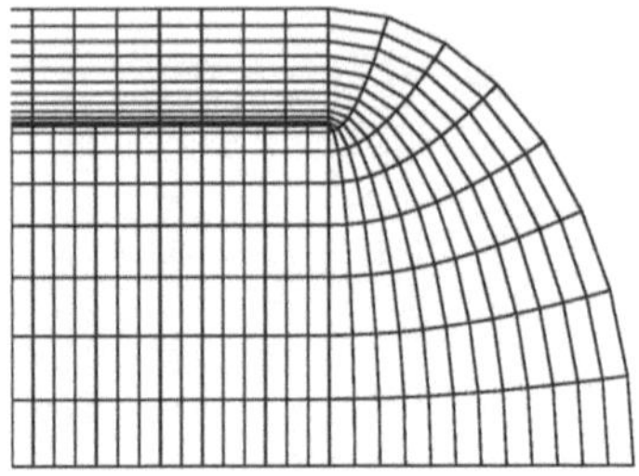

(b) FE Mesh.

(c) Computed near field acoustic pressure; mode $(26, 1)$, Mach number $= -0.5$, $ka = 35$.

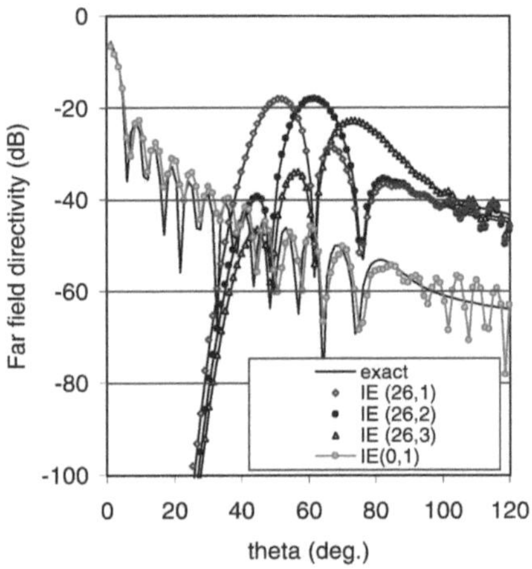

(d) Exact and computed far field directivities. Mach number $= -0.5$, $ka = 35$. Modes $(26, 1), (26, 2), (26, 3)$ and $(0, 1)$.

Fig. 7.9 Radiation from an unflanged duct with flow. A benchmark problem for acoustic radiation from turbofan intakes.

mon parameter against which to compare accuracy and computational effort between methods. [2]

The effectiveness of the conjugated schemes is illustrated in Figures 7.8 and 7.9. These show comparisons of computed and exact solutions for two benchmark problems. Figure 7.8 shows IE results for scattering of a plane wave by a rigid sphere of radius R. The mesh, shown in Figure 7.8(a) consists entirely of infinite elements of radial order 10 which are directly attached to the sphere. Computed and analytic instantaneous pressure contours are shown in Figure 7.8(b) for $kR = 10.0$, and the frequency response at two field points A and B within the IE domain are shown in

[2]This is not entirely true when the conditioning of the equations has an effect on solution time. If the condition number can be reduced to a level where indirect solvers can be used, as proposed by Dreyer and von Estorff [24] the solution time and memory requirements scale in quite a different way.

Figure 7.8(c) for frequencies in the range $kR \in [0, 20]$. The solution was generated by using a mapped code, but the elements are aligned in this case with spherical coordinates and are in effect indistinguishable from their separable equivalents. Further details are to be found in [14].

Solutions for a more challenging, high–frequency problem which involves propagation from an unflanged duct in the presence of mean flow is shown in Figure 7.9. The FE/IE model in this case has been formulated for the convected Helmholtz equation. Details of this modification to the current formulation are given elsewhere and will not be repeated here [25]. The geometry of the benchmark problem is shown in Figure 7.9(a) and the inner FE mesh in Figure 7.9(b). Conjugated Astley–Leis elements of high radial order are attached to the outer boundary of the FE mesh and a spinning mode of azimuthal mode order 26 propagates along the duct from left to right. The (adverse) Mach number of the mean flow is -0.5. Solutions are presented for a non–dimensional frequency $ka = 35.0$ where a is the duct radius. Contours of instantaneous near field pressure are shown in Figure 7.9(c) and a comparison of exact and computed far field directivity for the first thee radial modes and for the fundamental plane mode are shown in Figure 7.9(d). Further details of this benchmark problem are given in [12]. The parameters chosen are typical of acoustic tones propagating in a turbofan intake at maximum power. It can be seen that the predicted and analytic far field radiation patterns are in very close agreement. The conjugated schemes are used quite extensively for such problems and have been implemented in commercial codes such as the ACTRAN[3] range of products, and LMS SYSNOISE[4] for applications of this type.

The effectiveness of the unconjugated formulations has been demonstrated by Bettess for many two–dimensional problems involving surface wave propagation on shallow water [43]. Validation cases for axisymmetric or three–dimensional problems are less numerous. One such test case is presented in Figure 7.10 [22]. Here the accuracy of the spheroidal, separable, unconjugated formulation is demonstrated for a low–frequency acoustic–structural problem. A spheroidal FE/IE mesh of the type used for this calculation is shown in Figure 7.10(a). The mesh consists of a single layer of finite elements, indicated by solid lines, matched at an intermediate spheroidal surface to a layer of infinite elements of the Burnett type. These are aligned with a prolate spheroidal coordinate system. Figure 7.10(b) shows the close correspondence achieved between exact and computed solutions for the far–field directivity due to prescribed structural excitation of the thin cylindrical shell.

7.6 Conditioning

The results shown in the previous section were obtained by using elements of varying radial orders. The results of Figure 7.10, for example were obtained by using elements of radial order 2, whereas in the case of Figure 7.8 conjugated elements

[3] see http://www.fft.be
[4] see http://www.lmsintl.com

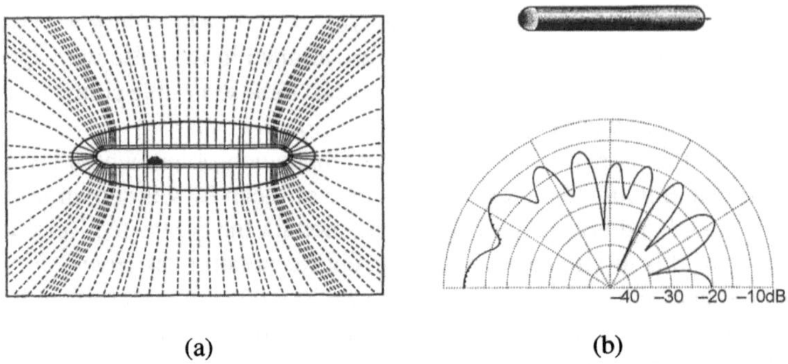

Fig. 7.10 Performance of unconjugated spheroidal elements of the Burnett type. Acoustic radiation from a cylindrical shell due to structural excitation. (a) Spheroidal FE/IE mesh. (b) Computed and exact far field sound pressure level plotted against polar angle measured from the cylinder axis, $ka = 1.96$ (– computed, ... exact) (Reused with permission from David S. Burnett, *The Journal of the Acoustical Society of America,* 96, 2798 (1994). Copyright 1994, Acoustical Society of America).

of radial order 10 were used. In the latter case, however, it was reported that when elements of radial order 11 were applied to the same problem, the accuracy of the solution deteriorated rapidly [14]. This degradation of the computed solution is associated with the conditioning of the coefficient matrix.

Numerical studies have shown that in all the formulations described so far, the condition number of the coefficient matrix is relatively constant with frequency, but varies significantly with radial element order [7]. This variation is illustrated in Figure 7.11 where the condition number is shown for a simple benchmark problem of radiation from a vibrating sphere. Condition number is plotted against radial element order for conjugated and unconjugated schemes. The axisymmetric FE/IE mesh used for these calculations is shown in Figure 7.11(a). A single layer of finite elements extends for one twentieth of a wavelength from the sphere and is attached to an infinite element domain of variable radial order. The variation of condition number with radial order is shown in Figure 7.11(b). The formulations are labelled "BB( )" and "AL( )", to distinguish between the *unconjugated* Bettess–Burnett (BB) and the *conjugated* Astley–Leis (AL) schemes. The descriptor in parentheses ("Lagrange" or "S.Legendre") indicates the type of radial shape function that has been used. The "Lagrange" schemes use Lagrange interpolation in $1/r$ as described in Sections 7.4.5 and 7.4.6. The "S.Legendre" (Shifted Legendre) schemes use a different approach which will be discussed shortly.

The condition numbers of the two Lagrange formulations, the *conjugated* ("AL–Lagrange") and the *unconjugated* ("BB–Lagrange") schemes, both increase rapidly with element order reaching values in excess of 10^{14} for element orders in excess of 10. This is an indication that numerical rounding errors may corrupt these solutions for $q > 10$ since the condition number then approaches machine precision.

In the case of the conjugated formulations, however, the conditioning can be greatly improved by using a more "orthogonal" class of shape functions. This was

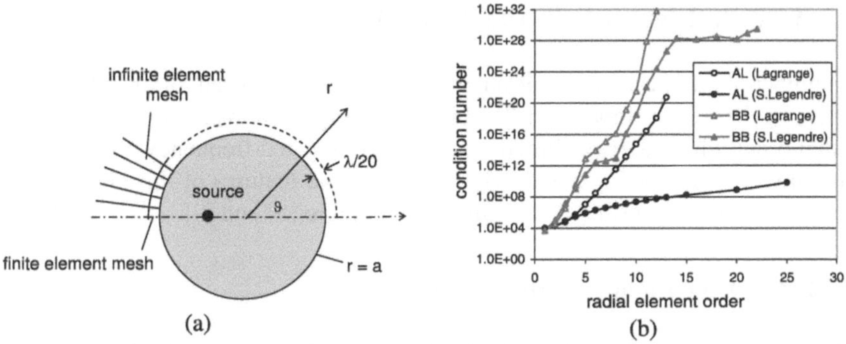

Fig. 7.11 Vibrating sphere test problem [7]. (a) Geometry and FE/IE mesh. (b)Variation of condition number with radial element order (ka=10π). Reprinted from the *Communications in Numerical Methods in Engineering*, **17**, RJ Astley and JP Coyette "Conditioning of infinite element schemes for wave problems", 31–34, (2001), with permission from John Wiley and Sons Ltd.

reported first by Shirron [40] who used shifted Legendre polynomials in $1/r$ as a trial basis rather than Lagrange interpolation. While this means abandoning nodal values of pressure as degrees of freedom and replacing them by polynomial coefficients, the use of such a basis greatly improves the conditioning of the conjugated schemes. This is demonstrated in Figure 7.11(b) where the "AL(S.Legendre)" values of condition number remain within acceptable limits to high orders (25 and beyond). The conditioning of the unconjugated scheme however is not significantly altered, as indicated by the "BB(S. Legendre)" curve on the same figure.

Recent studies [24] have shown that the conditioning of the conjugated schemes can be further improved by the use of different sets of orthogonal polynomials. This is discussed at greater length elsewhere in this volume, see Chapter 8. In practice therefore, the conditioning of conjugated infinite element schemes is not a significant limitation on their use. Certainly elements of radial order in the range $10 - -30$ are routinely used for real–world applications [12].

No remedy exists currently for the high condition numbers of unconjugated schemes, which become unstable for quite low radial orders. Paradoxically however, this is not a pressing issue in practice since very low radial orders appear to produce "superconvergent" near field results [40]. However, such elements give very poor — or indeed completely unstable — far field results within the IE domain. In practice, this also is not a major issue since the highly accurate near field solution can be used to generate equally accurate far field data by integration over a Kirchhoff surface, bypassing any inaccuracy in the IE representation.

In summary therefore, conditioning and stability both of the conjugated and the unconjugated schemes appear in theory to limit their application. In practice they pose few constraint on the use of these elements for the reasons mentioned above. It is true however that these issues are still not well understood. They are discussed at somewhat greater length in references [7] and [8].

7.7 Infinite Elements in the Time Domain

A number of infinite element formulations have been proposed for time domain problems. The most developed and flexible of these is the transient element of variable radial order which derives directly from the conjugated frequency–domain element presented in sections 7.4.5 and 7.4.6. The essential features of this element will be discussed here, but further details are to be found in [5] and [9].

7.7.1 The Time Domain Formulation

The development of frequency domain infinite elements was based on an assumption that the acoustic pressure field had achieved a steady time–harmonic state given by expression (7.2). Alternatively, the complex pressure amplitude $p(\boldsymbol{x},\omega)$ of Section 7.2, can be regarded as the complex Fourier transform of a time domain solution $\tilde{p}(\boldsymbol{x},t)$ which satisfies an initial value problem in time. The relationship between these two quantities is governed by the Fourier transform pair;

$$p(\boldsymbol{x},\omega) = F\left\{\tilde{p}(\boldsymbol{x},t)\right\} = \int_{-\infty}^{\infty} \tilde{p}(\boldsymbol{x},t)\mathrm{e}^{-\mathrm{i}\omega t}dt\ , \tag{7.76}$$

$$\tilde{p}(\boldsymbol{x},t) = F^{-1}\left\{p(\boldsymbol{x},\omega)\right\} = \frac{1}{2\pi}\int_{-\infty}^{\infty} p(\boldsymbol{x},\omega)\mathrm{e}^{\mathrm{i}\omega t}d\omega. \tag{7.77}$$

The same transform can be applied to nodal degrees of freedom and to the surface acceleration on Γ, in which case we can regard the degrees of freedom $q_j(\omega)$ and surface accelerations $a_n(\boldsymbol{x},\omega)$ as Fourier transforms of their transient equivalents $\tilde{q}_j(t)$ and $\tilde{a_n}(\boldsymbol{x},t)$. By the same argument, if $q_j(\omega)$ represents a nodal value of pressure amplitude, its inverse Fourier transform, $\tilde{q}_j(t)$, represents the time history of the acoustic pressure at that point. In principle, we can take the inverse Fourier transform of the discrete equations in the frequency domain — Equation (7.46) — to obtain an equivalent discretized relationship in the time domain of the form

$$\int_{-\infty}^{\infty} \tilde{\boldsymbol{A}}\,(t-\tau)\,\tilde{\boldsymbol{q}}\,(\tau)\,d\tau \;=\; \tilde{\boldsymbol{f}}\,(t), \tag{7.78}$$

where $\tilde{\boldsymbol{A}}\,(t)$ is the inverse Fourier transform of $\boldsymbol{A}(\omega)$. This equation is of little use in practice, unless the convolution integral on the left hand side can be evaluated simply. This is definitely *not* the case for the unconjugated schemes since the dependence of the coefficient matrix on frequency is complex owing to the presence of oscillatory terms, e^{-2ikr}, in the integrands for $\boldsymbol{A}(\omega)$.

In the case of the conjugated formulations however the frequency dependence of $\boldsymbol{A}(\omega)$ is much simpler. From expression (7.72) we see that the coefficient matrix for the conjugated element can *almost* be written in terms of frequency–independent mass, stiffness and damping matrices. This is impeded only by the constant factor $\mathrm{e}^{-ik(r_\alpha - r_\beta)}$ which occurs outside each integral and which conceals a dependence on ω through the wavenumber $k = \omega/c$. This can be removed quite simply however by a small modification to the trial expansion. Instead of the unknown coefficients $q_j(\omega)$ being defined as nodal values of the acoustic pressure amplitude at node j, they are normalized with respect to the phase of the solution at the corresponding innermost node; i.e we take as our unknown quantities in the IE region

$$q_\alpha(\omega) = p_\alpha(\omega)\mathrm{e}^{\mathrm{i}k(r_\alpha - r_1)}, \tag{7.79}$$

where r_1 is the radius of the appropriate inner node of the element in the notation of Figure 7.6. This removes the offending exponential term from the expression for $\Delta A^{ie}_{\alpha\beta}$ while preserving compatibility with the FE mesh on the inner boundary of the infinite element domain. The resulting coefficient matrix, $\boldsymbol{A}^{ie}(\omega)$, for the infinite element region then has components

$$A^{ie}_{\alpha\beta} = K^{ie}_{\alpha\beta} + \mathrm{i}\omega C^{ie}_{\alpha\beta} - \omega^2 M^{ie}_{\alpha\beta} \tag{7.80}$$

where $\boldsymbol{K}^{ie}$, $\boldsymbol{C}^{ie}$ and $\boldsymbol{M}^{ie}$ are constant matrices. When the finite element contributions $\boldsymbol{K}^{fe}$ and $\boldsymbol{M}^{fe}$ are added, this gives a set of equations in the frequency domain of the form

$$\boldsymbol{A}(\omega)\,\boldsymbol{q}(\omega) = \boldsymbol{K} + \mathrm{i}\omega\boldsymbol{C} - \omega^2\boldsymbol{M}\boldsymbol{q}(\omega) = \boldsymbol{f}(\omega), \tag{7.81}$$

where $\boldsymbol{K} = \boldsymbol{K}^{ie} + \boldsymbol{K}^{fe}$, etc. Performing the inverse Fourier transform of the above equation is then trivial, provided that acoustic pressure and velocity are initially zero at all points. This gives an equivalent discrete set of equations in the time domain

$$\boldsymbol{K}\tilde{\boldsymbol{q}} + \boldsymbol{C}\frac{d\tilde{\boldsymbol{q}}}{dt} + \boldsymbol{M}\frac{d^2\tilde{\boldsymbol{q}}}{dt^2} = \tilde{\boldsymbol{f}}\,(t). \tag{7.82}$$

These mimic the discrete equations of structural dynamics which are routinely encountered in the FE dynamic analysis. Many schemes exist for integrating these equations in time, and these can be used for the acoustic problem with the *proviso* that care must be taken to ensure that the transient discrete system is stable in time. Numerical studies have shown that this is the case provided that the interface between the finite and infinite element meshes is spherical or spheroidal [11]. Since the time domain equations derive directly from the infinite element model in the frequency domain, they exhibit the same characteristics with regard to radial element order and multipole accuracy. That is to say, by increasing the element order, the transparency of the FE/IE interface to near–field multipole components can be increased indefinitely.

The application of such a model to a simple test problem is illustrated in Figure 7.12(a). A segment of a hard sphere of radius a vibrates to form a spherical piston in a hard spherical baffle. The motion starts at time $t = 0$ and is driven at a steady radian frequency ω. The radiated sound field settles to a steady harmonic state after a short transient phase. Both the transient and steady state solutions can be calculated analytically [32]. The exterior acoustic domain is modeled by a FE mesh which extends to a radius of $3a$, as shown in Figure 7.12(b). The remainder of the exterior region is modelled by transient infinite elements of radial order 10. The exact

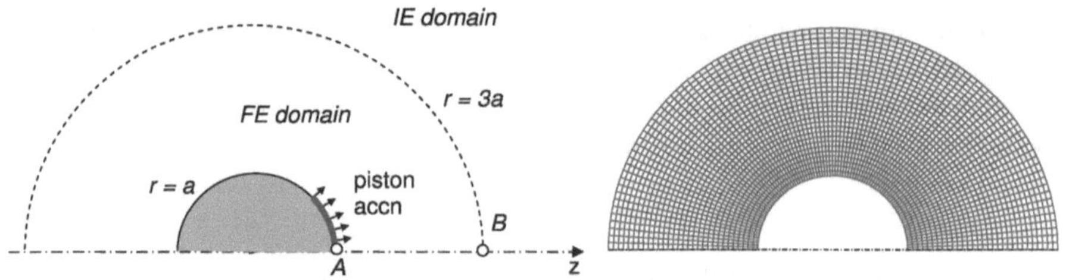

(a) (b)

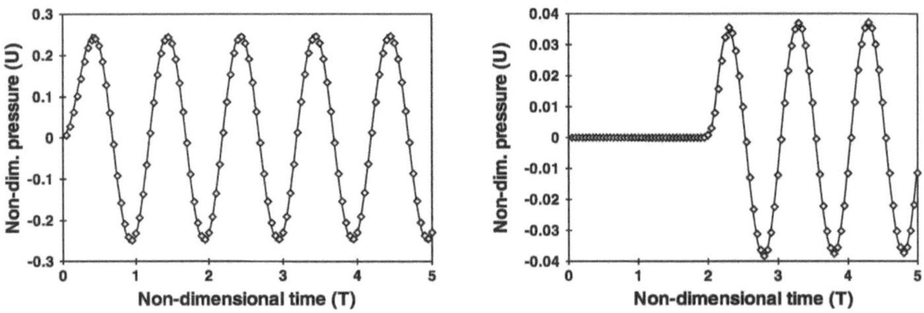

(a) (b)

Fig. 7.13 Transient test problem [32]. (a) Acoustic response at A. (b) Acoustic response at $B.(T = ct/a)$.

and computed acoustic pressure response at a point A on the surface of the radiator, and at point B at the FE/IE interface, are shown in Figures 7.13(a) and 7.13(b). The exact and predicted contours of instantaneous acoustic pressure throughout the domain are shown in Figure 7.14 at times $T = 1, T = 2$ and $T = 4$, where $T = tc/a$. An implicit Newmark scheme is used to integrate the equations.

Clearly, the transient and steady parts of the solution are resolved to a high degree of accuracy. This is a demanding problem in the sense that the solution in the near field contains creeping waves which propagate around the surface of the sphere to form a weak radiation lobe behind the piston. The magnitude of the acoustic pressure in this region is almost two orders of magnitude below that of the primary lobe but is well resolved by the FE/IE model as is the associated phase. The fact that the FE solution is not contaminated even at these low levels by spurious reflections from the IE domain, demonstrates the effectiveness of the infinite element model in providing an accurate termination for the inner model. Although not shown here, it is also possible to extract the pressure history at any point in the unbounded region from the transient FE/IE solution. The accuracy at such points is comparable to that in the FE portion of the mesh. The reader is referred to [9] and [4] for details of this procedure and evidence of its effectiveness.

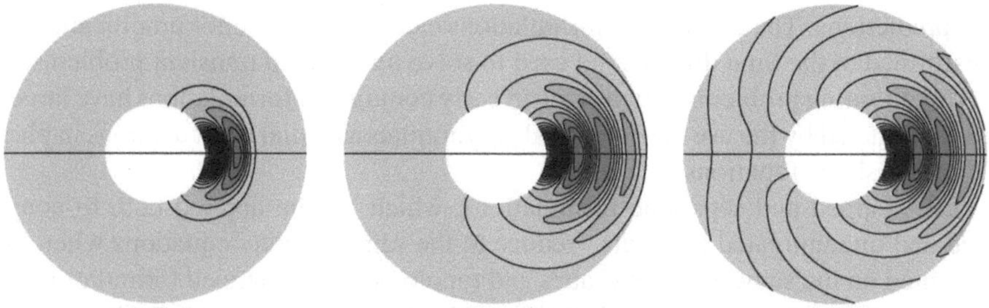

(a) T=1 (b) T=2 (c) T=4

Fig. 7.14 Test problem [32]. Radiation from a spherical piston in a spherical baffle. Contours of exact (lower half) and FE/IE (upper half) solutions at $T = 1, 2$ and 3 ($T = ct/a$).

Since Equations (7.82) are linear, an alternative to direct time integration of this equation, would be to first expand the acoustic field in terms of the "modes" of the homogeneous problem – Equation (7.81) with the right hand side set equal to zero. The structure of the coefficient matrix tells us that such modes exist for the exterior problem, albeit with complex eigenfrequencies when the damping matrix $\boldsymbol{C}$ is included. This arises exclusively from the infinite element contributions. The resulting modes therefore correspond to natural damped acoustic solutions for the coupled near and far fields taking into account radiation damping. Astley has used such modes to study the stability of the discrete time domain equations [11] and Marburg has looked at their general characteristics with a view to using them as the basis for computing the response of exterior problems [37–39].

7.8 Conclusions

The Infinite element method is one of many numerical techniques which have been developed to simulate anechoic conditions for unbounded wave problems. The method has been implemented both in the frequency and time domains but is more highly developed in the frequency domain. Its practical advantages include;

- Its general robustness,
- Its similarity to conventional finite element models, and ease of integration with existing finite element codes,
- Its preservation of the non–local nature of such models on the radiation boundary
- Its ability — in the case of conjugated elements — to give far field solutions without additional postprocessing.

The conjugated and unconjugated schemes have different strengths and weaknesses. The unconjugated schemes converge rapidly on the surface of a scatterer for very low element orders but are unable to resolve the far field directly. The conjugated schemes require higher radial orders to resolve accurately the near field, but the solution is then uniformly convergent over the entire exterior domain and gives accurate far field predictions. The conjugated formulations can also, with minor modification, be transformed to the time domain and used to solve unbounded transient problems.

Conditioning problems which arose in early conjugated formulations have largely been eliminated by the use of orthogonal polynomials as radial trial functions in place of "nodal" shape functions.

The mapped and separable formulations, which can be applied both to conjugated and unconjugated elements, reduce to the same discrete equations when implemented on the same element shapes and topologies. The mapped formulation can be more convenient in real world applications because of the additional freedom it confers in placing the Finite/Infinite element boundary.

This chapter has been able to present only the most basic introduction to infinite element formulations for acoustics. No attempt has been made to touch upon applications of the method to other types of wave propagation, and, even within acoustics, it has not been possible to deal in detail with important applications in flow acoustics and elastodynamics. For an appreciation of the wider applications of infinite elements in engineering mechanics and for a more thorough history of their development a good starting point is still Peter Bettess' 1992 monograph on the subject [21] supplemented by more recent reviews by the same author in the multi–volume texts of Zienkiewicz and Taylor [44].

In final conclusion, it is perhaps worth observing that infinite elements as described in this chapter can also be regarded as seminal members of a much broader class of finite element schemes for wave problems in which *multiple* "wave–type" solutions are used as the basis for a discrete numerical model. In the infinite element formulation a single (outward) wave direction is assumed, but the same concept can be applied to elements in which a finite number of plane waves is used within each element to describe a trial solution. The Partition of Unity Method (PUM) is perhaps the most straightforward of these [16, 29, 35] being an obvious extension of continuous, polynomial–based FE discretisations. However, the Ultra Weak Variational Formulation (UWVF) [33] whose theoretical basis is somewhat more esoteric [23] has also been shown to be effective as have a number of other discontinuous formulations [26, 27]. The reader is referred to [28] for an up–to–date review of these methods and of the linkages between them.

References

1. Astley RJ (1985) A finite element wave envelope formulation for acoustical radiation in moving flows. Journal of Sound and Vibration 103:471–485
2. Astley RJ (1996) FE mode matching schemes for the exterior Helmholtz problem and their relationship to the FE–DtN approach. Communications in Numerical Methods in Engineering 11:257–267
3. Astley RJ (1998) Mapped spheroidal elements for unbounded wave problems. International Journal of Numerical Methods in Engineering 41:1235–1254
4. Astley RJ (1998) Transient spheroidal elements for unbounded wave problems. Computer Methods in Applied Mechanics and Engineering 164:3–15

5. Astley RJ (1996) Transient wave envelope elements for wave problems. Journal of Sound and Vibration 192:245–261
6. Astley RJ, Bettess P, Clark PJ (1991) Letter to the editor. Re: Mapped infinite elements for exterior wave problems. International Journal of Numerical Methods in Engineering 32:207–209
7. Astley RJ, Coyette J–P (2001) Conditioning of infinite element schemes for wave problems. Communications in Numerical Methods in Engineering 17:31–34.
8. Astley RJ, Coyette J–P (2001) The performance of spheroidal infinite elements. International Journal of Numerical Methods in Engineering 52:1379–1396
9. Astley RJ, Coyette J–P, Cremers L (1998) Three dimensional wave envelope elements of variable order for acoustic radiation and scattering. Part II: Formulation in the time domain. Journal of the Acoustical Society of America 103:64–72
10. Astley RJ, Eversman W (1983) Finite element formulations for acoustical radiation. Journal of Sound and Vibration 88:47–64
11. Astley RJ, Hamilton JA (2006) The stability of infinite element schemes for transient wave problems. Computer Methods in Applied Mechanics and Engineering 195:3553–3571
12. Astley RJ, Hamilton JA, Baker N, Kitchen E (2002) Modelling tone propagation from turbofan inlets. The effect of extended lip liners. In: Proceedings of the 8th AIAA/CEAS Aeroacoustics Conference AIAA Paper 2002–2449, Breckenridge
13. Astley RJ, Macaulay GJ, Coyette J–P (1994) Mapped wave envelope elements for acoustical radiation and scattering. Journal of Sound and Vibration 170:97–118
14. Astley RJ, Macaulay GJ, Coyette J–P, Cremers L (1998) Three dimensional wave envelope elements of variable order for acoustic radiation and scattering. Part I: Formulation in the frequency domain. Journal of the Acoustical Society of America 103:49–63
15. Babuska I, Ihlenburg F, Strouboulis T, Gangaraj SK (1997) A posteriori error estimation for finite element solutions of Helmholtz' equation. Part II: estimation of the pollution error. International Journal of Numerical Methods in Engineering 40:3883–3900
16. Babuska I, Melenk JM (1997) The partition of unity method. International Journal of Numerical Methods in Engineering 40:727–758
17. Bayliss A, Gunzberger M, Turkel E (1982) Boundary conditions for the numerical solution of elliptic equations in exterior regions. SIAM Journal on Applied Mathematics 42:430–450
18. Bayliss A, Turkel E (1980) Radiation boundary conditions for wave–like equations. Communications in Pure and Applied Mathematics 33:707–725
19. Berenger J (1994) A perfectly matched layer for the absorbtion of electromagnetic waves. Journal of Computational Physics 114:185–200
20. Bettess P (1977) Diffraction and refraction of surface waves using finite and infinite elements. International Journal of Numerical Methods in Engineering 11:1271–1290
21. Bettess P (1992) Infinite elements, Penshaw Press, Sunderland
22. Burnett D (1994) A three–dimensional acoustic infinite element based on a prolate spheroidal multipole expansion. Journal of the Acoustical Society of America 96:2798–2816
23. Cessenat O, Despres B (1998) Application of an ultra–weak variational formulation of elliptic PDEs to the two–dimension Helmholtz problem. SIAM Journal on Numerical Analysis 35:255–299
24. Dreyer D, Estorff O von (2003) Improved conditioning of infinite elements for exterior acoustics. International Journal of Numerical Methods in Engineering 58:933-953
25. Eversman W (1999) Mapped infnite wave envelope elements for acoustic radiation in a uniformly moving medium. Journal of Sound and Vibration 224:665–687

26. Farhat C, Harari I, Franca L (2001) The discontinuous enrichment method. Computer Methods in Applied Mechanics and Engineering 190:6455–6479
27. Farhat C, Wiedemann–Goiran P, Tezaur R (2004) A discontinuous Galerkin method with plane waves and Lagrange multipliers for the solution of short wave exterior Helmholtz problems on unstructured meshes. Wave Motion 39:307–317
28. Gabard G (2007) Discontinuous Galerkin methods with plane waves for time–harmonic problems. Journal of Computational Physics 225:1961–1984
29. Gamallo P, Astley RJ (2006) The partition of unity finite element method for short wave acoustic propagation on non–uniform potential flows. International Journal of Numerical Methods in Engineering 65:425–444
30. Gerdes K, Demkowicz L (1996) Solution of 3d Laplace and Helmholtz equations in exterior domains using *hp*–infinite elements. Computer Methods in Applied Mechanics and Engineering 137:239–273
31. Gorranson P, Davidsson C F (1987) Three–dimensional infinite elements for wave propagation. Journal of Sound and Vibration 115:556–559
32. Hamilton JA (2003) Transient infinite element solutions for unbounded wave problems. PhD Thesis, University of Southampton, Southampton.
33. Huttunen T, Monk P, Kaipio JP (2002) Computational aspects of the ultra–weak variational formulation. Journal of Computational Physics 182:27–46
34. Ihlenburg F (1998) Finite element analysis of acoustic scattering. Springer–Verlag, New York
35. Laghrouche O, Bettess P, Astley RJ (2002) Modelling of short wave diffraction problems using approximating systems of plane waves. International Journal of Numerical Methods in Engineering 54:1501–1533
36. Leis R (1986) Initial boundary value problems in mathematical physics. John Wiley and Teubner Verlag, Stuttgart
37. Marburg S (2005) Normal modes in external acoustics. Part I: Investigation of the one–dimensional duct problem. Acta Acustica united with Acustica 91:1063–1078
38. Marburg S (2006) Normal modes in external acoustics. Part III: Sound power evaluation based on frequency–independent superposition of modes. Acta Acustica united with Acustica 92:296–311
39. Marburg S, Dienerowitz F, Horst T, Schneider S (2006) Normal modes in external acoustics. Part II: Eigenvalues and eigenvectors in 2d. Acta Acustica united with Acustica 92:97–111
40. Shirron J, Babuska I (1998) Comparison of approximate boundary conditions and infinite element methods for exterior Helmholtz problems. Computer Methods in Applied Mechanics and Engineering 164:121–139
41. Sommerfeld A (1949) Partial differential equations in physics. Academic Press, New York
42. Wilcox CH (1956) A generalisation of theorems of Rellich and Atkinson. Proceedings of the American Mathematical Society 7:271–276
43. Zienkiewicz OC, Bando K, Bettess P, Emson C, Chiam T (1985) Mapped infinite elements for exterior wave problems. International Journal of Numerical Methods in Engineering 21:1229–1252
44. Zienkiewicz OC, Taylor RL (2004) The finite element method. Vol 3 Fluid Dynamics, Butterworth Heinemann, London, 5th edition

8

Efficient Infinite Elements based on Jacobi Polynomials

Otto von Estorff[1], Steffen Petersen[2], and Daniel Dreyer[3]

[1] Institute of Modelling and Computation, Hamburg University of Technology, Denickestraße 17, D–21073 Hamburg, Germany
estorff@tu-harburg.de
[2] Department of Mechanical Engineering, Stanford University, 488 Escondido Mall, Mail code 3035, Stanford, CA 94305, USA
steffenp@stanford.edu
[3] Hanse Wohnbau GmbH, Birkenweg 15, D–22850 Norderstedt, Germany
daniel.dreyer@h-wb.de

Summary. In this contribution an optimized version of the so–called mapped wave envelope elements, also known as Astley–Leis elements, is presented and its practical usability is assessed. The elements are based on Jacobi polynomials in the direction of radiation, which leads to a low conditioning of the resulting system matrices and to a superior performance in conjunction with iterative solvers. This is shown for practically relevant simulations in the frequency as well as in the time domain.

8.1 Introduction

The finite element method (FEM) is a domain based approach and hence special care is necessary to simulate the response of an unbounded acoustic fluid coupled to a vibrating structure. In order to treat exterior problems while using the FEM an artificial boundary is generally introduced, that truncates the unbounded computational domain. Special techniques are then required to suppress spurious reflections on the artificial boundary. Up to now various approaches are available for this, where the best known are non–reflecting boundary conditions [34–36, 41], absorbing layers [14, 38, 51, 54], and infinite elements [2, 16, 27, 32, 50]. Surveys of these methods can be found in the recent reviews on the time–harmonic acoustic FEM in [37] and [52]. Though already a few years old, Shirron and Babuška give an excellent comparison of the advantages and weaknesses of absorbing boundary conditions and infinite elements in [49].

In contrast to approximate non–reflecting boundary conditions and absorbing layer approaches (such as the perfectly matched layer method), infinite elements do not terminate the computational domain, but actually represent the complete unbounded acoustic fluid. In the combined infinite/finite element method (IFEM) the absorbing boundary conditions are replaced by a layer of elements that extends to infinity. The Sommerfeld condition is fulfilled by means of special approximation functions in the radial (infinite) direction.

Today, primarily two types of infinite elements are in common use, that mainly differ in the definition of weighting functions. These are the Burnett elements [19] (also referred to as Bettess–Burnett elements) and the Astley–Leis or mapped wave envelope elements [9]. Depending on the choice of the weighting functions, the formulations may be further distinguished between conjugated and unconjugated elements. It is also noted, that infinite elements have been formulated for different separable coordinate systems, cf. [4, 20, 21]. Various aspects such as accuracy, convergence and stability as well as conditioning of the various element types have been studied in numerous publications [3,4,6,24,27,31,33,49]. Comparing the two aforementioned element types, it is generally acknowledged that the unconjugated Burnett elements provide most accurate results within the conventional finite element domain. On the other hand they lack true stability in the exterior [40, 49], where the Astley–Leis elements are most effective. In this contribution emphasis is placed on a variant of the Astley–Leis element that is based on Jacobi polynomials and leads to highly efficient iterative simulations procedures [27].

Compared to standard formulations of the boundary element method (BEM), a major advantage of the IFEM is the fact that the infinite elements preserve the banded structure of the system matrices, enabling for efficient iterative solution procedures. A direct cost comparison of the infinite elements presented here and commercial BEM software can be found in [26]. Additionally, the formulation employed here may be easily transformed to the time domain and with some slight modifications this method provides stable simulations of transient exterior acoustics, whereas transient boundary element formulations are known to behave rather unstable.

In the subsequent section, the general concept of acoustic computations with finite and infinite elements is described. This also includes some brief remarks on the iterative solution of the resulting system matrices by means of Krylov subspace methods. A crucial point of the acoustic computations in this contribution is the radial shape approximation in the Astley–Leis infinite element formulation. The derivation and identification of improved radial basis functions is given in Section 8.4. In order to highlight the advantages of the improved elements, their derivation is accompanied by a rather small sized numerical example with focus on the conditioning of the resulting system matrices. Practically relevant simulations in the frequency as well as in the time domain are then discussed in Section 8.5.1.

8.2 Acoustic Finite and Infinite Elements

Although first infinite element formulations originate from the 1970's [15, 17], their use remained a rather uncommon tool in practical applications for decades. Developments up to the beginning of the 1990's are given in the book by Bettess [16]. More recent reviews may be found in [2, 32].

For details on the governing relations it is referred to the introductory chapter of this book. And only major principals of the combined infinite/finite elements method

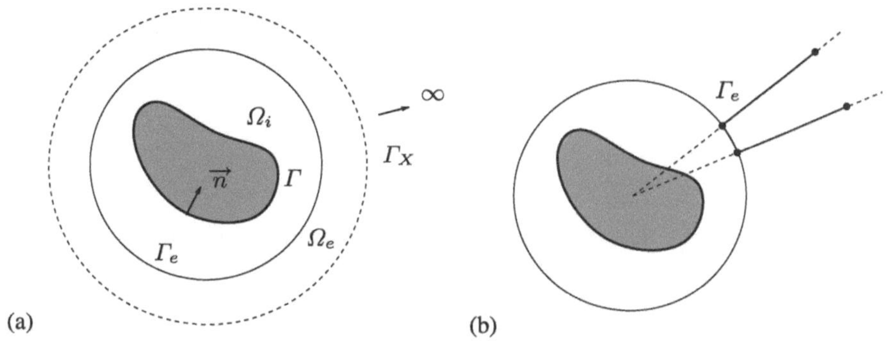

Fig. 8.1 (a) Geometry of the exterior problem; and (b) typical infinite element configuration.

(IFEM) are briefly described here. A detailed derivation of the Astley–Leis elements is also given in [9].

When using the IFEM to solve exterior boundary value problems the exterior domain Ω is decomposed into an interior part Ω_i and exterior part Ω_e through the envelope Γ_e as depicted in Figure 8.1. In the interior, conventional finite elements are employed for discretization, while the exterior is represented through infinite elements. These infinite elements serve to fulfill the Sommerfeld condition and suppress spurious reflections on the envelope Γ_e. This is achieved by means of approximation functions that are basically composed of an exponential term to represent the wave character of the numerical approximation, and an amplitude decay assuring that the sound pressure vanishes at infinity. Furthermore, the infinite elements are designed such that the solution of the Helmholtz equation is resembled with increasing order of the radial approximation.

The Astley–Leis formulation that is utilized in this contribution comprises the following fundamental properties:

- The shape approximation within an infinite element is obtained from a tensor product of conventional finite element shape functions on the element base (i.e. on the envelope) and special radial basis functions in the infinite direction as introduced in [19].
- The radial coordinate that extends to infinity is mapped on the reference element using the transformation described in [45].
- The variational formulation is based on a Petrov–Galerkin scheme, where conjugates of the basis functions are used as weighting functions.
- The test functions are taken from weighted Sobolev spaces and include an additional factor assuring that all inner products that appear in the weak form are finite, cf. [8,40,43]).
- The radial shape approximation is based on Jacobi polynomials that lead to favorable properties of the resulting system matrices [25,27]. More details on this aspect are given in Section 8.4.

Omitting terms that would arise from boundary conditions for ease of notation, the contribution from the exterior domain Ω_e to the weak form of the governing boundary value problem reads

$$\int_{\Omega_e} \left(\nabla\chi \cdot \nabla p - k^2 \chi p\right) \, d\Omega \,, \tag{8.1}$$

where the notation χ has been adopted for the weighting function, p is the sound pressure, and k is the wave number.

It will be shown in Section 8.4 that the discrete version of (8.1) matches the general form of the resulting system in terms of acoustic stiffness $\mathbf{K}$, damping $\mathbf{C}$, and mass matrix $\mathbf{M}$ given by

$$\mathbf{A}(\omega)\mathbf{p} := \left[\mathbf{K} + \mathrm{i}k\mathbf{C} - k^2\mathbf{M}\right] \mathbf{p} = \mathbf{f} \,, \tag{8.2}$$

where $\mathbf{f}$ is the right hand side vector due to prescribed Neumann boundary conditions. Note that the system in Equation (8.2) is based on time harmonic solutions of the form $p(\mathbf{x}, t) = \Re\{\hat{p}(\mathbf{x}, t)\mathrm{e}^{+\mathrm{i}\omega t}\}$.

8.2.1 Transformation to the Time Domain

A major advantage of the Astley–Leis formulation is the fact, that it can be straightforwardly transformed to the time domain [1, 5]. Since the system matrices $\mathbf{K}$, $\mathbf{C}$, and $\mathbf{M}$ are independent of the frequency an inverse Fourier transform of the form

$$p(\mathbf{x}, t) = F^{-1}(\hat{p}(\mathbf{x}, \omega)) = \frac{1}{2\pi} \int_{-\infty}^{+\infty} \hat{p}(\mathbf{x}, \omega) \, \mathrm{e}^{+\mathrm{i}\omega t} \, d\omega \tag{8.3}$$

may be applied to yield a second order system

$$\mathbf{K}\,\mathbf{p} + \frac{1}{c}\mathbf{C}\,\dot{\mathbf{p}} + \frac{1}{c^2}\,\mathbf{M}\,\ddot{\mathbf{p}} = \mathbf{f} \,. \tag{8.4}$$

The system can be solved employing common time integration schemes. Currently, no reliable concept to diagonalize the operators $\mathbf{C}$ and $\mathbf{M}$ is available, which would render explicit schemes particularly attractive. Hence, an implicit time integration, such as the Newmark algorithm, is generally used in order to solve Equation (8.4). A treatment of frequency dependent admittance boundary conditions in transient infinite element computations is described in [55].

It is noted here that, although the transformation to the time domain is straightforward, some problems may occur in the transient mapped wave envelope formulation. These problems are related to the approximation of the phase within the infinite elements. The phase approximation may lead to eigenvalues with negative real part in the mass matrix $\mathbf{M}$, rendering the time integration unstable. This issue has recently been addressed in various publications [7,22,23], and remedies have been suggested. A more straightforward approach to circumvent the instabilities is used in the simulations of Section 8.5.2. Namely, a discretization including a spherical envelope is chosen. This allows to explicitly set the mass matrix to zero, without deterioration of the numerical results [25].

8.3 Preconditioners and Krylov Subspace Methods

The matrices arising from finite element formulations for the Helmholtz equation are generally sparse, indefinite, complex valued and non–Hermitian. The most commonly employed iterative solution algorithms for the resulting systems are Krylov subspace methods such as BiCG–Stab [56], GMRES [48], QMR [30] or TFQMR [29].

Such Krylov solvers are prone to failure when the overall system matrix is ill–conditioned, and when the eigenvalue distribution of the matrix under consideration possesses certain unfavorable characteristics. Therefore, iterative solvers are commonly used together with preconditioners, that intend to improve the overall conditioning of the system matrix, so that the iterative solvers perform well. The infinite elements that are described in the subsequent section are designed to improve the conditioning of the system matrices. Hence, to some extent the improved elements may be considered as a preconditioning technique on the element level. Similar investigations in the framework of high order finite elements have been made in [46].

For the computations in Section 8.5.1, the the transpose–free variant of the quasi–minimal residual method, TFQMR, and the restarted generalized minimal residual, is employed. For preconditioning, a block incomplete LU–factorization is used, with zero level fill–in, as implemented in the software package PETSc [10]. This is a rather generic iLU(0) preconditioner. An efficiency increase may be expected when special preconditioning techniques [44], that decrease the indefiniteness of the Helmholtz operator, are used.

Derivations of the Krylov subspace methods mentioned above can be found in, e.g., [47,53]. The algorithms for of most commonly used solvers and preconditioners are given [11] and an extensive survey of preconditioning methods is given in [13].

8.3.1 Remarks on the Condition of a Matrix

To inspect in detail the improved infinite elements, the condition numbers of the resulting system matrices are considered. The most visible effect of severe condition numbers is the loss of accuracy, while the performance of iterative solvers also suffers from large conditioning. The condition number of a regular matrix $\mathbf{A} \in \mathbb{R}^{(n,n)}$ is defined based on some norm, say $\|\cdot\|_p$:

$$\kappa_p(\mathbf{A}) = \|\mathbf{A}^{-1}\|_p \, \|\mathbf{A}\|_p \,. \tag{8.5}$$

Assume, for example, $\kappa_p(\mathbf{A}) = 10^\gamma$, while the numerical representation of $\mathbf{A}$ employs finite precision of order 10^{-l}. Then the relative error inherent to the problem at hand, independent of the solution algorithm, is of order $10^{\gamma-l}$. It is apparent that orthogonal matrices perform best with respect to condition, such that $\kappa_p(\mathbf{A}) = 1$. Otherwise, the closer $\mathbf{A}$ is to a singular matrix, the higher is its condition number.

Concerning the solution speed of iterative solution techniques for symmetric and positive definite systems, it is a fact that not only the condition number, but especially the distribution of eigenvalues affects the performance [53]. Experience shows

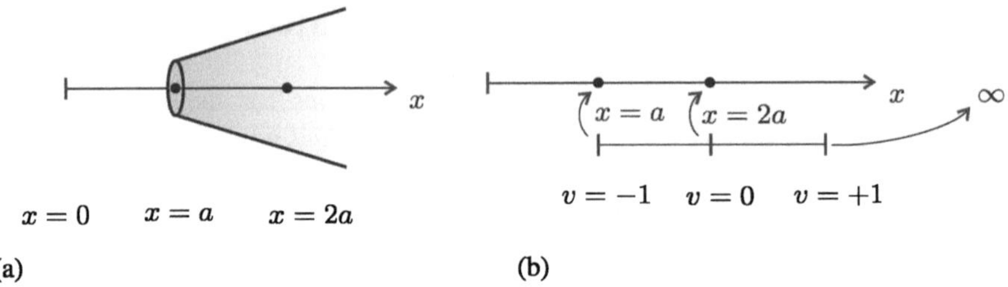

Fig. 8.2 (a) Geometry of the horn represented by one single infinite element with two mapping nodes. (b) Coordinate mapping $v \mapsto x$.

that clustering of eigenvalues is generally desirable. However, for the current case of non–symmetric indefinite systems, no clear concepts of preferable eigenvalue distributions have been proven yet.

8.4 Improved Astley–Leis Elements

In 1994, Astley et al. [8] derived the mapped wave envelope elements for the case of negligible circumferential variation, represented by just a single infinite element in the shape of a horn, as depicted in Figure 8.2(a). A similar approach is considered here, with particular emphasis on the integrands resulting from the weak form. Simple inspection leads to an optimized radial polynomial choice, which may easily be extended to three dimensions.

The cross section of the horn given in Figure 8.2(a) varies with x as $A(x) = \eta x^2$. This ensures similar behavior of the Helmholtz equation as in three dimensions. The elements are based on a two–noded non–isoparametric mapping. A deliberately chosen source location, here $x = 0$, defines the distance a between the source point and the base node, see Figure 8.2(b).

Employing a second node, generally located at $x = 2\,a$, the radial coordinate range $x \in [a, \infty)$ is transformed to a parent space $v \in [-1, +1)$ [45]. The two nodes at $x_1 = a$ and $x_2 = 2\,a$ define the geometry of the element. In two and three dimensions the element employs interpolants in the circumferential direction, perpendicular to v, while the coordinate transformation $v \mapsto r$ remains unchanged. Through

$$x = M_1(v)x_1 + M_2(v)x_2\,, \tag{8.6}$$

where

$$M_1(v) = \frac{-2v}{1-v}, \qquad M_2(v) = \frac{1+v}{1-v}\,, \tag{8.7}$$

the transformation is unique. In three dimensions the mapping in v remains valid, provided that rays of the element are diverging. Note that in two and three dimensions, this mapping reduces at the nodes to the inverse map $a/r = (1-v)/2$. Details may be found in [9].

The trial and test functions p and χ are approximated by $T(v)$ and $W(v)$, respectively. In indicial notation, using the summation convention, one obtains

$$p(\mathbf{x}) \approx \mathbf{p}_i\, T_i(v) \ = \mathbf{p}_i\, \Phi_i(v)\, \mathrm{e}^{-\mathrm{i}\, ka \frac{1+v}{1-v}}\,, \tag{8.8}$$

$$\chi(\mathbf{x}) \approx \boldsymbol{\chi}_j\, W_j(v) = \boldsymbol{\chi}_j \left(\frac{1-v}{2}\right)^2 \Phi_j(v)\, \mathrm{e}^{+\mathrm{i}\, ka \frac{1+v}{1-v}}\,, \tag{8.9}$$

where $\boldsymbol{\chi}$ is chosen such that the discrete weak form is tested independently at all degrees of freedom (reducing to the identity matrix), $\mathbf{p}_i$ is the i–th component of the vector of nodal pressure unknowns, and the polynomials $\Phi_i(v)$ and $\Phi_j(v)$ are weighted such that the Sommerfeld radiation condition is satisfied

$$\Phi_k(v) = \frac{1-v}{2}\, P_k(v)\,. \tag{8.10}$$

The choice of $P_k(v)$ is crucial for the performance of Astley–Leis elements. Note that in Equation (8.9) $W_j(v)$ includes an additional weight, $[(1-v)/2]^2$, providing finiteness in the associated modified norms, see e.g. [39] for details. Employing Equations (8.8, 8.9) in Equation (8.1), the exponential terms from p and χ cancel out. Including the coordinate transformation from Equation (8.6) leads to the discrete version of the weak form given in Equation (8.2).

The ij–th elements of the system matrices for a single infinite element are, in terms of polynomials defined in Equation (8.10):

$$\mathbf{K}_{ij} = \int_{-1}^{+1} \left[\left(\frac{1-v}{2}\right)^2 \frac{d\,\Phi_i(v)}{d\,v}\frac{d\,\Phi_j(v)}{d\,v} - \frac{1-v}{2}\Phi_i(v)\frac{d\,\Phi_j(v)}{d\,v} \right] 2\,a\,dv\,, \tag{8.11}$$

$$\mathbf{C}_{ij} = \int_{-1}^{+1} \left[\frac{2\,a^2}{1-v}\Phi_i(v)\Phi_j(v) - a^2\frac{d\,\Phi_i(v)}{d\,v}\Phi_j(v) + a^2\Phi_i(v)\frac{d\,\Phi_j(v)}{d\,v} \right] dv\,, \tag{8.12}$$

$$\mathbf{M}_{ij} = 0\,. \tag{8.13}$$

Note that in the current configuration $\mathbf{M}$ vanishes. The matrix expressions may be evaluated using conventional numerical integration.

Identification of Improved Polynomial Bases for Radial Expansion

The early concept of the Astley–Leis elements included standard Lagrange interpolation functions for the radial shape approximation. This led to extremely ill–conditioned system matrices for elements with high radial orders. Therefore infinite elements based on Legendre polynomials have been developed that lead to low condition numbers [3]. The approach is based on an infinite element concept modified by Shirron and Babuška [49]. In the following an optimized polynomial choice in the radial direction is presented which leads to even better performance of the infinite elements.

Consider now the discrete form of the simplified one–dimensional model, given in Equations (8.11) to (8.13), where $\Phi_i(v)$ and $\Phi_j(v)$ denote the trial and test functions, respectively, defined in Equation (8.10). It should be noted that the term $(1 - v)/2$ is crucial for decay towards infinity, otherwise the Sommerfeld radiation condition is not satisfied. In Equations (8.11) to (8.13) one may now attempt to find polynomial bases which are orthogonal with respect to the various products occurring in the element matrices. However, inspecting only the simple case of $\Phi_i(v)\Phi_j(v)$, one may rewrite this term using Equation (8.10) to obtain:

$$\Phi_i(v)\,\Phi_j(v) = \left(\frac{1-v}{2}\right)^2 P_i(v)\,P_j(v)\,. \tag{8.14}$$

This matches perfectly well with the orthogonality property of the Jacobi polynomials $P^{\alpha,\beta}$ given by

$$\int_{-1}^{+1} (1-v)^{\alpha}\,(1+v)^{\beta}\,P_i^{\alpha,\beta}(v)\,P_j^{\alpha,\beta}(v)\,dv = \mu_i\delta_{ij}\,, \tag{8.15}$$

and using $\alpha = 2$ and $\beta = 0$

$$\int_{-1}^{+1} (1-v)^2\,P_i^{2,0}(v)\,P_j^{2,0}(v)\,dv = \mu_i\delta_{ij}\,. \tag{8.16}$$

Clearly the choice of $\alpha = 2, \beta = 0$ yields an optimal basis for products of the type of Equation (8.14). Reconsidering the various types of products involved in the formulation of the discrete weak form raises thoughts of trying additional combinations of α and β. Therefore, in the following numerical examples, Jacobi polynomials with $\alpha \in \{1, 2, 3, 4\}$ and $\beta = 0$ are employed.

Numerical Results

Six different polynomial bases are employed to form the system matrices $\mathbf{M}$, $\mathbf{C}$, and $\mathbf{K}$, namely the still widely–used Lagrange polynomials, and shifted variants of Legendre and Jacobi polynomials using the four parameter pairs $(1, 0)$, $(2, 0)$, $(3, 0)$, $(4, 0)$ for (α, β). In all that follows, condition estimates are reported with respect to the matrix 1–norm.

Although computations with solely infinite elements are possible, they are most commonly used when coupled to conventional finite elements, which are employed to discretize the region in between a radiating structure or scatterer and the envelope (eventually refer to a previous figure). Two such discretizations are presented in Figure 8.3, namely one spherical radiator and one non–spherical radiator. In both cases symmetry is exploited.

In the discretization of the first example, one layer of linear quadrilaterals represents the inner fluid region, see Figure 8.3(a). The envelope is located at $r = 15$,

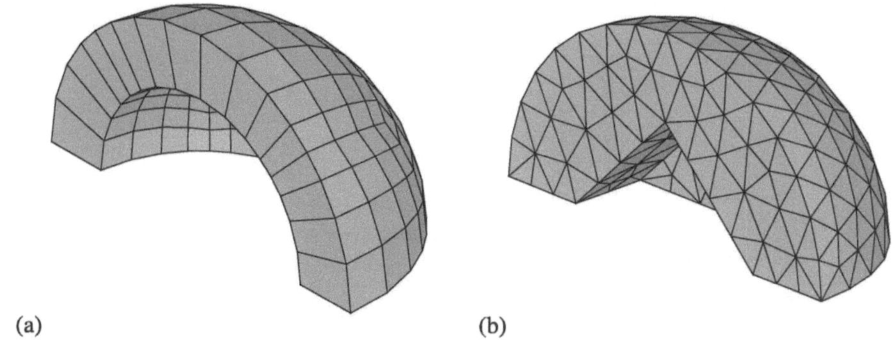

Fig. 8.3 Combined FEM–IFEM discretizations with (a) linear quadrilateral finite elements and (b) linear tetrahedral finite elements directly connecting to infinite elements on the outer face.

while the radius of the vibrating structure is $a = 10$. The second example, depicted in Figure 8.3(b), employs infinite elements enclosing a cube, tilted 45°. The typical length a is taken to be one half of the diagonal extending parallel to the symmetry planes, such that $a = 4\sqrt{2}$. Note that the envelope is located at the radial distance $r = 10$.

The condition of the overall system matrix $\mathbf{A}$ is computed for the maximum allowable frequency of the given discretization assuming that six elements are used to represent one wavelength. This relates to a non–dimensional wave number of $ka = 0.6\pi$. In Figure 8.4 the condition is given using infinite elements with increasing radial order m.

The Jacobi(2,0)–based infinite elements achieve, as expected, certainly low condition numbers, similar to the conditioning obtained using Jacobi(3,0)–based elements, while the Legendre–based elements exhibit slightly increased condition numbers.

From both figures showing the overall condition of the system matrices for settings including either infinite elements only or finite and infinite elements, it appears that Jacobi polynomials with $\alpha = 2$, $\beta = 0$ and $\alpha = 3$, $\beta = 0$ behave similar. Both outperform the remaining four different polynomial bases, but none of the two seems clearly more advantageous than the other one. In the following section, tests evaluating the performance of the new infinite elements in conjunction with iterative solvers will be considered. Prior to this, some aspects concerning the accuracy of the improved infinite elements will be discussed.

Similar results, see Figure 8.5, are obtained for the case of the non–spherical radiator, depicted in Figure 8.3(b). Here, the condition of the overall system matrix $\mathbf{A}$ is reported for $ka = 0.5\pi$, which corresponds to six elements per wavelength.

Comments on the Accuracy of the Improved Infinite Elements

As seen in the previous numerical examples, the use of Jacobi(2,0) polynomials lead to reduced condition numbers, compared to the formerly used Legendre and

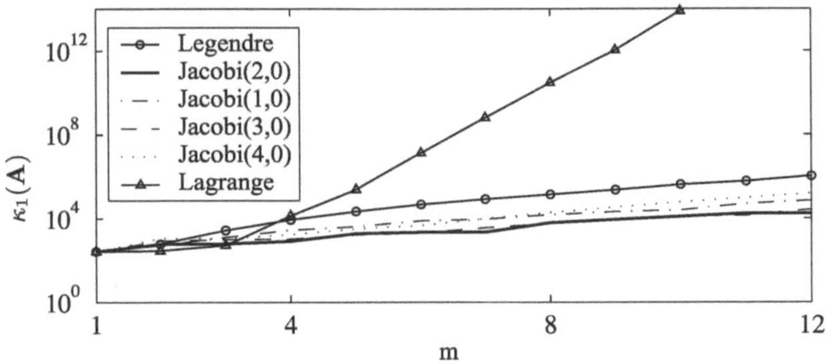

Fig. 8.4 1–Condition of **A** over m for finite elements and infinite elements at $ka = 0.6\pi$.

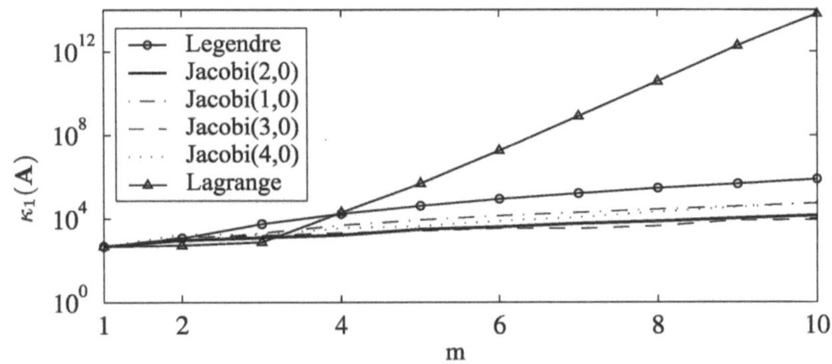

Fig. 8.5 1–Condition of **A** over m for finite elements and infinite elements at $ka = 0.5\pi$.

Lagrange polynomial bases. Now, one may ask how this improvement relates to the overall accuracy of the underlying numerical method. Using computers, and therefore finite arithmetic, the total accuracy of the numerical results are reduced to a finite number of digits, say double precision. Here the ill–conditioning of the Lagrange–based infinite elements becomes important, since the operator **A** notoriously decreases the accuracy of the input data, as already seen in several publications. However, the slight discrepancy in conditioning between Legendre and Jacobi polynomial–based infinite elements will only become visible when the remaining sources of error, like discretization, boundary conditions etc., are negligible in comparison to the difference in the condition numbers. Therefore, the new polynomial bases *never* reduce accuracy, compared with Legendre–based Astley–Leis elements, and may only increase accuracy marginally. The major benefit of the improved conditioning is the fact that it generally results in a superior performance in conjunction with iterative solution algorithms.

8.5 Numerical Examples

The following examples intend to show the applicability and efficiency of the improved infinite elements. Therefore numerical simulations are presented, which are comparable with typical industrial applications. It should be mentioned that a multitude of further examples have been computed, see e.g. [26]. These results are not shown here, but exhibit the same beneficial effects of the improved infinite elements. Application of the element in the framework of tire road noise simulations can be found in [18].

Each numerical example consists of discretizations, constructed out of trilinear tetrahedral elements Tet4. These discretizations are generally of sufficient complexity to reap the benefits of scaling on multiple compute nodes. Simulations are performed with a varying number of processors np, varying boundary conditions, different radial orders m, and three different polynomial families for radial approximation. Namely, Jacobi(2,0), Jacobi(3,0) and Legendre polynomials are used. In the following, these polynomial families are denoted J(2,0), J(3,0) and Leg, respectively.

A particularly interesting aspect of the tabulated summaries is the number of boundary nodes for each discretization. This is an indicator for the size of an equivalent boundary element (BEM) simulation, not accounting for additional burden due to suppressing irregular frequencies.

As a measure for the acoustic frequency compared with the spatial extension of the radiating structure under consideration, non–dimensional wave numbers ka are used, where a is a typical size of the structure. The acoustic frequencies correspond in case of linear elements to one wavelength being equal to 6 times the smallest element length in the mesh. Most of these wave numbers intentionally correspond to almost inadequate resolutions, since experience shows that iterative solvers are less prone to failure the higher the element resolution is. For all simulations, a heterogeneous cluster of Linux compute nodes with typically 1 GB RDRAM, and intel Pentium or Xeon processors was used. The computational investigations were performed using the open–source software package libMesh [42].

8.5.1 Radiation from an Engine Block

The automotive industry is one of the most prominent disciplines concerned with the acoustic behavior of structures. In the following, the radiated sound from an engine block, depicted in Figure 8.6, is computed.

Details of the engine block model are described by von Estorff [28]. There, possible speed–up of the simulation phase is obtained through coarsening the boundary element mesh. However, the validity of such an approach requires careful investigation. Instead of using the BEM, the following numerical examples are computed employing the improved Astley–Leis elements. This enables to rely on automatable discretizations of great detail, while still achieving low simulation times.

The modes of a structural finite element analysis serve as input for the simulation of the acoustic radiation. Namely, two different mode shapes are selected, corresponding to the non–dimensional wave numbers ka listed in Table 8.1. Figure 8.7 de-

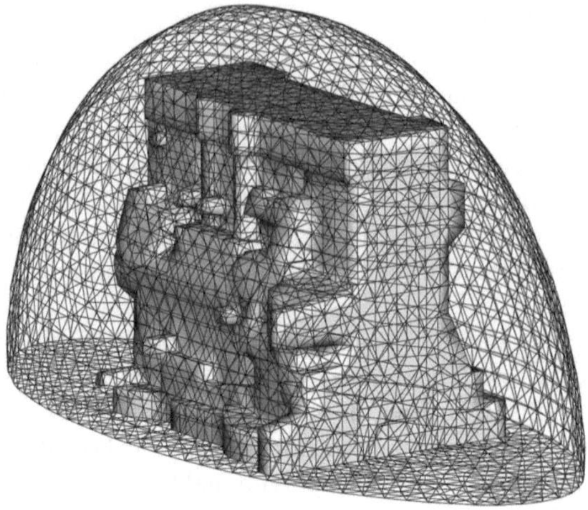

Fig. 8.6 Coarsest discretization of an engine block on a flat surface. Volume fluid elements are not shown, only the envelope Γ_e and the engine block surface.

Table 8.1 Simulation details for the engine block.

Mesh	Simulations using np	FE	ka	Boundary nodes	m	dof
mb7854	1, 2, 4	Tet4	3.38	2 569	6	18 450
					8	21 982
					10	25 514
					12	29 046
					14	32 578
					16	36 110
mb51416	1, 2, 4, 8	Tet4	6.76	10 185	6	93 278
					8	107 232
					10	121 186
					12	135 140
					14	149 094
					16	163 048

picts the radiated pressure field for the structural mode corresponding to $ka = 3.38$, simulated using the coarsest discretization mb7854, described in Table 8.1.

The block is surrounded by air with density $\rho = 1.225\,\mathrm{kg/m^3}$ and wave velocity $c = 340\,\mathrm{m/s}$. The typical length a for the non–dimensional wave number ka is taken to be the minimum edge length of the block, $a = 0.306\,\mathrm{m}$. Table 8.1 summarizes further details of the simulations performed, like, e.g., the required size of a Boundary Element simulation with similar accuracy.

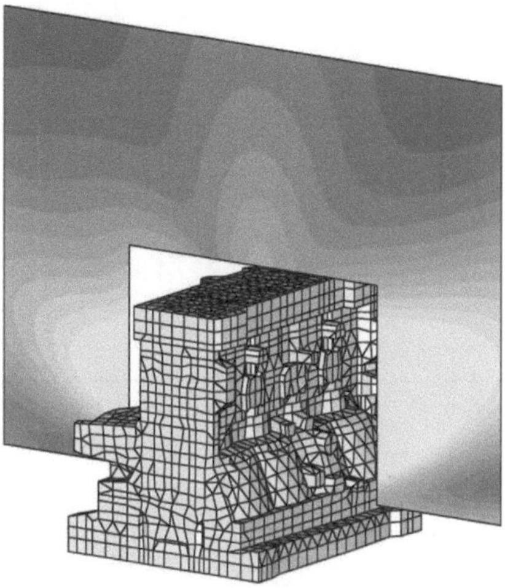

Fig. 8.7 Pressure field radiated by a structural mode of the engine block corresponding to $ka = 3.38$, computed using discretization mb7854.

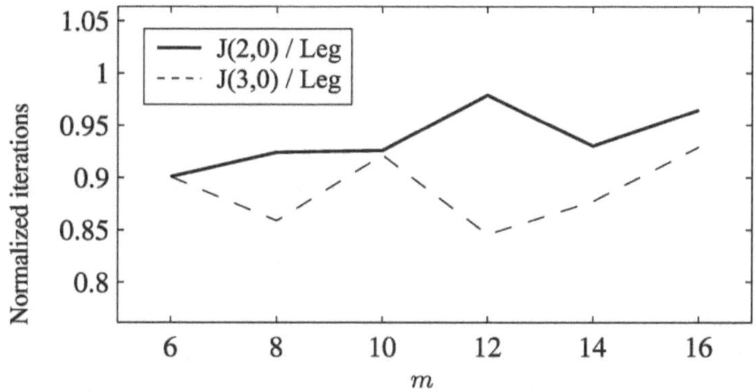

Fig. 8.8 Normalized iteration counts of TFQMR over radial order m, applied to mb7854, using $np = 4$ processors.

To assess the advantage of the improved infinite elements based on Jacobi polynomials over the elements using Legendre polynomials, relative iteration counts are presented in Figures 8.8 to 8.10.

The exterior acoustic problem is solved using Astley–Leis elements with Jacobi(2,0), Jacobi(3,0) or Legendre polynomials for radial approximation. The iterations of the Jacobi–based infinite elements are normalized with respect to the iterations needed by the Legendre–based Astley–Leis elements

$$\text{Normalized iterations}_{\,\mathrm{J}(n,0)} = \frac{\text{Iterations}_{\,\mathrm{J}(n,0)}}{\text{Iterations}_{\,\mathrm{Leg}}} \qquad \text{with } n \in \{2,3\}\,.$$

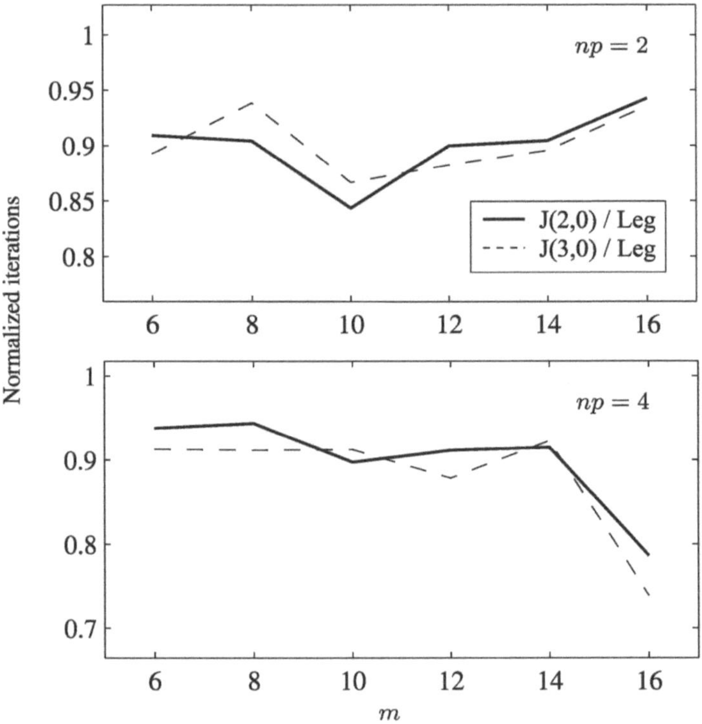

Fig. 8.9 Normalized iteration counts of TFQMR over radial order m, applied to `mb51416`, using $np = 2$ and 4 processors.

This implies that 1.0 corresponds to identical performance of Jacobi and Legendre polynomials, while ratios less than one correspond to a better performance of the Jacobi polynomials. Figure 8.8 depicts the relative iteration counts for simulations performed using discretization `mb7854` on $np = 2$ processors. Both Jacobi polynomials obviously require less iterations than Legendre polynomials. Such better performance of the Jacobi–based formulation comes at no cost, i.e., the overall flop count of the matrix assembly phase is identical, no matter whether Legendre– or Jacobi–based elements are used. However, the performance of the Jacobi(2,0) and Jacobi(3,0) polynomials differs, the latter ones seem to perform slightly better when employed in small–scale simulations. Yet for larger, truly demanding discretizations, the two Jacobi polynomial families perform rather similar, see Figures 8.9 and 8.10.

Even on $np = 8$ processors, the improved infinite elements are able to provide matrices highly suitable for the iterative solution in parallel. It appears that Jacobi polynomials decrease the number of iterations of TFQMR with block incomplete factorization preconditioning by a few percent to up to 25 percent, compared to Legendre polynomials. This is in agreement with other simulations, not depicted here.

To conclude this aspect, Figure 8.11 depicts the overall number of iteration counts and elapsed time for the discretization `mb51416`. Obviously, the small num-

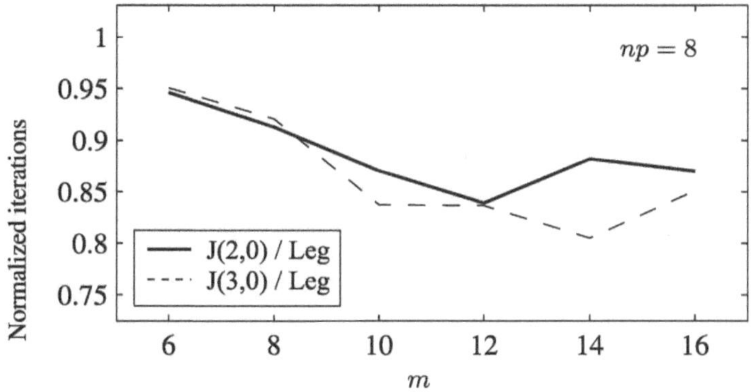

Fig. 8.10 Normalized iteration counts of TFQMR over radial order m, applied to `mb51416`, using $np = 8$ processors.

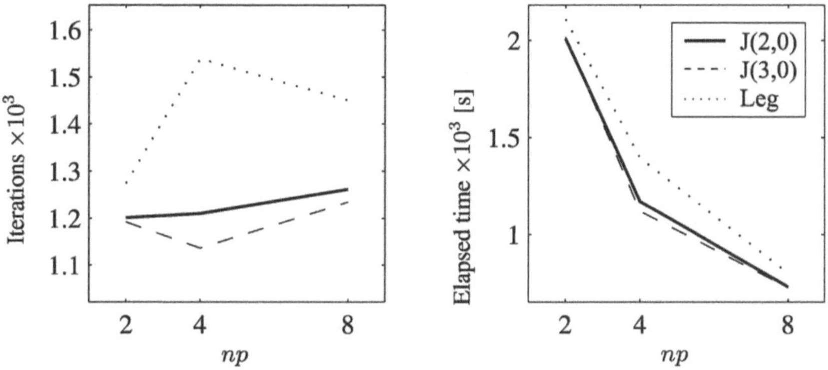

Fig. 8.11 Iteration counts and elapsed time for the discretization `mb51416` over np, solved with TFQMR, using radial order $m = 16$.

ber of iterations due to Jacobi polynomials fully translates into decreased elapsed time. However, this decrease is only visible when the solution is the dominant phase during the complete simulation process. For example, the decrease in iterations for the small discretization `mb7854`, reported in Figure 8.8, is visible in the elapsed time, however not considerably. It should be noted that this is not a defect of the improved polynomials, but merely an order–of–magnitude issue. The complete simulation for one frequency of `mb7854` with $m = 16$ requires roughly 150 seconds on $np = 4$ processors, while the same simulation with radial order $m = 16$, but discretization `mb51416`, and $ka = 6.76$, requires slightly less than 1 200 seconds on $np = 4$ processors. So the true challenge lies in reducing the elapsed time of medium–sized and large problems. Such a decrease in elapsed time is clearly visible already for the medium–sized discretization `mb51416`. Figure 8.11 shows how elapsed time is less for the Jacobi polynomials than for the Legendre polynomials, in between 2 to 4 minutes.

8.5.2 Simulation of Transient Phenomena

For the case of time dependent exterior acoustic simulations, certainly less numerical schemes are available. Especially the dominance of boundary integral methods for harmonic simulations is not apparent in the time dependent case. Still, time dependent simulations are of particular importance, since, e.g., nonlinear phenomena may be taken into account. This is already common practice with the FEM applied to bounded domains [12].

In the following, the engine block already described before is used. However, for the reason stated in Section 8.2.1, this time with a spherical envelope Γ_e, not shown here. The engine block is excited once with a sinusoidal signal, exciting the structural eigenmode of the block for $ka = 3.38$. Note that the normal velocities of the eigenmode govern the pressure magnitude in the exterior. A Newmark iteration scheme with the same iterative solver TFQMR, and preconditioner iLU(0), as before, is used. This is in contrast to the common approach for constant time steps of using a direct solver for one single matrix decomposition. Astley and Hamilton [6] have already shown the performance gain and reduced memory consumption of iterative solvers employed in transient analyses. Namely, the fact that waves may only travel the distance $c\Delta t$ from one time step to the next, provides the Krylov solver with an excellent initial guess, so that only a few iterations are necessary for convergence. In the following examples, the time step $\Delta t = 5.8 \times 10^{-5}$ s is used. Much larger time steps could have been used, however the small time step is chosen in order to obtain smooth pressure time histories at the field points. Two field points, denoted A and B, are considered to evaluate the computed pressure field. Point A is centered on the forefront of the engine, at a normal distance of 22 mm, while point B is located close to the long side–face of the block, with a normal distance of approximately 40 mm. These points are located rather close to the vibrating surface, which is excited once with a sinusoidal signal, and subsequently comes to rest again.

Figure 8.12 depicts the pressure time history at point A with a normal distance of very close to the radiating block, and at point B, the one slightly further away from the engine block. As expected, the delay of the acoustic wave, due to the difference in normal distances of the two field points, is solely slightly visible in the pressure plots. It should be noted, that the differences in pressure amplitude is governed by the differences of the magnitude of the normal velocities, not by the distances. It can be seen clearly how the sinusoidal signal passes the field points once, as expected. Subsequently, the pressure comes to rest again. However, the infinite elements with $m = 0$ exhibit rather strong oscillations, after the wave passed the field point. It should be noted that infinite elements with $m = 0$ are identical with a Sommerfeld condition directly defined on Γ_e. Only poor accuracy is obtained with this simple boundary condition, the wave is not correctly traveling outwards. The infinite elements with $m = 5$, in turn, perform certainly well. The pressure comes to rest again, with only negligible artificial oscillations. Therefore, it is strongly recommended to employ Astley–Leis elements of sufficiently increased radial order. Such demands towards the order m may be most suitably realized through employing the improved

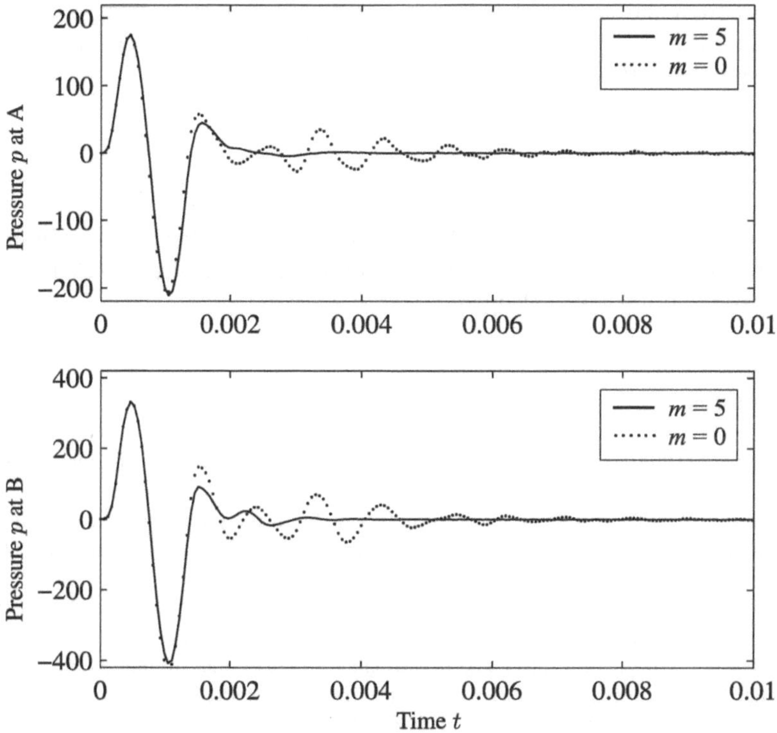

Fig. 8.12 Pressure time history of the engine block at field points A and B, using the improved infinite elements with Jacobi(2,0) polynomials.

infinite elements, because the advantage of the new elements grows with radial order m.

8.6 Conclusions

An improved version of the Astley–Leis elements has been presented and its applicability has been demonstrated and discussed in detail. The major improvements lead to lower conditioning of the system matrices and, as a consequence, to lower computation times, compared to other equivalently versatile formulations using iterative solvers. For industrial–type simulations, in between 5 to 25 percent less iteration counts are gained, compared to the best competitor. This translates directly into reduced computation time. The improved infinite elements also modify the system matrices in such a way that the Krylov solver TFQMR could solve the system of equations in an extremely reliable way.

Through the improved infinite elements suggested here, the use of suitable iterative solvers achieves unprecedented stability and efficiency. This enables to perform small, medium and large scale exterior acoustic simulations on single stand–alone workstations, and also multiple distributed or shared–memory parallel computers with tremendous efficiency.

Acknowledgments

The work has been partly supported by a fellowship within the post–doc program of the German Academic Exchange Service (DAAD) under contract D/06/44695. This support is greatfully acknowledged.

References

1. Astley RJ (1996) Transient wave–envelope elements for wave problems. Journal of Sound and Vibration 192:245–261
2. Astley RJ (2000) Infinite element formulations for wave problems: a review of current formulations and an assessment of accuracy. International Journal for Numerical Methods in Engineering 49:951–976
3. Astley RJ, Coyette JP (2001) Conditioning of infinite element schemes for wave problems. Communications in Numerical Methods in Engineering 17:31–41
4. Astley RJ, Coyette JP (2001) The performance of spheroidal infinite elements. International Journal for Numerical Methods in Engineering 52:951–976
5. Astley RJ, Coyette JP, Cremers L (1998) Three–dimensional wave–envelope elements of variable order for acoustic radiation and scattering Part II. Formulation in the time domain. Journal of the Acoustical Society of America 103:64–72
6. Astley RJ, Hamilton JA (2000) Numerical studies of conjugated infinite elements for acoustic radiation. Journal of Computational Acoustics 8:1–24
7. Astley RJ, Hamilton JA (2006) The stability of infinite element schemes for transient wave problems. Computer Methods in Applied Mechanics and Engineering 195:3553–3571
8. Astley RJ, Macaulay GJ, Coyette JP (1994) Mapped wave envelope elements for acoustic radiation and scattering. Journal of Sound and Vibration 170:97–118
9. Astley RJ, Macaulay GJ, Coyette JP Cremers L (1998) Three–dimensional wave–envelope elements of variable order for acoustic radiation and scattering Part I. Formulation in the frequency domain. Journal of the Acoustical Society of America 103:49–63
10. Balay S, Buschelman K, Gropp WD, Kaushik D, Knepley M, McInnes LC, Smith BF, Zhang H (2004) PETSc users manual (Portable, Extensible Toolkit for Scientific Computation). Technical Report ANL–95/11 – Revision 2.1.5, Argonne National Laboratory
11. Barrett R, Berry M, Chan TF, Demmel J, Donato J, Dongarra J, Eijkhout V, Pozo R, Romine C, Van der Vorst H (1994) Templates for the solution of linear systems: Building blocks for iterative methods. SIAM, Philadelphia, 2. edition
12. Belytschko T, Liu WK, Moran B (2000) Nonlinear finite elements for continua and structures. Wiley, New York
13. Benzi M (2002) Preconditioning techniques for large linear systems: A survey. Journal of Computational Physics 182:418–477
14. Bérenger JP (1994) A perfectly matched layer for the absorption of electromagnetic waves. Journal of Computational Physics 114:185–200

15. Bettess P (1977) Infinite elements. International Journal for Numerical Methods in Engineering 11:53–64
16. Bettess P (1992) Infinite elements. Penshaw Press, Sunderland
17. Bettess P, Zienkiewicz OC (1977) Diffraction and refraction of surface waves using finite and infinite elements. International Journal for Numerical Methods in Engineering 11:1271–1290
18. Brinkmeier M, Nackenhorst U, Petersen S, Estorff O von (2007) A numerical model for the simulation of tire rolling noise. Journal of Sound and Vibration, accepted for publication
19. Burnett DS (1994) A 3–d acoustic infinite element based on a prolate spheroidal multipole expansion. Journal of the Acoustical Society of America 96:2798–2816
20. Burnett DS, Holford RL (1998) An ellipsoidal acoustic infinite element. Computer Methods in Applied Mechanics and Engineering 164:49–76
21. Burnett DS, Holford (1998) Prolate and oblate spheroidal acoustic infinite elements. Computer Methods in Applied Mechanics and Engineering 158:117–141
22. Cipolla J (2002) Acoustic infinite elements with improved robustness. In: Sas P, Van Hal B (eds) Proceedings of ISMA 2002, Katholieke Universiteit Leuven, 2181–2187
23. Coyette JP, Meerbergen K, Robbé M (2005) Time integration for spherical acoustic finite–infinite element models. International Journal for Numerical Methods in Engineering 64:1752–1768
24. Demkowicz L, Gerdes K (1998) Convergence of the infinite element methods for the Helmholtz equation in separable domains. Numerische Mathematik 79:11–42
25. Dreyer D (2004) Efficient infinite elements for exterior acoustics. PhD Thesis, Technical University of Hamburg–Harburg
26. Dreyer D, Petersen S, Estorff O von (2006) Effectiveness and robustness of improved infinite elements for exterior acoustics. Computer Methods in Applied Mechanics and Engineering 195:3591–3607
27. Dreyer D, Estorff O von (2003) Improved conditioning of infinite elements for exterior acoustics. International Journal for Numerical Methods in Engineering 58:933–953
28. Estorff O von (2003) Efforts to reduce computation time in numerical acoustics – an overview. Acta Acustica united with Acustica 89:1–13
29. Freund RW (1993) A transpose–free quasi–minimal residual algorithm for non–hermitian linear systems. SIAM Journal on Scientific Computing 14:470–482
30. Freund RW, Nachtigal NM (1991) QMR: a quasi–minimal residual method for non–hermitian linear systems. Numerische Mathematik 60:315–339
31. Gerdes K (1998) The conjugate vs. the unconjugate infinite element method for the Helmholtz equation in exterior domains. Computer Methods in Applied Mechanics and Engineering 152:125–145
32. Gerdes K (2000) A review of infinite element methods for exterior Helmholtz problems. Journal of Computational Acoustics 8:43–62
33. Gerdes K, Demkowicz L (1996) Solution of 3d–Laplace and Helmholtz equation in exterior domains using hp–infinite elements. Computer Methods in Applied Mechanics and Engineering 137:239–273
34. Givoli D (2004) High–order local non–reflecting boundary conditions: a review. Wave Motion 39:319–326
35. Grote MJ, Keller JB (1995) On nonreflecting boundary conditions. Journal of Computational Physics 122:231–243
36. Guddati MN, Lim KW (2006) Continued fraction absorbing boundary conditions for convex polygonal domains. International Journal for Numerical Methods in Engineering 66:949–977

37. Harari I (2006) A survey of finite element methods for time–harmonic acoustics. Computer Methods in Applied Mechanics and Engineering 195:1594–1607
38. Harari I, Slavutin M, Turkel E (2006) Studies of FE/PML for exterior problems of time–harmonic elastic waves. Computer Methods in Applied Mechanics and Engineering 195:3854–3879
39. Ihlenburg F (1998) Finite element analysis of acoustic scattering. Springer–Verlag, New York
40. Ihlenburg F (2000) On fundamental aspects of exterior approximations with infinite elements. Journal of Computational Acoustics 8:63–80
41. Keller JB, Givoli D (1989) Exact non–reflecting boundary conditions. Journal of Computational Physics 82:172–192
42. Kirk BS, Peterson JW, Stogner RH, Carey GF (2006) libMesh: a C++ library for parallel adaptive mesh refinement/coarsening simulations. Engineering with Computers 22:237–254
43. Leis R (1986) Initial boundary value problems in mathematical physics. Wiley & Teubner, Stuttgart
44. Magolu monga Made M (2001) Incomplete factorization–based preconditionings for solving the Helmholtz equation. International Journal for Numerical Methods in Engineering 50:1077–1101
45. Marques JMMC, Owen DRJ (1984) Infinite elements in quasi–static materially nonlinear problems. Computers & Structures 18:739–751
46. Petersen S, Dreyer D, Estorff O von (2006) Assessment of finite and spectral element shape functions for efficient iterative simulations of interior acoustics. Computer Methods in Applied Mechanics and Engineering 195:6463–6478
47. Saad Y (2003) Iterative methods for sparse linear systems. SIAM, Philadelphia, 2nd edition
48. Saad Y, Schultz MH (1986) GMRES: a generalized minimal residual algorithm for solving nonsymmetric linear systems. SIAM Journal on Scientific and Statistical Computing 7:856–869
49. Shirron JJ, Babuška I (1998) A comparison of approximate boundary conditions and infinite element methods for exterior Helmholtz problems. Computer Methods in Applied Mechanics and Engineering 164:121–139
50. Shirron JJ, Dey S (2002) Acoustic infinite elements for non–separable geometries. Computer Methods in Applied Mechanics and Engineering 191:4123–4139
51. Shirron JJ, Giddings TE (2006) A finite element model for acoustic scattering from objects near a fluid–fluid interface. Computer Methods in Applied Mechanics and Engineering 195:279–288
52. Thompson LL, (2006) A review of finite–element methods for time–harmonic acoustics. Journal of the Acoustical Society of America 119:1315–1330
53. Trefethen LN, Bau D (1997) Numerical linear algebra. SIAM, Philadelphia
54. Turkel E, Yefet A (1998) Absorbing PML boundary layers for wave–like equations. Applied Numerical Mathematics 27:533–557
55. Van den Nieuwenhof B, Coyette JP (2001) Treatment of frequency–dependent admittance boundary conditions in transient acoustic finite/infinite–element models. Journal of the Acoustical Society of America 110:1743–1751
56. Van der Vorst HA (1992) Bi–CGSTAB: a fast and smoothly converging variant of Bi–CG for the solution of nonsymmetric linear systems. SIAM Journal on Scientific and Statistical Computing 13:631–644

Part III

FEM: Related Problems

9

Fluid–Structure Acoustic Interaction

Alfredo Bermúdez[1], Pablo Gamallo[2], Luis Hervella–Nieto[3], Rodolfo Rodríguez[4], and Duarte Santamarina[5]

[1] Departamento de Matemática Aplicada, Universidade de Santiago de Compostela, 15782 Santiago de Compostela, Spain
mabermud@usc.es
[2] Departamento de Matemática Aplicada II, Universidade de Vigo, 36310 Vigo, Spain
pgamallo@uvigo.es
[3] Departamento de Matemáticas, Universidade da Coruña, 15071 A Coruña, Spain
luisher@udc.es
[4] GI^{2}MA, Departamento de Ingeniería Matemática, Universidad de Concepción, Casilla 160–C, Concepción, Chile
rodolfo@ing-mat.udec.cl
[5] Departamento de Matemática Aplicada, Universidade de Santiago de Compostela, 15782 Santiago de Compostela, Spain
duartesr@usc.es

Summary. This is a survey of numerical methods to compute a particular fluid–solid interaction: the so called "elastoacoustic problem". It concerns the determination of the motion of an elastic structure in contact with a compressible fluid. In this case displacements are small and, then, we can assume a linear response of the structure. We neglect gravity effects and consider a homogeneous fluid for which its reference density is constant. Other usual simplifications for this kind of problems are that viscous effects are not relevant in the fluid and that velocities are small enough for convective effects to be neglected.

We review several alternative formulations of the elastoacoustic vibration problem, which differ from each other in the variables used to describe the fluid: pressure, a displacement potential or both. In all these cases, standard Lagrange finite elements are used for the discretization. We compare the results obtained with all these methods and with the pure displacement formulation, which has to be discretized by Raviart–Thomas elements. Next, we show how to apply a modal synthesis approach based on the displacement potential formulation. Finally, we report how the pure displacement formulation can be used to deal with thin layers of interface acoustic damping material and to solve the elastoacoustic problem in the time–domain.

9.1 Introduction

The need of computing fluid–solid interactions arises in many important engineering problems. A large amount of work has been devoted to this subject during the last years. A general overview can be found in the monographs by Morand and Ohayon [24], Conca et al. [18], and Ohayon and Soize [25], where numerical methods and further references are also given.

This chapter deals with one interaction of this kind: the *elastoacoustic problem*. It concerns the determination of the motion of an elastic structure in contact with a compressible fluid. In this case displacements are small and, then, we can assume a linear response of the structure. We neglect gravity effects and consider a homogeneous fluid for which its reference density is constant. Other usual simplifications for this kind of problems are that viscous effects are not relevant in the fluid and that velocities are small enough for convective effects to be neglected, see, for instance, Zienkiewicz and Taylor [31].

We will mainly focus on determining the vibrations of the coupled fluid–structure system. The acoustic fluid is usually treated by choosing pressure as primary variable, [31]. However, for coupled systems, such a choice leads to non–symmetric eigenvalue problems. To avoid this, the fluid has been described using different variables: displacements, Kiefling and Feng [23], displacement potential, Morand and Ohayon [24], velocity potential, Everstine [19], or combinations of some of them.

Since the solid is generally described in terms of displacements, to choose the same variable for the fluid presents one important advantage: compatibility and equilibrium through the fluid–solid interface are satisfied automatically. This approach could be in principle applied to the solution of a broad range of problems and, after discretization, leads to sparse symmetric matrices. Nevertheless, it is well known that the displacement formulation for the fluid suffers from the presence of zero–frequency spurious circulation modes with no physical meaning, see [23]. After discretization by standard finite elements, these modes are approximated by others with nonzero frequencies interspersed among those of the physical modes.

Several approaches have been proposed to circumvent this drawback. For instance Hamdi et al. [21] introduced an irrotational constraint, taken into account by a penalty term. This technique does not attain complete elimination of spurious modes, but they are pushed towards higher frequencies. Olson and Bathe [26] showed that this penalty formulation "locks up" for non–convex fluid domains and that although reduced integration yields some improvement, it does not assure convergence in the general case. Later on, Bathe et al. [3] and Wang and Bathe [30] proposed a mixed fluid displacement–pressure formulation, and Gastaldi [20] analyzed a finite element discretization, proving convergence for two–dimensional fluid domains without reentrant corners.

An alternative approach was introduced by Bermúdez and Rodríguez [12]. It consists of using standard piecewise linear elements in the solid combined with triangular or tetrahedral *Raviart–Thomas* elements for the fluid displacements. The degrees of freedom of the latter correspond to the mean fluxes of the displacements through the element faces. The method has been proved to be free of spurious modes and to attain optimal–order convergence for any kind of fluid and solid domains, see [4, 28]. Numerical experiments showing its performance were reported in [12] for two–dimensional (2D) triangular meshes and in [10] for three–dimensional (3D) tetrahedral ones.

In what follows we will review this approach as well as other alternative formulations of the elastoacoustic vibration problem, which differ from each other in the variables used to describe the fluid: pressure [6], a displacement potential [11] or both [15]. In all these cases, standard Lagrange finite elements are used for the discretization. We compare the results obtained with all these methods and with the pure displacement formulation [4]. Next, we show how to apply a modal synthesis approach based on the displacement potential formulation [11, 24]. Finally, we report how the pure displacement formulation can be used to deal with thin layers of interface acoustic damping material [13, 14] and to solve the elastoacoustic problem in the time–domain [16].

9.2 Notation

Throughout this chapter we will use the following notation for Sobolev spaces:

- $\mathrm{L}_2(\Omega)$ is the space of square–integrable functions,

$$\mathrm{L}_2(\Omega) := \left\{ f : \Omega \to \mathbb{R} : \int_\Omega |f|^2 < \infty \right\}.$$

- We use boldface to represent vector spaces (and variables). For instance,

$$\mathbf{L}_2(\Omega) := \left\{ \boldsymbol{f} : \Omega \to \mathbb{R}^n : \int_\Omega |\boldsymbol{f}|^2 < \infty \right\}.$$

- $\mathrm{H}^1(\Omega)$ is the space of functions in $\mathrm{L}_2(\Omega)$ with derivatives (in the sense of distributions) also in $\mathrm{L}_2(\Omega)$,

$$\mathrm{H}^1(\Omega) := \left\{ v \in \mathrm{L}_2(\Omega) : \boldsymbol{\nabla} v := \left(\frac{\partial v}{\partial x_1}, \cdots, \frac{\partial v}{\partial x_n} \right) \in \mathbf{L}_2(\Omega) \right\}.$$

- If Γ is a part of the boundary of Ω, $\mathrm{H}^1_\Gamma(\Omega)$ is the space of functions in $\mathrm{H}^1(\Omega)$ with vanishing trace on Γ,

$$\mathrm{H}^1_\Gamma(\Omega) := \left\{ v \in \mathrm{H}^1(\Omega) : v|_\Gamma = 0 \right\}.$$

- $\mathring{\mathrm{H}}^1(\Omega_\mathrm{F})$ is the subspace of zero–mean functions in $\mathrm{H}^1(\Omega_\mathrm{F})$,

$$\mathring{\mathrm{H}}^1(\Omega_\mathrm{F}) := \left\{ v \in \mathrm{H}^1(\Omega) : \int_{\Omega_\mathrm{F}} v = 0 \right\}.$$

- $\mathbf{H}(\mathrm{div}, \Omega)$ is the space of vector fields in $\mathbf{L}_2(\Omega)$ with divergence in $\mathrm{L}_2(\Omega)$,

$$\mathbf{H}(\mathrm{div}, \Omega) := \left\{ \boldsymbol{v} \in \mathbf{L}_2(\Omega) : \mathrm{div}\, \boldsymbol{v} := \sum_{i=1}^n \frac{\partial v_i}{\partial x_i} \in \mathrm{L}_2(\Omega) \right\}.$$

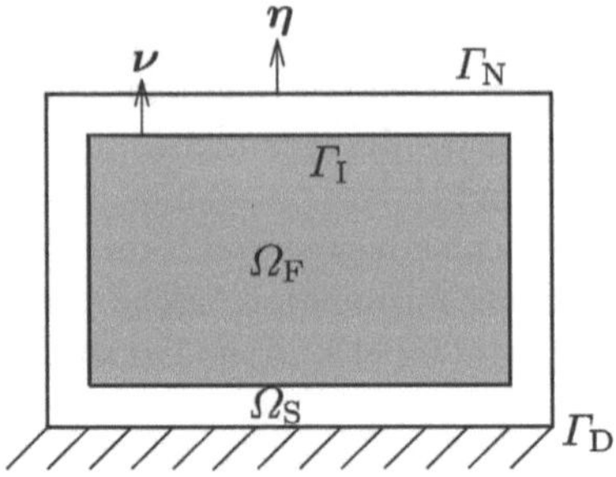

Fig. 9.1 Geometrical notation.

Let $\Omega_{\rm F}$ and $\Omega_{\rm S}$ denote polygonal or polyhedral bounded domains in $\mathbb{R}^n$ ($n = 2$ or 3) occupied by the fluid and the solid, respectively, as shown in Figure 9.1 in a 2D case. Let $\Gamma_{\rm I} = \partial\Omega_{\rm F}$ be the interface between the two media, and $\boldsymbol{\nu}$ its unit normal vector pointing outwards $\Omega_{\rm F}$. The exterior boundary of the solid domain is the union of $\Gamma_{\rm D}$ and $\Gamma_{\rm N}$: the structure is fixed along $\Gamma_{\rm D}$ and free of stress along $\Gamma_{\rm N}$; we assume that the measure of $\Gamma_{\rm D}$ is strictly positive. Finally, $\boldsymbol{\eta}$ denotes the outwards unit normal vector along $\Gamma_{\rm N}$. We use the following notation for the physical magnitudes; in the fluid:

- $\boldsymbol{u}^{\rm F}$: the displacement vector,
- p: the pressure fluctuation with respect to a reference pressure,
- ϕ: a displacement potential ($\boldsymbol{u}^{\rm F} = \boldsymbol{\nabla}\phi$),
- $\rho_{\rm F}$: the density,
- c: the acoustic speed,

and in the solid:

- $\boldsymbol{u}^{\rm S}$: the displacement vector,
- $\rho_{\rm S}$: the density,
- $\lambda_{\rm S}$ and $\mu_{\rm S}$: the Lamé's coefficients,
- $\varepsilon(\boldsymbol{u}^{\rm S})$: the infinitesimal strain tensor defined by $\varepsilon_{ij}(\boldsymbol{u}^{\rm S}) := \frac{1}{2}\left(\frac{\partial u_i^{\rm S}}{\partial x_j} + \frac{\partial u_j^{\rm S}}{\partial x_i}\right)$, $i, j = 1, \dots, n$,
- $\sigma(\boldsymbol{u}^{\rm S})$: the stress tensor which is related to the strain tensor by Hooke's law: $\sigma_{ij}(\boldsymbol{u}^{\rm S}) = \lambda_{\rm S} \sum_{k=1}^{n} \varepsilon_{kk}(\boldsymbol{u}^{\rm S})\delta_{ij} + 2\mu_{\rm S}\varepsilon_{ij}(\boldsymbol{u}^{\rm S}), \quad i, j = 1, \dots, n.$

9.3 Spectral PDE Problems in Elastoacoustics

The study presented here focus on elastoacoustic systems, which consist of a homogeneous inviscid barotropic ideal fluid contained in an isotropic homogeneous linear elastic structure. A typical setting will be a vessel completely filled with fluid, as shown in Figure 9.1. In particular, we are interested in the problem of determining the free harmonic motions of the coupled system. In this situation, the governing equations are, see, for instance [24]:

- in the fluid domain:

$$\nabla p - \omega^2 \rho_F \boldsymbol{u}^F = \boldsymbol{0} \qquad \text{in } \Omega_F, \tag{9.1}$$

$$p + \rho_F c^2 \operatorname{div} \boldsymbol{u}^F = 0 \qquad \text{in } \Omega_F; \tag{9.2}$$

- in the solid domain:

$$\operatorname{div}\left[\boldsymbol{\sigma}(\boldsymbol{u}^S)\right] + \omega^2 \rho_S \boldsymbol{u}^S = \boldsymbol{0} \qquad \text{in } \Omega_S; \tag{9.3}$$

- on the interface (coupling conditions):

$$\boldsymbol{\sigma}(\boldsymbol{u}^S)\boldsymbol{\nu} + p\boldsymbol{\nu} = \boldsymbol{0} \qquad \text{on } \Gamma_I, \tag{9.4}$$

$$\boldsymbol{u}^S \cdot \boldsymbol{\nu} - \boldsymbol{u}^F \cdot \boldsymbol{\nu} = 0 \qquad \text{on } \Gamma_I, \tag{9.5}$$

  where the first equation is a kinetic condition relating the solid stress on the interface to the pressure exerted by the fluid (action–reaction principle), while the second one is a kinematic condition which states that fluid and solid are in contact at the interface without friction (slippery condition);
- boundary conditions in the solid:

$$\boldsymbol{\sigma}(\boldsymbol{u}^S)\boldsymbol{\eta} = \boldsymbol{0} \qquad \text{on } \Gamma_N, \tag{9.6}$$

$$\boldsymbol{u}^S = \boldsymbol{0} \qquad \text{on } \Gamma_D, \tag{9.7}$$

  where the first equation states that the solid is free on Γ_N and the second one that it is clamped on Γ_D.

Equations (9.1) and (9.2) in the fluid domain can be reduced to a single one by eliminating either $\boldsymbol{u}^F$ or p to obtain, respectively, a coupled system in terms of p and $\boldsymbol{u}^S$, or another one in terms of $\boldsymbol{u}^F$ and $\boldsymbol{u}^S$.

In the first case, the unknown $\boldsymbol{u}^F$ can be eliminated by taking divergence of (9.1),

$$\Delta p - \omega^2 \rho_F \operatorname{div} \boldsymbol{u}^F = 0$$

and then substituting $\operatorname{div} \boldsymbol{u}^F$ in terms of p from Equation (9.2),

$$\operatorname{div} \boldsymbol{u}^F = -\frac{p}{\rho_F c^2}.$$

Thus, the so–called pressure/displacement formulation for the elastoacoustic problem is obtained:

PDE1.–

$$
\begin{aligned}
\Delta p+\frac{\omega^2}{c^2}p&=0 && \text{in } \Omega_{\mathrm{F}},\\
\operatorname{div}\left[\boldsymbol{\sigma}(\boldsymbol{u}^{\mathrm{S}})\right]+\omega^2\rho_{\mathrm{S}}\boldsymbol{u}^{\mathrm{S}}&=\mathbf{0} && \text{in } \Omega_{\mathrm{S}},\\
\boldsymbol{\sigma}(\boldsymbol{u}^{\mathrm{S}})\boldsymbol{\nu}+p\boldsymbol{\nu}&=\mathbf{0} && \text{on } \Gamma_{\mathrm{I}},\\
\frac{1}{\rho_{\mathrm{F}}}\frac{\partial p}{\partial \boldsymbol{\nu}}-\omega^2\boldsymbol{u}^{\mathrm{S}}\cdot\boldsymbol{\nu}&=0 && \text{on } \Gamma_{\mathrm{I}},\\
\boldsymbol{\sigma}(\boldsymbol{u}^{\mathrm{S}})\boldsymbol{\eta}&=\mathbf{0} && \text{on } \Gamma_{\mathrm{N}},\\
\boldsymbol{u}^{\mathrm{S}}&=\mathbf{0} && \text{on } \Gamma_{\mathrm{D}}.
\end{aligned}
$$

The fourth equation of this system has been obtained after rewriting (9.5) from Equation (9.1), as follows:

$$
\boldsymbol{u}^{\mathrm{F}}\cdot\boldsymbol{\nu}=\frac{1}{\omega^2\rho_{\mathrm{F}}}\boldsymbol{\nabla}p\cdot\boldsymbol{\nu}=\frac{1}{\omega^2\rho_{\mathrm{F}}}\frac{\partial p}{\partial\boldsymbol{\nu}}.
$$

Analogously, in order to eliminate p, we consider Equation (9.1) in the form

$$
\boldsymbol{u}^{\mathrm{F}}=\frac{1}{\omega^2\rho_{\mathrm{F}}}\boldsymbol{\nabla}p.
$$

Then, by taking the gradient of Equation (9.2), it is straightforward to obtain the pure displacement formulation for the elastoacoustic problem:

PDE2.–

$$
\begin{aligned}
\rho_{\mathrm{F}}c^2\boldsymbol{\nabla}\left(\operatorname{div}\boldsymbol{u}^{\mathrm{F}}\right)+\omega^2\rho_{\mathrm{F}}\boldsymbol{u}^{\mathrm{F}}&=\mathbf{0} && \text{in } \Omega_{\mathrm{F}},\\
\operatorname{div}\left[\boldsymbol{\sigma}(\boldsymbol{u}^{\mathrm{S}})\right]+\omega^2\rho_{\mathrm{S}}\boldsymbol{u}^{\mathrm{S}}&=\mathbf{0} && \text{in } \Omega_{\mathrm{S}},\\
\boldsymbol{\sigma}(\boldsymbol{u}^{\mathrm{S}})\boldsymbol{\nu}-\rho_{\mathrm{F}}c^2\operatorname{div}\boldsymbol{u}^{\mathrm{F}}\boldsymbol{\nu}&=\mathbf{0} && \text{on } \Gamma_{\mathrm{I}},\\
\boldsymbol{u}^{\mathrm{F}}\cdot\boldsymbol{\nu}-\boldsymbol{u}^{\mathrm{S}}\cdot\boldsymbol{\nu}&=0 && \text{on } \Gamma_{\mathrm{I}},\\
\boldsymbol{\sigma}(\boldsymbol{u}^{\mathrm{S}})\boldsymbol{\eta}&=\mathbf{0} && \text{on } \Gamma_{\mathrm{N}},\\
\boldsymbol{u}^{\mathrm{S}}&=\mathbf{0} && \text{on } \Gamma_{\mathrm{D}}.
\end{aligned}
$$

As will be shown in the next section, the formulation **PDE1** leads to a non–symmetric weak problem. To avoid this drawback, Morand and Ohayon introduced in [24] an alternative procedure which consists in using a redundant formulation in which the fluid is described by means of two variables: a fluid displacement potential ϕ and the pressure p.

To obtain such formulation let us introduce a new variable, $\widetilde{\phi}:=\frac{1}{\rho_{\mathrm{F}}\omega^2}p$. Actually, $\widetilde{\phi}$ is a fluid displacement potential field; indeed, from (9.1) we deduce $\boldsymbol{u}^{\mathrm{F}}=\boldsymbol{\nabla}\widetilde{\phi}$ in Ω_{F}.

In this case, when p and $\widetilde{\phi}$ are coupled directly (i.e., not by means of their derivatives), we obtain the so–called *mass coupling* redundant formulation for the elastoacoustic problem:

PDE3.–

$$\begin{aligned}
p - \omega^2 \rho_F \widetilde{\phi} &= 0 && \text{in } \Omega_F,\\
p + \rho_F c^2 \Delta \widetilde{\phi} &= 0 && \text{in } \Omega_F,\\
\operatorname{div}\left[\boldsymbol{\sigma}(\boldsymbol{u}^S)\right] + \omega^2 \rho_S \boldsymbol{u}^S &= \mathbf{0} && \text{in } \Omega_S,\\
\boldsymbol{\sigma}(\boldsymbol{u}^S)\boldsymbol{\nu} + \rho_F \omega^2 \widetilde{\phi}\,\boldsymbol{\nu} &= \mathbf{0} && \text{on } \Gamma_I.\\
\frac{\partial \widetilde{\phi}}{\partial \boldsymbol{\nu}} - \boldsymbol{u}^S \cdot \boldsymbol{\nu} &= 0 && \text{on } \Gamma_I,\\
\boldsymbol{\sigma}(\boldsymbol{u}^S)\boldsymbol{\eta} &= \mathbf{0} && \text{on } \Gamma_N,\\
\boldsymbol{u}^S &= \mathbf{0} && \text{on } \Gamma_D.
\end{aligned}$$

Remark 1. It is also possible to couple p and $\widetilde{\phi}$ by means of their derivatives, which leads to the so–called *stiffness coupling* formulation for the elastoacoustic problem, see [24]. We do not include this formulation in the present work, because it has been proved in [6] that the corresponding continuous and discrete problems are equivalent to those of the pressure/displacement formulation **PDE1**. The mass coupling formulation **PDE3** turns out to be equivalent at continuous and discrete level to **PDE1**, as well; the latter will be proved in Remark 5 below.

Finally, it is also possible to derive a formulation where the fluid is described only by means of the displacement potential. With this aim, we consider Equations (9.1) and (9.2) and a fluid displacement potential ϕ, normalized in such a way that $\int_{\Omega_F} \phi = 0$. Then,

$$p = \rho_F \omega^2 \phi - \frac{\rho_F c^2}{|\Omega_F|} \int_{\Gamma_I} \boldsymbol{u}^F \cdot \boldsymbol{\nu},$$

where $|\Omega_F|$ denotes the measure (area) of Ω_F (note that this potential differs in an additive constant from that introduced for problem **PDE3**, $\widetilde{\phi}$). Therefore, p and $\boldsymbol{u}^F$ can be eliminated in the set of equations (9.1)–(9.7) to obtain the following potential/displacement formulation:

PDE4.–

$$\begin{aligned}
\rho_F \Delta \phi - \frac{\rho_F}{|\Omega_F|} \int_{\Gamma_I} \boldsymbol{u}^S \cdot \boldsymbol{\nu} + \omega^2 \frac{\rho_F}{c^2} \phi &= 0 && \text{in } \Omega_F,\\
\operatorname{div}\left[\boldsymbol{\sigma}\left(\boldsymbol{u}^S\right)\right] + \omega^2 \rho_S \boldsymbol{u}^S &= \mathbf{0} && \text{in } \Omega_S,\\
\boldsymbol{\sigma}\left(\boldsymbol{u}^S\right)\boldsymbol{\nu} + \left(\rho_F \omega^2 \phi - \frac{\rho_F c^2}{|\Omega_F|} \int_{\Gamma_I} \boldsymbol{u}^S \cdot \boldsymbol{\nu}\right)\boldsymbol{\nu} &= \mathbf{0} && \text{on } \Gamma_I,\\
\frac{\partial \phi}{\partial \boldsymbol{\nu}} - \boldsymbol{u}^S \cdot \boldsymbol{\nu} &= 0 && \text{on } \Gamma_I,\\
\boldsymbol{\sigma}\left(\boldsymbol{u}^S\right)\boldsymbol{\eta} &= \mathbf{0} && \text{on } \Gamma_N,\\
\boldsymbol{u}^S &= \mathbf{0} && \text{on } \Gamma_D.
\end{aligned}$$

9.4 Weak Formulations of the Spectral Problems

Next, in order to implement finite element discretizations, we introduce weak formulations for the four PDE problems defined in the previous section.

To obtain that corresponding to **PDE1**, we multiply the first equation in this system by $\frac{1}{\rho_{\mathrm{F}}}$ times a test function $q \in \mathrm{H}^1(\Omega_{\mathrm{F}})$ and we use a Green's formula to obtain

$$\int_{\Omega_{\mathrm{F}}} \frac{1}{\rho_{\mathrm{F}}} \nabla p \cdot \nabla q - \int_{\Gamma_{\mathrm{I}}} \frac{1}{\rho_{\mathrm{F}}} \frac{\partial p}{\partial \boldsymbol{\nu}} q = \omega^2 \int_{\Omega_{\mathrm{F}}} \frac{1}{\rho_{\mathrm{F}} c^2} pq. \tag{9.8}$$

Next, multiplying the second equation in **PDE1** by a test function $\boldsymbol{v}^{\mathrm{S}} \in \mathbf{H}^1_{\Gamma_{\mathrm{D}}}(\Omega_{\mathrm{S}})$ and using a Green's formula, we obtain

$$\int_{\Omega_{\mathrm{S}}} \boldsymbol{\sigma}\left(\boldsymbol{u}^{\mathrm{S}}\right) : \boldsymbol{\varepsilon}\left(\boldsymbol{v}^{\mathrm{S}}\right) + \int_{\Gamma_{\mathrm{I}}} \boldsymbol{\sigma}\left(\boldsymbol{u}^{\mathrm{S}}\right) \boldsymbol{\nu} \cdot \boldsymbol{v}^{\mathrm{S}} = \omega^2 \int_{\Omega_{\mathrm{S}}} \rho_{\mathrm{S}} \boldsymbol{u}^{\mathrm{S}} \cdot \boldsymbol{v}^{\mathrm{S}}. \tag{9.9}$$

Finally, adding together (9.8) and (9.9) and taking into account the coupling conditions (third and fourth equations in **PDE1**), we are led to the following weak formulation for the spectral problem

WP1.– *Find* $\omega \in \mathbb{R}$ *and* $\mathbf{0} \neq \left(\boldsymbol{u}^{\mathrm{S}}, p\right) \in \boldsymbol{\mathcal{V}}_1$ *such that*

$$\begin{aligned} &\int_{\Omega_{\mathrm{S}}} \boldsymbol{\sigma}\left(\boldsymbol{u}^{\mathrm{S}}\right) : \boldsymbol{\varepsilon}\left(\boldsymbol{v}^{\mathrm{S}}\right) + \int_{\Omega_{\mathrm{F}}} \frac{1}{\rho_{\mathrm{F}}} \nabla p \cdot \nabla q - \int_{\Gamma_{\mathrm{I}}} p \boldsymbol{v}^{\mathrm{S}} \cdot \boldsymbol{\nu} \\ &\quad = \omega^2 \left(\int_{\Omega_{\mathrm{S}}} \rho_{\mathrm{S}} \boldsymbol{u}^{\mathrm{S}} \cdot \boldsymbol{v}^{\mathrm{S}} + \int_{\Omega_{\mathrm{F}}} \frac{1}{\rho_{\mathrm{F}} c^2} pq + \int_{\Gamma_{\mathrm{I}}} \boldsymbol{u}^{\mathrm{S}} \cdot \boldsymbol{\nu} q \right) \qquad \forall \left(\boldsymbol{v}^{\mathrm{S}}, q\right) \in \boldsymbol{\mathcal{V}}_1, \end{aligned}$$

where $\boldsymbol{\mathcal{V}}_1 := \mathbf{H}^1_{\Gamma_{\mathrm{D}}}(\Omega_{\mathrm{S}}) \times \mathrm{H}^1(\Omega_{\mathrm{F}})$.

Regarding the pure displacement formulation **PDE2**, multiplying the first equation by a test function $\boldsymbol{v}^{\mathrm{F}} \in \mathbf{H}(\mathrm{div}, \Omega_{\mathrm{F}})$ and using a Green's formula we have

$$\int_{\Omega_{\mathrm{F}}} \rho_{\mathrm{F}} c^2 \operatorname{div} \boldsymbol{u}^{\mathrm{F}} \operatorname{div} \boldsymbol{v}^{\mathrm{F}} - \int_{\Gamma_{\mathrm{I}}} \rho_{\mathrm{F}} c^2 \operatorname{div} \boldsymbol{u}^{\mathrm{F}} \boldsymbol{\nu} \cdot \boldsymbol{v}^{\mathrm{F}} = \omega^2 \int_{\Omega_{\mathrm{F}}} \rho_{\mathrm{F}} \boldsymbol{u}^{\mathrm{F}} \cdot \boldsymbol{v}^{\mathrm{F}}.$$

If we add this equation to (9.9), take into account the coupling conditions and impose the continuity of the normal traces across Γ_{I} for the test functions, we obtain the following weak problem:

WP2.– *Find* $\omega \in \mathbb{R}$ *and* $\mathbf{0} \neq \left(\boldsymbol{u}^{\mathrm{S}}, \boldsymbol{u}^{\mathrm{F}}\right) \in \boldsymbol{\mathcal{V}}_2$ *such that*

$$\begin{aligned} &\int_{\Omega_{\mathrm{S}}} \boldsymbol{\sigma}\left(\boldsymbol{u}^{\mathrm{S}}\right) : \boldsymbol{\varepsilon}\left(\boldsymbol{v}^{\mathrm{S}}\right) + \int_{\Omega_{\mathrm{F}}} \rho_{\mathrm{F}} c^2 \operatorname{div} \boldsymbol{u}^{\mathrm{F}} \operatorname{div} \boldsymbol{v}^{\mathrm{F}} \\ &\quad = \omega^2 \left(\int_{\Omega_{\mathrm{S}}} \rho_{\mathrm{S}} \boldsymbol{u}^{\mathrm{S}} \cdot \boldsymbol{v}^{\mathrm{S}} + \int_{\Omega_{\mathrm{F}}} \rho_{\mathrm{F}} \boldsymbol{u}^{\mathrm{F}} \cdot \boldsymbol{v}^{\mathrm{F}} \right) \qquad \forall \left(\boldsymbol{v}^{\mathrm{S}}, \boldsymbol{v}^{\mathrm{F}}\right) \in \boldsymbol{\mathcal{V}}_2, \end{aligned}$$

where $\boldsymbol{\mathcal{V}}_2 := \left\{ \left(\boldsymbol{v}^{\mathrm{S}}, \boldsymbol{v}^{\mathrm{F}}\right) \in \mathbf{H}^1_{\Gamma_{\mathrm{D}}}(\Omega_{\mathrm{S}}) \times \mathbf{H}(\mathrm{div}, \Omega_{\mathrm{F}}) \,:\, \boldsymbol{v}^{\mathrm{S}} \cdot \boldsymbol{\nu} = \boldsymbol{v}^{\mathrm{F}} \cdot \boldsymbol{\nu} \text{ on } \Gamma_{\mathrm{I}} \right\}$.

To obtain the weak formulation of problem **PDE3**, we multiply the first equation by $\frac{1}{\rho_{\mathrm{F}} c^2}$ times a test function $q \in \mathrm{L}_2(\Omega_{\mathrm{F}})$. Then we integrate in Ω_{F}, add the resulting equation to (9.9) and take into account the coupling condition. Thus we obtain

$$\int_{\Omega_S} \boldsymbol{\sigma}(\boldsymbol{u}^S) : \boldsymbol{\varepsilon}(\boldsymbol{v}^S) + \int_{\Omega_F} \frac{1}{\rho_F c^2} pq = \omega^2 \left(\int_{\Omega_S} \rho_S \boldsymbol{u}^S \cdot \boldsymbol{v}^S + \int_{\Omega_F} \frac{1}{c^2} \widetilde{\phi} q + \int_{\Gamma_I} \rho_F \widetilde{\phi}\, \boldsymbol{v}^S \cdot \boldsymbol{\nu} \right). \tag{9.10}$$

Similarly, multiplying the second equation in **PDE3** by $\frac{1}{c^2}$ times a test function $\psi \in \mathrm{H}^1(\Omega_F)$, integrating in Ω_F and using a Green's formula and the appropriate coupling condition (fifth equation in **PDE3**), we obtain

$$\int_{\Omega_F} \rho_F \nabla \widetilde{\phi} \cdot \nabla \psi = \int_{\Omega_F} \frac{1}{c^2} p\psi + \int_{\Gamma_I} \rho_F \boldsymbol{u}^S \cdot \boldsymbol{\nu}\, \psi. \tag{9.11}$$

Now we multiply (9.11) by ω^2 and add the resulting equation to (9.10). Therefore we obtain the following weak problem:

WP3.– *Find* $\omega \in \mathbb{R}$ *and* $\mathbf{0} \neq (\boldsymbol{u}^S, p, \widetilde{\phi}) \in \boldsymbol{\mathcal{V}}_3$ *such that*

$$\int_{\Omega_S} \boldsymbol{\sigma}(\boldsymbol{u}^S) : \boldsymbol{\varepsilon}(\boldsymbol{v}^S) + \int_{\Omega_F} \frac{1}{\rho_F c^2} pq = \omega^2 \left(\int_{\Omega_S} \rho_S \boldsymbol{u}^S \cdot \boldsymbol{v}^S + \int_{\Omega_F} \frac{1}{c^2} \widetilde{\phi} q + \int_{\Gamma_I} \rho_F \widetilde{\phi}\, \boldsymbol{v}^S \cdot \boldsymbol{\nu} + \int_{\Omega_F} \frac{1}{c^2} p\psi + \int_{\Gamma_I} \rho_F \boldsymbol{u}^S \cdot \boldsymbol{\nu}\, \psi - \int_{\Omega_F} \rho_F \nabla \widetilde{\phi} \cdot \nabla \psi \right) \quad \forall (\boldsymbol{v}^S, q, \psi) \in \boldsymbol{\mathcal{V}}_3,$$

where $\boldsymbol{\mathcal{V}}_3 := \mathbf{H}^1_{\Gamma_D}(\Omega_S) \times \mathrm{L}^2(\Omega_F) \times \mathrm{H}^1(\Omega_F)$.

Finally, to obtain a weak formulation of problem **PDE4**, we multiply the first two equations in this system by appropriate test functions and use Green's formulas, which leads to the following weak problem:

WP4.– *Find* $\omega \in \mathbb{R}$ *and* $\mathbf{0} \neq \left(\boldsymbol{u}^S, \phi\right) \in \boldsymbol{\mathcal{V}}_4$ *such that*

$$\int_{\Omega_S} \boldsymbol{\sigma}\left(\boldsymbol{u}^S\right) : \boldsymbol{\varepsilon}\left(\boldsymbol{v}^S\right) + \int_{\Omega_F} \rho_F \nabla \phi \cdot \nabla \psi - \int_{\Gamma_I} \rho_F \boldsymbol{u}^S \cdot \boldsymbol{\nu}\, \psi + \frac{\rho_F c^2}{|\Omega_F|} \int_{\Gamma_I} \boldsymbol{u}^S \cdot \boldsymbol{\nu} \int_{\Gamma_I} \boldsymbol{v}^S \cdot \boldsymbol{\nu} = \omega^2 \left(\int_{\Omega_F} \frac{\rho_F}{c^2} \phi\psi + \int_{\Omega_S} \rho_S \boldsymbol{u}^S \cdot \boldsymbol{v}^S + \int_{\Gamma_I} \rho_F \phi \boldsymbol{v}^S \cdot \boldsymbol{\nu} \right) \quad \forall \left(\boldsymbol{v}^S, \psi\right) \in \boldsymbol{\mathcal{V}}_4,$$

where $\boldsymbol{\mathcal{V}}_4 := \mathbf{H}^1_{\Gamma_D}(\Omega_S) \times \mathring{\mathrm{H}}^1(\Omega_F)$.

Let us emphasize that the bilinear forms arising in formulations **WP1** and **WP4** are not symmetric. Thus, the eigenvalues of these formulations could be, in principle, complex numbers. However, it is proved in [29] and in [11], respectively, that these spectra are real. The bilinear forms arising in formulations **WP2** and **WP3** are symmetric.

The results from [4, 6, 11, 15] for the weak formulations introduced above are summarized in the next theorem. It states that the spectrum is the same for all the formulations of the elastoacoustic problem, except for the eigenvalue $\omega^2 = 0$, which appears with multiplicity one in the pressure/displacement formulation and with infinite multiplicity in the pure displacement and in the mass coupling formulations.

Theorem 1.

- *The eigenfrequencies of* $\mathbf{WP1}$ *consist of a sequence of positive real numbers diverging to* $+\infty$*. All of them have finite multiplicity and their ascent is one (namely, algebraic and geometric multiplicities coincide). Moreover,* $\omega = 0$ *is an eigenfrequency of this problem with associated eigenfunction* $(\mathbf{u}_C^{\mathrm{S}}, C)$*, where* C *is any nonzero constant function and* $\mathbf{u}_C^{\mathrm{S}} \in \mathbf{H}^1_{\Gamma_{\mathrm{D}}}(\Omega_{\mathrm{S}})$ *is defined by*

$$\int_{\Omega_{\mathrm{S}}} \boldsymbol{\sigma}(\boldsymbol{u}_C^{\mathrm{S}}) : \boldsymbol{\varepsilon}(\boldsymbol{v}^{\mathrm{S}}) = C \int_{\Gamma_{\mathrm{I}}} \boldsymbol{v}^{\mathrm{S}} \cdot \boldsymbol{\nu} \qquad \forall \boldsymbol{v}^{\mathrm{S}} \in \mathbf{H}^1_{\Gamma_{\mathrm{D}}}(\Omega_{\mathrm{S}}).$$

- *For* $\omega \neq 0$*,* ω *is an eigenfrequency of* $\mathbf{WP2}$ *if and only if it is an eigenfrequency of* $\mathbf{WP1}$*, with the same ascent and multiplicity. Moreover,* $\omega = 0$ *is an eigenfrequency of* $\mathbf{WP2}$ *with infinite multiplicity, having as eigenfunctions all the rotational fluid motions, of the form* $(\mathbf{0}, \mathbf{curl}\,\psi)$*, with* $\psi \in \mathrm{H}^1_{\Gamma_{\mathrm{I}}}(\Omega_{\mathrm{F}})$*. (Recall that* $\mathbf{curl}\,\psi := (-\partial\psi/\partial x_2, \partial\psi/\partial x_1)$*.)*
- *For* $\omega \neq 0$*,* $(\omega, (\boldsymbol{u}^{\mathrm{S}}, p))$ *is an eigenpair of* $\mathbf{WP1}$ *if and only if there exists* $\widetilde{\phi} \in \mathrm{H}^1(\Omega_{\mathrm{F}})$ *such that* $(\omega, (\boldsymbol{u}^{\mathrm{S}}, p, \widetilde{\phi}))$ *is an eigenpair of* $\mathbf{WP3}$*. The corresponding multiplicities and ascent also coincide. Moreover,* $\omega = 0$ *is an eigenfrequency of* $\mathbf{WP3}$ *with infinite–multiplicity eigenspace* $\{\mathbf{0}\} \times \{0\} \times \mathrm{H}^1(\Omega_{\mathrm{F}})$*.*
- *For* $\omega \neq 0$*,* ω *is an eigenfrequency of* $\mathbf{WP4}$ *if and only if it is an eigenfrequency of* $\mathbf{WP1}$ *and its ascent and multiplicity are the same in both formulations. Moreover,* $\omega = 0$ *is not an eigenfrequency of* $\mathbf{WP4}$*.*

9.5 Finite Element Approximation

In this section, we introduce suitable finite element methods to solve efficiently the weak problems $\mathbf{WP1}$, $\mathbf{WP2}$ and $\mathbf{WP3}$. In the following section we show a modal reduction method to solve $\mathbf{WP4}$.

For simplicity, we present the methods in a two–dimensional setting. In all cases, the extension to 3D domains is straightforward, see [6, 7, 10].

Let $\mathcal{T}_h^{\mathrm{S}}$ and $\mathcal{T}_h^{\mathrm{F}}$ be regular partitions in triangles of Ω_{F} and Ω_{S}, respectively, where h denotes the typical mesh size in both domains. In order to simplify the exposition, we assume that these triangulations are compatible on the common interface Γ_{I}, see Figure 9.2, although this requirement is not essential, as can be seen in [7, 8].

The equations in Ω_{S} are common to the three formulations, so it is natural to use the same finite elements for $\boldsymbol{u}^{\mathrm{S}}$ in all cases. We will consider a standard choice

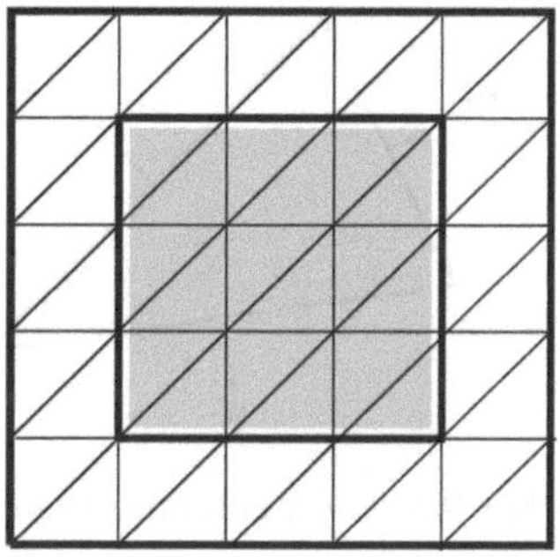

Fig. 9.2 Compatible meshes

for structural problems: each component of $\boldsymbol{u}_h^{\mathrm{S}}$ is discretized by classical piecewise linear Lagrange finite elements (three nodes in each triangle located at its vertices). Moreover, the Dirichlet boundary conditions are imposed as essential conditions, which leads to the following finite element space:

$$\boldsymbol{\mathcal{L}}_h^{\Gamma_{\mathrm{D}}}(\Omega_{\mathrm{S}}) := \left\{\boldsymbol{v}_h^{\mathrm{S}} \in \boldsymbol{\mathcal{C}}(\Omega_{\mathrm{S}}) : \boldsymbol{v}_h^{\mathrm{S}}|_T \in \boldsymbol{\mathcal{P}}_1(T) \quad \forall T \in \mathcal{T}_h^{\mathrm{S}} \quad \text{and} \quad \boldsymbol{v}_h^{\mathrm{S}}\big|_{\Gamma_{\mathrm{D}}} = \mathbf{0}\right\}.$$

Here and thereafter, $\mathcal{P}_k(T)$ and $\boldsymbol{\mathcal{P}}_k(T)$ denotes respectively scalar and vector valued polynomial functions defined in T of degree not greater than k.

Different discretizations will be used in the fluid for vector and scalar variables. For the scalar magnitudes (pressure or potential) we choose again piecewise linear Lagrange finite elements,

$$\mathcal{L}_h(\Omega_{\mathrm{F}}) := \left\{q_h \in \mathcal{C}(\Omega_{\mathrm{F}}) : q_h|_T \in \mathcal{P}_1(T) \quad \forall T \in \mathcal{T}_h^{\mathrm{F}}\right\}.$$

With these spaces, we state the finite element approximation of problem $\mathbf{WP1}$:

$\mathbf{WP1_h}$.– *Find* $\omega_{1h} \in \mathbb{R}$ *and* $\mathbf{0} \neq \left(\boldsymbol{u}_h^{\mathrm{S}}, p_h\right) \in \boldsymbol{\mathcal{V}}_{1h}$ *such that*

$$\int_{\Omega_{\mathrm{S}}} \boldsymbol{\sigma}\left(\boldsymbol{u}_h^{\mathrm{S}}\right) : \boldsymbol{\varepsilon}\left(\boldsymbol{v}_h^{\mathrm{S}}\right) + \int_{\Omega_{\mathrm{F}}} \frac{1}{\rho_{\mathrm{F}}} \boldsymbol{\nabla} p_h \cdot \boldsymbol{\nabla} q_h - \int_{\Gamma_{\mathrm{I}}} p_h \boldsymbol{v}_h^{\mathrm{S}} \cdot \boldsymbol{\nu}$$
$$= (\omega_{1h})^2 \left(\int_{\Omega_{\mathrm{S}}} \rho_{\mathrm{S}} \boldsymbol{u}_h^{\mathrm{S}} \cdot \boldsymbol{v}_h^{\mathrm{S}} + \int_{\Omega_{\mathrm{F}}} \frac{1}{\rho_{\mathrm{F}} c^2} p_h q_h + \int_{\Gamma_{\mathrm{I}}} \boldsymbol{u}_h^{\mathrm{S}} \cdot \boldsymbol{\nu} q_h\right) \quad \forall \left(\boldsymbol{v}_h^{\mathrm{S}}, q_h\right) \in \boldsymbol{\mathcal{V}}_{1h},$$

where $\boldsymbol{\mathcal{V}}_{1h} := \boldsymbol{\mathcal{L}}_h^{\Gamma_{\mathrm{D}}}(\Omega_{\mathrm{S}}) \times \mathcal{L}_h(\Omega_{\mathrm{F}})$.

Concerning $\mathbf{WP2}$, the fluid is represented by the displacement field $\boldsymbol{u}^{\mathrm{F}} \in \mathbf{H}(\mathrm{div}, \Omega_{\mathrm{F}})$, so that a finite element space to approximate vector fields in this space is needed. The simplest method that comes to mind would be to use again piecewise linear Lagrange finite elements for each component. However, this choice is unsuitable because the rotational fields associated to $\omega = 0$, see Theorem 1, are not

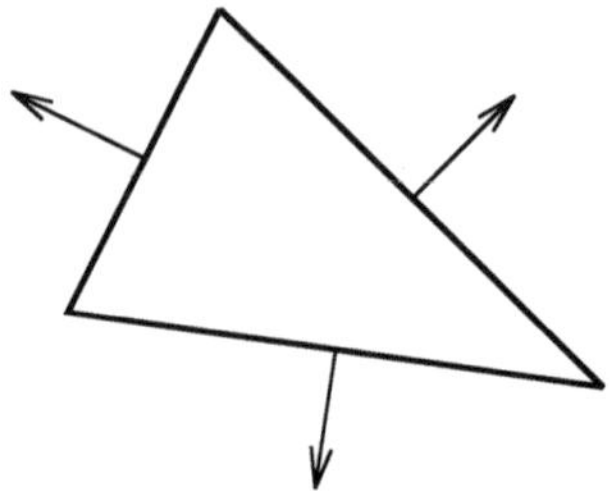

Fig. 9.3 Raviart–Thomas triangle

correctly represented at the discrete level. So, spurious modes appear polluting the numerical results, see [21,23].

Several alternatives to approximate correctly functions in $\mathbf{H}(\mathrm{div}, \Omega_{\mathrm{F}})$ are described and analyzed in the book by Brezzi and Fortin [17]. In the present work only the lowest–order Raviart–Thomas (RT) finite elements will be considered. This finite elements are vector fields that, for each triangle T, are incomplete linear polynomials of the form

$$(a + cx, b + cy), \qquad a, b, c \in \mathbb{R}.$$

Thus, the finite element space is defined by

$$\mathcal{R}_h(\Omega_{\mathrm{F}}) := \left\{ \boldsymbol{v}_h^{\mathrm{F}} \in \mathbf{H}(\mathrm{div}, \Omega_{\mathrm{F}}) : \boldsymbol{v}_h^{\mathrm{F}}\big|_T (x, y) = (a + cx, b + cy), \ a, b, c \in \mathbb{R} \right\}.$$

It is easy to check that these vector fields have constant normal components along the edges of the triangle T. Furthermore, these three values define a unique polynomial of the type above. Therefore, the constant normal components on all the edges of the triangulation can be regarded as the degrees of freedom of the lowest–order RT discretization, see Figure 9.3.
We emphasize that for this RT discretization

- there will be as many degrees of freedom as edges in $\mathcal{T}_h^{\mathrm{F}}$,
- vector fields in $\mathcal{R}_h(\Omega_{\mathrm{F}})$ have continuous normal components in Ω_{F} but they may have discontinuous tangential components on the edges of $\mathcal{T}_h^{\mathrm{F}}$.

Prior to presenting the discrete formulation corresponding to the pure displacement formulation **WP2** there is another point to address: the kinematic coupling condition, which for the continuous space $\mathcal{V}_2$ reads

$$\boldsymbol{v}^{\mathrm{S}} \cdot \boldsymbol{\nu} = \boldsymbol{v}^{\mathrm{F}} \cdot \boldsymbol{\nu} \qquad \text{on } \Gamma_{\mathrm{I}}. \tag{9.12}$$

Due to the different nature of the fluid and solid displacement discretizations, this condition cannot be imposed exactly in the finite element space $\mathcal{L}_h^{\Gamma_{\mathrm{D}}}(\Omega_{\mathrm{S}}) \times \mathcal{R}_h(\Omega_{\mathrm{F}})$, because it would be too restrictive. Indeed, since the functions in $\mathcal{R}_h(\Omega_{\mathrm{F}})$ have constant normal components on any edge of Γ_{I}, forcing (9.12) to be satisfied exactly would result in $\boldsymbol{v}_h^{\mathrm{S}} \cdot \boldsymbol{\nu}$ being constant all over Γ_{I}. Thus, to avoid this too rigid constraint, the coupling condition (9.12) is prescribed in a weak sense as follows:

$$\int_\ell \boldsymbol{v}_h^{\mathrm{S}} \cdot \boldsymbol{\nu} = \int_\ell \boldsymbol{v}_h^{\mathrm{F}} \cdot \boldsymbol{\nu} \qquad \forall \text{ edge } \ell \subset \Gamma_{\mathrm{I}}. \tag{9.13}$$

From this equation, it is clear that, in general, the normal components $\boldsymbol{v}_h^{\mathrm{S}} \cdot \boldsymbol{\nu}$ and $\boldsymbol{v}_h^{\mathrm{F}} \cdot \boldsymbol{\nu}$ do not match at every point on Γ_{I}. In fact, it is easy to check that both are equal only at the midpoints of the edges on Γ_{I}, see [10].

A simple numerical way to impose the constraint (9.13) is by means of a Lagrange multiplier. With this purpose, consider a new finite element space defined only on Γ_{I}:

$$\mathcal{Q}_h(\Gamma_{\mathrm{I}}) := \left\{ q_h \in \mathrm{L}_2(\Gamma_{\mathrm{I}}) : q_h|_\ell \in \mathcal{P}_0(\ell) \quad \forall \text{ edge } \ell \subset \Gamma_{\mathrm{I}} \right\}.$$

Equation (9.13) can be equivalently written

$$\int_{\Gamma_{\mathrm{I}}} \left(\boldsymbol{v}_h^{\mathrm{S}} \cdot \boldsymbol{\nu} - \boldsymbol{v}_h^{\mathrm{F}} \cdot \boldsymbol{\nu} \right) q_h = 0 \qquad \forall q_h \in \mathcal{Q}_h(\Gamma_{\mathrm{I}}).$$

Thus, we have all the ingredients to state a discrete version of **WP2**:

WP2$_\mathbf{h}$.- *Find* $\omega_{2h} \in \mathbb{R}$, $\mathbf{0} \neq \left(\boldsymbol{u}_h^{\mathrm{S}}, \boldsymbol{u}_h^{\mathrm{F}}\right) \in \boldsymbol{\mathcal{V}}_{2h}$ *and* $p_h^{\mathrm{I}} \in \mathcal{Q}_h(\Gamma_{\mathrm{I}})$ *such that*

$$\begin{aligned}
&\int_{\Omega_{\mathrm{S}}} \boldsymbol{\sigma}\left(\boldsymbol{u}_h^{\mathrm{S}}\right) : \boldsymbol{\varepsilon}\left(\boldsymbol{v}_h^{\mathrm{S}}\right) + \int_{\Omega_{\mathrm{F}}} \rho_{\mathrm{F}} c^2 \operatorname{div} \boldsymbol{u}_h^{\mathrm{F}} \operatorname{div} \boldsymbol{v}_h^{\mathrm{F}} + \int_{\Gamma_{\mathrm{I}}} p_h^{\mathrm{I}} \left(\boldsymbol{v}_h^{\mathrm{S}} \cdot \boldsymbol{\nu} - \boldsymbol{v}_h^{\mathrm{F}} \cdot \boldsymbol{\nu} \right) \\
&\qquad = (\omega_{2h})^2 \left(\int_{\Omega_{\mathrm{S}}} \rho_{\mathrm{S}} \boldsymbol{u}_h^{\mathrm{S}} \cdot \boldsymbol{v}_h^{\mathrm{S}} + \int_{\Omega_{\mathrm{F}}} \rho_{\mathrm{F}} \boldsymbol{u}_h^{\mathrm{F}} \cdot \boldsymbol{v}_h^{\mathrm{F}} \right) \qquad \forall \left(\boldsymbol{v}_h^{\mathrm{S}}, \boldsymbol{v}_h^{\mathrm{F}}\right) \in \boldsymbol{\mathcal{V}}_{2h}, \\
&\int_{\Gamma_{\mathrm{I}}} q_h \left(\boldsymbol{u}_h^{\mathrm{S}} \cdot \boldsymbol{\nu} - \boldsymbol{u}_h^{\mathrm{F}} \cdot \boldsymbol{\nu} \right) = 0 \qquad \forall q_h \in \mathcal{Q}_h(\Gamma_{\mathrm{I}}),
\end{aligned}$$

where $\boldsymbol{\mathcal{V}}_{2h} := \boldsymbol{\mathcal{L}}_h^{\Gamma_{\mathrm{D}}}(\Omega_{\mathrm{S}}) \times \boldsymbol{\mathcal{R}}_h(\Omega_{\mathrm{F}})$.

It is shown in [5] that the Lagrange multiplier p_h^{I} is a discretization of the pressure on the fluid–solid interface.

Remark 2. In Section 9.9.2, we show an alternative equivalent way to impose the coupling condition by means of a static condensation.

Finally, to discretize the redundant formulation **WP3**, we use the same Lagrange finite element spaces used in **WP1$_\mathbf{h}$**. The discrete problem reads as follows:

WP3$_\mathbf{h}$.– *Find* $\omega_{3h} \in \mathbb{R}$ *and* $\mathbf{0} \neq (\boldsymbol{u}_h^{\mathrm{S}}, p_h, \widetilde{\phi}_h) \in \boldsymbol{\mathcal{V}}_{3h}$ *such that*

$$\begin{aligned}
&\int_{\Omega_{\mathrm{S}}} \boldsymbol{\sigma}(\boldsymbol{u}_h^{\mathrm{S}}) : \boldsymbol{\varepsilon}(\boldsymbol{v}_h^{\mathrm{S}}) + \int_{\Omega_{\mathrm{F}}} \frac{1}{\rho_{\mathrm{F}} c^2} p_h q_h \\
&= (\omega_{3h})^2 \left(\int_{\Omega_{\mathrm{S}}} \rho_{\mathrm{S}} \boldsymbol{u}_h^{\mathrm{S}} \cdot \boldsymbol{v}_h^{\mathrm{S}} + \int_{\Omega_{\mathrm{F}}} \frac{1}{c^2} \widetilde{\phi}_h q_h + \int_{\Gamma_{\mathrm{I}}} \rho_{\mathrm{F}} \widetilde{\phi}_h \, \boldsymbol{v}_h^{\mathrm{S}} \cdot \boldsymbol{\nu} + \int_{\Omega_{\mathrm{F}}} \frac{1}{c^2} p_h \psi_h \right. \\
&\qquad \left. + \int_{\Gamma_{\mathrm{I}}} \rho_{\mathrm{F}} \boldsymbol{u}_h^{\mathrm{S}} \cdot \boldsymbol{\nu} \, \psi_h - \int_{\Omega_{\mathrm{F}}} \rho_{\mathrm{F}} \boldsymbol{\nabla} \widetilde{\phi}_h \cdot \boldsymbol{\nabla} \psi_h \right) \qquad \forall (\boldsymbol{v}_h^{\mathrm{S}}, q_h, \psi_h) \in \boldsymbol{\mathcal{V}}_{3h},
\end{aligned}$$

where $\boldsymbol{\mathcal{V}}_{3h} := \boldsymbol{\mathcal{L}}_h^{\Gamma_{\mathrm{D}}}(\Omega_{\mathrm{S}}) \times \mathcal{L}_h(\Omega_{\mathrm{F}}) \times \mathcal{L}_h(\Omega_{\mathrm{F}})$.

The main results derived in [4, 6, 15, 28] are presented in the following theorem, which can be summarized as follows: all the discrete problems are well posed and the approximate eigenfrequencies and eigenfunctions converge with optimal order.

It is convenient to introduce some notation before settling the theorem. Let $0 < \omega_1 \leq \ldots \leq \omega_n \leq \ldots$ be the strictly positive eigenfrequencies of problems $\mathbf{WP1} - \mathbf{WP3}$, repeated according to their respective multiplicities (let us recall that, according to Theorem 1, except for $\omega = 0$, the spectra of all these problems are the same). Let $(\boldsymbol{u}_n^{\mathrm{S}}, p_n)$, $(\boldsymbol{u}_n^{\mathrm{S}}, \boldsymbol{u}_n^{\mathrm{F}})$, and $\left(\boldsymbol{u}_n^{\mathrm{S}}, p_n, \widetilde{\phi}_n\right)$ denote normalized eigenfunctions of problems $\mathbf{WP1}$, $\mathbf{WP2}$, and $\mathbf{WP3}$, respectively, associated to the eigenfrequency ω_n. In what follows, C will be used to denote a strictly positive constant, not necessarily the same at each instance.

Theorem 2. *The finite element problems* $\mathbf{WP1_h} - \mathbf{WP3_h}$ *are well posed and all of them have real non–negative spectra. Moreover, there exists a constant* $r \in (0,1]$, *which depends on the geometry of the domains* Ω_{F} *and* Ω_{S} *and on the Lamé's coefficients, such that the following properties hold:*

1. *Let* $\omega_{1h,1} \leq \cdots \leq \omega_{1h,N_1}$ *be the strictly positive eigenfrequencies of* $\mathbf{WP1_h}$, *repeated according to their multiplicities. There exist normalized eigenfunctions* $(\boldsymbol{u}_{1h,n}^{\mathrm{S}}, p_{1h,n})$ *associated to* $\omega_{1h,n}$ *such that*

$$
\begin{aligned}
|\omega_n - \omega_{1h,n}| &\leq Ch^{2r}, \\
\left\| \left(\boldsymbol{u}_n^{\mathrm{S}}, p_n\right) - \left(\boldsymbol{u}_{1h,n}^{\mathrm{S}}, p_{1h,n}\right) \right\|_{\mathbf{H}^1(\Omega_{\mathrm{S}}) \times \mathrm{H}^1(\Omega_{\mathrm{F}})} &\leq Ch^r.
\end{aligned}
$$

2. *Let* $\omega_{2h,1} \leq \cdots \leq \omega_{2h,N_2}$ *be the strictly positive eigenfrequencies of* $\mathbf{WP2_h}$, *repeated according to their multiplicities. There exist normalized eigenfunctions* $(\boldsymbol{u}_{2h,n}^{\mathrm{S}}, \boldsymbol{u}_{2h,n}^{\mathrm{F}})$ *associated to* $\omega_{2h,n}$ *such that*

$$
\begin{aligned}
|\omega_n - \omega_{2h,n}| &\leq Ch^{2r}, \\
\left\| \left(\boldsymbol{u}_n^{\mathrm{S}}, \boldsymbol{u}_n^{\mathrm{F}}\right) - \left(\boldsymbol{u}_{2h,n}^{\mathrm{S}}, \boldsymbol{u}_{2h,n}^{\mathrm{F}}\right) \right\|_{\mathbf{H}^1(\Omega_{\mathrm{S}}) \times \mathbf{H}(\mathrm{div}, \Omega_{\mathrm{F}})} &\leq Ch^r.
\end{aligned}
$$

3. *Let* $\omega_{3h,1} \leq \cdots \leq \omega_{3h,N_3}$ *be the strictly positive eigenfrequencies of* $\mathbf{WP3_h}$, *repeated according to their multiplicities. There exist normalized eigenfunctions* $(\boldsymbol{u}_{3h,n}^{\mathrm{S}}, p_{3h,n}, \widetilde{\phi}_{3h,n})$ *associated to* $\omega_{3h,n}$ *such that*

$$
\begin{aligned}
|\omega_n - \omega_{3h,n}| &\leq Ch^{2r}, \\
\left\| \left(\boldsymbol{u}_n^{\mathrm{S}}, p_n\right) - \left(\boldsymbol{u}_{3h,n}^{\mathrm{S}}, p_{3h,n}\right) \right\|_{\mathbf{H}^1(\Omega_{\mathrm{S}}) \times \mathrm{L}_2(\Omega_{\mathrm{F}})} &\leq Ch^r.
\end{aligned}
$$

Remark 3. The error bound for the eigenfunctions in the previous theorem can be extended to distances between eigenspaces. In fact, for instance the following result holds for problem $\mathbf{WP1}$ (analogous results hold for problems $\mathbf{WP2}$ and $\mathbf{WP3}$, as well): For a given interval $I \subset \mathbb{R}$, let $\boldsymbol{\mathcal{E}}_I^{\mathrm{P}}$ be the direct sum of the eigenspaces for the continuous problem $\mathbf{WP1}$ associated with the eigenvalues in I and let $\boldsymbol{\mathcal{E}}_{I,1h}^{\mathrm{P}}$

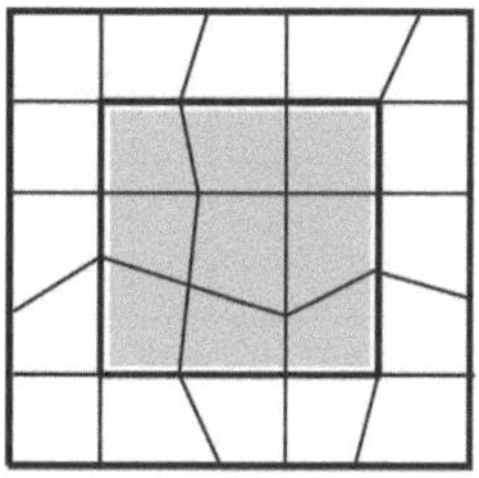

Fig. 9.4 Quadrilateral compatible meshes

the analogous space for the discrete problem $\mathbf{WP1_h}$. Then $\forall(\boldsymbol{u}^{\mathrm{S}}_{1h,n}, p_{1h,n}) \in \boldsymbol{\mathcal{E}}^{\mathrm{P}}_{I,1h}$ such that $\|(\boldsymbol{u}^{\mathrm{S}}_{1h,n}, p_{1h,n})\|_{\mathbf{H}^1(\Omega_{\mathrm{S}})\times \mathrm{H}^1(\Omega_{\mathrm{F}})} = 1$,

$$\operatorname{dist}\left(\left(\boldsymbol{u}^{\mathrm{S}}_{1h,n}, p_{1h,n}\right), \boldsymbol{\mathcal{E}}^{\mathrm{P}}_{I}\right) \leq Ch^r,$$

where dist denotes the distance in the $\mathbf{H}^1(\Omega_{\mathrm{S}}) \times \mathrm{H}^1(\Omega_{\mathrm{F}})$ norm.

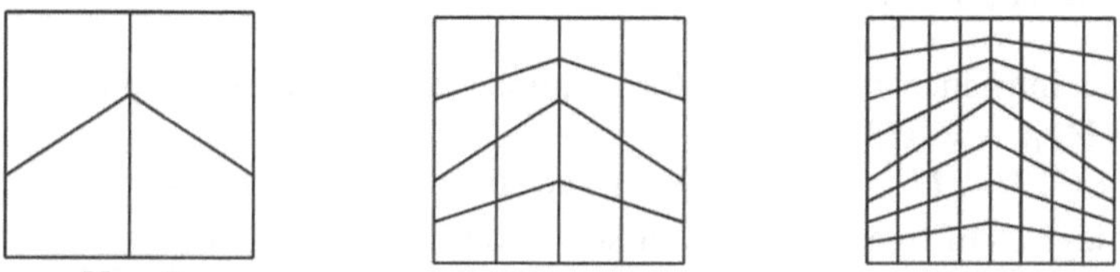

Fig. 9.5 Asymptotically parallelogram nonuniform meshes.

Remark 4. Regarding the discrete spectral problem $\mathbf{WP2_h}$, it can also be proved that it is free of spurious modes. Moreover, $\omega_h = 0$ is an eigenfrequency of the discrete problem with eigenspace equal to the set of discrete rotational fluid motions of the form $(\mathbf{0}, \mathbf{curl}\, \psi_h)$, with $\psi_h \in \mathcal{L}_h(\Omega_{\mathrm{F}})$ such that $\psi_h|_{\Gamma_{\mathrm{I}}} = 0$.

We have presented so far finite element discretizations based on triangular meshes for the different methods above. Certainly, this is not essential and quadrilateral meshes for Ω_{F} and Ω_{S}, as those shown in Figure 9.4, can be used too by introducing the corresponding Lagrange and Raviart–Thomas finite element spaces.

In this case, Lagrange elements preserve their good behavior, so it can be expected that the estimates of the previous theorem for the pressure and for the mass coupling formulations, $\mathbf{WP1}$ and $\mathbf{WP3}$, respectively, will also hold (although, to the authors' knowledge, they have not been proved yet).

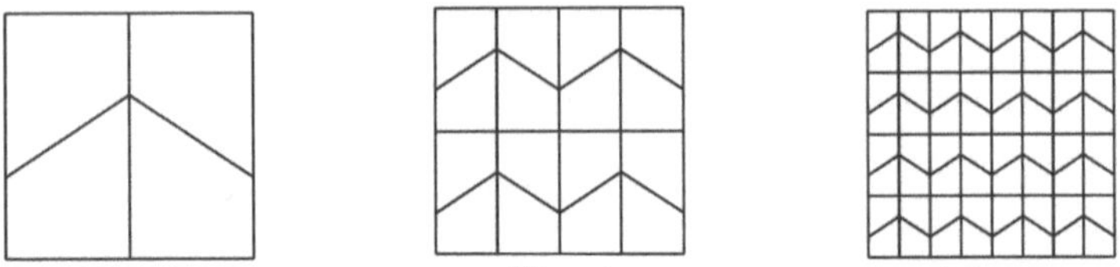

Fig. 9.6 Non–asymptotically parallelogram meshes (trapezoidal elements).

However, for the Raviart–Thomas elements, these estimates hold true only if the quadrilateral (hexahedral) meshes are at least asymptotically parallelogram (parallelepiped in 3D), see Figure 9.5.

In particular, in [7], see also [1,2], an example is shown where, although the optimal order was achieved, the numerical eigenfrequencies converge to wrong values. In that test, the pure displacement formulation of the spectral problem for an uncoupled fluid was solved by using the non asymptotically parallelogram (trapezoidal) meshes shown in Figure 9.6.

9.5.1 Matrix Formulations

Next, we introduce the matrix formulations corresponding to the discrete problems $\mathbf{WP1_h} - \mathbf{WP3_h}$.

Let $\boldsymbol{U}^{\mathrm{S}}$, $\boldsymbol{V}^{\mathrm{S}}$, $\boldsymbol{U}^{\mathrm{F}}$, $\boldsymbol{V}^{\mathrm{F}}$, $\boldsymbol{P}$, $\boldsymbol{Q}$, $\boldsymbol{P}_{\mathrm{I}}$ and $\widetilde{\boldsymbol{\Phi}}$ be column vectors whose components are the degrees of freedom (d.o.f.) associated to $\boldsymbol{u}_h^{\mathrm{S}}$, $\boldsymbol{v}_h^{\mathrm{S}}$, $\boldsymbol{u}_h^{\mathrm{F}}$, $\boldsymbol{v}_h^{\mathrm{F}}$, p_h, q_h, p_h^{I}, and $\widetilde{\phi}_h$, respectively.

The different matrices that will be used in the sequel are defined by the following relations:

$$\boldsymbol{V}^{\mathrm{S}^{\mathrm{t}}}\boldsymbol{K}^{\mathrm{S}}\boldsymbol{U}^{\mathrm{S}} = \int_{\Omega_{\mathrm{S}}} \boldsymbol{\sigma}\left(\boldsymbol{u}_h^{\mathrm{S}}\right) : \boldsymbol{\varepsilon}\left(\boldsymbol{v}_h^{\mathrm{S}}\right), \qquad \boldsymbol{V}^{\mathrm{S}^{\mathrm{t}}}\boldsymbol{M}^{\mathrm{S}}\boldsymbol{U}^{\mathrm{S}} = \int_{\Omega_{\mathrm{S}}} \rho_{\mathrm{S}}\boldsymbol{u}_h^{\mathrm{S}}\cdot\boldsymbol{v}_h^{\mathrm{S}},$$

$$\boldsymbol{Q}^{\mathrm{t}}\boldsymbol{K}_{\mathrm{P}}^{\mathrm{F}}\boldsymbol{P} = \int_{\Omega_{\mathrm{F}}} \boldsymbol{\nabla} p_h\cdot\boldsymbol{\nabla} q_h, \qquad \boldsymbol{Q}^{\mathrm{t}}\boldsymbol{M}_{\mathrm{P}}^{\mathrm{F}}\boldsymbol{P} = \int_{\Omega_{\mathrm{F}}} p_h q_h,$$

$$\boldsymbol{V}^{\mathrm{F}^{\mathrm{t}}}\boldsymbol{K}_{\mathrm{D}}^{\mathrm{F}}\boldsymbol{U}^{\mathrm{F}} = \int_{\Omega_{\mathrm{F}}} \rho_{\mathrm{F}}c^2 \operatorname{div}\boldsymbol{u}_h^{\mathrm{F}} \operatorname{div}\boldsymbol{v}_h^{\mathrm{F}}, \qquad \boldsymbol{V}^{\mathrm{F}^{\mathrm{t}}}\boldsymbol{M}_{\mathrm{D}}^{\mathrm{F}}\boldsymbol{U}^{\mathrm{F}} = \int_{\Omega_{\mathrm{F}}} \rho_{\mathrm{F}}\boldsymbol{u}_h^{\mathrm{F}}\cdot\boldsymbol{v}_h^{\mathrm{F}},$$

$$\boldsymbol{V}^{\mathrm{S}^{\mathrm{t}}}\boldsymbol{C}\boldsymbol{P} = \int_{\Gamma_{\mathrm{I}}} p_h\boldsymbol{v}_h^{\mathrm{S}}\cdot\boldsymbol{\nu}, \qquad \boldsymbol{V}^{\mathrm{S}^{\mathrm{t}}}\boldsymbol{D}\boldsymbol{P} = \int_{\Gamma_{\mathrm{I}}} p_h\boldsymbol{v}_h^{\mathrm{F}}\cdot\boldsymbol{\nu},$$

$$\boldsymbol{V}^{\mathrm{S}^{\mathrm{t}}}\boldsymbol{C}_{\mathrm{I}}\boldsymbol{P}_{\mathrm{I}} = \int_{\Gamma_{\mathrm{I}}} p_h^{\mathrm{I}}\boldsymbol{v}_h^{\mathrm{S}}\cdot\boldsymbol{\nu}, \qquad \boldsymbol{V}^{\mathrm{S}^{\mathrm{t}}}\boldsymbol{D}_{\mathrm{I}}\boldsymbol{P}_{\mathrm{I}} = \int_{\Gamma_{\mathrm{I}}} p_h^{\mathrm{I}}\boldsymbol{v}_h^{\mathrm{F}}\cdot\boldsymbol{\nu}.$$

Thus, the matrix formulations for the discrete weak problems introduced above can be written as follows:

$\mathbf{WP1_h}$.–

$$\begin{bmatrix} \boldsymbol{K}^{\mathrm{S}} & -\boldsymbol{C} \\ \boldsymbol{0} & \frac{1}{\rho_{\mathrm{F}}}\boldsymbol{K}_{\mathrm{P}}^{\mathrm{F}} \end{bmatrix} \begin{bmatrix} \boldsymbol{U}^{\mathrm{S}} \\ \boldsymbol{P} \end{bmatrix} = (\omega_{1h})^2 \begin{bmatrix} \boldsymbol{M}^{\mathrm{S}} & \boldsymbol{0} \\ \boldsymbol{C}^{\mathrm{t}} & \frac{1}{\rho_{\mathrm{F}} c^2}\boldsymbol{M}_{\mathrm{P}}^{\mathrm{F}} \end{bmatrix} \begin{bmatrix} \boldsymbol{U}^{\mathrm{S}} \\ \boldsymbol{P} \end{bmatrix}.$$

The discrete generalized eigenvalue problem written in this form is not suited to be solved with standard solvers, because the matrix on the right hand side is neither symmetric nor positive definite. As shown in [6], this difficulty can be overcome by rewriting the matrix equation in the following equivalent form:

$$\begin{bmatrix} \boldsymbol{K}^{\mathrm{S}} & -\boldsymbol{C} \\ \boldsymbol{0} & \delta\frac{1}{\rho_{\mathrm{F}}}\boldsymbol{K}_{\mathrm{P}}^{\mathrm{F}} \end{bmatrix} \begin{bmatrix} \boldsymbol{U}^{\mathrm{S}} \\ \boldsymbol{P} \end{bmatrix} = \lambda \begin{bmatrix} \boldsymbol{M}^{\mathrm{S}} - \delta\boldsymbol{K}^{\mathrm{S}} & \delta\boldsymbol{C} \\ \delta\boldsymbol{C}^{\mathrm{t}} & \delta\left(\frac{1}{\rho_{\mathrm{F}} c^2}\boldsymbol{M}_{\mathrm{P}}^{\mathrm{F}} - \delta\frac{1}{\rho_{\mathrm{F}}}\boldsymbol{K}_{\mathrm{P}}^{\mathrm{F}}\right) \end{bmatrix} \begin{bmatrix} \boldsymbol{U}^{\mathrm{S}} \\ \boldsymbol{P} \end{bmatrix},$$

with $\lambda := (\omega_{1h}) / [1 - \delta (\omega_{1h})^2]$. The matrix on the right hand side now is symmetric and also positive definite if $\delta > 0$ is small enough.

$\mathbf{WP2_h}$.–

$$\begin{bmatrix} \boldsymbol{K}^{\mathrm{S}} & \boldsymbol{0} & \boldsymbol{C}_{\mathrm{I}} \\ \boldsymbol{0} & \boldsymbol{K}_{\mathrm{D}}^{\mathrm{F}} & \boldsymbol{D}_{\mathrm{I}} \\ \boldsymbol{C}_{\mathrm{I}}^{\mathrm{t}} & \boldsymbol{D}_{\mathrm{I}}^{\mathrm{t}} & \boldsymbol{0} \end{bmatrix} \begin{bmatrix} \boldsymbol{U}^{\mathrm{S}} \\ \boldsymbol{U}^{\mathrm{F}} \\ \boldsymbol{P}_{\mathrm{I}} \end{bmatrix} = (\omega_{2h})^2 \begin{bmatrix} \boldsymbol{M}^{\mathrm{S}} & \boldsymbol{0} & \boldsymbol{0} \\ \boldsymbol{0} & \boldsymbol{M}_{\mathrm{D}}^{\mathrm{F}} & \boldsymbol{0} \\ \boldsymbol{0} & \boldsymbol{0} & \boldsymbol{0} \end{bmatrix} \begin{bmatrix} \boldsymbol{U}^{\mathrm{S}} \\ \boldsymbol{U}^{\mathrm{F}} \\ \boldsymbol{P}_{\mathrm{I}} \end{bmatrix}.$$

We recall that the unknown $\boldsymbol{P}_{\mathrm{I}}$ for the pressure in this formulation is defined only on the interface Γ_{I}.

$\mathbf{WP3_h}$.–

$$\begin{bmatrix} \boldsymbol{K}^{\mathrm{S}} & \boldsymbol{0} & \boldsymbol{0} \\ \boldsymbol{0} & \frac{1}{\rho_{\mathrm{F}} c^2}\boldsymbol{M}_{\mathrm{P}}^{\mathrm{F}} & \boldsymbol{0} \\ \boldsymbol{0} & \boldsymbol{0} & \boldsymbol{0} \end{bmatrix} \begin{bmatrix} \boldsymbol{U}^{\mathrm{S}} \\ \boldsymbol{P} \\ \widetilde{\boldsymbol{\Phi}} \end{bmatrix} = (\omega_{3h})^2 \begin{bmatrix} \boldsymbol{M}^{\mathrm{S}} & \boldsymbol{0} & \rho_{\mathrm{F}}\boldsymbol{C} \\ \boldsymbol{0} & \boldsymbol{0} & \frac{1}{c^2}\boldsymbol{M}_{\mathrm{P}}^{\mathrm{F}} \\ \rho_{\mathrm{F}}\boldsymbol{C}^{\mathrm{t}} & \frac{1}{c^2}\boldsymbol{M}_{\mathrm{P}}^{\mathrm{F}} & -\rho_{\mathrm{F}}\boldsymbol{K}_{\mathrm{P}}^{\mathrm{F}} \end{bmatrix} \begin{bmatrix} \boldsymbol{U}^{\mathrm{S}} \\ \boldsymbol{P} \\ \widetilde{\boldsymbol{\Phi}} \end{bmatrix}.$$

Remark 5. For non–zero eigenvalues, the spectral problems $\mathbf{WP1_h}$ and $\mathbf{WP3_h}$ are equivalent. In fact, from the second block in the matrix formulation of $\mathbf{WP3_h}$,

$$(\omega_{3h})^2 \widetilde{\boldsymbol{\Phi}} = \frac{1}{\rho_{\mathrm{F}}}\boldsymbol{P},$$

and substituting this in the first and third block we obtain the equations

$$\boldsymbol{K}^{\mathrm{S}}\boldsymbol{U}^{\mathrm{S}} = (\omega_{3h})^2 \boldsymbol{M}^{\mathrm{S}}\boldsymbol{U}^{\mathrm{S}} + \boldsymbol{C}\boldsymbol{P},$$

$$\boldsymbol{0} = (\omega_{3h})^2 \left(\rho_{\mathrm{F}}\boldsymbol{C}^{\mathrm{t}}\boldsymbol{U}^{\mathrm{S}} + \frac{1}{c^2}\boldsymbol{M}_{\mathrm{P}}^{\mathrm{F}}\boldsymbol{P}\right) - \boldsymbol{K}_{\mathrm{P}}^{\mathrm{F}}\boldsymbol{P},$$

which are clearly equivalent to $\mathbf{WP1_h}$.

9.6 A Modal Synthesis Method

The three techniques studied so far are based on different discretizations of the coupled fluid–structure equations. In this section we introduce another strategy based on solving, first, uncoupled problems in the solid and in the fluid and, finally, a smaller coupled problem. A description of such method can be found in the book by Morand and Ohayon [24], and a mathematical analysis in [11].

We illustrate this technique with problem **WP4**, which certainly could also be solved by means of the finite element spaces defined above. Instead, we consider the following modal synthesis method:

Algorithm 1 *1. Compute by finite elements the first*[1] N_{S} *eigenmodes of the solid in vacuo:*

$$\begin{cases} \operatorname{div}\left[\boldsymbol{\sigma}\left(\boldsymbol{u}^{\mathrm{S}}\right)\right] + \omega^2 \rho_{\mathrm{S}} \boldsymbol{u}^{\mathrm{S}} = \mathbf{0} & \text{in } \Omega_{\mathrm{S}}, \\ \boldsymbol{\sigma}\left(\boldsymbol{u}^{\mathrm{S}}\right) \boldsymbol{\nu} = \mathbf{0} & \text{on } \Gamma_{\mathrm{I}}, \\ \boldsymbol{\sigma}\left(\boldsymbol{u}^{\mathrm{S}}\right) \boldsymbol{\eta} = \mathbf{0} & \text{on } \Gamma_{\mathrm{N}}, \\ \boldsymbol{u}^{\mathrm{S}} = \mathbf{0} & \text{on } \Gamma_{\mathrm{D}}. \end{cases}$$

Specifically, solve the discrete eigenvalue problem:

Find $\omega_h^{\mathrm{S}} > 0$ *and* $\mathbf{0} \neq \boldsymbol{u}_h^{\mathrm{S}} \in \boldsymbol{\mathcal{L}}_h^{\Gamma_{\mathrm{D}}}(\Omega_{\mathrm{S}})$ *such that*

$$\int_{\Omega_{\mathrm{S}}} \boldsymbol{\sigma}\left(\boldsymbol{u}_h^{\mathrm{S}}\right) : \boldsymbol{\varepsilon}\left(\boldsymbol{v}_h^{\mathrm{S}}\right) = \left(\omega_h^{\mathrm{S}}\right)^2 \int_{\Omega_{\mathrm{S}}} \rho_{\mathrm{S}} \boldsymbol{u}_h^{\mathrm{S}} \cdot \boldsymbol{v}_h^{\mathrm{S}} \qquad \forall \boldsymbol{v}_h^{\mathrm{S}} \in \boldsymbol{\mathcal{L}}_h^{\Gamma_{\mathrm{D}}}(\Omega_{\mathrm{S}}),$$

to obtain the first N_{S} *approximate vibration frequencies* $\omega_{h,1}^{\mathrm{S}}, \dots, \omega_{h,N_{\mathrm{S}}}^{\mathrm{S}}$ *and the corresponding eigenfunctions* $\boldsymbol{u}_{h,1}^{\mathrm{S}}, \dots, \boldsymbol{u}_{h,N_{\mathrm{S}}}^{\mathrm{S}}$.

2. *Compute by finite elements the corresponding* N_{S} *static liftings in the fluid (i.e., the static response of the fluid to the normal displacement induced by each solid–eigenmode):*

$$\begin{cases} \Delta \phi^m = \dfrac{1}{|\Omega_{\mathrm{F}}|} \displaystyle\int_{\Gamma_{\mathrm{I}}} \boldsymbol{u}_m^{\mathrm{S}} \cdot \boldsymbol{\nu} & \text{in } \Omega_{\mathrm{F}}, \\ \dfrac{\partial \phi^m}{\partial \boldsymbol{\nu}} = \boldsymbol{u}_m^{\mathrm{S}} \cdot \boldsymbol{\nu} & \text{on } \Gamma_{\mathrm{I}}, \\ \displaystyle\int_{\Omega_{\mathrm{F}}} \phi^m = 0. \end{cases}$$

This amounts to solving, for $m = 1, \dots, N_{\mathrm{S}}$, *the discrete Neumann problem:*

Find $\phi_h^m \in \mathcal{L}_h^0(\Omega_{\mathrm{F}})$ *such that*

$$\int_{\Omega_{\mathrm{F}}} \boldsymbol{\nabla} \phi_h^m \cdot \boldsymbol{\nabla} \psi_h = \int_{\Gamma_{\mathrm{I}}} \boldsymbol{u}_{h,m}^{\mathrm{S}} \cdot \boldsymbol{\nu}\, \psi_h \qquad \forall \psi_h \in \mathcal{L}_h^0(\Omega_{\mathrm{F}}),$$

where $\mathcal{L}_h^0(\Omega_{\mathrm{F}}) := \left\{ \psi_h \in \mathcal{L}_h(\Omega_{\mathrm{F}}) : \int_{\Omega_{\mathrm{F}}} \psi_h = 0 \right\}$.

[1]From now on, we order the eigenmodes according to the corresponding vibration frequencies, so that, for instance, "first" actually means *eigenmodes with lowest vibration frequencies*.

3. *Compute by finite elements the first N_{F} non–constant eigenmodes of the fluid in a rigid cavity:*

$$\begin{cases} \rho_{\mathrm{F}} \Delta\phi + \omega^2 \dfrac{\rho_{\mathrm{F}}}{c^2}\phi = 0 & \text{in } \Omega_{\mathrm{F}}, \\ \dfrac{\partial\phi}{\partial\boldsymbol{\nu}} = 0 & \text{on } \Gamma_{\mathrm{I}}. \end{cases}$$

Specifically, solve the discrete eigenvalue problem:

Find $\omega_h^{\mathrm{F}} > 0$ and $0 \neq \phi_h \in \mathcal{L}_h(\Omega_{\mathrm{F}})$ such that

$$\int_{\Omega_{\mathrm{F}}} \rho_{\mathrm{F}} \boldsymbol{\nabla}\phi_h \cdot \boldsymbol{\nabla}\psi_h = \left(\omega_h^{\mathrm{F}}\right)^2 \int_{\Omega_{\mathrm{F}}} \frac{\rho_{\mathrm{F}}}{c^2}\phi_h\psi_h \qquad \forall\psi_h \in \mathcal{L}_h(\Omega_{\mathrm{F}}),$$

to obtain the first N_{F} approximate vibration frequencies $\omega_{h,1}^{\mathrm{F}}, \ldots, \omega_{h,N_{\mathrm{F}}}^{\mathrm{F}}$ and the corresponding eigenfunctions $\phi_{h,1}, \ldots, \phi_{h,N_{\mathrm{F}}}$.

4. *Find an approximate solution of problem* **WP4** *in the finite–dimensional space*

$$\boldsymbol{\mathcal{V}}_{4h,1} := \left\langle \left\{ \left(\boldsymbol{u}_{h,m}^{\mathrm{S}}, \phi_h^m\right) \right\}_{m=1}^{N_{\mathrm{S}}}, \{(\boldsymbol{0}, \phi_{h,i})\}_{i=1}^{N_{\mathrm{F}}} \right\rangle,$$

by taking the test functions in

$$\boldsymbol{\mathcal{V}}_{4h,2} := \left\langle \left\{ \left(\boldsymbol{u}_{h,m}^{\mathrm{S}}, 0\right) \right\}_{m=1}^{N_{\mathrm{S}}}, \{(\boldsymbol{0}, \phi_{h,i})\}_{i=1}^{N_{\mathrm{F}}} \right\rangle.$$

The reason to choose a space for the test functions, $\boldsymbol{\mathcal{V}}_{4h,2}$, different to that for the trial functions, $\boldsymbol{\mathcal{V}}_{4h,1}$, is that this choice leads to a symmetric matrix eigenvalue problem, as it is shown below.

We describe in more detail the last step of this algorithm. Let us denote the generalized masses of fluid and solid eigenfunctions respectively by

$$\mu_{h,i}^{\mathrm{F}} := \int_{\Omega_{\mathrm{F}}} \rho_{\mathrm{F}} \left|\boldsymbol{\nabla}\phi_{h,i}\right|^2 \qquad \text{and} \qquad \mu_{h,m}^{\mathrm{S}} := \int_{\Omega_{\mathrm{S}}} \rho_{\mathrm{S}} \left|\boldsymbol{u}_{h,m}^{\mathrm{S}}\right|^2.$$

The following orthogonality properties hold true:

$$\int_{\Omega_{\mathrm{F}}} \rho_{\mathrm{F}} \boldsymbol{\nabla}\phi_{h,i} \cdot \boldsymbol{\nabla}\phi_{h,j} = \delta_{ij}\mu_{h,i}^{\mathrm{F}},$$

$$\int_{\Omega_{\mathrm{F}}} \frac{\rho_{\mathrm{F}}}{c^2}\phi_{h,i}\phi_{h,j} = \delta_{ij}\frac{\mu_{h,i}^{\mathrm{F}}}{(\omega_{h,i}^{\mathrm{F}})^2},$$

$$\int_{\Omega_{\mathrm{S}}} \boldsymbol{\sigma}\left(\boldsymbol{u}_{h,m}^{\mathrm{S}}\right) : \boldsymbol{\varepsilon}\left(\boldsymbol{u}_{h,n}^{\mathrm{S}}\right) = \delta_{mn}\left(\omega_{h,m}^{\mathrm{S}}\right)^2 \mu_{h,m}^{\mathrm{S}},$$

$$\int_{\Omega_{\mathrm{S}}} \rho_{\mathrm{S}} \boldsymbol{u}_{h,m}^{\mathrm{S}} \cdot \boldsymbol{u}_{h,n}^{\mathrm{S}} = \delta_{mn}\mu_{h,m}^{\mathrm{S}},$$

where δ_{ij} is the Kronecker's delta ($\delta_{ij} = 1$, if $i = j$, and $\delta_{ij} = 0$, otherwise).

Then, it is easy to check, see [11], that the modal synthesis approximation of **WP4** (Step 4) can be written in matrix form as follows:

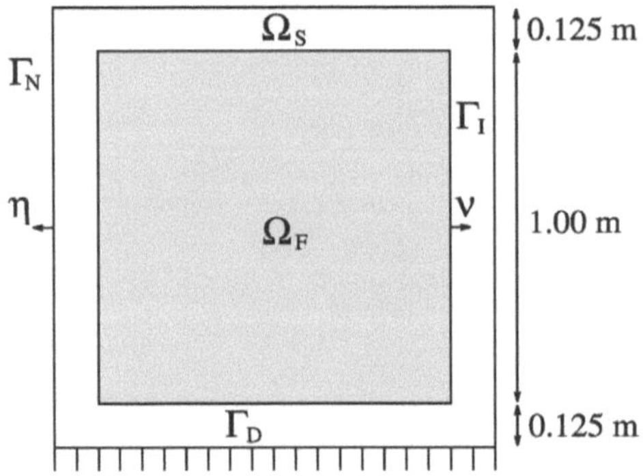

Fig. 9.7 Steel cavity with water inside.

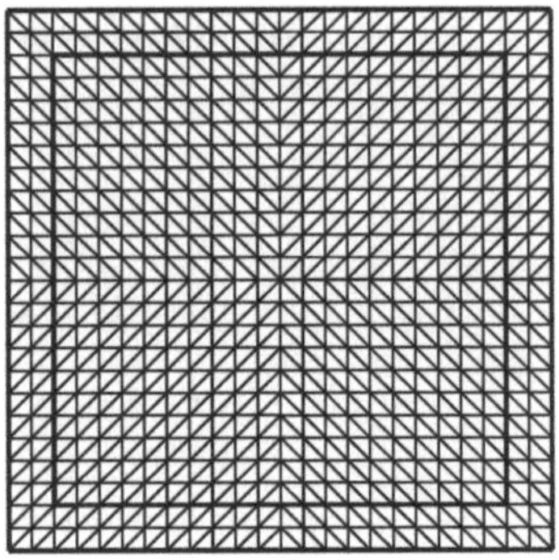

Fig. 9.8 Coarse mesh (N=2).

$$\begin{bmatrix} \boldsymbol{K}_{11} & 0 \\ 0 & \boldsymbol{K}_{22} \end{bmatrix} \begin{bmatrix} \boldsymbol{\alpha} \\ \boldsymbol{\beta} \end{bmatrix} = (\omega_{4,h})^2 \begin{bmatrix} \boldsymbol{M}_{11} & \boldsymbol{M}_{12} \\ \boldsymbol{M}_{12}^{\mathrm{t}} & \boldsymbol{M}_{22} \end{bmatrix} \begin{bmatrix} \boldsymbol{\alpha} \\ \boldsymbol{\beta} \end{bmatrix}, \tag{9.14}$$

where, for $1 \leq i, j \leq N_{\mathrm{F}}$ and $1 \leq m, n \leq N_{\mathrm{S}}$,

- $(\boldsymbol{K}_{11})_{ij} := \delta_{ij}(\omega_{h,i}^{\mathrm{F}})^2 \mu_{h,i}^{\mathrm{F}}$,
- $(\boldsymbol{K}_{22})_{mn} := \delta_{mn} \left(\omega_{h,m}^{\mathrm{S}}\right)^2 \mu_{h,m}^{\mathrm{S}} + \dfrac{\rho_{\mathrm{F}} c^2}{|\Omega_{\mathrm{F}}|} \displaystyle\int_{\Gamma_{\mathrm{I}}} \boldsymbol{u}_{h,m}^{\mathrm{S}} \cdot \boldsymbol{\nu} \int_{\Gamma_{\mathrm{I}}} \boldsymbol{u}_{h,n}^{\mathrm{S}} \cdot \boldsymbol{\nu}$,
- $(\boldsymbol{M}_{11})_{ij} := \delta_{ij} \mu_{h,i}^{\mathrm{F}}$,
- $(\boldsymbol{M}_{12})_{in} := \displaystyle\int_{\Omega_{\mathrm{F}}} \rho_{\mathrm{F}} \boldsymbol{\nabla} \phi_{hi} \cdot \boldsymbol{\nabla} \phi_h^n$,
- $(\boldsymbol{M}_{22})_{mn} := \delta_{mn} \mu_{h,m}^{\mathrm{S}} + \displaystyle\int_{\Omega_{\mathrm{F}}} \rho_{\mathrm{F}} \boldsymbol{\nabla} \phi_h^m \cdot \boldsymbol{\nabla} \phi_h^n$.

Finally, the entries of $\boldsymbol{\alpha} \in \mathbb{R}^{N_{\mathrm{S}}}$ and $\boldsymbol{\beta} \in \mathbb{R}^{N_{\mathrm{F}}}$ in (9.14), are the components in the basis of $\boldsymbol{\mathcal{V}}_{4h,1}$ of the approximate eigenmode, namely,

$$\begin{aligned} \left(\boldsymbol{u}_h^{\mathrm{S}}, \phi_h\right) &= \alpha_1 \left(\boldsymbol{u}_{h,1}^{\mathrm{S}}, \phi_h^1\right) + \cdots + \alpha_{N_{\mathrm{S}}} \left(\boldsymbol{u}_{h,N_{\mathrm{S}}}^{\mathrm{S}}, \phi_h^{N_{\mathrm{S}}}\right) \\ &\quad + \beta_1 \left(\boldsymbol{0}, \phi_{h,1}\right) + \cdots + \beta_{N_{\mathrm{F}}} \left(\boldsymbol{0}, \phi_{h,N_{\mathrm{F}}}\right). \end{aligned}$$

We notice that both block matrices in (9.14) are symmetric. Typically, N_{S} and N_{F} are small numbers. Hence, (9.14) is a low–dimension eigenproblem. Furthermore, it is proved in [11] that the matrix on the left hand side is positive definite, so that the numerical solution of (9.14) is simple.

9.7 Numerical Results for Spectral Problems

We report in this section the results computed for a test case with the different formulations introduced in the previous sections.

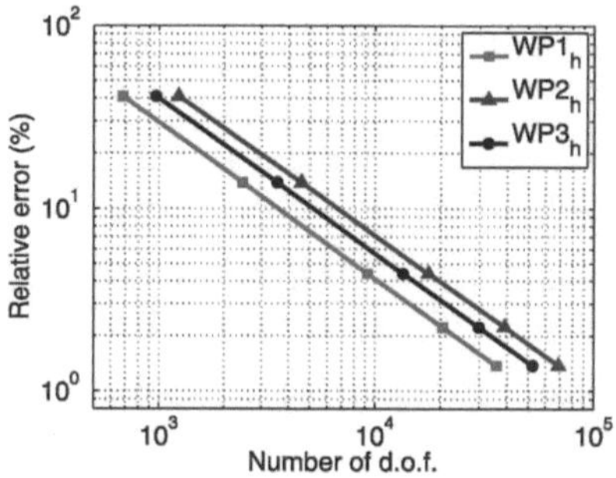

Fig. 9.9 First mode. Error versus d.o.f. (log–log scale).

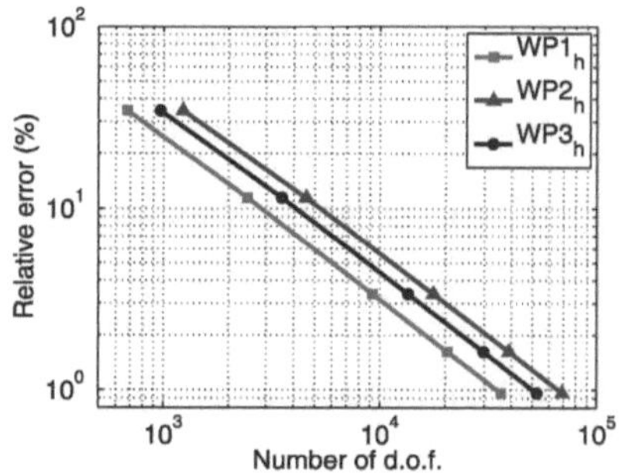

Fig. 9.10 Second mode. Error versus d.o.f. (log–log scale).

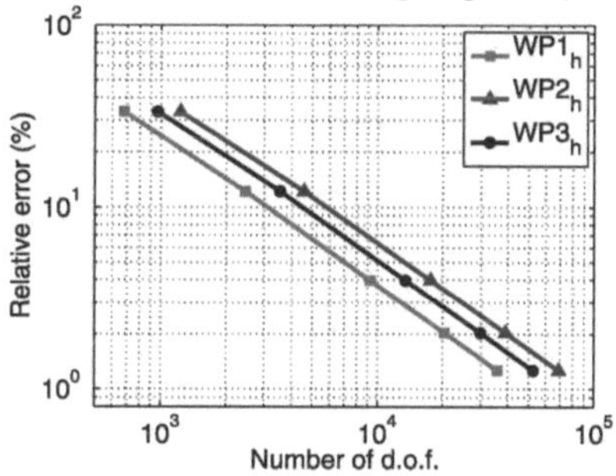

Fig. 9.11 Third mode. Error versus d.o.f. (log–log scale).

We consider that the solid is steel and the fluid is water, with the following physical parameters:

- $\nu \;=\; 0.35\,,$
- $E \;=\; 1.44 \times 10^{11}\,\mathrm{Pa}\,,$
- $\rho_{\mathrm{S}} \;=\; 7700\,\mathrm{kg/m^3}\,,$
- $\rho_{\mathrm{F}} \;=\; 1000\,\mathrm{kg/m^3}\,,$
- $c \;=\; 1430\,\mathrm{m/s}\,.$

The test we consider is the 2D fluid–solid coupled system depicted in Figure 9.7. The refinement parameter N denotes the number of layers of triangles in the solid. The mesh shown in Figure 9.8 corresponds to $N = 2$.

The error of each of the three first computed eigenfrequencies versus the number of degrees of freedom are shown in Figures 9.9, 9.10, and 9.11. The slopes of these curves are approximately -0.84. Since in 2D the number of d.o.f. for uniform meshes is asymptotically proportional to $\frac{1}{\sqrt{h}}$, we have $|\omega - \omega_h| \leq Ch^{1.68}$ in this problem.

Table 9.1 Number of degrees of freedom.

Method	Meshes	$N = 2$	$N = 4$	$N = 8$	$N = 16$
$\mathbf{WP1_h}$	Triangles and Quadrilaterals	679	2447	9247	35903
$\mathbf{WP2_h}$	Triangles	1232	4576	17600	68992
	Quadrilaterals	976	3552	13504	52608
$\mathbf{WP3_h}$	Triangles and Quadrilaterals	968	3536	13472	52544

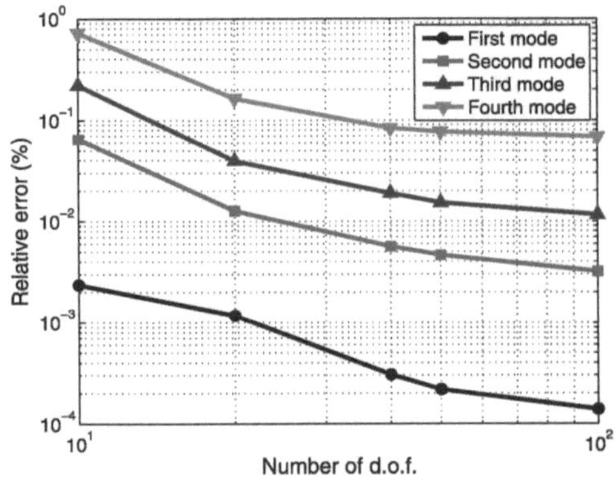

Fig. 9.12 Relative error of the modal synthesis method for varying values of $N_F + N_S$ and a fixed mesh with degree of refinement $N = 4$ (log–log scale).

The number of d.o.f. for each mesh are listed in Table 9.1. It is worth noting that, for a given mesh, this number is larger for the pure displacement formulation $\mathbf{WP2_h}$ than for the pressure/displacement formulation $\mathbf{WP1_h}$ (and even than for the redundant formulation $\mathbf{WP3_h}$), but this difference is smaller for quadrilateral meshes.

We show in Table 9.2 a comparison between the eigenfrequencies computed from $\mathbf{WP1_h}$, $\mathbf{WP2_h}$ and the modal synthesis method. We do not include results from $\mathbf{WP3_h}$, because they coincide exactly with those from $\mathbf{WP1_h}$. For the modal synthesis, we have taken a fixed number of fluid and solid eigenmodes $N_F = N_S = 10$. We have solved the uncoupled problems in the fluid and in the solid with the meshes corresponding to $N = 2, 4, 8, 16$ and, finally, a problem with only 20 degrees of freedom for each mesh. As can be seen and in spite of the low number of eigenfrequencies used, the results are very accurate.

On the other hand, in Figure 9.12, we fix the mesh ($N = 4$) and increase N_S and N_F. In this case, the convergence rate is very small. Indeed, the error goes to zero as $N_F, N_S \to \infty$, but very slowly; thus, it is not worth using a large number of uncoupled frequencies. However, it is important to remark that, even for the lowest values of N_F and N_S that we have considered, the relative errors are very small.

Table 9.2 First eigenfrequencies computed with the different methods.

Mode	Method	$N=2$	$N=4$	$N=8$	$N=16$	Extrapolated
	$\mathbf{WP1_h}$	87.930	71.042	65.136	63.263	62.404
1st	$\mathbf{WP2_h}$	87.915	71.039	65.136	63.263	62.401
	Modal Synthesis	87.931	71.043	65.138	63.264	62.387
	$\mathbf{WP1_h}$	299.963	248.741	230.717	225.351	223.203
2nd	$\mathbf{WP2_h}$	299.764	248.696	230.702	225.348	223.197
	Modal Synthesis	300.018	248.772	230.739	225.374	223.192
	$\mathbf{WP1_h}$	480.572	404.416	374.728	364.999	360.446
3rd	$\mathbf{WP2_h}$	480.186	404.333	374.710	364.994	360.431
	Modal Synthesis	480.863	404.575	374.855	365.116	360.466

9.8 Fluid–Structure Interaction with Interface Damping

In this section we show how the pure displacement formulation can be used to deal with damping. In particular, we consider a fluid–structure coupled problem, with a thin layer of a porous fabric separating both media. In practical applications, these viscoelastic materials are used to reduce the transmission of sound (passive control of noise). The damping effect of this material will be taken into account by means of a change in the coupling conditions (namely, as a *wall impedance*).

According to [9, 22], if the absorbing layer is thin enough and we assume a time dependence of the type $\mathrm{e}^{-\mathrm{i}\omega t}$, then the coupling condition (9.5) can be relaxed to model the effect of the viscoelastic material as follows:

$$\mathrm{i}\omega Z(\omega)\left(\boldsymbol{u}^{\mathrm{S}}\cdot\boldsymbol{\nu}-\boldsymbol{u}^{\mathrm{F}}\cdot\boldsymbol{\nu}\right)-p=0 \qquad \text{on } \Gamma_{\mathrm{I}},$$

where $Z(\omega)$ is the frequency–dependent, complex–valued impedance of the viscoelastic material. Consequently, Equation (9.4) can be rewritten

$$\boldsymbol{\sigma}\left(\boldsymbol{u}^{\mathrm{S}}\right)\boldsymbol{\nu}+\mathrm{i}\omega Z(\omega)\left(\boldsymbol{u}^{\mathrm{S}}\cdot\boldsymbol{\nu}-\boldsymbol{u}^{\mathrm{F}}\cdot\boldsymbol{\nu}\right)\boldsymbol{\nu}=\mathbf{0} \qquad \text{on } \Gamma_{\mathrm{I}}.$$

If an external harmonic load $\boldsymbol{f}$ acts on Γ_{N}, then the partial differential equation for the damping problem written in terms of the displacements reads:

$$\begin{aligned}
\rho_{\mathrm{F}}c^2\boldsymbol{\nabla}\left(\operatorname{div}\boldsymbol{u}^{\mathrm{F}}\right)+\omega^2\rho_{\mathrm{F}}\boldsymbol{u}^{\mathrm{F}}&=\mathbf{0} &&\text{in } \Omega_{\mathrm{F}},\\
\operatorname{div}\left[\boldsymbol{\sigma}(\boldsymbol{u}^{\mathrm{S}})\right]+\omega^2\rho_{\mathrm{S}}\boldsymbol{u}^{\mathrm{S}}&=\mathbf{0} &&\text{in } \Omega_{\mathrm{S}},\\
\boldsymbol{\sigma}\left(\boldsymbol{u}^{\mathrm{S}}\right)\boldsymbol{\nu}+\mathrm{i}\omega Z(\omega)\left(\boldsymbol{u}^{\mathrm{S}}\cdot\boldsymbol{\nu}-\boldsymbol{u}^{\mathrm{F}}\cdot\boldsymbol{\nu}\right)\boldsymbol{\nu}&=\mathbf{0} &&\text{on } \Gamma_{\mathrm{I}},\\
\mathrm{i}\omega Z(\omega)\left(\boldsymbol{u}^{\mathrm{S}}\cdot\boldsymbol{\nu}-\boldsymbol{u}^{\mathrm{F}}\cdot\boldsymbol{\nu}\right)+\rho_{\mathrm{F}}c^2\operatorname{div}\boldsymbol{u}^{\mathrm{F}}&=0 &&\text{on } \Gamma_{\mathrm{I}},\\
\boldsymbol{\sigma}(\boldsymbol{u}^{\mathrm{S}})\boldsymbol{\eta}&=\boldsymbol{f} &&\text{on } \Gamma_{\mathrm{N}},\\
\boldsymbol{u}^{\mathrm{S}}&=\mathbf{0} &&\text{on } \Gamma_{\mathrm{D}}.
\end{aligned}$$

By proceeding as in Section 9.4, we obtain the following weak formulation:

Find $\left(\boldsymbol{u}^{\mathrm{S}}, \boldsymbol{u}^{\mathrm{F}}\right) \in \mathbf{H}^1_{\Gamma_{\mathrm{D}}}(\Omega_{\mathrm{S}}) \times \mathbf{H}(\mathrm{div}, \Omega_{\mathrm{F}})$ *such that*

$$\begin{aligned}
&\int_{\Omega_{\mathrm{S}}} \boldsymbol{\sigma}\left(\boldsymbol{u}^{\mathrm{S}}\right) : \boldsymbol{\varepsilon}\left(\boldsymbol{v}^{\mathrm{S}}\right) + \int_{\Omega_{\mathrm{F}}} \rho_{\mathrm{F}} c^2 \operatorname{div} \boldsymbol{u}^{\mathrm{F}} \operatorname{div} \boldsymbol{v}^{\mathrm{F}} \\
&-\omega^2 \left(\int_{\Omega_{\mathrm{S}}} \rho_{\mathrm{S}} \boldsymbol{u}^{\mathrm{S}} \cdot \boldsymbol{v}^{\mathrm{S}} + \int_{\Omega_{\mathrm{F}}} \rho_{\mathrm{F}} \boldsymbol{u}^{\mathrm{F}} \cdot \boldsymbol{v}^{\mathrm{F}} \right) \\
&+\mathrm{i}\omega Z(\omega) \int_{\Gamma_{\mathrm{I}}} \left(\boldsymbol{u}^{\mathrm{S}} \cdot \boldsymbol{\nu} - \boldsymbol{u}^{\mathrm{F}} \cdot \boldsymbol{\nu}\right) \left(\boldsymbol{v}^{\mathrm{S}} \cdot \boldsymbol{\nu} - \boldsymbol{v}^{\mathrm{F}} \cdot \boldsymbol{\nu}\right) \\
&\quad = \int_{\Gamma_{\mathrm{N}}} \boldsymbol{f} \cdot \boldsymbol{v}^{\mathrm{S}} \qquad \forall \left(\boldsymbol{v}^{\mathrm{S}}, \boldsymbol{v}^{\mathrm{F}}\right) \in \mathbf{H}^1_{\Gamma_{\mathrm{D}}}(\Omega_{\mathrm{S}}) \times \mathbf{H}(\mathrm{div}, \Omega_{\mathrm{F}}).
\end{aligned}$$

It is proved in [13] that this problem has a unique solution for all $\omega \in \mathbb{R}$, except for very special geometric configurations. In these cases, there could exist particular pathological values of ω for which the problem above is not well–posed.

9.8.1 Numerical Approximation and Matrix Description

We use triangular compatible meshes in Ω_{S} and Ω_{F} and the finite element space introduced in Section 9.5: $\mathcal{V}_{2h} := \mathcal{L}_h^{\Gamma_{\mathrm{D}}}(\Omega_{\mathrm{S}}) \times \mathcal{R}_h(\Omega_{\mathrm{F}})$; namely, Lagrange piecewise linear elements for each component of $\boldsymbol{u}^{\mathrm{S}}$ and lowest–order Raviart–Thomas elements for $\boldsymbol{u}^{\mathrm{F}}$. It is proved in [13], under reasonable hypothesis, that the order of convergence of this approximation is optimal.

To introduce the matrix formulation of the problem above, let

$$\boldsymbol{F} := \int_{\Gamma_{\mathrm{N}}} \boldsymbol{f} \cdot \boldsymbol{v}_h^{\mathrm{S}}$$

represent the load vector and consider matrices $\boldsymbol{G}_1$, $\boldsymbol{G}_2$ and $\boldsymbol{G}_3$ defined by:

$$\begin{aligned}
\boldsymbol{V}^{\mathrm{S}^{\mathrm{t}}} \boldsymbol{G}_1 \boldsymbol{U}^{\mathrm{S}} &= \int_{\Gamma_{\mathrm{I}}} \left(\boldsymbol{u}_h^{\mathrm{S}} \cdot \boldsymbol{\nu}\right)\left(\boldsymbol{v}_h^{\mathrm{S}} \cdot \boldsymbol{\nu}\right), \\
\boldsymbol{V}^{\mathrm{F}^{\mathrm{t}}} \boldsymbol{G}_2 \boldsymbol{U}^{\mathrm{F}} &= \int_{\Gamma_{\mathrm{I}}} \left(\boldsymbol{u}_h^{\mathrm{F}} \cdot \boldsymbol{\nu}\right)\left(\boldsymbol{v}_h^{\mathrm{F}} \cdot \boldsymbol{\nu}\right), \\
\boldsymbol{V}^{\mathrm{S}^{\mathrm{t}}} \boldsymbol{G}_3 \boldsymbol{U}^{\mathrm{F}} &= \int_{\Gamma_{\mathrm{I}}} \left(\boldsymbol{u}_h^{\mathrm{F}} \cdot \boldsymbol{\nu}\right)\left(\boldsymbol{v}_h^{\mathrm{S}} \cdot \boldsymbol{\nu}\right).
\end{aligned}$$

Thus, the matrix formulation of the damped problem is written as follows:

$$\left\{ \begin{bmatrix} \boldsymbol{K}^{\mathrm{S}} & 0 \\ 0 & \boldsymbol{K}_{\mathrm{D}}^{\mathrm{F}} \end{bmatrix} + \mathrm{i}\omega Z(\omega) \begin{bmatrix} \boldsymbol{G}_1 & \boldsymbol{G}_3 \\ \boldsymbol{G}_3^{\mathrm{t}} & \boldsymbol{G}_2 \end{bmatrix} - \omega^2 \begin{bmatrix} \boldsymbol{M}^{\mathrm{S}} & 0 \\ 0 & \boldsymbol{M}_{\mathrm{D}}^{\mathrm{F}} \end{bmatrix} \right\} \begin{bmatrix} \boldsymbol{U}^{\mathrm{S}} \\ \boldsymbol{U}^{\mathrm{F}} \end{bmatrix} = \begin{bmatrix} \boldsymbol{F} \\ 0 \end{bmatrix},$$

where $\boldsymbol{K}^{\mathrm{S}}$, $\boldsymbol{K}_{\mathrm{D}}^{\mathrm{F}}$, $\boldsymbol{M}^{\mathrm{S}}$ and $\boldsymbol{M}_{\mathrm{D}}^{\mathrm{F}}$ are the matrices defined in Section 9.5.1.

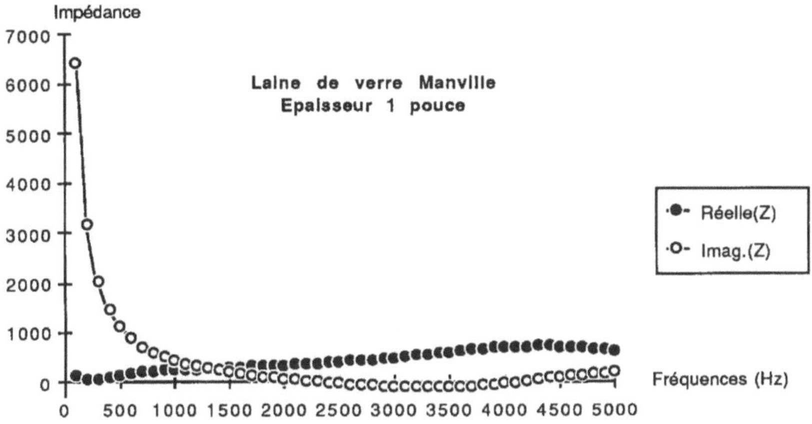

Fig. 9.13 Acoustic impedance for glasswool Manville (1").

9.8.2 Numerical Results

We report some numerical results corresponding to the geometry and physical data introduced in Section 9.7, except for the physical parameters of the fluid. In this test, we have used those of air: $\rho_F = 1\,\mathrm{kg/m^3}$ and $c = 340\,\mathrm{m/s}$. Further numerical tests can be found in [13, 14].

For the interface damping we have considered a frequency–dependent impedance corresponding to an insulating fabric: glasswool *Manville*. Figure 9.13 shows the measured real and imaginary parts of the acoustic impedance (taken from [22]) for the layer of this material (1 inch width) that we have used.

Using the mesh in Section 9.7 corresponding to $N = 4$, we have computed the response of the damped coupled system when one of the vertical walls of the vessel is subject to a periodical excitation force with varying frequency. For each frequency, the L^2 norm of the pressure of the air contained in the vessel, given by

$$\|p\|_{L^2} = \left(\int_{\Omega_F} p^2 \right)^{1/2}, \tag{9.15}$$

has been computed. Figure 9.14 shows the *response curve* of the coupled system. In this curve $\log_{10} \|p\|_{L^2}$ is plotted for frequencies $\frac{\omega}{2\pi}$ ranging from 70 to 350 Hz.

Six response peaks can be observed in this curve. The first and the fourth ones correspond to the fundamental vibration modes of the structure in this range of frequencies. The other peaks correspond to the natural vibration modes of the fluid excited by the external force. The modes of the structure have been only slightly damped, instead the damping was significantly stronger for those of the fluid. This agrees with the expected behavior of the damped coupled system.

Table 9.3 shows the computed resonance frequencies for both, the undamped and the damped coupled system. Notice that while the frequencies corresponding to the

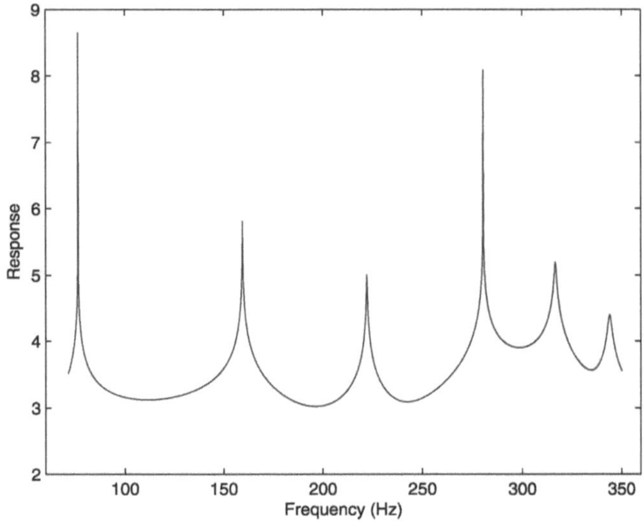

Fig. 9.14 Response curve.

solid modes remain practically unchanged, those associated with the fluid modes decrease between 6 and 10 % when the damping material is introduced. This table also includes in its last column the computed complex eigenfrequencies of the nonlinear spectral problem associated to the damped problem closest to this response peaks.

Table 9.3 Resonance vibration frequencies and complex eigenfrequencies

Undamped response peaks	Damped response peaks	Complex eigenfrequencies
76.514	76.514	$76.514 - 0.000\,i$
169.972	159.518	$159.505 - 0.056\,i$
240.476	222.230	$222.231 - 0.176\,i$
280.454	280.429	$280.429 - 0.001\,i$
339.961	316.611	$316.776 - 0.537\,i$
380.108	343.949	$343.953 - 1.057\,i$

9.9 A Pure Displacement Formulation for the Time Domain Problem

In this section we study a numerical method to solve the pure displacement formulation of the elastoacoustic problem in the time domain by using the same finite element approximation proposed to solve **WP2**.

We consider as a model problem the configuration depicted in Figure 9.1. When a time–dependent load $\boldsymbol{f}(t)$ is applied on Γ_{N}, the equations governing the motion of the coupled system are the following, see for instance [24]:

$$
\begin{aligned}
\rho_{\mathrm{S}} \frac{\partial^2 \boldsymbol{u}^{\mathrm{S}}}{\partial t^2} - \operatorname{div} \boldsymbol{\sigma}(\boldsymbol{u}^{\mathrm{S}}) &= \boldsymbol{0} && \text{in } \Omega_{\mathrm{S}}, \\
\rho_{\mathrm{F}} \frac{\partial^2 \boldsymbol{u}^{\mathrm{F}}}{\partial t^2} - \boldsymbol{\nabla}(\rho_{\mathrm{F}} c^2 \operatorname{div} \boldsymbol{u}^{\mathrm{F}}) &= \boldsymbol{0} && \text{in } \Omega_{\mathrm{F}}, \\
\boldsymbol{\sigma}(\boldsymbol{u}^{\mathrm{S}}) \boldsymbol{\eta} &= \boldsymbol{f} && \text{on } \Gamma_{\mathrm{N}}.
\end{aligned}
$$

The remaining boundary and interface conditions are as in **PDE2**. Initial conditions must be also given:

$$
\boldsymbol{u}^{\mathrm{S}}(0) = \boldsymbol{u}_0^{\mathrm{S}}, \quad \boldsymbol{u}^{\mathrm{F}}(0) = \boldsymbol{u}_0^{\mathrm{F}}, \quad \frac{\partial \boldsymbol{u}^{\mathrm{S}}}{\partial t}(0) = \boldsymbol{u}_1^{\mathrm{S}}, \quad \text{and} \quad \frac{\partial \boldsymbol{u}^{\mathrm{F}}}{\partial t}(0) = \boldsymbol{u}_1^{\mathrm{F}}.
$$

Analogous computations as those done for problem **WP2**, allow us to write the time domain problem in a weak sense as follows:

TDP.– *Given* $(\boldsymbol{u}_0^{\mathrm{S}}, \boldsymbol{u}_0^{\mathrm{F}}) \in \boldsymbol{\mathcal{V}}_2$, $(\boldsymbol{u}_1^{\mathrm{S}}, \boldsymbol{u}_1^{\mathrm{F}}) \in \mathbf{L}_2(\Omega_{\mathrm{S}}) \times \mathbf{L}_2(\Omega_{\mathrm{F}})$ *and* $\boldsymbol{f}$ *regular enough, find* $\boldsymbol{u}^{\mathrm{S}} : [0,T] \to \mathbf{H}^1_{\Gamma_{\mathrm{D}}}(\Omega_{\mathrm{S}})$ *and* $\boldsymbol{u}^{\mathrm{F}} : [0,T] \to \mathbf{H}(\operatorname{div}, \Omega_{\mathrm{F}})$ *such that,* $\forall t \in (0,T)$, $\big(\boldsymbol{u}^{\mathrm{S}}(t), \boldsymbol{u}^{\mathrm{F}}(t)\big) \in \boldsymbol{\mathcal{V}}_2$ *satisfies*

$$
\begin{aligned}
&\int_{\Omega_{\mathrm{S}}} \rho_{\mathrm{S}} \frac{\partial^2 \boldsymbol{u}^{\mathrm{S}}}{\partial t^2} \cdot \boldsymbol{v}^{\mathrm{S}} + \int_{\Omega_{\mathrm{F}}} \rho_{\mathrm{F}} \frac{\partial^2 \boldsymbol{u}^{\mathrm{F}}}{\partial t^2} \cdot \boldsymbol{v}^{\mathrm{F}} + \int_{\Omega_{\mathrm{S}}} \boldsymbol{\sigma}(\boldsymbol{u}^{\mathrm{S}}) : \boldsymbol{\varepsilon}(\boldsymbol{v}^{\mathrm{S}}) \\
&\quad + \int_{\Omega_{\mathrm{F}}} \rho_{\mathrm{F}} c^2 \operatorname{div} \boldsymbol{u}^{\mathrm{F}} \operatorname{div} \boldsymbol{v}^{\mathrm{F}} = \int_{\Gamma_{\mathrm{N}}} \boldsymbol{f} \cdot \boldsymbol{v}^{\mathrm{S}} \qquad \forall (\boldsymbol{v}^{\mathrm{S}}, \boldsymbol{v}^{\mathrm{F}}) \in \boldsymbol{\mathcal{V}}_2,
\end{aligned}
$$

and

$$
\boldsymbol{u}^{\mathrm{S}}(0) = \boldsymbol{u}_0^{\mathrm{S}}, \quad \boldsymbol{u}^{\mathrm{F}}(0) = \boldsymbol{u}_0^{\mathrm{F}}, \quad \frac{\partial \boldsymbol{u}^{\mathrm{S}}}{\partial t}(0) = \boldsymbol{u}_1^{\mathrm{S}}, \quad \textit{and} \quad \frac{\partial \boldsymbol{u}^{\mathrm{F}}}{\partial t}(0) = \boldsymbol{u}_1^{\mathrm{F}}.
$$

Under weak regularity hypothesis on the initial data, it is proved in [16] that this problem admits a unique solution.

9.9.1 Numerical Approximation

To obtain a numerical approximation of **TDP** it is necessary to combine a space discretization with a time discretization.

The finite element spaces introduced in Section 9.5 will be used for the space discretization: piecewise linear Lagrange finite elements for each component of the solid displacement and lowest–order Raviart–Thomas elements for the fluid displacement. The kinematic constraint is weakly imposed as in (9.13). However, we will do it this time by means of a static condensation instead of using Lagrange multipliers. Therefore, we include the constraint in the finite element space:

$$\widetilde{\mathcal{V}}_{2h} := \left\{ \left(\boldsymbol{u}^{\mathrm{S}}, \boldsymbol{u}^{\mathrm{F}}\right) \in \mathcal{V}_{2h} : \int_{\ell} \boldsymbol{u}_h^{\mathrm{S}} \cdot \boldsymbol{\nu} = \int_{\ell} \boldsymbol{u}_h^{\mathrm{F}} \cdot \boldsymbol{\nu} \quad \forall\, \text{edge}\ \ell \subset \Gamma_{\mathrm{I}} \right\}.$$

For the time discretization we apply Newmark's method. This is a second–order method (with respect to the time–step) frequently used to solve second–order in time partial differential equations, see for instance [27].

Consider a uniform partition of the time interval $[0, T]$ with step–size $\Delta t := \frac{T}{K}$:

$$t_k := k\Delta t, \qquad k = 0, 1, \ldots, K.$$

Then, given approximate initial data $(\boldsymbol{u}_{0,h}^{\mathrm{S}}, \boldsymbol{u}_{0,h}^{\mathrm{F}})$ and $(\boldsymbol{u}_{1,h}^{\mathrm{S}}, \boldsymbol{u}_{1,h}^{\mathrm{F}}) \in \widetilde{\mathcal{V}}_{2h}$ (for instance, the finite element interpolants of the exact initial data $(\boldsymbol{u}_0^{\mathrm{S}}, \boldsymbol{u}_0^{\mathrm{F}})$ and $(\boldsymbol{u}_1^{\mathrm{S}}, \boldsymbol{u}_1^{\mathrm{F}})$, respectively), the Newmark's method applied to our problem can be written as follows:

Algorithm 2 *Let* $\left(\boldsymbol{u}_h^{\mathrm{S},0}, \boldsymbol{u}_h^{\mathrm{F},0}\right) = \left(\boldsymbol{u}_{0,h}^{\mathrm{S}}, \boldsymbol{u}_{0,h}^{\mathrm{F}}\right)$,

- *Let* $\left(\boldsymbol{u}_h^{\mathrm{S},1}, \boldsymbol{u}_h^{\mathrm{F},1}\right) \in \widetilde{\mathcal{V}}_{2h}$ *be the solution of*

$$\begin{aligned}
&\int_{\Omega_{\mathrm{S}}} \rho_{\mathrm{S}} \frac{\boldsymbol{u}_h^{\mathrm{S},1} - \boldsymbol{u}_h^{\mathrm{S},0} - \Delta t \boldsymbol{u}_{1,h}^{\mathrm{S}}}{\Delta t^2} \cdot \boldsymbol{v}_h^{\mathrm{S}} + \int_{\Omega_{\mathrm{F}}} \rho_{\mathrm{F}} \frac{\boldsymbol{u}_h^{\mathrm{F},1} - \boldsymbol{u}_h^{\mathrm{F},0} - \Delta t \boldsymbol{u}_{1,h}^{\mathrm{F}}}{\Delta t^2} \cdot \boldsymbol{v}_h^{\mathrm{F}} \\
&+ \frac{1}{4}\left[\int_{\Omega_{\mathrm{S}}} \boldsymbol{\sigma}\left(\boldsymbol{u}_h^{\mathrm{S},1} + \boldsymbol{u}_h^{\mathrm{S},0}\right) : \boldsymbol{\varepsilon}\left(\boldsymbol{v}_h^{\mathrm{S}}\right) + \int_{\Omega_{\mathrm{F}}} \rho_{\mathrm{F}} c^2 \operatorname{div}\left(\boldsymbol{u}_h^{\mathrm{F},1} + \boldsymbol{u}_h^{\mathrm{F},0}\right) \operatorname{div} \boldsymbol{v}_h^{\mathrm{F}} \right] \\
&= \frac{1}{4} \int_{\Omega_{\mathrm{F}}} \left[\boldsymbol{f}(t_1) + \boldsymbol{f}(t_0)\right] \cdot \boldsymbol{v}_h^{\mathrm{S}} \qquad \forall \left(\boldsymbol{v}_h^{\mathrm{S}}, \boldsymbol{v}_h^{\mathrm{F}}\right) \in \widetilde{\mathcal{V}}_{2h}.
\end{aligned}$$

- *For* $k = 1, 2, \ldots, K-1$, *let* $\left(\boldsymbol{u}_h^{\mathrm{S},k+1}, \boldsymbol{u}_h^{\mathrm{F},k+1}\right) \in \widetilde{\mathcal{V}}_{2h}$ *be the solution of*

$$
\begin{aligned}
&\int_{\Omega_{\mathrm{S}}} \rho_{\mathrm{S}} \frac{\boldsymbol{u}_h^{\mathrm{S},k+1} - 2\boldsymbol{u}_h^{\mathrm{S},k} + \boldsymbol{u}_h^{\mathrm{S},k-1}}{\Delta t^2} \cdot \boldsymbol{v}_h^{\mathrm{S}} \\
&+ \int_{\Omega_{\mathrm{F}}} \rho_{\mathrm{F}} \frac{\boldsymbol{u}_h^{\mathrm{F},k+1} - 2\boldsymbol{u}_h^{\mathrm{F},k} + \boldsymbol{u}_h^{\mathrm{F},k-1}}{\Delta t^2} \cdot \boldsymbol{v}_h^{\mathrm{F}} \\
&+ \frac{1}{4} \left[\int_{\Omega_{\mathrm{S}}} \boldsymbol{\sigma} \left(\boldsymbol{u}_h^{\mathrm{S},k+1} + 2\boldsymbol{u}_h^{\mathrm{S},k} + \boldsymbol{u}_h^{\mathrm{S},k-1} \right) : \boldsymbol{\varepsilon} \left(\boldsymbol{v}_h^{\mathrm{S}} \right) \right. \\
&\qquad \left. + \int_{\Omega_{\mathrm{F}}} \rho_{\mathrm{F}} c^2 \operatorname{div} \left(\boldsymbol{u}_h^{\mathrm{F},k+1} + 2\boldsymbol{u}_h^{\mathrm{F},k} + \boldsymbol{u}_h^{\mathrm{F},k-1} \right) \operatorname{div} \boldsymbol{v}_h^{\mathrm{F}} \right] \\
&= \frac{1}{4} \int_{\Omega_{\mathrm{F}}} \left[\boldsymbol{f}(t_{k+1}) + 2\boldsymbol{f}(t_k) + \boldsymbol{f}(t_{k-1}) \right] \cdot \boldsymbol{v}_h^{\mathrm{S}} \qquad \forall \left(\boldsymbol{v}_h^{\mathrm{S}}, \boldsymbol{v}_h^{\mathrm{F}} \right) \in \widetilde{\boldsymbol{\mathcal{V}}}_{2h}.
\end{aligned}
$$

It is simple to show that this algorithm is well–posed for any $\Delta t > 0$, (see Remark 6 below for a matrix description from which this can be easily verified) which yields existence and uniqueness of the discrete solution.

It is proved in [16] that, for a time–step small enough, this scheme is stable and completely conservative.

9.9.2 Matrix Description

Let $\boldsymbol{K}_{\mathrm{D}}^{\mathrm{F}}$, $\boldsymbol{M}_{\mathrm{D}}^{\mathrm{F}}$, $\boldsymbol{K}^{\mathrm{S}}$ and $\boldsymbol{M}^{\mathrm{S}}$, be the matrices defined in Section 9.5.1. Let

$$
\boldsymbol{F}(t) := \int_{\Gamma_{\mathrm{N}}} \boldsymbol{f}(t) \cdot \boldsymbol{v}_h^{\mathrm{S}},
$$

be the time–dependent load vector.

As stated above, we impose the kinematic constraint on the fluid–solid interface by means of a static condensation over the nodes on Γ_{I}.

Let N_1 be the number of d.o.f. in the solid and N_2 in the fluid. Assume, for simplicity, that the first N_{I} d.o.f in the fluid correspond to the edges lying on Γ_{I}. Each of the nodal components $\boldsymbol{U}_1^{\mathrm{F}}, \ldots, \boldsymbol{U}_{N_{\mathrm{I}}}^{\mathrm{F}}$ can be statically condensed in terms of the nodal values of the solid displacements. In fact, for each edge ℓ_i of the fluid mesh lying on the interface, there holds

$$
\boldsymbol{U}_i^{\mathrm{F}} |\ell_i| = \int_{\ell_i} \boldsymbol{u}_h^{\mathrm{F}} \cdot \boldsymbol{\nu} = \int_{\ell_i} \boldsymbol{u}_h^{\mathrm{S}} \cdot \boldsymbol{\nu} = \sum_{j=1}^{N_1} \boldsymbol{U}_j^{\mathrm{S}} \int_{\ell_i} \boldsymbol{\phi}_j^{\mathrm{s}} \cdot \boldsymbol{\nu}.
$$

Therefore, if we write $\widetilde{\boldsymbol{U}}^{\mathrm{F}} := (\boldsymbol{U}_1^{\mathrm{F}}, \ldots, \boldsymbol{U}_{N_{\mathrm{I}}}^{\mathrm{F}})$, $\widehat{\boldsymbol{U}}^{\mathrm{F}} := (\boldsymbol{U}_{N_{\mathrm{I}}+1}^{\mathrm{F}}, \ldots, \boldsymbol{U}_{N_2}^{\mathrm{F}})$ and $\boldsymbol{U}^{\mathrm{S}} := (\boldsymbol{U}_1^{\mathrm{S}}, \ldots, \boldsymbol{U}_{N_1}^{\mathrm{S}})$, then we have

$$
\widetilde{\boldsymbol{U}}^{\mathrm{F}} = \boldsymbol{E}\boldsymbol{U}^{\mathrm{S}},
$$

where the matrix $\boldsymbol{E} := (E_{ij})$ is defined by

$$E_{ij} := \frac{1}{|\ell_i|}\int_{\ell_i} \boldsymbol{\phi}_j^{\mathrm{S}}\cdot\boldsymbol{\nu}, \qquad i=1,\dots,N_{\mathrm{I}}, \quad j=1,\dots,N_{\mathrm{I}}.$$

Hence, $\widetilde{\boldsymbol{U}}^{\mathrm{F}}$ can be eliminated. In fact, let us split $\boldsymbol{K}_{\mathrm{D}}^{\mathrm{F}}$ and $\boldsymbol{M}_{\mathrm{D}}^{\mathrm{F}}$ into blocks corresponding to the unknowns $\widetilde{\boldsymbol{U}}^{\mathrm{F}}$ and $\widehat{\boldsymbol{U}}^{\mathrm{F}}$:

$$\boldsymbol{K}_{\mathrm{D}}^{\mathrm{F}} = \begin{pmatrix} \widetilde{\boldsymbol{K}_{\mathrm{D}}^{\mathrm{F}}} & \boldsymbol{C}_K \\ \boldsymbol{C}_K^{\mathrm{t}} & \widehat{\boldsymbol{K}_{\mathrm{D}}^{\mathrm{F}}} \end{pmatrix} \qquad \text{and} \qquad \boldsymbol{M}_{\mathrm{D}}^{\mathrm{F}} = \begin{pmatrix} \widetilde{\boldsymbol{M}_{\mathrm{D}}^{\mathrm{F}}} & \boldsymbol{C}_M \\ \boldsymbol{C}_M^{\mathrm{t}} & \widehat{\boldsymbol{M}_{\mathrm{D}}^{\mathrm{F}}} \end{pmatrix},$$

and denote

$$\boldsymbol{\mathcal{K}} := \begin{pmatrix} \boldsymbol{K}^{\mathrm{S}} + \boldsymbol{E}^{\mathrm{t}}\widetilde{\boldsymbol{K}_{\mathrm{D}}^{\mathrm{F}}}\boldsymbol{E} & \boldsymbol{E}^{\mathrm{t}}\boldsymbol{C}_K \\ \boldsymbol{C}_K^{\mathrm{t}}\boldsymbol{E} & \widehat{\boldsymbol{K}_{\mathrm{D}}^{\mathrm{F}}} \end{pmatrix}, \qquad \boldsymbol{\mathcal{M}} := \begin{pmatrix} \boldsymbol{M}^{\mathrm{S}} + \boldsymbol{E}^{\mathrm{t}}\widetilde{\boldsymbol{M}_{\mathrm{D}}^{\mathrm{F}}}\boldsymbol{E} & \boldsymbol{E}^{\mathrm{t}}\boldsymbol{C}_M \\ \boldsymbol{C}_M^{\mathrm{t}}\boldsymbol{E} & \widehat{\boldsymbol{M}_{\mathrm{D}}^{\mathrm{F}}} \end{pmatrix},$$

$$\boldsymbol{\mathcal{F}}^k := \boldsymbol{F}(t_k), \qquad \boldsymbol{\mathcal{U}}_i := \begin{pmatrix} [\boldsymbol{U}^{\mathrm{S}}]_i \\ [\widehat{\boldsymbol{U}}^{\mathrm{F}}]_i \end{pmatrix} \qquad \text{and} \qquad \boldsymbol{\mathcal{U}}^k := \begin{pmatrix} [\boldsymbol{U}^{\mathrm{S}}]^k \\ [\widehat{\boldsymbol{U}}^{\mathrm{F}}]^k \end{pmatrix},$$

where $[\boldsymbol{U}^{\mathrm{S}}]_i$ $(i=0,1)$ and $[\boldsymbol{U}^{\mathrm{S}}]^k$ $(k=0,\dots,K)$ are the nodal components of $\boldsymbol{u}_{i,h}^{\mathrm{S}}$ and $\boldsymbol{u}_h^{\mathrm{S},k}$, respectively, whereas $[\widehat{\boldsymbol{U}}^{\mathrm{F}}]_i$ and $[\widehat{\boldsymbol{U}}^{\mathrm{F}}]^k$ are respectively those of $\boldsymbol{u}_{i,h}^{\mathrm{F}}$ and $\boldsymbol{u}_h^{\mathrm{F},k}$ corresponding to edges not lying on Γ_{I}.

Thus, Algorithm 2 can be written in terms of the genuine $N_{\mathrm{T}} := N_1 + (N_2 - N_{\mathrm{I}})$ d.o.f. as follows:

Algorithm 3 *For $\boldsymbol{\mathcal{U}}_0, \boldsymbol{\mathcal{U}}_1 \in \mathbb{R}^{N_{\mathrm{T}}}$ and $\boldsymbol{\mathcal{F}}^k \in \mathbb{R}^{N_{\mathrm{T}}}$, $k=0,1,\dots,K$, being given, let*

- $\boldsymbol{\mathcal{U}}^0 := \boldsymbol{\mathcal{U}}_0,$
- $\boldsymbol{\mathcal{U}}^1 \in \mathbb{R}^{N_{\mathrm{T}}}$ *be the solution of*

$$\boldsymbol{\mathcal{M}}\left(\frac{\boldsymbol{\mathcal{U}}^1 - \boldsymbol{\mathcal{U}}^0 - \Delta t\boldsymbol{\mathcal{U}}_1}{\Delta t^2}\right) + \frac{1}{4}\boldsymbol{\mathcal{K}}\left(\boldsymbol{\mathcal{U}}^1 + \boldsymbol{\mathcal{U}}^0\right) = \frac{1}{4}\left(\boldsymbol{\mathcal{F}}^1 + \boldsymbol{\mathcal{F}}^0\right),$$

- *and, for $k=1,2,\dots,K-1$, $\boldsymbol{\mathcal{U}}^{k+1} \in \mathbb{R}^{N_{\mathrm{T}}}$ be the solution of*

$$\boldsymbol{\mathcal{M}}\left(\frac{\boldsymbol{\mathcal{U}}^{k+1} - 2\boldsymbol{\mathcal{U}}^k + \boldsymbol{\mathcal{U}}^{k-1}}{\Delta t^2}\right) + \frac{1}{4}\boldsymbol{\mathcal{K}}\left(\boldsymbol{\mathcal{U}}^{k+1} + 2\boldsymbol{\mathcal{U}}^k + \boldsymbol{\mathcal{U}}^{k-1}\right)$$
$$= \frac{1}{4}\left(\boldsymbol{\mathcal{F}}^{k+1} + 2\boldsymbol{\mathcal{F}}^k + \boldsymbol{\mathcal{F}}^{k-1}\right).$$

Remark 6. In order to compute the new unknown $\boldsymbol{\mathcal{U}}^{k+1}$, it is necessary to solve at each step a system with associated matrix $\boldsymbol{\mathcal{M}} + \frac{\Delta t^2}{4}\boldsymbol{\mathcal{K}}$, which is symmetric and positive definite. If a direct method is used with this purpose, then the factorization of this matrix must be done only once and, at each time–step, it is only necessary to solve two triangular linear systems.

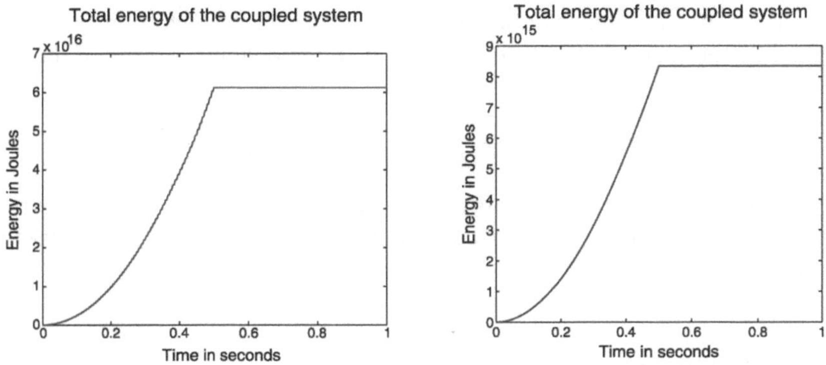

Harmonic force (first mode) Harmonic force (third mode)

Fig. 9.15 Total energy of the system for a resonant harmonic excitation.

9.9.3 Numerical Experiments

We report in this section numerical results obtained with a MATLAB code that we have developed to implement the above numerical method.

For our numerical experiments we have considered the same steel vessel clamped by its bottom and completely filled with water as in Section 9.7 and the mesh corresponding to $N = 4$.

Our first experiment consists in starting from rest and then exciting the system with a periodic force acting on the left edge of Γ_{N}, see Figure 9.7. As a first test, we compute the response during 1 second of the coupled problem excited during the initial 0.5 seconds with a normal harmonic load $F(t,x,y) = 10^{11} y \cos(\omega t)$. The frequency ω is chosen to coincide with one eigenfrequency of the coupled system, also computed with the mesh corresponding to $N = 4$; in particular we have used the frequencies of the first and the third vibration modes, see Table 9.2. We have used approximately ten time–steps per period; more precisely, 5000 time–steps for the first mode and 27500 time–steps for the third one.

Figure 9.15 shows the total energy of the coupled system excited with harmonic forces with the frequencies of the first (left) and the third modes (right).

In both cases, the energy shows the typical parabolic pattern while the force is acting. Indeed, it is well known that, when an effect of resonance occurs, the energy of the system has a "parabolic" shape of the form:

$$\mathrm{Energy}(t) \propto \left(A\cos^2(\omega t) + B\sin^2(\omega t)\right) t^2 + \text{ small periodic terms.}$$

On the other hand, once the force ceases, the energy becomes constant. This coincides with the theoretical assertion that the method is conservative, see [16].

As a second test we have excited the system, starting from rest, during $\frac{1}{10}$ seconds with a normal harmonic load $F(t,x,y) = 10^{11} y \cos(\omega t)$ of angular frequency $\omega =$ 600 rad/s, which is far from all the eigenfrequencies of the coupled problem. Again, we have used the mesh with degree of refinement $N = 4$.

For the time discretization we have used 500 time steps, which is approximately eight steps per period. We show in Figure 9.16 the potential energy of the system, namely,

$$E_{\mathrm{pot}}^{h,n} = \left(\frac{\mathcal{U}^{n+1}+\mathcal{U}^{n}}{2}\right)^{\mathrm{t}} \mathcal{K} \left(\frac{\mathcal{U}^{n+1}+\mathcal{U}^{n}}{2}\right).$$

In this case, the potential energy shows the typical response of a harmonic oscillator to a non resonant excitation: a fast oscillation with a slowly varying oscillating amplitude.

9.10 Conclusions

We have studied several problems related with the interaction between an elastic structure and an acoustic fluid. For all of these problems we have described numerical methods for their solution together with numerical tests to assess the performance of each method.

We have studied in more detail the fluid–structure vibration problem. In this case, we have considered four different formulations, which differ from each other in the variables used for the fluid equations. We have presented finite element numerical methods to solve each of them, and a modal synthesis method, as well. An overview of theoretical results concerning all the methods is shown, together with academic numerical tests which allow comparing the different methods. The conclusion of both, the theoretical analysis and the numerical experimentation is that all the approaches are similarly efficient, so that the choice of formulation can be done attending to other aspects of the global problem to solve.

Finally, we have chosen one of the formulations based on displacement variables for both media and we have shown how it can be applied to deal with other more complex fluid–structure problems. In particular we have analyzed a dissipative acoustics problem and a time–domain problem. The former consists of modeling a viscoelastic layer placed among the fluid and the structure as a wall impedance. For the latter we have combined the finite element method we propose in the space variables with a Newmark's method in time.

Acknowledgments

A. Bermúdez, L.M. Hervella–Nieto and D. Santamarina were partially supported by MEC (Spain) research project DPI2004–05504–C02–02 and Xunta de Galicia project PGIDIT05PXIC20705PN (Spain). L.M. Hervella–Nieto was also partially funded by MEC (Spain) research project MTM2004–05796–C02–01. P. Gamallo was supported by MEC (Spain) under grant number 18–08–463B–750. R. Rodríguez was partially funded by FONDAP in Applied Mathematics (Chile).

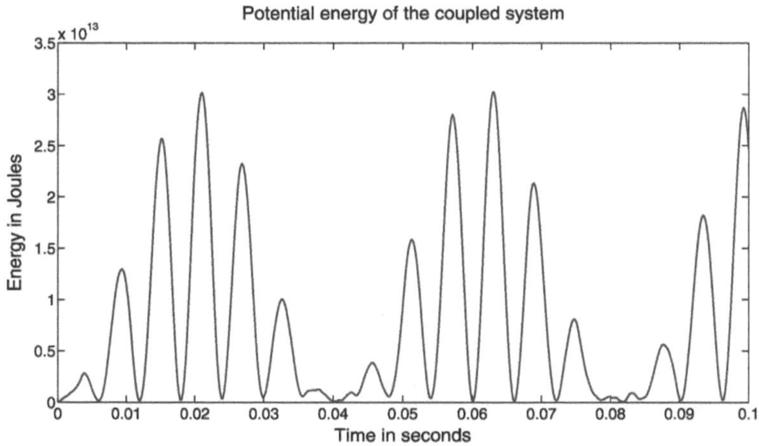

Fig. 9.16 Potential energy for a non–resonant harmonic excitation.

References

1. Arnold DN, Boffi D, Falk RS (2002) Approximation by quadrilateral finite elements. Mathematics of Computation 71:902–922
2. Arnold DN, Boffi D, Falk RS (2005) Quadrilateral $\boldsymbol{H}(\mathrm{div})$ finite elements. SIAM Journal on Numerical Analysis 42:2429–2451
3. Bathe KJ, Nitikitpaiboon C, Wang X (1995) A mixed displacement–based finite element formulation for acoustic fluid–structure interaction. Computer & Structures 56:225–237
4. Bermúdez A, Durán R, Muschietti MA, Rodríguez R, Solomín J (1995) Finite element vibration analysis of fluid–solid systems without spurious modes. SIAM Journal on Numerical Analysis 32:1280–1295
5. Bermúdez A, Durán RG, Rodríguez R (1998) Finite element analysis of compressible and incompressible fluid–solid systems. Mathematics of Computation 67:111–136
6. Bermúdez A, Gamallo P, Hervella–Nieto L, Rodríguez R (2003) Finite element analysis of pressure formulation of the elastoacoustic problem. Numerische Mathematik 95:29–51
7. Bermúdez A, Gamallo P, Nogueiras M, Rodríguez R (2006) Approximation of a structural vibration problem by hexahedral finite elements. IMA Journal on Numerical Analysis 26:391–421
8. Bermúdez A, Gamallo P, Rodríguez R (2001) A hexahedral face element for elastoacoustic vibration problems. Journal of the Acoustical Society of America 109:422–425
9. Bermúdez A, Hervella–Nieto L, Prieto A, Rodríguez R (2007) Validation of acoustic models for time harmonic dissipative scattering problems. Journal of Computational Acoustics 206:440–453.
10. Bermúdez A, Hervella–Nieto L, Rodríguez R (1999) Finite element computation of three–dimensional elastoacoustic vibrations. Journal of Sound and Vibration 219:279–306
11. Bermúdez A, Hervella–Nieto L, Rodríguez R (2002) A modal synthesis method for the elastoacoustic vibration problem. ESAIM: Mathematical Modelling and Numerical Analysis – Modélisation Mathématique et Analyse Numérique 36:121–142
12. Bermúdez A, Rodríguez R (1994) Finite element computation of the vibration modes of a fluid–solid system. Computer Methods in Applied Mechanics and Engineering 119:355–370

13. Bermúdez A, Rodríguez R (1998) Numerical computation of elastoacoustic vibrations with interface damping. In: Équations aux Dérivées Partielles et Applications, Gauthier–Villars, Paris 165–187
14. Bermúdez A, Rodríguez R (1999) Modeling and numerical solution of elastoacoustic vibrations with interface damping. International Journal for Numerical Methods in Engineering 46:1763–1799
15. Bermúdez A, Rodríguez R (2002) Analysis of a finite element method for pressure/potential formulation of elastoacoustic spectral problems. Mathematics of Computation 71:537–552
16. Bermúdez A, Rodríguez R, Santamarina D (2003) Finite element approximation of a displacement formulation for time–domain elastoacoustic vibrations. Journal of Computational and Applied Mathematics 152:17–34
17. Brezzi F, Fortin M (1991) Mixed and hybrid finite element methods. Series in Computational Mathematics 15, Springer–Verlag, New York
18. Conca C, Planchard J, Vanninathan M (1995) Fluids and periodic structures. Masson, Paris
19. Everstine GC (1981) A symmetric potential formulation for fluid–structure interaction. Journal of Sound and Vibration 79:157–160
20. Gastaldi L (1996) Mixed finite element methods in fluid structure systems. Numerische Mathematik 74:153–176
21. Hamdi M, Ousset Y, Verchery G (1978) A displacement method for the analysis of vibrations of coupled fluid–structure systems. International Journal for Numerical Methods in Engineering 13:139–150
22. Kehr–Candille V, Ohayon R (1992) Elastoacoustic damped vibrations. finite element and modal reduction methods. In: Zienkiewicz OC, Ladevèze P (eds) New advances in Computational Structural Mechanics. Elsevier 321–334
23. Kiefling L, Feng GC (1976) Fluid–structure finite element vibrational analysis. AIAA Journal 14:199–203
24. Morand HJP, Ohayon R (1995) Fluid Structure Interaction. John Wiley & Sons, Chichester
25. Ohayon R, Soize C (1998) Structural acoustics and vibration. Academic Press, New York
26. Olson LG, Bathe KJ (1983) A study of displacement–based fluid finite elements for calculating frequencies of fluid and fluid–structure systems. Nuclear Engineering and Design 76:137–151
27. Raviart PA, Thomas JM (1992) Introduction à L'Analyse Numérique des Équations aux Dérivées Partielles. Masson, Paris
28. Rodríguez R, Solomín J (1996) The order of convergence of eigenfrequencies in finite element approximations of fluid–structure interaction problems. Mathematics of Computation 65:1463–1475
29. Wandinger J (1994) Analysis of small vibrations of coupled fluid–structure systems. Zeitschrift für Angewandte Mathematik und Mechanik 74:37–42
30. Wang X, Bathe KJ (1997) Displacement/pressure based mixed finite element formulations for acoustic fluid–structure interaction problems. International Journal for Numerical Methods in Engineering 40:2001–2017
31. Zienkiewicz OC, Taylor RL (1991) The finite element method. Vol 2. McGraw–Hill, London

10

Energy Finite Element Method

Robert Bernhard[1] and Shuo Wang[2]

[1] Purdue University, Hovde Hall, 610 Purdue Mall, West Lafayette, IN 47907, USA
bernhard@purdue.edu
[2] Science Applications International Corporation (SAIC), 2450 NASA Road One, Houston, TX 77058, USA
shuowang_usa@yahoo.com

Summary. The energy finite element method was developed to predict the average response of built-up structural acoustic systems consisting of subsystems such as rods, beams, plates, and acoustical enclosures. The methodology for predicting the behavior in the subsystems is based on a diffuse energy field approximation that is most appropriate for high frequency analysis where traditional finite element approaches become expensive. Subsystems are coupled together using net energy flow and energy superposition principles [3].

10.1 Background and Motivation

At high frequency, acoustical and structural wavelengths are short relative to the dimensions of the systems of interest. The model size and cost of traditional finite element models become prohibitively large. In addition, system response is highly sensitive to environmental and dimensional variation and deterministic methods have less value than statistical methods for predicting performance. Also, many noise applications are described in terms of one–third octave band or octave band metrics. Such analyses are most efficiently done using methods that are able to accurately predict average results.

To obtain an accurate and simple mathematical model representing the average energy propagation in systems, significant effort has been made to develop equations that govern the average energy flow in continuous structures. Belov and Rybak developed transport equations utilizing the Green's function for energy flow in infinite vibrating plates [1], and formulated conduction–like equations for the energy flow in ribbed plates [2]. From these equations Nefske and Sung [10] developed a finite element method to predict the energy flow in homogeneous finite beams in terms of energy variables.

Wohlever and Bernhard [23] derived locally averaged energy governing equations using a method that is consistent with classical mechanics and obtained a second order differential equation, which governs the smoothed energy distribution in rods and beams. Coupling of the energy densities for rods and beams was resolved by Cho and Bernhard [6]. By this approach, energy conduction is simple to predict and can be implemented using standard finite element methods with relatively few elements. The technique is capable of predicting the spatial distribution of the energetics of built–up structures. The energy finite element method (EFEM) was developed to utilize available finite element geometric models for high frequency structural–acoustic analysis, based on the energy governing equations and the finite element formulation given by Wohlever and Bernhard and the coupling methods developed by Cho and Bernhard.

The EFEM techniques have been successfully applied to various engineering problems. Wang and Bernhard [16, 17] applied the EFEM techniques to the realistic case study of a heavy equipment cab. Vlahopoulos et al. [14] applied the EFEM to a complex double–hull ship structure. Wang et al. [20] used the EFEM formulation to compute the structural response of an aluminum cylinder shell structure with periodic stiffeners.

Wang and Bernhard [15, 18] also reformulated the numerical implementation of the equations of Wohlever and Cho using a simplified energy finite element method, referred to as $EFEM^0$ (the superscript "0" denotes that a zero–order finite element interpolation or the finite volume method is used). The resulting formulation results in less degree of freedom for a model and can be implemented using Statistical Energy Analysis (SEA) software. Wang et al. combined the $EFEM^0$ technique with SEA for sound package optimization of a trimmed van [22], and also for exterior acoustic modeling of a truck [19]. Klos applied the $EFEM^0$ technique for point–excited shells [8].

The EFEM is complementary to low frequency FEM models since it can use existing FEM databases. A finite element geometric model can be applied to both low and high frequency analyses. The prediction of a spatially varying energy level within a subsystem is predicted with the EFEM computation. The post–processing of EFEM results also provides straightforward visualization of the energy flow in a system, which is convenient for diagnosis and control of noise propagation. EFEM can be used to model relatively highly damped or non–uniformly damped materials, and to model distributed masses as well as multi–point power input. Due to the utilization of the finite element technique, EFEM also has the other advantages of traditional FEM. It can be easily applied to irregular domains and geometries that are composed of different materials or mixed boundary conditions.

The structural energy density and the power radiation obtained from an EFEM model can be used as boundary conditions of energy boundary element method (EBEM) to compute the radiated far-field sound pressure. An energy boundary element method was developed by Wang et al. [21] using numerical boundary element principles and is a useful complement to energy finite element methods for high frequency acoustical analysis in large or open spaces. Franzoni et al. have also formulated an acoustic boundary element method based on time–averaged energy and intensity variables [7].

Statistical energy analysis (SEA) is the most widely used energy-based methods for predicting average behavior at high frequencies [9]. SEA is based on the physical concept of power balance. However, the assumptions used to develop SEA limit application to lightly coupled, low damping applications. When a more detailed model is needed to include behavior such as the excitation location effects, spatial distribution of response for high–damping system, etc., SEA is not applicable. The assumptions of SEA limit the applicable range of the method.

10.2 Governing Equations

For steady state vibrational energy propagation within a control volume V, the principle of conservation of energy requires that the total power Π_{in} entering the control volume must be balanced by the summation of the power dissipated within the volume, Π_{diss}, and the energy flow through the boundary S

$$\Pi_{\text{diss}} + \int_S \boldsymbol{I} \cdot \boldsymbol{n} dS = \Pi_{\text{in}} \tag{10.1}$$

where $\boldsymbol{I}$ is the intensity of the field. For a general case, assuming the intensity function has continuous first partial derivatives, the divergence theorem can be applied such that

$$\int_S \boldsymbol{I} \cdot \boldsymbol{n} dS = \int_V \boldsymbol{\nabla} \cdot \boldsymbol{I} dV \, . \tag{10.2}$$

Writing each term in (10.1) in the format of volume integrals,

$$\int_V \pi_{\text{diss}} dV + \int_V \boldsymbol{\nabla} \cdot \boldsymbol{I} dV = \int_V \pi_{\text{in}} dV \, , \tag{10.3}$$

then, the energy balance equation is obtained for steady state vibrational energy propagation such that

$$\pi_{\text{diss}} + \boldsymbol{\nabla} \cdot \boldsymbol{I} = \pi_{\text{in}} \, , \tag{10.4}$$

where π_{in} is the input power density (power input per unit volume) and π_{diss} is the dissipated power density (power dissipated per unit volume).

For EFA implementations, a simple loss factor model of damping is used for power dissipation

$$\pi_{\text{diss}} = \eta \omega e \, , \tag{10.5}$$

where η is the damping loss factor, ω is the angular frequency, and e is the time–averaged and locally space–averaged energy density. To develop an EFA equation and express (10.4) in terms of energy density, a relationship is required between the divergence of intensity and the energy density. For an omni–directional source in an infinite medium, these relationships are known. In a two–dimensional medium this relationship is

$$\boldsymbol{\nabla} \cdot \boldsymbol{I} = \frac{1}{r} \frac{d}{dr} \left(r c_g e_d \right) \, , \tag{10.6}$$

where c_g is the group speed. In a three-dimensional medium, this relationship is

$$\boldsymbol{\nabla} \cdot \boldsymbol{I} = \frac{1}{r^2} \frac{d}{dr} \left(r^2 c_g e_d \right) \, , \tag{10.7}$$

where e_d is the time–averaged energy density in the direct field, and r is the radial distance from the excitation.

Thus, the governing differential equation for the energy density distribution in the direct field of a point–excited infinite plate is

$$\frac{1}{r}\frac{d}{dr}(rc_g e_d) + \eta\omega e = \pi_{\text{in}} \,. \tag{10.8}$$

The solution for the direct-field energy density due to a point source is

$$e_d = \frac{\Pi_{\text{in}}}{2\pi h r c_g}\,\mathrm{e}^{-\frac{\eta\omega r}{c_g}} \tag{10.9}$$

and the radial intensity is

$$I_{\text{rad}} = c_g e_d = \frac{\Pi_{\text{in}}}{2\pi h r}\,\mathrm{e}^{-\frac{\eta\omega r}{c_g}} \tag{10.10}$$

Equations (10.6) through (10.10) potentially could be basis of the simplest versions of the Energy Boundary Element Method.

For certain cases, such as in a reverberant field, the response can be assumed to be the superposition of moderately damped plane waves. Bouthier and Bernhard [4] showed that the smoothed intensity $\boldsymbol{I}$ is related to the energy density by

$$\boldsymbol{I} = -\left(\frac{c_g^2}{\eta\omega}\right)\boldsymbol{\nabla} e \tag{10.11}$$

where e is the time–averaged and locally space–averaged energy density, $\boldsymbol{\nabla} e$ is the gradient of the smoothed energy density, and c_g is the group speed. Using equations (10.4), (10.5), and (10.11), the general form of the differential equation governing the energy flow through an isotropic, homogeneous system can be written as

$$\boldsymbol{\nabla}\cdot\left[\left(\frac{c_g^2}{\eta\omega}\right)\boldsymbol{\nabla} e\right] - \eta\omega e + \pi_{\text{in}} = 0 \,. \tag{10.12}$$

Equation (10.12) is one of the governing equations of so–called energy flow analysis (EFA) and is applicable where the wave field can be modeled as a superposition of moderately damped plane waves.

10.3 Energy Finite Element Method (EFEM) Formulations

The EFA governing differential equation (10.12) for damped plane wave behavior can be rewritten as

$$\boldsymbol{\nabla}\cdot\left(\hat{D}\boldsymbol{\nabla} e\right) - \hat{G}e + \pi_{\text{in}} = 0 \tag{10.13}$$

where the coefficients are defined as $\hat{D} = c_g^2/(\eta\omega)$ and $\hat{G} = \eta\omega$. Finite element approximations of the EFA governing differential equation (10.13) can be developed by using the Galerkin's method or the finite volume method.

10.3.1 EFEM for Continuous Systems

For a continuous system, using a conventional finite element formulation, the residual integral is expressed on an element (e) as [11]

$$\boldsymbol{R}^{(e)} = -\int_V \boldsymbol{N}^T \left[\boldsymbol{\nabla}\cdot\left(\hat{D}\boldsymbol{\nabla}e\right) - \hat{G}e + \pi_{\text{in}}\right] dV \tag{10.14}$$

where $\boldsymbol{N}$ is the row vector of the shape functions (interpolation and test functions), and

$$e = \boldsymbol{N}\boldsymbol{e}_n \tag{10.15}$$

where $\boldsymbol{e}_n$ denotes the nodal value vector of the energy density. Substituting (10.15) into (10.14) and forcing the residual integral to vanish results in the scalar equation as

$$\boldsymbol{R}^{(e)} = \boldsymbol{q}^{(e)} + \left[\boldsymbol{k}_{\hat{D}}^{(e)} + \boldsymbol{k}_{\hat{G}}^{(e)}\right]\boldsymbol{e}_n^{(e)} - \boldsymbol{f}_\pi^{(e)} = 0\,. \tag{10.16}$$

For element (e), $\boldsymbol{q}^{(e)}$ represents the inter–element energy flow along the element boundary

$$\boldsymbol{q}^{(e)} = \int_S \boldsymbol{N}^T dq = \int_S \boldsymbol{N}^T\left(-\hat{D}\boldsymbol{n}\cdot\boldsymbol{\nabla}e\right) dS \tag{10.17}$$

The remaining element matrices are

$$\boldsymbol{k}_{\hat{D}}^{(e)} = \int_V \boldsymbol{B}^T\hat{\boldsymbol{D}}\boldsymbol{B}dV \tag{10.18}$$

and

$$\boldsymbol{k}_{\hat{G}}^{(e)} = \int_V \boldsymbol{N}^T\hat{G}\boldsymbol{N}dV\,. \tag{10.19}$$

The element input power vector is

$$\boldsymbol{f}_\pi^{(e)} = \int_V \boldsymbol{N}^T\pi_{\text{in}}dV\,. \tag{10.20}$$

In equation (10.18), $\boldsymbol{B}$ is a derivative vector and $\hat{\boldsymbol{D}}$ is a diagonal matrix as for 1-D systems

$$\boldsymbol{B} = \frac{\partial\boldsymbol{N}}{\partial x} \qquad \text{and} \qquad \hat{\boldsymbol{D}} = \hat{D}_x\ , \tag{10.21}$$

for 2-D systems

$$\boldsymbol{B} = \begin{bmatrix}\dfrac{\partial\boldsymbol{N}}{\partial x}\\[2mm] \dfrac{\partial\boldsymbol{N}}{\partial y}\end{bmatrix} \qquad \text{and} \qquad \hat{\boldsymbol{D}} = \begin{bmatrix}\hat{D}_x & 0\\ 0 & \hat{D}_y\end{bmatrix}\ , \tag{10.22}$$

and for 3-D systems

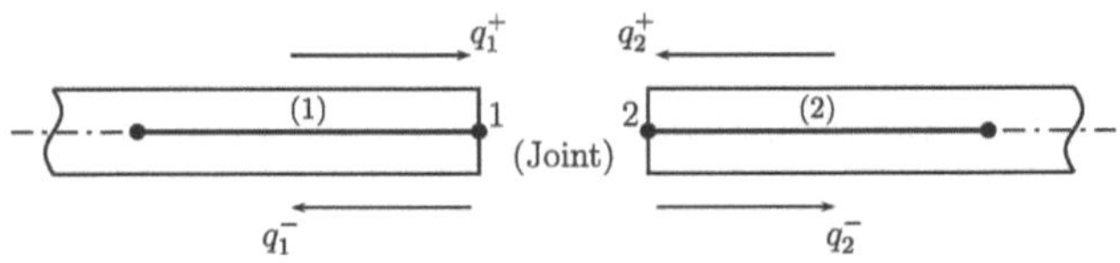

Fig. 10.1 Two collinear beam elements coupled by a joint.

$$\boldsymbol{B} = \begin{bmatrix} \dfrac{\partial \boldsymbol{N}}{\partial x} \\ \dfrac{\partial \boldsymbol{N}}{\partial y} \\ \dfrac{\partial \boldsymbol{N}}{\partial z} \end{bmatrix} \qquad \text{and} \qquad \hat{\boldsymbol{D}} = \begin{bmatrix} \hat{D}_x & 0 & 0 \\ 0 & \hat{D}_y & 0 \\ 0 & 0 & \hat{D}_z \end{bmatrix} . \tag{10.23}$$

For clamped, free, and simply supported boundaries, the energy flow is zero and $\hat{D}Ade/dn = 0$.

10.3.2 Coupling Relations for Joined Systems at Discontinuities

As described by Cho [5], the energy density discontinuities that occur when there is a discontinuity of material or geometric properties can be modeled using approximations that are consistent with the EFA equations. The joint relationships will be illustrated for a simple case of two coupled collinear beams as shown in Figure 10.1. The energy density and energy flow are expressed in terms of components associated with positive and negative traveling waves

$$e_i = e_i^+ + e_i^- \tag{10.24}$$

$$q_i = q_i^+ - q_i^- = c_{g_i} e_i^+ A_i - c_{g_i} e_i^- A_i \qquad \text{with} \quad i = 1, 2 . \tag{10.25}$$

At the joint position, the e_i are the nodal values of energy density on either side of the joint and q_i are the net energy flows out of the joint from beam i. The net energy flow away from the joint in each beam can be expressed as

$$q_2^- = \tau_{12} q_1^+ + r_{22} q_2^+ \tag{10.26}$$

$$q_1^- = \tau_{21} q_2^+ + r_{11} q_1^+ \tag{10.27}$$

where τ_{ij} is the power transmission coefficient from beam i to beam j with $i, j = 1, 2$, and r_{ii} is the power reflection coefficient in beam i. A_i represents the cross section area of beam i.

Substituting equation (10.25) into (10.26) and (10.27) gives

$$c_{g_2} e_2^- A_2 = \tau_{12} c_{g_1} e_1^+ A_1 + r_{22} c_{g_2} e_2^+ A_2 \tag{10.28}$$

$$c_{g_1} e_1^- A_1 = \tau_{21} c_{g_2} e_2^+ A_2 + r_{11} c_{g_1} e_1^+ A_1 \tag{10.29}$$

The values of $e_i^{\pm}$ are obtained by solving equations (10.24), (10.28) and (10.29). Substituting the values of $e_i^{\pm}$ into (10.25) yields the expression of energy flow through the joint from beam 1, which is also the energy flow from beam 1 to 2

$$q_{1\to 2} = q_1 = \frac{1}{r_{11} + r_{22}} \left[\tau_{12} c_{g_1} A_1 \ , \ \ - \tau_{21} c_{g_2} A_2 \right] \begin{bmatrix} e_1 \\ e_2 \end{bmatrix} \tag{10.30}$$

For steady–state energy conduction, it is required that the energy flow is continuous at the joint. The energy flow from beam 2 to beam 1 is $q_{2\to 1} = q2 = -q1$.

The derivation of q_1 and q_2 does not require the joint to be conservative. Thus, it is possible to model a dissipative joint using EFA methods. However, for conservative coupling of rods and beams $\tau_{12} = \tau_{21}$, $r_{11} = r_{22}$ and $\tau + r = 1$. For this discussion the joint is assumed to be conservative. Using these conservative coupling relationships, the relationship between the energy and energy flow at a joint can be reduced to

$$\begin{bmatrix} q_1 \\ q_2 \end{bmatrix} = \frac{1}{2 - \tau_{12} - \tau_{21}} \begin{bmatrix} \tau_{12} c_{g_1} A_1 & -\tau_{21} c_{g_2} A_2 \\ -\tau_{12} c_{g_1} A_1 & \tau_{21} c_{g_2} A_2 \end{bmatrix} \begin{bmatrix} e_1 \\ e_2 \end{bmatrix} \tag{10.31}$$

This approach has been generalized to cases of multiple structural members connected to the joint and multiple wave types in each structural member as well as line joints and area connections between plates and acoustical spaces. Cho [5] shows the extension of the approach for the joint matrix $\boldsymbol{K}_J$ for these more complex and general joints.

When all of the effects of energy dissipation and inter–element energy flow are accounted for, the EFEM formulation takes the form

$$\boldsymbol{K} \boldsymbol{e}_n = \left(\boldsymbol{K}_{\hat{D}} + \boldsymbol{K}_{\hat{G}} + \boldsymbol{K}_J \right) \boldsymbol{e}_n = \boldsymbol{F} \, . \tag{10.32}$$

The EFEM approach is essentially a simple conduction finite element program with a unique joint to account for the discontinuity of energy density at discontinuous joints, which are accounted for using the joint matrix $\boldsymbol{K}_J$.

10.3.3 Simplified Energy Finite Element Method (EFEM^0) Model

A simpler version of the EFEM formulation was developed by Wang [15] using the finite volume formulation of equation (10.12). To illustrate this, a one–dimensional form of the EFA equation (10.12) will be utilized

$$\frac{d}{dx} \left(\hat{D} \frac{de}{dx} \right) - \eta \omega e + \pi_{\text{in}} = 0 \, . \tag{10.33}$$

Using the finite volume method [13], the problem domain is discretized into a number of control volumes. The center of each control volume is treated as a node. A capital letter (P, W, or E) is used to represent both a volume and its center (node), as shown in Figure 10.2. The lower case letters w and e denote the *west* and *east*

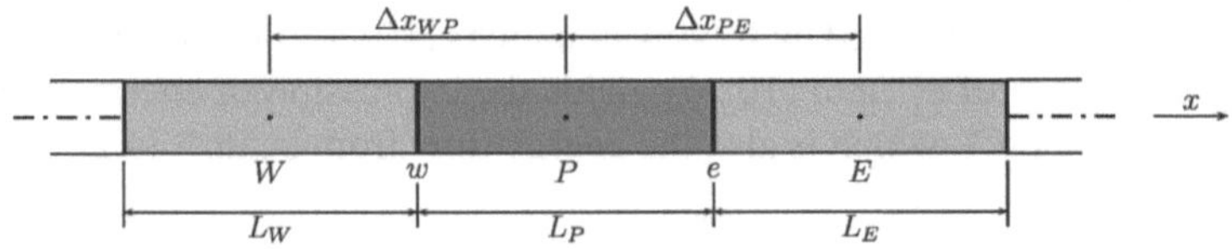

Fig. 10.2 One–dimensional finite volume grids.

boundaries of volume (or element) P. The spacing between the nodes is Δx_{WP} and Δx_{PE}. The length of element I (I is any of P, W or E) is identified by L_I.

Integration of the 1–D EFA governing equation (10.33) over the control volume P yields

$$\int_V \frac{d}{dx}\left(\hat{D}\frac{de}{dx}\right)dV - \int_V \eta\omega e dV + \int_V \pi_{\text{in}} dV = \hat{D}A\frac{de}{dx}\bigg|_{x=w}^{x=e} - \eta\omega e V + \pi_{\text{in}} V = 0 \tag{10.34}$$

where $dV = Adx$ and A is the cross–sectional area of the one–dimensional system. The volume of P is $V = AL_P$. Equation (10.34) is an energy balance equation, which requires that the sum of the energy leaving the control volume and the energy dissipated in the domain per unit time is equal to the power input.

The derivatives in the energy flow terms in (10.34) can be rewritten as

$$\frac{de}{dx}\bigg|_e = \frac{e_E - e_P}{\Delta x_{PE}} \quad \text{and} \quad \frac{de}{dx}\bigg|_w = \frac{e_P - e_W}{\Delta x_{WP}}\,. \tag{10.35}$$

Substitution of (10.35) into (10.34) yields

$$\hat{D}\frac{e_E - e_P}{\Delta x_{PE}} - \hat{D}\frac{e_P - e_W}{\Delta x_{WP}} - \eta\omega e_P L_P + \pi_{\text{in}} L_P = 0 \tag{10.36}$$

which can be rearranged as

$$\left(\eta\omega L_P + \frac{\hat{D}}{\Delta x_{PE}} + \frac{\hat{D}}{\Delta x_{WP}}\right)e_P - \frac{\hat{D}}{\Delta x_{WP}}e_W - \frac{\hat{D}}{\Delta x_{PE}}e_E = \pi_{\text{in}} L_P\,. \tag{10.37}$$

In order to solve the energy distribution within the system, an equation of this form can be developed for each nodal point. As demonstrated here, one of the major advantages of the finite volume method is that the numerical algorithm is closely related to the underlying physical conservation principle. This feature makes the method easy to apply and adapt to novel problems.

To illustrate the comparison of the EFEM^0 to SEA, the energy flow terms in (10.36) can be restated in terms of two types of *coupling* coefficients η_e^c and finite volume total energy as

$$\omega\left[\left(\eta^c_{e_{PE}} E_P - \eta^c_{e_{EP}} E_E\right) + \left(\eta^c_{e_{PW}} E_P - \eta^c_{e_{WP}} E_W\right)\right] + \omega\eta E_P = \Pi_{P_{\text{in}}}\,. \tag{10.38}$$

where $E_I = AL_I e_I$, and

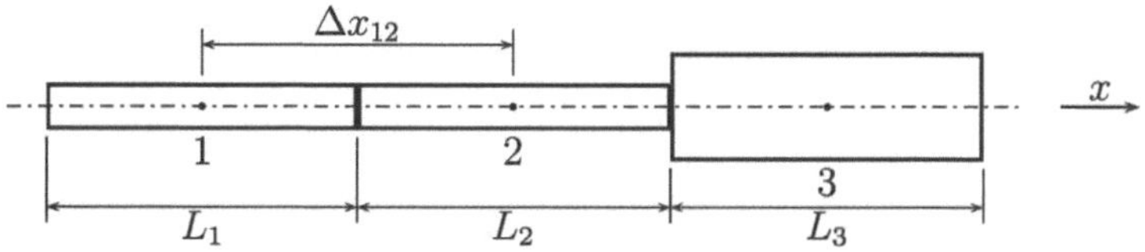

Fig. 10.3 Coupled three–element beam model with continuous and discontinuous joints.

$$\eta^{c}_{e_{IJ}} = \frac{\hat{D}A}{\omega V_I \Delta x_{IJ}} = \frac{\hat{D}}{\omega L_I \Delta x_{IJ}} = \frac{2c_g^2}{\eta\omega^2 L_I (L_I + L_J)} \tag{10.39}$$

and

$$\eta^{dc}_{e_{ij}} = \frac{c_{g_i}}{\omega L_i} \frac{\tau_{ij}}{r_{ii} + r_{jj}} \approx \frac{c_{g_i}}{\omega L_i} \frac{\tau_{ij}}{2 - \tau_{ij} - \tau_{ji}} \tag{10.40}$$

where I and J denote the central node numbers of adjacent control volumes, i.e. E, W, P. Equation (10.38) is similar to an SEA equation, which has the general form

$$\sum_{j=1,\neq i}^{m} \omega \left(\eta_{ij} E_i - \eta_{ji} E_j\right) + \omega \eta_i E_i = \Pi_{i_{\text{in}}} \,. \tag{10.41}$$

The EFEM^0 equations for the three–element system shown in Figure 10.3 can be written in matrix form as

$$\begin{bmatrix} \left(\eta_1 + \eta^{c}_{e_{12}}\right)\omega & -\eta^{c}_{e_{21}}\omega & 0 \\ -\eta^{c}_{e_{12}}\omega & \left(\eta_2 + \eta^{c}_{e_{21}} + \eta^{dc}_{e_{23}}\right)\omega & -\eta^{dc}_{e_{32}}\omega \\ 0 & -\eta^{dc}_{e_{23}}\omega & \left(\eta_3 + \eta^{dc}_{e_{32}}\right)\omega \end{bmatrix} \begin{bmatrix} E_1 \\ E_2 \\ E_3 \end{bmatrix} = \begin{bmatrix} \Pi_{\text{in}_1} \\ \Pi_{\text{in}_2} \\ \Pi_{\text{in}_3} \end{bmatrix} . \tag{10.42}$$

Equation (10.42), which has been derived using a finite volume implementation of an EFA equation, has the same structure as the SEA matrix equation, which has been derived for lightly damped, weakly coupled modal systems with high modal overlap. Many of the terms are the same, particularly those associated with power input and dissipation of energy within the element.

In this paper, we derived generic equations for 1–D, 2–D and 3–D problems based on structural–acoustic theories. The 3–D theory and equations included the fluid or acoustic problems, although some derivations and examples may start with 1–D problem. In EFEM^0 formulation one can change the coupling factors to account for acoustics based on radiation efficiency. In EFEM one can change the coupling terms in the joint matrix.

10.3.4 EFEM^0 Models for One–dimensional Systems

In order to illustrate the performance of the EFEM^0, a convergence study for a single uniform damped beam with simply supported boundary conditions is done using EFEM^0 theory. An analytical EFA solution is also obtained for validation. The beam

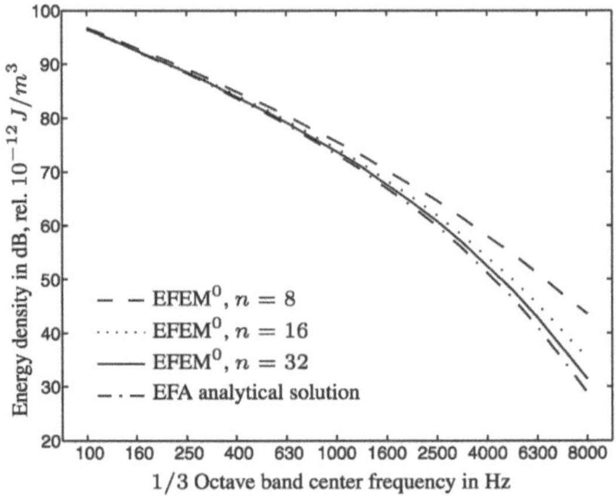

Fig. 10.4 Predictions of energy density at $x = 3$m using EFEM^0 when dividing one uniform damped beam into n subsystems. ($E = 7.1 \cdot 10^{10}$Pa, $L = 4$m, $\eta = 0.24$, $A = 4 \cdot 10^{-4}$m^2).

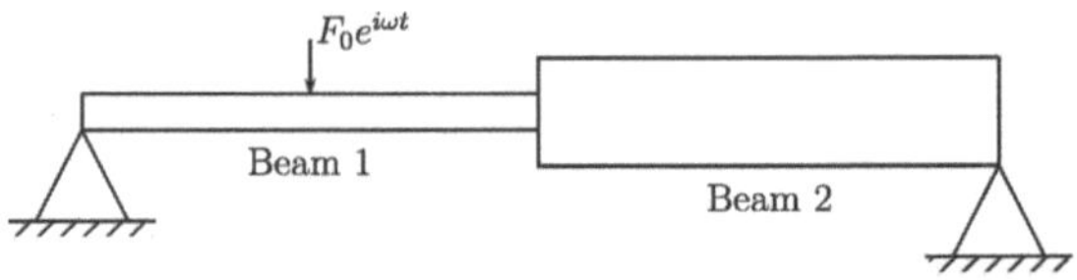

Fig. 10.5 Two coupled beams with simply supported boundary conditions.

is divided into an increasing number of elements. As shown in Figure 10.4, the accuracy of EFEM^0 prediction increases as more elements are used, and converges to the EFA analytical solution.

The coupled beams shown in Figure 10.5 with simply supported boundary conditions were also studied by Wang and Bernhard [18]. The physical properties are listed in Table 10.1. Analytical solutions were obtained using a wave solution for Euler–Bernoulli beam theory. As shown in Figure 10.6, the EFEM^0 predictions were compared to analytical solutions for three cases: strong coupling ($\tau_{12} = 0.999$), medium coupling ($\tau_{12} = 0.629$) and weak coupling ($\tau_{12} = 0.236$). The predictions using in equation (10.40) for discontinuous joint match the analytical solutions well for all three cases. In the study each beam was discretized into 100 elements to obtain a converged solution.

A simplified EFEM (EFEM^0) has been developed from the energy flow analysis equations using the finite volume method. The resulting systems of equations have a similar form to SEA equations. The formulation results in apparent *coupling loss factors* for both continuous and discontinuous systems for relatively heavily damped, strongly coupled systems where conventional SEA coupling loss factors are not suitable.

For a one–dimensional system, the prediction by EFEM^0 agrees very well with the analytical solution and is numerically shown to converge to the exact EFA solution as the number of elements is increased.

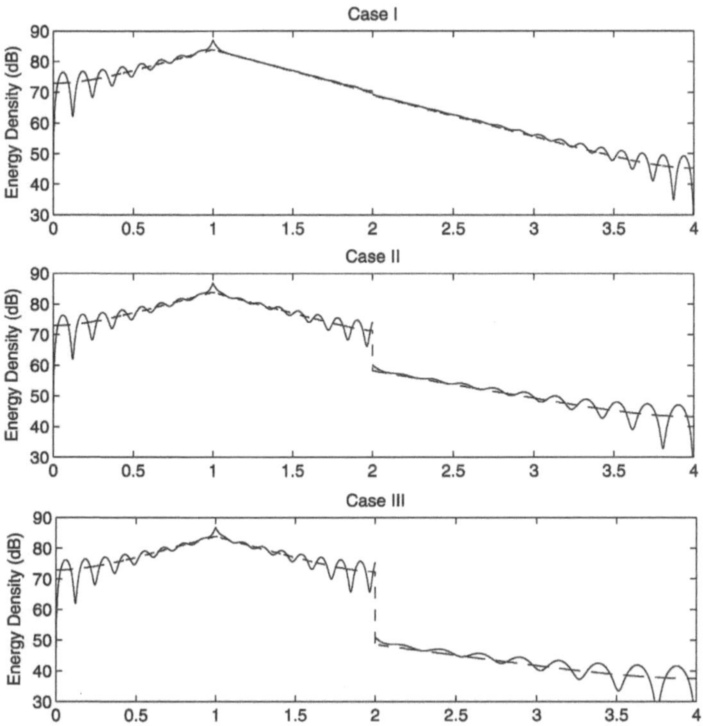

Fig. 10.6 Comparison of EFEM^0 predictions with analytical solutions, dashed line: EFEM^0, solid line: analytical wave equation solution.

Table 10.1 Physical parameters of the two beams, results for three cases: Case I for $\tau_{12} = 0.999$, Case II for $\tau_{12} = 0.629$, and Case III for $\tau_{12} = 0.236$.

	Beam 1	Beam 2		
		Case I	Case II	Case III
cross–sectional area $[m^2]$	$A_1 = 4 \cdot 10^{-4}$	$A_2 = 1.2A_1$	$A_2 = 6A_1$	$A_2 = 16A_1$
area moment of inertia $[m^4]$	$I_1 = 1.33 \cdot 10^{-8}$	$I_2 = 1.44I_1$	$I_2 = 36I_1$	$I_2 = 256I_1$
Young's modulus [Pa]	$E_1 = 7.1 \cdot 10^{10}$	$E_2 = E_1$	$E_2 = E_1$	$E_2 = E_1$
density [kg/m^3]	$\rho_1 = 2.7 \cdot 10^3$	$\rho_2 = \rho_1$	$\rho_2 = \rho_1$	$\rho_2 = \rho_1$
loss factor	$\eta_1 = 0.25$	$\eta_2 = \eta_1$	$\eta_2 = \eta_1$	$\eta_2 = \eta_1$
length [m]	$L_1 = 2$	$L_2 = L_1$	$L_2 = L_1$	$L_2 = L_1$

Generally, EFEM^0 includes the advantages of both EFEM and SEA. Its global matrix has a simple band matrix form similar to SEA. Thus, computational cost is low. The method can be used to predict the local response in each subsystem and to deal with the local effects such as patch damping treatments and partial joints. The coupling relations are also generalized to heavily coupled systems. The EFEM^0 potentially can be integrated into existing finite element model preparation software and will be able to utilize models that have been created for other purposes, such as

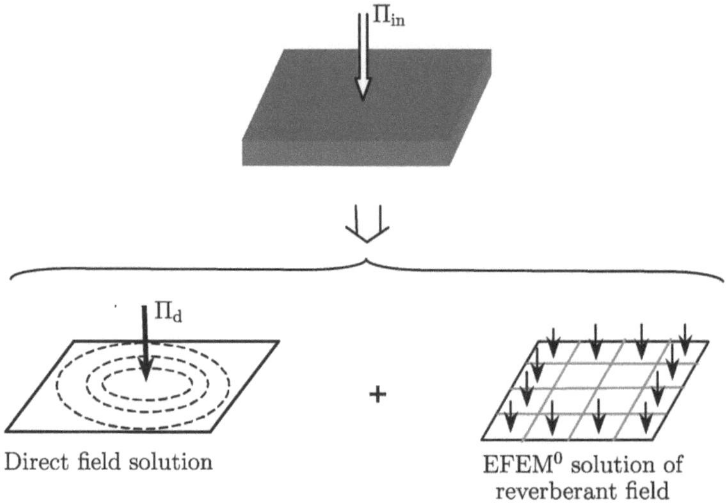

Fig. 10.7 Hybrid–EFEM^0 model for single plate.

low frequency structural acoustic models. EFEM^0 coupling coefficients can also be used in existing SEA programs to increase the capability of the programs. Therefore the EFEM^0 is a good combination of the two methods, and potentially will be quite useful to model built–up, complex systems.

10.3.5 Hybrid EFEM^0 Model for Single Plate

The EFA plane wave equation was derived using a damped plane wave assumption, and thus, represents well the energy distribution in a damped reverberant field. When the direct field of a source is dominant, the EFA equation based on superimposed damped plane waves is not a good model of the response of such systems. A hybrid energy modeling method was developed using the approached described by Smith [12] to address this problem. The approach is to superimpose the contributions of the direct field and the reverberant field. The direct field is calculated assuming the plate is infinitely large. The reverberant field is modeled assuming damped plane wave behavior and the input power from the direct field reflected at the boundary of the finite system. The overall response is the superposition of the two fields.

For numerical implementation, the two–dimensional problem domain is discretized into finite volumes. The power input to the EFEM^0 is modeled as the intensity at the boundary due to the direct field, which is assumed to be reflected into the reverberant field at the boundary. The direct field energy density for each volume is calculated from the energy density at the location of the center node of the finite volume relative to the source. The general strategy of this hybrid–EFEM^0 technique is demonstrated in Figure 10.7.

As a simple example of this implementation, a plate, is considered as shown in Figure 10.8. For S on the boundary surface of a region T in space, $\boldsymbol{n}$ is the outer unit normal vector of S. The direct field power flow from a point source to a small

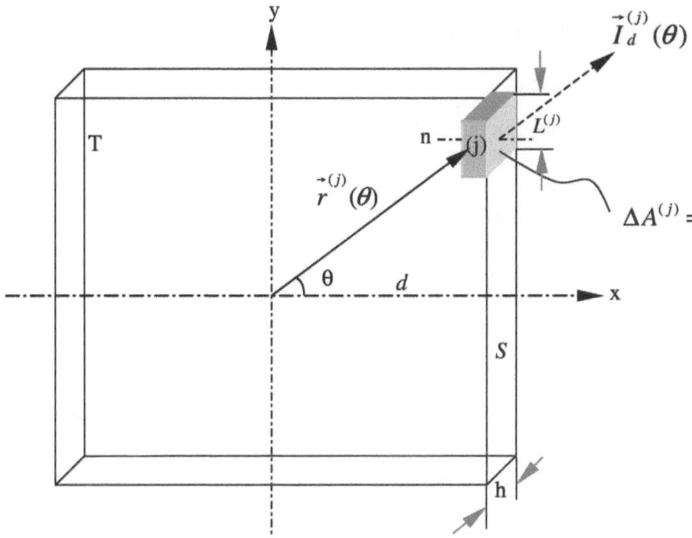

Fig. 10.8 One element on the plate boundary.

surface area ΔA is equal to $\boldsymbol{I} \cdot \boldsymbol{n} \Delta A$, where $\boldsymbol{I} \cdot \boldsymbol{n}$ is the normal component of $\boldsymbol{I}$ in the direction of $\boldsymbol{n}$. For the EFEM^0 numerical implementation, the magnitude of the intensity at a point (j) due to the direct field is

$$I_d^{(j)} = c_g e_d^j = \frac{\Pi_{\text{in}}}{2\pi h r^{(j)}} \mathrm{e}^{-\eta\omega r^{(j)}/c_g} \tag{10.43}$$

where $r^{(j)}$ is the distance from the excitation point (x_0, y_0) to the center of the element $(x^{(j)}, y^{(j)})$, and is calculated by

$$r^{(j)} = \sqrt{\left(x^{(j)} - x_0\right)^2 + \left(y^{(j)} - y_0\right)^2}\,, \tag{10.44}$$

$r^{(j)}$ can be expressed as a function of angle θ as shown in Figure 10.8

$$r^{(j)} = \frac{d}{\cos\theta}\,, \tag{10.45}$$

where d is the normal distance from the excitation point to the boundary.

The power reflected into the element (j) of the reverberant field along the boundary is

$$\Pi_\nu^{(j)}\Big|_\Gamma = r_{11}(\theta) I_d^{(j)} \cos\theta \Delta A^{(j)} \tag{10.46}$$

where $\Delta A^{(j)}$ is the boundary elemental area, $\Delta A^{(j)} = hL^{(j)}$, and θ is the angle between a radial vector from the source and the normal vector of the boundary. Substitution of equations (10.43)– (10.45) into (10.46) gives

$$\Pi_\nu^{(j)}(\theta)\Big|_\Gamma = r_{11}(\theta) D(\theta) \beta^{(j)} \Pi_{\text{in}} \tag{10.47}$$

Table 10.2 Physical properties of the square plate properties.

properties	case
Young's modulus [Pa]	$E = 2 \cdot 10^8$
Poisson's Ratio	$\mu = 0.5$
loss factor	$\eta = 0.02$
density [kg/m^3]	$\rho = 1100$
thickness [m]	$h = 0.01$
edge length [m]	$a = b = 1.5$

where $D(\theta)$ is a directivity function expressed as

$$D(\theta) = \cos^2\theta\, \mathrm{e}^{-\frac{\eta\omega d}{c_g \cos\theta}} \tag{10.48}$$

and is defined as

$$\beta^{(j)} = \frac{L^{(j)}}{2\pi d}\,. \tag{10.49}$$

The elemental reverberant power is a function of angle θ and can be incorporated into the input power vector of the global matrix of the EFEM^0 model.

In [15], Wang applied the hybrid–EFEM^0 models to the predictions for 8 square plates with different properties. One of the plates with moderate damping has the properties listed in Table 10.2. The predictions of the plate with simply supported boundaries are shown in Figure 10.9, compared with the frequency–averaged modal solutions for the plate. The results represent well the frequency– and space–averaged value of the modal solution, and are accurate in both the near field of the excitation point and at the far field of the boundary region.

10.4 Engineering Application

10.4.1 Applications to Cab Rear Window

The hybrid–EFEM^0 model was used to model two coupled plates at the rear window of a heavy equipment cab. The top panel is tempered glass and the bottom panel is steel. For the EFEM^0 model each panel is discretized into a 27 by 27 array of elements. This level of discretization is not necessary for typical systems but is used here to ensure that convergence occurred. The hybrid method is used for the prediction. The reflected and transmitted power at the joints is computed from the direct field using diffuse field transmission and reflection coefficients.

The experimental setup is shown in Figure 10.10. The rear window is divided into 12 subsystems. On each subsystem 5 points are picked at random. Acceleration responses to force excitation at the center of the glass plate are measured and converted to energy density. Both the average response and the response variation are computed.

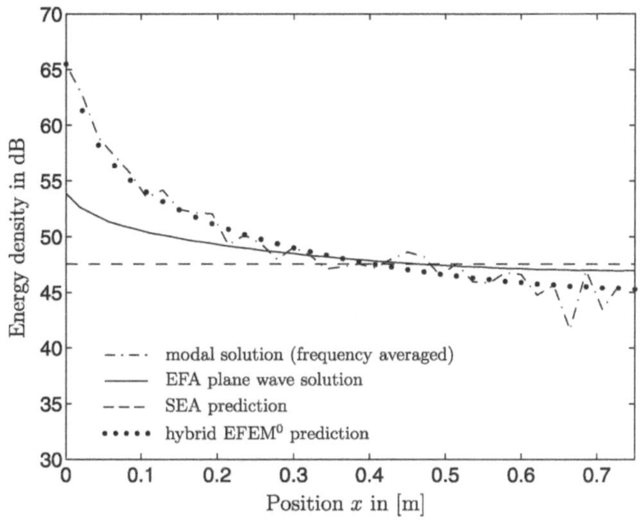

Fig. 10.9 Energy density predictions for square plate.

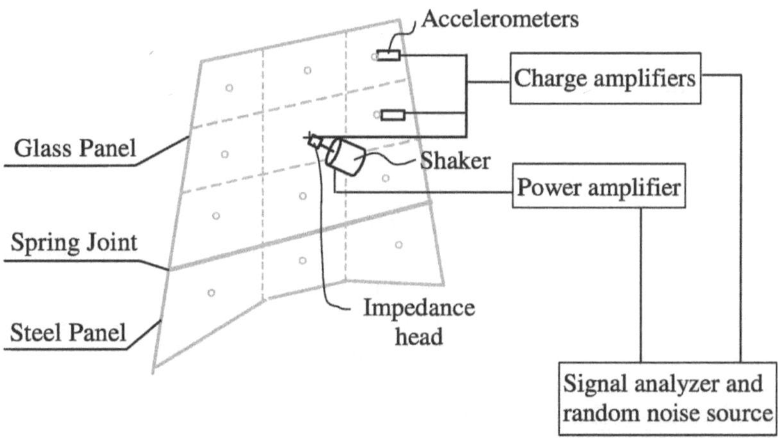

Fig. 10.10 Test setup to measure the responses of cab rear window.

The predicted values are averaged for each of the 12 experimental subsystems and compared to the test data. The comparisons for steel panel subsystem are shown in Figure 10.11. Generally the predicted energy density distributions lie in the region bounded by the experimental variation over the entire frequency domain.

The spatial distributions of the direct field, reverberant field and total field of the predicted responses of the glass panel are shown in Figure 10.12 for 630 Hz and 2500 Hz. Since the damping of the steel panel is light, its response is nearly uniform for most of the frequency range. It is not necessary to do a dense discretization of the steel plate. However, it is important to discretize the glass panel and use the hybrid method in order to account for the effects of damping and the direct field of

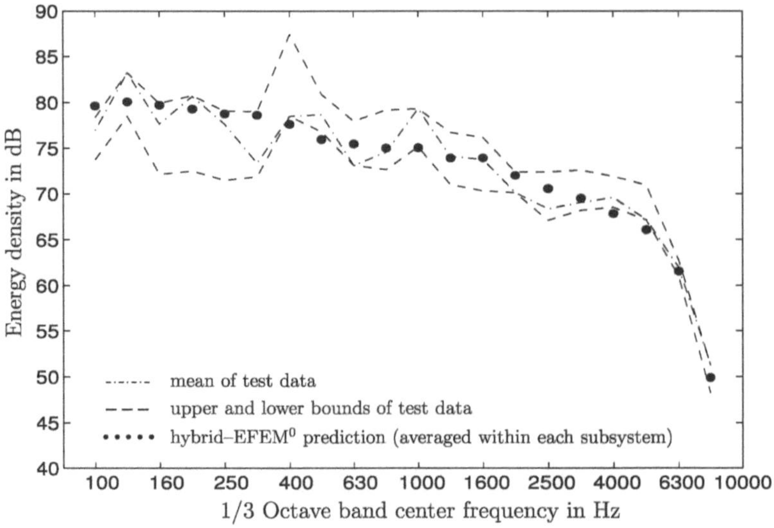

Fig. 10.11 Comparison of the EFEM^0 prediction versus experimental data for cab rear window subsystem 11.

the source as demonstrated in Figure 10.12. The reverberant field is more important at low frequency than at high frequency. However, in both cases, both fields are important.

10.4.2 EFEM Models for Cab Front Window

The cab front and rear windows are typical structures and contain representative joints. In [16], the power transmission and reflection coefficients for joints were calculated and incorporated into both EFEM and SEA models to obtain results for comparison with each other and with measured data. When the transmission and reflection coefficients are obtained, the key step for EFEM is to calculate the general joint matrix (GJM). The key step for SEA is to calculate the coupling loss factor (CLF). The calculated power transmission and reflection coefficients are substituted into the joint matrix of EFEM.

An existing cab NASTRAN model was imported into the EFEM preprocessor to generate a model with 20,225 elements and 19,405 nodes. The front window was divided into four subsystems, which are coupled with each other through five joints. The joint between the middle and side glass panels is a rubber–like sealant which is modeled as a spring joint between plates. Due to the unique materials of the glass plates and the sealant between them, it is only possible to estimate the stiffness and loss factors by measurement of typical samples. The boundaries of the entire front window system were modeled as simply supported boundaries. Thus, the energy flow boundary condition of the front window was zero. The power input was computed using the input force and the real part of the impedance of an infinite plate.

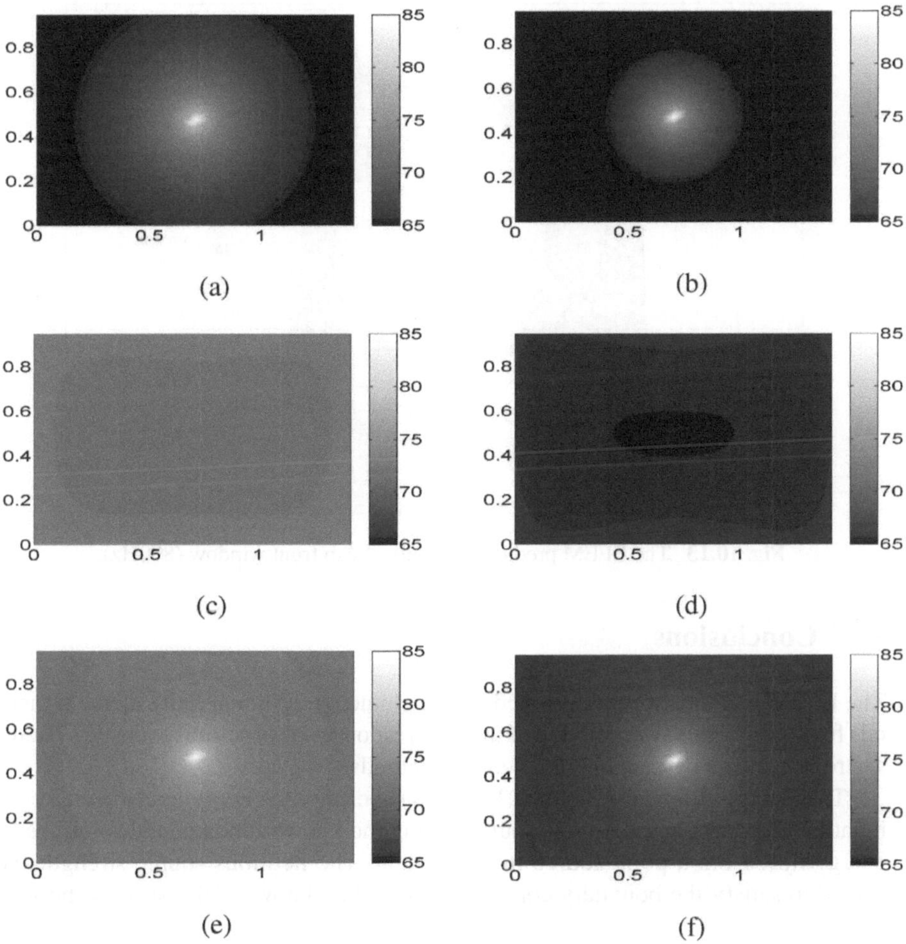

Fig. 10.12 The energy density fields predicted by EFEM^0 for the glass panel of the cab rear window: (a) the direct field at 630 Hz; (b) the direct field at 2500 Hz; (c) the reverberant field at 630 Hz; (d) the reverberant field at 2500 Hz; (e) the overall field at 630 Hz; (f) the overall field at 2500 Hz. The unit is in dB, re 10^{-12} J/m^3.

A test was conducted to validate the EFEM model of the cab front window. A mini–shaker was used to generate power input at the excitation point. Pretests were made to monitor the background noise and choose the excitation level to ensure good coherence. Coherence data was carefully monitored for each measurement. The acceleration responses of two points on each side of the spring joint were measured.

The EFEM results after post–processing are shown in Figure 10.13 for the front window excited at the center. The spatial variation of energy density within each subsystem is clearly demonstrated in the EFEM results [16].

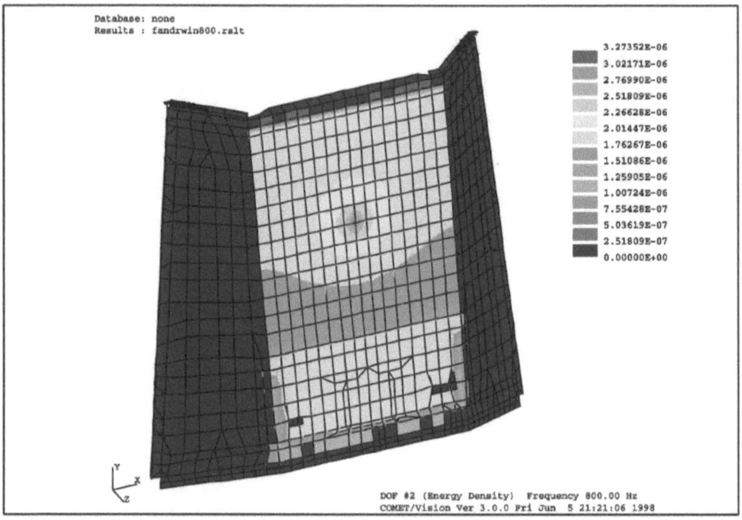

Fig. 10.13 The EFEM prediction results of cab front window (800Hz).

10.5 Conclusions

The EFEM methods were developed to complement traditional finite element methods for high frequency prediction of average responses of structural acoustic systems. There are several distinct energy element methods.

The Energy Boundary Element Methods superimpose energy fields from a combination of known and fictitious sources using the known fundamental solutions for energy flow from a point source in a free field. The fictitious source strengths are solved to satisfy the boundary conditions, and when known, fully specify the solution.

The EFEM methods are based on governing equations that were derived assuming the wave field can be modeled as a superposition of moderately damped plane waves. The governing equations under these conditions are conduction–like equations. Traditional finite element methods can be used to develop numerical approximations of these equations. Due to the simplicity of the behavior of the energy for such conditions, a finite volume approximation of the vibrational conductivity equations also gives a satisfactory approximation of the behavior. The finite volume approximation results in an SEA–like set of linear equations that have the advantage of relative simplicity.

For certain problems where a direct field is important, hybrid EFEM methods have been developed. For these implementations it is assumed that the energy field can be expressed as the superposition of a direct field, which is predicted using the fundamental solution for energy propagation from a point source, and a reverberant field, which is assumed to be modeled by an EFEM solution where the input is assumed to be the energy flow reflected at the boundary of the subsystem that is driven.

The entire family of EFEM and EBEM methods makes up a simple, but complete complementary set of methods to traditional acoustical FEM and BEM methods. With this family of methods, the analyst is able to make dual use of models and predict structural acoustic response across the spectrum for low frequency to high frequency at a level of detail that is appropriate and useful to engineering applications.

References

1. Belov VD, Rybak SA (1975) Applicability of the transport equation in the one–dimensional wave–propagation problem. Journal of Soviet Physics–Acoustics 21:110–114
2. Belov VD, Rybak SA, Tartakovskii BD (1977) Propagation of vibrational energy in absorbing structures. Journal of Soviet Physics–Acoustics 23:115–119
3. Bitsie F, Bernhard RJ (1998) Sensitivity calculations for structural–acoustic EFEM predictions. Noise Control Engineering Journal 46:91–96
4. Bouthier OM, Bernhard RJ (1995) Simple models of the energetics of transversely vibrating plates. Journal of Sound and Vibration 182:149–164
5. Cho PE (1993) Energy flow analysis of coupled structures. PhD Thesis, Purdue University, Herrick Laboratories, West Lafayette
6. Cho PE, Bernhard RJ (1998) Energy flow analysis of coupled beams. Journal of Sound and Vibration 211:593–605
7. Franzoni LP, Bliss DB, Rouse JW (2001) An acoustic boundary element method based on energy and intensity variables for predicting broadband sound fields. Journal of the Acoustical Society of America 110:3071–3080
8. Klos J (2001) Modeling cylindrical shell dynamics using energy flow methods: An analytical and experimental validation. MSME Thesis, School of Mechanical Engineering, Purdue University
9. Lyon RH, Dejong RG (1995) Statistical energy analysis. Butterworth–Heinemann, Boston, 2nd edition
10. Nefske DJ, Sung S (1989) Power flow finite element analysis of dynamic systems: Basic theory and application to beams. Journal of Vibration, Acoustics, Stress, and Reliability in Design 111:94–100
11. Segerlind, LJ (1984) Applied finite element analysis. John Wiley & Sons Inc., New York
12. Smith MJ (1997) A hybrid energy method for predicting high frequency vibrational response of point–loaded plates. Journal of Sound and Vibration 202:375–394
13. Versteeg HK, Malalasekera W (1995) An introduction to computational fluid dynamics: the finite volume method. John Wiley & Sons Inc., New York
14. Vlahopoulos N, Wang A, Wu K (2005) An EFEA formulation for computing the structural response of complex structures. In: Proceedings of 2005 ASME International Mechanical Engineering Congress and Exposition, Orlando
15. Wang S (2000) High frequency energy flow analysis methods: Numerical implementation, applications, and verification. PhD Thesis, Purdue University, Herrick Laboratories, West Lafayette
16. Wang S, Bernhard RJ (1999) Energy finite element method (EFEM) and statistical energy analysis (SEA) of a heavy equipment cab. In: Proceedings of SAE 99 Noise & Vibration Conference, Traverse City 443–450

17. Wang S, Bernhard RJ (1998) The energy finite element method applied to a heavy equipment cab. In: Proceedings of Internoise 98, Christchurch, Paper No. 229
18. Wang S, Bernhard RJ (2002) Theory and applications of a simplified energy finite element method and its similarity to SEA. Noise Control Engineering Journal 50:63–72
19. Wang S, Ebbitt G, Powell RE (2002) Development of a generic truck SEA model with EFEA–SEA hybrid exterior. In: Proceedings of the Internoise 2002, Dearborn
20. Wang A, Vlahopoulos N, Buehrle RD, Klos J (2005) Energy finite element analysis of the NASA aluminum test–bed cylinder. In: Proceedings of 2005 SAE Noise and Vibration Conference, Traverse City, SAE Paper No. 2005-01-2372
21. Wang A, Vlahopoulos N, Wu K (2004) Development of an energy boundary element formulation for computing high–frequency sound radiation from incoherent intensity boundary conditions. Journal of Sound and Vibration 278:413–436
22. Wang S, Xu H, Ebbitt G (2001) Sound package optimization using an $EFEM^0$ –SEA model for a trimmed van. In: Proceedings of SAE 01 Noise & Vibration Conference, Traverse City
23. Wohlever JC, Bernhard RJ (1992) Mechanical energy flow models of rods and beams. Journal of Sound and Vibration 153:1–19

Part IV

BEM: Numerical Aspects

11

Discretization Requirements: How many Elements per Wavelength are Necessary?

Steffen Marburg

Institut für Festkörpermechanik, Technische Universität Dresden,
01062 Dresden, Germany
marburg@ifkm.mw.tu-dresden.de

Summary. The commonly applied rule of thumb to use a fixed number of elements per wavelength in linear time–harmonic acoustics is discussed together with the question of using either continuous or discontinuous elements for collocation. Continuous interpolation of the sound pressure has been favored in most applications of boundary element methods for acoustics. Only a few papers are known where discontinuous elements are applied because they guarantee C^1 continuity of the geometry at element edges. In these cases, it was assumed that the same number of elements as for continuous elements is required for the same numeric error. Of course this implies a larger degree of freedom. An effect of superconvergence is known for boundary element collocation on discontinuous elements. This effect is observed if the collocation points are located at the zeros of orthogonal functions, e.g. at the zeros of the Legendre polynomials. We start with a review of continuous and discontinuous boundary elements using constant, linear and quadratic interpolation on triangular and quadrilateral elements. Major part of this contribution consists of the investigation of the computational example of a long duct. For that, the numeric solution is compared with the analytic solution of the corresponding one–dimensional problem. Error dependence in terms of frequency, element size and location of nodes on discontinuous elements is reported. It will be shown that the zeros of the Legendre polynomials account for an optimal position of nodes for this problem of interior acoustics. Similar results are observed for triangular elements. It can be seen that the error in the Euclidean norm changes by one or two orders of magnitude if the location of nodes is shifted over the element. The irregular mesh of a sedan cabin compartment accounts for the second example. The optimal choice of node position is confirmed for this example. It is one of the key results of this paper, that discontinuous boundary elements perform more efficiently than continuous ones, in particular for linear elements. This, however, implies that nodes are located at the zeros of orthogonal functions on the element. Furthermore, there is no indication of a similarity to the pollution effect which is known from finite element methods.

11.1 Introduction

It is widely accepted that the element size in element–based acoustic computations should be related to the wavelength. Often, the element size is measured in a certain (fixed) number of elements per wavelength. In many cases, this number of elements per wavelength is given for constant or linear/bilinear elements. It varies between six and ten. Obviously, this number is closely related to a certain desired accuracy. Often the error is of an acceptable magnitude which depends on the user and which meets certain technical requirements.

The idea of using a fixed number of elements per wavelength is most likely a consequence of Shannon's sampling theorem [14]. This theorem is of fundamental importance in vibration and acoustics for experimental measurements and frequency detection. It states that at least two points per wavelength (or period of oscillating function) are necessary to detect the corresponding frequency. However, a simple detection cannot be sufficient to approximate the function. Schmiechen [29] investigates discretization of axisymmetric structures for modal analysis. He states that two points per wavelength are strictly sufficient, but would still not lead to accurate mode shapes. Another factor of 3 to 5 is advised. This is equivalent to the number of six to ten nodes per wavelength. We mention that the number of nodes was given for bilinear shell elements. A similar result can be extracted from the author's paper [19]. There, the eigenvalues of a one–dimensional problem are shown for linear and quadratic finite elements. The eigenvalue distribution can be easily related to the Shannon frequency. For linear and quadratic elements, the largest eigenfrequency is slightly larger than the frequency which corresponds to two points per wavelength. For quadratic elements, a gap of eigenfrequencies occurs beyond frequencies of two (quadratic) elements per wavelength. This, however, is equivalent to four points per wave. Although done for finite elements these investigation will most likely hold for boundary element methods as well.

The common rule of using six linear boundary elements per wavelength (or, more general, six points per wavelength) are investigated in the author's papers [18, 22]. Therein, performance of constant elements, of linear and quadratic boundary elements are compared. The results of these papers are summarized in this article.

In boundary element methods, it is quite common to use Lagrangian elements [7, 8,15,36]. In contradiction to finite element methods, the basis functions of the boundary element method underly lower continuity requirements, i.e. there is no particular reason of demanding continuity at element edges for boundary element collocation methods. Consequently, the elements can be either continuous or discontinuous. This means that the physical quantity, i.e. the sound pressure may be either continuous or discontinuous at element edges but the geometry remains continuous. There are numerous examples of the use of discontinuous boundary elements in literature, cf. [1, 2, 9–12, 16, 17, 20, 23, 25–28, 30, 32, 33, 35, 37–40]. In many cases, discontinuous boundary elements have been used for collocation only because the normal derivative integral equation which requires $\mathcal{C}^1$–continuity in the collocation point. This is relevant for methods such as the one by Burton and Miller [4,24,27]. Wu and Seybert [38, 39] proposed a discontinuous variant of using continuous interpolation while the collocation points were located inside the elements. This resulted in an overdetermined system of equations for the hypersingular formulation. In their review paper on DtN–methods for acoustic scattering and radiation, Harari et al. [11] mentioned that the hypersingular formulation of the boundary element method for wave scattering and radiation would be inefficient since necessity of discontinuous elements would require much more memory and computational time than conventional boundary element methods that encounter the problem of the so–called irregular frequencies. This remark presupposes that the numeric error of discontinuous elements is about the same as for continuous elements while the degree of freedom is much higher for discontinuous interpolation. Similarly, same accuracy was presumed for continuous and discontinuous boundary elements in an older paper of Harari and Hughes [12] where efficiency of finite and boundary element methods was compared.

There are a number of papers in which discontinuous elements are compared with continuous elements. Often, continuous elements are considered advantageous, especially because they physical continuity condition at element edges is fulfilled [17,40]. For that reason, discontinuous elements were considered as an alternative to continuous elements if either higher continuity in the collocation point is required (see above) or if boundary conditions are discontinuous [17, 26]. In the paper [40] performance of discontinuous elements is evaluated based on the element–to–element jump of the physical quantity. Many authors (of engineering) conclude that they recommend the use of mixed continuous/discontinuous elements, i.e. discontinuous elements only where they are really required, but an overall mesh of discontinuous elements would supply too many degrees of freedom. The idea of using these mixed meshes is nicely worked out and presented in detail by do Rego Silva et al. [27, 28]. The element definitions given are will be reviewed here.

Herein, we will compare meshes of continuous and discontinuous elements based on the overall error arising due to a certain element size. It will be shown that discontinuous elements may give much smaller errors even for the same degree of freedom as a mesh of continuous elements. This is due to a superconvergence effect which has been well–known in mathematical literature.

Atkinson [2] reviewed this effect of superconvergence for error dependence upon element size for discontinuous boundary elements for the case that collocation points are located at the zeros of orthogonal functions for the standard interval. This effect had been discovered earlier by Chandler [5] and Chatelin and Lebbar [6]. Although this superconvergence effect was studied well, discontinuous boundary elements with nodes at zeros of orthogonal functions, in particular of Legendre polynomials as the simplest case, have hardly been used for practical applications. The authors have found very few papers that described this technique for the Helmholtz equation. A similar idea was proposed by Branski [3] who applied Čebyčev polynomials for acoustic source modeling. Tadeu and Antonio [32] found for the axisymmetric case that linear discontinuous elements are substantially influenced by position of collocation points whereas quadratic elements showed marginal dependence upon their locations. Extended investigations on continuous and discontinuous elements were presented in the aforementioned papers [18, 22]. The dependence on location of the collocation points on the element shows a clear optimum.

This paper is organized as follows: We start with the review of continuous and discontinuous Lagrangian boundary elements. These elements are compared with respect to two examples. The first considers traveling waves in a long duct. The numerical solution of the 3d–problem is compared with the simple analytic solution of the 1d–problem. A sedan cabin compartment is investigated as the second example.

11.2 Continuous and Discontinuous Boundary Elements

11.2.1 Quadrilateral Elements

Quadrilateral boundary elements are usually transformed from a standard interval $-1 \leq \eta_1 \leq +1$ and $-1 \leq \eta_2 \leq +1$ to an arbitrary piecewise smooth surface. Thus, the geometry is approximated by shape functions. It is not the subject of this paper to investigate the influence of these shape functions. It is assumed that all shape approximations in this paper are exact or include an error which may be neglected. In the practical application, this is ensured by using linear or quadratic polynomial shape functions. In general, an element with constant interpolation may be defined on a parabolic surface (superparametric approximation of the geometry). In other cases, a subparametric approximation is suitable, especially on plane surface areas.

Continuous elements account for the most commonly used types of boundary elements. Most likely, this is due to the experiences which users have made in the context of finite elements and, also, due to the misunderstanding that a continuous physical quantity, e.g. the sound pressure, should be approximated by continuous functions. Most popular elements are constructed by (bi)linear and (bi)quadratic Lagrangian interpolation functions.

Interpolation functions of continuous quadrilateral surface elements are easily constructed by multiplying two one–dimensional polynomials ψ_1 and ψ_2, cf. [27]. Introducing the notation of upper indices l and q for linear and quadratic polynomials, respectively, these linear polynomials are formulated as

$$\begin{aligned} \psi_1^l(\eta_k) &= \frac{1}{2}(1-\eta_k) \qquad \text{and} \\ \psi_2^l(\eta_k) &= \frac{1}{2}(1+\eta_k) \end{aligned} \tag{11.1}$$

whereas quadratic polynomials are given by

$$\begin{aligned} \psi_1^q(\eta_k) &= \frac{1}{2}\eta_k(1-\eta_k) \qquad , \\ \psi_2^q(\eta_k) &= \frac{1}{2}\eta_k(1+\eta_k) \qquad \text{and} \\ \psi_3^q(\eta_k) &= (1-\eta_k^2) \ . \end{aligned} \tag{11.2}$$

Interpolation functions of discontinuous quadrilateral elements are constructed in a similar way. The simplest discontinuous elements use constant interpolation. Hence we write constant interpolation function as

$$\psi_1^c(\eta_k) = 1 \ . \tag{11.3}$$

For linear and quadratic discontinuous elements, we assume that the distance between the element edge and the closest nodal point on the standard element is given by the value of a with $0 < a < 1$. Introducing the constant $\zeta = 1 - \alpha$, we write the

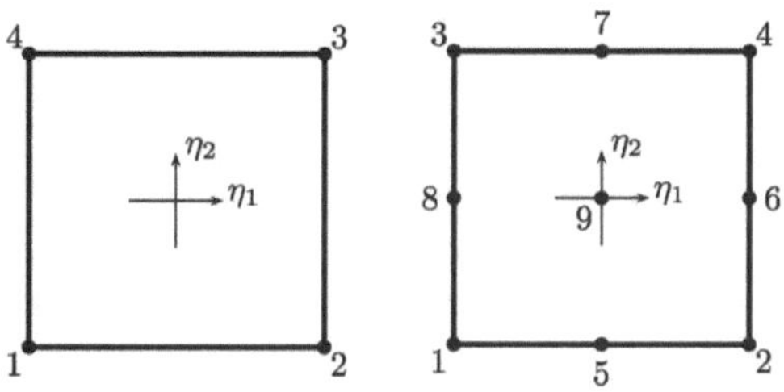

Fig. 11.1 Linear (left) and quadratic (right) continuous quadrilateral elements.

$$\begin{aligned} \psi_1^l(\eta_k) &= \frac{1}{2\zeta}\,(\zeta - \eta_k) \qquad \text{and} \\ \psi_2^l(\eta_k) &= \frac{1}{2\zeta}\,(\zeta + \eta_k) \end{aligned} \tag{11.4}$$

whereas quadratic polynomials are given by

$$\begin{aligned} \psi_1^q(\eta_k) &= \frac{1}{2\zeta^2}\,\eta_k\,(\eta_k - \zeta) \qquad , \\ \psi_2^q(\eta_k) &= \frac{1}{2\zeta^2}\,\eta_k\,(\eta_k + \zeta) \qquad \text{and} \\ \psi_3^q(\eta_k) &= \frac{1}{\zeta^2}\,(\zeta - \eta_k)\,(\zeta + \eta_k) \ . \end{aligned} \tag{11.5}$$

The actual interpolation functions φ_l on the quadrilateral element are evaluated by multiplying the two one–dimensional polynomials $\psi_i(\eta_1)$ and $\psi_k(\eta_2)$, cf. [27]. This gives the polynomials for constant quadrilaterals as

$$\varphi_1^c \; = \; \psi_1^c(\eta_1)\psi_1^c(\eta_2) \; = \; 1 \ , \tag{11.6}$$

for linear elements as

$$\begin{aligned} \varphi_1^l &= \psi_1^l(\eta_1)\,\psi_1^l(\eta_2)\,, & \varphi_2^l &= \psi_2^l(\eta_1)\,\psi_1^l(\eta_2)\,, \\ \varphi_3^l &= \psi_2^l(\eta_1)\,\psi_2^l(\eta_2)\,, & \varphi_4^l &= \psi_1^l(\eta_1)\,\psi_2^l(\eta_2)\,, \end{aligned} \tag{11.7}$$

and analogously for quadratic quadrilateral elements as

$$\begin{aligned} \varphi_1^q &= \psi_1^q(\eta_1)\,\psi_1^q(\eta_2)\,, & \varphi_2^q &= \psi_2^q(\eta_1)\,\psi_1^q(\eta_2)\,, & \varphi_3^q &= \psi_2^q(\eta_1)\,\psi_2^q(\eta_2)\,, \\ \varphi_4^q &= \psi_1^q(\eta_1)\,\psi_2^q(\eta_2)\,, & \varphi_5^q &= \psi_3^q(\eta_1)\,\psi_1^q(\eta_2)\,, & \varphi_6^q &= \psi_2^q(\eta_1)\,\psi_3^q(\eta_2)\,, \\ \varphi_7^q &= \psi_3^q(\eta_1)\,\psi_2^q(\eta_2)\,, & \varphi_8^q &= \psi_1^q(\eta_1)\,\psi_3^q(\eta_2)\,, & \varphi_9^q &= \psi_3^q(\eta_1)\,\psi_3^q(\eta_2)\,. \end{aligned} \tag{11.8}$$

Figures 11.1 and 11.2 show the locations of nodes on continuous and discontinuous quadrilateral elements, respectively. Throughout the computational examples we will indicate different polynomial degree and element types for quadrilaterals as follows:

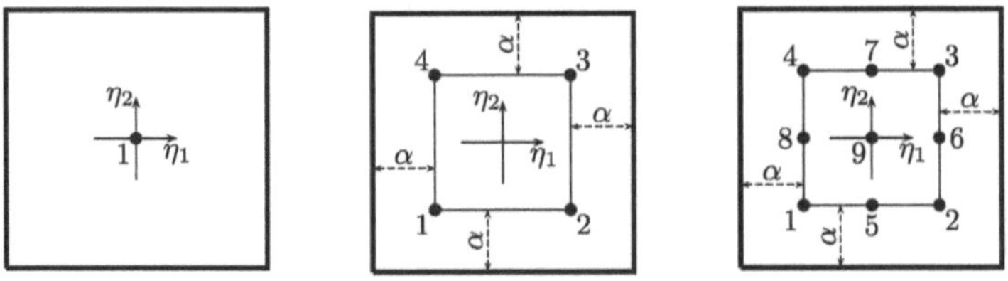

Fig. 11.2 Constant (left), linear (middle) and quadratic (right) discontinuous quadrilateral elements.

P_0	constant (discontinuous)
P_{1c}	linear continuous
P_{1e}	linear discontinuous, equidistant distribution of nodes ($\alpha = 0.5$)
P_{1L}	linear discontinuous, nodes at zeros of Legendre polynomials ($\alpha = 0.4226$)
P_{2c}	quadratic continuous
P_{2e}	quadratic discontinuous, equidistant distribution of nodes ($\alpha = 0.3333$)
P_{2L}	quadratic discontinuous, nodes at zeros of Legendre polynomials ($\alpha = 0.2254$)

Note that linear elements are actually bilinear elements since they involve a nonlinear term. Similarly, quadratic elements are actually biquadratic elements since higher order terms occur. The configurations of P_{1e} and P_{2e} produce an equidistantly distributed pattern of nodes on a regular mesh. In literature, this was recommended as a reasonable configuration for discontinuous elements, cf. [27,28].

11.2.2 Triangular Elements

For triangular elements we use transformation from the standard interval $0 \leq \gamma_1 \leq 1$ and $0 \leq \gamma_2 \leq \gamma_1$ to an arbitrary smooth triangular surface element. Coordinate transformation and interpolation are done in an analogous way to quadrilateral elements. The details are different though. The parameter which is used to define the position of nodes on the element is called β.

In a similar way as for the quadrilateral elements, we start with introduction of one–dimensional polynomials. In contradiction to quadrilateral elements, at this stage, we will not distinguish between linear and quadratic elements. We introduce $\gamma_3 = 1 - \gamma_1 - \gamma_2$ and

$$\psi_k(\gamma_k) = \frac{\gamma_k - \beta}{1 - 3\beta} \qquad \text{for} \quad k = 1, 2, 3 \, . \tag{11.9}$$

The interpolation functions, however, require distinction into constant elements

$$\varphi_1^c = 1 \ , \tag{11.10}$$

linear elements

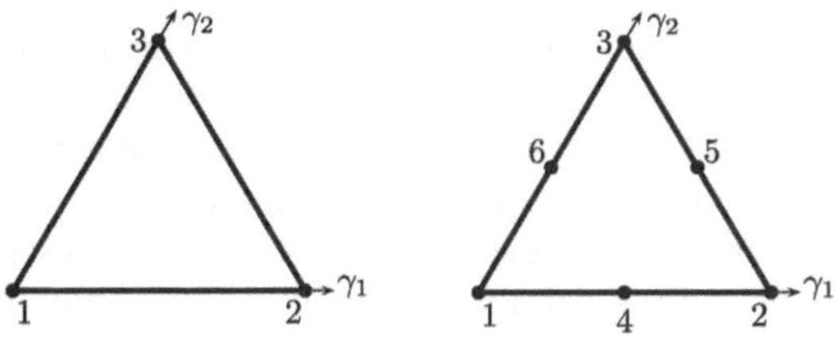

Fig. 11.3 Linear (left) and quadratic (right) continuous triangular elements.

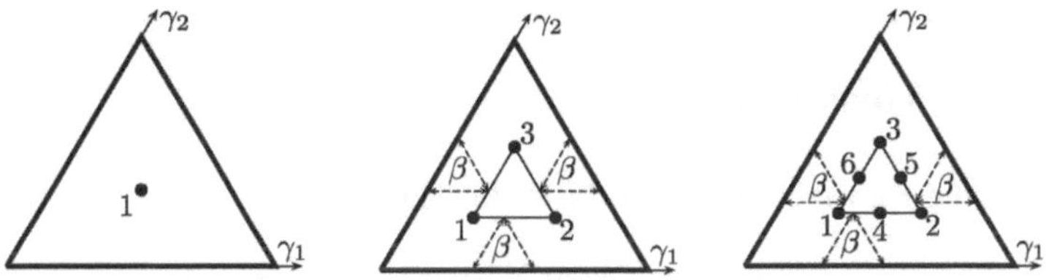

Fig. 11.4 Constant (left), linear (middle) and quadratic (right) discontinuous triangular elements.

$$\begin{aligned} \varphi_1^l &= \psi_3(\gamma_3)\,, \\ \varphi_2^l &= \psi_1(\gamma_1)\,, \\ \varphi_3^l &= \psi_2(\gamma_2) \end{aligned} \tag{11.11}$$

and quadratic elements as

$$\begin{aligned} \varphi_1^q &= \psi_3(\gamma_3)\ [2\psi_3(\gamma_3)-1]\,, & \varphi_4^q &= 4\,\psi_1(\gamma_1)\,\psi_3(\gamma_3)\,, \\ \varphi_2^q &= \psi_1(\gamma_1)\ [2\psi_1(\gamma_1)-1]\,, & \varphi_5^q &= 4\,\psi_1(\gamma_1)\,\psi_2(\gamma_2)\,, \\ \varphi_3^q &= \psi_2(\gamma_2)\ [2\psi_2(\gamma_2)-1]\,, & \varphi_6^q &= 4\,\psi_2(\gamma_2)\,\psi_3(\gamma_3)\,. \end{aligned} \tag{11.12}$$

The node of a constant element is defined at the centroid of triangle, i.e. $(\gamma_1, \gamma_2) = (1/3, 1/3)$. The nodes of linear elements are located at (β, β) (1), $(1-2\beta, \beta)$ (2) and $(\beta, 1-2\beta)$ (3), whereas for quadratic elements, nodes 4–6 are assumed half way between the two nodes of 1–3 respectively, e.g. node 4 half way between 1 and 3. The nodal configurations of continuous and discontinuous triangular elements are shown in Figures 11.3 and 11.4 respectively. In a similar way as for the quadrilateral elements we use P_0, P_{1c} and P_{2c}, whereas linear and quadratic discontinuous elements require further specifications as

P_{1e} linear discontinuous,
equidistant distribution of nodes ($\beta = 0.25$)

P_{1L} linear discontinuous,
nodes at zeros of Legendre polynomials ($\beta = 0.1667$)

P_{2e} quadratic discontinuous,
equidistant distribution of nodes ($\beta = 0.1667$)

P_{2L} quadratic discontinuous,
nodes at zeros of Legendre polynomials ($\beta = 0.0916$)

Note that orthogonal polynomials with respect to the weighting function 1 on a triangular domain are not called Legendre polynomials. We just keep this notation to compare with quadrilateral elements. Furthermore, note that the zeros of second order orthogonal polynomials with respect to the weighting function 1 on a triangular domain do not form a triangle. Actually, nodes 1–3 and 4–6 describe two triangles according to Figure 11.4. However, their location is well approximated by one triangle. To be exact [31], two values of β are required, i.e. for nodes 1–3 we have $\beta = 0.0916$ whereas nodes 4–6 require $\beta = 0.1081$.

11.3 Error Measures

We define an error function e^{Γ} for the surface error as

$$e^{\Gamma}(\boldsymbol{x}) = \bar{p}(\boldsymbol{x}) - p(\boldsymbol{x}) \qquad \boldsymbol{x} \in \Gamma \tag{11.13}$$

where $\bar{p}(\boldsymbol{x})$ represents the approximate solution yielded by using the boundary element formulation and $p(\boldsymbol{x})$ represents either the analytic solution of the one–dimensional duct problem or, in case of the sedan cabin, the solution which is obtained by the finest discretization using quadratic elements P_{2L}. An error e^{Ω} is defined analogously in the interior domain, i.e. for $\boldsymbol{x} \in \Omega$.

The discrete error function is evaluated in discrete points, i.e. all collocation points for the surface error and 33 internal points for the error in the cavity, i.e. all points located in centroid of the cross section and equidistantly distributed along the length. Then, the discrete surface error is determined as

$$||e^{\Gamma}||_m = \left(\frac{1}{N_n} \sum_{i=1}^{N_n} ||e(\boldsymbol{x}_i)||^m \right)^{\frac{1}{m}} \tag{11.14}$$

where N_n represents the number of nodes and m the specific norm, the Euclidean norm (rms) for $m = 2$ and the maximum norm for $m \rightarrow \infty$.

Further, we use relative errors e_m^{Γ} for the sound pressure error

$$e_m^{\Gamma} = \frac{||e^{\Gamma}||_m}{||p^{\Gamma}||_m} \tag{11.15}$$

where $||p^{\Gamma}||_m$ accounts for the discrete norm of the exact sound pressure. Analogously, e_m^{Ω} accounts for the sound pressure error at 33 equally spaced points inside the duct.

An iterative solver is used for the linear system of equations. The residuum of 10^{-8} which is demanded guarantees that the iterative solver does not essentially influence the error functions.

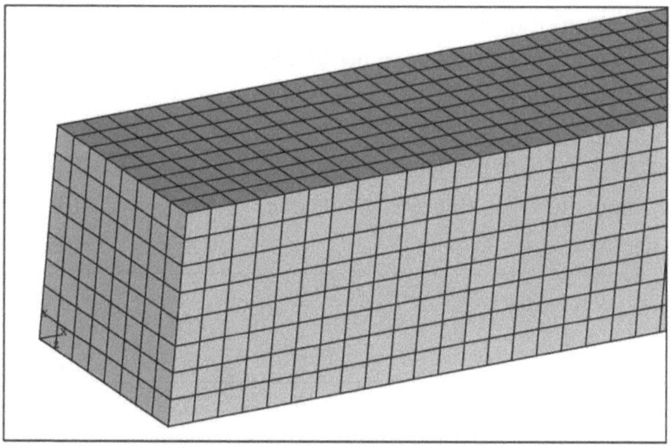

Fig. 11.5 BE mesh of duct end, h_8.

11.4 Computational Example: Long Duct

11.4.1 Model Description and Error Measures

We consider an air–filled duct of length $l = 3.4$ m and a 0.2m×0.2m square cross section. Material data of air being $\rho = 1.3\,\text{kg/m}^3$ and $c = 340\,\text{m/s}$, we expect one wave over the length at $f = 100\,\text{Hz}$. We assume $Y = 0$ on the entire surface with the exception of $Y(l) = (\rho c)^{-1}$. Furthermore, for the particle velocity we use $v_s = 0$ with the exception $v_s(x = 0) = 1\,\text{m/s}$ where the coordinate x is used in the interval of $0 \leq x \leq l$. Since we want to compare solutions of the three–dimensional method with the analytical solution of the corresponding one–dimensional problem, it is necessary to apply zero boundary conditions for the boundary admittance and the particle velocity at all other surfaces.

The exact solution of the corresponding one–dimensional problem is given by

$$p(x) \quad = \quad -\,v_s(0)\,\rho c\, e^{ikx} \quad . \tag{11.16}$$

The sound pressure magnitude is constant in the duct and over the entire frequency range. The solution may be considered as waves traveling through the duct. The boundary condition at $x = l$ ensures that the wave is fully absorbed.

Although, a smooth solution is expected over the entire frequency range, the numeric solution is unstable if modes perpendicular to the traveling waves occur. For the above given cross section, these modes occur for frequencies of 850, 1700, 2550, 3400 Hz and higher. To present a smooth solution for higher frequencies, we compare with the additional example of a particular thin duct of 0.025m×0.025m square cross section. There, perpendicular modes may be expected from 6800 Hz on. This corresponds to a wavenumber $kl = 136\pi$.

Several different meshes are considered. In what follows, we will use h_n to indicate element size. Subscript n counts the number of elements over the width of the duct. Figure 11.5 shows the h_8–mesh of the duct end. More detailed information

Table 11.1 Comparison of mesh sizes (number of nodes and elements) for different element sizes, for continuous and discontinuous quadrilateral elements and for different polynomial degree of interpolation.

number of elements per length	number of elements per width	ref. to	edge–length of element h in [m]	polynomial degree p	number of elements	number of nodes for elements contin.	number of nodes for elements discontin.
				0			70
17	1	h_1	0.2	1	70	72	280
				2		282	630
				0			280
34	2	h_2	0.1	1	280	282	1120
				2		1122	2520
				0			630
51	3	h_3	0.0667	1	630	632	2522
				2		2522	5670
				0			1120
68	4	h_4	0.05	1	1120	1122	4480
				2		4482	10080
102	6	h_6	0.0333	0	2520		2520
				1		2522	
136	8	h_8	0.025	0	4480		4480
				1		4482	
(thin)				0			546
136	1		0.025	1	546	548	2184
				2		2186	4914

about mesh size, number of nodes and elements for different types of elements can be found in Table 11.1.

11.4.2 Error in Terms of Location in the Duct

We start with investigating the error along the length of the duct. In Figure 11.6, we compare one line in models of P_0, P_{1c} and P_{2c} elements. Using constant elements, we observe that the approximate solution of the sound pressure magnitude underestimates the exact solution over the whole length of the duct. The deviation varies rather little. In contrast to the solution with constant elements, linear elements provide a solution where the sound pressure magnitude is somewhat overestimated at one end and clearly underestimated at the other. Obviously, the ratio between maximum norm and Euclidean norm is greater for linear elements compared to constant elements.

Moreover, Figure 11.6 illustrates the difference between edges and smooth surfaces. Although, the problem is essentially one–dimensional, the sound pressure

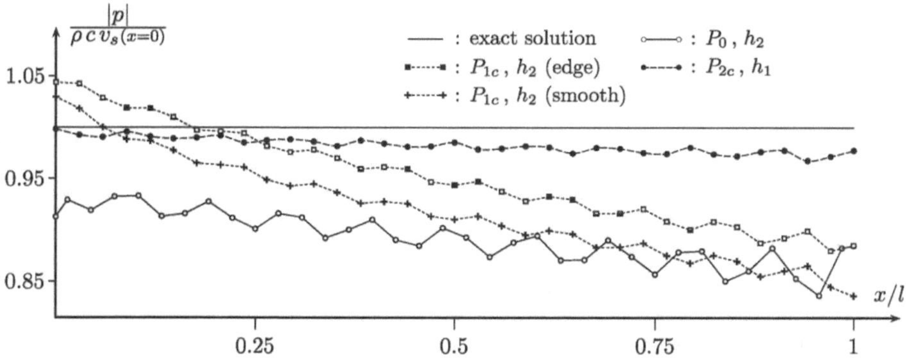

Fig. 11.6 Sound pressure magnitude along lines on surface over the length of duct (collocation points marked) at $f = 500\,\text{Hz}$: Comparison of exact and numerical solutions using P_0, P_{1c} and P_{2c} elements.

magnitude at collocation points on smooth surfaces is significantly smaller that at edge points. This observation holds for continuous elements only since discontinuous do not allow collocation points at edges. In Figure 11.6, this effect is only shown for linear elements since it is less significant for quadratic elements, cf. [18].

11.4.3 Error in Terms of Frequency

Figure 11.7 shows the error e_2^Γ for different meshes in terms of frequency. We observe similar errors and slopes for P_0 and P_{1c} elements. It is common to write the error in the form

$$e^\Gamma(h,k) \quad \sim \quad C(h,\ldots)\,k^\alpha\,. \tag{11.17}$$

For the P_{1L} elements, the error is much lower than for P_0 and P_{1c} elements. However, we observe that for P_{1L} the error functions are quite flat up to a certain frequency. Above that frequency, these curves increase drastically. Most likely, the error dependence cannot be written in the form of Equation (11.17).

For P_{2c} elements, the error curves in terms of k are much steeper. Functions of error show lower rise for elements P_{2L}, especially if the error is of low level. In general, the error level is very low for elements P_{1L} and P_{2L}. This indicates that the theoretical prediction that collocation at the zeros of Legendre polynomials causes an effect of superconvergence, cf. Atkinson [2], seems to hold. Actually, it is just possible to identify a very low error but the superconvergence effect is not confirmed at this point.

In the lower right subfigure of Figure 11.7 the smoother functions of error of the thin duct are collected. All of these models consist of the same number of elements while the number of nodes differs significantly. It can be seen that more or less straight lines represent the error dependence upon k for the elements $P_0, P_{1c}, P_{1e}, P_{2c}$ and P_{2e} provided that the error is less than $20\ldots30\%$. When looking at the error dependence for P_{1L} and P_{2L} straight lines are observed for low errors.

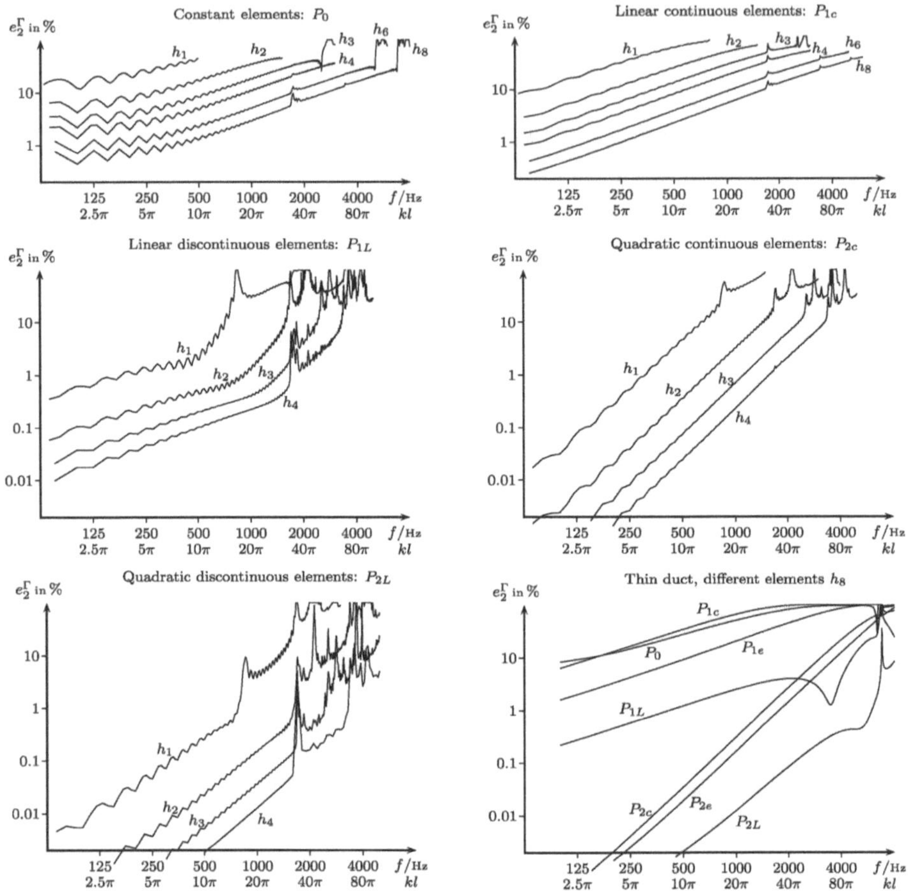

Fig. 11.7 Long duct: Surface error in Euclidean norm in terms of frequency and wave number; resp., Lower right subfigure: Surface error for thin duct and different element types.

To our surprise, the error decreases significantly beyond 2000 Hz in the case of linear elements P_{1L}, whereas the error stagnates in a frequency range between 3500 and 6000 Hz for elements P_{2L}. These phenomena will be discussed later in this section when looking for an optimal location of nodal points on the element.

Efficiency of element types is compared for surface error in Euclidean norm and in maximum norm in Figure 11.8 and Figure 11.9, respectively. Based on the assumption that the number of nodes mainly controls the computational costs (memory and CPU time), we compare different meshes of the same degree of freedom being 2520 or 2522, cf. Table 11.1. When considering the error in the Euclidean norm, quadratic elements P_{2L} prove to be most efficient. Quadratic elements P_{2c} and P_{2e} perform well too. However, linear elements P_{1L} prove to perform efficiently if very low errors are desired. Furthermore, it can be seen that constant elements are as efficient as discontinuous linear elements P_{1e} and provide lower errors than continuous linear

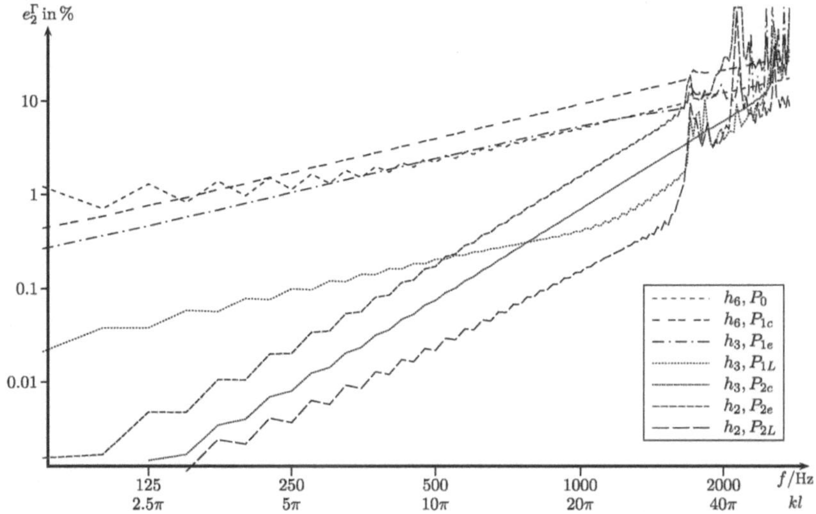

Fig. 11.8 Long duct, surface error in Euclidean norm: Comparison of different element types. Note, that all models have the same degree of freedom of approx. 2520.

elements P_{1c}. We observe a slightly different behaviour when the error is measured in the maximum norm, cf. Figure 11.9. Then, quadratic elements P_{2L} and P_{2c} give about the same error, whereas elements P_{2e} come out with larger errors. Again, linear elements P_{1L} give very low errors compared with linear elements P_{1e} and P_{1c}. The latter have larger errors than constant elements P_0.

Based on the results of this chapter there is no indication of a pollution effect for boundary element collocation method. We will review this problem in the subsequent subsection again.

In the papers [18, 22], the reader finds additional graphs on error in terms of frequency. These include comparisons of quadrilaterals and triangles, where hardly any difference could be observed. Further, the differences of errors in Euclidian and in maximum norms are compared. More detailed comparisons of surface error and field point errors are shown.

11.4.4 Error in Terms of Element Size

The computational example in Ref. [18] confirmed the effect of superconvergence for regular meshes and even polynomial degree of the interpolation functions. Following Atkinson's prediction [2] we should expect an error dependence as

$$e^\Gamma(h,k) \quad \sim \quad C(k,\ldots)\; h^{p+1} \tag{11.18}$$

where p is the polynomial degree. For a regular mesh of constant or continuous elements of even number of polynomial degree of the interpolation functions, we can expect an additional factor of h in the error dependence as

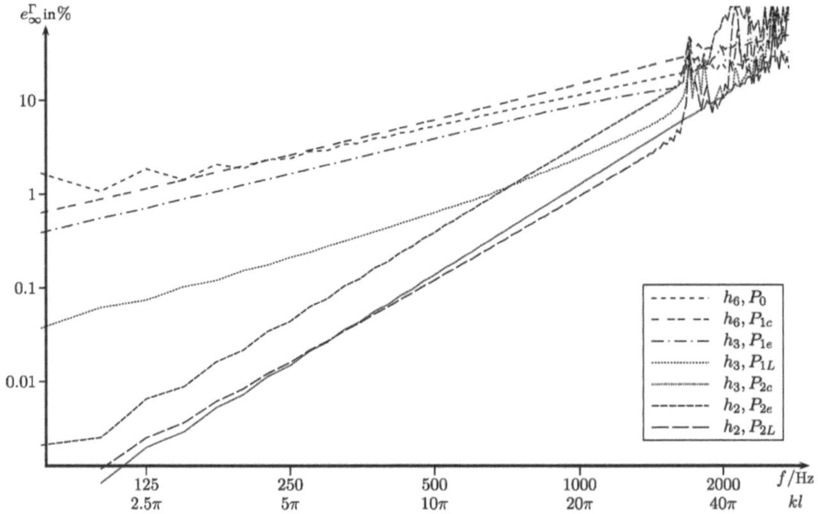

Fig. 11.9 Long duct, surface error in maximum norm: Comparison of different element types. Note, that all models have the same degree of freedom of approx. 2520.

$$e^{\Gamma}(k,h) \quad \sim \quad C_h(k,\ldots)\, h^{p+2} \quad , \qquad p = 0, 2, 4, \ldots \tag{11.19}$$

The higher exponents β for regular meshes of constant or continuous elements of even polynomial degree are valid for the Euclidean norm measured in the discrete collocation points only. If we consider the error measured in the Euclidean norm over the entire surface we would observe an error behaviour as given in (11.18).

The effect of superconvergence at collocation points is well documented in the literature, see for example Hackbusch [10]. There, the superconvergence effect is reported for arbitrary meshes of constant elements. This coincides with the observation [2, 5, 6] indicating that the effect of superconvergence may be achieved on arbitrary meshes of discontinuous Lagrange elements with nodal point at the zeros of orthogonal polynomials, e.g. Legenndre polynomials.

Error dependence in terms of the element size is presented for two different frequencies in Figure 11.10. We easily realize that different functions of error occur for different frequencies. For 500 Hz, lines for P_0, P_{1e}, P_{1c} and P_{1L} are almost parallel but on different levels. The remaining three functions are almost parallel too but much steeper. For 1500 Hz, the lines for P_0, P_{1e} are P_{1c} parallel. The error for elements P_{1L}, however, is now parallel to the lines of quadratic elements which indicates higher convergence rate for these higher frequencies. Note, that functions of error for P_{1L} and P_{2e} coincide!

So we realize that slopes of error for P_{1L} are about the same as for other linear or for constant elements at the low frequency of 500 Hz but much greater for the higher frequency of 1500 Hz. A similar behaviour is assumed for quadratic elements P_{2L}. The frequency is not large enough to confirm. (Solutions at higher frequencies are perturbed due to the ill–conditioning of perpendicular modes.)

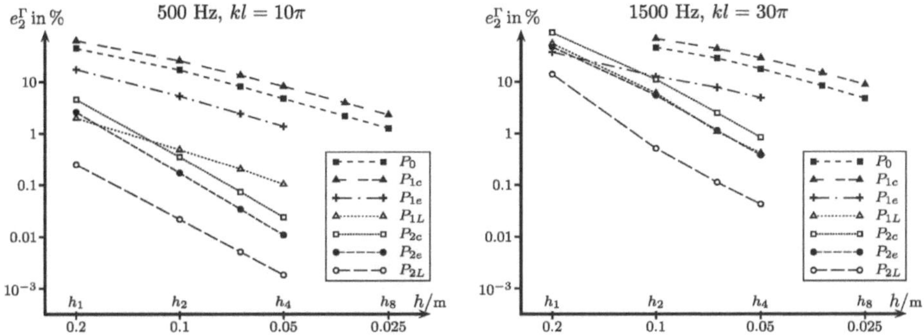

Fig. 11.10 Long duct, surface error in Euclidean norm, error in terms of element size.

Since the error functions in terms of mesh size show the same slopes at least for the $P_0, P_{1e}, P_{1c}, P_{2c}$ and P_{2e} elements, we assume that an occurrence of a pollution effect which is well–known from finite elements [13] is very unlikely. In the subsequent subsection, we will discuss why the slopes of error functions of P_{1L} and P_{2L} elements show or may show a dependence on frequency.

11.4.5 Error in Terms of Location of Collocation Points

In this subsection, the location of nodal points is investigated. It became obvious in the previous subsections that location of nodes at the zeros of Legendre polynomials provides lower errors compared to an equidistant distribution of nodes on the surface. According to Atkinson, it is not required to put nodes at the zeroes of Legendre polynomials. More general, nodes should be located at zeroes of orthogonal functions being defined on the standard interval $[-1, 1]$. Legendre polynomials account for the simplest selection of orthogonal functions since they are well–known and particularly designed for the interval $[-1, 1]$. In what follows, it will be investigated if the zeros of the Legendre polynomials actually account an optimal position of nodes.

Figure 11.11 contains the errors e_2^Γ and e_2^Ω in terms of α. Our test model h_2 consists of 280 elements. We expect the lowest error at $\alpha = 0.4226$ for linear elements and at $\alpha = 0.2254$ for quadratic elements. Although not exactly fulfilled we see that an optimal location of nodal values is very close to the zeros of the Legendre polynomials. The optimal value varies with frequency. For low frequencies and low error, lower values of α account for an optimal position, for higher frequencies and, consequently, higher errors an α greater than that providing the zeros of the Legendre polynomials is required for optimal elements. In between, a large frequency range is observed where nodal points are optimally placed as predicted [2, 5, 6].

Actually, the optimal location of nodes at the zeros of Legendre polynomials refers to pure Neumann problems using the double layer potential operator. Herein, a mixed problem is considered because a Robin boundary condition is applied at one end of the duct. Apparently, the choice of nodes at the zeros of the Legendre polynomials is a good approximation of the optimal location. In case of other operators,

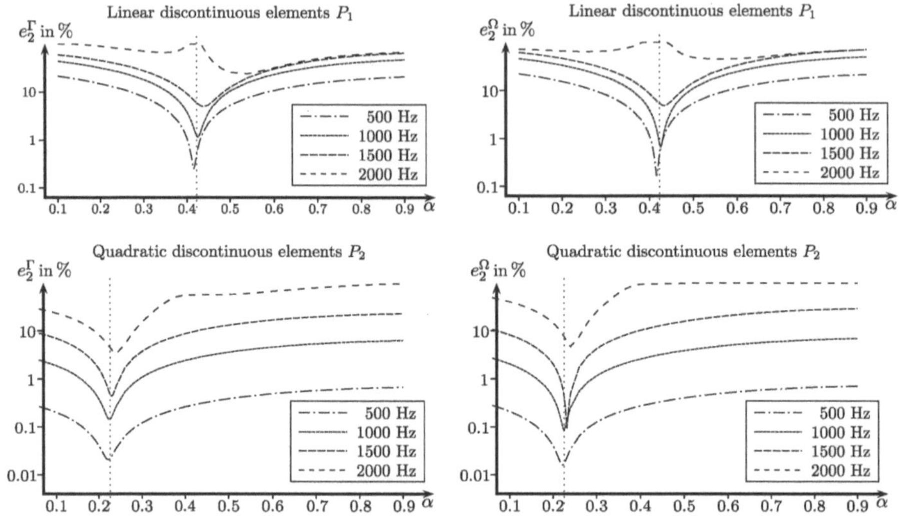

Fig. 11.11 Long duct, quadrilateral elements, surface error and error at internal points in Euclidean norm, error in terms of the position of nodal points on elements.

i.e. the hypersingular operator, and other boundary conditions, the optimal position of nodes may differ from the one identified here.

It shall be mentioned at this point that for certain frequencies extremely low errors are gained for field points compared to surface error. The most remarkable example is found for quadratic elements at 1500 Hz. However, significant differences can be found at 500 and 1000 Hz for both, linear and quadratic elements. The error of the solution at the surface and at internal points is almost the same for very low and for higher frequencies.

The fact that the optimal location of nodal points varies with frequency explains the error behaviour for the thin duct, cf. lower right subfigure of Figure 11.7. There, we have observed decreasing error for increasing frequency in certain ranges. It does further explain why error dependence upon element size changes with frequency.

In Table 11.2, the error e_2^Γ is compared for continuous and discontinuous elements. The third column contains the numeric error of the h_2 discretization of continuous elements which is the same element size as for the discontinuous elements. The fourth column contains the error of the h_4 discretization of continuous elements which results in (approximately) the same degree of freedom as for the discontinuous elements. These results confirm that the node location at the Legendre zeros and, even better, the optimal location give much more accuracy for linear elements and, mostly, better accuracy for quadratic elements.

By now, our considerations were limited to quadrilateral elements. When looking at the error in terms of node position it is necessary to investigate triangles separately. For that, we create a mesh of triangles simply by dividing each quadrilateral into two triangles. Starting with the element size $h = 0.1m$ we yield 560 triangles. Linear and quadratic discontinuous elements give 1680 and 3360 nodes, respectively.

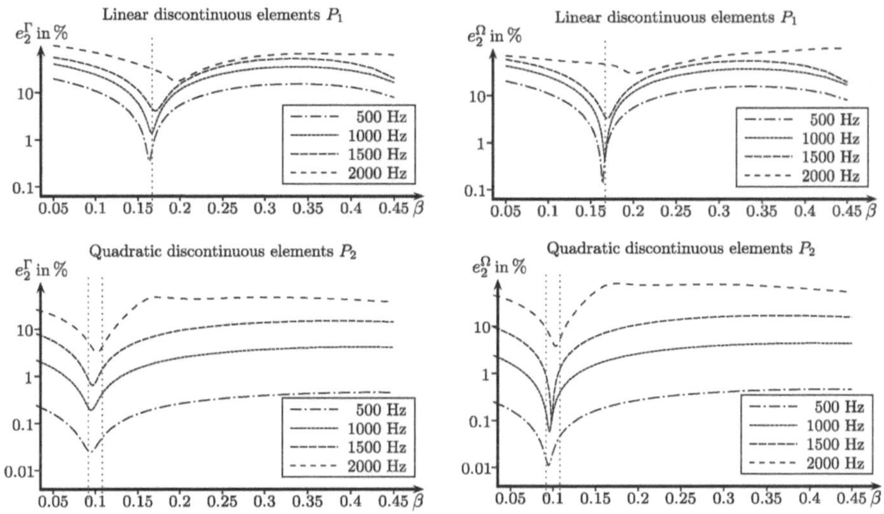

Fig. 11.12 Long duct, triangular elements, surface error and error at internal points in Euclidean norm, error in terms of the position of nodal points on elements.

Table 11.2 Long duct, quadrilateral elements, e_2^Γ in %, comparison between continuous (different element size) and discontinuous elements (different location of nodes).

p	f [Hz]	continuous elements h_2	continuous elements h_4	discontinuous elements h_2, location of nodes: equidistant	Legendre zeros	best solution
	100	4.1	1.2	0.82	0.11	0.032
	500	25.6	8.2	5.3	0.49	0.25
1	1000	51.3	18.5	10.5	1.24	1.16
	1500	68.7	28.8	12.7	6.04	5.0
	2000	—	39.3	26.0	100	24.2
	100	0.0024	0.0002	0.0017	0.0004	0.0003
	500	0.35	0.024	0.17	0.021	0.021
2	1000	3.2	0.23	1.55	0.15	0.145
	1500	11.1	0.85	5.49	0.51	0.446
	2000	37.3	2.1	28.7	4.66	3.37

Figure 11.12 supplies estimates for optimal location of nodal points in triangles. The results confirm the prediction that the optimal position is at $\beta = 0.1667$ on linear elements and in the interval $0.0916 \leq \beta \leq 0.1081$. A value of $\beta \approx 0.098$ is identified as the optimum. Similar to quadrilateral elements, we observe a particular gain in accuracy for internal points at certain frequencies. These frequencies coincide with those reported for quadrilateral elements.

As a conclusion for this subsection we want to emphasize that the position of nodal points on the element can influence the accuracy of the solution by one to two orders of magnitude.

Table 11.3 Comparison of mesh sizes (number of nodes and elements) for different element sizes, triangular and quadrilateral elements and for different polynomial degree of interpolation.

maximum elem. size h in [m]	polynom. degree p	number of elem.	number of triang.	number of quadril.	number of nodes elements contin.	discontin.
	0					258
0.4	1	258	104	154	208	928
	2				826	2010
	0					604
0.2	1	604	128	476	542	2288
	2				2162	5052
	0					1774
0.1	1	1774	74	1700	1739	7022
	2				6950	15744
0.05	0	5738	24	5714		5738
	1				5728	

11.5 Computational Example: Sedan Cabin Compartment

This example is chosen to examine an irregular mesh which is the result of an automatic mesh generation. Four meshes are investigated. The meshes contain quadrilaterals and triangles. Their detailed data are given in Table 11.3. A more vivid description of these models is shown in Figure 11.13. For discontinuous elements with nodes at zeroes of the Legendre polynomials, P_{1L} and P_{2L}, we distinguish between triangles and quadrilaterals as discussed in the previous section.

A fictitious excitation with uniform normal particle velocity $\bar{v}_s = 1$ mm/s at the lower left front area is applied, cf. Figure 11.13. A uniform boundary admittance of

$$Y = \frac{1}{\rho c} \frac{f}{f_0} \qquad (f_0 = 2800 \text{ Hz}) \tag{11.20}$$

is applied to simulate the absorbing behaviour of the surfaces inside the cabin [21]. This value corresponds to experimental measurements of the reverberation time and a corresponding average absorption coefficient. The sound pressure is computed at ten points inside the cabin.

It shall be mentioned that the author is aware that realistic calculations of cabin noise problems are done for frequencies up to (max.) 150...200 Hz. However, the major uncertainty of these calculations are structural transfer functions and realistic distributions of the boundary admittance values. We will show that even a coarse boundary element mesh for the fluid can give an excellent approximation of the sound pressure field over the entire frequency range.

Our reference solution is computed by using discontinuous quadratic elements of size $h \leq 0.1$ m. As indicated by Table 11.3, the associated system of equations has

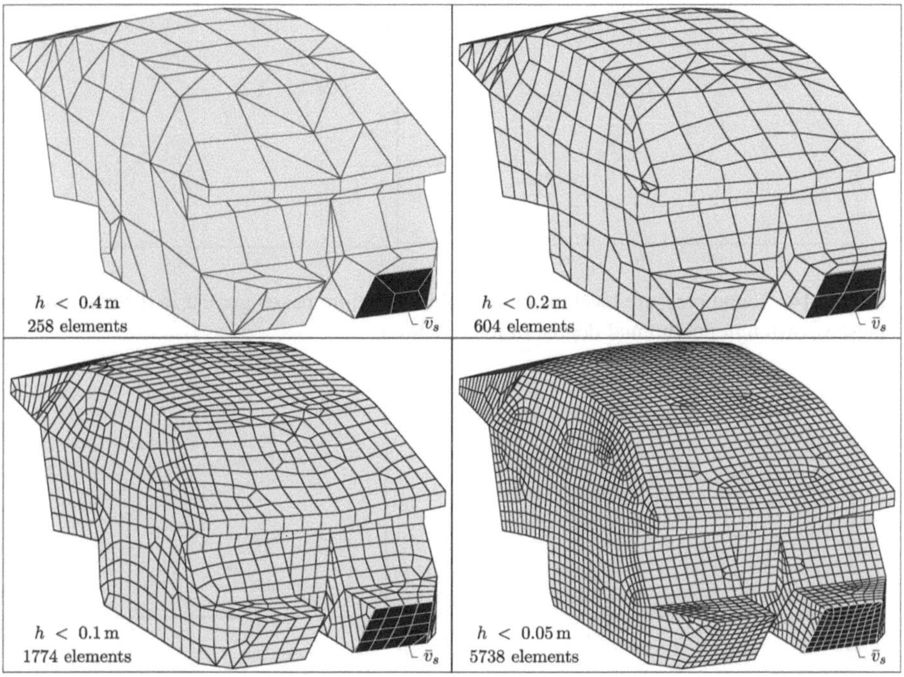

Fig. 11.13 Four meshes of sedan cabin compartment, element size with upper limits, indication of excitation area by particle velocity $\bar{v}_s$.

15744 unknowns. In what follows, we will call this solution our reference solution and all errors are evaluated with respect to this reference.

The left subfigure of Figure 11.14 shows the sound pressure level at an arbitrarily chosen internal point for different meshes. All transfer functions are in good agreement for frequencies lower than 100 Hz. Above that, the solution of $p = 0$ deviates from the others but maxima and minima of the transfer functions are found at correct frequencies up to about 350 Hz. Unlike the case of uniform mesh, linear elements provide a better approximation of the sound pressure level in the present application. However, even by using such a coarse mesh, the general approximation is good. Quadratic elements hardly allow us to find differences between the different solutions. In the right subfigure of Figure 11.14, we observe hardly any differences between the different sound pressure level curves.

Looking at the error at internal points in terms of element size, Figures 11.15 and 11.16 show the error functions for different types of elements and different frequencies. Most of these functions are virtually linear but not necessarily parallel for the same element type, cf. Figure 11.15. The comparison of different element types confirms excellent performance of discontinuous elements, cf. Figure 11.16. So, we realize that, in this example, constant elements give lower error than continuous linear elements. Furthermore, discontinuous linear elements give lower error

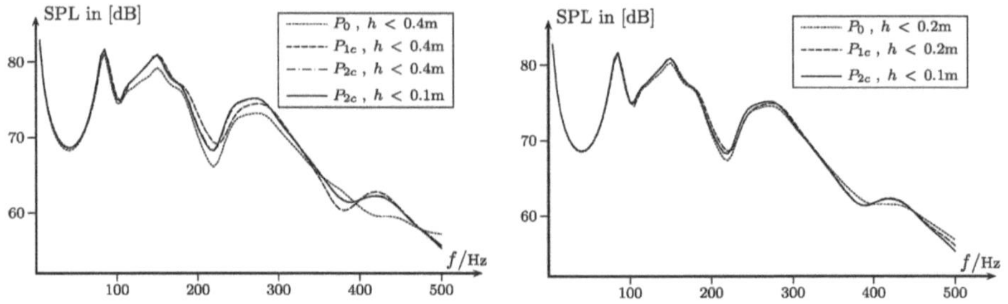

Fig. 11.14 Sound pressure level inside the cabin, location: passenger ear in the rear, different meshes, different polynomial degree of interpolation.

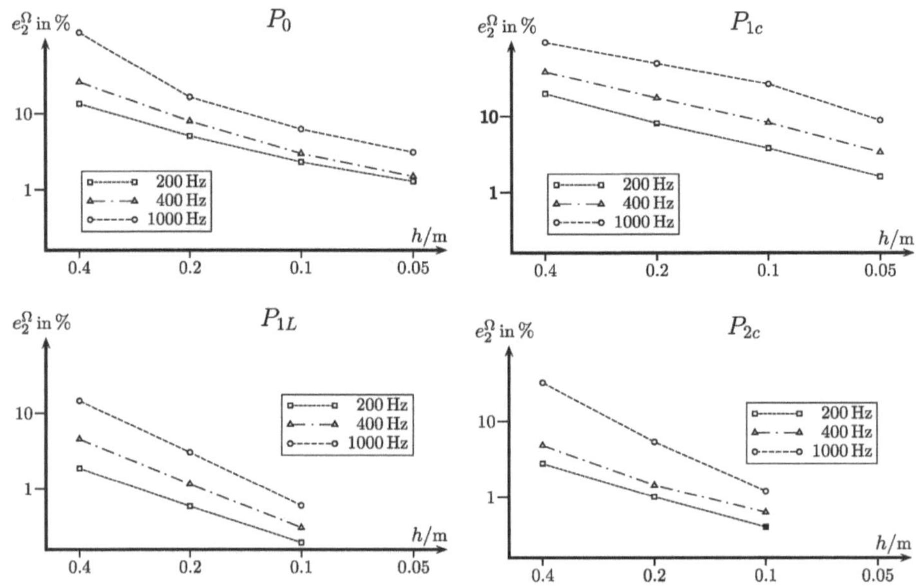

Fig. 11.15 Sedan cabin, error at ten internal points in terms of element size for different frequencies, subfigures show behaviour for different types of elements.

than continuous quadratic elements. Note, that this presumes that a solution using discontinuous quadratic elements provides the lowest error.

In the next step, we investigate the error in terms of location of nodal points. For that, we use the parameter α for quadrilaterals, Figure 11.2, and β for triangles, Figure 11.4. To use only one variable, it is assumed that $\alpha = 2\beta$. The model for $h \leq 0.2m$ consists of 476 quadrilaterals and 128 triangles. As we have seen for the duct's regular meshes either consisting of quadrilaterals or of triangles, optimal values of α and 2β are different. Figure 11.17 shows that significant improvements can be reported in the vicinity of $0.35 \leq \alpha \leq 0.42$ for linear and $\alpha \approx 0.2$ for quadratic elements.

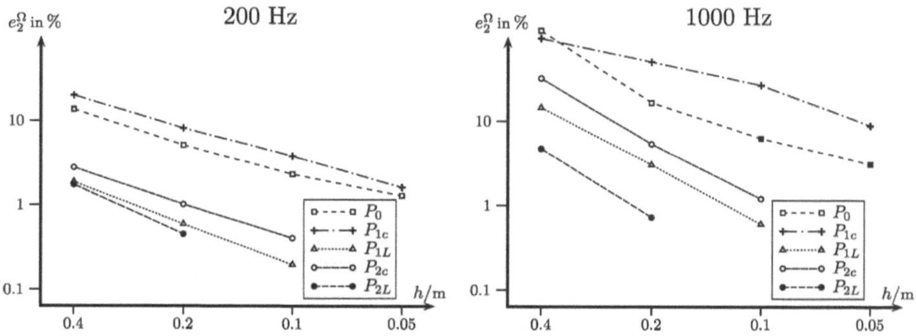

Fig. 11.16 Sedan cabin, error at ten internal points in terms of element size for different element types, subfigures show behaviour for different frequencies.

For linear elements, a discontinuity is observed at lower frequencies at about $\alpha = 0.35$. The author could not find a reasonable explanation for this phenomenon. Originally thought to be caused by the integration scheme, integration was performed extremely accurate, i.e. number of integration points for Gauss–Legendre quadrature controlled by distance between element and source point varied between 8×8 and 30×30 on single element. Polynom transformation [34] is applied for nearly singular integrals.

Another unexpected result is found for quadratic elements at 250 Hz. There, the error appears virtually insensitive with respect to the location of nodes.

11.6 Conclusions

This paper has reviewed the results of two former papers [18, 22]. Continuous and discontinuous Lagrangian boundary elements are compared for the collocation method.

It could be shown for these low order elements that discontinuous elements perform more efficient than continuous ones if nodal points are located at the zeros of the Legendre polynomials. To achieve very low errors, the use of discontinuous quadratic elements is recommended. If a larger error is accepted, discontinuous linear or even constant elements can be efficiently used. The most commonly used linear continuous elements seem to be the most unreliable and inefficient element type of those which have been tested here.

The author has presented tables for the long duct example. These tables show how many elements P_0, P_{1c} and P_{2c} are required to remain below a certain error. They confirm that six linear boundary elements correspond to 10...15 percent error whereas the same number of constant elements correspond to approximately 10 percent error. The same result is achieved when using approximately two P_{2c} elements per wavelength. However, the author recommends to use finer meshes in particular if the mesh is not regular and if it contains edges and corners with geometric or other singularities.

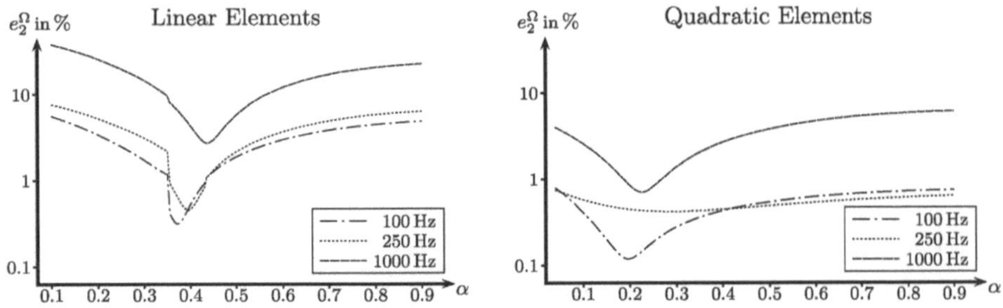

Fig. 11.17 Sedan cabin, error at ten internal points in terms of location of nodes.

In the example of the long duct and, in particular, in the example of the thin duct, we could not find any indication of a pollution effect in the frequency range of $0 \leq kl \leq 160\pi$. Consequently, a discretization rule of using a fixed number of elements per wavelength remains acceptable for the boundary element collocation method. When looking at the error in terms of location of the nodes on the elements, we found that this can influence the error by one or two orders of magnitude. This could even be shown for the irregular mesh of the sedan cabin compartment.

In addition to the remarkable gain in efficiency for discontinuous elements, it shall be mentioned that they possess a number of interesting features as

- they are well suited for adaptive mesh refinement,
- they fulfill C^1 continuity condition at collocation points which is required for the hypersingular formulation,
- they simplify construction of mesh dependent preconditioners for iterative solvers, and
- they are well suited for development of parallel codes.

Finally concluding, we strongly recommend the use of discontinuous boundary elements with nodes located at the zeros of the Legendre polynomials provided our problem is essentially related to inversion of the double layer potential operator. In case of mixed problems and when using the hypersingular operator, it is likely that other optimal locations of nodes will be found.

References

1. Araujo FC, Silva KI, Telles JCF (2006) Generic domain decomposition and iterative solvers for 3D BEM problems. International Journal for Numerical Methods in Engineering 68:448–472
2. Atkinson KE (1997) The Numerical Solution of Integral Equations of the Second Kind. Cambridge University Press, Cambridge
3. Branski A (1997) Model of an acoustic source with discontinuous optimal elements. Archives of Acoustics 22:383–395
4. Burton AJ, Miller GF (1971) The application of integral equation methods to the numerical solution of some exterior boundary–value problems. Proceedings of the Royal Society of London 323:201–220

5. Chandler G (1979) Superconvergence of numerical solutions to second kind integral equations. PhD Thesis, Australian National University, Canberra
6. Chatelin F, Lebbar R (1981) The iterated projection solution for the fredholm integral equation of the second kind. Journal of the Australian Mathematical Society, Series B 22:439–451
7. Ciskowski RD, Brebbia CA (eds) (1991) Boundary Elements in Acoustics. Computational Mechanics Publications and Elsevier Applied Science, Southampton–Boston
8. Estorff O von (ed) (2000) Boundary elements in acoustics: Advances and applications. WIT Press, Southampton
9. Florez WF, Power H (2001) Comparison between continuous and discontinuous boundary elements in the multidomain dual reciprocity method for the solution of the two–dimensional Navier–Stokes equation. Engineering Analysis with Boundary Elements 25:57–69
10. Hackbusch W (1995) Integral Equations: Theory and Numerical Treatment, Birkhäuser–Verlag, Basel–Boston–Berlin
11. Harari I, Grosh K, Hughes TJR, Malhotra M, Pinsky PM, Stewart JR, Thompson LL (1996) Recent development in finite element methods for structural acoustics. Archives of Computational Methods in Engineering 3:131–309
12. Harari I, Hughes TJR (1992) A cost comparison of boundary element and finite element methods for problems of time–harmonic acoustics. Computer Methods in Applied Mechanics and Engineering 97:77–102
13. Ihlenburg F (1998) Finite element analysis of acoustic scattering. Springer–Verlag, New York
14. Jerry AJ (1977) The Shannon sampling theorem – it various extensions and applications: A tutorial review. Proceedings of the IEEE 65:1565–1596
15. Kirkup SM (1998) The boundary element method in acoustics. Integrated Sound Software, Heptonstall
16. Makarov SN, Ochmann M (1998) An iterative solver for the Helmholtz integral equation for high frequency scattering. Journal of the Acoustical Society of America 103:742–750
17. Manolis GD, Banerjee PK (1986) Conforming versus non–conforming boundary elements in three–dimensional elastostatics. International Journal for Numerical Methods in Engineering 23:1885–1904
18. Marburg S (2002) Six elements per wavelength. Is that enough? Journal of Computational Acoustics 10:25–51
19. Marburg S (2005) Normal modes in external acoustics. Part I: Investigation of the one–dimensional duct problem. Acta Acustica united with Acustica 91:1063–1078
20. Marburg S, Amini S (2005) Cat's eye radiation with boundary elements: Comparative study on treatment of irregular frequencies. Journal of Computational Acoustics 13: 21–45
21. Marburg S, H.–J. Hardtke (1999) A study on the acoustic boundary admittance. Determination, results and consequences. Engineering Analysis with Boundary Elements 23:737–744
22. Marburg S, Schneider S (2003) Influence of element types on numeric error for acoustic boundary elements. Journal of Computational Acoustics 11:363–386
23. Marburg S, Schneider S (2003) Performance of iterative solvers for acoustic problems. Part I: Solvers and effect of diagonal preconditioning. Engineering Analysis with Boundary Elements 27:727–750
24. Meyer WL, Bell WA, Zinn BT, Stallybrass MP (1978) Boundary Integral Solutions of Three Dimensional Acoustic Radiation Problems. Journal of Sound and Vibration 59:245–262

25. Natalini B, Popov V (2007) On the optimal implementation of the boundary element dual reciprocity method–multi domain approach for 3d problems. Engineering Analysis with Boundary Elements 31:275–287
26. Patterson C, Sheikh (1984) Interelement continuity in the boundary element method. In: Brebbia CA (ed) Topics in Boundary Element Research Vol 1: Basic Principles and Applications. Chapter 6. Springer–Verlag, Berlin–Heidelberg–New York
27. Rego Silva JJd (1993) Acoustic and elastic wave scattering using boundary elements. Computational Mechanics Publications, Southampton–Boston
28. Rego Silva JJd, Wrobel LC, Telles JCF (1993) A new family of continuous/discontinuous three–dimensional boundary elements with application to acoustic wave propagation. International Journal for Numerical Methods in Engineering 36:1661–1679
29. Schmiechen P (1997) Travelling wave speed coincidence. PhD Thesis, Imperial College of Science, Technology and Medicine, University of London
30. Schneider S (2003) Application of fast methods for acoustic scattering and radiation problems. Journal of Computational Acoustics 11:387–401
31. Stroud AH (1971) Approximate calculation of multiple integrals. Prentice–Hall, Englewood Cliffs
32. Tadeu A, Antonio J (2000) Use of constant, linear and quadratic boundary elements in 3d wave diffraction analysis. Engineering Analysis with Boundary Elements 24:131–144
33. Tadeu AJB, Godinho L, Santos P (2001) Performance of the BEM solution in 3D acoustic wave scattering. Advances in Engineering Software 32:629–639
34. Telles JCF (1987) A self–adaptive coordinate transformation for efficient numerical evaluation of general boundary element integrals. International Journal for Numerical Methods in Engineering 24:959–973
35. Tsinopoulos SV, Agnantiaris JP, Polyzos D (1999) An advanced boundary element/fast Fourier transform axisymmetric formulation for acoustic radiation and wave scattering problems. Journal of the Acoustical Society of America 105:1517–1526
36. Wu TW (ed) (2000) Boundary element acoustics: Fundamentals and computer codes. WIT Press, Southampton
37. Wu TW, Cheng CYR (2003) Boundary element analysis of reactive mufflers and packed silencers with catalyst converters. Electronic Journal of Boundary Elements 1:218–235
38. Wu TW, Seybert AF, Wan GC (1991) On the numerical implementation of a cauchy principal value integral to insure a unique solution for acoustic radiation and scattering. Journal of the Acoustical Society of America 90:554–560
39. Wu TW, Wan GC (1992) Numerical modeling of acoustic radiation and scattering from thin bodies using a Cauchy principal integral equation. Journal of the Acoustical Society of America 92:2900–2906
40. Zhang XS, Ye TQ, Ge SL (2001) Several problems associated with discontinuous boundary elements. Chinese Journal of Computational Mechanics 18:331–334

12

Fast Solution Methods

Tetsuya Sakuma[1], Stefan Schneider[2], and Yosuke Yasuda[3]

[1] Graduate School of Frontier Sciences, University of Tokyo, 5–1–5 Kashiwanoha, Kashiwa, Chiba 277–8563, Japan
sakuma@k.u-tokyo.ac.jp
[2] CNRS/LMA, 31 chemin Joseph–Aiguier, 13402 Marseille Cedex 20, France
schneider@lma.cnrs-mrs.fr
[3] Graduate School of Frontier Sciences, University of Tokyo, 5–1–5 Kashiwanoha, Kashiwa, Chiba 277–8563, Japan
yyasuda@k.u-tokyo.ac.jp

Summary. The standard boundary element method applied to the time harmonic Helmholtz equation yields a numerical method with $\mathcal{O}(N^3)$ complexity when using a direct solution of the fully populated system of linear equations. Strategies to reduce this complexity are discussed in this paper. The $\mathcal{O}(N^3)$ complexity issuing from the direct solution is first reduced to $\mathcal{O}(N^2)$ by using iterative solvers. Krylov subspace methods as well as strategies of preconditioning are reviewed. Based on numerical examples the influence of different parameters on the convergence behavior of the iterative solvers is investigated. It is shown that preconditioned Krylov subspace methods yields a boundary element method of $\mathcal{O}(N^2)$ complexity. A further advantage of these iterative solvers is that they do not require the dense matrix to be set up. Only matrix–vector products need to be evaluated which can be done efficiently using a multilevel fast multipole method. Based on real life problems it is shown that the computational complexity of the boundary element method can be reduced to $\mathcal{O}(N \log^2 N)$ for a problem with N unknowns.

12.1 Introduction

The standard boundary element method, that is the calculation and storage of the dense matrices $\boldsymbol{H}$ and $\boldsymbol{G}$ and the direct solution of the system of linear equations, requires $\mathcal{O}(N^2)$ memory locations and $\mathcal{O}(N^3)$ arithmetic operations. This makes the method hardly applicable to the solution of problems of practical interest. However, the fact that the BEM requires only a discretization of the boundary Γ of the fluid domain Ω is advantageous when solving radiation or scattering problems involving complex shaped radiators or scatterers.

The following chapter discusses possibilities to increase the efficiency of the BEM such that it will be applicable to problems of practical interest. Two directions are issued. Firstly, the iterative solution of the system of linear equations and secondly the efficient evaluation of the product of the matrices $\boldsymbol{H}$ and $\boldsymbol{G}$ with a vector as needed within the iterative solution process.

12.2 Iterative Solution

Iterative methods have been widely used for solving linear systems in various fields. For the linear system $\boldsymbol{Ax} = \boldsymbol{b}$, an iterative method gives successive approximations $\boldsymbol{x}_i$ at each step i from an initial approximation $\boldsymbol{x}_0$ in a systematic way, until the residual norm $||\boldsymbol{r}_i|| = ||\boldsymbol{b} - \boldsymbol{Ax}_i||$ sufficiently decreases. If the iterative process rapidly converges, iterative methods are much faster than direct methods, such as Gaussian elimination.

Iterative methods are classified into two groups: stationary methods and nonstationary methods. In the former methods, the coefficients for renewal of the approximation $\boldsymbol{x}_i$ do not depend on the iteration count. Jacobi, Gauss–Seidel, Successive Over–Relaxation (SOR), and Symmetric Successive Over–Relaxation (SSOR) methods, etc. belong to this group. The latter methods use iteration–dependent coefficients so that the computations can involve renewal information obtained at each iteration. The Krylov subspace methods are most popular in this group.

The Krylov subspace methods, which were selected as one of the top 10 algorithms of the 20th century in *SIAM News* [19], are very effective for large systems of linear equations. A Krylov method successively gives the approximations $\boldsymbol{x}_i$ in the process of increasing the dimensionality of the Krylov subspace, which is spanned by the Krylov sequence generated by the initial residual $\boldsymbol{r}_0$ and the system matrix $\boldsymbol{A}$. The Krylov subspace methods become more powerful when they are used with preconditioning technique for improvement of iterative convergence. One special feature of the methods is that the system matrix $\boldsymbol{A}$ is not explicitly necessary, and only the products of the matrix $\boldsymbol{A}$ with vectors (matrix–vector products) are required. Therefore, the Krylov methods are applicable with fast evaluation of the products without generating $\boldsymbol{A}$.

In the field of computational acoustics, there exist many examples using the Krylov methods: FE [45, 46] and BE [43, 55] analyses for interior problems, BE analyses for exterior problems [6, 54, 57, 60], structure–acoustic problems [17], etc.

This chapter describes the outline of the Krylov subspace iterative methods and its application to the BEM. For the details about iterative methods, see References [8, 10, 53, 66], for example.

12.2.1 The Krylov Subspace Methods

Consider that the linear system of equations

$$\boldsymbol{Ax} = \boldsymbol{b} \tag{12.1}$$

is solved with an iterative method. The initial residual $\boldsymbol{r}_0$ is expressed as:

$$\boldsymbol{r}_0 = \boldsymbol{b} - \boldsymbol{Ax}_0, \tag{12.2}$$

where $\boldsymbol{x}_0$ denotes the initial approximate solution of $\boldsymbol{x}$. The Krylov subspace methods are iterative methods in which the approximate solution $\boldsymbol{x}_n$ is taken as satisfying the next condition,

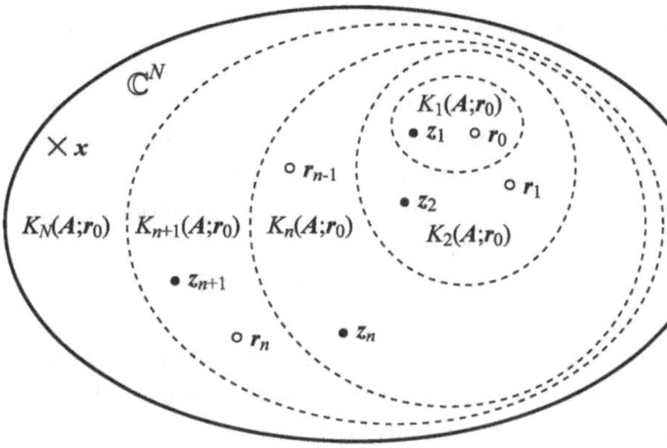

Fig. 12.1 Relationship among subspaces.

$$\boldsymbol{x}_n = \boldsymbol{x}_0 + \boldsymbol{z}_n, \quad \boldsymbol{z}_n \in K_n(\boldsymbol{A}; \boldsymbol{r}_0), \tag{12.3}$$

where $K_n(\boldsymbol{A}; \boldsymbol{r}_0)$ denotes the n–dimensional Krylov subspace spanned by a Krylov sequence $\{\boldsymbol{A}^i \boldsymbol{r}_0\}_{i \geq 0}$:

$$K_n(\boldsymbol{A}; \boldsymbol{r}_0) = \text{span}\{\boldsymbol{r}_0, \boldsymbol{A}\boldsymbol{r}_0, \ldots, \boldsymbol{A}^{n-1}\boldsymbol{r}_0\}. \tag{12.4}$$

The iteration residual $\boldsymbol{r}_n$ is expressed as follows:

$$\boldsymbol{r}_n = \boldsymbol{b} - \boldsymbol{A}\boldsymbol{x}_n = \boldsymbol{r}_0 - \boldsymbol{A}\boldsymbol{z}_n \in K_{n+1}(\boldsymbol{A}; \boldsymbol{r}_0). \tag{12.5}$$

Figure 12.1 illustrates relationship among subspaces. Extension of the subspace with each iteration enables one to express better approximations of $\boldsymbol{x}_n$.

Krylov subspace methods require an approach for identification of the approximate solution $\boldsymbol{x}_n$ as follows.

- The minimum norm residual approach: identify $\boldsymbol{r}_n$ for which the norm $||\boldsymbol{b} - \boldsymbol{A}\boldsymbol{x}||$ is minimal over $K_n(\boldsymbol{A}; \boldsymbol{r}_0)$.

$$||\boldsymbol{r}_n|| = \min_{\boldsymbol{x} \in \boldsymbol{x}_0 + K_n(\boldsymbol{A}; \boldsymbol{r}_0)} ||\boldsymbol{b} - \boldsymbol{A}\boldsymbol{x}||. \tag{12.6}$$

- The Ritz–Galerkin approach: identify $\boldsymbol{r}_n$ that $\boldsymbol{r}_n$ is orthogonal to the current subspace.

$$\boldsymbol{w}^H \boldsymbol{r}_n = 0, \quad \boldsymbol{w} \in K_n(\boldsymbol{A}; \boldsymbol{r}_0), \tag{12.7}$$

  where $\boldsymbol{w}^H$ denotes the conjugate transpose of $\boldsymbol{w}$.

Classification

Lanczos, Arnoldi and Bi–Lanczos Types

In the Krylov subspace methods, an orthogonal basis for the Krylov subspace is required for stability of computation. The Krylov subspace methods are classified by the algorithms for production of an orthogonal basis and by the approaches for identification of the approximate solution $\boldsymbol{x}_n$ mentioned above.

If $\boldsymbol{A}$ is a Hermitian matrix, one can compute an orthogonal basis for the Krylov subspace $K_n(\boldsymbol{A}; \boldsymbol{r}_0)$ based on only three–term reccurence relations. This process is called the Lanczos process. The methods based on the Lanczos process have the advantage that the operation count and the required memory per iteration step do not increase with the number of iteration.

If $\boldsymbol{A}$ is a non–Hermitian matrix such as that obtained with the BEM, one cannot compute an orthogonal basis with short reccurence relations. One have to keep all elements of the orthogonal basis computed in the iterative process, to obtain the next element. This process is called the Arnoldi process. The methods based on the Arnoldi process can be applied any regular matrices, whereas they have a major disadvantage that the operation count and required memory per iteration step rise linearly with the number of iteration.

There is an inexpensive process for computation of bases for non–Hermitian matrices, where bi–orthogonal bases are computed instead of an orthogonal basis. In addition to the Krylov subspace $K_n(\boldsymbol{A}; \boldsymbol{r}_0)$, one can consider another Krylov subspace $K_n(\boldsymbol{A}^H; \boldsymbol{r}_0^*)$, and construct bi–orthogonal bases $\{\boldsymbol{v}_i\}$ and $\{\boldsymbol{u}_j\}$ for the two subspaces, satisfying the following relation

$$\boldsymbol{v}_i^H \boldsymbol{u}_j = 0, \quad i \neq j, \tag{12.8}$$

where $\boldsymbol{r}_n^*$ denotes the shadow residual, satisfying $\boldsymbol{r}_0^{*H} \boldsymbol{r}_0 \neq 0$. This process is called the Bi–Lanczos process, where one can compute the bi–orthogonal bases with only three–term reccurence relations, like the Lanczos process. The main drawback of the methods based on this process is the possibility of break down in the iterative process, which does not occur in methods based on the Lanczos process. Since the above approaches Equations (12.6) and (12.7) for identification of $\boldsymbol{x}_n$ cannot be directly used with the Bi–Lanczos algorithm, the following approach is used instead.

- The Petrov–Galerkin approach: find $\boldsymbol{r}_n$ so that $\boldsymbol{r}_n$ is orthogonal to the other subspace $K_n(\boldsymbol{A}^H; \boldsymbol{r}_0^*)$

$$\boldsymbol{w}^H \boldsymbol{r}_n = 0, \quad \boldsymbol{w} \in K_n(\boldsymbol{A}^H; \boldsymbol{r}_0^*). \tag{12.9}$$

Table 12.1 shows popular Krylov subspace methods classified by the process for computation of the basis and the approach for identification of the approximate solution.

There is a short description of the representative methods below.

Conjugate Gradient (CG) [38] : The most basic method based on the Lanczos process. This method is used for linear systems with positive–definite Hermitian matrices, and extremely effective. This cannot be directly used for the BEM.

Generalized Minimal Residual (GMRes) [52] : This method is based on the Arnoldi process and known as one of the robust methods. This is applicable to the linear systems with non–Hermitian matrices and often used for sound field analyses using the

Table 12.1 Classification of Krylov subspace methods.

	Lanczos	Arnoldi	Bi–Lanczos
Ritz–Galerkin	CG	FOM	–
Minimum norm residual	MinRes	GMRes	–
Petrov–Galerkin	–	–	BiCG

BEM. The major drawback is that the amount of operation and memory required per iteration step rise linearly with the number of iteration. Thus, one needs to restart the process every l iterations to overcome this limitation (GMRes(l)). It is quite difficult to choose an appropriate value for l, since GMRes may converge slowly, or completely fail to converge if l is too small.

Bi–Conjugate Gradient (BiCG) [26, 41] : This method is based on the Bi–Lanczos process and can be applied to the systems with non–Hermitian matrices. Unlike the methods based on the Arnoldi process, the amount of operation and required memory per iteration step do not rise with the number of iteration. However, BiCG requires not only the original system matrix but also its conjugate transpose. Another drawback is that the method might be break down, or converge very irregular.

BiCG–based Methods

BiCG is an alternative of the methods based on the Arnoldi process in efficiency, whereas it has some disadvantages mentioned above. A lot of variants of BiCG, here we call the BiCG–based methods, have been developed to improve these disadvantages. All of the methods shortly described below do not need the conjugate transpose of the system matrix. The convergence behavior of the methods of this group including BiCG is hardly known.

Conjugate Gradient Squared (CGS) [63] : CGS often converges or diverges much faster than BiCG, because the residual polynomial of CGS is the square of that of BiCG. The convergence behavior may be irregular for large problems, due to round–off errors.

Biconjugate Gradient Stabilized (BiCGStab) [65] : This method has smoother convergence behavior than CGS, keeping its rapid convergence. This is regarded as the product of BiCG and GMRes(1).

Generalized Product Type of Biconjugate Gradient (GPBiCG) [75] : This method is derived from generalization of BiCG–based methods. This is effective when the eigenvalues of the system matrix are complex.

BiCGStab2 [33], BiCGStab(l) [61] and TFQMR [28], etc. are also well known.

Methods Based on Normal Equations

There is another class of Krylov subspace methods for non–Hermitian matrices. These methods are based on the application of the methods for Hermitian matrices to the normal equations. Popular methods of this class are CGNE and CGNR, in which CG is applied to the systems $(\boldsymbol{A}^H\boldsymbol{A})\boldsymbol{x} = \boldsymbol{A}^H\boldsymbol{b}$ and $(\boldsymbol{A}\boldsymbol{A}^H)(\boldsymbol{A}^{-H}\boldsymbol{x}) = \boldsymbol{b}$, respectively. These methods have a disadvantage that the matrices of normal equations are more ill–conditioned than the original matrices. The convergence may be slow.

12.2.2 Preconditioning

Preconditioning is a well known technique for improvement of convergence when one attempts to solve linear systems using iterative methods. A matrix $\boldsymbol{M} = \boldsymbol{M}_1\boldsymbol{M}_2$, which approximates the system matrix $\boldsymbol{A}$ in some way, transforms the original system into the following system with more favorable properties for iterative solution:

$$\boldsymbol{M}_1^{-1}\boldsymbol{A}\boldsymbol{M}_2^{-1}(\boldsymbol{M}_2\boldsymbol{x}) = \boldsymbol{M}_1^{-1}\boldsymbol{b}. \tag{12.10}$$

In general, a good preconditioner $\boldsymbol{M}$ should meet the following requirements: the preconditioned system should be easy to solve, and the preconditioner should be cheap to construct and apply. For the detail review on preconditioning techniques, an intensive review paper [10] can be available. We describe below some preconditioning techniques, which can be used for the BEM.

Diagonal Preconditioning

The preconditioner $\boldsymbol{M}$ consists of just the diagonal of the matrix $\boldsymbol{A}$. This preconditioning is applicable without extra memory and time, and easy to apply to almost all kinds of the BEM. This preconditioning was reported to be effective for the BEM for some fields, for example, the thermal and elastic BEM [47], whereas reported not so effective for the acoustic BEM [43, 72].

Incomplete Factorizations

Preconditioners based on incomplete factorizations of the system matrix, such as incomplete LU (ILU) and incomplete Cholesky (IC) factorizations, have been widely used. In ILU preconditioning, the system matrix $\boldsymbol{A}$ is incompletely factorized as a preconditioner $\boldsymbol{M} = \boldsymbol{L}\boldsymbol{U} \approx \boldsymbol{A}$, where $\boldsymbol{L}$ and $\boldsymbol{U}$ are lower and upper triangular matrices, respectively. If the system matrix is Hermitian and positive definite, IC factorization can be applied, which is a special case of ILU factorizations.

It is better to make ILU factorization near to complete factorization for good convergence, but resulting in high cost. Actually, complete LU factorization is equivalent to Gaussian elimination, and $\boldsymbol{L}$ and $\boldsymbol{U}$ obtained by complete factorization generally have many fill–in (nonzero) entries. There are some dropping strategies for discarding fill–ins to make ILU preconditioners. ILU(k) allows fill–ins only to a certain level k. ILU(0) corresponds to allowance of fill–ins only at positions for which the corresponding entries of $\boldsymbol{A}$ are nonzero. ILU(k) has some disadvantages; for example, the computational cost rapidly increases with k, and the storage and computational cost cannot be predicted in advance. ILUT(τ, p) [51] is a threshold–based ILU factorization. This employs a dual dropping strategy and more powerful than ILU(k); the threshold τ for dropping elements having small magnitude and the maximum number p of fill–in elements in each row of $\boldsymbol{L}$ and $\boldsymbol{U}$ are used to control the computational costs for factorization and application of the preconditioner.

ILU preconditioners are usually applied to sparse linear systems. They are very expensive when this is directly applied to the system with a dense system matrix, typically obtained with the BEM. Some techniques have been proposed to overcome this limitation. Please see 12.2.3.

Sparse Approximate Inverses (SPAI)

The basic idea of this class is that $\boldsymbol{M}^{-1} \approx \boldsymbol{A}^{-1}$ is explicitly computed and used as a preconditioner. The computational cost for naive construction and application of $\boldsymbol{M}^{-1}$ is very high, since $\boldsymbol{M}^{-1}$ is usually dense. In SPAI preconditioning, a *sparse* matrix $\boldsymbol{M}^{-1} \approx \boldsymbol{A}^{-1}$ is computed as the solution of the constrained minimization problem

$$\min_{\boldsymbol{M}^{-1} \in \mathcal{S}} ||\boldsymbol{A}\boldsymbol{M}^{-1} - \boldsymbol{I}||_F, \tag{12.11}$$

where $\mathcal{S}$ is a set of sparse matrices and $||\cdot||_F$ denotes the Frobenius norm of a matrix. This results in n independent least–squares problems for each columns of $\boldsymbol{M}^{-1}$. The computational cost mainly depends on how to give the sparsity pattern $\mathcal{S}$. To use this for dense linear systems, $\mathcal{S}$ must be sufficiently sparse.

12.2.3 Application to the BEM

The Krylov subspace methods require repeated calculation of matrix–vector products from the system matrix $\boldsymbol{A}$ and vectors, which occupies quite a large amount of operations in the iterative process. If $\boldsymbol{A}$ is a $N \times N$ dense matrix as generally obtained with the BEM, the operation count for a matrix–vector product is $\mathcal{O}(N^2)$. Thus, if the number of iteration is sufficiently smaller than N, the Krylov methods remarkably reduce the total operation count for solving the system compared with direct methods, which require $\mathcal{O}(N^3)$ operations. In addition, more reduction of the operation count can be achieved by using methods for efficient evaluation of the matrix–vector products, such as the fast multipole method (FMM). The required memory is also reduced sharply by using such methods, because it is not necessary to store the entire system matrix $\boldsymbol{A}$. For more details on the efficient evaluation of matrix–vector products, please see Section 12.3.

Iterative Methods for the BEM

The BEM gives a linear system with a non–Hermitian matrix, to which the iterative methods based on the Arnoldi or Bi–Lanczos process and based on normal equations are applicable. For an advanced BEM that does not store the entire system matrix, it is difficult to apply the methods that require the conjugate transpose of the system matrix, such as BiCG, QMR and the methods for the normal equations.

Preconditioners for the BEM

There are not many preconditioners effective for non–Hermitian dense matrices, because preconditioners usually have been proposed for the sparse matrices. For the preconditioners for dense linear systems, see References [14,15,67,70] and the references therein. Moreover, fewer preconditioners are applicable to the advanced BEM that does not store the entire system matrix.

One class of preconditioners for the BEM is based on the splitting of the boundary integral operators [5,14]. The system matrix $\boldsymbol{A}$ is splitted into two matrices $\boldsymbol{A}_{\text{near}}$ and $\boldsymbol{A}_{\text{far}}$, the former of which is sparse and composed of the influence parts between near elements. ILU factorization is done for the matrix $\boldsymbol{A}_{\text{near}}$, and the results are used for the preconditioning to the original system with $\boldsymbol{A}$. This technique is very suitable for the advanced BEM that does not store the entire $\boldsymbol{A}$, because this type of the BEM requires to generate $\boldsymbol{A}_{\text{near}}$ in the same manner as the standard BEM. This technique has been proved to be very effective through some investigations, where practical acoustic problems were calculated with the advanced BEM using the regular grid method (RGM) [59] and the FMM [59,72]. For the standard BEM, a simple way of making a substitute of $\boldsymbol{A}_{\text{near}}$ is to take the tri–diagonal band of $\boldsymbol{A}$ together with the anti–diagonal corner elements a_{1n} and a_{n1} [5].

12.2.4 Convergence Behavior for the BEM

The rate at which an iterative method converges relates directly to its computational time. Regarding non–Hermitian matrices, the convergence of iterative methods is not clear, and depends greatly on the properties of the matrices [8].

There are some studies on the convergence for the acoustic BEM [2–5,14–17,37, 43,44,59,72,74]. Many of them are especially for exterior problems with Burton–Miller formulation [3,16,37,59,74], because iterative methods do not converge well in this case. Preconditioning techniques were mainly discussed in References [5,14–16,37]. Detailed comparison through numerical experiments has been done for a variety of practical problems [43,59,72].

The following refers to general tendencies on the convergence, giving the examples using the collocation BEM with constant elements. The multilevel fast multipole algorithm (MLFMA) is used for efficient evaluation of matrix–vector products. For more details on MLFMA, see Section 12.3. The following equation is used as a stopping criterion for the linear system,

$$\frac{|\boldsymbol{r}_i|}{|\boldsymbol{b}|} = \frac{|\boldsymbol{b} - \boldsymbol{A}\boldsymbol{x}_i|}{|\boldsymbol{b}|} \leq \varepsilon, \tag{12.12}$$

where $|\cdot|$ is the Euclidean norm. $\varepsilon = 10^{-6}$ is used, unless noted otherwise. Instead of the number of iteration, the number of matrix–vector products is counted. This is because the operation of matrix–vector products accounts for the greater part of the iterative process, and iterative methods have different numbers of matrix–vector products per iteration. For example, CGS requires two matrix–vector multiplications per iteration, whereas GMRes requires one.

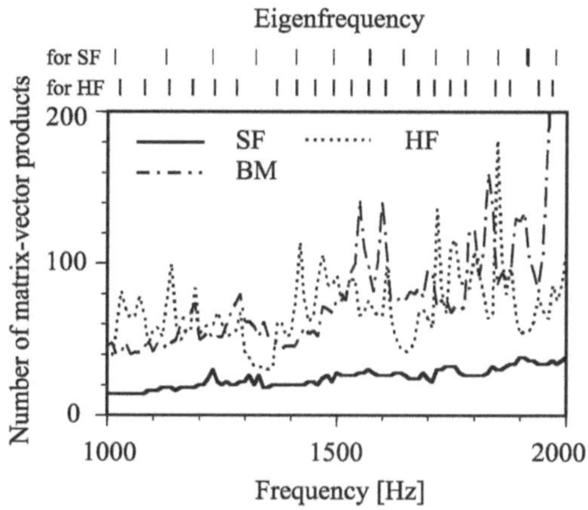

Fig. 12.2 Effect of different kinds of formulation on the number of matrix–vector products for an exterior problem with a vibrating cube. The iterative method is unpreconditioned BiCGStab, DOF is 6,144, and $\varepsilon = 10^{-3}$.

Effect of Formulation

The convergence behavior is affected by the formulation of the BEM: basic formulation (singular formulation: SF), normal derivative formulation (hypersingular formulation: HF), and formulation by Burton and Miller (BM) [13] to avoid fictitious eigenfrequency difficulties. Figure 12.2 shows the history of the residual norms for an exterior problem with a vibrating cube. The fictitious eigenfrequencies for SF and for HF are also shown in the upper part of the figure. The convergence with HF and with BM is slower than that with SF at all frequencies. This is because the matrices obtained with HF generally have eigenvalues which are not clustered compared to those with SF (i.e., HF matrices are ill–conditioned), and BM matrices inherit the ill condition from HF matrices.

It has been reported that the hypersingular equation can be reduced to weakly singular one by using the high–order Galerkin BEM, resulting in improvement of the matrix condition and the convergence behavior with BM [37, 74].

Effect of Boundary Shapes

The shape of boundary can greatly affect the convergence behavior. There is a tendency that the convergence is more rapid with smoother surface [43, 72]. Figure 12.4 shows the history of the residual norms for two interior models (Figure 12.3), which have nearly the same DOF and ratio of the element width to the wavelength. Slower convergence is seen for the problem with a complex shape (auditorium) than with a simple shape (cube), when both the problems are under the same boundary condition.

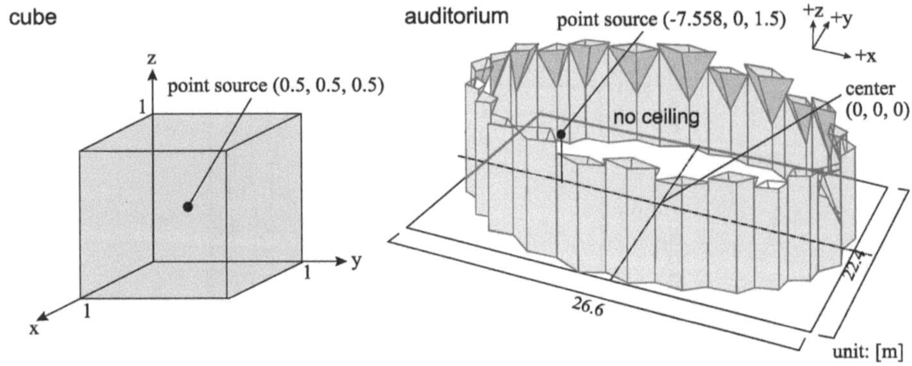

Fig. 12.3 Analysis models for interior problems: a cube (simple shape) and an auditorium (complex shape).

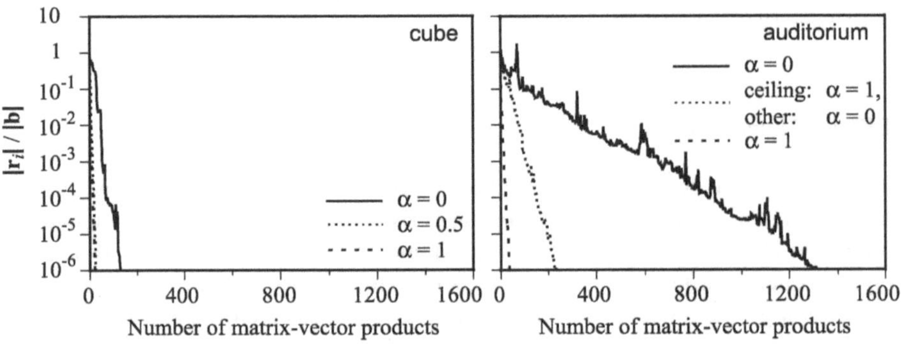

Fig. 12.4 Effect of shapes and boundary conditions on the iteration residual. α is the absorption coefficient on the surfaces. The formulation is SF, the iterative method is unpreconditioned GPBiCG, DOF is 24,576 (cube) and 24,514 (auditorium).

Effect of Boundary Conditions

The larger the absorption coefficient α is, the faster the convergence is [43, 72], as is seen in Figure 12.4. This is the general tendency, independent of calculated problems and iterative methods. As shown in Figure 12.5, a little absorption greatly improves the convergence, compared to the case of perfectly rigid boundaries.

Effect of Preconditioning

The following refers to the effect of ILUT(τ, p) [51], which is one of the powerful preconditioners, on the convergence behavior with Burton–Miller formulation. In particular, it focuses on the effect of the parameter p, which greatly affects memory

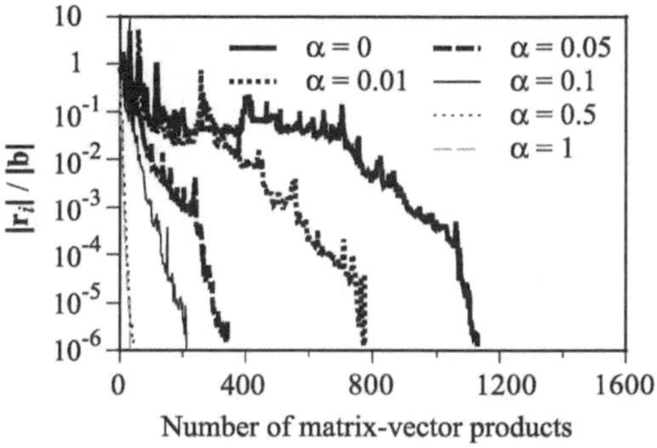

Fig. 12.5 Effect of absorption coefficients α on the iteration residual for an auditorium at 63 Hz. The formulation is SF, the iterative method is unpreconditioned BiCGStab, and DOF is 6,110.

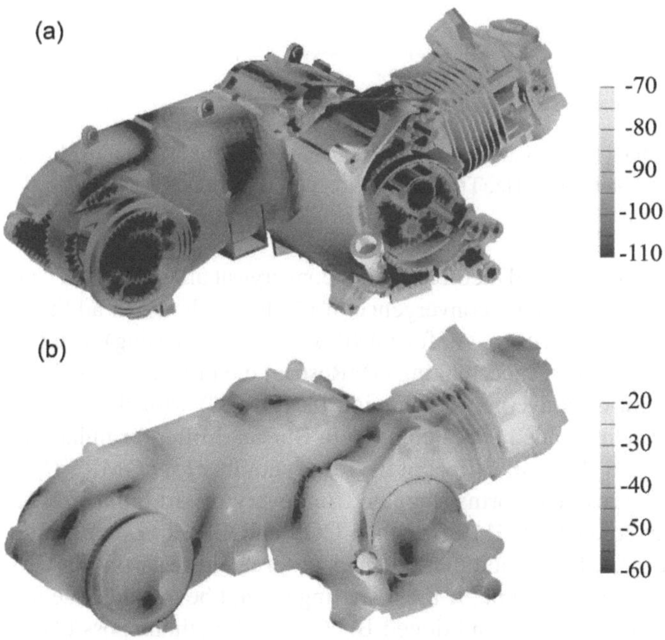

Fig. 12.6 An engine model: (a) relative vibration velocity level distribution, and (b) relative SPL distribution at 1,977 Hz. The formulation is BM, and DOF is 42,152.

requirements. This preconditioner is referred to as ILUT(p) with $\tau = 10^{-5}$ in the below.

Figure 12.7 shows the history of the residual norms with three kinds of iterative methods, for an exterior problem with a vibrating engine, Figure 12.6. In this case,

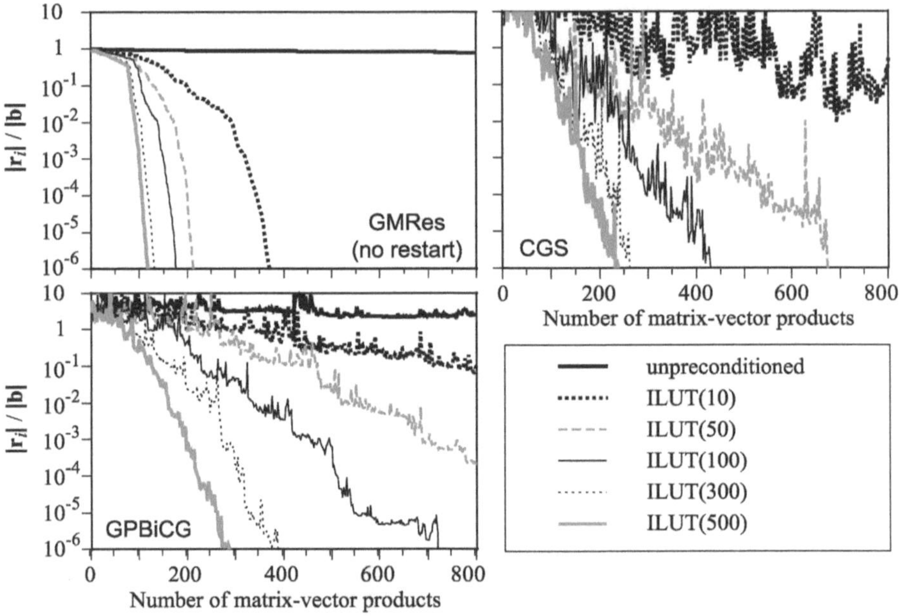

Fig. 12.7 Effect of the number of fill–in elements p for ILUT(p) on the iteration residual for the engine problem at 3 kHz. The formulation is BM, and DOF is 42,152.

all the unpreconditioned methods do not converge at all, while ILUT preconditioning remarkably improves the convergence of all the methods. In addition, the effect of ILUT(p) increases with p. As for GMRes(∞) (no restarting), even small p greatly improves the convergence, making GMRes(∞) the fastest among the methods. From the viewpoint of convergence behavior, GMRes(∞) with ILUT preconditioning is generally recommended for calculation in Burton–Miller formulation.

Figure 12.8 shows the effect of the restart number l for GMRes(l) on the iteration residual. The residual norms stagnate from every restart point [59, 72], resulting in slower convergence of GMRes(l) than CGS and GPBiCG, depending on ILUT(p), see Figure 12.7. It is stated that one had better avoid restarting in GMRes(l) for reducing computational time. If the restarting cannot be avoided due to the restriction of memory storage, preconditioned BiCG–based methods possibly converge faster than GMRes(l). For fast convergence of GMRes(l) with ILUT(p), one should consider the trade–off between the two parameters l and p in the restriction of memory storage.

12.3 Efficient Evaluation of the Matrix–Vector Product

Within the iterative solution of the dense linear system arising from a solution of the Helmholtz equation using Boundary Element Method, namely, when calculating

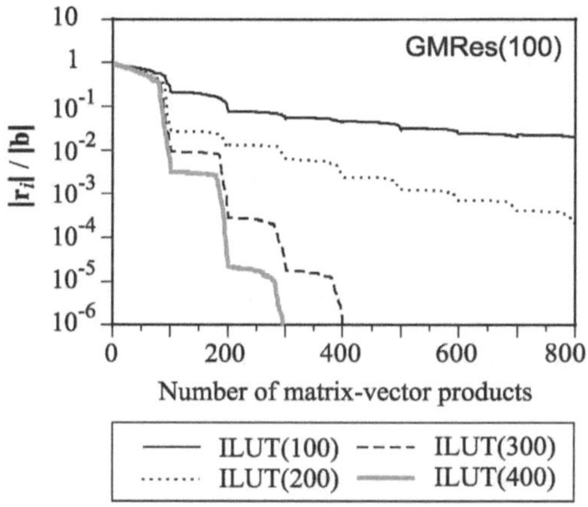

Fig. 12.8 Effect of the restart number l for GMRes(l) on the iteration residual for the engine problem at 3 kHz. The formulation is BM, $l = 100$, and DOF is 42,152.

the matrix–vector product $\boldsymbol{z} = (\boldsymbol{I} - \boldsymbol{A})\boldsymbol{u}$ one of the following quantities has to be evaluated

$$z_j = \int_\Gamma \varphi_j(\mathbf{x})\varphi(\mathbf{x})\, d\Gamma{\cdot}\boldsymbol{u} - \int_\Gamma \varphi_j(\mathbf{y}) \int_\Gamma k(\mathbf{x}-\mathbf{y})\varphi(\mathbf{x})\, d\Gamma\, d\Gamma{\cdot}\boldsymbol{u} \quad (12.13)$$

when using a Galerkin method,

$$z_j = u_j - \int_\Gamma k(\mathbf{x}-\mathbf{y}_j)\varphi(\mathbf{x})\, d\Gamma{\cdot}\boldsymbol{u} \quad (12.14)$$

when using a Collocation method or

$$z_j = u_j - w_i k(\mathbf{x}_i - \mathbf{y}_j)\varphi(\mathbf{x}_i){\cdot}\boldsymbol{u} \quad (12.15)$$

when using a Nyström method for all $j = 1, \ldots, N$ with N being the number of unknowns of the given problem. Due to the fact that the kernel

$$k(\mathbf{x}-\mathbf{y}) = \alpha G(\mathbf{x},\mathbf{y}) + \beta \nabla_{\mathbf{x}} G(\mathbf{x},\mathbf{y}){\cdot}\nu_{\mathbf{x}} \quad (12.16)$$

with the fundamental solution $G(\mathbf{x},\mathbf{y})$ has non–local support this operation costs $\mathcal{O}(N^2)$ arithmetic operations.

Several possibilities to achieve a reduction of the complexity have been proposed in the literature.

One approach consists in using a suitable wavelet basis together with a drooping strategy to sparsify the dense matrix $\boldsymbol{A}$ [12,21,27,39]. The most challenging part of this approach is the direct evaluation of the matrix $\boldsymbol{A}$ in the wavelet basis.

A second approach consists in a low rank approximation of $\boldsymbol{A}$ together with efficient $\mathcal{H}$–matrix techniques [9,35]. Such an approach is based only on smoothness properties of the kernel function without necessarily knowing it explicitly. This is a very important point when considering numerical implementation. A possible change of the kernel function, as long as the smoothness properties are satisfied, does not require a complete recoding of the numerical algorithms.

A third approach is based on a suitable approximation of a specific kernel function in order to separate the $\mathbf{x}$ and $\mathbf{y}$ dependency of the kernel function. For two points $\mathbf{x}$ and $\mathbf{y}$ with $|\mathbf{x}-\mathbf{y}| > \varrho > 0$ the kernel Equation (12.16) is a smooth function, thus an approximation $k(\mathbf{x},\mathbf{y}) \approx h(\mathbf{y}-\mathbf{y}_c)\mu(\mathbf{y}_c-\mathbf{x}_c)g(\mathbf{x}-\mathbf{x}_c)$ can be used to replace the original kernel function. Such an approximation is in general linked to a specific kernel $k(\cdot)$. Hence, changing the kernel function in general requires a complete recoding of the numerical algorithms. The use of such a kernel approximation forms the basis of methods like regular grid method [11], panel clustering [36,56] or the fast multipole method [18,31]. The latter method, especially its multilevel variant, seems to be the most widely accepted method as it covers the very low to high frequency range. An overview of the state–of–the–art of the fast multipole method can be found in the textbook [32].

12.3.1 Basic Concept of the Fast Multipole Method

The usage of the Fast Multipole Method for the solution of boundary value problems for the Laplace equation goes back to Greengard and Rohklin [31]. The application of this method for two–dimensional scattering problems can be found in [49]. It was again Rohklin [50] who extended it to the three–dimensional case. The Fast Multipole Method for the three–dimensional Helmholtz equation will be derived related to the work done by Anderson [7] and Rahola [48] which is often referred to as the "fast multipole method without multipoles".

The Fast Multipole Method was first used to calculate particle interactions. From this point of view (integration replaced by summation) Equation (12.13) to Equation (12.15) can be regarded as the communication of every point $\mathbf{y}_j$ on the surface with all other points $\mathbf{x}_i$ on the surface, hence the direct evaluation of these equations has $\mathcal{O}(N^2)$ complexity. Avoiding the evaluation of the interaction of all $(\mathbf{y}_j, \mathbf{x}_i)$ pairs will reduce the complexity of the algorithm. This can be achieved by splitting the direct path $\mathbf{y}_j - \mathbf{x}_i$. Therefore sets of points $\mathbf{z}_i$ are introduced where the information of assigned points $\mathbf{x}_i$ is aggregated. Thereafter, only the interaction between aggregation points is evaluated. Information is redistributed from the points $\mathbf{z}_j$ to the associated points $\mathbf{y}_j$ on the surface Γ after all interactions have been calculated. This situation is depict in Figure 12.9.

The idea behind the Fast Multipole Method is to replace (approximate) the kernel function $k(\mathbf{x}-\mathbf{y})$ for all well separated point $\mathbf{x}$ and $\mathbf{y}$ with $|\mathbf{x}-\mathbf{y}| > \varrho$ by

$$\begin{aligned} k(\mathbf{x}-\mathbf{y}) &\approx g(\mathbf{y}-\mathbf{z}_2)\mu(\mathbf{z}_2-\mathbf{z}_1)h(\mathbf{z}_1-\mathbf{x}) \\ &\approx g(\mathbf{c}_2)\mu(\mathbf{a})h(\mathbf{c}_1)\,. \end{aligned} \tag{12.17}$$

Such a factorization of the kernel can be obtained using the truncation of the series expansion of the fundamental solution of the Helmholtz equation. The use of a kernel Equation (12.17) instead of the original kernel Equation (12.16) leads to an approximate factorization of the matrix $\boldsymbol{A}$ in Equation (12.13) to Equation (12.15) such that

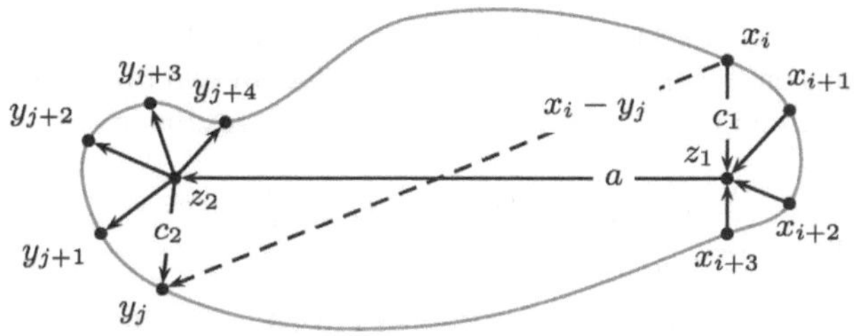

Fig. 12.9 Splitting of the $\mathbf{y}_j - \mathbf{x}_i$ path such that $\mathbf{y}_j - \mathbf{x}_i = (\mathbf{y}_j - \mathbf{z}_2) + (\mathbf{z}_2 - \mathbf{z}_1) + (\mathbf{z}_1 - \mathbf{x}_i)$.

$\boldsymbol{A} \approx (\boldsymbol{I} - \boldsymbol{A}_{\text{near}} - \boldsymbol{V}\boldsymbol{B}\boldsymbol{W})$ where $\boldsymbol{A}_{\text{near}}$, $\boldsymbol{V}$, $\boldsymbol{B}$ and $\boldsymbol{W}$ are sparse matrices. In the following the application of the Fast Multipole Method to Equation (12.14) will be considered. Its application to the Galerkin–Method can be found in [25].

12.3.2 Series Expansion of the Fundamental Solution

The fundamental solution $G(\mathbf{x}, \mathbf{y})$

$$G(\mathbf{x}, \mathbf{y}) = \frac{e^{\mathrm{i}k|\mathbf{x}-\mathbf{y}|}}{4\pi|\mathbf{x}-\mathbf{y}|} = \frac{e^{\mathrm{i}kr}}{4\pi r}, \quad \mathbf{r} = |\mathbf{x}-\mathbf{y}|, \quad \mathbf{x} \neq \mathbf{y} \tag{12.18}$$

of the Helmholtz equation in three dimensions can be expressed in terms of spherical Hankel and Bessel functions, Legendre polynomials and spherical harmonics. For a detailed description we refer to [1].

The derivation of the desired formula Equation (12.17) is mainly based on the truncation of the series expansion of the fundamental solution Equation (12.18). For $|\mathbf{a}| > |\mathbf{c}|$ the fundamental solution can be written as, see [1, 10.1.45],

$$\frac{e^{\mathrm{i}k|\mathbf{a}+\mathbf{c}|}}{4\pi|\mathbf{a}+\mathbf{c}|} = \frac{\mathrm{i}k}{4\pi} \sum_{l=0}^{\infty} (2l+1)(-1)^l h_l^{(1)}(k|\mathbf{a}|) j_l(k|\mathbf{c}|) P_l(\hat{a} \cdot \hat{c}) \tag{12.19}$$

with the spherical Hankel function $h_l^{(1)}$, the spherical Bessel function j_l and the Legendre polynomial P_l of order l. The notation $\hat{x} = \mathbf{x}/|\mathbf{x}|$, $|\mathbf{x}| \neq 0$ has been used for vectors on the unit sphere. Using the partial wave expansion of the plane wave (see [1, 10.1.47])

$$e^{\mathrm{i}k\hat{x}\cdot\mathbf{z}} = \sum_{l=0}^{\infty} \mathrm{i}^l (2l+1) j_l(k|\mathbf{z}|) P_l(\hat{x} \cdot \hat{z}) \tag{12.20}$$

and making use of the orthogonality of the Legendre polynomials yields the following identity

$$\begin{aligned} \int_{\mathbb{S}^2} e^{\mathrm{i}k\mathbf{y}\cdot\hat{s}} P_m(\hat{s} \cdot \hat{x})\, do(\hat{s}) &= \sum_{l=0}^{\infty} \mathrm{i}^l (2l+1) j_l(k|\mathbf{y}|) \int_{\mathbb{S}^2} P_m(\hat{y} \cdot \hat{s}) P_l(\hat{s} \cdot \hat{x})\, do(\hat{s}) \\ &= 4\pi \mathrm{i}^m j_m(k|\mathbf{y}|) P_m(\hat{x} \cdot \hat{y}) \,. \end{aligned} \tag{12.21}$$

Equation (12.21) enables to replace the function $j_l(k|\mathbf{c}|)P_l(\hat{a}\cdot\hat{c})$ in Equation (12.19) which depends on the product of $\hat{a}\cdot\hat{c}$ by a product of two functions each depending on $\hat{a}$ or $\hat{c}$. Combining Equations (12.21) and (12.19) results in

$$\frac{e^{\mathrm{i}k|\mathbf{a}+\mathbf{c}|}}{4\pi|\mathbf{a}+\mathbf{c}|} = \frac{\mathrm{i}k}{4\pi}\sum_{l=0}^{\infty}(2l+1)\mathrm{i}^l h_l^{(1)}(k|\mathbf{a}|)\frac{1}{4\pi}\int_{\mathbb{S}^2} e^{\mathrm{i}k\mathbf{c}\cdot\hat{\boldsymbol{s}}} P_l(\hat{\boldsymbol{s}}\cdot\hat{a})\, do(\hat{\boldsymbol{s}})\,. \tag{12.22}$$

By setting $\mathbf{a} = \mathbf{z}_1 - \mathbf{z}_2$ and $\mathbf{c} = (\mathbf{z}_2 - \mathbf{y}) + (\mathbf{x} - \mathbf{z}_1)$ the fundamental solution Equation (12.18) can be expressed as follows

$$\begin{aligned}\frac{e^{\mathrm{i}k|\mathbf{x}-\mathbf{y}|}}{4\pi|\mathbf{x}-\mathbf{y}|} &= \frac{\mathrm{i}k}{4\pi}\sum_{l=0}^{\infty}(2l+1)\mathrm{i}^l h_l^{(1)}(k|\mathbf{z}_2-\mathbf{z}_1|)\\ &\quad\times\frac{1}{4\pi}\int_{\mathbb{S}^2} e^{\mathrm{i}k\,(\mathbf{y}-\mathbf{z}_2)\cdot\hat{\boldsymbol{s}}} e^{\mathrm{i}k\,(\mathbf{z}_1-\mathbf{x})\cdot\hat{\boldsymbol{s}}} P_l(\hat{\boldsymbol{s}}\cdot(\widehat{z_2-z_1}))\, do(\hat{\boldsymbol{s}})\end{aligned} \tag{12.23}$$

for all point $\mathbf{x}$ and $\mathbf{y}$ satisfying the admissibility condition

$$|(\mathbf{z}_2-\mathbf{y})+(\mathbf{x}-\mathbf{z}_1)| < |\mathbf{z}_2-\mathbf{z}_1|\,. \tag{12.24}$$

In the same manner the series expansion for the gradient of the fundamental solution

$$\begin{aligned}\nabla_{\mathbf{x}}G(\mathbf{x},\mathbf{y}) &= \frac{\mathrm{i}k}{4\pi}\sum_{l=0}^{\infty}(2l+1)\mathrm{i}^l h_l^{(1)}(k|\mathbf{z}_2-\mathbf{z}_1|)\\ &\quad\times\frac{1}{4\pi}\int_{\mathbb{S}^2} e^{\mathrm{i}k\,(\mathbf{y}-\mathbf{z}_2)\cdot\hat{\boldsymbol{s}}}(-\mathrm{i}k\hat{\boldsymbol{s}}) e^{\mathrm{i}k\,(\mathbf{z}_1-\mathbf{x})\cdot\hat{\boldsymbol{s}}} P_l(\hat{\boldsymbol{s}}\cdot(\widehat{z_2-z_1}))\, do(\hat{\boldsymbol{s}})\end{aligned} \tag{12.25}$$

is obtained. Thus, for all pairs $(\mathbf{x},\mathbf{y})$ with $|(\mathbf{z}_2-\mathbf{y})+(\mathbf{x}-\mathbf{z}_1)| < |\mathbf{z}_1-\mathbf{z}_2|$ the kernel function $k(\mathbf{x}-\mathbf{y})$ can be replaced by

$$\begin{aligned}k(\mathbf{x}-\mathbf{y}) &= \alpha G(\mathbf{x},\mathbf{y}) + \beta\nabla_{\mathbf{x}}G(\mathbf{x},\mathbf{y})\cdot\boldsymbol{\nu}_x = \\ &= \frac{\mathrm{i}k}{4\pi}\sum_{l=0}^{\infty}(2l+1)\mathrm{i}^l h_l^{(1)}(k|\mathbf{z}_2-\mathbf{z}_1|)\frac{1}{4\pi}\int_{\mathbb{S}^2} e^{\mathrm{i}k\,(\mathbf{y}-\mathbf{z}_2)\cdot\hat{\boldsymbol{s}}}(\alpha\mathrm{i}k + \beta k^2\hat{\boldsymbol{s}}\cdot\boldsymbol{\nu}_{\mathbf{x}})\\ &\quad\times P_l(\hat{\boldsymbol{s}}\cdot(\widehat{z_2-z_1}))e^{\mathrm{i}k\,(\mathbf{z}_1-\mathbf{x})\cdot\hat{\boldsymbol{s}}}\, do(\hat{\boldsymbol{s}})\end{aligned} \tag{12.26}$$

This rather technical representation of the kernel function is not yet in the desired form of Equation (12.17) as the infinite summation and the integration can not be interchanged. But after truncation of the infinite summation Equation (12.26) becomes

$$\begin{aligned}k(\mathbf{x}-\mathbf{y}) &= \alpha G(\mathbf{x},\mathbf{y}) + \beta\nabla_{\mathbf{x}}G(\mathbf{x},\mathbf{y})\cdot\boldsymbol{\nu}_{\mathbf{x}}\\ &\approx \int_{\mathbb{S}^2} \underbrace{\frac{e^{\mathrm{i}k\,(\mathbf{y}-\mathbf{z}_2)\cdot\hat{\boldsymbol{s}}}}{4\pi}}_{g(\mathbf{y}-\mathbf{z}_2,\hat{s})}\ \underbrace{\mu^M(\mathbf{z}_2-\mathbf{z}_1,\hat{s})}_{\mu^M(\mathbf{z}_2-\mathbf{z}_1,\hat{s})}\ \underbrace{(\alpha\mathrm{i}k + \beta k^2\hat{\boldsymbol{s}}\cdot\boldsymbol{\nu}_{\mathbf{x}})e^{\mathrm{i}k\,(\mathbf{z}_1-\mathbf{x})\cdot\hat{\boldsymbol{s}}}}_{h(\mathbf{z}_1-\mathbf{x},\hat{s})}\, do(\hat{\boldsymbol{s}})\\ &\approx \int_{\mathbb{S}^2} g(\mathbf{y}-\mathbf{z}_2,\hat{s})\mu^M(\mathbf{z}_2-\mathbf{z}_1,\hat{s})h(\mathbf{z}_1-\mathbf{x},\hat{s})\, do(\hat{\boldsymbol{s}})\end{aligned}$$

with

$$\mu^M(\mathbf{z}_2 - \mathbf{z}_1, \hat{s}) = \frac{1}{4\pi}\sum_{l=0}^{M}(2l+1)\mathrm{i}^l h_l^{(1)}(k|\mathbf{z}_2 - \mathbf{z}_1|)P_l(\hat{s}\cdot(\widehat{z_2 - z_1})). \quad (12.27)$$

To apply this expansion to the fast evaluation of Equation (12.14) for all points $\mathbf{y}_j$ on Γ, the surface Γ has to be splitted according to the admissibility condition Equation (12.24). Therefore, the elements of the surface triangulation $\mathcal{T}$ are grouped to clusters τ_j with the radius ρ_j and the center $\mathbf{z}_j$. Thereafter, the far and near field of a cluster τ_j are defined as follows

$$\mathcal{F}(\tau_j) := \{\tau_i |\, \rho_i + \rho_j < \eta|\mathbf{z}_j - \mathbf{z}_i|\} \quad \text{far field of } \tau_j$$

$$\mathcal{N}(\tau_j) := \{\tau_i |\, \tau_i \notin \mathcal{F}(\tau_j)\} \quad \text{near field of } \tau_j\,.$$

The parameter $\eta \in (0,1)$ defines the number of buffered clusters. The series expansion of the kernel $k(\mathbf{x}-\mathbf{y}_j)$ for a point $\mathbf{y}_j$ with $\mathbf{y}_j \in \tau_j$ can now be used for all $\mathbf{x} \in \tau_i$ with $\tau_i \in \mathcal{F}(\tau_j)$. The interaction of the cluster τ_j with the remaining clusters $\mathcal{N}(\tau_j)$ – the near field of τ_j – has to be calculated directly using standard boundary element techniques.

Using the splitting of the elements of the surface triangulation with respect to $\mathbf{y}_j \in \tau_j$ Equation (12.14) becomes

$$z_j = u_j - \underbrace{\int_{\Gamma\cap\mathcal{N}(\tau_j)} k(\mathbf{x}-\mathbf{y}_j)\varphi(\mathbf{x})\, d\Gamma\cdot\boldsymbol{u}}_{=:\, \boldsymbol{A}_{\mathrm{near}<j,:>}} - \underbrace{\int_{\Gamma\cap\mathcal{F}(\tau_j)} k(\mathbf{x}-\mathbf{y}_j)\varphi(\mathbf{x})\, d\Gamma\cdot\boldsymbol{u}}_{=:\, \boldsymbol{A}_{\mathrm{far}<j,:>}}$$

The second term forms the j–th row of the sparse matrix $\boldsymbol{A}_{\mathrm{near}}$. The contribution of all clusters τ_i belonging to the far field of τ_j can be approximated using Equation (12.26)

$$\begin{aligned}
&\int_{\Gamma\cap\mathcal{F}(\tau_j)} k(\mathbf{x}-\mathbf{y}_j)\varphi(\mathbf{x})\, d\Gamma\cdot\boldsymbol{u} = \\
&\quad = \sum_{\tau_i\in\mathcal{F}(\tau_j)} \int_{\tau_i} k(\mathbf{x}-\mathbf{y}_j)\varphi(\mathbf{x})\, do(\mathbf{x})\cdot\boldsymbol{u} \\
&\quad \approx \sum_{\tau_i\in\mathcal{F}(\tau_j)} \int_{\tau_i}\int_{\mathbb{S}^2} \frac{e^{\mathrm{i}k\,(\mathbf{y}-\mathbf{z}_j)\cdot\hat{s}}}{4\pi}\mu^M(\mathbf{z}_j-\mathbf{z}_i,\hat{s})(\alpha\mathrm{i}k+\beta k^2\hat{s}\cdot\boldsymbol{\nu}_{\mathbf{x}}) \\
&\qquad\qquad \times e^{\mathrm{i}k\,(\mathbf{z}_i-\mathbf{x})\cdot\hat{s}}\, do(\hat{s})\varphi(\mathbf{x})\, do(\mathbf{x})\cdot\boldsymbol{u}\,.
\end{aligned}$$

Changing the order of integration and summation yields

$$\int_{\Gamma\cap\mathcal{F}(\tau_j)} k(\mathbf{x}-\mathbf{y}_j)\varphi(\mathbf{x})\,d\Gamma\cdot\boldsymbol{u} =$$

$$\approx \int_{\mathbb{S}^2} \frac{e^{ik\,(\mathbf{y}-\mathbf{z}_j)\cdot\hat{s}}}{4\pi} \sum_{\tau_i\in\mathcal{F}(\tau_j)} \mu^M(\mathbf{z}_j-\mathbf{z}_i,\hat{s}) \times \underbrace{\int_{\tau_i} (\alpha ik+\beta k^2 \hat{s}\cdot\boldsymbol{\nu}_{\mathbf{x}}) e^{ik\,(\mathbf{z}_i-\mathbf{x})\cdot\hat{s}} \varphi(\mathbf{x})\,do(\mathbf{x})\cdot\boldsymbol{u}}_{=:\,\Psi^{\tau_i}(\mathbf{z}_i,\hat{s})}\, do(\hat{s})\,.$$

The function $\Psi^{\tau_i}(\mathbf{z}_i, \hat{s})$ is often referred to as far field pattern of the cluster τ_i. Using this notation gives

$$\int_{\Gamma\cap\mathcal{F}(\tau_j)} k(\mathbf{x}-\mathbf{y}_j)\varphi(\mathbf{x})\,d\Gamma\cdot\boldsymbol{u} \approx \int_{\mathbb{S}^2} \frac{e^{ik\,(\mathbf{y}-\mathbf{z}_j)\cdot\hat{s}}}{4\pi} \times \underbrace{\sum_{\tau_i\in\mathcal{F}(\tau_j)} \mu^M(\mathbf{z}_j-\mathbf{z}_i,\hat{s})\Psi^{\tau_i}(\mathbf{z}_i,\hat{s})}_{=:\,\Upsilon^{\tau_j}(\mathbf{z}_j,\hat{s})}\, do(\hat{s})\,.$$

All far field pattern $\Psi^{\tau_i}(\mathbf{z}_i, \hat{s})$ with $\tau_i \in \mathcal{F}(\tau_j)$ have been converted to the near field pattern $\Upsilon^{\tau_j}(\mathbf{z}_j, \hat{s})$ of the cluster τ_j. Thus, the contribution of the far field of the cluster τ_j to the j–th row of the matrix–vector–product can be approximated via

$$\int_{\Gamma\cap\mathcal{F}(\tau_j)} k(\mathbf{x}-\mathbf{y}_j)\varphi(\mathbf{x})\,d\Gamma\cdot\boldsymbol{p} \approx \int_{\mathbb{S}^2} \frac{e^{ik\,(\mathbf{y}-\mathbf{z}_j)\cdot\hat{s}}}{4\pi} \Upsilon^{\tau_j}(\mathbf{z}_j,\hat{s})\,do(\hat{s}) \tag{12.28}$$

using the near field pattern $\Upsilon^{\tau_j}(\mathbf{z}_j, \hat{s})$ of the cluster τ_j. Finally Equation (12.14), j–th row of $\boldsymbol{z} = (\boldsymbol{I} - \boldsymbol{A})\boldsymbol{u}$, becomes

$$z_j \approx u_j - \boldsymbol{A}_{\mathrm{near}\,j}\cdot\boldsymbol{u} - \int_{\mathbb{S}^2} \frac{e^{ik\,(\mathbf{y}-\mathbf{z}_j)\cdot\hat{s}}}{4\pi} \Upsilon^{\tau_j}(\mathbf{z}_j,\hat{s})\,do(\hat{s})\,. \tag{12.29}$$

To use Equation (12.29) to evaluate a matrix–vector product in a first step the far field pattern $\Psi^{\tau_i}(\mathbf{z}_i, \hat{s})$

$$\Psi^{\tau_i}(\mathbf{z}_i,\hat{s}) = \int_{\tau_i} (\alpha ik+\beta k^2 \hat{s}\cdot\boldsymbol{\nu}_{\mathbf{x}}) e^{ik\,(\mathbf{z}_i-\mathbf{x})\cdot\hat{s}} \varphi(\mathbf{x})\,do(\mathbf{x})\cdot\boldsymbol{p} \tag{12.30}$$

$$= \sum_{i=1}^{N} p_i \int_{\tau_i\cap\mathrm{supp}\varphi_i} (\alpha ik+\beta k^2 \hat{s}\cdot\boldsymbol{\nu}_{\mathbf{x}}) e^{ik\,(\mathbf{z}_i-\mathbf{x})\cdot\hat{s}} \varphi_i(\mathbf{x})\,do(\mathbf{x})$$

must be calculated for all $\tau \in \mathcal{T}$. This corresponds to the evaluation of a surface integral over a smooth function. Hence, standard Gaussian quadrature can be applied. In a second step the near field pattern $\Upsilon^{\tau_j}(\mathbf{z}_j, \hat{s})$ of a cluster τ_j

$$\Upsilon^{\tau_j}(\mathbf{z}_j,\hat{s}) = \sum_{\tau_i \in \mathcal{F}(\tau_j)} \mu^M(\mathbf{z}_j - \mathbf{z}_i,\hat{s}) \Psi^{\tau_i}(\mathbf{z}_i,\hat{s}) \tag{12.31}$$

is the translation of the previously calculated far field pattern of all η–admissible clusters of τ_j. Therefore, the operator $\mu^M(\mathbf{z}_j - \mathbf{z}_i, \hat{s})$ is referred to as translation operator. In a third step the sound pressure at the point $\mathbf{y}_j$ on the surface Γ is the results of the integration over the unit sphere

$$p(\mathbf{y}_j) = \int_{\mathbb{S}^2} \frac{e^{\mathrm{i}k\,(\mathbf{y}-\mathbf{z}_j)\cdot\hat{s}}}{4\pi} \Upsilon^{\tau_j}(\mathbf{z}_j,\hat{s})\, do(\hat{s})\,. \tag{12.32}$$

For a numerical implementation of the above equations a suitable truncation of the infinite sum in Equation (12.23) and numerical integration over the unit sphere as well as the error introduced by doing so requires some attentions.

12.3.3 Truncation of the Series Expansion and Integration on the Unit Sphere

For the numerical implementation of the Fast Multipole Method the infinite sum over l in Equation (12.23) has to be truncated for a certain integer M

$$\mu(\mathbf{a},\hat{s}) \approx \mu^M(\mathbf{a},\hat{s}) = \frac{1}{4\pi} \sum_{l=0}^{M} (2l+1)\mathrm{i}^l h_l^{(1)}(k|\mathbf{a}|) P_l(\hat{s}\cdot\hat{a})\,. \tag{12.33}$$

One would expect that for increasing M the Fast Multipole Method gives greater accuracy. This holds true only in exact arithmetic as the sum over l diverges as M tends to infinity. In exact arithmetic the growth of $h_l^{(1)}(k|\mathbf{a}|)$ is implicitly balanced by the decay of $j_l(k|\mathbf{c}|)P_l(\hat{c}\cdot\hat{a})$ but which has been replaced by

$$j_l(k|\mathbf{c}|)P_l(\hat{c}\cdot\hat{a}) = \frac{1}{4\pi\mathrm{i}^l} \int_{\mathbb{S}^2} e^{\mathrm{i}k\mathbf{c}\cdot\hat{s}} P_l(\hat{s}\cdot\hat{a})\, do(\hat{s})\,.$$

As the integration over the unit sphere has to be undertaken numerically

$$\int_{\mathbb{S}^2} f(\hat{s})\, do(\hat{s}) \approx \sum_{i=1}^{K_{\mathrm{I}}} w_i f(\hat{\xi}_i)$$

using a quadrature rule with the nodes $\hat{\xi}_i$ and the weights w_i ($i = 1,\ldots,K_{\mathrm{I}}$) the values of $j_l(k|\mathbf{c}|)P_l(\hat{c}\cdot\hat{a})$ appear only to a finite precision and the index of truncation M has to be chosen according to the accuracy of the numerical integration. As the value of K_{I} determines the efficiency of the algorithm the smallest possible value should be used. A detailed analysis of the truncation error was presented by Darve [22]. The author shows (Proposition 1 and 2) that for $\eta \leq 2/\sqrt{5}$ there exist constants C_1, C_2, C_3 and D_1, D_2, D_3 such that if $M \geq C_1 + C_2 k|\mathbf{c}| + C_2 \log(k|\mathbf{c}|) + C_3 \log \varepsilon^{-1}$ terms are kept in the sum in Equation (12.33) and the quadrature rule integrates exactly the first $K \geq D_1 + D_2 k|\mathbf{c}| + D_2 \log(k|\mathbf{c}|) + D_3 \log \varepsilon^{-1}$ spherical harmonics on $\mathbb{S}^2$, then the error introduced by truncation and numerical integration when replacing the fundamental solution by the truncated series expansion Equation (12.22) is bounded by ε. Darve further states "*Practically a good choice for* K *is* $K \geq 2M$". Out of the numerical tests [20, 34, 62] the authors empirically gave formulas for the value of M of the following type

$$M = k|\mathbf{c}| + \frac{d}{1.6}\log(k|\mathbf{c}| + \pi) \tag{12.34}$$

where the value of d controls the number of significant digits in the approximation. For numerical implementation the value of K is often fixed to $K = 2M$. Consequently, the quadrature rule has to be exact up to the first K spherical harmonics. Using a Gaussian quadrature rule $\widehat{Q}_M$ which integrates exactly the first $2M$ spherical polynomials – see [64, Theorem 2.7-1] for details – $K_{\mathrm{I}} = (M+1)(2M+1)$ sample points of each far field and near field pattern as well for each translation operator need to be stored.

12.3.4 Aspects of the Numerical Implementation

To evaluate

$$z_j = (\boldsymbol{I} - \boldsymbol{A})_{j\cdot}\boldsymbol{u} = u_j - \sum_{i=1}^{N} u_i \int_{\Gamma \cap \mathrm{supp}\varphi_i} k(\mathbf{x} - \mathbf{y}_j)\varphi_i(\mathbf{x})\, d\Gamma \tag{12.35}$$

at each collocation point $\mathbf{y}_j, j = 1, \ldots, N$ with less than $\mathcal{O}(N^2)$ arithmetic complexity using the Fast Multipole Method the elements of the surface triangulation must be grouped into clusters. It is assumed that each element Δ_i of the surface triangulation $\mathcal{T}$ can be uniquely assigned to a cluster τ_i with the radius ρ_i and the center $\mathbf{z}_i$. The maximal cluster radius is defined as

$$\rho := \max_{i=1,\ldots,N_{\mathrm{C}}} \rho_i$$

with the number of clusters N_{C}. Thereafter the order of expansion of the fundamental solution M in Equation (12.33) is set to

$$M = \max(M_0, 2k\rho + \frac{d}{1.6}\log(2k\rho + \pi)) \,. \tag{12.36}$$

The constant M_0 ensures that a sufficient number of coefficients is kept even at small wavenumbers k. Commonly $M_0 = 4$ is a good choice. Further details on the treatment of the low frequency range using the Fast Multipole Method we refer to [23, 24, 30, 40, 69]. Integration on the unit sphere is performed using a quadrature rule $\widehat{Q}_M$ that integrates exactly the first $2M$ spherical polynomials. The partition of the surface Γ is based on Equation (12.24). Two clusters τ_i and τ_j are called η–admissible with respect to the parameter $\eta \in (0, 1)$ if

$$\rho_j + \rho_i < \eta|\mathbf{z}_i - \mathbf{z}_j| \tag{12.37}$$

holds true, i.e. the sum of their radii is smaller than the distance between their cluster centers. Thereafter, the interaction of all η–admissible clusters can be calculated using Equation (12.26) and for the collocation point $\mathbf{y}_j \in \tau_j$ Equation (12.35) becomes

$$
\begin{aligned}
z_j &= (\boldsymbol{I} - \boldsymbol{A})_{j\cdot}\boldsymbol{u} \\
&\approx u_j - \sum_{q=1}^{N} u_q \underbrace{\sum_{\tau_i \in \mathcal{N}(\tau_j)} \int_{\tau_i \cap \operatorname{supp}\varphi_q} k(\mathbf{x}-\mathbf{y}_j)\varphi_q \, do(\mathbf{x})}_{=: \boldsymbol{A}_{\text{near}\, j,q}} \\
&\quad - \sum_{\alpha} \underbrace{\frac{1}{4\pi} w_\alpha e^{\mathrm{i}k(\mathbf{y}_j - \mathbf{z}_j)\cdot\hat{\xi}_\alpha}}_{=: \boldsymbol{V}_{j,j_\alpha}} \sum_{\tau_i \in \mathcal{F}(\tau_j)} \underbrace{\mu^M(\mathbf{z}_j - \mathbf{z}_i, \hat{\xi}_\alpha)}_{=: \boldsymbol{B}_{j_\alpha, i_\alpha}} \Psi^{\tau_i}(\hat{\xi}_\alpha)
\end{aligned}
\tag{12.38}
$$

with

$$
\Psi^{\tau_i}(\hat{\xi}_\alpha) = \sum_{q=1}^{N} u_q \underbrace{\int_{\operatorname{supp}\varphi_q \cap \tau_i} \frac{1}{4\pi} e^{\mathrm{i}k(\mathbf{z}_i - \mathbf{x})\cdot\hat{\xi}_\alpha} \left(\alpha \mathrm{i}k + \beta k^2 \hat{\xi}_\alpha \cdot \boldsymbol{\nu}_{\mathbf{x}}\right) \varphi_q(\mathbf{x}) \, do(\mathbf{x})}_{=: \boldsymbol{W}_{i_\alpha, q}}
$$

Using the notations introduced above the matrix–vector–product now reads as

$$
\boldsymbol{z} = (\boldsymbol{I} - \boldsymbol{A})u \approx (\boldsymbol{I} - \boldsymbol{A}_{\text{near}} - \boldsymbol{V}\boldsymbol{B}\boldsymbol{W})\boldsymbol{u}
$$

where $\boldsymbol{A}_{\text{near}}$, $\boldsymbol{V}$, $\boldsymbol{B}$, and $\boldsymbol{W}$ are sparse matrices. The matrix $\boldsymbol{A}_{\text{near}}$ contains all nearby interactions that must still be calculated using standard boundary element techniques and the original kernel function Equation (12.16). The remaining interactions are evaluated using the Fast Multipole Method.

It can be shown by counting the non–zero elements in $\boldsymbol{A}_{\text{near}}$, $\boldsymbol{V}$, $\boldsymbol{B}$ and $\boldsymbol{W}$ that the effort for evaluating one matrix–vector product is $\mathcal{O}(N^{3/2})$, cf. [22]. Therefore, a further reduction of the complexity of the algorithm is highly desirable. Obviously still $\mathcal{O}(N_{\mathrm{C}}^2)$ interactions of clusters have to be calculated. However, one will find that many of the η–admissible clusters would stay η–admissible if their radii were larger. In other words the interaction of larger parts of the surface could have been calculated using the series expansion. Hence, to increase the efficiency of the Fast Multipole Method the size of the clusters has to be enlarged as long as they stay η–admissible and evaluating of their interactions must be performed when they have reached the largest possible radius. This leads to the so called Multilevel Fast Multipole Algorithm (MLFMA).

12.3.5 The Multilevel Fast Multipole Algorithm

When looking at the Fast Multipole Method described in Section 12.3.1 it turns out that there are pairs of clusters τ_i and τ_j which would satisfy the admissible condition

Equation (12.37) even with larger radii ρ_i and ρ_j or in other words more interactions of points on the surface could have been evaluated at once using the Fast Multipole Method. The main idea of the Multilevel Fast Multipole Algorithm is to enlarge the radii of the clusters adaptively using a hierarchy of clusters, a so–called cluster tree. The admissibility condition Equation (12.37) is then applied at every level l of the tree. Only these clusters interact on level l which are η–admissible and their fathers on the next higher level are not η–admissible. The interaction of the remaining clusters will be calculated at a higher level. The shift of the far and near field pattern Ψ^{τ} and Υ^{τ} through the different levels of the cluster tree is undertaken using the following relation

$$\Psi^{\tau_1}_{\mathbf{z}_2}(\hat{z}) = \Psi^{\tau_1}_{\mathbf{z}_1}(\hat{z})e^{\mathrm{i}k(\mathbf{z}_2-\mathbf{z}_1)\cdot\hat{z}} \tag{12.39}$$

$$\Upsilon^{\tau_1}_{\mathbf{z}_2}(\hat{z}) = \Upsilon^{\tau_1}_{\mathbf{z}_1}(\hat{z})e^{\mathrm{i}k(\mathbf{z}_2-\mathbf{z}_1)\cdot\hat{z}}\,, \tag{12.40}$$

following directly from Equation (12.23). Analyzing the complexity of the multilevel version of the Fast Multipole Method shows that the application of the Fast Multipole Method on $\mathcal{O}(\log N)$ levels of a cluster tree leads to an algorithm of $\mathcal{O}(N\log^2 N)$ complexity [18, 22, 62]. Two important differences of the multilevel and the single–level algorithm must be addressed. Firstly, the splitting of the elements of the surface triangulation must be replaced by a hierarchy of such splittings based on a cluster tree. Secondly, as cluster radii differ on different levels the expansion length M in Equation (12.36) must be adapted and in addition to the shift of the near and far field pattern interpolation and filtering of the data will be necessary when passing information between different levels. Using the notation $\boldsymbol{B}^l$ for the translation operator on level l, $\boldsymbol{I}_l^{l+1}$ for the shift and interpolation operator from level l to level $l+1$ and $\boldsymbol{F}_l^{l+1}$ for the shift and filter operator from level $l+1$ to level l the approximation of the product of a vector $\boldsymbol{u}$ with the system matrix $\boldsymbol{A}$ can be written in the following form

$$\boldsymbol{A}\boldsymbol{u} \approx \boldsymbol{A}_{\text{near}}\boldsymbol{u} + \boldsymbol{V}(\boldsymbol{F}_0^1(\boldsymbol{F}_1^2(\boldsymbol{F}_2^3(\ldots)\boldsymbol{I}_2^3 + \boldsymbol{B}^2)\boldsymbol{I}_1^2 + \boldsymbol{B}^1)\boldsymbol{I}_0^1 + \boldsymbol{B}^0)\boldsymbol{W}\boldsymbol{u}\,.$$

Similar to the single–level version first the far field pattern of the vector $\boldsymbol{u}$ is evaluated. But in the multilevel version the translation operator $\boldsymbol{B}^l$ translates far field pattern to the near field pattern on level l only for admissible clusters on level l whose fathers on level $l+1$ are not admissible. Then the far field pattern from level l is shifted to level $l+1$. Once the highest level of the cluster tree is reached the near field pattern from level l is shifted to level $l-1$ and accumulated to the pattern of level $l-1$. When the lowest level of the cluster tree is reached the accumulated near field pattern is converted to a sound pressure on the surface Γ.

12.3.6 Construction of the Cluster Tree

A cluster tree is a hierarchy of sets containing the elements of the surface triangulation. Starting on the coarsest level where the elements of the surface triangulation

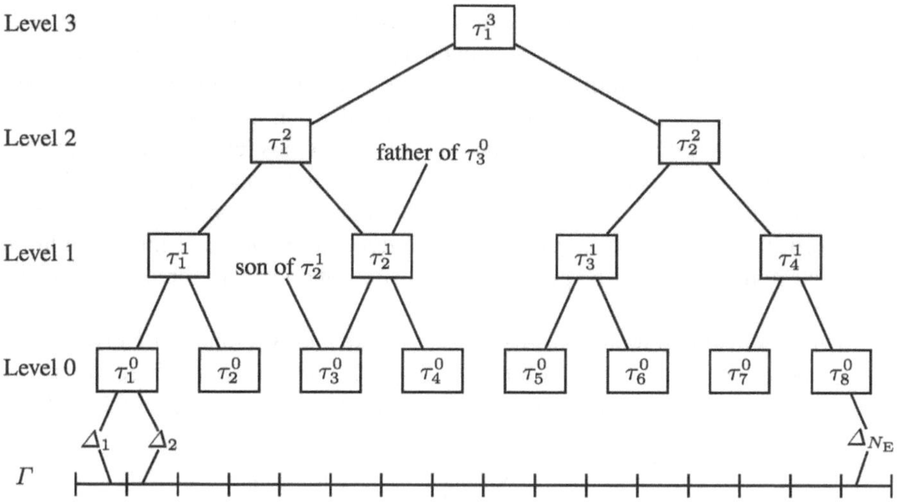

Fig. 12.10 Binary cluster tree based on the elements Δ_i on the triangulation $\mathcal{T}$ of the boundary Γ.

form a single set, sets of clusters on the next finer level are obtained by a successive subdivision of the previous sets. The newly obtained sets are referred to as sons and will become the father sets for the corresponding sets of the next finer level. As each set is subdivided into two subsets, see Figure 12.10, a binary cluster tree is obtained [29]. The process is stopped when the number of elements in a set is smaller than a given value. A different strategy where the subdivision is obtained by a successive subdivision of a cube in $\mathbb{R}^3$ yielding an oct–tree [22, 55] will not be considered here.

Once the cluster tree is build up, for each of its levels the radius ρ and the center z of a cluster is defined as the radius and center, respectively, of the smallest open ball $B(\rho, \mathbf{z})$ containing τ entirely. Further, the clusters have to satisfy $\tau_i \cap \tau_j = \emptyset$ if $i \neq j$ and $\Gamma = \cup_i \tau_i$. It is pointed out that the elements of $\mathcal{T}$ within a cluster do not necessarily have to be adjacent elements.

12.3.7 Interpolation and Filtering on the Sphere

The proper handling of the shifting of near and far field patterns through various levels of the cluster tree is of crucial importance for the efficiency of the multilevel fast multipole algorithm. As the radius of the clusters changes from level to level of the cluster tree, the number of coefficients to be kept according to Equation (12.34) is also different for each level of the cluster tree. To obtain the complexity estimates of the algorithm given above the number of coefficients M in Equation (12.33) must be adapted according to the radius of the clusters on each level. Hence, on lower levels fewer coefficients are needed than on higher levels. Consequently, the number of sample points on the unit sphere needed for numerical integration also changes. Thus, the values at the new sample points needed when moving up and down the tree must be interpolated of the values from the previous level. As a consequence of [22, Proposition 1 and 2] this operation must not influence the number of spherical harmonics, which are required to have, so that the interpolation error is in the same order as the truncation error. Hence, interpolation must not introduce higher order harmonics. This can be guaranteed by using spherical harmonic analysis when interpolating sample points.

The spherical harmonic analysis consists of expanding a function f given at a set of points (φ_i, θ_j), $i = 0, \ldots, N$ and $j = 0, \ldots, 2N$ on the unit sphere in terms of spherical harmonics Y_n^m, that is calculating a set of coefficients α_{nm} such that

$$f(\varphi_i, \theta_j) = \sum_{n=0}^{N} \sum_{m=-n}^{n} \alpha_{nm} Y_n^m(\varphi_i, \theta_j) \tag{12.41}$$

holds true. The required function values at a new set of points $(\tilde{\varphi}_i, \tilde{\theta}_j)$, $i = 0, \ldots, \tilde{N}$ and $j = 0, \ldots, 2\tilde{N}$ are now obtained using the above expansion

$$f(\tilde{\varphi}_i, \tilde{\theta}_j) = \sum_{n=0}^{\min(N,\tilde{N})} \sum_{m=-n}^{n} \alpha_{nm} Y_n^m(\tilde{\varphi}_i, \tilde{\theta}_j)\,. \tag{12.42}$$

The values of N and $\tilde{N}$ in Equation (12.42) determine interpolation or filtering. In the case where $\tilde{N}$ is larger than N Equation (12.42) is referred to as interpolation otherwise it is referred to as filtering. Using the orthogonality properties of the spherical polynomials the coefficients α_{nm} are given by

$$\alpha_{nm} = \int_{\mathbb{S}^2} f(\varphi, \theta) \overline{Y}_n^m(\varphi, \theta)\, do(\hat{s}) = \int_{-1}^{1} \int_{0}^{2\pi} f(\varphi, t) \overline{Y}_n^m(\varphi, t)\, dt\, d\varphi \tag{12.43}$$

with the substitution $t = \cos(\theta)$. The numerical implementation of the interpolation and filtering consist of several steps. First, the integration over the φ–direction can be seen as a Fourier–transformation

$$\beta_{im} = \frac{2\pi}{2M+1} \sum_{j=0}^{2M} f(\varphi_j, \theta_i) e^{-\mathrm{i}\frac{2\pi}{2M+1} jm}\,.$$

Integration with respect to t in Equation (12.43) and summation over n in Equation (12.42) yield the new Fourier coefficients $\tilde{\beta}_{im}$

$$\tilde{\beta}_{im} = \sum_{n=0}^{M'} C_n^m C_n^m \sum_{j=0}^{M} w_j P_n^{|m|}(t_j) \beta_{jm} P_n^{|m|}(\tilde{t}_i)\,.$$

An algorithm for large M, $M^{'}$ that is more efficient than the direct evaluation of the double summation above was presented in [71]. It is base on the Christoffel–Darboux formula [1] and the application of a one dimensional Fast Multipole Method to evaluate the arising matrix–vector product in an efficient manner.

Data at the new sample points $(\tilde{\varphi}_i, \tilde{\theta}_j)$ is obtained by an inverse Fourier–transformation of $\beta^{'}_{im}$

$$f(\tilde{\varphi}_j, \tilde{\theta}_i) = \sum_{m=-M'}^{M^{'}} \tilde{\beta}_{im} e^{\mathrm{i}\frac{2\pi}{2M^{'}+1}jm} . \tag{12.44}$$

With a proper choice of M and $M^{'}$ the above required Fourier–transformations can be implemented efficiently using fast algorithms.

12.3.8 Implementation of the Algorithm

The application of the Multilevel Fast Multipole Algorithm for the Helmholtz equation is divided into two parts. In the first part commonly referred to as *setup*–step the following operations have to be performed

- definition of the number of elements of the surface triangulation n_{elem} contained in a leave of the cluster tree
- construction of the cluster tree and definition of number of levels $n_{\text{level}} = c\log(N)$ that will be used,
- definition of expansion order on each level, $M^l = \max(M^0, 2k\rho^l + \frac{d}{1.6}\log(2k\rho^l + \pi))$,
- adjustment of M^l such that fast Fourier–transformation can be applied in the filter and interpolation step,
- definition of the parameter η and construction of an admissible–list of each cluster at each level according to Equation (12.37),
- calculation of the matrix $\boldsymbol{A}_{\text{near}}$ of the interactions of non–η–admissible clusters on level zero and
- calculation of the sparse matrices $\boldsymbol{W}_{i_\alpha,q}$ and $\boldsymbol{B}_{j_\alpha,i_\alpha}$.

Details on the choice of the values for n_{elem}, n_{level} and M^l can be found in [55,73]. Often the matrix $\boldsymbol{V}$ in Equation (12.38) is not explicitly calculated and the numerical integration over the unit sphere, see Equation (12.28) is performed directly at each matrix–vector–product.

The evaluation of a matrix–vector–product $\boldsymbol{v} = (\boldsymbol{I} - \boldsymbol{A})\boldsymbol{u}$ using the Multilevel Fast Multipole Algorithm, often referred to as *apply*–step, is realized using the following algorithm:

Algorithm *apply*:

input: $\boldsymbol{u}$
output: $\boldsymbol{z} = (\boldsymbol{I} - \boldsymbol{A}_{\text{near}})\boldsymbol{u} - \boldsymbol{A}_{\text{far}}\boldsymbol{u}$

$\Psi = \boldsymbol{W}\boldsymbol{u}$! far field pattern of sound pressure distribution $\boldsymbol{u}$ on Γ
do $l = 0, n_{\text{level}}$! up tree
  `for` all clusters i on level l `do`
    `for` all clusters j on level l `do`
      `if` (i,j) η–admissible $\wedge$ $(f(i), f(j))$ `not` η–admissible
        $\Upsilon^{\tau_i}(\Xi_{M^l}) = \mu^{M^l}(\mathbf{z}_i - \mathbf{z}_j, \Xi_{M^l})\Psi^{\tau_j}(\Xi_{M^l})$
      `end`
    `end`
    $\Psi^{\tau_i}(\Xi_{M^{l+1}})$=`interpol`$(\Psi^{\tau_i}, M^l, M^{l+1})$
    $\Psi^{\tau_{f(i)}}(\Xi_{M^{l+1}}) = \Psi^{\tau_{f(i)}}(\Xi_{M^{l+1}}) + \Psi^{\tau_i}(\Xi_{M^{l+1}})e^{ik(\mathbf{z}_{f(i)} - \mathbf{z}_i)\cdot \Xi_{M^{l+1}}}$
  `end`
`end`
do $l = n_{\text{level}} - 1, 0$! down tree
  `for` all boxes i on level l `do`
    $\Upsilon^{\tau_i}(\Xi_{M^{l+1}}) = \Upsilon^{\tau_{f(i)}}(\Xi_{M^{l+1}})e^{ik(\mathbf{z}_i - \mathbf{z}_{f(i)})\cdot \Xi_{M^{l+1}}}$
    $\Upsilon^{\tau_i}(\Xi_{M^l}) = \Upsilon^{\tau_i}(\Xi_{M^l}) + \texttt{filter}(\Upsilon^{\tau_i}(\Xi_{M^{l+1}}), M^{l+1}, M^l)$
  `end`
`end`
`for` all clusters i on level 0 `do`
  $z_{2_j} = \sum_\alpha \frac{1}{4\pi} w_\alpha e^{ik(\mathbf{y}_j - \mathbf{z}_j)\cdot\hat{\xi}_\alpha}\Upsilon^{\tau_i}(\hat{\xi}_\alpha)$
`end`
$\boldsymbol{z}_1 = \boldsymbol{A}_{\text{near}}\boldsymbol{u}$
$\boldsymbol{z} = \boldsymbol{u} - \boldsymbol{z}_1 - \boldsymbol{z}_2$

The functions `interpol` and `filter` transfer data given on a set of points Ξ_{M^l} to a set of points $\Xi_{M^{l+1}}$ and vice versa as discussed in Section 12.3.7.

The implementation of the algorithm above shows that a large part of the memory required for the multipole part is needed to store the data $\mu^M(\mathbf{z}_j - \mathbf{z}_i, \Xi_{M^l}) = \mu^M(\mathbf{z}, \Xi_{M^l})$ with $\mathbf{z} = [z^1, z^2, z^3]$ which is the translation operator that translates far field pattern to near field pattern. This is a drawback of the use of a binary cluster tree over an oct–tree. Taking a closer look at the structure of the data shows the dependence on the absolute values of the components[4] of z only. Hence, using a suitable permutation P the values $\mu^M([\pm|z^{1,2}|, \pm|z^{2,1}|, \pm|z^3|], \Xi_{M^l})$ can be calculated out of the values of $\mu^M([|z^1|, z^2|, |z^3|], \Xi_{M^l})$ yielding to the situation that the memory requirement is now negligible. To increase the probability that the difference of two cluster centers $\mathbf{z} = \mathbf{z}_j - \mathbf{z}_i$ differs only by the signs of the components the position of the cluster centers can be restricted to points on a regular grid $[ih, jh, kh]$. This yields a situation quite similar to that of an oct–tree obtained by a successive subdivision of a cube. If however necessary, a further compression of the data μ^M can be achieved by the recalculation of $\mu^M(\mathbf{z}, \Xi_{M^l})$ as it is needed. An efficient algorithm

[4]The components 1 and 2 of z may be interchanged.

for this task is given in [68]. There the authors use a one–dimensional Fast Multipole Method to evaluate

$$\mu^M(\mathbf{z}, \hat{s}) = \sum_{i=0}^{M} \alpha_i P_i(\hat{z} \cdot \hat{s}) \tag{12.45}$$

with $\alpha_l = (2l+1)\mathrm{i}^l h_l^{(1)}(k|z|)$ in an efficient manner. The evaluation of μ^M at $\mathcal{O}(M^2)$ locations needs only $\mathcal{O}(M^2 \log(1/\epsilon))$ arithmetic operations where ϵ determines the accuracy of the approximation. Thereafter, only the coefficients α_i which depend on the modulus of $\mathbf{z}$ need be stored.

12.3.9 An Example

The Multilevel Fast Multipole Algorithm described above will be used to demonstrate that the boundary element method can be applied efficiently to large scale problems.

The objective of the following numerical example is the prediction of the quality of an anechoic chamber in the low frequency range. The fact that the acoustic lining of such a chamber is not sufficiently absorbing at low frequencies creates several inconveniences. First, from a practical point of view, remaining reflections of the walls perturb experimental results. Secondly, from a numerical point of view, neither the geometry of the lining nor sound propagation within the absorbing material can be neglected. This excludes the use of the model of a rectangular cavity where walls are equipped with a local admittance condition to account for the acoustic lining. Therefore the numerical model must respect the real geometry of the lining and modelling of the absorbing material requires special attention. Here an admittance matrix – instead of a scalar value –, taking into account for sound propagation within a specific part of the lining and its vicinity, was used to represent the behaviour of the acoustical treatment. Hence, on the air lining interface the sound pressure p at a specific part of the surface is coupled with the surface velocity $v_{\boldsymbol{\nu}}$ on that part and its vicinity via

$$v_{\boldsymbol{\nu}} = \boldsymbol{Y} \boldsymbol{p} \tag{12.46}$$

with the dense and frequency dependent matrix $\boldsymbol{Y}$. The approximation, that only the surface velocity at the vicinity is considered, can been seen as a localisation of the non–local behavior of an absorbing material.

Numerical results will be compared with experimental data obtained from measurements carried out in the large anechoic chamber of the LMA, see Figure 12.11. The 1.5 dB region in the 20 to 200 Hz frequency range was obtained by measuring the sound pressure radiated by a bass–reflex box. For a detailed description of the experiment we refer to [58]. The anechoic chamber at the LMA has inner dimensions of $5.4 \times 6.3 \times 11.4$ m^3 measured from wedge tip to wedge tip. Each of the 3720 wedges consists of a rectangular parallelepiped measuring $.3 \times .3 \times .4$ m^3 forming the base of the wedge and a tapering section of .7 m in length. Wedges are made of melamine foam. To calculate the admittance matrix $\boldsymbol{Y}$ a central wedge and its 8 adjacent wedges have been used. As all of the wedges are identical only a single

Fig. 12.11 Interior of the large anechoic chmaber of the LMA (left sub–figure) and a photo of a sample of 3 × 3 elements of the acoustic lining (right sub–figure).

admittance matrix $\boldsymbol{Y}(\omega) \in \mathbb{C}^{324\times 36}$ is needed. However, to take into account for the anisotropy of the melamine foam three different matrices $\boldsymbol{Y}^i$, representing the three different material orientations, have actually been used. These matrices were precalculated in the frequency range of 20 to 200 Hz with a frequency resolution of 1 Hz using a finite element method. The geometry of the wedges has been modelled using 9 elements resulting in a mesh size of ≈.3 m. Using linear discontinuous basis functions [42] yields a linear system with N=138 280 unknowns and standard boundary element methods, requiring ≈ 300 Gb of memory to hold the system matrix, are hardly applicable. Therefore a six–level fast multipole method was applied. The leaves of the cluster tree contained up to $n_{\mathrm{elem}} = 4$ surface elements. The parameter η in Equation (12.37) was set to $\eta = .7$. The memory requirement of the algorithm at 200 Hz is given in the left sub–table of Table 12.2. It can be seen that the matrices of the near field interactions occupy almost all of the required memory. The matrix named "*work*" represents the working space needed for the multipole algorithm to hold the far and near field patterns on the different levels. The expansion length M in Equation (12.36) varied from 5 to 16 depending on the level. The memory required to hold the translation operator μ^{M}, matrix $\boldsymbol{B}$ in Table 12.2, is indeed negligible, as stated in Section 12.3.8, due to the performed compression. Each stored operator has be reused ≈ 1500 times. Computational resources, the time and the number of floating point operations required to perform a matrix–vector–product, are given in the right sub–table of Table 12.2. It is pointed out that due to the admittance boundary condition the product $\boldsymbol{A}_{\mathrm{near}}\boldsymbol{u}$ reads as $\boldsymbol{A}_{\mathrm{near}}\boldsymbol{u} = \boldsymbol{H}_{\mathrm{near}}\boldsymbol{u} - \boldsymbol{G}_{\mathrm{near}}\boldsymbol{Y}\boldsymbol{u}$. Therefore the near field part of the matrices $\boldsymbol{H}$ and $\boldsymbol{G}$ must be stored seperately. Performing a

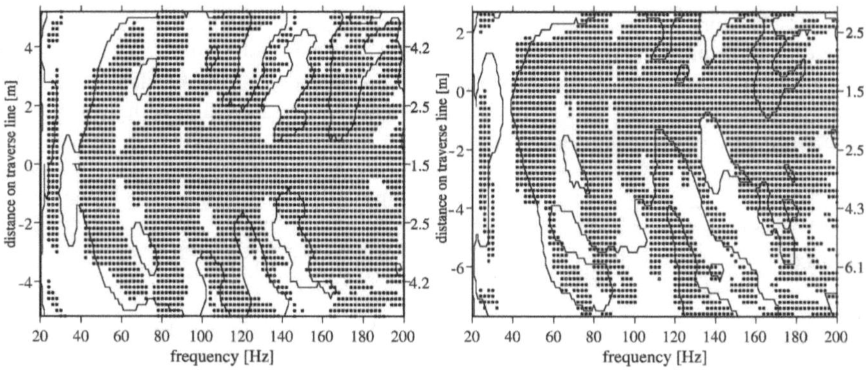

Fig. 12.12 1.5 dB regions for two source positions in the chamber obtained using numerical simulations.

single product $(\boldsymbol{I}-\boldsymbol{A})\boldsymbol{u}$ took 72.0 s on a SGI Origin3800 and required 6.4e9 floating point operations. In contrast the standard BEM would require at least N^2 $\approx$19.1e9 floating point operations. The above given computational resources have been measured using a *performance counter library*[5].

The linear system was solved using the GMRes [52] solver. Depending on frequency 80...120 iterations were necessary to obtain a residual of $\varepsilon = 10^{-6}$. The total solution time was 2.5...4 hours per frequency on a SGI Origin3800 of the Center for Information Services and High Performance Computing at the Technische Universität Dresden, Germany. Numerical results are compared with experimental data in Figure 12.12 for two different source positions. A solid dot represents the numerical result. The bounds of the 1.5 dB region obtained from experimental data are represented by a solid line. For both source positions numerical and experimental results agree well for frequencies higher than 100 Hz.

The example shows that the BEM can be applied to large scale problems. Especially when the boundary of the fluid domain has a complex shape the boundary element method is competitive to the finite element method as meshing the complex geometry of the fluid domain can be avoided.

12.4 Conclusion

Fast solution methods have been discussed to overcome the $\mathcal{O}(N^3)$ complexity of the standard boundary element method when using a direct solution of the system of linear equations. By the use of an iterative solver the complexity can be reduced to $\mathcal{O}(N^2)$, if the number of required iterations is much less than the number of unknowns N. Krylov subspace methods are the most suitable class of iterative solvers in the context here. The convergence of the iterative solvers is fairly accelerated

[5] http://www.fz-juelich.de/zam/PCL

Table 12.2 Memory requirement (left table) and computational resources needed to perform one matrix–vector product (right table) when using a six–level fast multipole algorithm. The first line, labeled "Equation (12.46)", represents the application of the boundary condition. The last column of the right table gives the performance of the algorithm with respect to the peak–performance of the computer's processor.

Matrix	Memory [Mb]
$\boldsymbol{H}_{\text{near}}$	2148
$\boldsymbol{G}_{\text{near}}$	2148
$\boldsymbol{W}$	111
$\boldsymbol{B}$	5
work	107
total	4519

line of *apply*	CPU–time [s]	CPU–time [%]	Flops [Mflops]	Flops [%]	Flop rate [%]
Eq. (12.46)	5.3	7.4	1569.5	24.3	73.5
1	0.6	0.8	34.6	0.5	14.3
22	1.3	1.8	16.2	0.3	3.2
20	1.1	1.5	214.6	3.3	48.1
6	23.5	32.7	2807.5	43.5	29.8
9–10	7.2	10.0	371.6	5.8	12.8
15–16	7.0	9.8	365.9	5.7	13.0
22	25.9	35.9	1074.1	16.6	10.4
total	72.0	100.0	6454.0	100.0	22.4

by strategies of preconditioning. Preconditioners based on the splitting of the system matrix are effective for the BEM that generates the dense system matrix. The sparse matrix corresponding to the near field interaction is factorized using an incomplete LU decomposition. The use of the hyper singular formulation leads to ill–conditioned matrices that cause slow convergence. Furthermore the properties of the boundary of the domain have a significant influence on the required number of iterations. In general it can be stated that complex shapes and low absorption will cause slow convergence.

A further reduction can be achieved through the use of fast BEMs which avoid the explicit set–up of the dense system matrix. Especially the multilevel fast multipole method seems to be the most widely accepted as such fast method. Based on the truncation of the series expansion of the fundamental solution, the fast multipole method yields an approximate factorization of the system matrix. The application of this method on multiple levels of a cluster tree enables the evaluation of a matrix–vector product with $\mathcal{O}(N \log^2 N)$ complexity. Therefore, the iterative solution of the linear system together with the multilevel fast multipole method yields a very efficient numerical method. The obtained higher efficiency of the accelerated BEM makes this numerical method applicable to real life problems.

References

1. Abramowitz M, Stegun IA (1965) Handbook of mathematical functions. Dover Publications, New York
2. Amini S (1987) An iterative method for the boundary element solution of the exterior acoustic problem. Journal of Computational and Applied Mathematics 20:109–117

3. Amini S (1999) On boundary integral operators for the Laplace and the Helmholtz equations and their discretisations. Engineering Analyis with Boundary Elements 23:327–337
4. Amini S, Chen K (1989) Conjugate gradient method for second kind integral equations–applications to the exterior acoustic problem. Engineering Analyis with Boundary Elements 6:72–77
5. Amini S, Maines ND (1998) Preconditioned Krylov subspace methods for boundary element solution of the Helmholtz equation. International Journal for Numerical Methods in Engineering 41:875–898
6. Amini S, Profit ATJ (2003) Multi–level fast multipole solution of the scattering problem. Engineering Analysis with Boundary Elements 27:547–564
7. Anderson CR (July 1992). An implementation of the fast multipole method without multipoles. SIAM Journal on Scientific and Statistical Computing 13:923–947
8. Barret R, Berry M, Chan TF, Demmel J, Donato J, Dongarra J, Eijkhout V, Pozo R, Romine C, Van der Vorst H (1994) Templates for the solution of linear systems: building blocks for iterative methods. Society for Industrial and Applied Mathematics, Philadelphia, USA
9. Bebendorf M, Rjasanov S (2003) Adaptive low–rank approximation of collocation matrices. Computing 70:1–24
10. Benzi M (2002) Preconditioning techniques for large linear systems: A survey. Journal of Computational Physics 182:418–477
11. Bespalov A (2000) On the usage of a regular grid for implementation of boundary integral methods for wave problems. Russian Journal of Numerical Analysis and Mathematical Modelling 15:469–488
12. Beylkin G, Coifman R, Rokhlin V (1991) Fast wavelet transforms and numerical algorithms. Communications on Pure and Applied Mathematics 44:141–183
13. Burton AJ, Miller GF (1971) The application of integral equation methods to the numerical solution of some exterior boundary value problems. Proceedings of the Royal Society of London A 323:201–210
14. Chen K (1998) On a class of preconditioning methods for dense linear systems from boundary elements. SIAM Journal of Scientific Computing 20:684–698
15. Chen K (2001) An analysis of sparse approximate inverse preconditioners for boundary integral equations. SIAM Journal on Matrix Analysis and Applications 22:1058–1078
16. Chen K, Harris PJ (2001) Efficient preconditioners for iterative solution of the boundary element equations for the three–dimensional Helmholtz equation. Applied Numerical Mathematics 36:475–489
17. Chen S, Liu Y (2000) A unified boundary element method for the analysis of sound and shell–like structure interactions. II. Efficient solution techniques. Journal of the Acoustical Society of America 108:2738–2745
18. Chew WC, Jin JM, Lu CC, Michelssen E, Song JM (1997) Fast solution methods in electromagnetics. IEEE Transaction on Antennas Propagation 45:533–543
19. Cipra BA (2000) The best of the 20th century: editors name top 10 algorithms. SIAM News 33:1–2
20. Coifman R, Rokhlin V, Wandzura S (June 1993) The fast multipole method for the wave equation: a pedestrian prescription. IEEE Antennas and Propagation Magazine 35:7–12
21. Dahmen W, Kleemann B, Prössdorf S, Schneider R (1997) Multiscale methods for the solution of the Helmholtz and Laplace equations. In: Wendland W (ed) Boundary Element Methods. Reports from the Final Conference of the Priority Research Program 1989–1995 of the German Research Foundation, Stuttgart, Springer–Verlag
22. Darve E (Feb. 2001) The fast multipole method I: Error analysis and asymptotic complexity. SIAM Journal on Numerical Analysis 38:98–128

23. Darve E, Havé P (2004) Efficient fast multipole method for low–frequency scattering. Journal on Computational Physics 197:341–363
24. Darve E, Havé P (2004) A fast multipole method for maxwell equations stable at all frequencies. Philosophical Transaction of the Royal Society of London A 362:1–27
25. Fischer M, Gauger U, Gaul L (2004) A multipole Galerkin boundary element method for acoustics. Engineering Analysis with Boundary Elements 28:155–162
26. Fletcher R (1976) Conjugate gradient methods for indefinite systems. In: Watson GA (ed) Dundee Conference on Numerical Analysis. 1975, Lecture Notes in Mathematics 506:73–89, Springer–Verlag, Berlin
27. Freeden W, Schneider F (1999) Runge–Walsh–wavelet approximation for the Helmholtz equation. Journal of Mathematical Analysis and Application 235:533–566
28. Freund RW (1993) A transpose–free quasi–minimum residual algorithm for non–Hermitian linear systems. SIAM Journal on Scientific Computing 14:470–482
29. Giebermann K (2001) Multilevel approximation of boundary integral operators. Computing 67:183–207
30. Greengard L, Huang J, Rokhlin V, Wandzura S (1998) Accelerating fast multipole methods for the Helmholtz equation at low frequencies. IEEE Computational Science and Engineering 5:32–38
31. Greengard L, Rokhlin V (1987) A fast algorithm for particle simulations. Journal of Computational Physics 73:325–348
32. Gumerov N, Duraiswami R (2005) Fast multipole methods for the Helmholtz equation in three dimensions. Elsevier, Oxford
33. Gutknecht MH (1993) Variations of BiCGSTAB for matrices with complex spectrum. SIAM Journal on Scientific Computing 14:1020–1033
34. Gyure MF, Stalzer MA (1998) A prescription for the multilevel Helmholtz FMM. IEEE Computanional Science & Engineering 5:39–47
35. Hackbusch W (1999) A sparse matrix arithmetic based on H–matrices. Part I: Introduction to H–matrices. Computing 62:89–108
36. Hackbusch W, Nowak Z (1989) On the fast matrix multiplication in the boundary element method by panel clustering. Numerische Mathmatik 54:463–491
37. Harris PJ, Chen K (2003) On efficient preconditioners for iterative solution of a Galerkin boundary element equation for the three–dimensional exterior Helmholtz problem. Journal of Computation and Applied Mathematics 156:303–318
38. Hestenes MR, Stiefel E (1952) Methods of conjugate gradients for solving linear systems. Journal of Research of the National Bureau of Standards 49:409–436
39. Huybrechs D, Simoens J, Vandewalle S (2004) A note on wave number dependence of wavelet matrix compression for integral equations with oscillatory kernel. Journal of Computation and Applied Mathematics 172:233–246
40. Zhao J–S, Chew WC (2000) Three-dimensional multilevel fast multipole algorithm from static to electrodynamic. Microwave and Optical Technology Letters 26:43–48
41. Lanczos C (1952) Solution of systems of linear equations by minimized iterations. Journal of Research of the National Bureau of Standards 49:33–53
42. Marburg S, Schneider S (2003) Influence of element types on numeric error for acoustic boundary elements. Journal of Computational Acoustics 11:363–386
43. Marburg S, Schneider S (2003) Performance of iterative solvers for acoustic problems. Part I. Solvers and effect of diagonal preconditioning. Engineering Analysis with Boundary Element 27:727–750
44. Ochmann M, Homm A, Makarov S, Semenov S (2003) An iterative GMRES–based boundary element solver for acoustic scattering. Engineering Analysis with Boundary Element 27:717–725

45. Okamoto N, Otsuru T, Tomiku R, Yasuda Y (2007) Numerical analysis of large–scale sound fields using iterative methods. Part II: Application of Krylov subspace methods to finite element analysis. Journal of Computational Acoustics 15, accepted for publication
46. Otsuru T, Uchinoura Y, Tomiku R, Okamoto N, Takahashi Y (2004) Basic concept, accuracy and application of large–scale finite element sound field analysis of rooms. In: Proceedings of the 18th International Congress on Acoustics (ICA), Kyoto I 479–482
47. Prasad KG, Kane JH, Keyes DE, Balakrishna C (1994) Preconditioned Krylov solvers for BEA. International Journal for Numerical Methods in Engineering 37:1651–1672
48. Rahola J (1996) Diagonal forms of the translation operators in the fast multipole algorithm for scattering problems. BIT Numerical Mathematics 36:333–358
49. Rokhlin V (Feb. 1990) Rapid solution of integral equations of scattering theory in two dimensions. Journal of Computational Physics 86:414–439
50. Rokhlin V (1993) Diagonal forms of the translation operators for the Helmholtz equation in three dimensions. Applied and Computational Harmonic Analysis 1:82–93
51. Saad Y (1994) ILUT: a dual threshold incomplete LU factorization. Numer. Linear Algebra 1:387–402
52. Saad Y, Schultz MH (1986) GMRES: a generalized minimal residual algorithm for solving nonsymmetric linear systems. SIAM Journal on Scientific and Statistical Computing 7:856–869
53. Saad Y, van der Vorst HA (2000) Iterative solution of linear systems in the 20th century. Journal of Computational and Applied Mathematics 123:1–33
54. Sakuma T, Takamura N, Yasuda Y, Sakamoto S (2005) Numerical analysis of the additional attenuation due to the tops of edge modified barriers. In: Proceedings of the Inter–Noise 2005, Rio de Janeiro, No. 1956 (CD)
55. Sakuma T, Yasuda Y (2002) Fast multipole boundary element method for large–scale steady–state sound field analysis, Part I: Setup and validation. Acta Acustica united with Acustica 88:513–525
56. Sauter S (2000) Variable order panel clustering. Computing 64:223–261
57. Schneider S (2003) Application of fast methods for acoustic scattering and radiation problems. Journal of Computational Acoustics 11:387–401
58. Schneider S, Kern C (2007) Acoustical behavior of the large anechoic chamber at the Laboratoire de Mécanique et d'Acoustique in the low frequency range. Acta Acustica united with Acustica, submitted for publication.
59. Schneider S, Marburg S (2003) Performance of iterative solvers for acoustic problems. Part II. Acceleration by ILU–type preconditioner. Engineering Analysis with Boundary Elements 27:751–757
60. Shen L, Liu YJ (2006) An adaptive fast multipole boundary element method for solving large–scale three–dimensional acoustic wave problems based on the Burton–Miller formulation. Computational Mechanics, accepted for publication
61. Sleijpen GLC, Fokkema DR (1993) BiCGStab(l) for linear matrices involving unsymmetric matrices with complex spectrum. Electronic Transactions on Numerical Analysis 1:11–32.
62. Song J, Lu CC, Chew WC (Okt. 1997) Multilevel fast multipole algorithm for electromagnetic scattering by large complex objects. IEEE Transactions on Antennas and Propagation 45:1488–1493
63. Sonneveld P (1989) CGS, a fast Lanczos–type solver for nonsymmetric linear systems. SIAM Journal on Scientific Statistical Computing 10:36–52
64. Stroud AH (1971) Approximate calculation of multiple integrals. Prentice–Hall, New York

65. van der Vorst HA (1992) Bi–CGSTAB: a fast and smoothly converging variant of Bi–CG for the solution of nonsymmetric linear systems. SIAM Journal on Scientific Statistical Computing 13:631–644
66. van der Vorst HA (2003) Iterative Krylov methods for large linear systems. Cambridge University Press, Cambridge
67. Vavasis SA (1992) Preconditioning for boundary integral equations. SIAM Journal on Matrix Analysis and Applications 13:905–925
68. Velamparambil S, Chew WC (2001) A fast polynomial representation for the translation operators of an MLFMA. Microwave and Optical Technology Letters 28:298–303
69. Wallén H, Sarvas J(2005) Translation procedures for broadband MLFMA. Progress in Electromagnetics Research (PIER) 55:47–78
70. Yan Y (1994) Sparse preconditioned iterative methods for dense linear systems. SIAM Journal on Scientific Computing 15:1190–1200
71. Yarvin N, Rokhlin V (1999) An improved fast multipole algorithm for potential fields on the line. SIAM Journal on Numerical Analysis 36:629–666
72. Yasuda Y, Sakamoto S, Kosaka Y, Sakuma T, Okamoto N, Oshima T (2007) Numerical analysis of large–scale sound fields using iterative methods. Part I: Application of Krylov subspace methods to boundary element analysis. Journal of Computational Acoustics 15, accepted for publication
73. Yasuda Y, Sakuma T (2003) Fast multipole boundary element method for large–scale steady–state sound field analysis. Part II: Examination of numerical items. Acta Acustica united with Acustica 89:28–38
74. Ylä–Oijala P, Järvenpää S (2006) Iterative solution of high–order boundary element method for acoustic impedance boundary value problems. Journal of Sound and Vibration 291:824–843
75. Zhang SL (1997) GPBi–CG: Generalized product–type methods based on Bi–CG for solving nonsymmetric linear systems. SIAM Journal on Scientific Computing 18:537–551

13

Multi–domain Boundary Element Method in Acoustics

Ting–Wen Wu

Department of Mechanical Engineering, University of Kentucky, Lexington, KY 40506, USA
timwu@engr.uky.edu

Summary. This chapter reviews the development and applications of the multi–domain boundary element method in acoustics. Early development of the method originated from the need for solving problems involving two or more acoustic media, as well as radiation and scattering from thin bodies. The concept of the multi–domain method may be simple, but real implementation could be tedious when one has to define many subdomains and match many interface boundary conditions. As new techniques for regularizing the hypersingular integral equation became more available in the early 1990's, thin–body problems no longer required the multi–domain boundary element method. A so–called direct *mixed–body boundary element method* was also developed to handle a mix of regular and thin bodies. Later, problems involving two or more acoustic media could also be solved by an extension of the mixed–body concept. Nevertheless, the multi–domain boundary element still remains to be a viable choice today. When implemented in a modular way using the impedance matrix approach, the multi–domain boundary element method is very efficient and requires only a small amount of computer memory. Instead of being totally eliminated, the multi–domain boundary element method has actually become more powerful and easier to use when used in conjunction with the direct mixed–body boundary element method. Recent examples in muffler and silencer analysis are used to demonstrate the idea.

13.1 Introduction

The multi–domain boundary element method (BEM) has been used in potential problems and elasticity for decades [1]. It has also been applied to a variety of acoustic problems since the early 1990's [2,7,8,11,12]. The most common application of the multi–domain BEM in acoustics is to solve problems with two or more acoustic media, such as an acoustic cavity with bulk–reacting sound absorbing materials [11]. It may also be applied to coupled interior/exterior problems [7] and problems involving thin bodies [2,8,12]. The basic idea of the multi–domain BEM is to divide the acoustic domain into a few subdomains so that each subdomain is homogeneous and has a well–defined boundary. As demonstrated in Figure 13.1, a two-medium problem can be easily divided into two homogeneous subdomains according to their material properties division. Two surface meshes are generated for the two subdomains, re-

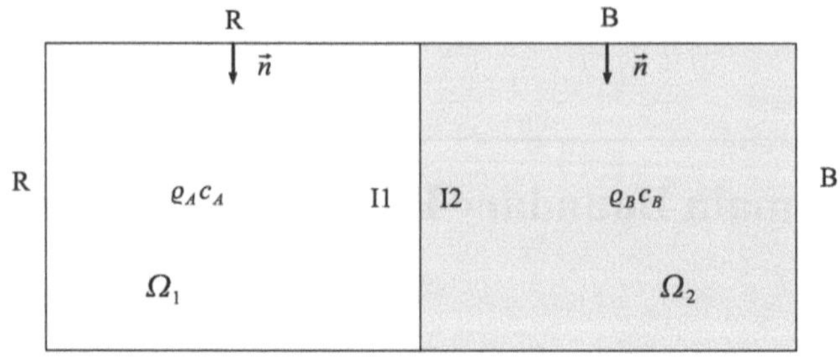

Fig. 13.1 Multi-domain BEM for the two-medium problem.

spectively. Continuity of sound pressure and the normal particle velocity is enforced at the interface. To match the interface boundary conditions, it may be desirable for two subdomains to have an identical mesh along the interface except that their normal directions are opposite to each other. Figure 13.2 shows another application of the multi–domain BEM in which a muffler has some thin bodies (such as the extended inlet tube and the thin baffle) inside its interior. It is noted that imaginary interfaces, as marked by the dotted line segments in Figure 13.2, have to be constructed in order to create a well–defined boundary for each subdomain. Although the concept of the multi–domain BEM may be simple, real implementation could become very tedious when one has to define many subdomains or create a lot of imaginary interfaces for problems with a complex geometry.

It has been well known that radiation and scattering from thin bodies can be solved by the hypersingular integral equation without using the multi–domain BEM [10]. The derivation actually begins with a multi–domain BEM formulation, but the Helmholtz integral equations of the subdomains are added together [14]. The integrals over any imaginary interfaces cancel out with each other due to continuity of sound pressure and the normal particle velocity. The numerical system is then supplemented by adding the hypersingular normal–derivative integral equation. In 1980, Terai [10] was the first to derive a simple formula for evaluating the hypersingular integral equation on constant planar elements. Since 1990, new techniques for regularizing the hypersingular integration equation on curvilinear elements have also become more available [4,5]. By then, it was clear that the classical thin–body problem no longer required the tedious multi–domain approach. To put the hypersingular integral equation into more practical use, Wu and his co–workers [13, 18, 19] have come up with a so–called *direct mixed–body BEM* to handle problems involving a mix of regular, thin, and perforated surfaces. Since 2002, the mixed–body concept has even been extended to problems with bulk–reacting sound absorbing materials [16, 17].

If the Helmholtz integral equations for all subdomains can be added together, and the hypersingular integral equation is provided to make up for any missing equations, there seems to be no need to define any subdomains. As far as the end users are concerned, defining surfaces may be part of their routine pre-processing job, but defining subdomains could be an extra burden, especially when many subdomains are intricately connected to one another. The direct mixed–body BEM does provide

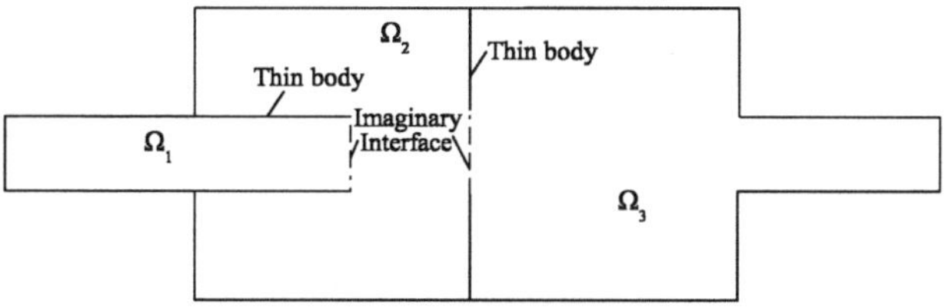

Fig. 13.2 Multi-domain BEM for a muffler with thin bodies.

a relief to that end. However, one advantage of the multi–domain BEM for interior problems is its numerical efficiency. When implemented in a modular way using the impedance matrix, a subdomain should produce a much smaller BEM matrix than the entire one-domain approach. For real-life problems consisting of more than thousands of elements, it may be necessary to divide the entire domain into smaller subdomains. Lou et al. [6] have presented a substructuring technique that uses the direct mixed–body BEM in each subdomain for a multi–domain silencer analysis. With the help of the direct mixed–body BEM, each subdomain does not need to be homogeneous or thin–body free anymore. The breakup of the entire domain into subdomains can be done more freely without worrying about the material properties division or thin bodies placement. It turns out that the direct mixed–body BEM, originally meant to totally eliminate the tedious multi–domain BEM, has actually facilitated the multi–domain BEM operation in the end.

In this chapter, the conventional multi–domain BEM is first reviewed. Impedance matrix is introduced as an efficient way to solve the coupled system of equations. The direct mixed–body BEM and the substructuring technique are then reviewed. Recent examples in muffler and silencer analysis are used to demonstrate the idea.

13.2 Conventional Multi–domain BEM

We will use the simple two-medium problem as shown in Figure 13.1 to demonstrate the conventional multi–domain BEM formulation. Without loss of generality, let's assume the first subdomain Ω_1 to be air, and the second subdomain Ω_2 to be a bulk–reacting sound absorbing material. Let ϱ_A and c_A denote the mean density and speed of sound of air, and ϱ_B and c_B denote the mean density and speed of sound of the bulk–reacting material, respectively. The governing differential equations for this two-medium problem in the frequency domain are

$$\nabla^2 p + k_A^2 p = 0 \qquad \text{in air} \tag{13.1}$$

$$\nabla^2 p + k_B^2 p = 0 \qquad \text{in the bulk–reacting material} \tag{13.2}$$

where p is the sound pressure, and k_A and k_B are the wavenumbers in air and the bulk–reacting material, respectively. The boundary of the air subdomain comprises two parts denoted by R and I1, where R is the non–interface part and I1 is on the interface between the two subdomains. Similarly, the boundary of the bulk–reacting material subdomain comprises B and I2. Let $\mathbf{n}$ be the unit normal vector pointing into each subdomain. If the $\mathrm{e}^{\mathrm{i}\omega t}$ convention is adopted, where $\mathrm{i}^2 = -1$ and ω is the angular frequency, the Helmholtz integral equation can be written for each subdomain as follows:

$$\int_{\mathrm{R+I1}} \left(p\frac{\partial \psi_A}{\partial n} + \mathrm{i}\varrho_A \omega v \psi_A \right) \mathrm{d}S = \begin{cases} 4\pi p(P) & P \in \Omega_1 & (13.3\mathrm{a}) \\ 2\pi p(P) & P \in \mathrm{R+I1} & (13.3\mathrm{b}) \\ 0 & P \in \Omega_2 + \mathrm{B} & (13.3\mathrm{c}) \end{cases}$$

$$\int_{\mathrm{R+I2}} \left(p\frac{\partial \psi_B}{\partial n} + \mathrm{i}\varrho_B \omega v \psi_B \right) \mathrm{d}S = \begin{cases} 4\pi p(P) & P \in \Omega_2 & (13.4\mathrm{a}) \\ 2\pi p(P) & P \in \mathrm{R+I2} & (13.4\mathrm{b}) \\ 0 & P \in \Omega_1 + \mathrm{B} & (13.4\mathrm{c}) \end{cases}$$

where v is the normal particle velocity, P is the collocation point, ψ_A and ψ_B are the free–space Green's functions in air and the bulk–reacting material, respectively. For simplicity, we use discontinuous elements, such as the constant element, in the derivation, although continuous elements may also be used in the multi–domain BEM. As a result, the solid angle at the collocation point P on the surface is always 2π. The explicit expressions for the two Green's functions are

$$\psi_A = \frac{\mathrm{e}^{-\mathrm{i}k_A r}}{r} \qquad (13.5\mathrm{a})$$

$$\psi_B = \frac{\mathrm{e}^{-\mathrm{i}k_B r}}{r} \qquad (13.5\mathrm{b})$$

where $r = |P - Q|$, and Q is any integration point on the boundary.

For a well–posed boundary value problem, either p or v or a linear combination of them is specified on R and B. To simplify the discussion, we will assume that v is specified as the boundary condition on R and B, and the unknown is p. At the interface, both p and v are unknowns. If the two subdomains are in direct contact with each other, both p and v are continuous across the interface. In other words,

$$p_{\mathrm{I1}} = p_{\mathrm{I2}} \qquad (13.6\mathrm{a})$$

$$v_{\mathrm{I1}} = -v_{\mathrm{I2}} \qquad (13.6\mathrm{b})$$

where the subscripts denote the boundary sides the variables belong to. The negative sign in Equation (13.6b) is due to the opposite normal directions at the interface. There are other types of interface conditions that may be applied at the interface in the multi–domain BEM [12], for example, the transfer impedance for perforated tubes [9]. After discretization and collocation, Equations (13.3b) and (13.4b) may be assembled into a big global matrix equation, with the unknowns being p on $\mathrm{R} + \mathrm{B}$, and p_{I1} and v_{I1} at the interface. Equations (13.3b) and (13.4b) are coupled by the interface conditions in Equations (13.6a) and (13.6b).

A more efficient way to solve the coupled system of equations is to find the impedance matrix for each subdomain first. In other words, regardless of the boundary conditions specified, Equations (13.3b) and (13.4b) can be converted into

$$\begin{bmatrix} p_R \\ p_{I1} \end{bmatrix} = \begin{bmatrix} z_{11} & z_{12} \\ z_{21} & z_{22} \end{bmatrix} \begin{bmatrix} v_R \\ v_{I1} \end{bmatrix} \tag{13.7}$$

and

$$\begin{bmatrix} p_B \\ p_{I2} \end{bmatrix} = \begin{bmatrix} z_{31} & z_{32} \\ z_{41} & z_{42} \end{bmatrix} \begin{bmatrix} v_B \\ v_{I2} \end{bmatrix} \tag{13.8}$$

respectively. In each of the above two equations, the matrix relating the boundary pressures to the boundary normal velocities for each subdomain is called the *impedance matrix*. A matrix inverse operation (usually handled by the LU decomposition instead of direct matrix inverse) is involved in computing the impedance matrix for each subdomain, but the size of the matrix is limited to the number of nodes of each subdomain only, instead of the total number of nodes of the coupled system. Therefore, this approach is considered to be more efficient than solving the big global matrix directly. With the individual impedance matrices ready, we can equate the lower rows of Equations (13.7) and (13.8) *via* the interface continuity conditions, to get

$$z_{21} v_R + z_{22} v_{I1} = z_{41} v_B + z_{42} \left(-v_{I1} \right) . \tag{13.9}$$

The above equation can be solved for v_{I1}:

$$v_{I1} = \left(z_{22} + z_{42} \right)^{-1} \left(z_{41} v_B - z_{21} v_R \right) \tag{13.10}$$

where the matrix inverse operation is done on an even smaller matrix that involves the number of nodes on the interface surface only. Since v_R and v_B are already specified as the boundary conditions, the solution of v_{I1} can then be substituted back into Equations (13.7) and (13.8) to find the sound pressures on any boundary surfaces.

The impedance matrices are easy to define if discontinuous elements are used. In case that the more popular continuous elements are used, the particle normal velocity will not have a unique value at corners and edges. Then, it is desirable to assume a constant normal velocity on each element, and define the vector of the particle normal velocities based on the number of elements, instead of the number of nodes. Under such circumstances, since the sound pressure is still defined at each node, the impedance matrix would be a rectangular matrix, rather than a square matrix. Assuming a constant normal velocity on each continuous element will require taking an average of the nodal normal velocities for each element on R and B. This may have a minor negative effect on accuracy when compared to the big global matrix approach.

13.3 Direct Mixed–body BEM

Defining different subdomains and matching interfaces could become a very tedious job, especially when there are too many subdomains intricately connected to one another. To partially alleviate the difficulty, the direct mixed–body BEM was developed with the original goal to eliminate the tedious multi–domain approach. In the following, we will demonstrate the idea for the simple two–medium problem in Figure 13.1 first and then for the thin–body problem in Figure 13.2 as well. There are also many other types of interfaces that can be handled by the direct mixed–body BEM approach, and we will summarize the complete integral formulation at the end of this section.

13.3.1 Direct Mixed–body BEM for Two–medium Problem

Instead of defining two coinciding interface surfaces, I1 and I2 as shown in Figure 13.1, we will just keep I1 and simply call it "I". The unit normal vector on interface I is pointing into the air subdomain. Since I2 is replaced by I, a negative sign needs to be applied to the integral over the interface portion in Equation (13.4) due to the opposite normal direction of I as compared to I2. We then add Equation (13.3) to Equation (13.4) to get [17]

$$\int_{\mathrm{R}}\left(p\frac{\partial\psi_A}{\partial n}+\mathrm{i}\varrho_A\omega v\psi_A\right)\mathrm{d}S+\int_{\mathrm{B}}\left(p\frac{\partial\psi_B}{\partial n}+\mathrm{i}\varrho_A\omega v\psi_B\right)\mathrm{d}S+$$
$$+\int_{\mathrm{I}}\left[p\left(\frac{\partial\psi_A}{\partial n}-\frac{\partial\psi_B}{\partial n}\right)+\mathrm{i}\omega v\left(\varrho_A\psi_A-\varrho_B\psi_B\right)\right]\mathrm{d}S=$$
$$=\begin{cases}4\pi p(P) & P\in\Omega_1+\Omega_2 \qquad (13.11a)\\ 2\pi p(P) & P\in\mathrm{R}+\mathrm{B} \qquad (13.11b)\\ 4\pi p(P) & P\in\mathrm{I} \qquad (13.11c)\end{cases}$$

Note that the subtraction signs in the integral over I are due to the opposite normal directions from the two different subdomains. On the R and B surfaces, either p or v or a linear combination of them is specified as the boundary condition, and there is only one unknown variable at each node. Equation (13.11b) is sufficient for providing the necessary equation to solve the unknown variable at each node. However, at interface I, neither p nor v is known. Equation (13.11c) itself would not be sufficient to solve the two unknowns at each node on I. An additional equation is needed to supplement Equation (13.11c).

To provide the additional equation when P is on the interface, the hypersingular normal–derivative integral equations corresponding to Equations (13.3b) and (13.4b), respectively, are first written. They are

$$\int_{\mathrm{R}}\left(p\frac{\partial^2\psi_A}{\partial n\partial n^P}+\mathrm{i}\varrho_A\omega v\frac{\partial\psi_A}{\partial n^P}\right)\mathrm{d}S+\int_{\mathrm{I}}\left(p\frac{\partial^2\psi_A}{\partial n\partial n^P}+\mathrm{i}\varrho_A\omega v\frac{\partial\psi_A}{\partial n^P}\right)\mathrm{d}S$$
$$=2\pi\frac{\partial p}{\partial n}(P)\,,\qquad P\in\mathrm{I}\ \text{ from the air side} \qquad (13.12)$$

$$\int_{\mathrm{B}}\left(p\frac{\partial^2\psi_B}{\partial n\partial n^P}+\mathrm{i}\varrho_B\omega v\frac{\partial\psi_B}{\partial n^P}\right)\mathrm{d}S-\int_{\mathrm{I}}\left(p\frac{\partial^2\psi_B}{\partial n\partial n^P}+\mathrm{i}\varrho_B\omega v\frac{\partial\psi_B}{\partial n^P}\right)\mathrm{d}S$$
$$=2\pi\frac{\partial p}{\partial n}(P)\,,\qquad P\in\mathrm{I}\ \text{ from the bulk–reacting side} \qquad (13.13)$$

where n^P is the unit normal vector at P and the normal derivative is taken with respect to the coordinates of P. It should be noted that although both p and v are continuous across interface I, $\partial p/\partial n$ is not. Using the momentum equation $\partial p/\partial n = -\mathrm{i}\varrho\omega v$ in each of Equations (13.12) and (13.13) and then summing up the two equations produce

$$\int_{\mathrm{R}} \left(p \frac{\partial^2 \psi_A}{\partial n \partial n^P} + \mathrm{i}\varrho_A \omega v \frac{\partial \psi_A}{\partial n^P} \right) \mathrm{d}S + \int_{\mathrm{B}} \left(p \frac{\partial^2 \psi_B}{\partial n \partial n^P} + \mathrm{i}\varrho_B \omega v \frac{\partial \psi_B}{\partial n^P} \right) \mathrm{d}S +$$

$$+ \int_{\mathrm{I}} \left[p \left(\frac{\partial^2 \psi_A}{\partial n \partial n^P} - \frac{\partial^2 \psi_B}{\partial n \partial n^P} \right) + \mathrm{i}\omega v \left(\varrho_A \frac{\partial \psi_A}{\partial n^P} - \varrho_B \frac{\partial \psi_B}{\partial n^P} \right) \right] \mathrm{d}S =$$

$$= -2\pi \mathrm{i}\omega \left(\varrho_A + \varrho_B \right) v(P) \qquad P \in \mathrm{I}\,. \tag{13.14}$$

Equation (13.14) provides the additional equation to supplement Equation (13.11c) for the two unknowns at each node on I. Note that although both Equations (13.12) and (13.13) are hypersingular, Equation (13.14) is only weakly singular because the strong singularity in the two Green's functions cancel out in subtraction. Equations (13.11b), (13.11c), and (13.14) are then solved simultaneously for the boundary unknowns on R, B and I. It is noted that there is no need to define any subdomains because all the subdomain integral equations are added together in one single equation.

13.3.2 Direct Mixed–body BEM for Thin Bodies

Similarly, instead of defining two coinciding surfaces along any thin body or any imaginary interface in Figure 13.2, we will just define a single layer of surface called "T" for any thin body, and a single layer of surface called "I" for any imaginary interface. Since both sides of T or I are air, it does not matter which side the unit normal vector is pointing into. On any T surface, the side into which the normal is pointing is called the positive side, while the opposite side is called the negative side. Sound pressure on the positive side is denoted by p^+, and on the negative side p^-. The normal particle velocity v is continuous across T. The Helmholtz integral equation identical to Equation (13.3) is first written for each subdomain. All the subdomain integral equations are then added together. This summing procedure produces [13, 14]

$$\int_{\mathrm{R}} \left(p \frac{\partial \psi_A}{\partial n} + \mathrm{i}\varrho_A \omega v \psi_A \right) \mathrm{d}S + \int_{\mathrm{T}} \frac{\partial \psi_A}{\partial n} \left(p^+ - p^- \right) \mathrm{d}S =$$

$$= \begin{cases} 4\pi p(P) & P \in \Omega \quad (13.15\mathrm{a}) \\ 2\pi p(P) & P \in \mathrm{R} \quad (13.15\mathrm{b}) \\ 2\pi \left[p^+(P) - p^-(P) \right] & P \in \mathrm{T} \quad (13.15\mathrm{c}) \end{cases}$$

It should be noted that the integral contributions over any imaginary surface I are canceled out due to continuity of p and v, and the opposite normal directions from

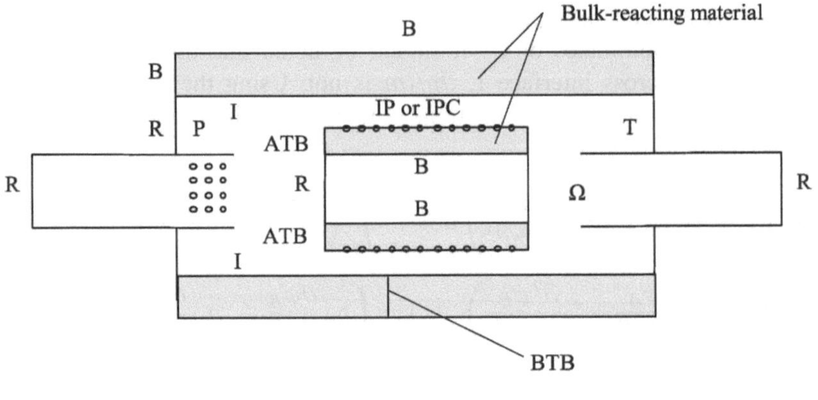

Fig. 13.3 Types of surface modeled in the direct mixed–body BEM.

both sides. Equation (13.15c) is not sufficient to solve for the two unknowns, p^+ and p^-, on T. A companion hypersingular normal–derivative integral equation is needed to supplement Equation (13.15c). The equation is

$$\int_{\mathrm{R}} \left(p \frac{\partial^2 \psi_A}{\partial n \partial n^P} + \mathrm{i} \varrho_A \omega v \frac{\partial \psi_A}{\partial n^P} \right) \mathrm{d}S + \int_{\mathrm{T}} \frac{\partial^2 \psi_A}{\partial n \partial n^P} (p^+ - p^-) \, \mathrm{d}S =$$

$$= -4\pi \mathrm{i} \varrho_A \omega v(P) \qquad P \in \mathrm{T} \tag{13.16}$$

Equations (13.15b), (13.15c), and (13.16) can be solved simultaneously for all the unknown variables on R and T. Normally, only Equation (13.15b) and (13.16) are needed to solve for p on R and $(p^+ - p^-)$ on T. Then the solutions are substituted back into Equations (13.15a) and (13.15c) to find p in Ω and $(p^+ + p^-)$ on T. With the pressure sum and the pressure difference known, it is then easy to find the individual pressure on either side of T. It is noted again that there is no need to define any subdomains in this direct mixed–body BEM approach. The biggest advantage is perhaps the total elimination of any imaginary interfaces.

13.3.3 Complete Mixed–body BEM Formulation for Silencer Analysis

In addition to the interface I between any two different media and the thin body surface T, there are other types of surfaces that can be modeled by the direct mixed–body BEM. Figure 13.3 shows a typical packed silencer configuration that can be solved by the mixed–body approach. With reference to the figure, Ω denotes the interior acoustic domain of the silencer. The symbols R, T, P, B, I, IP, IPC, ATB and BTB represent *Regular, Thin, Perforated, Bulk–reacting, Interface, Interface with Perforated surface, Interface with Perforated surface and Cloth, Air–Thin–Bulk–reacting and Bulk–reacting–Thin–Bulk–reacting* surfaces, respectively. The R surfaces include the exterior silencer surfaces (with no bulk–reacting packing), the external inlet/outlet tubes, and the inlet/outlet ends. The T surfaces are the thin components inside the silencer, such as the extended inlet/outlet tubes, thin baffles, thin flow plugs, and internal connecting tubes. Note that a T surface has air on both sides of the thin surface. The P surfaces are designated for perforated tubes or any perforated thin plates with air on both sides. The R surfaces are the exterior boundary surfaces of the bulk–reacting lining. The I surfaces are the interfaces between the bulk–reacting material and air. The IP surfaces are the perforated interfaces between the bulk–reacting material and air. The IPC interfaces are the same as the IP interfaces except that a layer of protective cloth is inserted between the perforated plate and the bulk–reacting material to protect the bulk–reacting material from being blown out. The ATB surfaces represent any rigid thin plates between the bulk–reacting material and air. The BTB surfaces represent any embedded rigid thin plates with the bulk–reacting material on both sides.

Imagine how many subdomains we would need to define and how many different interface conditions we would have to match for a typical silencer problem in Figure 13.3. The direct mixed–body BEM does provide a relief to that end. Without going through the derivation details, the complete boundary integral equations are given in the following [6, 15–17]:

$$
\begin{aligned}
&\int_{\mathrm{R}} \left(p \frac{\partial \psi_A}{\partial n} + \mathrm{i} \varrho_A \omega v \psi_A \right) \mathrm{d}S + \int_{\mathrm{T+P}} \frac{\partial \psi_A}{\partial n} \left(p^+ - p^- \right) \mathrm{d}S + \\
&+ \int_{\mathrm{B}} \left(p \frac{\partial \psi_B}{\partial n} + \mathrm{i} \varrho_B \omega v \psi_B \right) \mathrm{d}S + \int_{\mathrm{BTB}} \frac{\partial \psi_B}{\partial n} \left(p^+ - p^- \right) \mathrm{d}S + \\
&+ \int_{\mathrm{I}} \left[p \left(\frac{\partial \psi_A}{\partial n} - \frac{\partial \psi_B}{\partial n} \right) + \mathrm{i} \omega v \left(\varrho_A \psi_A - \varrho_B \psi_B \right) \right] \mathrm{d}S + \\
&+ \int_{\mathrm{IP+IPC}} \left[p_A \left(\frac{\partial \psi_A}{\partial n} - \frac{\partial \psi_B}{\partial n} \right) - \varrho_A c_A \xi^* v \frac{\partial \psi_B}{\partial n} + \right. \\
&\qquad\qquad \left. + \mathrm{i} \omega v \left(\varrho_A \psi_A - \varrho_B \psi_B \right) \right] \mathrm{d}S + \\
&+ \int_{\mathrm{ATB}} \left(p_A \frac{\partial \psi_A}{\partial n} - p_B \frac{\partial \psi_B}{\partial n} \right) \mathrm{d}S =
\end{aligned}
$$

$$
= \begin{cases}
4\pi p(P) & P \in \Omega & (13.17\mathrm{a}) \\
2\pi p(P) & P \in \mathrm{R+B} & (13.17\mathrm{b}) \\
2\pi \left[p^+(P) - p^-(P) \right] & P \in \mathrm{T+P+BTB} & (13.17\mathrm{c}) \\
4\pi p(P) & P \in \mathrm{I} & (13.17\mathrm{d}) \\
4\pi p_A(P) + 2\pi \varrho_A c_A \xi^* v(P) & P \in \mathrm{IP+IPC} & (13.17\mathrm{e}) \\
2\pi \left[p_A(P) + p_B(P) \right] & P \in \mathrm{ATB} & (13.17\mathrm{f})
\end{cases}
$$

and

$$\int_{\mathrm{R}} \left(p \frac{\partial^2 \psi_A}{\partial n \partial n^P} + \mathrm{i} \varrho_A \omega v \frac{\partial \psi_A}{\partial n^P} \right) \mathrm{d}S + \int_{\mathrm{T+P}} \frac{\partial^2 \psi_A}{\partial n \partial n^P} \left(p^+ - p^- \right) \mathrm{d}S +$$

$$+ \int_{\mathrm{B}} \left(p \frac{\partial^2 \psi_B}{\partial n \partial n^P} + \mathrm{i} \varrho_B \omega v \frac{\partial \psi_B}{\partial n^P} \right) \mathrm{d}S + \int_{\mathrm{BTB}} \frac{\partial^2 \psi_B}{\partial n \partial n^P} \left(p^+ - p^- \right) \mathrm{d}S +$$

$$+ \int_{\mathrm{I}} \left[p \left(\frac{\partial^2 \psi_A}{\partial n \partial n^P} - \frac{\partial^2 \psi_B}{\partial n \partial n^P} \right) + \mathrm{i} \omega v \left(\varrho_A \frac{\partial \psi_A}{\partial n^P} - \varrho_B \frac{\partial \psi_B}{\partial n^P} \right) \right] \mathrm{d}S +$$

$$+ \int_{\mathrm{IP+IPC}} \left[p_A \left(\frac{\partial^2 \psi_A}{\partial n \partial n^P} - \frac{\partial^2 \psi_B}{\partial n \partial n^P} \right) - \varrho_A c_A \xi^* v \frac{\partial^2 \psi_B}{\partial n \partial n^P} + \right.$$
$$\left. + \mathrm{i} \omega v \left(\varrho_A \frac{\partial \psi_A}{\partial n^P} - \varrho_B \frac{\partial \psi_B}{\partial n^P} \right) \right] \mathrm{d}S +$$

$$+ \int_{\mathrm{ATB}} \left(p_A \frac{\partial^2 \psi_A}{\partial n \partial n^P} - p_B \frac{\partial^2 \psi_B}{\partial n \partial n^P} \right) \mathrm{d}S =$$

$$= \begin{cases} 0 & P \in \mathrm{T} + \mathrm{BTB} + \mathrm{ATB} \quad (13.18\mathrm{a}) \\ 4\pi \dfrac{\mathrm{i} k_A}{\xi} \left[p^+(P) - p^-(P) \right] & P \in \mathrm{P} \quad (13.18\mathrm{b}) \\ -2\pi \mathrm{i} \omega \left(\varrho_A + \varrho_B \right) v(P) & P \in \mathrm{I} + \mathrm{IP} + \mathrm{IPC} \quad (13.18\mathrm{c}) \end{cases}$$

where ξ is the dimensionless transfer impedance for perforated tubes or perforated surfaces [9], and ξ^* is the dimensionless transfer impedance for IP or IPC interfaces. On an IP interface, ξ^* is identical to ξ. On an IPC interface, ξ^* is the sum of ξ and the dimensionless transfer impedance of the protective cloth [16].

13.4 Substructuring: Multi–domain BEM Revisited

Although the direct mixed–body BEM is easy to use, it does produce a big global matrix because all the subdomain integral equations are added together. For a large–size model consisting of thousands of elements, it may become necessary to divide the large domain into several smaller subdomains first so that each subdomain can fit into the limited amount of computer memory available. Therefore, the multi–domain BEM using the impedance matrix solution procedure as demonstrated in Equations (13.7)–(13.10) needs to be revisited for large–size problems. Fortunately, the concept of the direct mixed–body BEM is not totally against the multi–domain BEM. In fact, with the help of the direct mixed–body BEM, the multi–domain BEM can be done more flexibly because each subdomain does not need to be homogeneous or thin–body free anymore. The decision to divide a large domain into smaller subdomains is based purely on the matrix size consideration, rather than on material properties distribution or thin–body locations. As a result, the idea of using the multi–domain BEM is very similar to the substructuring technique used in the structural finite element analysis. In the following, we will continue to use the silencer analy-

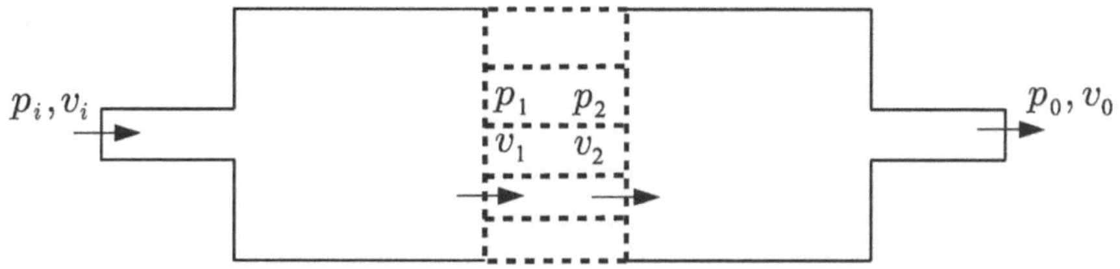

Fig. 13.4 Two substructures connected by an acoustic filter element.

sis as an example to demonstrate the substructuring idea, although the same concept may be applied to other types of acoustic analysis as well.

To begin with, we assume that a silencer is divided into two substructures as shown in Figure 13.4. It is possible that the two substructures are connected to each other by a built–in acoustic filter element, such as a catalyst converter (CC) or a diesel particulate filter (DPF), represented by the dotted lines in Figure 13.4. Let p_i and v_i denote the sound pressure and the particle velocity in the longitudinal direction, respectively, at the inlet of the first substructure, and p_o and v_o denote the corresponding variables at the outlet of the second substructure. Also, p_1 and v_1 are the variables at the outlet of the first substructure, and p_2 and v_2 are at the inlet of the second substructure. It should be noted that sound pressure and particle velocity may not be uniform at any cross section. Therefore, each p and v variable used in this section actually represents a vector. The length of each vector depends on the number of elements used at each cross section.

For substructure 1, the sound pressures at the inlet and the outlet are related to the corresponding particle velocities by an impedance matrix [3, 6]

$$\begin{bmatrix} p_i \\ p_1 \end{bmatrix} = \begin{bmatrix} Z_{11} & Z_{12} \\ Z_{21} & Z_{22} \end{bmatrix} \begin{bmatrix} v_i \\ v_1 \end{bmatrix} \tag{13.19}$$

It is important to note that this *partial* impedance matrix relationship is similar to the *full* impedance matrix defined in Equations (13.7) and (13.8) except that only the variables at the inlet and the outlet are included in the relationship. This is valid only when the rest of the boundary is either rigid or just having a *passive* boundary condition. In other words, p and v on the rest of the boundary will not be independently activated. Instead, they will be proportional to p and v at the inlet and the outlet only. Under such circumstances, there is no need to include them in the impedance matrix relationship. To obtain the impedance matrix, one will need to run the BEM on substructure 1 with a multiple of velocity boundary conditions. For example, to obtain the first column of the impedance matrix, $v = 1$ is applied to the first element at the inlet of substructure 1, and $v = 0$ is applied elsewhere. The sound pressure solutions at the inlet and outlet will become the first column of the impedance matrix. Similarly, by making $v = 1$ on each element at the inlet and outlet one at a time, the whole impedance matrix in Equation (13.19) can be obtained. Similarly, one can create the impedance matrix for substructure 2. The impedance matrix relationship is

$$\begin{bmatrix} p_2 \\ p_0 \end{bmatrix} = \begin{bmatrix} Z_{31} & Z_{32} \\ Z_{41} & Z_{42} \end{bmatrix} \begin{bmatrix} v_2 \\ v_0 \end{bmatrix} . \tag{13.20}$$

For the acoustic filter element that connects the two substructures, a four–pole transfer matrix is used to describe the acoustic property of the filter element. That is,

$$\begin{bmatrix} p_1 \\ v_1 \end{bmatrix} = \begin{bmatrix} A & B \\ C & D \end{bmatrix} \begin{bmatrix} p_2 \\ v_2 \end{bmatrix} \tag{13.21}$$

Note that for a through–flow type catalytic monolith, each capillary tube can be assumed to have a rigid wall. Since the diameter of each capillary tube is small, a plane–wave motion that consists of a forward traveling wave and a backward reflective wave is sufficient to describe the behavior in each capillary tube. Due to the rigid–wall assumption, there is no interaction between any two tubes. Under such circumstances, the transfer matrix coefficients A, B, C and D are each a diagonal matrix. In other words,

$$A = \begin{bmatrix} A_1 & & & \\ & A_1 & & \\ & & \cdots & \\ & & & A_1 \end{bmatrix}, \quad B = \begin{bmatrix} B_1 & & & \\ & B_1 & & \\ & & \cdots & \\ & & & B_1 \end{bmatrix}$$
$$C = \begin{bmatrix} C_1 & & & \\ & C_1 & & \\ & & \cdots & \\ & & & C_1 \end{bmatrix}, \quad D = \begin{bmatrix} D_1 & & & \\ & D_1 & & \\ & & \cdots & \\ & & & D_1 \end{bmatrix} \tag{13.22}$$

where A_1, B_1, C_1 and D_1 are the four–pole parameters of a straight tube. With the diagonal matrices in Equations (13.22), Equation (13.21) actually represents an *element–to–element* connection between the two substructures. To account for resistance and friction loss, the four–pole parameters are expressed in terms of a complex mean density ϱ^* and a complex speed of sound c^*. Both material properties can be measured by the two–cavity method on a sample of the CC or DPF with the same length (as in real applications). The four–pole parameters for a straight tube are

$$A_1 = \cos k^* L \tag{13.23a}$$

$$B_1 = \mathrm{i}\varrho^* c^* \sin k^* L \tag{13.23b}$$

$$C_1 = \frac{\mathrm{i}}{\varrho^* c^*} \sin k^* L \tag{13.23c}$$

$$D_1 = \cos k^* L \tag{13.23d}$$

where the complex wavenumber $k^* = \omega / c^*$ and L is the length of the CC or DPF.

It should be noted that when the two substructures are in direct contact (without a CC or DPF in between), the transfer matrix reduces to the identity matrix. In other words, $A = D = I$ and $B = C = 0$, where I is the identity matrix.

The interface variables, p_1, v_1, p_2 and v_2, can be eliminated from Equations (13.19)–(13.21) so that p_i and v_i are directly related to p_0 and v_0. The result is [6]

$$\begin{bmatrix} p_i \\ p_0 \end{bmatrix} = \begin{bmatrix} Z_{51} & Z_{52} \\ Z_{61} & Z_{62} \end{bmatrix} \begin{bmatrix} v_i \\ v_0 \end{bmatrix}, \tag{13.24}$$

where

$$Z_{51} = Z_{11} + Z_{12}\left[C\left(Z_{31}K^{-1}Z_{21}\right) + DK^{-1}Z_{21}\right] \tag{13.25a}$$

$$Z_{52} = Z_{12}C\left[Z_{32} - Z_{31}K^{-1}\left(A - Z_{22}C\right)Z_{32}\right] - Z_{12}D\left[K^{-1}\left(A - Z_{22}C\right)Z_{32}\right] \tag{13.25b}$$

$$Z_{61} = Z_{41}K^{-1}Z_{21} \tag{13.25c}$$

$$Z_{62} = Z_{42} - Z_{41}K^{-1}\left(A - Z_{22}C\right)Z_{32} \tag{13.25d}$$

with

$$K = \left[\left(A - Z_{22}C\right)Z_{31} + \left(B - Z_{22}D\right)\right]. \tag{13.26}$$

Note that sound pressure and particle velocity are usually uniform across any cross section in the inlet tube and in the outlet tube as well. That means each of the vectors p_i, v_i, p_0 and v_0 can be lumped into one single variable, and the impedance matrix in Equation (13.24) can be further reduced to a 4×4 impedance matrix. With the lumped 4×4 impedance matrix ready, the transmission loss (TL) of the silencer can be easily obtained [19].

As mentioned earlier, when the two substructures are in direct contact, $A = D = I$ and $B = C = 0$. Equation (13.14) reduces to

$$K = Z_{31} - Z_{22} \tag{13.27}$$

and Equations (13.25a)–(13.25d) become

$$Z_{51} = Z_{11} + Z_{12}\left(Z_{31} - Z_{22}\right)^{-1}Z_{21} \tag{13.28a}$$

$$Z_{52} = -Z_{12}\left(Z_{31} - Z_{22}\right)^{-1}Z_{32} \tag{13.28b}$$

$$Z_{61} = Z_{41}\left(Z_{31} - Z_{22}\right)^{-1}Z_{21} \tag{13.28c}$$

$$Z_{62} = Z_{42} - Z_{41}\left(Z_{31} - Z_{22}\right)^{-1}Z_{32} \tag{13.28d}$$

The procedure described above is sometimes referred to as the impedance matrix synthesis (IMS) [9]. If a silencer is divided into multiple substructures (more than two) in series connection, the IMS can be first applied to the first two substructures to produce a resulting impedance matrix. Then the resulting impedance matrix is combined with the impedance matrix of the third substructure by the same synthesis

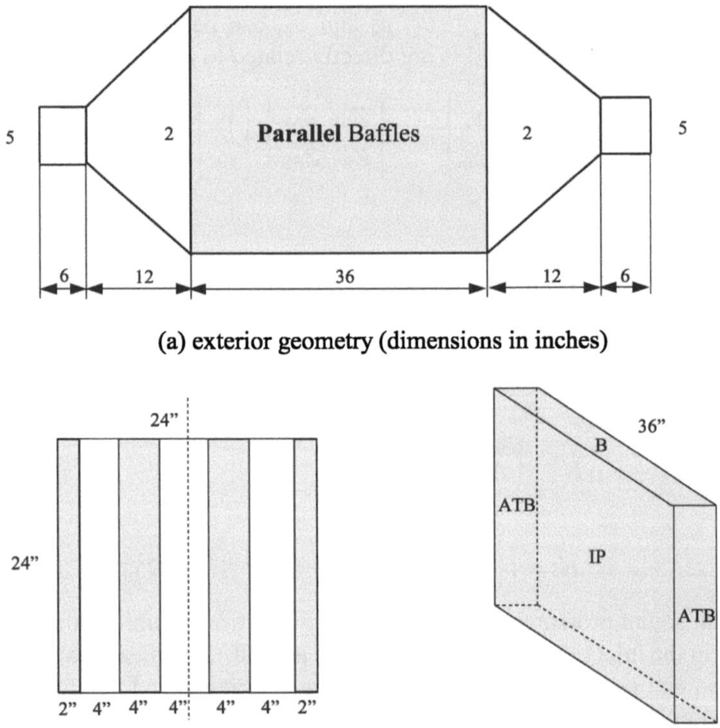

(a) exterior geometry (dimensions in inches)

(b) cross section of the parallel baffles (c) splitter geometry

Fig. 13.5 Parallel–baffle silencer with a lining on both side walls.

method. The procedure is repeated until the last substructure is reached. It is noticed that many packed silencers in practical use may have an identical cross section in a large portion of the silencer structure. For those kinds of silencers, the impedance matrix of a small section of the identical geometry can be used as a *template*, which can be repeatedly used in the IMS procedure.

13.5 Silencer Test Case

The first test case is a parallel–baffle silencer [6] as shown in Figure 13.5. Both the inlet and the outlet tubes have a diameter of 5". The two transition ducts that connect the inlet/outlet tubes to the parallel–baffle section have a square cross section of varying dimensions.

Figure 13.5b shows the cross section of the parallel–baffle section. Two 4" × 24" × 36" Polyester parallel baffles are used as center splitters in a 24" × 24" × 36" rectangular duct, and two 2" × 24" × 36" baffles are attached to the two side walls, respectively, as linings. This creates three 4" air gaps in the center. As shown in Figure 13.5c, each Polyester splitter is covered by a perforated metal sheet (designated

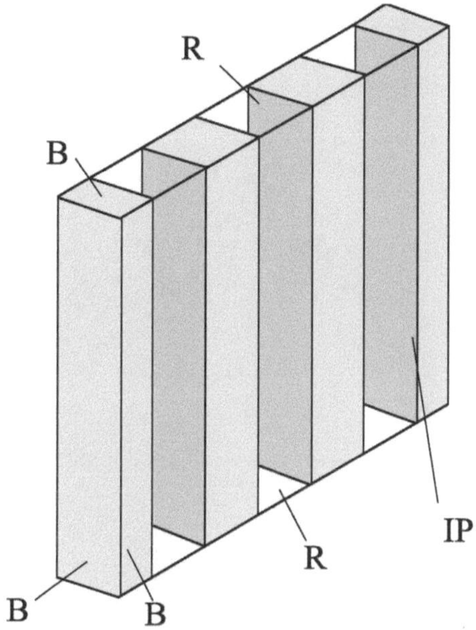

Fig. 13.6 A substructure template created from the identical center portion of the parallel–baffle silencer..

as IP surface) on its sides and a rigid plate (designated as ATB surface) at both ends. The porosity of the IP interfaces is 23%.

Since the model has two planes of symmetry, only a quarter of the silencer geometry has to be modeled. Even so, the whole model still cannot fit in a PC with 1 GB memory at high frequencies without dividing the silencer into smaller substructures. It is noticed the parallel–baffle section has an identical cross section over its length except at the two ATB ends (covering plates). Thus, a template as shown in Figure 13.6 can be created and its impedance matrix, once computed for each frequency, can be repeatedly used in the synthesis procedure. Figure 13.7 shows the comparison of the BEM predictions using four and ten substructures and the experimental TL curve. It is seen that the BEM results compare fairly well with the measured TL curve, although some small discrepancies do exist between models with different number of substructures. Dividing a large structure into smaller substructures not only reduces the computer memory requirement, but it also speeds up the computation. In Reference [6], Lou et al. have shown that the CPU time speedup is roughly proportional to the number of substructures used.

The second test case is a catalytic converter (CC) consisting of a ceramic monolith as shown in Figure 13.8 [15]. The catalytic monolith is made up of capillaries with square cross–sections and has a cell density of 400 cells per square inch (CPSI). The monolith used in this study is uncoated. A circular substrate sample is cored out from the catalytic monolith in full length for measuring its characteristic impedance and propagation constant using the two–cavity method. There are two ways to make

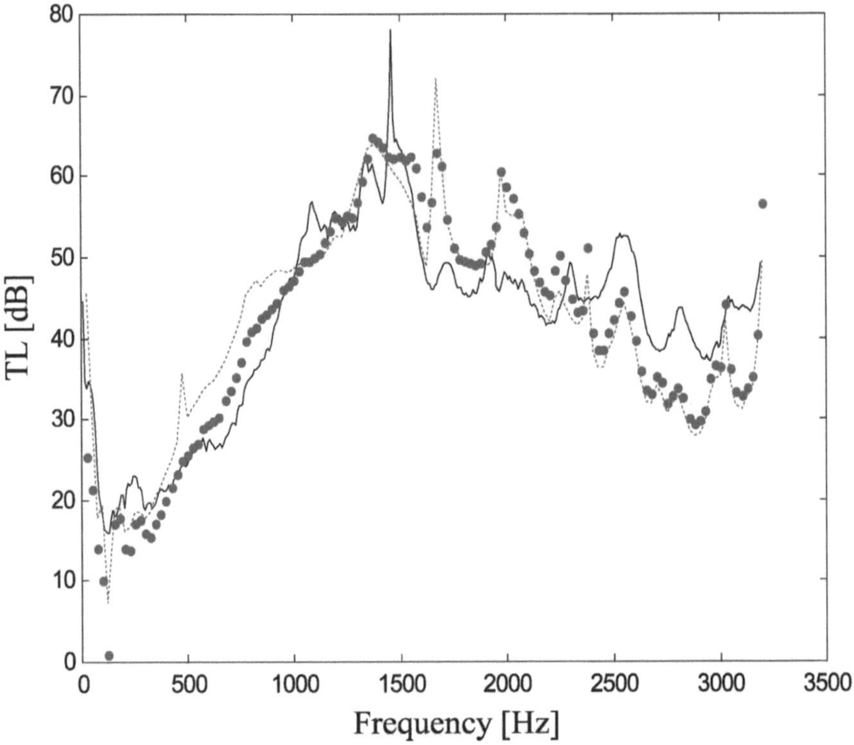

Fig. 13.7 Transmission loss for the parallel–baffle silencer. Solid line: experiment; solid dots: BEM using four substructures; dotted line: BEM using ten substructures.

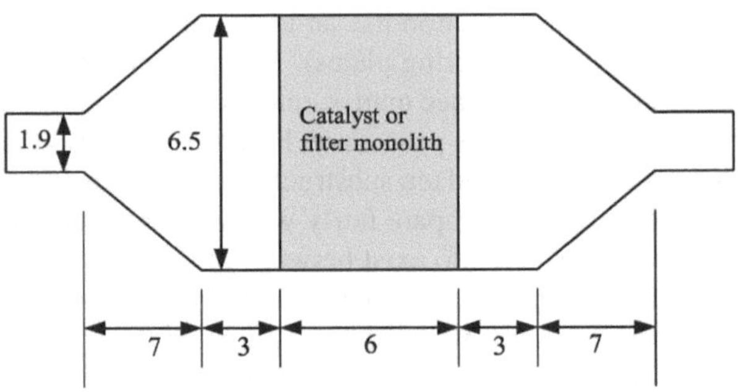

Fig. 13.8 Emission control device test case (dimension in inches).

BEM predictions. One is to directly model the CC as a block of homogeneous and isotropic bulk–reacting material (BRM), and the other is to model the CC by the *element–to–element* four–pole transfer matrix as described earlier in the impedance matrix synthesis (IMS) procedure. As shown in Figure 13.9, both BEM predictions

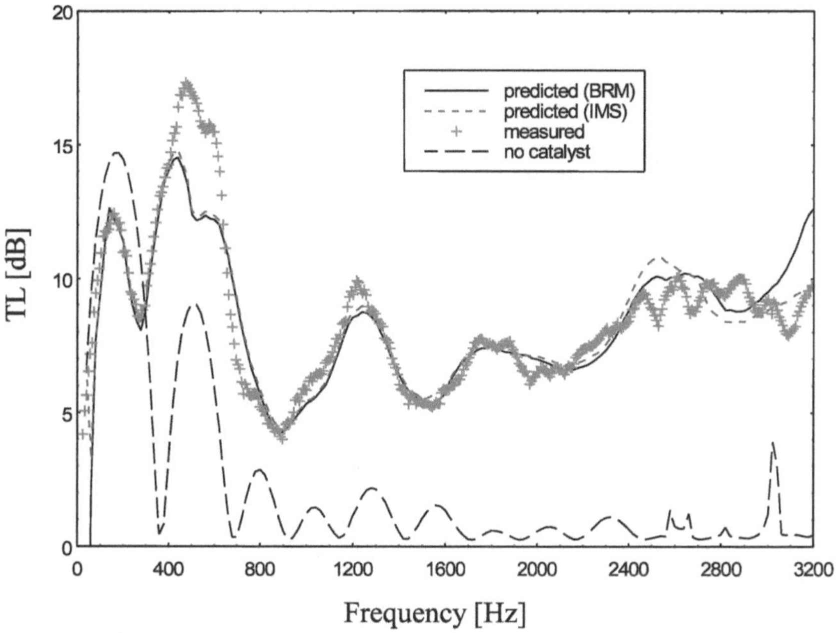

Fig. 13.9 Comparison of the predicted and measured TL of the CC test case.

compare reasonably well with the experimental data. Both numerical approaches produce very similar results. The IMS approach, however, generates numerical results much faster than the direct BRM modeling method. Since it is assumed that the acoustic propagation in the catalytic monolith is predominantly one–dimensional, a simple four–pole transfer matrix would be adequate to represent the catalytic monolith. The predicted TL curve for an empty housing without the catalytic monolith is also plotted in the same graph for comparison. The results reveal that the first two domes of the TL curves are the reactive effect caused by wave reflections due to area changes in the housing. Beyond the first two domes, the noise attenuation is dominated by the dissipative effect inside the capillaries of the monolith. The catalytic monolith increases the overall TL by more than 5 dB except for the first dome.

Using the same test configuration, the catalytic monolith is replaced by a ceramic wall–flow filter monolith (a DPF) in the third test case [15]. The DPF monolith has a cell density of 100 CPSI and is also uncoated. Figure 13.10 shows the BEM predictions versus the experimental data. It can be seen that both numerical approaches under–predict the TL values though the trend of the curves agrees well with the experimental results. The discrepancy is due to the fact that sound propagation in the wall–flow monolith is not limited to only the axial direction along the capillaries. There is significant amount of sound transmission through the walls of the filter monolith.

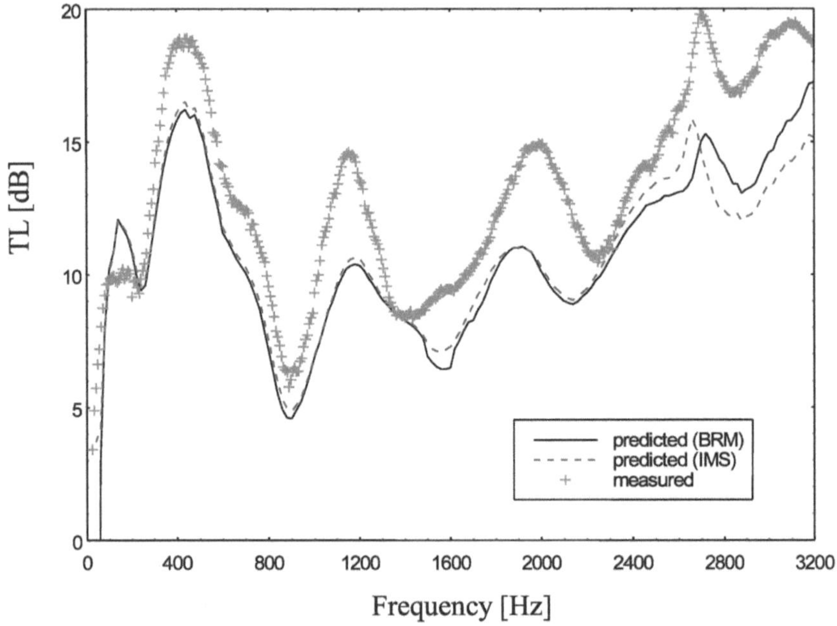

Fig. 13.10 Comparison of the predicted and measureed TL of the DPF test.

13.6 Conclusions

In this chapter, the conventional multi–domain BEM in acoustics is first reviewed. It is noted that computing the impedance matrix for each subdomain may be a more efficient way to solve the coupled system of equations.

Recent developments of the direct mixed–body BEM for muffler and silencer analysis are also reviewed. A key ingredient in the direct mixed–body BEM is the hypersingular integral equation. It is well known that the hypersingular integral equation is ideally suited to thin–body problems so that a multi–domain approach can be avoided. We have also demonstrated that even a two-medium problem can benefit from the direct mixed–body BEM concept.

When the direct mixed–body BEM for muffler analysis was first presented in 1996, it was meant to eliminate the tedious multi–domain BEM zoning and interface matching so that a muffler model with complicated internal components could be created and solved quickly in a single computation. However, as frequency and the number of elements increase, a large silencer model cannot be computed on a PC anymore due to the physical limitation of available computer memory. Dividing a large structure into smaller substructures becomes a requirement rather than an option. Therefore, the multi–domain BEM has to be revisited in the form of impedance matrix synthesis. Fortunately, the concept of the direct mixed–body BEM is not totally against the multi–domain BEM. In this reformulation of the multi–domain BEM, the direct mixed–body BEM actually facilitates the substructuring operations.

Acknowledgements

The research of the direct mixed–body BEM and the substructuring technique has been supported by Cummins, Inc. (the Nelson Industries division).

References

1. Brebbia CA, Walker S (1980) Boundary element technique in engineering. Butterworth, London
2. Cheng CYR, Seybert AF, Wu TW (1991) A multidomain boundary element solution for silencer and muffler performance prediction. Journal of Sound and Vibration 151:119–129
3. Ji Z, Ma Q, Zhang Z (1994) Application of the boundary element method to predicting acoustic performance of expansion chamber mufflers with mean flow. Journal of Sound and Vibration 173:57–71
4. Krishnasamy G, Schmerr LW, Rudolphi TJ, Rizzo FJ (1990) Hypersingular boundary integral equations: some applications in acoustic and elastic wave scattering. Journal of Applied Mechanics 57:404–414
5. Liu YJ, Rizzo FJ (1992) A weakly singular form of the hypersingular boundary integral equation applied to 3–D acoustic wave problems. Computer Methods in Applied Mechanics and Engineering 96:271–287
6. Lou G, Wu TW, Cheng CYR (2003) Boundary element analysis of packed silencers with a substructuring technique. Engineering Analysis with Boundary Elements 27:643–653
7. Seybert AF, Cheng CYR, Wu TW (1990) Measurement and model of the cat middle ear: Evidence of tympanic membrane acoustic delay. Journal of the Acoustical Society of America 88:1612–1618
8. Seybert AF, Jia ZH, Wu TW (1993) Solving knife–edge scattering problems using singular boundary elements. Journal of the Acoustical Society of America 91:1278–1283
9. Sullivan JW, Crocker MJ (1978) Analysis of concentric–tube resonators having unpartitioned cavities. Journal of the Acoustical Society of America 64:207–215
10. Terai T (1980) On calculation of sound fields around three dimensional objects by integral equation methods. Journal of Sound and Vibration 69:71–100
11. Utsuno H, Wu TW, Seybert AF, Tanakav T (1990) Prediction of sound fields in cavities with sound absorbing materials. AIAA Journal 28:1870–1876
12. Wang CN, Tse CC, Chen YN (1993) A boundary element analysis of a concentric–tube resonator. Engineering Analysis with Boundary Elements 12:21–27
13. Wu TW (1995) A direct boundary element method for acoustic radiation and scattering from mixed regular and thin bodies. Journal of the Acoustical Society of America 97:84–91
14. Wu TW (ed) (2000) Boundary element acoustics, fundamentals and computer codes. WIT Press, Southampton
15. Wu TW, Cheng CYR (2003) Boundary element analysis of reactive mufflers and packed silencers with catalyst converters. Electronic Journal of Boundary Elements 1:218–235
16. Wu TW, Cheng CYR, Tao Z (2003) Boundary element analysis of packed silencers with protective cloth and embedded thin surfaces. Journal of Sound and Vibration 261:1–15
17. Wu TW, Cheng CYR, Zhang P (2002) A direct mixed–body boundary element method for packed silencers. Journal of the Acoustical Society of America 111:2566–2572

18. Wu TW, Wan GC (1996) Muffler studies using a direct mixed–body boundary element method and a three–point method for evaluating transmission loss. Journal of Vibration and Acoustics 96:479–484
19. Wu TW, Zhang P, Cheng CYR (1998) Muffler studies using a direct mixed–body boundary element method and a three–point method for evaluating transmission loss. Journal of Sound and Vibration 217:767–779

14

Waveguide Boundary Spectral Finite Elements

Andrew Peplow

Marcus Wallenberg Laboratory for Sound & Vibration Research, Department of Aeronautics & Vehicle Engineering, Kungl Tekniska Högskolan, S–10044 Stockholm, Sweden
atpeplow@kth.se

Summary. A waveguide boundary spectral finite element method (SFEM) is developed for the study of acoustical wave propagation in non–uniform waveguide–like geometries. The formulation is based on a variational approach using a mixture of non–internal node element shape functions and wave solutions. The numerical method provides solutions to acoustic duct or fluid waveguide environments which may be divided into uniform cross–sectional regions. Trial functions are determined by solution of an eigenvalue problem defined in the cross–section, which in turn, depends upon the boundary data. Illustration of the method through demonstration of transmission loss of acoustic energy through two–dimensional dissipative mufflers is presented and solutions of a three–dimensional elliptical duct problem are shown.

14.1 Introduction

The spectral element method is an advanced implementation of the finite element method in which the solution over each element is expressed in terms of a priori unknown values at carefully selected spectral nodes. The advantage of the spectral element method is that stable solution algorithms and high accuracy can be achieved with a low number of elements under a broad range of conditions. Spectral element techniques are high order methods which allow for either obtaining very accurate results or reducing the number of degrees of freedom for fixed standard precision. As explained by Gottlieb and Orszag [20] in the 1970's spectral methods for boundary value problems rely on high degree polynomial approximations on square and cubic domains. Trefethen [31] also describes the theory behind spectral methods and the connection between Fourier series, Chebyshev polynomials and includes MATLAB code. Both Boyd [7] and Fornberg [17] books are also excellent constructive editions on spectral methods on regular finite domains. Recently though the article by Philipps and Davies [28] proposes a general spectral method for Poissons equation in rectangularly decomposable regions including semi–infinite regions. Eigenfunctions of a differential operator are chosen as trial functions within each element and are nonconforming. The solution is determined by matching the solution and its derivative across element interfaces. A similar approach of using eigenfunctions of a differential operator in the trial function space is adopted in here where the principal field equation is the Helmholtz equation. This is the underpinning principle for the waveguide spectral element method.

Handling complex geometries by spectral finite element methods are now an established alternative to finite difference and finite element methods to solve elliptic Partial Differential Equations (PDE). Spectral methods are naturally chosen to solve problems in regular rectangular, cylindrical or spherical regions. However in a general irregular region it would be unwise to turn away from the finite element method since models defined in such regions are extremely difficult to implement and solve with a spectral method. The examples here show how the spectral method uses boundary data to devise trial functions so that, in essence, no discretization of the interior domain is necessary. Thus the method maybe viewed as a boundary element method as the method is "meshless" as possible but unlike most boundary element methods it is not derived from Greens theorem. Hence for a complex *waveguide* the method uses the *boundary* data to devise an efficient and accurate *spectral* method and is combined with the flexibility of *finite elements* to produce a high–performance engineering tool.

The work described here is concerned with the development of the waveguide boundary spectral finite element method (SFEM). Its beginnings in structures as a combination of the dynamic stiffness method and the finite element method based on a variational formulation for a non-conservative motion. A dynamic stiffness approach for frame structures has been developed originally by Richard and Leung [29]. However, a major advance, was made by Gavric [19], where the cross-sectional motion of a given waveguide was approximated by standard finite element polynomials. Wave propagation along the waveguide could then be studied by finding eigenvalues, corresponding to propagating wavenumbers, from a system of differential equations. This innovation inspired the study for beam and plate structures where Finnveden used the spectral finite element method in [11, 13]. Orrenius and Finnveden [24] and more recently Nilsson and Finnveden [23] used this approach for rib–stiffened plate structures used in train wagons. For two–dimensional modelling, one–dimensional finite element shape polynomial functions describe the motion's z–dependence where, without loss of generality, it is assumed that the waveguide is aligned with the x–axis. It follows that nodal displacements, vertical and longitudinal, are functions of the x variable and may be found by the elastic waveguide boundary spectral finite element method. The underlying is through the solution of a matrix polynomial eigenvalue problem. This novel approach has been used to describe the dynamic motion of sandwich composite structures Bonfiglio et al. [6] and ground vibration in layered geomechanic media Peplow and Finnveden [25] modelled masses lying within bedrock layers of infinite extent.

Turning now to complex acoustic systems the underlying polynomial eigenvalue problem for acoustic problems are straightforward, however, see section 14.2 and 14.5.1 for large–scale three–dimensional problems. Waveguides with uniform cross sectional properties will be described here using the spectral finite element method. For such an analysis, a natural starting point is the variational formulation. In this context, this type of finite element analysis first appeared in Finnveden [14] and [12] for fluid–filled flexible pipes , and was used to compare with experimental techniques by Finnveden and Pinnington [16]. The effect of including distributed loading by turbulent boundary layers on pipes can be found in [3]. Peplow and Finnveden developed the method to study sound transmission in various acoustic waveguide configurations in [26]. Recently the SFEM has been used for the study of and comparison with measurements, by Birgersson et al. to predict the response of a structure by turbulence excitation [4, 5, 15]. Applications of the method can be found in the theses by Birgersson [2] and Nilsson [22] on applications in fluid–structure interaction and Fraggstedt [18] on modelling wave propagation in car tyres. For three–dimensional problems, two–dimensional polynomial shape functions describe the motion's y– and z– dependence. Kirby used collocation to determine transmission loss for elliptic dissipative silencers with and without mean flow [21] and [9]. Although a specific three–dimensional finite element was not constructed the efficiency of the method is clearly described.

In general the SFEM technique is an important development in engineering analysis. The flexibility of the waveguide boundary spectral finite element and the versatility of the modelling applications give new solutions to engineering dynamics and acoustic problems.

14.2 Weak Formulation for the Spectral Finite Element

A weak formulation is based on introducing the weight function $\chi(\boldsymbol{x})$, where $\boldsymbol{x} = (x, z)$, and testing it with the Helmholtz operator such that integrating by parts gives

$$\begin{aligned} L_\Omega = & \int_\Gamma \chi(\boldsymbol{x})\, a\, v_f(\boldsymbol{x}) d\Gamma(\boldsymbol{x}) - \int_\Omega \Big[\boldsymbol{\nabla}\chi(\boldsymbol{x}) \cdot \boldsymbol{\nabla} p(\boldsymbol{x}) - k^2 \chi(\boldsymbol{x}) p(\boldsymbol{x}) \Big] d\Omega(\boldsymbol{x}) \\ = & \; k^2 \int_\Omega \chi(\boldsymbol{x}) p(\boldsymbol{x}) d\Omega(\boldsymbol{x}) - \int_\Omega \chi_z(\boldsymbol{x}) p_z(\boldsymbol{x}) d\Omega(\boldsymbol{x}) \\ & + \int_\Gamma \chi(\boldsymbol{x})\, a\, v_f(\boldsymbol{x}) d\Gamma(\boldsymbol{x}) - \int_\Omega \chi_x(\boldsymbol{x}) p_x(\boldsymbol{x}) d\Omega(\boldsymbol{x}) \; = \; 0. \end{aligned} \quad (14.1)$$

Often, equation (14.1) represents the starting point for conventional finite element discretizations, e.g. Galerkin method. The second part (lower row) consists of a domain integral and a boundary integral.

The variational statement (14.1) is used to obtain *wave influence* basis functions $W(x)$. An approximate solution to the original problem is also found by selecting a solution $p_j(x, z)$ from a discrete set of trial functions determined by a finite element discretization of a region with a uniform cross–section Ω_j.

14.2.1 Approximation Functions

We approximate the sound pressure $p(\boldsymbol{x})$ as

$$p(\boldsymbol{x}) \; = \; \sum_{i=1}^{N} \phi_i(\boldsymbol{x})\, p_i \; = \; \boldsymbol{\phi}^T(\boldsymbol{x}) \boldsymbol{p} \quad (14.2)$$

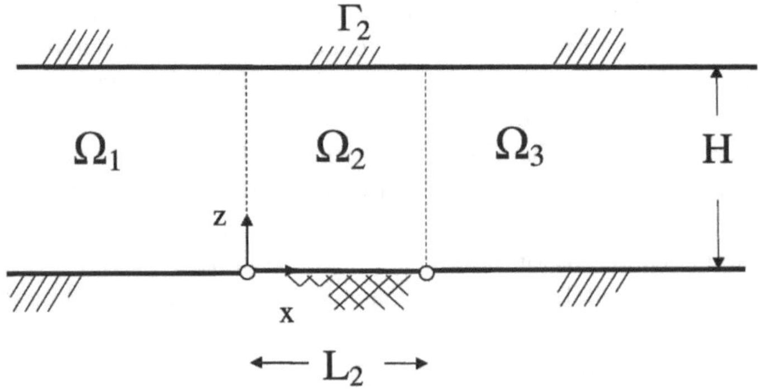

Fig. 14.1 Two dimensional acoustic duct with expansion chamber.

where p_i represents the discrete sound pressure at point $\boldsymbol{x}_i$ and ϕ_i is the i–th basis function for our approximation. First we start the analysis with a two–dimensional formulation and then followed by an three–dimensional example.

Figure 14.1 shows a typical two–dimensional waveguide section where $\Omega = \Omega_1 \cup \ldots \cup \Omega_3$. In the super–spectral method this section will define one *waveguide finite element*. Within one *super–spectral finite element*, $\Omega_2 := D \times C$, $D := [0, L_2]$, $C := [0, H]$ for example, the approximate solution $p(.)$ may be represented by an expression of the form

$$p(\boldsymbol{x}) = \sum_{i=1}^{N} p_i \phi_i(z) W_i(x) \tag{14.3}$$

using piecewise quadratic polynomial shape functions. Shape functions, $\phi(s)$, are defined sub–locally over the cross–sections. Elemental shape functions $\boldsymbol{W}(x)$ are defined later in this section. The complete set of quadratic polynomials defined over $[-1, 1]$ is $\boldsymbol{\phi}(s) = [\phi^1(s)\ \phi^2(s)\ \phi^3(s)\ ]$ where

$$\phi^1(s) = \frac{1}{4}\left(1 - 2s + s^2\right), \quad \phi^2(s) = \frac{1}{4}\left(1 + 2s + s^2\right), \quad \phi^3(s) = \frac{-1}{\sqrt{2}}\left(1 - s^2\right). \tag{14.4}$$

Consider the functional L_Ω, (14.1), for a single arbitrary region Ω_2 say. Substitution of expression (14.3) into the resulting form, where subscript $_x$ denotes x–derivative, yields the approximation

$$\begin{aligned} L_{\Omega_j} &= k^2 \int_{\Omega_i} \boldsymbol{W}^T \boldsymbol{K}_1 \boldsymbol{W}\, dx - \int_{\Omega_i} \boldsymbol{W}^T \boldsymbol{K}_2 \boldsymbol{W}\, dx \\ &\quad + \mathrm{i}\, k \int_{\Omega_i} \boldsymbol{W}^T \boldsymbol{K}_3 \boldsymbol{W}\, dx - \int_{\Omega_i} \boldsymbol{W}_x^T \boldsymbol{K}_4 \boldsymbol{W}_x\, dx, \end{aligned} \tag{14.5}$$

where the matrices are defined by

$$\boldsymbol{K}_1 = \int_C \boldsymbol{\phi}^T(z)\boldsymbol{\phi}(z)\,dz, \tag{14.6}$$

$$\boldsymbol{K}_2 = \int_C \frac{d\boldsymbol{\phi}}{dz}^T \frac{d\boldsymbol{\phi}}{dz}\,dz, \tag{14.7}$$

$$\boldsymbol{K}_3 = \left[\frac{1}{\zeta}\right]_0^H, \tag{14.8}$$

$$\boldsymbol{K}_4 = \int_C \boldsymbol{\phi}^T(z)\boldsymbol{\phi}(z)\,dz. \tag{14.9}$$

Explicitly the *mass* and *stiffness* matrices in (14.6) and (14.7), defined for the interval $C = [0, H]$, are given by

$$\boldsymbol{K}_1 = H \int_{-1}^{+1} \boldsymbol{\phi}^T(s)\boldsymbol{\phi}(s)\,ds = \frac{H}{15}\begin{bmatrix} 6 & 1 & -6\sqrt{2} \\ 1 & 6 & -6\sqrt{2} \\ -6\sqrt{2} & -6\sqrt{2} & 8 \end{bmatrix}, \tag{14.10}$$

$$\boldsymbol{K}_2 = H \int_{-1}^{+1} \frac{d\boldsymbol{\phi}}{ds}^T \frac{d\boldsymbol{\phi}}{ds}\,ds = \frac{H}{3}\begin{bmatrix} 2 & -1 & \sqrt{2} \\ -1 & 2 & \sqrt{2} \\ \sqrt{2} & \sqrt{2} & 4 \end{bmatrix}. \tag{14.11}$$

For each geometrical sector Ω in Figure 14.1 the finite element cross–sectional matrices $\boldsymbol{K}_1, \ldots, \boldsymbol{K}_4$ are frequency independent, real valued and are fairly small in size (for a two–dimensional problem at least) and hence may be stored. In the case here these are clearly 3×3 real–valued symmetric matrices. For a full problem a dynamic stiffness matrix requires assemblage. To do this wave influence functions $\boldsymbol{W}$ for each sector Ω are required.

14.3 Wave Influence Functions & the Dynamic Stiffness Matrix

Construction of the wave influence functions follows by consideration of the ordinary differential equations which correspond to Equation (14.5) found by taking an appropriate first variation and *ignoring* any boundary conditions:

$$\boldsymbol{K}_4 \frac{d^2}{dx^2}\boldsymbol{W}(x) + k^2\,\boldsymbol{K}_1\boldsymbol{W}(x) - \boldsymbol{K}_2\boldsymbol{W}(x) + \mathrm{i}\,k\boldsymbol{K}_3\boldsymbol{W}(x) = 0. \tag{14.12}$$

Crucial to the fundamental principle of determining waveguide boundary spectral finite elements is that the system of Equation (14.12) are **not** defined over a specific region. It could be argued that Equation (14.12) is defined over a region of infinite length. The differential equations have constant coefficients in the form of symmetric positive and semi–definite $(N \times N)$ real–valued matrices $\boldsymbol{K}_1, \ldots, \boldsymbol{K}_4$. Hence, the solutions of the linear homogeneous system may be written as :

$$\boldsymbol{W}_m(x) = \boldsymbol{\Phi}_m\,\mathrm{e}^{\mathrm{i}\lambda_{m-1}x}, \quad m = 1, \ldots, N \tag{14.13}$$

Table 14.1 Computed eigenvalues associated with wave propagation in the positive direction for a uniform rigid duct; $f = 740$ Hz, and $H = 0.25$ m against increasing number of $h-element$ quadratic polynomial cross–section shape functions. Relative errors for 4 element model also shown.

λ_m (m^{-1})	1 element	2 elements	3 elements	4 elements	Rel. error (%) 4 elements
λ_0	-1.3675E+1	-1.3675E+1	-1.3675E+1	-1.3675E+1	2.6E−14
λ_1	-2.2337E+0i	-5.2828E+0	-5.3709E+0	-5.3866E+0	0.14
λ_2	-2.7802E+1i	-2.4103E+1i	-2.1412E+1i	-2.1199E+1i	0.53
λ_3		-4.3273E+1i	-3.9255E+1i	-3.5798E+1i	1.90
λ_4		-6.0440E+1i	-5.5832E+1i	-5.3712E+1i	11.05
λ_5			-7.8141E+1i	-6.8884E+1i	12.32
λ_6			-9.1940E+1i	-8.9728E+1i	21.01
λ_7				-1.1193E+2i	28.82
λ_8				-1.2317E+2i	23.68

where $\boldsymbol{\Phi}_m$ is a vector representing the cross–sectional mode shapes. Under this assumption Equation (14.13) reduces to a linear eigenvalue problem, $\boldsymbol{K}$:

$$\boldsymbol{K}(\lambda)\boldsymbol{\Phi} = \{k^2\,\boldsymbol{K}_1 - \boldsymbol{K}_2 + \mathrm{i}k\boldsymbol{K}_3 - \lambda^2\boldsymbol{K}_4\}\,\boldsymbol{\Phi} = 0 \tag{14.14}$$

of order N for the parameters λ^2. The solutions of Equation (14.14) yield values for λ that occur in pairs, $\lambda^{\pm} = \pm\lambda$, indicating that pairs of eigenmodes result with the same phase speed propagating in the positive and negative axial directions. The dimension of the eigenvalue problem is $(N \times N)$ and a finite $2N$ number of propagating wavenumbers are obtained.

The resolution of the matrix eigenvalue problem (14.14) itself may be achieved by a number of standard computational routines. In the present analysis a *QZ* algorithm was used as implemented in Matlab 7.0.2 and requires 46 N^3 operations for to determine all eigenvalues and right eigenvectors. For *large* problems this numerical analysis procedure can dominate the total computation time for a single problem. This produces a complete set of cross–sectional mode shapes and corresponding eigenvalues λ_m^2, $m = 0, \ldots, N-1$. Table 14.1 shows eigenvalues for the problem with a duct of width $H = 0.25$ m using one finite element. The size of the matrix is 3×3 giving $N = 3$ eigenvalues $(\lambda_0^2, \ldots, \lambda_2^2)$ in the first column. At the given frequency 740 Hz the acoustic wavenumber takes the value $k = 13.675$ m^{-1}. Doubling the number of elements and enforcing pressure continuity increases the size of problem to $N = 5$ but reduces the error considerably by comparing with exact values $\lambda^2 = k^2 - m^2\pi^2/H^2$ in the final column.

Now the finite element trial functions have been constructed the dynamic stiffness for a general spectral waveguide element Ω is described. The local dynamic stiffness matrix defined over a region Ω shown in Figure 14.1 will now be described.

A simple translation to $T := \{x : -D \leq x \leq +D\}$, where $2D$ is the length of the sector, is a key element in defining the wave influence trial functions. By consideration of the eigensolutions in Equation (14.13) it is clear that each wave influence function may be written as

$$W_{jk}(x) = \sum_{l=1}^{2N} \overline{\Phi}_{jl} E_{ll}(x) A_{lk}\, p_k, \qquad j = 1, \ldots, N, \quad k = 1, \ldots, 2N, \qquad (14.15)$$

where entries $\overline{\Phi}_{jl}$ and E_{ll}, a diagonal matrix, take the values of the eigenmodes and wavefunctions respectively. Coefficients A_{lk} are determined by appropriate scaling of the set of wave influence functions, to be described shortly, and p_k are the unknown coefficients. Now, the set of eigenvectors, from Equation (14.13), $\boldsymbol{\Phi}_m$ have been cast as a partitioned matrix

$$\overline{\boldsymbol{\Phi}} = \left[\boldsymbol{\Phi} \,\vdots\, \boldsymbol{\Phi} \right]. \qquad (14.16)$$

Also the vector $\boldsymbol{\Lambda}$, formed by the negative and positive direction propagating wavenumber eigenvalues, is given by

$$\boldsymbol{\Lambda} = \left(\mathrm{i}\lambda_0^-, \ldots, \mathrm{i}\lambda_{N-1}^-, \mathrm{i}\lambda_0^+, \ldots, \mathrm{i}\lambda_{N-1}^+ \right), \qquad (14.17)$$

and introducing the vector $\boldsymbol{\Lambda}^p$ defined by

$$\Lambda_m^p = \begin{cases} \Lambda_m, & \Re\{\Lambda_m\} \geq 0, \\ -\Lambda_m, & \Re\{\Lambda_m\} < 0, \end{cases} \qquad (14.18)$$

$m = 1, \ldots, 2N$ the basis functions may be scaled in the longitudinal direction taking the form:

$$E_{ll}(x) = \mathrm{e}^{\Lambda_l x - \Lambda_l^p D}, \qquad |x| \leq D, \;\; l = 1, \ldots, 2N. \qquad (14.19)$$

Scaling is an important part of the super–spectral method as this allows "very long" elements, in terms of wavelengths, to be considered. Without scaling, it is clear that entries in the matrix could grow exponentially with length. Projecting the basis functions, (to form the trial functions) Equation (14.15), onto the end–points of the region T:

$$\begin{aligned} W_{jk}(-T) &= \delta_{jk} p_k, \quad j = 1, \ldots, N, k = 1, \ldots, N, \\ &\text{and} \\ W_{(j-N)k}(+T) &= \delta_{jk} p_k, \quad j = N+1, \ldots, 2N, k = N+1, \ldots, 2N, \end{aligned} \qquad (14.20)$$

gives a system of $2N \times 2N$ equations determining the basis functions coefficients A_{lk},

$$\boldsymbol{X}\,\boldsymbol{A} = \boldsymbol{I}, \qquad (14.21)$$

where

$$\boldsymbol{X} = \begin{bmatrix} \overline{\boldsymbol{\Phi}} \lceil E(-T) \rfloor \\ \cdots\cdots\cdots\cdots \\ \overline{\boldsymbol{\Phi}} \lceil E(+T) \rfloor \end{bmatrix}, \tag{14.22}$$

where $\lceil . \rfloor$ represents diagonal matrix and $\boldsymbol{I}$ is the identity matrix. Substitution of Equations (14.16) and (14.19) into Equation (14.22), yields a system of $2N \times 2N$ linear equations. Since we are finding all the entries of an $2N \times 2N$ inverse matrix the cost for generating the coefficient matrix $\boldsymbol{A}$ is substantial, to third order a little over $5N^3$. Hence the local dynamic stiffness matrix for a certain element may be written in matrix form as:

$$\boldsymbol{L} = \boldsymbol{B}^T \left(k^2 \boldsymbol{K}_1 - \boldsymbol{K}_2 \right) \boldsymbol{B} - \lceil \boldsymbol{\Lambda} \rfloor \boldsymbol{B}^T \boldsymbol{K}_4 \boldsymbol{B} \lceil \boldsymbol{\Lambda} \rfloor + \mathrm{i} k \boldsymbol{B}^T \boldsymbol{K}_3 \boldsymbol{B}, \tag{14.23}$$

where the matrix $\boldsymbol{B}$, of order $2N \times 2N$, combines wave influence function matrices

$$\boldsymbol{B} = \overline{\boldsymbol{\Phi}} \lceil E \rfloor \boldsymbol{A}. \tag{14.24}$$

The *ansatz* dynamic stiffness matrix above may seem a little contrived compared to direct FEM but it can be constructed very simply from Equation (14.23) using matrix algebra operations.

Within the computation of local dynamic stiffness matrix entries use is made of an important matrix generating function for the combination of matrix diagonal exponential terms, $\lceil E \rfloor$, in (14.23)

$$\boldsymbol{E}_I(\boldsymbol{\Lambda}, D) = \int_{-D}^{D} \left[\mathrm{e}^{\boldsymbol{\Lambda} x - \boldsymbol{\Lambda}^p D} \right]^T \mathrm{e}^{\boldsymbol{\Lambda} x - \boldsymbol{\Lambda}^p D} \, dx, \tag{14.25}$$

which has an analytic form.

The global dynamic stiffness matrix is generated by calculating local dynamic stiffness matrices for each region and enforcing continuity of pressure (and velocity) across neighbouring interfaces. For example, consider a uniform waveguide geometry consisting of three spectral elements, as in Figure 14.1, with N degrees of freedom across each element cross–section. A $N \times N$ matrix eigenvalue problem is solved and a $2N \times 2N$ local dynamic stiffness matrix generated for each element. Enforcing continuity across neighbours results in a $4N \times 4N$ global dynamic stiffness matrix. Acoustic sources may be modelled as volume point sources but in the following examples normal accelerations are applied on the left–hand boundary. The total number of operations for solving this problem for an arbitrary length waveguide are around $57N^3$ for all the Gaussian elimination operations and $138N^3$ operations for all eigenvalue and eigenfunction computations. Note that any code written can be made efficient by re–using elements and finite element matrices. The estimates above represent over–estimates of real computations.

Construction of dynamic stiffness matrices for layered media follows exactly as with construction of stiffness matrices for a single layer waveguide as above. Hence it is possible to solve multi–layered waveguide problems using super–spectral elements bearing in mind the large generalised matrix eigenvalue problems to be solved. This is performed in Example 2 for multi–layered dissipative elements.

14.4 Examples for Two–dimensional Spectral Finite Elements

One area in acoustical engineering that is well–suited to spectral waveguide finite element methods is the design of silencer systems for noise control. There is much work that has been done for smaller systems such as those used in automobiles and small engines, however, the design of much larger systems (such as the parallel baffle type used for gas turbines and other large industrial machines) is still largely guesswork and empirical extensions of previous results. Due to the large size, difficulties in testing and high costs of these silencer systems, the ability to accurately predict the performance before construction and commissioning would be very beneficial. To properly predict the performance of a silencer system, many factors need to be involved in the calculation. Geometrical concerns, absorptive material characteristics, flow effects (turbulence), break out noise, self–generated noise, and source impedance all need to be included in the design calculations of insertion loss (IL). It is very important to note that the method derivations, and their use with the numerical methods are based on plane wave propagation sound sources (i.e. the entire face of the inlet section moving in unison) and an anechoic termination, i.e. $\zeta = \rho c$ at the left and right hand end sections of regions Ω_1 and Ω_4 respectively. Anechoic termination at the outlet and the inlet pipes are assumed.

For both two–dimensional examples cubic polynomials corresponding to $N = 4$ for SFEM and cubic polynomials defined over 9–noded triangular elements for FEMLAB 3.1 were used. The total number of degrees of freedom (DOF) for the spectral waveguide finite element totalled 68 and using FEMLAB 3.1 [10] 2210 DOF over 472 triangles were required. The length of the inlet and outlet pipes for the SFEM and FEMLAB geomtries were 1.0 m. Note that the pipe lengths for SFEM could have been considerably longer in their design but for the purposes of sensible computations it was necessary to keep the DOF for FEMLAB 3.1 to a minimum. The characteristic impedance $\zeta = \rho c$ was applied to both inlet and outlet pipes outer boundaries. The height of the outer pipes was fixed at $H = 0.05$ m and chamber extended by an extra $H_1 = 0.2$ m depth. In keeping with Belawchuk [1] the length of the chamber was fixed at $L_3 = 1.2$ m. To compare efforts of the two methods the CPU timing for 55 frequencies was 3.9 s for SFEM and 26.3 s CPU expenditure for FEMLAB [10]. It seems that the profit speedup for SFEM is rather conservative. However, if the chamber was $L_3 = 2.4$ m long the CPU time would remain 3.9 s for the waveguide finite element method, but FEMLAB 3.1 would now require 849 elements with 3928 DOF and using UMFPACK CPU expenditure would increase to 46.5 s. Both a substantial increase in both storage and computation time compared to the spectral element method.

14.4.1 Theory for Transmission Loss

The definition of transmission loss is the ratio of the incident sound power to the transmitted sound power. As long as the inlet and outlet regions of the silencer are of the same cross section, and the properties of the fluid (density, temperature) do not change, then the TL can be expressed as:

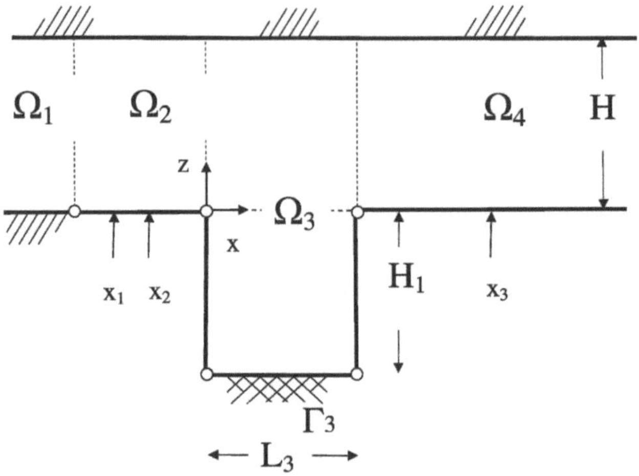

Fig. 14.2 Geometry used for expansion chamber examples. Inlet and outlet pipes are regions $\Omega_{1,2}$ and Ω_4 respectively and Ω_3 is the expansion chamber. Absorbing material may be placed along the boundary Γ_3.

$$TL = 20 \log 10 \left| \frac{P_i}{P_t} \right| \tag{14.26}$$

where P_i is the rms pressure of the incident wave without silencer in place, P_t is the rms pressure of the transmitted wave with silencer in place. This can be simplified to the following equation: $TL = SPL_i - SPL_t$ where it is understood that SPL_i is obtained without the silencer in place, and SPL_t is obtained with the silencer in place, on the exhaust side of the silencer. Figure 14.2 illustrates the geometry used to calculate SPL_i in region Ω_2 and SPL_t in Ω_4. The SPL_i is calculated with the straight pipe (no expansion chamber) and the SPL_t is calculated with the expansion chamber (no straight pipe). The inlet and outlet sections have the characteristic impedance ($\zeta = \rho c$) boundary condition applied. This models a completely anechoic source and termination. Also the inlet section is given a unit acceleration amplitude to model a sound source.

14.4.2 Example 1. Absorbing Boundary Material Lining Silencer Chamber

Starting from the one-dimensional wave equation, the so called 3-point method can be derived as Bilawchuk [1]:

$$p_i = \frac{p_1 - p_2 e^{ikx_{12}}}{1 - e^{2ikx_{12}}} \tag{14.27}$$

where referring to Figure 14.2: p_i is the incoming contribution of rms sound pressure wave; p_1 is the rms sound pressure at location x_1; p_2 =rms sound pressure at location x_2; $x_{12} = x_2 - x_1$ (microphone spacing). Now that the incoming rms pressure values

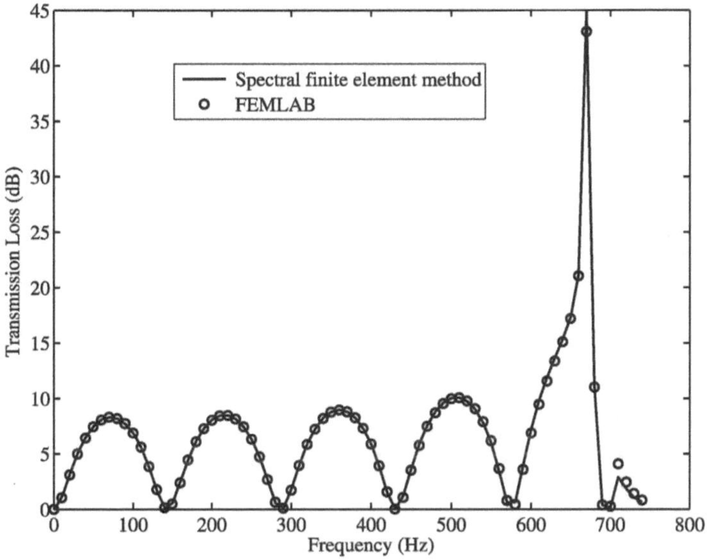

Fig. 14.3 Transmission loss results for expansion chamber shown in Figure 14.2, height and length of chamber $H_3 = 0.2$, $L_3 = 1.2$m respectively and width of pipes $H = 0.05$m.

are known, the exiting rms pressure can be obtained and the TL can be calculated simply as follows:

$$TL = 20 \log 10 \left| \frac{p_i}{p_3} \right| \tag{14.28}$$

where p_3 is the rms sound pressure at point x_3. The rms pressure this point can obtained directly since the termination at the exit is given the characteristic impedance ($Z = \rho c$). The FEM and/or BEM calculations can then be started. In the post–processing stage, the pressures at points x_1, x_2, and x_3 can be calculated and, knowing the distances $x_1 = -0.12$ m and $x_2 = -0.1$, $x_3 = 1.3$ m and the wave number, k, the transmission loss can be determined. Figure 14.3 shows results from SFEM and FEMLAB. Accuracy of the waveguide finite element method is clearly evident. Figure 14.4 shows the change in transmission loss where the chamber lining, Γ_3, is lined with absorbing material. Flow resistivity values taken were from Delany and Bazley formula [8].

14.4.3 Example 2. Dissipative Fibrous Material Chambers

The acoustic performance of a dissipative expansion chamber lined with two layers of fibrous material with different resistances is investigated as a two–dimensional version of Selamet et al. [30]. A two–dimensional numerical approach is used to determine the transmission loss of this dissipative silencer. The flow resistivity of the fibre in the dissipative chamber greatly influences the acoustic performance. The

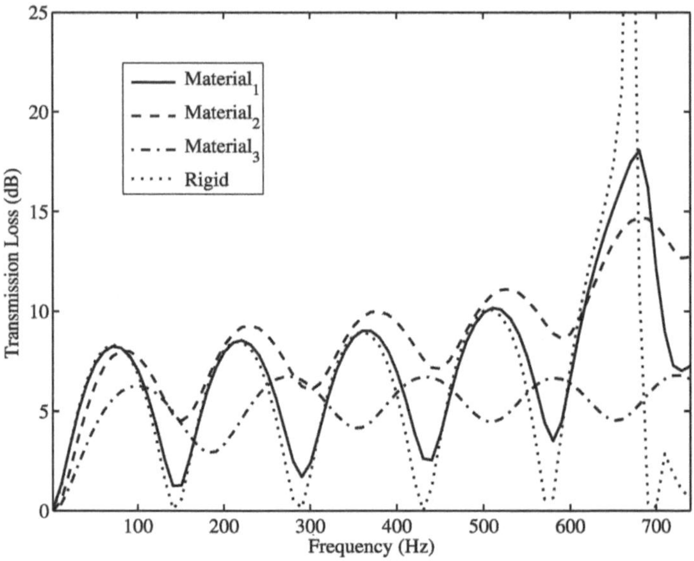

Fig. 14.4 Transmission loss for rigid expansion chamber and chamber with three types of absorbing lining along Γ_3. Flow resistivity values, from [8], for materials are $\sigma_{1,2,3} = 400, 50, \&1 \times 10^3$ Nsm^{-4}.

model used describes complex valued characteristic impedance and wavenumber leading to complex values of wavespeed and density to input into SFEM scheme:

$$\hat{Z} = \rho c \left(1 + 0.0855(f/R)^{-0.754}\right) - 0.0765(f/R)^{-0.732}\mathrm{i}, \qquad (14.29)$$

$$\hat{k} = \omega \left(1 + 0.1472(f/R)^{-0.577}\right) - 0.1734(f/R)^{-0.595}\mathrm{i} \qquad (14.30)$$

Generally, the increasing resistance of fibre in the dissipative chamber improves the sound attenuation in the mid to high frequency range, while deteriorating to a degree at low frequencies.

Thus, to improve the sound attenuation performance at all frequencies, it is a paradox to design a dissipative expansion chamber filled completely with a unique fibre. The present study considers a layered dissipative silencer to investigate the potential trade-offs. Thus a single–pass expansion chamber lined with two fibre layers of different fibre resistance is examined primarily by the SFEM approach, in Figure 14.5. Dissipative material be included in the regions of height $H_1 = 0.05$ m and $H_2 = 0.1$ m, see Figure 14.6. Respectively Dissipative material$_1$ comprised fibrous materials R_1 and R_1, Material$_2$ comprised fibrous materials R_2 and R_1, Material$_3$ comprised materials R_3 and R_1, and Material$_4$ comprised materials R_4 and R_1.

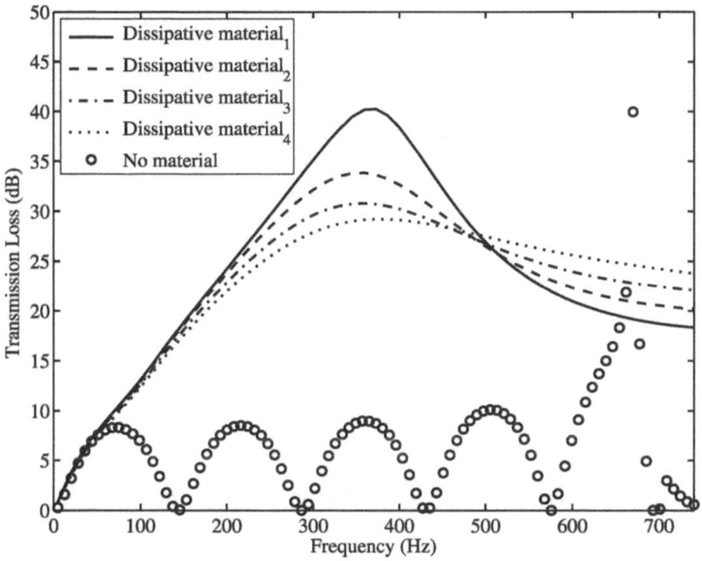

Fig. 14.5 Transmission loss for rigid expansion chamber and chamber with four types of dissipative lining material configurations. Resistivity values used $R_1 = 5000$ rayls/m, $R_2 = 10,000$ rayls/m, $R_3 = 17,000$ rayls/m, and $R_4 = 25,000$ rayls/m. See text for nomenclature.

14.5 Results from a Three–dimensional Analysis

A numerical technique has been developed for the analysis of rigid and absorbent lined silencers of arbitrary, but axially uniform, cross–section. The analysis begins by employing the spectral finite element method to extract the eigenvalues and associated eigenvectors for a silencer chamber. It is demonstrated also that the technique presented offers a considerable reduction in the computational expenditure when compared to a three-dimensional finite element analysis. The method for determining transmission loss predictions from Section 14.4.2 may be used to compare with experimental measurements taken for automotive dissipative silencers with elliptical cross sections.

The dissipative silencer consists of a concentric tube of arbitrary cross section is surrounded by absorbent lined material, on S_2 see Figure 14.8(b). The silencer chamber, which has a length L, is assumed to be uniform along its length, the outer walls of which are assumed to be absorbent except for the final example. The inlet and outlet pipes regions Ω_1 and Ω_3 are identical, each having a circular cross section radius R with rigid walls.

14.5.1 Dispersion Relations for Three-dimensional Examples

If the outcome of an analysis is a dispersion relation between frequency and propagating wavenumber it is prudent to use a sparse eigensolver, such as **eigs** in MATLAB 7.0.2. The finite element mesh for the silencer chamber cross–section consisted

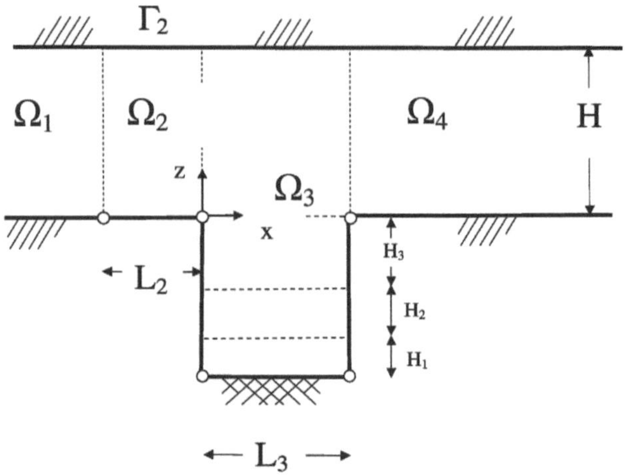

Fig. 14.6 Geometry used for expansion chamber example including dissipative lining in chamber. Lining material material may be included in the regions of height $H_1 = 0.05$ m and $H_2 = 0.1$ m. Region of height $H_3 = 0.05$m is air.

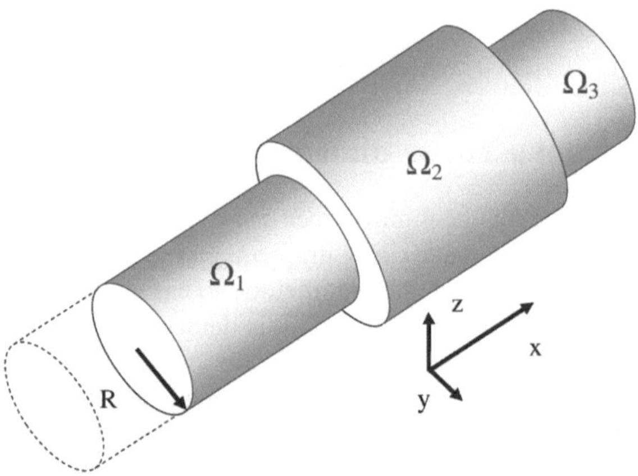

Fig. 14.7 Geometry for the three–dimensional waveguide problem. The system is finite in length with circular inlet and outlet pipes with radius R. The geometry of the silencer is arbitrary, but has an axially uniform cross–section.

of three-noded triangular elements and was generated using *distmesh* by Persson and Strang [27], see Figure 14.8(a). The elliptical cross–section had major-axis radius 0.25 m and minor–axis radius 0.15 m. For the silencer, 470 elements equating to 310 nodes (DOFs) were used to mesh the chamber. Finite element meshes, not shown here, for a square cross–section of width 0.5 m and circular cross–section, radius 0.25 m, were also constructed.

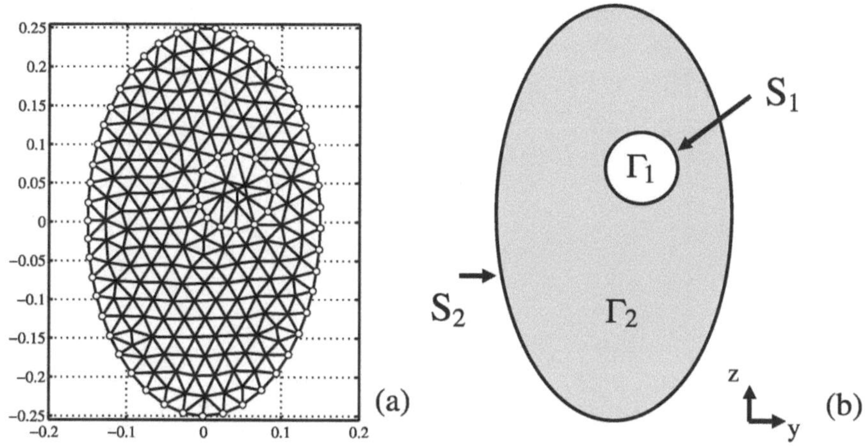

Fig. 14.8 Typical mesh for expansion chambers and cross–sections Γ_2, Γ_1 clearly shown in the right–hand figure.

The mesh (including inlet and outlet circular section) with 470 triangles and 310 nodes is shown in Figure 14.8. The acoustic pressure in the three–dimensional problem is approximated by piecewise linear triangular elements in the cross–section and wave influence functions in the axial direction, similar to the two–dimensional case, Equation (14.3)

$$p(\boldsymbol{x}) = \sum_{J=1}^{N} p_J \phi_J(y,z) W_J(x). \tag{14.31}$$

To arrive at the eigenvalue problem, in order to derive the wave trial functions, the following matrix entries are assembled across the elliptic and cylindrical cross–sections, see Equation (14.6)

$$\boldsymbol{K}_{\mathbf{1}(IJ)} = \int_{\Gamma_{1,2,3}} \boldsymbol{\phi}_I^T(y,z)\boldsymbol{\phi}_J(y,z)\,dy\,dz, \tag{14.32}$$

$$\boldsymbol{K}_{\mathbf{2}(IJ)} = \int_{\Gamma_{1,2,3}} \nabla\boldsymbol{\phi}_I^T(y,z)\cdot\nabla\boldsymbol{\phi}_J(y,z)\,dy\,dz, \tag{14.33}$$

$$\boldsymbol{K}_{\mathbf{3}(IJ)} = \int_{S_{1,2,3}} \frac{1}{\zeta(y,z)}\boldsymbol{\phi}_I^T(y,z)\boldsymbol{\phi}_J(y,z)\,dy\,dz, \tag{14.34}$$

$$\boldsymbol{K}_{\mathbf{4}(IJ)} = \int_{\Gamma_{1,2,3}} \boldsymbol{\phi}_I^T(y,z)\boldsymbol{\phi}_J(y,z)\,dy\,dz \tag{14.35}$$

where $\zeta(y,z)$ is the function defining specific surface impedance on the walls of the pipes and the chamber, that is the boundaries of $S_{1,2,3}$. The corresponding eigenvalue problem for the problem becomes

$$\boldsymbol{K}(\lambda)\boldsymbol{\Phi} = \left\{k^2\,\boldsymbol{K}_1 - \boldsymbol{K}_2 + \mathrm{i}\,k\boldsymbol{K}_3 - \lambda^2\boldsymbol{K}_4\right\}\boldsymbol{\Phi} = 0 \tag{14.36}$$

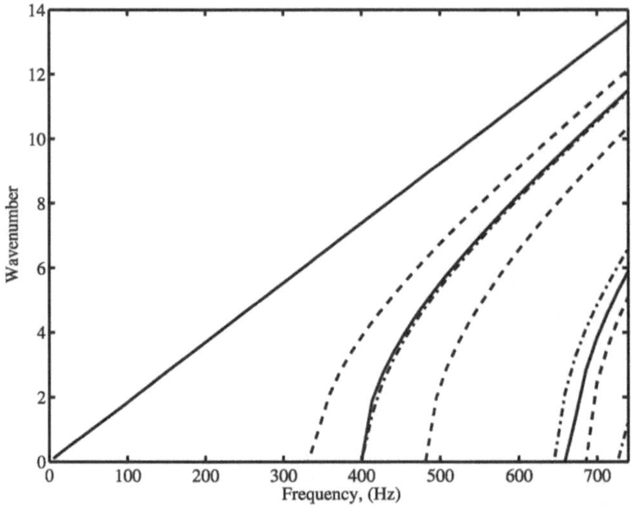

Fig. 14.9 Dispersion curves for various silencer chambers. Square cross–section of width 0.5 m (dashed line), circular cross–section radius 0.25 m (solid line), and elliptical cross–section major-axis radius 0.25 m minor–axis radius 0.15 m. Mesh (including inlet and outlet circular section) used with 470 triangles and 310 nodes shown in Figure 14.8.

Energy transmission through a waveguide system is possible when the propagating wavenumber λ, in Equation (14.37) has zero imaginary component. For a given frequency or wavenumber k it is possible to solve the eigenvalue problem below seeking real–valued wavenumbers and their corresponding mode shapes

$$\boldsymbol{W}_m(x) = \boldsymbol{\Phi}_m \, \mathrm{e}^{\mathrm{i}\lambda_{m-1}x}, \quad m = 1, \ldots, N \tag{14.37}$$

giving a dispersion relation between frequency $\omega = 2\pi f$ and wavenumber λ. The eigenvalue problem corresponding to the elliptic expansion chamber equated to solving a 310 $\times$ 310 *sparse* eigenvalue problem. For 55 frequencies the sparse eigensolver, for the first seven eigenvalues λ^2 with smallest imaginary part, CPU time costing around 15 s for the ellipse. For the circular and square cross–sections the CPU time increased to around 20 s due to the increased number of DOFs. Figure 14.9 shows dispersion relations for the three square, circular and elliptical configurations. The curves for the square and circular cross–sections could be easily compared with known solutions. Note that four propagating modes, $\boldsymbol{\Phi}$, exist for the elliptic geometry at 740 Hz. Omitting the plane–wave mode, corresponding to $\lambda_0 = 0$, these are shown in Figure 14.10 from lowest wavenumber to highest wavenumber respectively.

14.5.2 Solutions for Elliptic Silencer Problem

This section illustrates computations for a finite length combination of pipes and chamber assuming unit normal acceleration at the end of the inlet pipe. The mesh

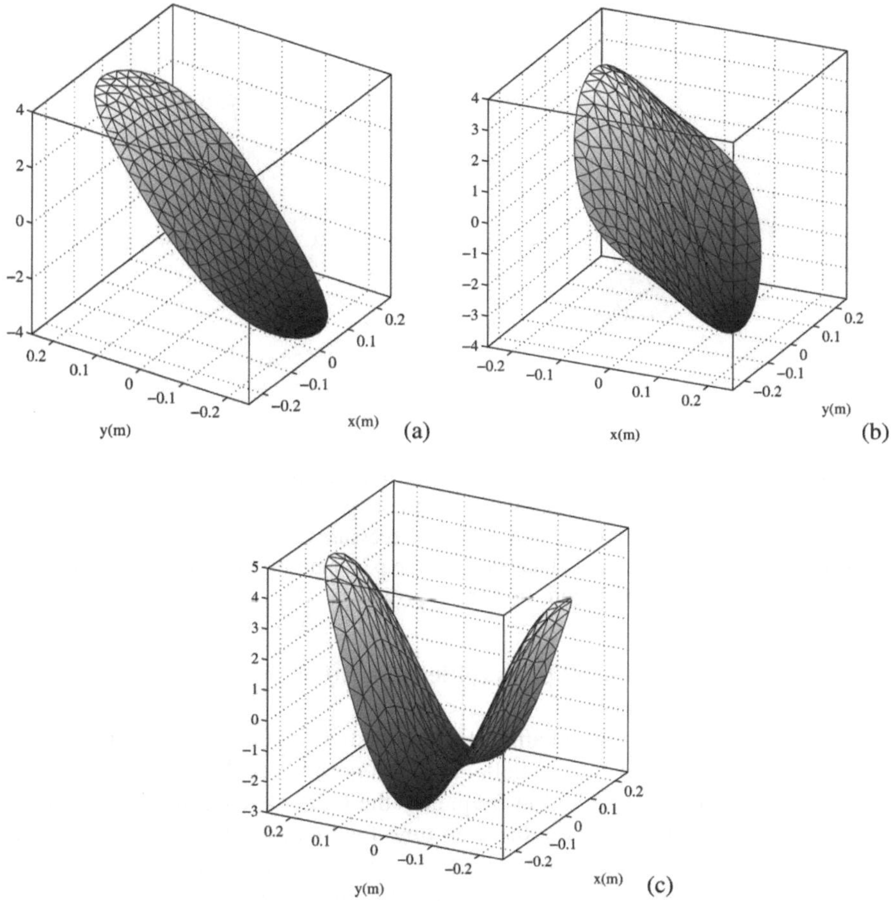

Fig. 14.10 Propagating modes for elliptic cross–section expansion chamber, at excitation frequency $f = 740$ Hz for geometry in Figure 14.8.

used, Figure 14.8, shows the finite elements for the chamber and the cross–section mesh for the rigid inlet and outlet pipes, located just above the centre. The computation of acoustic pressure for the silencer problem is dominated by the assemblage of the wave influence functions (14.36) and the corresponding dynamic stiffness matrix for the silencer chamber (14.23). All the frequency independent matrices $\boldsymbol{K}_1, \ldots, \boldsymbol{K}_4$ are stored as sparse matrices. However, all the eigenvectors and eigenvalues are required for the elliptic chamber problem of size $N_2 = 310$, requiring $46N_2^3$ operations. The dynamic stiffness matrix for the chamber costs a little over $5N_2^3$ operations, due to re–use of LU decomposition, from Equation (14.21). For the inlet and outlet pipes the numbers of degrees of freedom are somewhat lower, $N_{1,3} = 20$. Computing the wave influence functions and the dynamic stiffness ma-

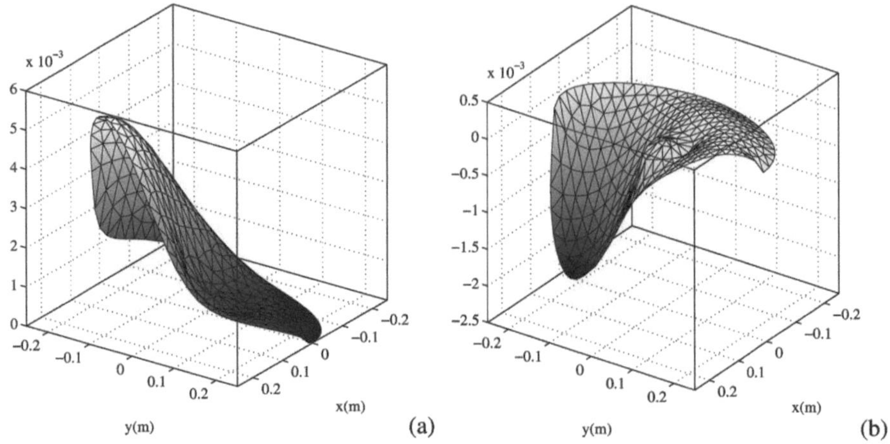

Fig. 14.11 Acoustic pressure at two cross–sections of lined expansion chamber (length $L = 0.4$ m) due to unit normal acceleration at inlet pipe (length $L = 1.0$ m) at excitation frequency 740 Hz. Left hand plot shows acoustic pressure at entry of chamber and right hand plot acoustic pressure at exit of chamber. Outlet pipe length of 4.0 m with rigid termination, geometry of pipes shown in Figure 14.7 and Figure 14.8.

trices for the inlet and outlet pipes (of lengths 1.0 m and 4.0 m) are negligible in comparison to the chamber.

If one assumes the full dynamic stiffness matrix for this problem the total numbers of DOFs amount to $N_{tot} = 660$. For a unit normal acceleration at the left–hand end of Ω_1 and a given excitation frequency the CPU expenditure time for solving the system of equation is 8.1 s, the total time for finding wave functions for chamber takes 30.3 s, and the total CPU time for solving the problem in MATLAB 7.0.2 on Pentium M machine took 54.8 seconds. Although the number of operations are proportional to $50N_2^3$ as discussed in the section on two–dimensional analysis the CPU expenditure is generally higher than for an equivalent analytic matching procedure. However, CPU expenditure compares favourably with alternative fully three-dimensional treatments.

Figures 14.11–14.14 show solutions at cross–sections of the silencer assembly. Specifically, solutions are shown at the beginning of the chamber (a) and at the midway point (b) for elliptic chambers of lengths $L = 0.8, 1.0$ and 1.2 m respectively with absorbent liner material (flow resistivity $\sigma = 400 \times 10^3$ Nsm^{-4}) covering the entire outer surface. The final figure shows acoustic pressure solutions, similar to Figure 14.13 but for rigid silencer chamber. Note that the size of the computational problem is not changed for each expansion chamber length.

14.6 Conclusions

A new spectral method in the form of a boundary finite element scheme has been derived to treat the problem of sound transmission in non–uniform waveguides or

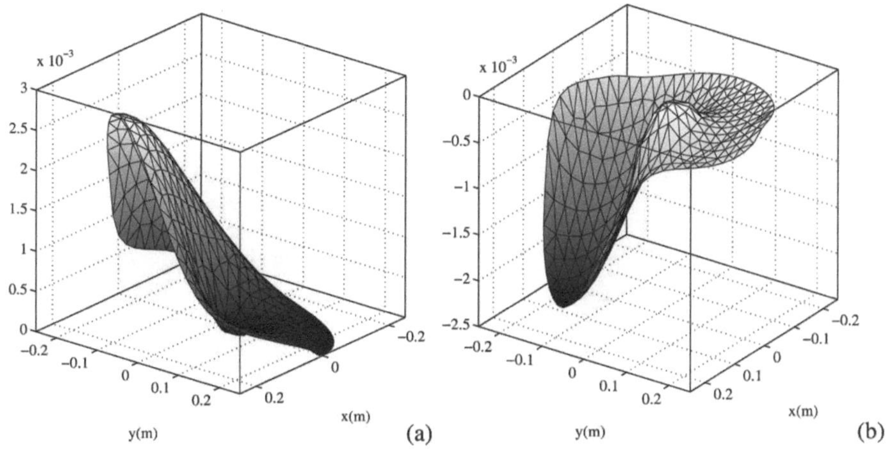

Fig. 14.12 Acoustic pressure at two cross–sections of lined expansion chamber (length $L = 0.8$ m) due to unit normal acceleration at inlet pipe (length 1.0 m) at excitation frequency 740 Hz. Left hand plot shows acoustic pressure at entry of chamber and right hand plot acoustic pressure at exit of chamber. Outlet pipe length of 4.0 m with rigid termination, geometry of pipes shown in Figure 14.7 and Figure 14.8.

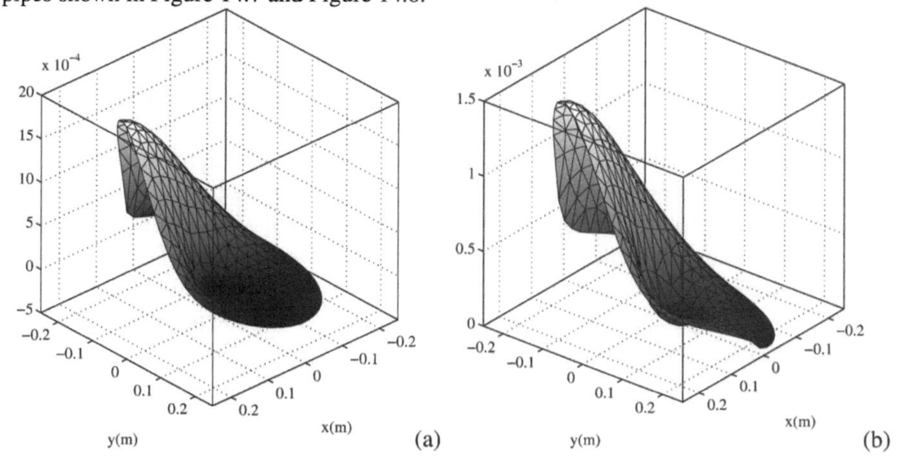

Fig. 14.13 Acoustic pressure at two cross–sections of lined expansion chamber (length $L = 1.2$ m) due to unit normal acceleration at inlet pipe (length 1.0 m) at excitation frequency 740 Hz. Left hand plot shows acoustic pressure at entry of chamber and right hand plot acoustic pressure at exit of chamber. Outlet pipe length of 4.0 m with rigid termination, geometry of pipes shown in Figure 14.7 and Figure 14.8.

ducts. A unique feature of the waveguide boundary spectral finite element approach is the use of basis functions generated from linear eigenvalue calculations. The basis functions, themselves solutions to the homogeneous reduced wave equations, may be defined over regions of arbitrary length with sound absorbing sides and dissipative material.

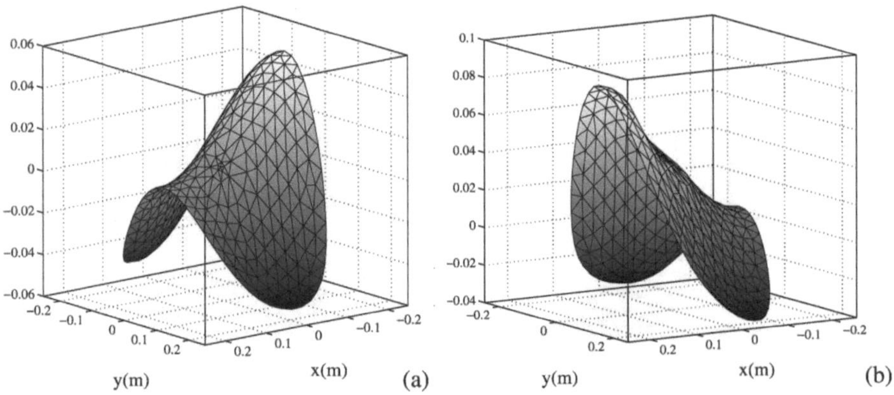

Fig. 14.14 Acoustic pressure at two cross–sections of rigid expansion chamber (length L = 1.2 m) to be compared with Figure 14.13.

The waveguide geometries that have been given the most attention in the present study are that of a rectangular duct partially lined on one side or a geometrical non–uniform rectangular duct. These configurations are not easily amenable to analytical treatment and the use of a numerical method is appropriate here. The finite element formulation in this investigation is, however, sufficiently general that it can be extended to any non–uniform rectangular waveguide with varying linings, fluid densities and a flexible structure.

Future work in this area could be dedicated to coupling the method to standard finite element codes especially for problems defined in built–up waveguide domains. Developers will profit on a huge saving on computational expenditure For three dimensional problems. On the other hand, for exterior problems, use of the Astley–Leis conjugated infinite element on the upper domain may be employed for sound propagation over half–spaces. This may lead to a singular eigenvalue problem but work in this area may lead to a novel solution.

References

1. Bilawchuk S, Fyfe KR (2003) Comparison and implementation of the various numerical methods used for calculating transmission loss in silencer systems. Applied Acoustics 64:903–916
2. Birgersson F (2004) Prediction of random vibration using spectral methods. PhD Thesis, Trita–AVE, ISSN 1651-7660, KTH, Stockholm
3. Birgersson F, Ferguson NS, Finnveden S (2003) Application of the spectral finite element method to turbulent boundary layer induced vibration of plates. Journal of Sound and Vibration 259:873–891
4. Birgersson F, Finnveden S (2005) A spectral super element for modelling of plate vibration. Part II: Turbulence excitation. Journal of Sound and Vibration, 287:315–328
5. Birgersson F, Finnveden S, Nilsson C–M (2005) A spectral super element for modelling of plate vibration. Part I: General theory. Journal of Sound and Vibration 287:297–314

6. Bonfiglio P, Pompoli F, Peplow AT, Nilsson AC (2007) Aspects of computational vibration transmission for sandwich panels. Journal of Sound and Vibration 303:780–797
7. Boyd JP (2000) Chebyshev and Fourier spectral methods. Dover Publications, New York, 2nd edition
8. Delany ME, Bazley EN (2003) Acoustical properties of fibrous absorbent materials. Applied Acoustics 3:105–116
9. Denia FD, Selamet A, Fuenmayor FJ, Kirby R (2007) Acoustic attenuation performance of perforated dissipative mufflers with empty inlet/outlet extensions. Journal of Sound and Vibration 302:1000–1017
10. FEMLAB 3.1 (2004) Users manual
11. Finnveden S (1994) Exact spectral finite element analysis of stationary vibrations in a railway car structure. Acta Acustica 2:461–482
12. Finnveden S (1997) Simplified equations of motion for the radial–axial vibrations of fluid–filled pipes. Journal of Sound and Vibration 208:685–703
13. Finnveden S (1996) Spectral finite element analysis of stationary vibrations in a beam–plate structure. Acustica / Acta Acustica 82:479–497
14. Finnveden S (1997) Spectral finite element analysis of the vibration of straight fluid–filled pipes with flanges. Journal of Sound and Vibration 199:125–154
15. Finnveden S, Birgersson F, Ross U, Kremer T (2005) A model of wall pressure correlation for prediction of turbulence–induced vibration. Journal of Fluids and Structures, 20:1127–1143
16. Finnveden S, Pinnington RJ (2000) A velocity method for estimating dynamic strain and stress in pipes. Journal of Sound and Vibration 229:147–182
17. Fornberg B (1996) A practical guide to pseudospectral methods. Cambridge University Press, Cambridge
18. Fraggstedt M (2006) Power dissipation in car tyres. PhD Thesis, KTH, Trita–AVE, ISSN 1651–7660, KTH, Stockholm
19. Gavric L (1994) Finite element computation of dispersion relations of thin–walled waveguides. Journal of Sound and Vibration 173:113–124
20. Gottlieb D, Orszag SA (1977) Numerical analysis of spectral methods: Theory and applications. SIAM, Philadelphia
21. Kirby R (2003) Transmission loss predictions for dissipative silencers of arbitrary cross section in the presence of mean flow. Journal of the Acoustical Society of America 114:200–209
22. Nilsson C–M (2004) Waveguide finite elements applied on a car tyre. PhD Thesis, Trita–AVE, ISSN 1651–7660, KTH, Stockholm
23. Nilsson C–M, Finnveden S (2007) Input power to waveguides calculated by a finite element method. Journal of Sound and Vibration 305:641–658
24. Orrenius U, Finnveden S (1996) Calculation of wave propagation in rib–stiffened plate structures. Journal of Sound and Vibration 198:203–224
25. Peplow AT, Finnveden S (2007) Calculation of vibration transmission over bedrock using a waveguide finite element mode. International Journal for Numerical and Analytical Methods in Geomechanics, DOI: 10.1002/nag.643
26. Peplow AT, Finnveden S (2004) A super–spectral finite element method for sound transmission in waveguides. Journal of the Acoustical Society of America 116:1389–1400
27. Persson PO, Strang G (2004) A simple mesh generator in MATLAB. SIAM Review 46:329–345
28. Philipps TN, Davies AR (1988) On semi–infinite spectral elements for Poisson problems with reentrant boundary singularities. Journal of Comutational and Applied Mathematics 21:173–188

29. Richard TH, Leung AYT (1979) An accurate method in structural vibrations. Journal of Sound and Vibration 55:363–376
30. Selamet A, Xu MB, Lee IJ, Huff NT (2005) Dissipative expansion chambers with two concentric layers of fibrous material. International Journal of Vehicle Noise and Vibration 1:341–357
31. Trefethen LN (2000) Spectral methods in Matlab. SIAM, Philadelphia

Part V

BEM: External Problems

15

Treating the Phenomenon of Irregular Frequencies

Steffen Marburg[1] and Ting–Wen Wu[2]

[1] Institut für Festkörpermechanik, Technische Universität Dresden, 01062 Dresden, Germany
marburg@ifkm.mw.tu-dresden.de

[2] Department of Mechanical Engineering, University of Kentucky Lexington, KY 40506, USA
timwu@engr.uky.edu

Summary. This chapter reviews a number of techniques developed to overcome the well–known non–uniqueness difficulty in the boundary element method for acoustic radiation and scattering in an exterior domain. The non–uniqueness difficulty occurs at a set of irregular frequencies associated with the eigenfrequencies of the corresponding interior problem. The chapter focuses on the comparison of two commonly used techniques, the CHIEF method and the Burton and Miller method, along with their variations. After briefly revisiting the example of the pulsating sphere, a cat's eye radiation problem is used as a test case for evaluating the effectiveness of different techniques. Numerical results confirm that the Burton and Miller method is a very reliable technique at all frequencies, while the CHIEF method is effective only at low frequencies. One potential drawback of Burton and Miller method is the requirement of the C^1 continuity condition at collocation points, which may rule out the use of any C^0 continuous elements. Modified versions of the Burton and Miller method, which compute the normal–derivative integral equation only at the center of each C^0 element, have been proposed in the past to partially alleviate the strict C^1 requirement. It is still uncertain that any of these modified versions is as theoretically robust as the original Burton and Miller method. Nonetheless, the cat's eye radiation test case demonstrates that a modified version that uses 9–node quadrilateral elements is as effective as the original Burton and Miller method in practical use. The paper is completed by applying the Burton and Miller method to the industrial problem of a radiating diesel engine for which the radiated sound power is evaluated.

15.1 Introduction

Boundary element methods (BEM) have been used in acoustic radiation and scattering for decades [6,9,18,45]. The major advantage of boundary element methods over other numerical techniques is that only the surface of the body needs to be modeled. The Sommerfeld radiation condition at infinity is automatically satisfied. However, one potential shortcoming is that the exterior boundary integral formulation, either direct or indirect, fails to produce a correct solution at a set of irregular frequencies associated with the eigenfrequencies of the corresponding interior domain. For the direct BEM formulation, the problem is referred to as "the non–uniqueness difficulty," while for the indirect BEM formulation, the problem is often referred to as "the non–existence difficulty."

Since the indirect BEM, cf. [45], solves the exterior and the interior problems (even though the interior domain may be just a solid) at the same time in a combined boundary integral equation, it is understandable that the exterior solution will fail when the interior domain (imaginary or not) becomes acoustically resonant at its eigenfrequencies associated with the type of boundary condition specified. The common practice in the indirect BEM is to always add some damping or sound absorbing surfaces in the interior domain to damp out any potential resonance. On the other hand, the non–uniqueness difficulty associated with the direct BEM, which is the focus of this chapter, is harder to explain from the physical ground because the integral equation used is for the exterior problem only, instead of a combined interior/exterior problem. Although the exterior domain and the corresponding (imaginary) interior domain share the same boundary, the direct boundary integral equations for the exterior and the interior problems are still slightly different in two aspects

1. their normal directions are opposite to each other, and
2. their solid angles are different at corners and edges.

It is hard to just directly compare the interior and exterior boundary integral equations to explain why the non–uniqueness difficulty would occur. Advanced mathematical explanations have been presented in papers more than half a century ago, cf. Kupradze [20] and Weyl [43]. A simple mathematical explanation of this phenomenon can be found in Wu and Seybert [51, 52] and later in Wu [47] as well. It has been shown that regardless of the type of boundary conditions prescribed for the exterior problem (Neumann, Dirichlet, or impedance), the Kirchhoff–Helmholtz integral equation will always fail to yield a unique solution at the eigenfrequencies of the corresponding interior Dirichlet problem. In real-world applications, knowing the exact locations of the irregular frequencies is actually not that important because it is impractical to solve an interior Dirichlet problem first just to find the eigenfrequencies. A more reasonable approach is to always apply some kind of treatment in the direct BEM at every frequency to prevent the non–uniqueness from happening. Actually, at high frequencies, the eigenfrequencies are so closely spaced that it is impossible to distinguish the regular frequencies from the irregular frequencies.

Over the last four or five decades, many different approaches were proposed to create a unique solution. There are countless publications on this topic. In this chapter, we will review a number of different methods attempting to overcome the non–uniqueness difficulty at irregular frequencies. Among them, the two most popular categories are

1. the Combined Helmholtz Integral Equation Formulation (CHIEF) method originally proposed by Schenck [33] and its variations, and
2. the linear combination of the Kirchhoff–Helmholtz integral equation and its normal derivative originally proposed by Panič [29] and Brakhage and Werner [3]. This approach was adopted to the Neumann problem by Burton and Miller [4]. We will consider the Burton and Miller method and its variations.

There are a couple of other methods which will be briefly discussed thereafter.

In the examples section, we will discuss results of different methods for a spherical radiator, for a cat's eye structure and finally apply the Burton and Miller method to solve for an engine radiation problem. The cat's eye radiation problem [24] is used as a test case to evaluate the effectiveness of a number of different methods and to show effects which are not produced by a spherical radiator because the cat's eye structure is complex enough to produce many irregular frequencies as would be encountered in real–world applications, and its smooth solution will make it easy to identify solution failures at irregular frequencies.

15.2 Boundary Element Formulation

Herein, we consider the three–dimensional linear time–harmonic problems of acoustics. The governing differential equation is the Helmholtz equation

$$\Delta p(\boldsymbol{x}) + k^2 p(\boldsymbol{x}) = 0 \qquad \boldsymbol{x} \in \Omega \subset \mathbb{R}^3 \,. \tag{15.1}$$

Since Ω represent the domain in which our equations are valid, Ω_c denotes the complementary domain and Γ stands for the boundary. We will limit our analysis to the Neumann problem in external acoustics. Hence, the boundary condition becomes

$$\frac{\partial p(\boldsymbol{x})}{\partial n(\boldsymbol{x})} = s\,k\,v_f(\boldsymbol{x}) = s\,k\,v_s(\boldsymbol{x})\,, \tag{15.2}$$

where k is the wavenumber ω/c and $s = \mathrm{i}\rho c$. With this boundary condition, the Kirchhoff–Helmholtz boundary integral equation of the first chapter simplifies to

$$c(\boldsymbol{y})p(\boldsymbol{y}) + \int_\Gamma \frac{\partial G(\boldsymbol{x},\boldsymbol{y})}{\partial n(\boldsymbol{x})}\, p(\boldsymbol{x}) d\Gamma(\boldsymbol{x}) = sk \int_\Gamma G(\boldsymbol{x},\boldsymbol{y})\, v_s(\boldsymbol{x}) d\Gamma(\boldsymbol{x}) \tag{15.3}$$

with $\boldsymbol{y} \in \Gamma$. This integral equation may also be referred to as the first boundary integral equation. In addition to this, the normal derivative of the Kirchhoff–Helmholtz integral equation, also referred to as the second boundary integral equation, is required

$$c(\boldsymbol{y})\frac{\partial p(\boldsymbol{y})}{\partial n(\boldsymbol{y})} + \int_\Gamma \frac{\partial^2 G(\boldsymbol{x},\boldsymbol{y})}{\partial n(\boldsymbol{x})\partial n(\boldsymbol{y})}\, p(\boldsymbol{x}) d\Gamma(\boldsymbol{x}) = sk \int_\Gamma \frac{\partial G(\boldsymbol{x},\boldsymbol{y})}{\partial n(\boldsymbol{y})}\, v_s(\boldsymbol{x}) d\Gamma(\boldsymbol{x})\,. \tag{15.4}$$

This equation is often referred to as the hypersingular integral equation since the integral on the left hand–side involves a hypersingular integrand. For the harmonic time–dependence of $\mathrm{e}^{-\mathrm{i}\omega t}$, we can write the Green's function G and its first and second–order derivatives in $\mathbb{R}^3$ as

$$
\begin{aligned}
G(\boldsymbol{x},\boldsymbol{y}) &= \frac{1}{4\pi}\frac{\mathrm{e}^{\mathrm{i}kr}}{r}\,, \\
\frac{\partial G(\boldsymbol{x},\boldsymbol{y})}{\partial n(\boldsymbol{\xi})} &= -\frac{1}{4\pi r^2}\,(1-\mathrm{i}kr)\,\frac{\partial r}{\partial n(\boldsymbol{\xi})}\,\mathrm{e}^{\mathrm{i}kr} \quad , \qquad \boldsymbol{\xi}=\boldsymbol{x},\boldsymbol{y} \\
\frac{\partial^2 G(\boldsymbol{x},\boldsymbol{y})}{\partial n(\boldsymbol{x})\partial n(\boldsymbol{y})} &= \frac{1}{4\pi r^3}\left[(3-3\mathrm{i}kr-k^2r^2)\frac{\partial r}{\partial n(\boldsymbol{x})}\frac{\partial r}{\partial n(\boldsymbol{y})}+\right. \\
&\qquad\left. +\,(1-\mathrm{i}kr)\,\boldsymbol{n}(\boldsymbol{x})\cdot\boldsymbol{n}(\boldsymbol{y})\right]\mathrm{e}^{\mathrm{i}kr}
\end{aligned}
\tag{15.5}
$$

where r represents the Euclidean distance between field point $\boldsymbol{x}$ and source point $\boldsymbol{y}$ as $r = r(\boldsymbol{x},\boldsymbol{y}) = |\boldsymbol{x}-\boldsymbol{y}|$.

Discretization of Equations (15.3) and (15.4) by collocation allows to formulate matrices $\boldsymbol{G}$ and $\boldsymbol{H}$ (15.3) as

$$
\begin{aligned}
g_{lj} &= s\,k\int_{\Gamma} G(\boldsymbol{x},\boldsymbol{z}_l)\,\bar{\phi}_j(\boldsymbol{x})\,d\Gamma(\boldsymbol{x}) \\
h_{lj} &= c(\boldsymbol{z}_l)\,\delta_{lj} + \bar{h}_{lj} \;= \\
&= c(\boldsymbol{z}_l)\,\delta_{lj} + \int_{\Gamma}\frac{\partial G(\boldsymbol{x},\boldsymbol{z}_l)}{\partial n(\boldsymbol{x})}\,\phi_j(\boldsymbol{x})\,d\Gamma(\boldsymbol{x})
\end{aligned}
\tag{15.6}
$$

and matrices $\boldsymbol{F}$ and $\boldsymbol{E}$ (15.4) as

$$
\begin{aligned}
f_{lj} &= -s\,k\,c(\boldsymbol{z}_l)\,\delta_{lj} + s\,k\int_{\Gamma}\frac{\partial G(\boldsymbol{x},\boldsymbol{z}_l)}{\partial n(\boldsymbol{z}_l)}\,\bar{\phi}_j(\boldsymbol{x})\,d\Gamma(\boldsymbol{x}) \\
e_{lj} &= \int_{\Gamma}\frac{\partial^2 G(\boldsymbol{x},\boldsymbol{y})}{\partial n(\boldsymbol{x})\partial n(\boldsymbol{y})}\,\phi_j(\boldsymbol{x})\,d\Gamma(\boldsymbol{x})\,.
\end{aligned}
\tag{15.7}
$$

Thus we can write two systems of equations

$$
\boldsymbol{H}\,\boldsymbol{p} \;=\; \boldsymbol{G}\,\boldsymbol{v}_s \tag{15.8}
$$

and

$$
\boldsymbol{E}\,\boldsymbol{p} \;=\; \boldsymbol{F}\,\boldsymbol{v}_s\,. \tag{15.9}
$$

Theoretically, both systems can be solved for $\boldsymbol{p}$. However, system matrices $\boldsymbol{H}$ and $\boldsymbol{E}$ suffer from an ill–conditioning at certain frequencies. This will be further discussed in what follows.

15.3 Irregular Frequencies

The exterior Helmholtz problem with appropriate boundary and radiation conditions has a unique solution for all frequencies. However, the boundary integral Equation (15.3), will fail to have a unique solution for a countable set $\mathcal{I}_\Gamma$ of values of k; see for example [20, 27, 43, 51]. This problem of irregular frequencies can be explained in various equivalent ways. For example, it can be seen as the consequence of using a free–space Green's function, which by nature, takes no account of the boundary Γ. A number of methods and formulations have been proposed over the last 4–5 decades for overcoming this problem. An excellent survey on these methods is presented by Rego Silva [30]. More recently, Ochmann [28] and Visser [42] have discussed various aspects of different solution methods. Herein, we discuss two categories of these remedies in detail and briefly outline some others.

The problem of irregular frequencies in (15.3) arises because, for $k \in \mathcal{I}_\Gamma$, both the left and the right hand–side of Equation (15.3) will be singular and so will be the system of equations, Equation (15.9). For values of k close to the countable set $\mathcal{I}_\Gamma$ the operators and therefore the matrix approximations to them will be ill–conditioned. Furthermore, it can be shown that the number of elements in $\mathcal{I}_\Gamma$ which are less than a value K grows with $\mathcal{O}(K^3)$. Therefore, the density of these irregular wavenumbers or frequencies increases as k increases. For example for a unit sphere $\mathcal{I}_\Gamma = \{k | j_n(k) = 0,\ n = 0, 1, 2, ...\} = \{...., 20.12, 20.18, 20.20, 20.37, 20.54, 20.98, ...\}$, where j_n is the spherical Bessel function of order n; cf. [1]. In general, the set $\mathcal{I}_\Gamma$ corresponds to the eigenvalues of the interior Dirichlet problem.

Similarly, if we choose to use the boundary integral equation (15.4) and the system of equations, Equation (15.9), there will also be a countable set, $\mathcal{J}_\Gamma$, of irregular frequencies, but these now correspond to the eigenvalues of the interior Neumann problem.

Practical aspects of solution behaviour close to irregular frequencies were described and discussed by Juhl [16]. In the following sections we briefly discuss some of the methods for alleviating the problem with irregular frequencies.

15.4 CHIEF and its Variations

15.4.1 CHIEF Originally Proposed by Schenck

First we examine the Combined Helmholtz Integral Equation Formulations (CHIEF) originally proposed by Schenck [33]. The basic idea is to add to Equation (15.8), i.e. the n point discretization of (15.3), an m point discretization of the interior Kirchhoff–Helmholtz relation (15.3), that is with $\boldsymbol{x} \in \Omega_c$ for which $c(\boldsymbol{x}) = 0$. We need to choose m collocation points located in the enclosed cavity. These points are usually referred to as CHIEF points. The system matrix $\boldsymbol{H}$ corresponding to the CHIEF method takes the form

$$
\boldsymbol{H} = \left[\begin{array}{cccc}
c_1 + \bar{h}_{11} & \bar{h}_{12} & \cdots & \bar{h}_{1n} \\
\bar{h}_{21} & c_2 + \bar{h}_{22} & \cdots & \bar{h}_{2n} \\
\vdots & \vdots & & \vdots \\
\bar{h}_{n1} & \bar{h}_{n2} & \cdots & c_n + \bar{h}_{nn} \\
\hline
\bar{h}_{n+1\,1} & \bar{h}_{n+1\,2} & \cdots & \bar{h}_{n+1\,n} \\
\vdots & \vdots & & \vdots \\
\bar{h}_{n+m\,1} & \bar{h}_{n+m\,2} & \cdots & \bar{h}_{n+m\,n}
\end{array}\right] . \qquad (15.10)
$$

The entries $\bar{h}_{lj}$ have been given in Equation (15.6). The rectangular system matrix reflects the fact that we now have an over–determined linear system of algebraic equations for the n–dimensional vector of unknowns, $\boldsymbol{p}$. This system is solved in a least squares sense, where the unknown solution is formally given by

$$
\boldsymbol{p} = \left(\boldsymbol{H}^H \boldsymbol{H}\right)^{-1} \boldsymbol{H}^H \boldsymbol{G}\, \boldsymbol{v}_s , \qquad (15.11)
$$

with superscript ^{H} denoting Hermitian, i.e. transposed conjugate complex matrix.

There are a large number of publications on the use and variations of this popular method; see for example Wu and Seybert [51] and references therein.

15.4.2 CHIEF with a Square Matrix [31]

At this point we will consider and test the CHIEF variation due to Rosen et al. [31], and, even earlier, Seybert and Rengarian [38]. They proposed to solve a square system of linear equations, having the same right hand side as CHIEF but with the new matrix $\boldsymbol{H}$ below:

$$
\boldsymbol{H} = \left[\begin{array}{cccc|cccc}
c_1 + \bar{h}_{11} & \bar{h}_{12} & \cdots & \bar{h}_{1n} & \bar{h}_{1\,n+1} & \bar{h}_{1\,n+2} & \cdots & \bar{h}_{1\,n+m} \\
\bar{h}_{21} & c_2 + \bar{h}_{22} & \cdots & \bar{h}_{2n} & \bar{h}_{2\,n+1} & \bar{h}_{2\,n+2} & \cdots & \bar{h}_{2\,n+m} \\
\vdots & \vdots & & \vdots & \vdots & \vdots & & \vdots \\
\bar{h}_{n1} & \bar{h}_{n2} & \cdots & c_n + \bar{h}_{nn} & \bar{h}_{n\,n+1} & \bar{h}_{n\,n+2} & \cdots & \bar{h}_{n\,n+m} \\
\hline
\bar{h}_{n+1\,1} & \bar{h}_{n+1\,2} & \cdots & \bar{h}_{n+1\,n} & 1 & 0 & \cdots & 0 \\
\bar{h}_{n+2\,1} & \bar{h}_{n+2\,2} & \cdots & \bar{h}_{n+2\,n} & 0 & 1 & \cdots & 0 \\
\vdots & \vdots & & \vdots & \vdots & \vdots & & \vdots \\
\bar{h}_{n+m\,1} & \bar{h}_{n+m\,2} & \cdots & \bar{h}_{n+m\,n} & 0 & 0 & \cdots & 1
\end{array}\right] . \qquad (15.12)
$$

The new linear system is now solved for the $(n+m)$–vector $(\boldsymbol{p}, \boldsymbol{0})$. In [31], the entries of the upper right block matrix are chosen as

$$
h_{lk} = h^*_{kl} \quad \text{with} \quad l = 1, \ldots, n \quad \text{and} \quad k = n+1, \ldots, n+m . \qquad (15.13)
$$

Hence, the upper right block matrix is the complex conjugate of the lower left block. We will refer to this formulation as the conjugated variant of the Rosen et al. method. We also test a modified version of this where

$$h_{lk} \quad = \quad h_{kl} \qquad \text{with} \quad l = 1, \ldots, n \qquad \text{and} \quad k = n+1, \ldots, n+m, \tag{15.14}$$

thus, the upper right block matrix is just the transpose of the lower left block. This will be referred to as the unconjugated variant of the Rosen et al. method which is equivalent to the method proposed by Seybert and Rengarian [38].

15.4.3 CHIEF Enhanced by Taking Derivatives

It is well known that when a CHIEF point falls on any of the internal nodal surfaces of the corresponding interior Dirichlet problem, that particular CHIEF point will not provide any constraint effect because sound pressure on any internal nodal surface is automatically zero by definition. For a general radiation/scattering problem, it is unlikely to know the exact locations of the internal nodal surfaces unless the corresponding interior Dirichlet problem is solved first. The problem is compounded by the fact that the internal nodal surfaces are clustered together at high frequencies in such a way that it is almost impossible for a CHIEF point not to fall on any nodal surface. To partially alleviate the difficulty, first–order [51] and second–order [11,36,37] derivatives of the original CHIEF equation have been taken at each CHIEF point along with the original CHIEF equation. The idea is based on the fact that even though sound pressure is zero on nodal surfaces, its derivatives may not be zero unless the CHIEF point also falls on the intersection of two or more nodal surfaces. This will greatly improve the "survival rate" of a randomly selected CHIEF point. If first–order derivatives (with respect to the three coordinate directions) are taken in addition to the original CHIEF equation, a total of four constraint equations are provided at each "enhanced CHIEF point." In principle, one enhanced CHIEF point (with four constraint equations) may still be better than four original CHIEF points. If second–order derivatives are also taken, the method is referred to as "SuperCHIEF method." A total of 7 constraint equations are provided at each "Super CHIEF point."

It should be noted that either the Enhanced CHIEF method or the SuperCHIEF method only extends the application range of the CHIEF method from low frequencies to intermediate frequencies. The rank deficiency of the original BEM matrix (without using any CHIEF) is usually greater than one. It has been found in numerical experiments that a single CHIEF point that does not fall on any nodal surfaces still may not be able to provide enough constraint effect. In practice, a number of Enhanced or SuperCHIEF points are still needed to produce decent results. How many CHIEF points (original, Enhanced, or Super) are needed for a particular problem may become an issue. In practice, it is usually based on the frequency range, user's experience, and a trial–and–error procedure. To provide some feedback to the user, Wu [44,46,48] has used an estimated condition number in a least–squares solver to monitor the effectiveness of the CHIEF method (or its enhanced versions).

15.4.4 Weighted Residual CHIEF

Wu and Seybert [52] also proposed doing a weighted residual integration of the CHIEF equation over a small interior volume to partially alleviate the annoying nodal–surface difficulty. Like the Enhanced CHIEF or SuperCHIEF, the Weight Residual CHIEF only extends the application range of the original CHIEF to intermediate frequencies. At high frequencies, the nodal surfaces are clustered together, and any small interior volume will most likely be filled with nodal surfaces. The weak weighted residual integration of the CHIEF equation over a small volume will not be strong enough to provide necessary constraint effects to yield a unique solution at high frequencies.

15.5 Superimposing First and Second Integral Equation

15.5.1 Burton and Miller Method

The second category of methods for overcoming the irregular frequencies consist of taking a linear combination of the first boundary integral equation (15.3) and its differentiated form, i.e. the second boundary integral equation (15.4) to give

$$c(\boldsymbol{y})p(\boldsymbol{y}) + \int_\Gamma \frac{\partial G(\boldsymbol{x},\boldsymbol{y})}{\partial n(\boldsymbol{x})} p(\boldsymbol{x}) d\Gamma(\boldsymbol{x}) + \mathrm{i}\eta \int_\Gamma \frac{\partial^2 G(\boldsymbol{x},\boldsymbol{y})}{\partial n(\boldsymbol{x}) \partial n(\boldsymbol{y})} p(\boldsymbol{x}) d\Gamma(\boldsymbol{x}) =$$

$$= sk \left\{ \int_\Gamma G(\boldsymbol{x},\boldsymbol{y}) v_s(\boldsymbol{x}) d\Gamma(\boldsymbol{x}) - \mathrm{i}\eta \left[c(\boldsymbol{y}) v_s(\boldsymbol{x}) - \int_\Gamma \frac{\partial G(\boldsymbol{x},\boldsymbol{y})}{\partial n(\boldsymbol{y})} v_s(\boldsymbol{x}) d\Gamma(\boldsymbol{x}) \right] \right\} \tag{15.15}$$

The coupling parameter η should be a real number, i.e. $\eta \in \mathbb{R}$, but must not be purely imaginary. A similar approach was originally proposed by Panich [29] and by Brakhage and Werner [3] in the context of indirect boundary integral equations for the Helmholtz equation; see also Kussmaul [21]. The direct formulation (15.15) which we use in this paper is due to Burton and Miller [4].

In a paper by Meyer et al. [27], based on computational experiments, the authors recommended the choice of the coupling parameter as $\eta = 1/k$. It was shown later, using rigorous mathematical analysis, that this choice of the coupling parameter *almost minimises* the condition number of the operators on the left and the right hand sides of (15.15), when Γ is a sphere; cf. [1] and references therein.

The hypersingular operator is a pseudo–differential operator of order $+1$; meaning that it behaves essentially as a first order differentiation operator. Because of this, in Equation (15.15), the unknown function $p(\boldsymbol{x})$ requires higher continuity than that for Equation (15.3). More specifically, the hypersingular operator requires C^1 continuity of the function at collocation points. As discussed already [23,25], it is quite common and indeed advantageous to use discontinuous boundary elements. Alternatively, it is possible to use Galerkin discretization and continuous elements.

15.5.2 Modified Burton & Miller Methods for Continuous Elements

To bypass the strict C^1 continuity requirement on every collocation point, Ingber and Hickox [13], and later Marburg and Amini [24] proposed a modified implementation of the Burton and Miller method on continuous quadratic quadrilateral elements, in which the Burton and Miller linear combination is taken only at the center node of each element. The conventional surface integral equation, cf. Equation (15.3) is still applied to all the other nodes on the surface where there is only C^0 continuity. The same concept could also be applied to 7–node triangular elements with a center node. It is uncertain if such a "reduced" implementation will be as theoretically robust as the original Burton and Miller method or not. Nevertheless, all test cases so far have shown very promising results, see also References [10, 12] which show similar schemes but not the same as the one which will be tested here.

A similar "reduced" implementation has also been applied to even more popular continuous elements, such as the 8–node continuous quadrilateral element (Serendipity element) and the 6–node continuous quadratic triangular element [14], in which the Burton and Miller linear combination is taken at the center of each element where no physical node exists. In other words, the Burton and Miller linear combination is used to produce constraint equations in addition to the conventional surface Helmholtz integral equation at physical nodes. The resulting matrix is then solved by a least–square solver. This approach is expected to produce very similar results as its 9–node counterpart.

To take the "reduced" implementation even further, Wu and Jia [49] later proposed a so–called "Hyper–CHIEF method," in which users are allowed to select only a number of elements to do the Burton and Miller linear combination at their centers. This is actually a Burton and Miller method implemented in a CHIEF way. Compared to the original CHIEF or its variations, this Hyper–CHIEF method is completely free of the annoying nodal–surface difficulty. Compared to the 9–node or 8–node modified Burton and Miller methods mentioned above, this method is more efficient at low frequencies. However, rank deficiency is always an uncertainty. At high frequencies, users will end up selecting most, if not all, of the surface elements to do the Burton and Miller linear combination in order to ensure reliability.

15.6 Other Methods

There are many other attempts in the literature in eliminating the irregular frequencies. An important category is what is generally referred to as modified Green's function methods. These were first investigated by Ursell [41] and Jones [15], where they addressed the problem by modifying the choice of the free–space Green's function G. The choice of the function G is somewhat arbitrary, as there are an infinite number of functions satisfying

$$\Delta G(\boldsymbol{x},\boldsymbol{y}) + k^2 G(\boldsymbol{x},\boldsymbol{y}) \quad = \quad -\,\delta(\boldsymbol{x},\boldsymbol{y}) \ . \tag{15.16}$$

Ursell [41] suggested the use of a modified Green's function G_1, with

$$G_1 \;=\; G \,+\, W \ , \tag{15.17}$$

where G is the standard fundamental solution (15.5) and $W(\boldsymbol{x},\boldsymbol{y})$ is an infinite series of spherical wave functions. Ursell was then able to show that using this modified

Green's function the boundary integral equation (15.3) would not suffer from singularities at any frequencies.

In order to reduce the burden of computation of G_1 and make this result of some practical use, Jones [15] suggested replacing W in G_1 by W_M, where W_M is simply the partial sum of the first M terms in W. Jones proved that, for any given Γ and a real value K, it is possible to choose $M = M(\Gamma, K)$ such that there are no irregular frequencies in the wavenumber range of $0 < kR < KR$. As mentioned earlier, the number of irregular frequencies below a given value of K grows as $\mathcal{O}(K^3)$. Unfortunately, it can be shown that even for modest values of K, we still require a large number of terms in W_M, specially for boundaries which have a large aspect ratio. These modified Green's function methods are computationally expensive for moderate to high frequency radiation and scattering problems. This is the main reason why these elegant methods have not gained favor with practitioners.

Finally, the interesting paper by Cremers et al. [7] proposes a multi–domain boundary–element technique utilizing infinite boundary elements. The basic intention of the authors was to investigate a substructure technique with infinite subdomains separated by infinite boundary elements. The method proposed suffers from inefficient integration techniques for the infinite elements but there is a positive side–effect that when using infinite boundary elements no effect from irregular frequencies is observed. It is likely that future developments may overcome the integration difficulties and render the scheme efficient.

15.7 Test Cases

In this section, we report on the performance of the two most popular categories of methods, namely two variants of CHIEF, the original by Schenck [33] and two variations of Rosen's method [31], i.e. the conjugated and the unconjugated, and two variants of the Burton and Miller method, namely the original [4] and the "reduced" variant [13,24]. For the cat's eye, we present results for different boundary elements.

15.7.1 Pulsating Sphere

In a number of papers on boundary element methods for acoustic radiation the pulsating sphere is used as an example. Most likely, one of the main reasons is that a simple analytic solution is available for this example. Some authors have remarked that very coarse discretization is sufficient to achieve accurate numerical solutions in the low frequency range, i.e. $kR < 10$, where R is the radius of sphere; see for example [5, 8, 30, 39, 50]. Discretization of this problem, having a constant solution on the boundary, should give the exact solution, even with one constant element, provided the boundary surface and boundary integrals are calculated exactly. Consequently, we should expect an excellent solution by using constant approximation of the sound pressure and the particle velocity. This, however presupposes high degree of geometry approximation, such as a nine–noded (quadratic) quadrilateral elements, and the accurate computation of the matrix elements.

The pulsating sphere has also been used as an example to show the effect of irregular frequencies. Because of the simple nature of the problem, many of the higher modes are not excited. Often, the only irregular frequencies observed in this case correspond to zeros of $j_0(kR)$, that is $kR = n\pi$, with $n = 1, 2, 3, \ldots$. However, we know that the set of irregular frequencies for a sphere is $\mathcal{I}_\Gamma = \{k | j_n(kR) = 0,\ n = 0, 1, 2, ...\}$, where the density of irregular frequencies increases with increasing wavenumber k. Therefore, the pulsating sphere is perhaps a good example for an initial testing of a new code but should not be used to check convergence rates or stability with respect to irregular frequencies.

This is confirmed by our first example. It considers the pulsating sphere. The entire sphere is discretized into 24 elements, i.e. three per octant. These quadrilateral elements approximate the geometry by using quadratic polynomials. We use the data $R = 1m$, $\rho = 1.3kg/m^3$, and $c = 340m/s$, for sphere radius, fluid density, and speed of sound, respectively. We assume unit particle velocity ($1m/s$) independent of the location on the surface and frequency. We consider the frequency range up to 2000 Hz. This is equivalent to the relative wavenumber $kR \approx 11.75\pi$. Since the integrand is a highly oscillatory function for high frequencies, we need a high order of integration. Here, all integrals were evaluated using Gauss–Legendre quadrature rule with 30 integration points per direction, i.e. 900 points per element. The analytic expression for the sound pressure magnitude at the surface can be written as

$$\bar{p}(R) = \rho\, c\, v_s \,\frac{k\, R}{\sqrt{1 + k^2 R^2}} \tag{15.18}$$

where v_s stands for the uniform particle velocity at the surface of the sphere.

In Figure 15.1, the numerical solutions are compared for different boundary element methods. The surface value of the sound pressure (left subfigures) is an average over all the nodes. However, as expected, we observed very little difference in the nodal values. The relative error (right subfigures) of the solution is based on this average. The solution in the top subfigure is calculated using the Kirchhoff–Helmholtz integral equation (15.3). The graphs in the middle subfigure correspond to the CHIEF solution [33] where only one CHIEF point was used. The lowest subfigures present the results for the uniquely solvable method of Burton and Miller [4,27,30].

In the upper subfigures of Figure 15.1 we can clearly identify irregular frequencies since poles are observed for these wavenumbers. The addition of only one CHIEF point in the centre of the sphere is sufficient to get rid of all spurious modes. The results show also a small decrease in the error as the wavenumber is increased. When looking at the solution by the Burton and Miller formulation we note an error of one order of magnitude higher than that for the CHIEF solution. As mentioned before, for the pulsating sphere, there is no interpolation error, even with our piecewise constant approximation. Therefore, any error in the solution (except near the irregular frequencies for the CHIEF and ordinary BEM) is due to numerical integration and surface approximation. We note that the Burton and Miller solution is continuous over the frequency range, showing its validity for all wavenumbers.

As mentioned before, to investigate the convergence of a code and effectiveness of the underlying method for overcoming the irregular frequencies, it will be better

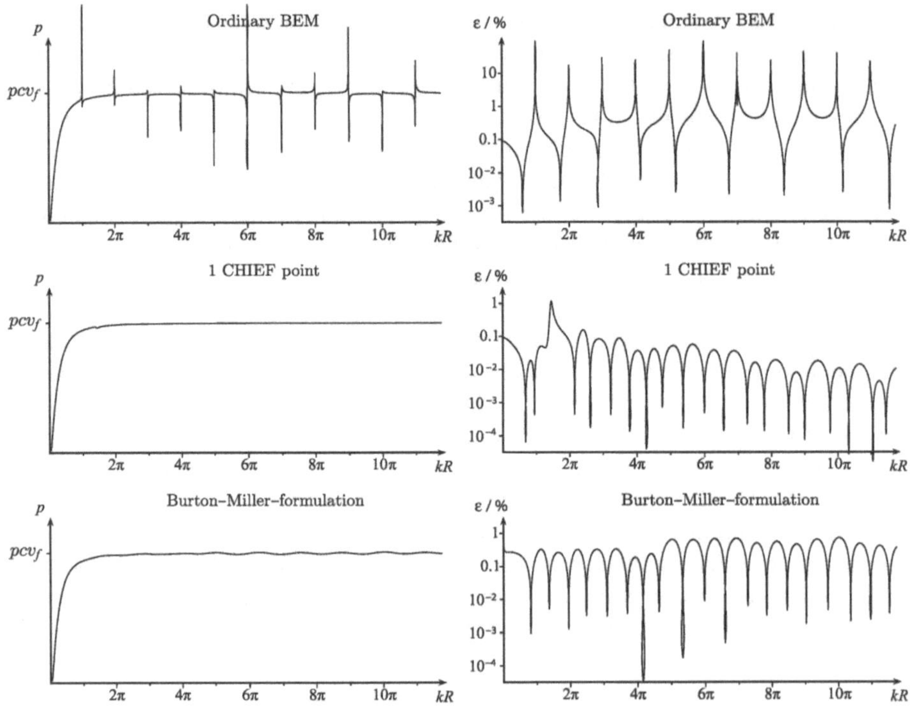

Fig. 15.1 Pulsating sphere: 24 elements P_0, surface solution (left) and its relative error (right), comparison of solutions by ordinary BEM, by CHIEF and by method of Burton and Miller (highest frequency equivalent to $kR \approx 37$).

to use a spherical scatterer than a pulsating sphere. It has the same benefit that an analytic solution is available, but because the solution is now oscillatory, one can test the effect of changing the order of interpolation too. Furthermore, we are likely to observe many more irregular frequencies for this example as more modes are excited. Examples of rigid scattering from spheres were given in [10, 17, 40].

15.7.2 Cat's Eye Radiator

After having seen that the sphere might not be a representative example, the idea of investigation of a cat's eye structure is essentially based on two considerations:

- As will be shown later, this radiator allows construction of a smooth solution that will make it easy to identify solution failures caused by the ill–conditioning of the integral operator, i.e. the irregular frequencies.
- The cat's eye structure is a more complicated shaped than a sphere. Hence, we expect more irregular frequencies.

The cat's eye has been analyzed in a number of papers in recent years, cf. [22, 26, 34, 35]. As shown in Figure 15.2, it is a sphere with the positive octant cut out.

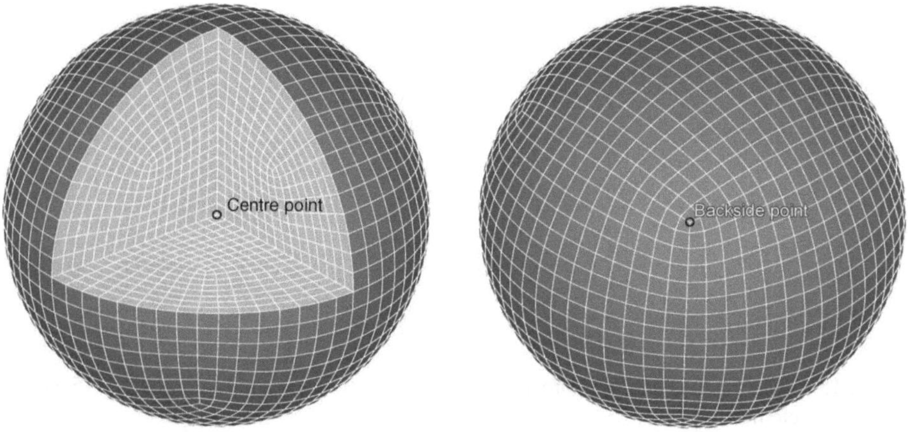

Fig. 15.2 Cat's eye: boundary element model (1920 elements), identification of observation points.

Herein, we investigate the radiation problem with a vibrating surface where it coincides with the spherical one. The plain surfaces of the missing octant remain calm. In Figure 15.2, the vibrating and the non–vibrating domains are distinguished by their brightness. The idea is that the sound pressure at points in the centre of the backside should behave asymptotically (as frequency increases) like the one at the surface of the sphere. This implies that the noise transfer function, i.e. sound pressure divided by surface velocity in terms of frequency, returns a very smooth function similar to the solution of the radiating sphere where it is easy to identify frequencies where the solution fails.

Again, we assume material data of air as density $\rho = 1.3kg/m^3$ and speed of sound $c = 340m/s$. The spherical radius is taken to be $R = 1.0m$ and the particle velocity for the vibrating surface as $v = 1m/s$. We use different types of boundary elements. They consist of discontinuous constant elements (P_0), discontinuous linear ones with collocation points at zeros of Legendre polynomials (P_{1L}), continuous linear P_{1c}, and continuous quadratic elements P_{2c}. These elements have been discussed in the paper [25]. For the P_{1L} and P_{2c} interpolations we use 1920 elements and for P_0 and P_{1c} we use 7680 elements. For degree of freedom for continuous and discontinuous elements is 7682 and 7680 respectively. We consider a frequency range of up to 1700 Hz which corresponds to a normalized wavenumber of $kR = 10\pi$. A frequency step size of 0.5 Hz is applied.

In case of CHIEF and Rosen et al. we used 240 CHIEF points. Considering that we use 7680/7682 collocation points on the surface nodes, the number of CHIEF points used are relatively small.

The arising linear system of equations is solved by using the Generalized Minimum Residual algorithm, cf. Saad [32]. Iterative techniques have become a reliable tool for the efficient solution of large scale acoustic problems as shown in a couple of papers, cf. [2, 22, 26, 35]. For the cat's eye investigation, we use an unpreconditioned

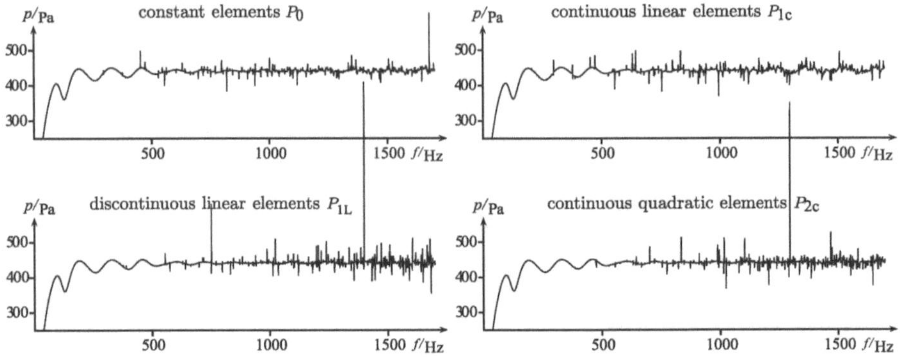

Fig. 15.3 Cat's eye: sound pressure at backside point evaluated solving Equation (15.3) using different element types.

Table 15.1 Cat's eye: coordinates of the four points under consideration.

name	x	y	z
centre point	0	0	0
backside point	$-1/\sqrt{3}$	$-1/\sqrt{3}$	$-1/\sqrt{3}$
point at $2R$	$-2/\sqrt{3}$	$-2/\sqrt{3}$	$-2/\sqrt{3}$
point at $10R$	$-10/\sqrt{3}$	$-10/\sqrt{3}$	$-10/\sqrt{3}$

tioned GMRes. In case of the least squares solution of the over–determined system arising in conventional CHIEF, the GMRes is applied to the normal equations. We set a stopping criterion on the residual of 10^{-8}. This is, at least, two to three orders of magnitude smaller than the level of discretization error. Thus, the additional error introduced by the iterative solver is negligible.

Although the solution is available at all surface nodes, we limit our interest to the sound pressure at four discrete points. These points are identified in Table 15.1 and in Figure 15.2. The sound pressure in terms of frequency will be called a noise transfer function.

We start our investigation by testing the performance of Equation (15.3). Figure 15.3 contains noise transfer functions for the backside point and the four different element types. Basically, we expect a smooth solution. This smooth solution can be guessed. However, it is often interrupted. The solution fails at irregular frequencies. This phenomenon starts at about 200 Hz. Hardly any smooth regions are identified above 600 Hz. Consequently, it appears impossible to actually distinguish between regular and irregular frequencies. We note that even at most irregularities, the solution seems to remain in the vicinity of the correct solution. This is because our linear systems often do not become singular at the frequencies considered, but simply ill–conditioned, where errors are merely magnified. Nevertheless, the performance of the method is clearly not acceptable.

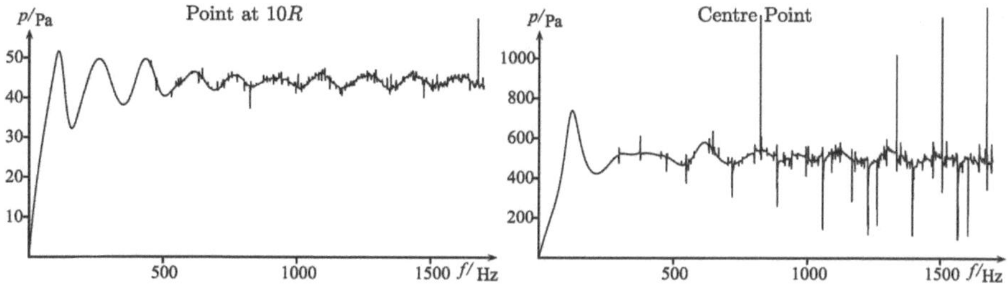

Fig. 15.4 Cat's eye: sound pressure at different points solving Equation (15.3) using elements P_0.

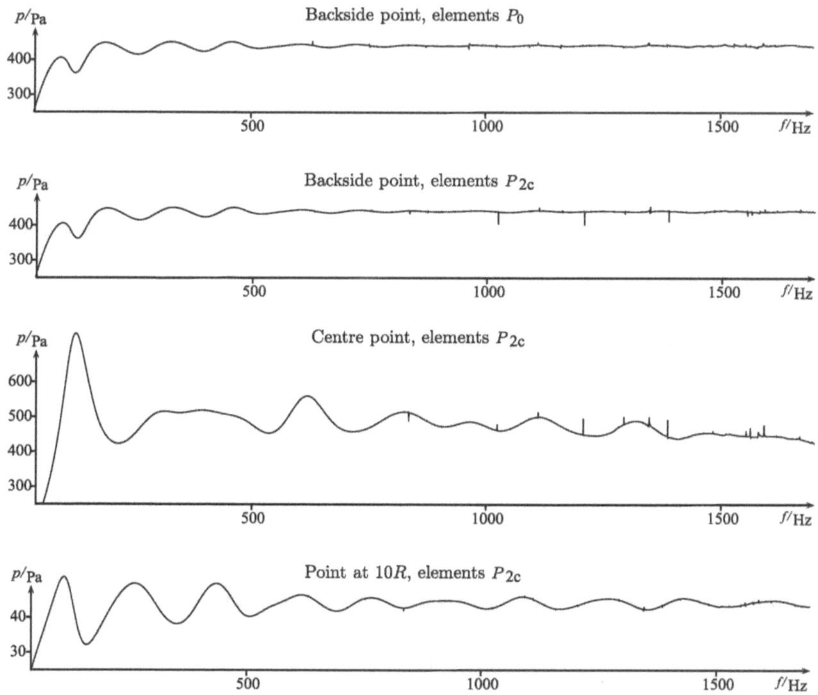

Fig. 15.5 Cat's eye: sound pressure at three points evaluated by CHIEF using different element types.

A comparison between solutions for different element types indicates more moderate failures in case of constant and continuous linear elements, especially in the higher frequency range. Discontinuous linear and continuous quadratic elements seem to produce smoother solution in the lower frequency range.

Figure 15.4 contains the noise transfer functions for constant elements at two other points, namely, Centre and at $10R$. Clearly, the solution at field points above the backside point (left subfigure) fails analogously to the backside point, cf. Figure 15.3. It was shown in [24] that the point at $2R$ behaves in the same way. How-

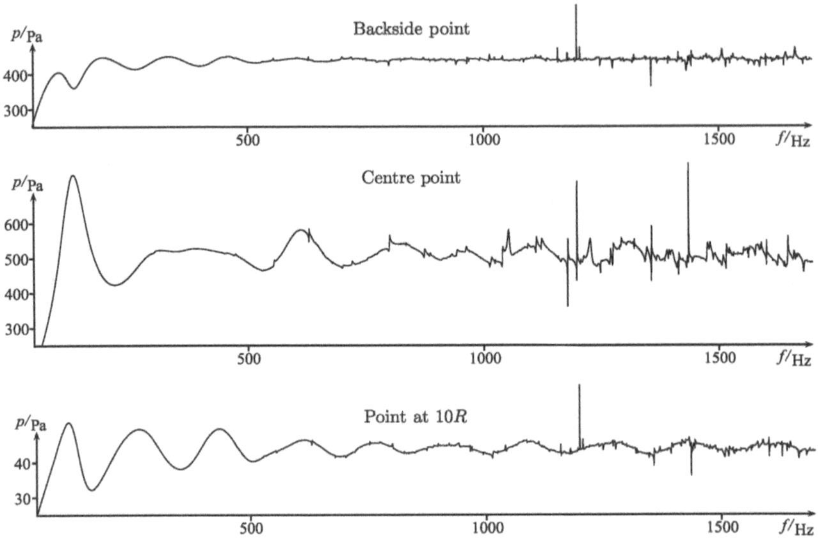

Fig. 15.6 Cat's eye: sound pressure at different locations evaluated by Rosen's method (conjugated) using elements P_0.

ever, the centre point solution appears much worse. This is likely to be caused by the geometric singularity and certainly less acceptable than the moderately failing solutions at the other points.

Noise transfer functions for the conventional CHIEF [33], i.e. solution in a least squares sense, are plotted in Figure 15.5. Solution is chosen for three points again, namely the backside point, the point at $10R$ and for the centre point. For the backside point, solutions of elements P_0 and P_{2c} are compared. As expected, the solution becomes much smoother when applying the CHIEF method. However, it still contains numerous discontinuities. We recall that we are using 240 CHIEF points here. Theoretically, only one CHIEF point (at the correct position) is sufficient. In practice many more are required to obtain a reliable solution.

We can state that the solution is smooth up to a frequency of about 600 Hz. Furthermore, as one might expect the solution at the field point ($10R$) is somewhat smoother than the surface solution. Use of quadratic element seems to magnify the effect of failures at certain frequencies. This is to be expected. The reason is that the more accurate our approximation is the more closely it behaves like the continuous boundary integral operator. That is, the severity of failure at the irregular frequencies increases but the width around these frequencies, where the failure occurs, narrows. The effect of irregular frequencies is magnified for the geometric singularity in the centre point.

Two variants of the Rosen et al. [31] modification of CHIEF are tested here; namely, the conjugated and the unconjugated versions. The resulting noise transfer functions for the original conjugated version are shown in Figure 15.6. Although smooth up to about 500 Hz, the solution in the upper frequency range appears hardly

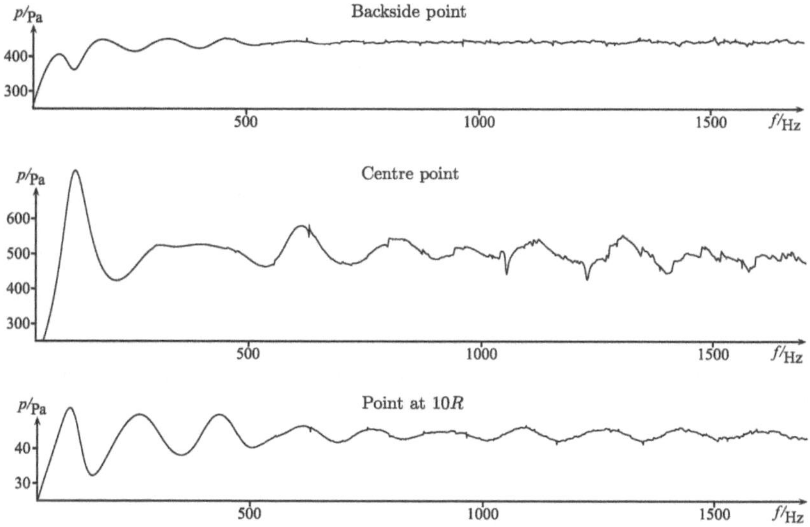

Fig. 15.7 Cat's eye: sound pressure at different locations evaluated by Rosen's method (unconjugated) using elements P_0.

better than those without any CHIEF points, cf. Figure 15.5. In contrast to the conjugated version, the unconjugated version returns much smoother curves, cf. Figure 15.7. Nevertheless, their smoothness and reliability is clearly worse than that for the conventional CHIEF, in particular at the geometric singularity at the centre point. The potential gain in efficiency obtained by solving a square system of equations instead of a least square problem should not be at the expense of reliability and accuracy.

Situation is completely different when using the alternative approach originally proposed by Burton and Miller [4]. Although Equation (15.15) is computationally more challenging it yields a smooth solution, as can be seen in Figure 15.8. Noise transfer functions are plotted for both types of discontinuous elements. Both solutions coincide well except for the centre point. Deviations are likely to have been caused by different quality of approximation close to the geometric singularity which is beyond the scope of this paper.

It was pointed out earlier that by setting the coupling parameter to zero at all but the centre nodes where the C^1 continuity is assured we are able to apply the Burton and Miller method to continuous quadratic elements [12, 13, 24]. Consequently, the hypersingular formulation appears for approximately one quarter of all collocation points. This simple and efficient modification of the original method produces a surprisingly smooth solution for the backside point and for the field points, cf. Figure 15.9. Some traces of discontinuities appear for the centre point. Compared to the variants of CHIEF that have been discussed above, they are much less pronounced even at this geometric singularity.

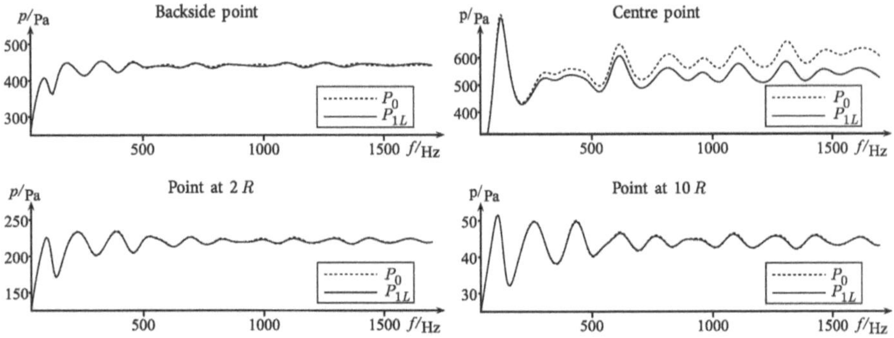

Fig. 15.8 Cat's eye: sound pressure at different locations evaluated by Burton and Miller method using different discontinuous elements.

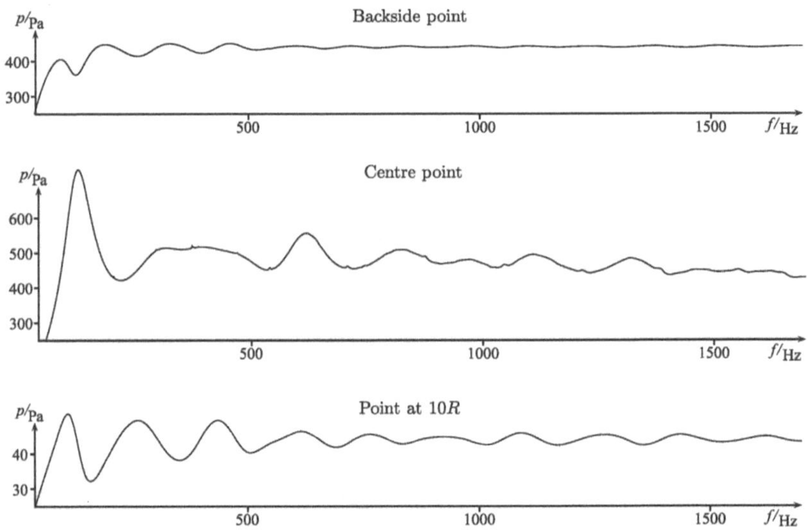

Fig. 15.9 Cat's eye: sound pressure at different locations evaluated by "reduced" Burton and Miller method using continuous elements P_{2c}.

In the paper [24], additional studies were presented. There, it is shown how the accuracy of integration in the far field may affect the pronunciation of spurious resonances. Furthermore, the number of iterations for solution of the linear system of equations is presented. It clearly shows that the number increases significantly with frequency.

15.7.3 Diesel Engine

The third example presents the sound radiation of a diesel engine. The fluid surface model contains 20172 nodes and 21497 elements P_0. The number of elements determines the number of collocation points and, thus, the degree of freedom. The

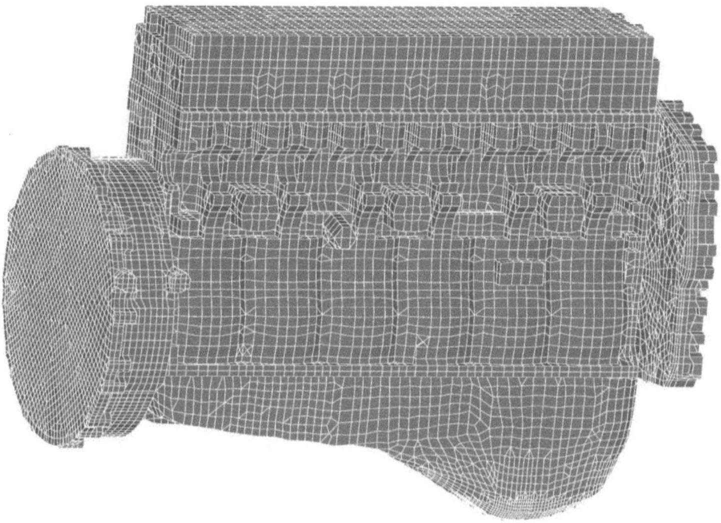

Fig. 15.10 Diesel engine: boundary element model.

problem is solved for the frequency range up to 3000 Hz. The size of the elements does not require to use higher order elements. The boundary element model is shown in Figure 15.10. Again we use the material data of air. The excitation of the acoustic field is applied by defining the particle velocity over the surface at each investigated frequency.

The particle velocity distribution over the engine's surface for a certain operations condition was computed and provided by the AVL / ACC Graz (Austria). In Figure 15.10, the geometry and the mesh topology of the surface are shown. Originally, the particle velocity was given on the mesh of linear continuous elements. The piecewise constant particle velocity data which is used for our simulations can be understood as an average of the normal velocity on each element.

To provide the reader with a vivid impression and a comparison of the level distributions of the particle velocity and the sound pressure, these data are visualized for two specified frequencies, i.e. approximately 500 Hz and approximately 2200 Hz, cf. Figure 15.11. It can be realized at the lower frequency of 500 Hz that the sound pressure level contribution does not match with the velocity contribution. Actually, the sound pressure level distribution is much coarser than the particle velocity level distribution which appear more detailed. At 2200 Hz, the surface plots of particle velocity and sound pressure look more similar in such a way that the sound pressure level distribution appears almost as detailed as the distribution of the particle velocity.

Clearly, the boundary element mesh is even too fine for the highest frequency of 3000 Hz. However, problems introduced by the coupling between the structure and the fluid mesh are avoided, since the outer surface of the structure is directly used as the fluid BE mesh.

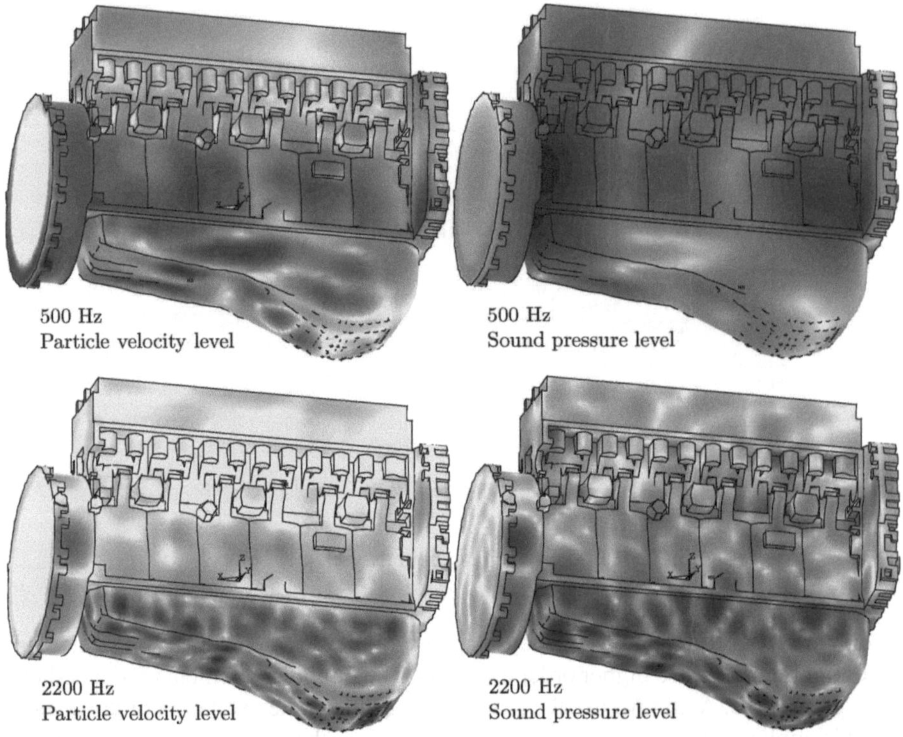

Fig. 15.11 Diesel engine: surface distribution of particle velocity level and sound pressure level at approximately 500 Hz and 2200 Hz, range between dark and light regions approximately 35 dB.

Based on the data of sound pressure and particle velocity, it is easy to evaluate the sound intensities. The integral of the sound intensity over the entire surface provides us with the radiated sound power

$$P = \frac{1}{2} \Re \left\{ \int p(x)\, v_s^*(x)\, d\Gamma(x) \right\} \tag{15.19}$$

where * denotes conjugate complex.

Alternatively, the radiated sound power can be estimated by just setting the radiation efficiency equal 1 which is an asymptotic high frequency solution, cf. Koopmann and Fahnline [19]. This solution is also referred to as the Equivalent Radiated Power (ERP). For this, the sound power is very easily estimated by

$$P_{ERP} = \frac{1}{2} \varrho c \int v_s^*(x)\, v_s(x)\, d\Gamma(x) = \frac{1}{2} \varrho c \int |v_s(x)|^2\, d\Gamma(x)\,. \tag{15.20}$$

The radiated sound power evaluated based on the Burton and Miller solution and Equation (15.19) is compared with the ERP solution (15.20) for the engine in Figure 15.12.

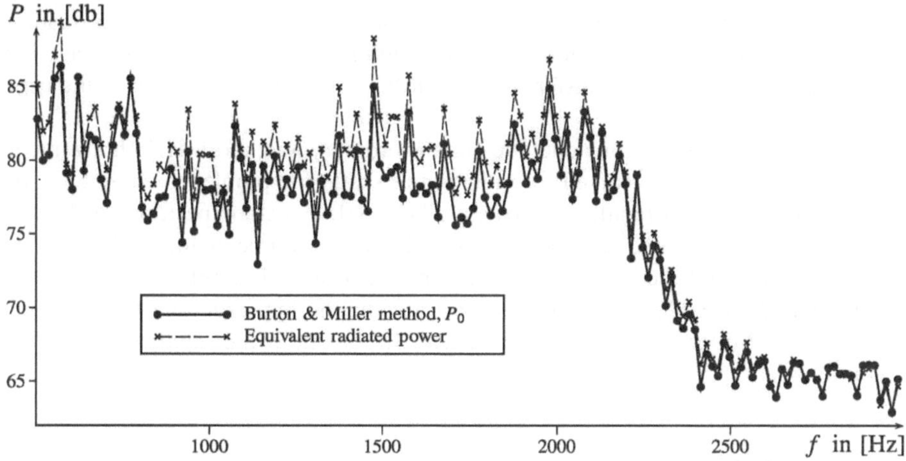

Fig. 15.12 Diesel engine: sound power in terms of frequency for realistic loadcase, solution by Burton and Miller method, elements P_0 vs. ERP solution, Equation (15.20).

There, the ERP values agree with the BEM solution better and better the higher the frequencies are. Actually, it is quite surprising that the simple approximation of ERP catches the behaviour of the Burton and Miller solution with this accuracy. In the lower frequency range, the difference of the approximate solution and the BEM reference are a couple of decibel. However, peaks and valleys in the curves are found at the same frequencies. Even this result can be taken as an indicator that the Burton and Miller technique provides reliable solutions even for complex industrial applications.

15.8 Conclusion

This comparative study has been motivated by the observation that radiation from a pulsating sphere can be accurately approximated by very few constant elements. The effect of irregular frequencies in this case can be removed by using either one CHIEF point or the method of Burton and Miller. Irregular frequencies which are excited are equidistantly distributed at normalized wave numbers $kR = n\pi$.

A more complicated shape such as the cat's eye shows many more irregular frequencies. In general a reasonable solution can not be obtained without careful treatment of that phenomenon. The most commonly used method for removing the effects of irregular frequencies is the CHIEF method where additional internal collocation points are used to create an over–determined system of equations. In our example, with a smooth noise transfer function, even with 240 CHIEF points, still numerous discontinuities remain. The method of Rosen et al. gives rise to a square system matrix. However, the solution is even less reliable than the conventional CHIEF. Smooth noise transfer functions are obtained when using the method of Burton and Miller.

A modified Burton and Miller formulation for application with continuous quadratic elements also provides us with a smooth solution. Some discontinuities have been discovered for the noise transfer function at the geometric singularity of the centre point.

More generally, we can conclude that the ordinary Kirchhoff–Helmholtz integral equation should not be solved for radiation or scattering problems without any particular treatment of irregular frequencies. The CHIEF method performs well at low frequencies, but in general an accurate solution can never be guaranteed. Method of Rosen et al. is not recommended since it performs even less reliably than the conventional CHIEF. Hence, the authors recommend the hypersingular formulation of Burton and Miller. Results of the modified Burton and Miller method are very encouraging.

Acknowledgement

The authors wish to acknowledge that the computation was run on the SGI Origin 3800 at the Zentrum für Hochleistungsrechnen of the Technische Universität Dresden. We further thank the AVL / ACC in Graz (Austria) for the data of the engine's model.

References

1. Amini S (1990) On the choice of the coupling parameter in boundary integral formulations of the exterior acoustics problem. Applicable Analysis 35:75–92
2. Amini S, Maines ND (1998) Preconditioned Krylov subspace methods for boundary element solution of the Helmholtz equation. International Journal for Numerical Methods in Engineering 41:875–898
3. Brakhage H, Werner P (1965) Über das Dirichlet'sche Außenraumproblem für die Helmholtz'sche Schwingungsgleichung. Archiv der Mathematik 16:325–329
4. Burton AJ, Miller GF (1971) The application of integral equation methods to the numerical solution of some exterior boundary–value problems. Proceedings of the Royal Society of London 323:201–220
5. Chien CC, Rajiyah H, Atluri SN (1990) An effective method for solving the hypersingular integral equations in 3–d acoustics. Journal of the Acoustical Society of America 88:918–937
6. Ciskowski RD, Brebbia CA (eds) (1991) Boundary elements in acoustics. Computational Mechanics Publications and Elsevier Applied Science, Southampton–Boston
7. Cremers L, Fyfe KR, Sas P (2000) A variable order infinite element for multi–domain boundary element modelling of acoustic radiation and scattering. Applied Acoustics 59:185–220
8. Cunefare KA, Koopmann GH, Brod K (1989) A boundary element method for acoustic radiation valid for all wavenumbers. Journal of the Acoustical Society of America 85:39–48
9. Estorff O von (ed) (2000) Boundary elements in acoustics: Advances and applications. WIT Press, Southampton

10. Francis DTI (1993) A gradient formulation of the Helmholtz integral equation for acoustic radiation and scattering. Journal of the Acoustical Society of America 93:1700–1709
11. Hahn SR, Ferri AA, Ginsberg JH (1991) A review and evaluation of methods to alleviate non-uniqueness in the surface Helmholtz integral equation. In: Keltie RF, Seybert AF, Kang DS, Olson L, Pinsky PM (eds) Structural Acoustics. Winter Annual Meeting Symposium Proceedings of ASME. NCA–Vol 12/AMD–Vol 128:223–234
12. Harris PJ, Amini S (1992) On the Burton and Miller boundary integral formulation of the exterior acoustic problem. ASME Journal of Vibration and Acoustics 114:540–546
13. Ingber MS, Hickox CE (1992) A modified Burton–Miller algorithm for treating the uniqueness of representation problem for exterior acoustic radiation and scattering problems. Engineering Analysis with Boundary Elements 9:323–329
14. Jia ZH, Wu TW, Seybert AF (1992) Simplified hypersingular boundary integral equations for acoustic scattering and radiation. In: Lee D (ed) Proceedings of the 3rd IMACS Symposium on Computational Acoustics. North–Holland
15. Jones DS (1974) Integral equations for the exterior acoustic problem. Quarterly Journal of Mechanics and Applied Mathematics 27:129–142
16. Juhl P (1994) A numerical study of the coefficient matrix of the boundary element method near characteristic frequencies. Journal of Sound and Vibration 175:39–50
17. Juhl P (1998) A note on the convergence of the direct collocation boundary element method. Journal of Sound and Vibration 212:703–719
18. Kirkup SM (1998) The boundary element method in acoustics. Integrated Sound Software, Heptonstall
19. Koopmann GH, Fahnline JB (1997) Designing quiet structures: A sound power minimization approach. Academic Press, San Diego–London
20. Kupradze VD (1956) Boundary value problems in vibrational theory and integral equations. Deutscher Verlag der Wissenschaften, Berlin (1st Russian edition 1950)
21. Kussmaul R (1969) Ein numerisches Verfahren zur Lösung des Neumannschen Außenraumproblems für die Helmholtzsche Schwingungsgleichung. Computing 4:246–273
22. Makarov SN, Ochmann M (1998) An iterative solver for the Helmholtz integral equation for high frequency scattering. Journal of the Acoustical Society of America 103:742–750
23. Marburg S (2002) Six elements per wavelength. Is that enough? Journal of Computational Acoustics 10:25–51
24. Marburg S, Amini S (2005) Cat's eye radiation with boundary elements: Comparative study on treatment of irregular frequencies. Journal of Computational Acoustics 13:21–45
25. Marburg S, Schneider S (2003) Influence of element types on numeric error for acoustic boundary elements. Journal of Computational Acoustics 11:363–386
26. Marburg S, Schneider S (2003) Performance of iterative solvers for acoustic problems. Part I: Solvers and effect of diagonal preconditioning. Engineering Analysis with Boundary Elements 27:727–750
27. Meyer WL, Bell WA, Zinn BT, Stallybrass MP (1978) Boundary integral solutions of three dimensional acoustic radiation problems. Journal of Sound and Vibration 59:245–262
28. Ochmann M, Mechel FP (2002) Analytical and numerical methods in acoustics. In: Mechel FP (ed) Formulas of Acoustics. Chapter O. Springer–Verlag, Berlin Heidelberg 930–1023
29. Panič OI (1965) K voprosu o razrešimosti vnešnich kraevich zadač dlja volnovogo uravnenija i dlja sistemi uravnenij MAXWELLa. Uspechi Mathematičkich Nauk 20:221–226
30. Rego Silva JJ (1993) Acoustic and elastic wave scattering using boundary elements. Computational Mechanics Publications, Southampton–Boston

31. Rosen EM, Canning FX, Couchman LS (1995) A sparse integral equation method for acoustic scattering. Journal of the Acoustical Society of America 98:599–610
32. Saad Y, Schultz MH (1986) GMRES: A generalized minimal residual algorithm for solving non–symmetric linear systems. SIAM Journal of Scientific and Statistical Computing 7:856–869
33. Schenck HA (1968) Improved integral formulation for acoustic radiation problems. Journal of the Acoustical Society of America 44:41–58
34. Schneider S (2003) Application of fast methods for acoustic scattering and radiation problems. Journal of Computational Acoustics 11:387–401
35. Schneider S, Marburg S (2003) Performance of iterative solvers for acoustic problems. Part II: Acceleration by ilu–type preconditioner. Engineering Analysis with Boundary Elements 27:751–757,
36. Segalman DJ, Lobitz DW (1990) SuperCHIEF: A modified CHIEF method. Sandia National Laboratories Report, SAND90-1266, Albuquerque
37. Segalman DJ, Lobitz DW (1992) A method to overcome computational difficulties in the exterior acoustic problem. Journal of the Acoustical Society of America 91:1855-1861
38. Seybert AF, Rengarian, TK (1987) The use of CHIEF to obtain unique solutions for acoustic radiation using boundary integral equation. Journal of the Acoustical Society of America 81:1299–1306
39. Seybert AF, Soenarko B, Rizzo FJ, Shippy FJ (1985) An advanced computational method for radiation and scattering of acoustic waves in three dimensions. Journal of the Acoustical Society of America 77:362–368
40. Tobocman W (1986) Extensions of the Helmholtz integral equation method to shorter wavelengths. Journal of the Acoustical Society of America 80:1828–1837
41. Ursell F (1973) On the exterior problems of acoustic. Proceedings of Cambridge Philosophers Society 74:117–125
42. Visser R (2004) A boundary element approach to acoustic radiation and source identification. PhD Thesis, University of Twente, Enschede
43. Weyl H (1952) Kapazität von Strahlungsfeldern. Mathematische Zeitschrift 55:187–198
44. Wu TW (1995) A direct boundary element method for acoustic radiation and scattering from regular and thin bodies. Journal of the Acoustical Society of America 97:84–91
45. Wu TW (ed) (2000) Boundary element acoustics: Fundamentals and computer codes. WIT Press, Southampton
46. Wu TW (1994) On computational aspects of the boundary element method for acoustic radiation and scattering in a perfect waveguide. Journal of the Acoustical Society of America 96:3733–3743
47. Wu TW (2000) The Helmholtz integral equation. Chapter 2 of [45] 9–28
48. Wu TW (2000) Two–dimensional Problems. Chapter 3 of [45] 29–49
49. Wu TW, Jia ZH (1992) A choice of practical approaches to overcome the nonuniqueness problem of the BEM in acoustic radiation and acattering. Boundary Element Technology VII, Ed CA Brebbia, MS Ingber, Computational Mechanics Publications, 501–510
50. Wu TW, Li WL, Seybert AF (1993) An efficient boundary element algorithm for multi–frequency acoustical analysis. Journal of the Acoustical Society of America 94:447–452
51. Wu TW, Seybert AF (1991) Acoustic radiation and scattering. Chapter 3 of [6] 61–76
52. Wu TW, Seybert AF (1991) A weighted residual formulation for the CHIEF method in acoustics. Journal of the Acoustical Society of America 90:1608–1614

16

A Galerkin–type BE–formulation for Acoustic Radiation and Scattering of Structures with Arbitrary Shape

Zhensheng Chen[1], Günter Hofstetter[2], and Herbert Mang[3]

[1] Acoustics Research Institute, Austrian Academy of Sciences, Wohllebengasse 12–14, A–1040 Vienna, Austria
zhensheng.chen@oeaw.ac.at

[2] Institute for Basic Sciences in Civil Engineering, University of Innsbruck, Technikerstraße 13, A–6020 Innsbruck, Austria
Guenter.Hofstetter@uibk.ac.at

[3] Institute for Mechanics of Materials and Structures, Vienna University of Technology, Karlsplatz 13, A–1040 Vienna, Austria
Herbert.Mang@tuwien.ac.at

Summary. This chapter deals with a Galerkin–type boundary element formulation for acoustic radiation and scattering of structures with arbitrary shape. The integral equations for radiation and scattering of sound are derived for three dimensional closed structures, thin–walled open structures and for combinations of both. For the numerical solution of these integral equations a Galerkin–type numerical solution scheme is described and programming aspects related to the evaluation of the hyper–singular integrals, determination of the number of unknowns at a particular nodal point and prevention of the singularity of the coefficient matrix at so–called irregular frequencies are addressed. Finally, the proposed method is applied to the numerical solution of some benchmark problems for acoustic radiation and scattering.

16.1 Introduction

The boundary element method (BEM) is very well suited for the numerical simulation of acoustic radiation of vibrating structures and scattering of sound by structures. Especially for exterior acoustic problems, the boundary element method offers the possibility to consider radiation and scattering of sound in an infinite fluid domain in a very efficient manner.

The standard discretization procedure for the integral equations, formulated for the boundary of the domain under consideration, is the collocation method. It is characterized by enforcing the residuals of the integral equations, which result from the approximation of the unknown functions on the boundary of the fluid domain under consideration, to be zero at the nodal points of the employed boundary element mesh. This discretization method results in a system of algebraic equations for the unknown nodal values of the sound field with a fully populated, unsymmetric coefficient matrix.

An alternative discretization procedure is the Galerkin method. It is characterized by enforcing the weighted residuals of the respective integral equations to be zero. To this end, the integral equations are multiplied by test functions and then are integrated over the boundary of the fluid domain under consideration. Thus, by contrast to the collocation method, the Galerkin method requires two integrations over the boundary and, consequently, the computational effort for the generation of the algebraic system of equations for determining the unknown nodal values of the sound field is higher than for the collocation method. However, if one starts from a self–adjoint system of integral equations, the Galerkin method results in a system of algebraic equations for the unknown nodal values of the sound field with a symmetric coefficient matrix. This feature is advantageous especially for boundary element meshes with a large number of degrees of freedom, since the solution of a symmetric algebraic system of equations requires considerably less time and storage capacity than the solution of an unsymmetric system. Boundary element formulations for acoustic problems, resulting in a symmetric system of algebraic equations, can be found, e.g., in [4, 5, 11].

Furthermore, the Galerkin formulation of the boundary element method offers advantages for coupled elasto–acoustic problems, in which the interaction between structural vibrations and the acoustic response have to be taken into account in the numerical model. In this context, coupling of the finite element discretization of an elastic structure with the boundary element discretization of a fluid domain will preserve the symmetry property of the finite element stiffness matrix [2, 3].

In this chapter a symmetric Galerkin formulation of the boundary element method for acoustic radiation and scattering of sound is presented, which offers the following features:

(i) The formulation allows consideration of three–dimensional (3D) closed structures, 3D thin–walled closed structures, 3D thin–walled open structures, and of structures, consisting of combinations of a 3D closed part with thin–walled open parts;
(ii) Different fluids may be present in the interior and exterior domain of a 3D closed structure with attached thin–walled open parts;
(iii)Different acoustic admittances can be prescribed on the two surfaces of thin–walled parts of a structure and a velocity boundary condition as well as an acoustic admittance boundary condition can be prescribed simultaneously;

In the subsequent section, the integral equations for radiation and scattering of sound are derived for 3D closed structures, thin–walled open structures and for combinations of both. Then, for these integral equations the Galerkin type numerical solution scheme is described and programming aspects related to the evaluation of the hyper–singular integrals, determination of the number of unknowns at a particular nodal point and prevention of the singularity of the coefficient matrix at so–called irregular frequencies are addressed. Finally, the proposed method is applied to the

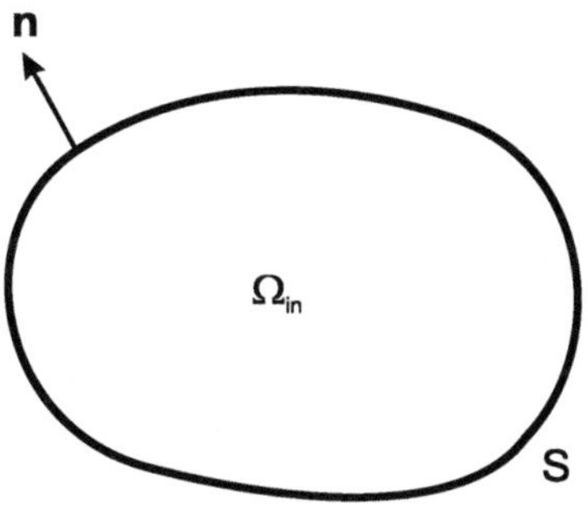

Fig. 16.1 Section of a 3D closed domain Ω_{in} with volume V, surface S and normal vector $\mathbf{n}$.

numerical solution of some benchmark problems for acoustic radiation and scattering.

16.2 Derivation of the Integral Equation

Consider a 3D closed domain Ω_{in} with the volume V bounded by the surface S, a section of which is shown in Figure 16.1. It is occupied by a fluid, characterized by the density ρ and the speed of sound c.

The basic differential equation for harmonic vibrations at a particular angular frequency ω is the Helmholtz differential equation [10]

$$\Delta\tilde{\Phi}(\mathbf{y}) + k^2\,\tilde{\Phi}(\mathbf{y}) = \tilde{\Psi}(\mathbf{y})\,. \tag{16.1}$$

In (16.1), Δ is the Laplace operator and $\tilde{\Phi}(\mathbf{y})$ denotes the amplitude of the time–harmonic velocity potential of the sound field

$$\Phi(\mathbf{y},t) = \tilde{\Phi}(\mathbf{y})\,\mathrm{e}^{-\mathrm{i}\omega t} \tag{16.2}$$

at a particular point $\mathbf{y} \in V$; $\mathrm{i} = \sqrt{-1}$ is the imaginary unit and t represents the time; $k = \omega/c$ is the wave number and $\tilde{\Psi}(\mathbf{y})$ is the amplitude of the time–harmonic intensity $\Psi(\mathbf{y},t) = \tilde{\Psi}(\mathbf{y})\,\mathrm{e}^{-i\omega t}$ of an acoustic source.

The relationship between the time–harmonic velocity potential $\Phi(\mathbf{y},t)$ and the time–harmonic sound pressure $p(\mathbf{y},t) = \tilde{p}(\mathbf{y})\,\mathrm{e}^{-\mathrm{i}\omega t}$ is given as [10]

$$p(\mathbf{y},t) = -\rho\frac{\partial\Phi(\mathbf{y},t)}{\partial t} - \rho\,\Psi(\mathbf{y},t)\,. \tag{16.3}$$

All subsequent equations will be formulated in terms of the amplitudes of the pressure and the velocity potential of the sound field at a particular angular frequency ω. Hence, in the following, omitting the tilde, the amplitudes of the time–harmonic pressure and the velocity potential will simply be referred to as sound pressure $p(\mathbf{y})$ and sound potential $\Phi(\mathbf{y})$, respectively.

From the differential equation (16.1) an integral equation can be derived by weighting this equation by a function $\bar{\Phi}(\mathbf{y})$, which is a known solution of (16.1), and integrating over the volume V, yielding

$$\int_V \bar{\Phi}(\mathbf{y})\, \Delta\Phi(\mathbf{y})\, dV(\mathbf{y}) + k^2 \int_V \bar{\Phi}(\mathbf{y})\, \Phi(\mathbf{y})\, dV(\mathbf{y}) = \int_V \bar{\Phi}(\mathbf{y})\, \Psi(\mathbf{y})\, dV(\mathbf{y})\,. \tag{16.4}$$

Exchanging in (16.4) Φ and $\bar{\Phi}$ and Ψ and $\bar{\Psi}$ (the latter represents the intensity of an acoustic source which gives $\bar{\Phi}$) and subtracting the so obtained equation from (16.4) yields

$$\int_V \left[\bar{\Phi}(\mathbf{y})\, \Delta\Phi(\mathbf{y}) - \Phi(\mathbf{y})\, \Delta\bar{\Phi}(\mathbf{y})\right] dV(\mathbf{y}) = \int_V \left[\bar{\Phi}(\mathbf{y})\, \Psi(\mathbf{y}) - \bar{\Psi}(\mathbf{y})\, \Phi(\mathbf{y})\right] dV(\mathbf{y})\,. \tag{16.5}$$

Application of Green's integral theorem to the left hand side of (16.5) results in

$$\int_S \left[\bar{\Phi}(\mathbf{y})\, \frac{\partial\Phi(\mathbf{y})}{\partial n_y} - \Phi(\mathbf{y})\, \frac{\partial\bar{\Phi}(\mathbf{y})}{\partial n_y}\right] dS(\mathbf{y}) = \int_V \left[\bar{\Phi}(\mathbf{y})\, \Psi(\mathbf{y}) - \bar{\Psi}(\mathbf{y})\, \Phi(\mathbf{y})\right] dV(\mathbf{y})\,, \tag{16.6}$$

where

$$\frac{\partial\Phi(\mathbf{y})}{\partial n_y} = \frac{\partial\Phi(\mathbf{y})}{\partial\mathbf{y}} \cdot \mathbf{n}(\mathbf{y}) = \mathbf{v}(\mathbf{y}) \cdot \mathbf{n}(\mathbf{y}) = v_n(\mathbf{y}) \tag{16.7}$$

denotes the derivative of the sound potential Φ at $\mathbf{y} \in S$ in the direction of the vector $\mathbf{n}$ normal to S. $v_n(\mathbf{y})$ is the amplitude of the time–harmonic particle velocity of the fluid in the direction of $\mathbf{n}$ at $\mathbf{y} \in S$.

For the known function $\bar{\Phi}(\mathbf{y})$ in (16.6) the solution of the Helmholtz equation (16.1) at an arbitrary field point $\mathbf{y}$ resulting from an acoustic point source at the source point $\mathbf{x}$ can be chosen. Hence,

$$\bar{\Psi}(\mathbf{y}) = -\delta(\mathbf{y}, \mathbf{x}) \tag{16.8}$$

is chosen, where

$$\delta(\mathbf{y}, \mathbf{x}) = \begin{cases} 0\,, & \mathbf{y} \neq \mathbf{x} \\ \infty\,, & \mathbf{y} = \mathbf{x} \end{cases} \tag{16.9}$$

represents the Dirac Delta function which for an arbitrary function $f(\mathbf{y})$ is characterized by

$$\int_V f(\mathbf{y})\, \delta(\mathbf{y}, \mathbf{x})\, dV(\mathbf{y}) = f(\mathbf{x}) \quad \forall\, \mathbf{x} \in V\,. \tag{16.10}$$

Inserting (16.8) into the Helmholtz equation (16.1) gives the so–called fundamental solution [10]

$$\bar{\Phi}(\mathbf{y}) \equiv G(\mathbf{y}, \mathbf{x}) = \frac{1}{4\pi r} \mathrm{e}^{ikr} \tag{16.11}$$

with

$$r = \sqrt{(\mathbf{x}-\mathbf{y})\cdot(\mathbf{x}-\mathbf{y})} \tag{16.12}$$

denoting the distance between the source point $\mathbf{x}$ and the field point $\mathbf{y}$. The derivative of $G(\mathbf{y},\mathbf{x})$ in the direction of $\mathbf{n}$ at $\mathbf{y}$ is obtained as

$$\frac{\partial \bar{\Phi}(\mathbf{y},\mathbf{x})}{\partial n_y} \equiv H(\mathbf{y},\mathbf{x}) = \frac{\mathrm{e}^{ikr}}{4\pi r}\left(\mathrm{i}k - \frac{1}{r}\right)\frac{\partial r}{\partial n_y} \,. \tag{16.13}$$

Substituting (16.7), (16.8), (16.11) and (16.13) into (16.6), assuming $\Psi(\mathbf{y}) = 0$ (i.e., neglecting acoustic sources in the domain V) and considering (16.10), yields the integral equation

$$\Phi(\mathbf{x}) = \int_S \Big[G(\mathbf{y},\mathbf{x})\, v_n(\mathbf{y}) - H(\mathbf{y},\mathbf{x})\,\Phi(\mathbf{y})\Big]\; dS(\mathbf{y})\,, \quad \forall\, \mathbf{x} \in \Omega_{in}\backslash S\,, \tag{16.14}$$

which holds for an arbitrary point $\mathbf{x}$ located within the domain Ω_{in}. For a point $\mathbf{x}$ on the boundary S, for the special case $\mathbf{x} = \mathbf{y}$ in (16.12) $r = 0$ follows and, thus, because of (16.13), strongly singular integrals have to be dealt with in (16.14). They can be evaluated in the sense of a Cauchy principal value by assuming a small sphere with radius ε at the singular point $\mathbf{x}$, such that the singularity is excluded from the integration domain, and solving the resulting integrals for the limiting case $\varepsilon \to 0$. This procedure yields

$$[1-\beta(\mathbf{x})]\;\Phi(\mathbf{x}) = \int_S \Big[G(\mathbf{y},\mathbf{x})\, v_n(\mathbf{y}) - H(\mathbf{y},\mathbf{x})\,\Phi(\mathbf{y})\Big]\; dS(\mathbf{y}) \tag{16.15}$$

with

$$\beta(\mathbf{x}) = \begin{cases} 0\,, & \mathbf{x} \in \Omega_{in}\backslash S \\ \frac{1}{2}\,, & \mathbf{x} \in S \;\; (S \text{ is smooth at } \mathbf{x}) \\ 1\,, & \mathbf{x} \in \Omega_{ex}\backslash S \end{cases} \,. \tag{16.16}$$

(16.15) holds for interior acoustic problems, for which the radiation and the scattering of sound are described in the interior domain Ω_{in}, bounded by the surface S as shown in Figure 16.1. If the amplitude of the velocity v_n in the direction of $\mathbf{n}$ and the sound potential Φ are known at the boundary S, then by means of (16.15) the sound potential Φ and, hence, the sound pressure p can be computed at any point $\mathbf{x}$.

For exterior acoustic problems, for which the radiation and the scattering of sound are described in the infinite exterior domain Ω_{ex}, bounded by the surface S, the respective integral equation is obtained from (16.15) by making use of the Sommerfeld radiation condition [10]. Maintaining the definition of $\mathbf{n}$ according to Figure 16.1, for exterior acoustic problems in (16.15) the sign has to be changed for those terms which contain derivatives with resepct to $\mathbf{n}$, yielding

$$\beta(\mathbf{x})\,\Phi(\mathbf{x}) = \int_S \Big[-G(\mathbf{y},\mathbf{x})\, v_n(\mathbf{y}) + H(\mathbf{y},\mathbf{x})\,\Phi(\mathbf{y})\Big]\; dS(\mathbf{y})\,. \tag{16.17}$$

The integral equations (16.15) and (16.17) can be generalized as

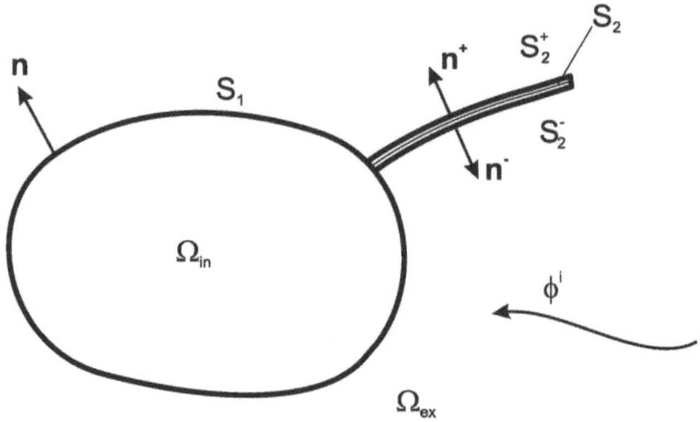

Fig. 16.2 Structure consisting of a 3D closed part and a thin–walled open part.

$$\alpha(\mathbf{x})\,\Phi(\mathbf{x}) = \int_S \tau \left[-G(\mathbf{y},\mathbf{x})\,v_n(\mathbf{y}) + H(\mathbf{y},\mathbf{x})\,\Phi(\mathbf{y})\right]\,dS(\mathbf{y}) \tag{16.18}$$

with

$$\alpha(\mathbf{x}) = \begin{cases} \beta(\mathbf{x}) & \text{for exterior acoustic problems} \\ [1-\beta(\mathbf{x})] & \text{for interior acoustic problems} \end{cases} \tag{16.19}$$

and

$$\tau = \begin{cases} 1 & \text{for exterior acoustic problems} \\ -1 & \text{for interior acoustic problems} \end{cases}\;. \tag{16.20}$$

By contrast to the fundamental solution (16.11), where $\mathbf{x}$ is the position vector of the acoustic point source and $\mathbf{y}$ is an arbitrary field point, in the integral equation (16.18) $\mathbf{x}$ can be referred to as a field point, at which the sound potential Φ is computed from known values of v_n and Φ at the boundary S of the domain under consideration.

Consider now a structure, consisting of a 3D closed part and a thin–walled part. A section of such a structure is shown in Figure 16.2. In this case the integral equation (16.18) can be written as

$$\begin{aligned}\alpha(\mathbf{x})\,\Phi(\mathbf{x}) = \Phi^i(\mathbf{x}) &+ \int_{S_1} \tau \left[-G(\mathbf{y},\mathbf{x})\,v_n(\mathbf{y}) + H(\mathbf{y},\mathbf{x})\,\Phi(\mathbf{y})\right]\,dS(\mathbf{y}) \\ &+ \int_{S_2^+\cup S_2^-} \left[-G(\mathbf{y},\mathbf{x})\,v_n(\mathbf{y}) + H(\mathbf{y},\mathbf{x})\,\Phi(\mathbf{y})\right]\,dS(\mathbf{y})\;,\end{aligned} \tag{16.21}$$

where, in addition, the sound potential Φ^i of an incident sound wave is taken into account. The integrals in (16.21) extend over the complete boundary S of the structure which consists of the boundary S_1 of the 3D closed part and the boundary $S_2^- \cup S_2^+$ of the thin–walled part. However, because of the small distance of adjacent points on the surfaces S_2^- and S_2^+ of the thin–walled part of the structure, direct application of (16.21) will result in numerical problems. Hence, the respective surface integrals are specialized for the limiting case of zero thickness. Consequently, the integration over the boundary $S_2^- \cup S_2^+$ of the thin–walled part is replaced by the integration over its middle surface S_2.

If the position vector $\mathbf{x}$ is located on the boundary S_1 of the 3D closed part, then one obtains from (16.21)

$$\begin{aligned}\alpha(\mathbf{x})\,\Phi(\mathbf{x}) = \Phi^i(\mathbf{x}) + \int_{S_1} \tau \Big[-G(\mathbf{y},\mathbf{x})\,v_n(\mathbf{y}) + H(\mathbf{y},\mathbf{x})\,\Phi(\mathbf{y})\Big]\,dS(\mathbf{y}) \\ + \int_{S_2} \Big[-G(\mathbf{y},\mathbf{x})\,v_n^d(\mathbf{y}) + H(\mathbf{y},\mathbf{x})\,\Phi^d(\mathbf{y})\Big]\,dS(\mathbf{y})\,, \qquad (16.22)\end{aligned}$$

where

$$\begin{aligned}\Phi^d(\mathbf{y}) &= \Phi^+(\mathbf{y}) - \Phi^-(\mathbf{y}) \\ v_n^d(\mathbf{y}) &= v_n^+(\mathbf{y}) - v_n^-(\mathbf{y}) = \frac{\partial \Phi^+(\mathbf{y})}{\partial n_y^+} - \frac{\partial \Phi^-(\mathbf{y})}{\partial n_y^+} \qquad (16.23)\end{aligned}$$

are the differences of the potential and the velocity of the sound field between both sides of S_2. Since v_n^- is also defined in the direction of $\mathbf{n}^+$, $\mathbf{n}^-$ will not be used in the following and, hence, for $\mathbf{n}^+$ the superscript + can be omitted, i.e. $\mathbf{n}^+ \equiv \mathbf{n}$.

If the position vector $\mathbf{x}$ is located on the middle surface S_2 of the thin–walled part, then one obtains from (16.21)

$$\begin{aligned}\frac{1}{2}\Big[\Phi^+(\mathbf{x}) + \Phi^-(\mathbf{x})\Big] = \Phi^i(\mathbf{x}) + \int_{S_1} \tau \Big[-G(\mathbf{y},\mathbf{x})\,v_n(\mathbf{y}) + H(\mathbf{y},\mathbf{x})\,\Phi(\mathbf{y})\Big]\,dS(\mathbf{y}) \\ + \int_{S_2} \Big[-G(\mathbf{y},\mathbf{x})\,v_n^d(\mathbf{y}) + H(\mathbf{y},\mathbf{x})\,\Phi^d(\mathbf{y})\Big]\,dS(\mathbf{y})\,. \qquad (16.24)\end{aligned}$$

If $v_n^d(\mathbf{y})$ is known, then on S_2 there are still the two unknowns Φ^+ and Φ^-. Hence, an additional equation is required, which is obtained by the derivative of (16.24) with respect to $\mathbf{n}$ at $\mathbf{x} \in S_2$ as

$$\begin{aligned}\frac{1}{2}\Big[v_n^+(\mathbf{x}) + v_n^-(\mathbf{x})\Big] = v_n^i(\mathbf{x}) + \int_{S_1} \tau \Big[-H^T(\mathbf{y},\mathbf{x})\,v_n(\mathbf{y}) + E(\mathbf{y},\mathbf{x})\,\Phi(\mathbf{y})\Big]\,dS(\mathbf{y}) \\ + \int_{S_2} \Big[-H^T(\mathbf{y},\mathbf{x})\,v_n^d(\mathbf{y}) + E(\mathbf{y},\mathbf{x})\,\Phi^d(\mathbf{y})\Big]\,dS(\mathbf{y})\,, \\ (16.25)\end{aligned}$$

with

$$v_n^+(\mathbf{x}) = \frac{\partial \Phi^+(\mathbf{x})}{\partial n_x}\,, \quad v_n^-(\mathbf{x}) = \frac{\partial \Phi^-(\mathbf{x})}{\partial n_x}\,, \quad v_n^i(\mathbf{x}) = \frac{\partial \Phi^i(\mathbf{x})}{\partial n_x}\,, \qquad (16.26)$$

$$H^T(\mathbf{y},\mathbf{x}) = \frac{\partial G(\mathbf{y},\mathbf{x})}{\partial n_x} = \frac{\mathrm{e}^{\mathrm{i}kr}}{4\pi r}\left(\mathrm{i}k - \frac{1}{r}\right)\frac{\partial r}{\partial n_x} \qquad (16.27)$$

and

$$E(\mathbf{y},\mathbf{x}) = \frac{\partial^2 G(\mathbf{y},\mathbf{x})}{\partial n_x \partial n_y}$$
$$= \frac{\mathrm{e}^{\mathrm{i}kr}}{4\pi r}\left[\left(\frac{3}{r^2} - \frac{\mathrm{i}3k}{r} - k^2\right)\frac{\partial r}{\partial n_x}\frac{\partial r}{\partial n_y} + \frac{1}{r}\left(\frac{1}{r} - \mathrm{i}k\right)\mathbf{n}(\mathbf{x})\cdot\mathbf{n}(\mathbf{y})\right]. \tag{16.28}$$

If the vibrating surface of a structure is lined by a sound absorbing material, then, the normal velocity v_n of the fluid at the interface between the structure and the fluid differs from the respective velocity component v_n^S of the structural vibrations. In this case the relationship between v_n and v_n^S can be described as

$$\begin{aligned} v_n(\mathbf{y}) &= v_n^S(\mathbf{y}) - \tau A(\mathbf{y})\,\Phi(\mathbf{y}), & \mathbf{y} &\in S_1, \\ v_n^-(\mathbf{y}) &= v_n^S(\mathbf{y}) + A^-(\mathbf{y})\,\Phi^-(\mathbf{y}), & \mathbf{y} &\in S_2^-, \\ v_n^+(\mathbf{y}) &= v_n^S(\mathbf{y}) - A^+(\mathbf{y})\,\Phi^+(\mathbf{y}), & \mathbf{y} &\in S_2^+, \end{aligned} \tag{16.29}$$

where

$$A = \mathrm{i}\omega\rho a, \quad A^- = \mathrm{i}\omega\rho a^-, \quad A^+ = \mathrm{i}\omega\rho a^+, \tag{16.30}$$

with a, a^+ and a^- as the acoustic admittance of the respective surface. The velocity difference between both surfaces of the thin–walled part of the structure can be expressed as

$$v_n^d = v_n^+ - v_n^- = -A^+\Phi^+ - A^-\Phi^- = -A^m\Phi^m - A^d\Phi^d \tag{16.31}$$

with

$$\begin{aligned} A^m &= \frac{1}{2}(A^+ + A^-), \\ A^d &= \frac{1}{2}(A^+ - A^-), \\ \Phi^m &= \Phi^+ + \Phi^- . \end{aligned} \tag{16.32}$$

With respect to the thin–walled part of a structure the integral equations (16.22), (16.24) and (16.25) are formulated in terms of the differences Φ^d and v_n^d of the potential and the velocity of the sound field. However, prescribed acoustic admittances on the surfaces S_2^+ and S_2^- of the thin–walled part may be discontinuous and, consequently, v_n^d may be discontinuous. For this reason it is convenient to use the continuous function Φ^m instead of v_n^d in the respective integral equations.

Considering (16.31) together with (16.32) in the integral equations (16.22), (16.24) and (16.25) yields the three integral equations

$$\begin{aligned} &\alpha(\mathbf{x})\,\Phi(\mathbf{x}) - \int_{S_1}\Big[G(\mathbf{y},\mathbf{x})\,A(\mathbf{y}) + \tau H(\mathbf{y},\mathbf{x})\Big]\,\Phi(\mathbf{y})\,dS(\mathbf{y}) \\ &\quad - \int_{S_2}\Big\{\Big[G(\mathbf{y},\mathbf{x})\,A^d(\mathbf{y}) + H(\mathbf{y},\mathbf{x})\Big]\,\Phi^d(\mathbf{y}) + G(\mathbf{y},\mathbf{x})\,A^m(\mathbf{y})\,\Phi^m(\mathbf{y})\Big\}\,dS(\mathbf{y}) \\ &\quad = \Phi^i(\mathbf{x}) - \tau\int_{S_1} G(\mathbf{y},\mathbf{x})\,v_n^S(\mathbf{y})\,dS(\mathbf{y}), \quad \mathbf{x}\in S_1; \end{aligned} \tag{16.33}$$

$$
\begin{aligned}
&\frac{1}{2}\,\Phi^m(\mathbf{x}) - \int_{S_1} \Big[G(\mathbf{y},\mathbf{x})\,A(\mathbf{y}) + \tau H(\mathbf{y},\mathbf{x})\Big]\,\Phi(\mathbf{y})\,dS(\mathbf{y})\\
&\quad - \int_{S_2} \Big\{\Big[G(\mathbf{y},\mathbf{x})\,A^d(\mathbf{y}) + H(\mathbf{y},\mathbf{x})\Big]\,\Phi^d(\mathbf{y}) + G(\mathbf{y},\mathbf{x})\,A^m(\mathbf{y})\,\Phi^m(\mathbf{y})\Big\}\,dS(\mathbf{y})\\
&\quad = \Phi^i(\mathbf{x}) - \tau \int_{S_1} G(\mathbf{y},\mathbf{x})\,v_n^S(\mathbf{y})\,dS(\mathbf{y})\,, \quad \mathbf{x}\in S_2\,;
\end{aligned}
\tag{16.34}
$$

$$
\begin{aligned}
&\frac{1}{2}\Big[A^d(\mathbf{x})\,\Phi^m(\mathbf{x}) + A^m(\mathbf{x})\,\Phi^d(\mathbf{x})\Big]\\
&\quad + \int_{S_1} \Big[H^T(\mathbf{y},\mathbf{x})\,A(\mathbf{y}) + \tau E(\mathbf{y},\mathbf{x})\Big]\,\Phi(\mathbf{y})\,dS(\mathbf{y})\\
&\quad + \int_{S_2} \Big\{\Big[H^T(\mathbf{y},\mathbf{x})\,A^d(\mathbf{y}) + E(\mathbf{y},\mathbf{x})\Big]\,\Phi^d(\mathbf{y})\\
&\qquad\qquad + H^T(\mathbf{y},\mathbf{x})\,A^m(\mathbf{y})\,\Phi^m(\mathbf{y})\Big\}\,dS(\mathbf{y})\,,\\
&\quad = v_n^S(\mathbf{x}) - v_n^i(\mathbf{x}) + \tau \int_{S_1} H^T(\mathbf{y},\mathbf{x})\,v_n^S(\mathbf{y})\,dS(\mathbf{y})\,, \quad \mathbf{x}\in S_2\,.
\end{aligned}
\tag{16.35}
$$

In order to obtain a selfadjoint system of integral equations from (16.33), (16.34) and (16.35), (16.33) is differentiated with respect to $\mathbf{n}(\mathbf{x})$, which gives

$$
\begin{aligned}
&\alpha(\mathbf{x})\Big[v_n^S(\mathbf{x}) - \tau A(\mathbf{x})\,\Phi(\mathbf{x})\Big] - \int_{S_1} \Big[H^T(\mathbf{y},\mathbf{x})\,A(\mathbf{y}) + \tau E(\mathbf{y},\mathbf{x})\Big]\,\Phi(\mathbf{y})\,dS(\mathbf{y})\\
&\quad - \int_{S_2} \Big\{\Big[H^T(\mathbf{y},\mathbf{x})\,A^d(\mathbf{y}) + E(\mathbf{y},\mathbf{x})\Big]\,\Phi^d(\mathbf{y}) + H^T(\mathbf{y},\mathbf{x})\,A^m(\mathbf{y})\,\Phi^m(\mathbf{y})\Big\}\,dS(\mathbf{y})\\
&\quad = v_n^i(\mathbf{x}) - \tau \int_{S_1} H^T(\mathbf{y},\mathbf{x})\,v_n^S(\mathbf{y})\,dS(\mathbf{y})\,, \quad \mathbf{x}\in S_1\,;
\end{aligned}
\tag{16.36}
$$

Multiplying (16.33) by $A(\mathbf{x})$ and (16.36) by τ, and adding the resulting equations yields

$$
\begin{aligned}
&\int_{S_1} \Big[G(\mathbf{y},\mathbf{x})\,A(\mathbf{x})\,A(\mathbf{y}) + \tau H(\mathbf{y},\mathbf{x})\,A(\mathbf{x}) + \tau H^T(\mathbf{y},\mathbf{x})\,A(\mathbf{y})\\
&\qquad + E(\mathbf{y},\mathbf{x})\Big]\,\Phi(\mathbf{y})\,dS(\mathbf{y}) + \int_{S_2} \Big[G(\mathbf{y},\mathbf{x})\,A(\mathbf{x})\,A^d(\mathbf{y}) + H(\mathbf{y},\mathbf{x})\,A(\mathbf{x})\\
&\qquad + \tau H^T(\mathbf{y},\mathbf{x})\,A^d(\mathbf{y}) + \tau E(\mathbf{y},\mathbf{x})\Big]\,\Phi^d(\mathbf{y})\,dS(\mathbf{y})\\
&\qquad + \int_{S_2} \Big[G(\mathbf{y},\mathbf{x})\,A(\mathbf{x}) + \tau H^T(\mathbf{y},\mathbf{x})\Big]\,A^m(\mathbf{y})\,\Phi^m(\mathbf{y})\,dS(\mathbf{y})\\
&\quad = \tau\alpha(\mathbf{x})v_n^S(\mathbf{x}) - A(\mathbf{x})\,\Phi^i(\mathbf{x}) - \tau v_n^i(\mathbf{x})\\
&\qquad + \int_{S_1} \Big[\tau G(\mathbf{y},\mathbf{x})\,A(\mathbf{x}) + H^T(\mathbf{y},\mathbf{x})\Big]\,v_n^S(\mathbf{y})\,dS(\mathbf{y})\,, \quad \mathbf{x}\in S_1\,.
\end{aligned}
\tag{16.37}
$$

Multiplying (16.34) by $A^d(\mathbf{x})$ and subtracting the resulting equation from (16.35) gives

$$\begin{aligned}
&\frac{1}{2}A^m(\mathbf{x})\,\Phi^d(\mathbf{x}) + \int_{S_1}\Big[G(\mathbf{y},\mathbf{x})\,A(\mathbf{y})\,A^d(\mathbf{x}) + \tau H(\mathbf{y},\mathbf{x})\,A^d(\mathbf{x}) \\
&\quad + H^T(\mathbf{y},\mathbf{x})\,A(\mathbf{y}) + \tau E(\mathbf{y},\mathbf{x})\Big]\,\Phi(\mathbf{y})\,dS(\mathbf{y}) + \int_{S_2}\Big[G(\mathbf{y},\mathbf{x})\,A^d(\mathbf{y})\,A^d(\mathbf{x}) \\
&\quad + H(\mathbf{y},\mathbf{x})\,A^d(\mathbf{x}) + H^T(\mathbf{y},\mathbf{x})\,A^d(\mathbf{y}) + E(\mathbf{y},\mathbf{x})\Big]\,\Phi^d(\mathbf{y})dS(\mathbf{y}) \\
&\quad + \int_{S_2}\Big[G(\mathbf{y},\mathbf{x})\,A^d(\mathbf{x}) + H^T(\mathbf{y},\mathbf{x})\Big]\,A^m(\mathbf{y})\,\Phi^m(\mathbf{y})\,dS(\mathbf{y}) \\
&\quad = -\Phi^i(\mathbf{x})\,A^d(\mathbf{x}) + v_n^S(\mathbf{x}) - v_n^i(\mathbf{x}) \\
&\quad + \tau\int_{S_1}\Big[G(\mathbf{y},\mathbf{x})\,A^d(\mathbf{x}) + H^T(\mathbf{y},\mathbf{x})\Big]\,v_n^S(\mathbf{y})\,dS(\mathbf{y})\,, \quad \mathbf{x}\in S_2\,.
\end{aligned} \tag{16.38}$$

The third integral equation is obtained by multiplying (16.34) by $-A^m(\mathbf{x})$ as

$$\begin{aligned}
&-\frac{1}{2}A^m(\mathbf{x})\,\Phi^m(\mathbf{x}) + \int_{S_1}\Big[G(\mathbf{y},\mathbf{x})\,A(\mathbf{y}) + \tau H(\mathbf{y},\mathbf{x})\Big]\,A^m(\mathbf{x})\,\Phi(\mathbf{y})\,dS(\mathbf{y}) \\
&\quad + \int_{S_2}\Big[G(\mathbf{y},\mathbf{x})\,A^d(\mathbf{y}) + H(\mathbf{y},\mathbf{x})\Big]\,A^m(\mathbf{x})\,\Phi^d(\mathbf{y}) \\
&\quad + \int_{S_2} G(\mathbf{y},\mathbf{x})\,A^m(\mathbf{y})\,A^m(\mathbf{x})\,\Phi^m(\mathbf{y})\,dS(\mathbf{y}) \\
&\quad = -\Phi^i(\mathbf{x})A^m(\mathbf{x}) + \tau\int_{S_1} G(\mathbf{y},\mathbf{x})\,A^m(\mathbf{x})\,v_n^S(\mathbf{y})\,dS(\mathbf{y})\,, \quad \mathbf{x}\in S_2\,.
\end{aligned} \tag{16.39}$$

The self–adjoint system of integral equations (16.37), (16.38) and (16.39), which is the basis for the Galerkin–type numerical solution procedure to be described in the following section, can be written concisely as

$$\mathbf{T}(\mathbf{y},\mathbf{x})\{\mathbf{z}(\mathbf{y})\} = \mathbf{f}(\mathbf{x}) \tag{16.40}$$

with

$$\mathbf{T}(\mathbf{y},\mathbf{x})\{\bullet\} = \begin{bmatrix} T_{11}\{\bullet\} & T_{12}\{\bullet\} & T_{13}\{\bullet\} \\ T_{21}\{\bullet\} & T_{22}\{\bullet\} & T_{23}\{\bullet\} \\ T_{31}\{\bullet\} & T_{32}\{\bullet\} & T_{33}\{\bullet\} \end{bmatrix} \tag{16.41}$$

and

$$\mathbf{z}(\mathbf{y}) = \begin{Bmatrix} \Phi(\mathbf{y}) \\ \Phi^d(\mathbf{y}) \\ \Phi^m(\mathbf{y}) \end{Bmatrix}, \qquad \mathbf{f}(\mathbf{x}) = \begin{Bmatrix} f_1(\mathbf{x}) \\ f_2(\mathbf{x}) \\ f_3(\mathbf{x}) \end{Bmatrix}. \tag{16.42}$$

The components of $\mathbf{T}$ and $\mathbf{f}$ follow from (16.37), (16.38) and (16.39) as

$$
\begin{aligned}
T_{11}\{\bullet\} &= \int_{S_1} \Big[G(\mathbf{y},\mathbf{x})\,A(\mathbf{x})\,A(\mathbf{y}) + \tau H(\mathbf{y},\mathbf{x})\,A(\mathbf{x}) + \tau H^T(\mathbf{y},\mathbf{x})\,A(\mathbf{y}) \\
&\qquad + E(\mathbf{y},\mathbf{x})\Big]\,\{\bullet\}\,dS(\mathbf{y})\,, \\
T_{12}\{\bullet\} &= \int_{S_2} \Big[G(\mathbf{y},\mathbf{x})\,A(\mathbf{x})\,A^d(\mathbf{y}) + H(\mathbf{y},\mathbf{x})\,A(\mathbf{x}) + \tau H^T(\mathbf{y},\mathbf{x})\,A^d(\mathbf{y}) \\
&\qquad + \tau E(\mathbf{y},\mathbf{x})\Big]\,\{\bullet\}\,dS(\mathbf{y})\,, \\
T_{13}\{\bullet\} &= \int_{S_2} \Big[G(\mathbf{y},\mathbf{x})\,A(\mathbf{x}) + \tau H^T(\mathbf{y},\mathbf{x})\Big]\,A^m(\mathbf{y})\,\{\bullet\}\,dS(\mathbf{y})\,, \\
T_{21}\{\bullet\} &= \int_{S_1} \Big[G(\mathbf{y},\mathbf{x})\,A(\mathbf{y})\,A^d(\mathbf{x}) + \tau H(\mathbf{y},\mathbf{x})\,A^d(\mathbf{x}) + H^T(\mathbf{y},\mathbf{x})\,A(\mathbf{y}) \\
&\qquad + \tau E(\mathbf{y},\mathbf{x})\Big]\,\{\bullet\}\,dS(\mathbf{y})\,, \\
T_{22}\{\bullet\} &= \frac{1}{2}\,A^m(\mathbf{x})\,\delta(\mathbf{y},\mathbf{x})\{\bullet\} + \int_{S_2} \Big[G(\mathbf{y},\mathbf{x})\,A^d(\mathbf{y})\,A^d(\mathbf{x}) + H(\mathbf{y},\mathbf{x})\,A^d(\mathbf{x}) \\
&\qquad + H^T(\mathbf{y},\mathbf{x})\,A^d(\mathbf{y}) + E(\mathbf{y},\mathbf{x})\Big]\,\{\bullet\}dS(\mathbf{y})\,, \\
T_{23}\{\bullet\} &= \int_{S_2} \Big[G(\mathbf{y},\mathbf{x})\,A^d(\mathbf{x}) + H^T(\mathbf{y},\mathbf{x})\Big]\,A^m(\mathbf{y})\,\{\bullet\}\,dS(\mathbf{y})\,, \\
T_{31}\{\bullet\} &= \int_{S_1} \Big[G(\mathbf{y},\mathbf{x})\,A(\mathbf{y}) + \tau H(\mathbf{y},\mathbf{x})\Big]\,A^m(\mathbf{x})\,\{\bullet\}\,dS(\mathbf{y})\,, \\
T_{32}\{\bullet\} &= \int_{S_2} \Big[G(\mathbf{y},\mathbf{x})\,A^d(\mathbf{y}) + H(\mathbf{y},\mathbf{x})\Big]\,A^m(\mathbf{x})\,\{\bullet\}\,dS(\mathbf{y})\,, \\
T_{33}\{\bullet\} &= -\frac{1}{2}\,A^m(\mathbf{x})\,\delta(\mathbf{y},\mathbf{x})\{\bullet\} + \int_{S_2} G(\mathbf{y},\mathbf{x})\,A^m(\mathbf{y})\,A^m(\mathbf{x})\,\{\bullet\}\,dS(\mathbf{y})
\end{aligned}
\tag{16.43}
$$

and

$$
\begin{aligned}
f_1(\mathbf{x}) &= \tau\alpha(\mathbf{x})v_n^S(\mathbf{x}) - A(\mathbf{x})\,\Phi^i(\mathbf{x}) - \tau v_n^i(\mathbf{x}) \\
&\qquad + \int_{S_1} \Big[\tau G(\mathbf{y},\mathbf{x})\,A(\mathbf{x}) + H^T(\mathbf{y},\mathbf{x})\Big]\,v_n^S(\mathbf{y})\,dS(\mathbf{y})\,, \\
f_2(\mathbf{x}) &= -\Phi^i(\mathbf{x})\,A^d(\mathbf{x}) + v_n^S(\mathbf{x}) - v_n^i(\mathbf{x}) \\
&\qquad + \tau \int_{S_1} \Big[G(\mathbf{y},\mathbf{x})\,A^d(\mathbf{x}) + H^T(\mathbf{y},\mathbf{x})\Big]\,v_n^S(\mathbf{y})\,dS(\mathbf{y})\,, \\
f_3(\mathbf{x}) &= -\Phi^i(\mathbf{x})\,A^m(\mathbf{x}) + \tau \int_{S_1} G(\mathbf{y},\mathbf{x})\,A^m(\mathbf{x})\,v_n^S(\mathbf{y})\,dS(\mathbf{y})\,.
\end{aligned}
\tag{16.44}
$$

If acoustically hard reflecting boundaries are prescribed for the whole boundary, then it follows from $a = 0$ that $A = 0$, $A^+ = 0$ and $A^- = 0$ and from (16.31) that

$v_n^d = 0$. Hence, (16.43) and (16.44) degenerate to

$$T_{11}\{\bullet\} = \int_{S_1} E(\mathbf{y},\mathbf{x})\{\bullet\}\, dS(\mathbf{y}) \,, \qquad T_{12}\{\bullet\} = \int_{S_2} \tau E(\mathbf{y},\mathbf{x})\{\bullet\}\, dS(\mathbf{y}) \,,$$

$$T_{21}\{\bullet\} = \int_{S_1} \tau E(\mathbf{y},\mathbf{x})\{\bullet\}\, dS(\mathbf{y}) \,, \qquad T_{22}\{\bullet\} = \int_{S_2} E(\mathbf{y},\mathbf{x})\, \{\bullet\} dS(\mathbf{y}) \,,$$

$$T_{13}\{\bullet\} = T_{31}\{\bullet\} = 0 \,, \qquad T_{23}\{\bullet\} = T_{32}\{\bullet\} = 0 \,, \qquad T_{33}\{\bullet\} = 0 \quad (16.45)$$

and

$$f_1(\mathbf{x}) = \tau\alpha(\mathbf{x}) v_n^S(\mathbf{x}) - \tau v_n^i(\mathbf{x}) + \int_{S_1} H^T(\mathbf{y},\mathbf{x}) v_n^S(\mathbf{y})\, dS(\mathbf{y}) \,,$$

$$f_2(\mathbf{x}) = v_n^S(\mathbf{x}) - v_n^i(\mathbf{x}) + \tau \int_{S_1} H^T(\mathbf{y},\mathbf{x})\, v_n^S(\mathbf{y})\, dS(\mathbf{y}) \,,$$

$$f_3(\mathbf{x}) = 0 \,. \qquad (16.46)$$

Hence, in this special case all terms associated with Φ^m vanish.

16.3 Galerkin–type Numerical Solution Scheme

Within the framework of the BEM the boundary of the domain under consideration is discretized by means of surface elements of finite size, which are commonly known as boundary elements. In the present case, the boundary S_1 of the 3D closed part of the structure is discretized by means of isoparametric boundary elements. In the following, the latter are denoted as S_1–elements. On the boundary S_1 the potential $\Phi(\mathbf{y})$ is the unknown function, which within a particular boundary element e is approximated by

$$\Phi(\mathbf{y}) = [\mathbf{N}^{(e)}]^T\, \mathbf{\Phi}^{(e)} \,. \qquad (16.47)$$

In (16.47)

$$[\mathbf{N}^{(e)}]^T = \lfloor\, N_1^{(e)}(\boldsymbol{\xi})\,, \ldots\,, N_i^{(e)}(\boldsymbol{\xi}), \ldots\,, N_k^{(e)}(\boldsymbol{\xi})\, \rfloor \qquad (16.48)$$

denotes a vector of shape functions, formulated in local coordinates $\boldsymbol{\xi} = \lfloor \xi_1\ \xi_2 \rfloor^T$ and

$$(\mathbf{\Phi}^{(e)})^T = \lfloor\, \Phi_1^{(e)}\,, \ldots\,, \Phi_i^{(e)}\,, \ldots\,, \Phi_k^{(e)}\, \rfloor \qquad (16.49)$$

is the vector of unknown values of the sound potential $\Phi(\mathbf{y})$ at the nodes of element e with $N_i^{(e)}(\boldsymbol{\xi})$ and $\Phi_i^{(e)}$ as the shape function and the nodal value of the sound potential for node i of boundary element e and k as the total number of nodes of this boundary element. Since isoparametric boundary elements are employed, the position vector $\mathbf{y}$ within a particular boundary element is approximated by

$$\mathbf{y} = \begin{bmatrix} N_1^{(e)} & 0 & 0 & \ldots & N_i^{(e)} & 0 & 0 & \ldots & N_k^{(e)} & 0 & 0 \\ 0 & N_1^{(e)} & 0 & \ldots & 0 & N_i^{(e)} & 0 & \ldots & 0 & N_k^{(e)} & 0 \\ 0 & 0 & N_1^{(e)} & \ldots & 0 & 0 & N_i^{(e)} & \ldots & 0 & 0 & N_k^{(e)} \end{bmatrix} \mathbf{y}^{(e)} \quad (16.50)$$

with

$$(\mathbf{y}^{(e)})^T = \lfloor\, (y_1)_1, (y_2)_1, (y_3)_1, \ldots, (y_1)_i, (y_2)_i, (y_3)_i, \ldots, (y_1)_k, (y_2)_k, (y_3)_k \,\rfloor \tag{16.51}$$

denoting the vector of nodal coordinates of boundary element e with $\mathbf{y}_i^T = \lfloor (y_1)_i,\ (y_2)_i,\ (y_3)_i \rfloor$ as the position vector of node i.

Since in the present formulation for the thin–walled part of the structure the limiting case of zero thickness is considered, the discretization of the thin–walled part just extends over its middle surface S_2. In the following, the respective isoparametric elements will be denoted as S_2–elements. In the middle surface S_2 of the thin–walled part of the structure the difference $\Phi^d(\mathbf{y})$ and the sum $\Phi^m(\mathbf{y})$ of the sound potential of both sides of the middle surface are the unknown functions. By analogy to (16.47), they are approximated within a particular boundary element e by

$$\Phi^d(\mathbf{y}) = [\mathbf{N}^{(e)}]^T\, \mathbf{\Phi}^d\,, \qquad \Phi^m(\mathbf{y}) = [\mathbf{N}^{(e)}]^T\, [\mathbf{\Phi}^m]^{(e)}\,, \tag{16.52}$$

where $[\mathbf{N}^{(e)}]^T$ is given according to (16.48) and $[\mathbf{\Phi}^d]^{(e)}$ and $[\mathbf{\Phi}^{(e)}]^m$ by analogy to (16.49) denote the unknown values of the respective functions at the nodes of the boundary element e.

In the following, the notation is simplified by omitting reference to the individual boundary elements. The approximation of the unknown function $\Phi(\mathbf{y})$ on the boundary S_1 is written as

$$\Phi(\mathbf{y}) = \mathbf{N}^T\, \mathbf{\Phi}\,, \tag{16.53}$$

where

$$\mathbf{N}^T = \lfloor\, N_1(\mathbf{y})\,, \ldots\,, N_i(\mathbf{y}), \ldots\,, N_{n_1}(\mathbf{y})\,\rfloor \tag{16.54}$$

denotes a vector of shape functions and

$$\mathbf{\Phi}^T = \lfloor\, \Phi_1\,, \ldots\,, \Phi_i\,, \ldots\,, \Phi_{n_1}\,\rfloor \tag{16.55}$$

is the vector of unknown nodal values of the sound potential $\Phi(\mathbf{y})$ for the boundary S_1 of the 3D closed part with n_1 being the total number of nodes of the boundary element mesh for the boundary S_1. If a particular boundary element is addressed, then in (16.54) and (16.55) all entries, which do not belong to nodes of this boundary element, will be set to zero.

By analogy to (16.53), the approximations for the thin–walled part of the structure are written as

$$\Phi^d(\mathbf{y}) = \mathbf{N}^T\, \mathbf{\Phi}^d\,, \qquad \Phi^m(\mathbf{y}) = \mathbf{N}^T\, \mathbf{\Phi}^m\,, \tag{16.56}$$

where

$$\mathbf{N}^T = \lfloor\, N_1(\mathbf{y})\,, \ldots\,, N_i(\mathbf{y}), \ldots\,, N_{n_2}(\mathbf{y})\,\rfloor \tag{16.57}$$

with n_2 being the total number of nodes of the boundary element mesh for the middle surface S_2 and $\mathbf{\Phi}^d$ and $\mathbf{\Phi}^m$ denoting the unknown nodal values of the respective functions for the middle surface S_2.

If the collocation method was applied for the numerical solution of the integral equations (16.37), (16.38) and (16.39), then the position vector $\mathbf{x}$ would be put successively to all nodal points and the integral equations would be enforced at these nodal points. Hence, the residuals of the integral equations (16.37), (16.38) and (16.39), which result from replacing the unknown functions Φ, Φ^d and Φ^m by the approximations, given in (16.53) and (16.56), would be required to be zero at all nodal points.

In this work, however, the Galerkin method is employed. It is characterized by requiring the weighted residuals of the integral equations (16.37), (16.38) and (16.39) to vanish. The discretized weight functions $\delta\Phi(\mathbf{y})$, $\delta\Phi^d(\mathbf{y})$ and $\delta\Phi^m(\mathbf{y})$ are given by analogy to (16.53) and (16.56) as

$$\delta\Phi = (\delta\mathbf{\Phi})^T\,\mathbf{N}\,, \quad \delta\Phi^d = (\delta\mathbf{\Phi}^d)^T\,\mathbf{N}\,, \quad \delta\Phi^m = (\delta\mathbf{\Phi}^m)^T\,\mathbf{N}. \tag{16.58}$$

The integral equation (16.37) is multiplied by $\delta\Phi(\mathbf{x})$ and integrated over S_1, the integral equation (16.38) is multiplied by $\delta\Phi^d(\mathbf{x})$ and integrated over S_2 and the integral equation (16.39) is multiplied by $\delta\Phi^m(\mathbf{x})$ and integrated over S_2.

Considering the so–obtained integral equations to hold for arbitrary nodal values of the weighting functions, the symmetric system of algebraic equations

$$\mathbf{K}\,\mathbf{q} = \mathbf{f} \tag{16.59}$$

is obtained with the coefficient matrix

$$\mathbf{K} = \begin{bmatrix} \mathbf{K}_{11} & \mathbf{K}_{12} & \mathbf{K}_{13} \\ \mathbf{K}_{21} & \mathbf{K}_{22} & \mathbf{K}_{23} \\ \mathbf{K}_{31} & \mathbf{K}_{32} & \mathbf{K}_{33} \end{bmatrix}, \tag{16.60}$$

the right–hand–side vector

$$\mathbf{f} = \begin{Bmatrix} \mathbf{f}_1 \\ \mathbf{f}_2 \\ \mathbf{f}_3 \end{Bmatrix} \tag{16.61}$$

and the vector of unknown nodal values

$$\mathbf{q} = \begin{Bmatrix} \mathbf{\Phi} \\ \mathbf{\Phi}^d \\ \mathbf{\Phi}^m \end{Bmatrix}. \tag{16.62}$$

The latter consists of the subvector $\mathbf{\Phi}$, representing the unknown nodal values of the sound potential Φ on S_1, and of the subvectors $\mathbf{\Phi}^d$ and $\mathbf{\Phi}^m$, representing the unknown nodal values of the difference and of the sum of the sound potential on S_2.

The submatrices of $\mathbf{K}$ and the subvectors of $\mathbf{f}$ are given as

$$\mathbf{K}_{11} = \int_{S_1}\int_{S_1} \mathbf{N}(\mathbf{x}) \Big[G(\mathbf{y},\mathbf{x})\,A(\mathbf{x})\,A(\mathbf{y}) + \tau H(\mathbf{y},\mathbf{x})\,A(\mathbf{x}) + \tau H^T(\mathbf{y},\mathbf{x})\,A(\mathbf{y}) + E(\mathbf{y},\mathbf{x})\Big]\,\mathbf{N}^T(\mathbf{y})\,dS(\mathbf{y})\,dS(\mathbf{x})\,,$$

$$\mathbf{K}_{12} = \int_{S_1}\int_{S_2} \mathbf{N}(\mathbf{x}) \Big[G(\mathbf{y},\mathbf{x})\,A(\mathbf{x})\,A^d(\mathbf{y}) + H(\mathbf{y},\mathbf{x})\,A(\mathbf{x}) + \tau H^T(\mathbf{y},\mathbf{x})\,A^d(\mathbf{y}) + \tau E(\mathbf{y},\mathbf{x})\Big]\,\mathbf{N}^T(\mathbf{y})\,dS(\mathbf{y})\,dS(\mathbf{x})\,,$$

$$\mathbf{K}_{13} = \int_{S_1}\int_{S_2} \mathbf{N}(\mathbf{x}) \Big[G(\mathbf{y},\mathbf{x})\,A(\mathbf{x}) + \tau H^T(\mathbf{y},\mathbf{x})\Big]\,A^m(\mathbf{y})\,\mathbf{N}^T(\mathbf{y})\,dS(\mathbf{y})\,dS(\mathbf{x})\,,$$

$$\mathbf{K}_{21} = \int_{S_2}\int_{S_1} \mathbf{N}(\mathbf{x}) \Big[G(\mathbf{y},\mathbf{x})\,A(\mathbf{y})\,A^d(\mathbf{x}) + \tau H(\mathbf{y},\mathbf{x})\,A^d(\mathbf{x}) + H^T(\mathbf{y},\mathbf{x})\,A(\mathbf{y}) + \tau E(\mathbf{y},\mathbf{x})\Big]\,\mathbf{N}^T(\mathbf{y})\,dS(\mathbf{y})\,dS(\mathbf{x})\,,$$

$$\mathbf{K}_{22} = \int_{S_2}\int_{S_2} \mathbf{N}(\mathbf{x}) \Big[\frac{1}{2}\,A^m(\mathbf{x})\,\delta(\mathbf{y},\mathbf{x}) + G(\mathbf{y},\mathbf{x})\,A^d(\mathbf{y})\,A^d(\mathbf{x}) + H(\mathbf{y},\mathbf{x})\,A^d(\mathbf{x}) + H^T(\mathbf{y},\mathbf{x})\,A^d(\mathbf{y}) + E(\mathbf{y},\mathbf{x})\Big]\,\mathbf{N}^T(\mathbf{y})\,dS(\mathbf{y})\,dS(\mathbf{x})\,, \tag{16.63}$$

$$\mathbf{K}_{23} = \int_{S_2}\int_{S_2} \mathbf{N}(\mathbf{x}) \Big[G(\mathbf{y},\mathbf{x})\,A^d(\mathbf{x}) + H^T(\mathbf{y},\mathbf{x})\Big]\,A^m(\mathbf{y})\,\mathbf{N}^T(\mathbf{y})\,dS(\mathbf{y})\,dS(\mathbf{x})\,,$$

$$\mathbf{K}_{31} = \int_{S_2}\int_{S_1} \mathbf{N}(\mathbf{x}) \Big[G(\mathbf{y},\mathbf{x})\,A(\mathbf{y}) + \tau H(\mathbf{y},\mathbf{x})\Big]\,A^m(\mathbf{x})\,\mathbf{N}^T(\mathbf{y})\,dS(\mathbf{y})\,dS(\mathbf{x})\,,$$

$$\mathbf{K}_{32} = \int_{S_2}\int_{S_2} \mathbf{N}(\mathbf{x}) \Big[G(\mathbf{y},\mathbf{x})\,A^d(\mathbf{y}) + H(\mathbf{y},\mathbf{x})\Big]\,A^m(\mathbf{x})\,\mathbf{N}^T(\mathbf{y})\,dS(\mathbf{y})\,dS(\mathbf{x})\,,$$

$$\mathbf{K}_{33} = \int_{S_2}\int_{S_2} \mathbf{N}(\mathbf{x}) \Big[-\frac{1}{2}\,\delta(\mathbf{y},\mathbf{x}) + G(\mathbf{y},\mathbf{x})A^m(\mathbf{y})\Big]\,A^m(\mathbf{x})\mathbf{N}^T(\mathbf{y})dS(\mathbf{y})dS(\mathbf{x})$$

and

$$\mathbf{f}_1(\mathbf{x}) = \int_{S_1} \mathbf{N}(\mathbf{x}) \Big\{\tau\alpha(\mathbf{x})v_n^S(\mathbf{x}) - A(\mathbf{x})\,\Phi^i(\mathbf{x}) - \tau v_n^i(\mathbf{x}) + \int_{S_1}\Big[\tau G(\mathbf{y},\mathbf{x})\,A(\mathbf{x}) + H^T(\mathbf{y},\mathbf{x})\Big]\,v_n^S(\mathbf{y})\,dS(\mathbf{y})\Big\}\,dS(\mathbf{x})\,,$$

$$\mathbf{f}_2(\mathbf{x}) = \int_{S_2} \mathbf{N}(\mathbf{x}) \Big\{-\Phi^i(\mathbf{x})\,A^d(\mathbf{x}) + v_n^S(\mathbf{x}) - v_n^i(\mathbf{x}) + \tau\int_{S_1}\Big[G(\mathbf{y},\mathbf{x})\,A^d(\mathbf{x}) + H^T(\mathbf{y},\mathbf{x})\Big]\,v_n^S(\mathbf{y})\,dS(\mathbf{y})\Big\}\,dS(\mathbf{x})\,, \tag{16.64}$$

$$\mathbf{f}_3(\mathbf{x}) = \int_{S_2} \mathbf{N}(\mathbf{x}) \Big[-\Phi^i(\mathbf{x})\,A^m(\mathbf{x}) + \tau\,A^m(\mathbf{x})\int_{S_1} G(\mathbf{y},\mathbf{x})\,v_n^S(\mathbf{y})\,dS(\mathbf{y})\Big]\,dS(\mathbf{x})\,.$$

If acoustically hard reflecting boundaries are prescribed for the whole boundary, then, because of $A = 0$, $A^+ = 0$ and $A^- = 0$, the system of algebraic equations (16.59) degenerates to

$$\mathbf{K} = \begin{bmatrix} \mathbf{K}_{11} & \mathbf{K}_{12} \\ \mathbf{K}_{21} & \mathbf{K}_{22} \end{bmatrix} , \qquad \mathbf{q} = \begin{Bmatrix} \boldsymbol{\Phi} \\ \boldsymbol{\Phi}^d \end{Bmatrix} , \qquad \mathbf{f} = \begin{Bmatrix} \mathbf{f}_1 \\ \mathbf{f}_2 \end{Bmatrix} \tag{16.65}$$

with

$$\begin{aligned} \mathbf{K}_{11} &= \int_{S_1} \int_{S_1} \mathbf{N}(\mathbf{x})\, E(\mathbf{y},\mathbf{x})\, \mathbf{N}^T(\mathbf{y})\, dS(\mathbf{y})\, dS(\mathbf{x}) \,, \\ \mathbf{K}_{12} &= \tau \int_{S_1} \int_{S_2} \mathbf{N}(\mathbf{x})\, E(\mathbf{y},\mathbf{x})\, \mathbf{N}^T(\mathbf{y})\, dS(\mathbf{y})\, dS(\mathbf{x}) \,, \\ \mathbf{K}_{21} &= \tau \int_{S_2} \int_{S_1} \mathbf{N}(\mathbf{x})\, E(\mathbf{y},\mathbf{x})\, \mathbf{N}^T(\mathbf{y})\, dS(\mathbf{y})\, dS(\mathbf{x}) \,, \\ \mathbf{K}_{22} &= \int_{S_2} \int_{S_2} \mathbf{N}(\mathbf{x})\, E(\mathbf{y},\mathbf{x})\, \mathbf{N}^T(\mathbf{y})\, dS(\mathbf{y})\, dS(\mathbf{x}) \,, \\ \mathbf{f}_1(\mathbf{x}) &= \int_{S_1} \mathbf{N}(\mathbf{x}) \left[\tau\alpha(\mathbf{x}) v_n^S(\mathbf{x}) - \tau v_n^i(\mathbf{x}) + \int_{S_1} H^T(\mathbf{y},\mathbf{x}) v_n^S(\mathbf{y}) dS(\mathbf{y}) \right] dS(\mathbf{x}) \,, \\ \mathbf{f}_2(\mathbf{x}) &= \int_{S_2} \mathbf{N}(\mathbf{x}) \left[v_n^S(\mathbf{x}) - v_n^i(\mathbf{x}) + \tau \int_{S_1} H^T(\mathbf{y},\mathbf{x})\, v_n^S(\mathbf{y})\, dS(\mathbf{y}) \right] dS(\mathbf{x}) \,. \end{aligned} \tag{16.66}$$

16.4 Programming Aspects

16.4.1 Hyper–singular Integrals

The integrals in (16.63) or (16.66), which contain the kernel $E(\mathbf{y}, \mathbf{x})$ are hyper–singular integrals requiring special attention. In case of 3D problems such integrals are characterized by an integrand of the type r^{-3}. According to [11] they can be transformed into weakly singular integrals by the relation

$$\begin{aligned} \int_S \int_S \Phi(\mathbf{x})\, E(\mathbf{y},\mathbf{x})\, \Phi(\mathbf{y})\, dS(\mathbf{y})\, dS(\mathbf{x}) &= \int_S \int_S \{k^2\, \mathbf{n}(\mathbf{x}) \cdot \mathbf{n}(\mathbf{y})\, \Phi(\mathbf{x})\, \Phi(\mathbf{y}) \\ &- \left[\mathbf{n}(\mathbf{x}) \times \boldsymbol{\nabla}\Phi(\mathbf{x})\right] \cdot \left[\mathbf{n}(\mathbf{y}) \times \boldsymbol{\nabla}\Phi(\mathbf{y})\right] \} \, G(\mathbf{y},\mathbf{x})\, dS(\mathbf{y})\, dS(\mathbf{x}) \,. \end{aligned} \tag{16.67}$$

In Equation (16.67), S is a closed surface and $\boldsymbol{\nabla}\Phi(\mathbf{x}) = \partial\Phi(\mathbf{x})/\partial\mathbf{x}$ and $\boldsymbol{\nabla}\Phi(\mathbf{y}) = \partial\Phi(\mathbf{y})/\partial\mathbf{y}$ are the gradients of $\Phi(\mathbf{x})$ and $\Phi(\mathbf{y})$, respectively. By means of (16.67) the derivatives of the fundamental solution, contained in the kernel $E(\mathbf{y}, \mathbf{x})$, are transferred to the potential functions $\Phi(\mathbf{x})$ and $\Phi(\mathbf{y})$.

The remaining weakly singular integrals, which for 3D problems are characterized by a singular integrand of the type r^{-1}, can be determined by means of a suitable coordinate transformation. It consists of subdivision of the respective boundary element into triangular subelements (Figure 16.3) such that the singular point, which is one of the integration points of the boundary element, becomes a corner point of these subelements. The triangular subelements are then mapped onto a square of unit side length by means of triangle polar coordinates such that the singular corner point corresponds to a edge of the unit square and the integration is performed on the unit square [6].

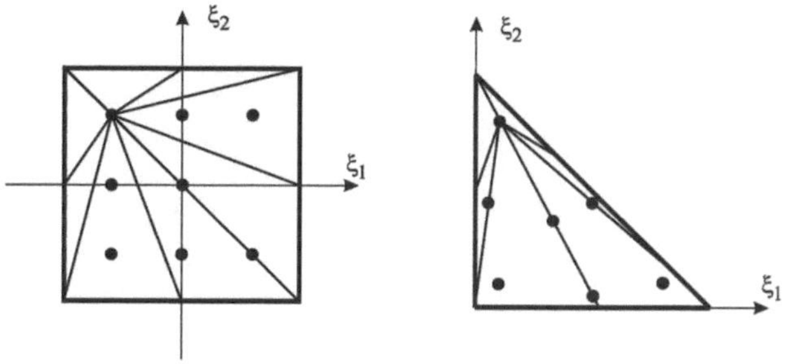

Fig. 16.3 Subdivision of boundary elements into triangular subelements.

16.4.2 Number of Unknowns at Nodal Points

The number of unknowns at a nodal point depends on the location of the nodal point and on the prescribed boundary conditions. With respect to the location of a node one has to distinguish between

(i) nodes on the boundary of the 3D closed part of a structure,
(ii) nodes on the middle surface of the thin–walled part of a structure,
(iii) nodes on a free edge of the thin–walled part of a structure,
(iv) nodes on a line of intersection between the 3D closed part and the thin–walled part of a structure,
(v) nodes located on a line of intersection or on a point of intersection of thin–walled parts of a structure.

The respective number of unknowns is given as follows:

(i) If the velocity on the boundary of the 3D closed part of a structure is known, then the potential Φ of the sound field is unknown;
(ii) If the surfaces of a thin–walled part of a structure are acoustically hard reflecting, then Φ^d is unknown. Otherwise, also Φ^m is unknown;
(iii) On a free edge of a thin–walled part of a structure, there are no unknowns, if both surfaces of the thin–walled part are acoustically hard reflecting. Otherwise, Φ^m is unknown;
(iv) For nodes on a line of intersection between the 3D closed part and a thin–walled part of a structure the number of unknowns is equal to the number m of subdomains of the fluid, which are generated by the intersecting parts of the structure, since for each of these subdomains the potential Φ of the sound field is unknown;
(v) For nodes on a line of intersection or on a point of intersection of acoustically hard reflecting thin–walled parts of a structure the number of unknowns is equal to $m - 1$ with m denoting the number of subdomains of the fluid, generated by the intersecting thin–walled parts of a structure. In the example shown in Figure 16.4 nodes between points A and B are characterized by four unknown values of Φ^d, which follows from the five subdomains of the fluid, generated by the intersecting thin–walled parts. Because of $\sum_{i=1}^{m} \Phi_i^d = 0$ there are only $m-1$ unknowns. For node A, because of $m = 8$, the number of unknowns is equal to

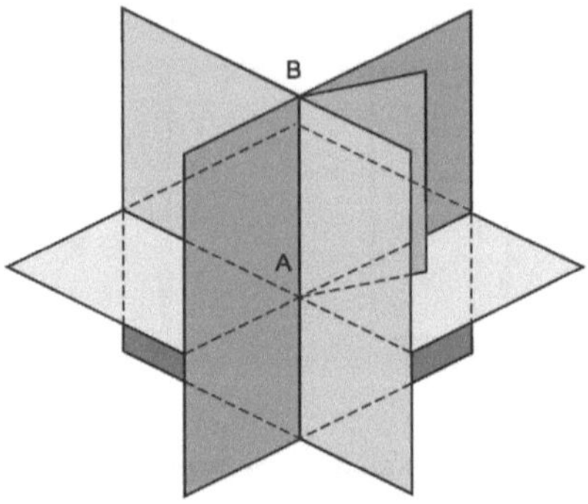

Fig. 16.4 Intersecting thin–walled parts of a structure.

seven. If, however, an admittance boundary condition is prescribed for a surface of any of these thin–walled parts, then the number of unknowns is equal to m, i.e., for each of the subdomains Φ is unknown. In the BE–program the number of unknowns at nodal points on lines or points of intersection of thin–walled parts of a structure is set equal to m. Hence, if no admittance boundary conditions are prescribed, then an additional equation is required at such nodes. This equation is obtained by replacing in (16.34) $\Phi^m/2 = \Phi^+/2 + \Phi^-/2$ by $\sum_{i=1}^m \alpha_i(\mathbf{x})\,\Phi_i(\mathbf{x})$ as

$$\begin{aligned}
&\sum_{i=1}^{m} \alpha_i(\mathbf{x})\,\Phi_i(\mathbf{x}) - \int_{S_1} \Big[G(\mathbf{y},\mathbf{x})\,A(\mathbf{y}) + \tau H(\mathbf{y},\mathbf{x})\Big]\,\Phi(\mathbf{y})\,dS(\mathbf{y}) \\
&- \int_{S_2} \Big\{\Big[G(\mathbf{y},\mathbf{x})\,A^d(\mathbf{y}) + H(\mathbf{y},\mathbf{x})\Big]\,\Delta\Phi(\mathbf{y}) + G(\mathbf{y},\mathbf{x})\,A^m(\mathbf{y})\,\Phi^m(\mathbf{y})\Big\}dS(\mathbf{y}) \\
&= \Phi^i(\mathbf{x}) - \tau \int_{S_1} G(\mathbf{y},\mathbf{x})\,v_n^S(\mathbf{y})\,dS(\mathbf{y})\,, \quad \mathbf{x} \in S_2\,,
\end{aligned} \tag{16.68}$$

where $\alpha_i(\mathbf{x})$ represents the ratio of the part of the surface of a sphere of infinitesimal radius with the center at $\mathbf{x}$ within the respective subdomain i over the total surface of this sphere. The determination of $\alpha_i(\mathbf{x})$ is described in [1].

16.4.3 Treatment of Irregular Frequencies

For exterior acoustic problems, which are characterized by the radiation of sound of a 3D closed body into the surrounding exterior fluid domain, the coefficient matrix of the system of algebraic equations becomes singular at certain frequencies, which are known as irregular frequencies. For the present formulation this drawback can be overcome by two different methods, which depend on the chosen type of elements.

In case of a 3D closed body, discretized by S_1–elements, which radiates sound into the exterior fluid domain, only this domain is considered in the mathematical model. Hence, additional points can be chosen in the interior of the body, at which the potential of the sound field is required to be zero. This method, which originally was proposed in [7] and is known as CHIEF–method, results in an over–determined system of algebraic equations. However, if these additional points coincidentally are located on nodal surfaces of the eigenforms for the potential of the sound field of the related interior problem, then this methods fails. Hence, the CHIEF–method was improved by forcing also the derivatives of the sound potential with respect to the directions of the coordinate axes to be zero at these additional points in the interior domain [8, 12].

In case of a sound radiating 3D closed body, discretized by S_2–elements, the body is viewed as a thin–walled closed body considering both the interior and the exterior fluid domain in the mathematical model. Hence, the irregular frequencies are the resonance frequencies of the interior domain, enclosed by the thin–walled body. The singularity of the coefficient matrix at the resonance frequencies can be avoided by prescribing suitable admittance boundary conditions on the interior surface, resulting in complex eigenfrequencies of the interior domain. This method is in line with the method proposed in [9] to prevent the mentioned singularities by using a linear combination of the Helmholtz integral equation with its normal derivative, multiplied by a complex number. The integrals over S_2 in (16.38) and (16.39) are linear combinations of different integral operators with A^m and A^d playing the role of multipliers. Because of $A = \mathrm{i}\omega\rho a$ according to (16.30), the prescribed acoustic admittance a should contain a real part.

16.5 Numerical Study

The numerical study consists of the numerical simulation of a popular benchmark example for acoustic radiation, which allows a comparison between the numerical results and analytical solutions, and of the numerical simulation of an experiment for acoustic radiation, which allows a comparison of the numerical results with measurement data. The examples demonstrate the versatility of the proposed Galerkin–type boundary element formulation for acoustic radiation and scattering.

For the numerical examples, the speed of sound and the density of the fluid are assumed as $c = 340$ m/s and $\rho = 1.225$ kg/m^3.

16.5.1 Sound Radiation of a Pulsating Sphere

The numerical simulation of the sound radiation of a pulsating sphere is a popular academic benchmark example for acoustic codes. Within the context of the proposed Galerkin–type boundary element formulation for acoustic radiation and scattering two different numerical models of the pulsating sphere are available. The pulsating sphere can be viewed as a 3D closed body, which radiates sound either in the interior spherical domain or in the unbounded exterior domain. Alternatively, the pulsating sphere can be viewed as a thin–walled spherical shell, radiating sound simultaneously in the interior spherical domain and in the unbounded exterior domain. In the

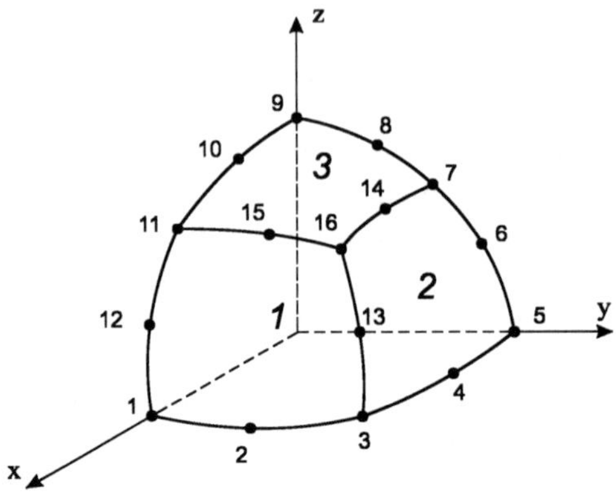

Fig. 16.5 Boundary element mesh for one octant of the sphere.

Table 16.1 Sound pressure on the surface of a pulsating spherical cavity for different values of prescribed acoustic admittance.

admittance a [m/(Pa·s)]	$0.001 + 0.001\mathrm{i}$	$0.01 + 0.01\mathrm{i}$	$0.1 + 0.1\mathrm{i}$
p [Pa] analytical solution	$-113.9 + 317.7\mathrm{i}$	$-41.84 + 49.33\mathrm{i}$	$-4.911 + 4.999\mathrm{i}$
p [Pa] numerical solution	$-116.6 + 318.4\mathrm{i}$	$-43.02 + 49.69\mathrm{i}$	$-4.972 + 5.018\mathrm{i}$
error [%]	0.5	1.6	0.8

former case the surface of the sphere is discretized by S_1–elements and the distinction between an interior and exterior acoustic problem only depends on the choice of τ according to (16.20). In the latter case the middle surface of the spherical shell is discretized by S_2–elements.

Because of symmetry only one octant of the sphere needs to be discretized. Both types of discretization consist of three isoparametric boundary elements with eight nodes each, see Figure 16.5.

The analytical solution for the sound pressure in a spherical cavity with radius r_a and prescribed acoustic admittance a on its surface, which is pulsating with an amplitude of the normal velocity v_n at the wave number k, is given as

$$p(r) = \frac{-\mathrm{i}\rho\omega k\, r_a^2\, v_n}{\sin(kr_a) - kr_a\cos(kr_a) + \mathrm{i}\rho\omega a\, r_a \sin(kr_a)}\,\frac{\sin(kr)}{kr}\,, \qquad r < r_a\,, \quad (16.69)$$

where r is the distance between the respective point and the center of the sphere. Table 16.1 contains a comparison of numerical results for $k = 1.25\pi$, $v_n = 1$ m/s, $r_a = 1$ m and different values of prescribed acoustic admittance a on the surface of the spherical cavity, which is discretized by S_1–elements, with the respective analytical solutions.

The analytical solution for the sound pressure in the unbounded exterior domain of a sphere with radius r_a and prescribed acoustic admittance a, which is pulsating with an amplitude of the normal velocity of v_n at the wave number k, is given as

Table 16.2 Sound pressure on the surface of a pulsating sphere for different values of prescribed acoustic admittance.

admittance a [m/(Pa·s)]	$0.001 + 0.001\mathrm{i}$	$0.01 + 0.01\mathrm{i}$	$0.1 + 0.1\mathrm{i}$
p [Pa] analytical solution	$231.7 - 120.2\mathrm{i}$	$45.99 - 39.92\mathrm{i}$	$4.69 - 4.881\mathrm{i}$
p [Pa] numerical solution	$235.3 - 116.3\mathrm{i}$	$46.68 - 41.00\mathrm{i}$	$5.031 - 4.924\mathrm{i}$
error [%]	0.6	1.6	4.0

Table 16.3 Difference and sum of the sound pressure on the outer and inner surface of the pulsating spherical shell for different values of prescribed acoustic admittance on the outer surface.

admittance a^+ [m/(Pa·s)]	$0.001 + 0.001\mathrm{i}$	$0.01 + 0.01\mathrm{i}$	$0.1 + 0.1\mathrm{i}$
$p^+ - p^-$ [Pa] analytical solution	$240.1 - 672.6\mathrm{i}$	$46.55 - 598.6\mathrm{i}$	$4.969 - 563.7\mathrm{i}$
$p^+ - p^-$ [Pa] numerical solution	$245.7 - 674.8\mathrm{i}$	$46.42 - 608.4\mathrm{i}$	$3.267 - 575.1\mathrm{i}$
error [%]	0.6	1.6	2.0
$p^+ + p^-$ [Pa] analytical solution	$240.1 + 445.0\mathrm{i}$	$46.55 + 518.9\mathrm{i}$	$4.969 + 553.9\mathrm{i}$
$p^+ + p^-$ [Pa] numerical solution	$234.5 + 457.3\mathrm{i}$	$46.20 - 531.4\mathrm{i}$	$6.579 - 565.6\mathrm{i}$
error [%]	1.6	2.4	2.1

$$p(r) = \frac{\mathrm{i}\rho\omega\, r_a^2\, v_n \mathrm{e}^{\mathrm{i}k(r-r_a)}}{r(\mathrm{i}\omega\rho a\, r_a + \mathrm{i}k\, r_a - 1)}\,, \qquad r > r_a\,. \tag{16.70}$$

Table 16.2 contains a comparison of numerical results for a pulsating sphere with $k = \pi$, $v_n = 1$ m/s, $r_a = 1$ m and different values of prescribed acoustic admittance a on the outer surface, discretized by S_1–elements, with the respective analytical solutions.

Alternatively, the pulsating sphere can be viewed as a thin–walled spherical shell, the middle surface of which is discretized by S_2–elements. Table 16.3 contains a comparison of numerical results for $k = 1.25\pi$, $v_n = 1$ m/s, $r_a = 1$ m and different values of prescribed acoustic admittance a^+ on the outer surface of the spherical shell. The inner surface is assumed as a hard reflecting surface, i.e. $a^- = 0$.

16.5.2 Sound Radiation of a Vibrating Box with a Rib

The sound radiation of a vibrating steel box with a rib welded on the top surface was measured by the Austrian company AVL in order to obtain measurement data from a simple structure for the validation of numerical results. The dimensions of the steel box and the rib and the distance from the bottom surface of the steel box to the floor are shown in Figure 16.6.

All surfaces of the box, the rib and the floor can be assumed as acoustically hard reflecting surfaces. The top surface and the rib were excited to vibrations at selected frequencies by means of an electrodynamic shaker, whereas all other surfaces of the

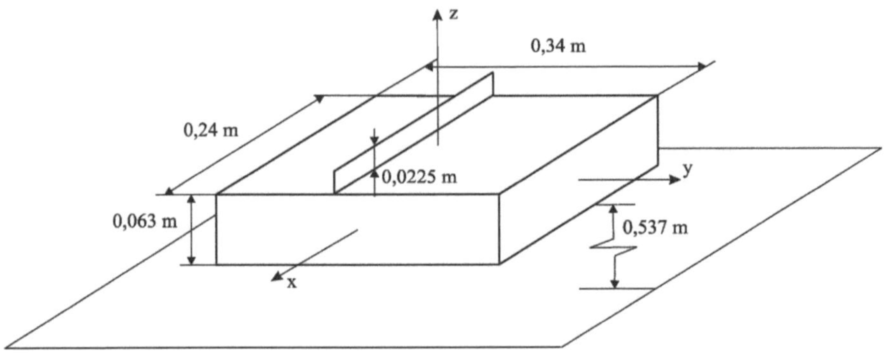

Fig. 16.6 Steel box with a rib welded on the top surface.

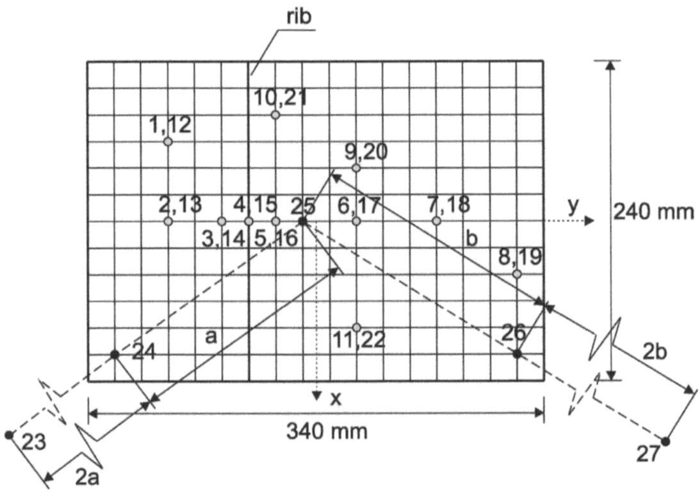

Fig. 16.7 Measurement points for the air–borne noise.

box were rigid surfaces. The amplitude and phase of the generated velocity on the top surface and the rib were measured at the points of a square grid of 20 × 20 mm. The sound pressure of the air–borne noise resulting from the excited vibrations of the top surface and the rib of the steel box was measured at 27 points, located at different distances from the steel box. The plan view of the measurement points is shown in Figure 16.7.

For measurement points 1 to 3 and 5 to 11 the normal distance from the top surface of the steel box was 30 mm and for measurement point 4, located above the rib, it was 55 mm. For measurement points 12 to 22 the normal distance from the top surface of the steel box was 100 mm and for measurement points 23 to 27 it was 1000 mm.

For the discretization with boundary elements the steel box can be viewed either as a 3D closed body or as a thin–walled structure. Hence, it can be discretized either

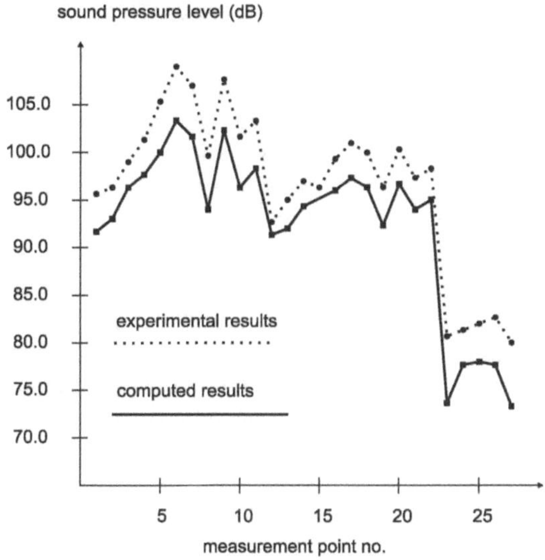

Fig. 16.8 Comparison of measured and computed values of the sound pressure level.

by S_1–elements or by S_2–elements. In the former case, a single CHIEF point, located at the center of the box, is used to avoid a singular coefficient matrix at irregular frequencies. In the latter case, an acoustic admittance boundary condition of a = 0.01 m/(Pa·s) is prescribed on the inner surface of the box in order to avoid the possibility of acoustic resonance in the box. The rib is discretized by S_2 elements. All boundary elements are elements with bilinear shape functions with 4 nodes each. The mesh width of the top and bottom surface is identical with the measurement grid of 20 × 20 mm, the element sizes for the rib are 20 × 22.5 mm and 20 × 21 mm for the lateral surfaces. Hence, for each node on the top surface and on the rib the measured normal velocity can be directly prescribed as a boundary condition.

Figure 16.8 contains a comparison of the numerical and the measured results for the sound pressure level of the air–borne noise at the 27 measurement points, resulting from structural vibrations at 682 Hz. Apart from a systematic difference between measured and computed sound pressure levels of about 3 dB, the computed sound pressure levels agree well with the measured ones. The mentioned systematic difference is the consequence of an identified difference between the measured velocity of the top surface of the steel box and the measured particle velocity of the adjacent fluid at the points on the measurement grid at the top surface of the steel box.

16.6 Conclusion

The symmetric Galerkin formulation of the Boundary Element Method, presented in this chapter, allows the numerical simulation of acoustic radiation and scattering of three–dimensional (3D) closed structures, 3D thin–walled closed structures, 3D thin–walled open structures, and of structures, consisting of combinations of a 3D closed part with thin–walled open parts. In addition, different fluids may be present in the interior and exterior domain of a 3D closed structure with attached thin–walled open parts and different acoustic admittances can be prescribed on the two surfaces of thin–walled parts of a structure and a velocity boundary condition as well as an acoustic admittance boundary condition can be prescribed simultaneously.

Since the Galerkin method requires two integrations over the boundary, the computational effort for generating the algebraic system of equations is higher than for the collocation method. However, starting from a self–adjoint system of integral equations, the Galerkin method results in a system of algebraic equations for the unknown nodal values of the sound field with a symmetric coefficient matrix. This feature is advantageous especially for boundary element meshes with a large number of degrees of freedom and for coupled elasto–acoustic problems [2].

References

1. Chen ZS, Hofstetter G, Mang HA (1997) A symmetric Galerkin formulation of the boundary element method for acoustic radiation and scattering. Journal of Computational Acoustics 5:219–241
2. Chen ZS, Hofstetter G, Mang HA (1998) A Galerkin–type BE–FE formulation for elasto–acoustic coupling. Computer Methods in Applied Mechanics and Engineering 152:147–155
3. Coyette JP, Fyfe KR (1989) Solution of elasto–acoustic problems using a variational finite/boundary element technique. In: Bernhard RJ, Keltie RF (eds) Numerical Techniques in Acoustic Radiation, ASMENCA – American Society of Mechanical Engineers, Noise Control and Acoustics Division 6:15–25
4. Freund DE, Farrell R (1990) A variational principle for the scattered wave. Journal of the Acoustical Society of America 87:1847–1860
5. Hamdi MA (1981) Une formulation variationelle par équations intégrales pour la résolution de l' équation de Helmholtz aves des conditions aux limites mixtes. Comptes Rendus l'Académie de Sciences, Paris, Séries II, 292:17–20.
6. Li HB, Han GM, Mang HA (1985) A new method for evaluating singular integrals in stress analysis of solids by the direct boundary element method. International Journal for Numerical Methods in Engineering 21:2071–2098
7. Schenk HA (1968) Improved integral formulation for acoustic radiation problems. Journal of the Acoustical Society of America 44:41–58
8. Segalman DJ, Lobitz DW (1992) A method to overcome computational difficulties in the exterior acoustics problem. Journal of the Acoustical Society of America 91:1855–1861
9. Seybert AF, Soenarko B, Rizzo FJ, Shippy DJ (1985) An advanced computational method for radiation and scattering of acoustic waves in three dimensions. Journal of the Acoustical Society of America 77:362–368
10. Skudrzyk E (1971) The foundation of acoustics. Springer–Verlag, New York
11. SYSNOISE Rev. 5.0 (1993) Theoretical manual. Numerical Integration Technologies N.V. B–3001 Leuven Belgium
12. Wu TW, Seybert AF (1991) A weighted residual formulation for the CHIEF method in acoustics. Journal of the Acoustical Society of America 90:1608–1614

17

Acoustical Radiation and Scattering above an Impedance Plane

Martin Ochmann and Haike Brick

TFH Berlin – University of Applied Sciences, Department of Mathematics, Physics and Chemistry, Luxemburger Str. 10, 13353 Berlin, Germany;
ochmann@tfh-berlin.de, brick@tfh-berlin.de

Summary. Half–space acoustical problems are of great importance, since sound fields caused by radiation or scattering from complex structures are seldom found in the boundless infinite three–dimensional space. In many cases, a flat ground with a certain surface impedance confines the acoustical space as it encounters for example when treating problems of outdoor sound propagation. Whereas the boundary element method (BEM) is a mighty tool for predicting sound fields in unlimited space, the occurrence of an infinite plane and the associated additional discretization effort weakens the advantages of the classical BEM. One possible remedy is to use a Green's function as a building block for the BEM which automatically satisfies the boundary condition at the infinite plane and to incorporate it into the BEM. In the present article, such a half–space Green's function will be constructed, inserted into a direct BEM formulation, and positively tested by solving numerically several different problems of acoustical sound propagation over impedance planes.

17.1 Introduction

The purpose of the present work is to calculate the sound pressure distribution caused by a one or more finite structures, which are located over an infinite plane. The structure is vibrating and radiates sound or it is hit by an incident wave and scatters sound waves into the surrounding half–space. The infinite plane is characterized by a normal surface impedance, and hence it is considered as locally reacting. Such an exterior half–space problem is well suited for the application of a boundary element method (BEM), since the method only requires solving the central integral equation at the boundaries of the problem, i. e. at the boundary of the structure and at the reflecting plane. However, the plane itself is infinite, and therefore computation time can be reduced drastically by using a Green's function as a cornerstone of the BEM, which takes into account the influence of the infinite plane automatically, so that it is not necessary to model and discretize the infinite plane.

Such a half–space Green's function can be constructed by a superposition of elementary acoustical point sources which was shown by Sommerfeld [44, Ch 3] for a related problem of heat conduction. Unfortunately, this formulation is only valid for masslike impedances. The important case of a plane with a springlike normal surface impedance leads to a diverging line integral of point sources. For obtaining a half–space Green's function which is valid for all kind of surface impedances, the source locations of the point sources in the line integral has to be shifted to the complex domain. Based on this "complex–source–point–Green's function", a direct "complex–source–point–BEM" (acronym: CBEM) was formulated, developed and implemented by using Matlab. Several numerical studies are presented and show the effectiveness and reliability of the CBEM.

The present work is organized as follows: First the half–space problem will be defined and the corresponding integral equations will be formulated. Then, the mirror source method of Sommerfeld will be represented and applied to problems with masslike impedances. If the impedance of the plane becomes springlike, the method of Sommerfeld leads to the appearance of an additional Hankel function, which destroys the validity of the method for perpendicular incidence. Therefore, the half–space Green's function will be constructed by a superposition of sources with complex source points. The properties of such "complex–source–point–solutions" will be considered briefly.

In the numerical part, a corresponding Matlab–based BEM–code is presented. The implementation of the solution for a masslike impedance plane as well as of the CBEM is described. Furthermore, the convergence behavior of the line integral in the half–space Green's function, the compilation of the different normal derivatives, and methods of integration for the infinite integral – the Gauss–Laguerre quadrature and an adaptive multigrid quadrature – are discussed with respect to the CBEM formulation. Finally, several numerical test cases are investigated including the solution for perpendicular incidence. All these calculations show an excellent agreement with corresponding reference solutions.

17.2 Theory

17.2.1 Definition of the Half–space Problem

We consider the following half–space problem: The three–dimensional space is separated by an infinite plane S_p into two half–spaces. For simplicity, we assume that the plane S_p is located at $z = 0$. Hence, as shown in Figure 17.1 the plane separates the whole space into the upper half–space $V_+ := \{(x, y, z)|z > 0\}$ and the lower half–space $V_- := \{(x, y, z)|z < 0\}$. The plane S_p should be characterized by a normal impedance

$$\frac{p}{v_n} = Z \qquad \text{on } S_p \,. \tag{17.1}$$

Here, p is the complex sound pressure, v_n the normal velocity, and Z the normal impedance, where the unit normal $\boldsymbol{n} = (0,\ 0,\ -1)$ at the plane is directed into the exterior of the acoustical domain V_+. Therefore, in this work, we only consider the sound propagation above a flat ground with a locally reacting boundary condition. Rough surfaces or boundaries with bulk reacting planes will be not investigated, and sound waves are not transmitted into the non–physical lower half–space V_-. The

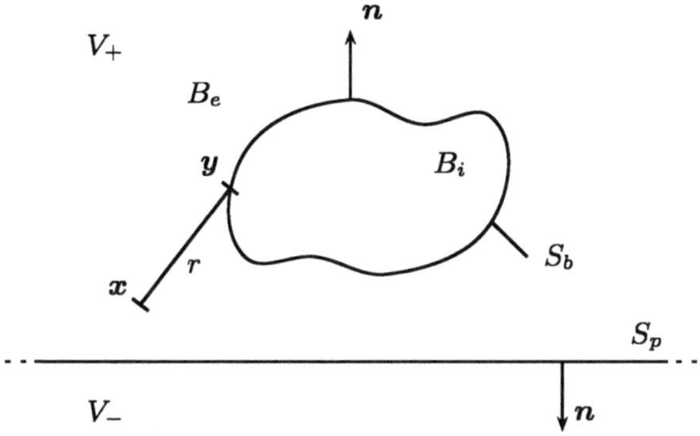

Fig. 17.1 Geometry of the radiation and scattering problem for a half–space above a flat impedance plane.

transmission problem, where both half–spaces V_+ and V_- have a physical meaning, is described in [28, Ch O.3.5].

As shown in Figure 17.1, a bounded structure is located in the physical half–space V_+. The structure is vibrating and radiates sound into the surrounding half–space (radiation problem) or it is hit by an incident wave and scatters sound waves into the upper half–space (scattering problem). The radiated or scattered sound field is characterized by the sound pressure p, sound velocity $\boldsymbol{v}$, and derived quantities such as the sound intensity $\boldsymbol{I}$, the radiated sound power P, the radiated efficiency σ etc., which shall be calculated by the boundary element method (BEM). The bounded volume of the radiating or scattering structure in three–dimensional half–space is denoted by B (like Body) with $B = B_i \cup S_b$. As presented in Figure 17.1, the interior of B is called B_i, the exterior B_e, and the structural surface S_b. The surface normal $\boldsymbol{n}$ of the body should be directed into the exterior B_e.

Again, the boundary condition on the surface of the body is restricted to the locally reacting case. Transmission of sound into the body or fluid–structure interaction are excluded.

17.2.2 The Boundary Integral Equation for the Half–space Problem

The most frequently used integral equation formulation in acoustics is the well–known Helmholtz integral equation (abbreviation HIE) for exterior field problems. The HIE is obtained by applying Green's second theorem to the Helmholtz equation, see for example [12, 45]. Depending on the location of the field point $\boldsymbol{x}$, the HIE takes the form

$$\iint_S \left[p(\boldsymbol{y}) \frac{\partial g(\boldsymbol{x}, \boldsymbol{y})}{\partial n(\boldsymbol{y})} - \frac{\partial p(\boldsymbol{y})}{\partial n(\boldsymbol{y})} g(\boldsymbol{x}, \boldsymbol{y}) \right] \mathrm{d}s(\boldsymbol{y}) = C(\boldsymbol{x}) \tag{17.2}$$

where

$$C(\boldsymbol{x}) = \begin{cases} p(\boldsymbol{x}), & \boldsymbol{x} \in B_e \\ \frac{1}{2} p(\boldsymbol{x}), & \boldsymbol{x} \in S = S_b \cup S_p \\ 0, & \boldsymbol{x} \in B_i \end{cases} \tag{17.3}$$

and

$$g(\boldsymbol{x}, \boldsymbol{y}) = \frac{1}{4\pi r} \mathrm{e}^{ikr} \quad \text{with } r = r(\boldsymbol{x}, \boldsymbol{y}) = \|\boldsymbol{r}\|,\ \boldsymbol{r} = \boldsymbol{x} - \boldsymbol{y} \tag{17.4}$$

is the free–space Green's function. $\boldsymbol{y} = (x_s, y_s, z_s)$ is a spatial point on the structural surface S_b or on the plane S_p. $k = \omega/c$ is the wave number with angular frequency ω and c is the speed of sound. All time–varying quantities should obey the time dependence $\mathrm{e}^{-\mathrm{i}\omega t}$ with $\mathrm{i} = \sqrt{-1}$. The geometrical notations are chosen as shown in Figure 17.1. Depending on the location of the point $\boldsymbol{x}$ in Equation (17.3), Equation (17.2) is called exterior HIE, surface HIE, or interior HIE, respectively. The Helmholtz formula (17.2) is valid if the surface S_p is assumed to be closed and both boundaries S_p and S_b are sufficiently smooth, i.e. there is a unique tangent to $S = S_p \cup S_b$ at every $\boldsymbol{x} \in S$. For the general case, where no unique tangent plane exists at $\boldsymbol{x} \in S$ (for example, when $\boldsymbol{x}$ is lying on a corner or an edge), the surface HIE has to be modified slightly [40]. Integral equation (17.2) is the basis equation for the direct BEM, see for example [12, 45], which solves the discretized version of the integral equation (17.2).

Since this formulation uses the free–space Green's function, it is necessary to integrate not only over the structural surface, but also over the infinite impedance plane. This procedure has serious drawbacks: First, one has to replace the surface integration about the infinite plane by a finite integration. Thus, depending on the parameters of the concrete problem, the corresponding truncation error must be estimated, which seems to be not an easy task. Secondly, the direct boundary element approach based on the HIE (17.2) can only handle boundaries S consisting of a finite number of disjoint, closed, bounded surfaces, cf. [5, p 68ff]. Hence, in order to avoid the so-called thin-body problem, cf. [45], we have to substitute the infinite plane by a finite plate containing a non-zero volume, which increases the number of boundary elements for the discretization of the finite plane by more than a factor of two. Thus, the number of elements necessary to discretize the boundary integral equation may become huge, especially if problems in the high–frequency range are investigated. An example should illuminate the situation. If high frequency scattering at 11333 Hz from a cube with a lateral surface of 0.24 m× 0.24 m should be treated, we need about 14000 constant quadrilateral elements for modeling the cube, in order to ensure six points per wavelength (the wavelength λ is 0.03 m in air with $c = 340$ m/s corresponding to eight wavelength or 48 elements per side length of the cube) at 11333 Hz. However, if a square section of the plane of 2.4 m× 2.4 m is taken into account for the integration and enlarged to a plate with a small finite thickness, we need about half a million elements only for modeling the plate! For this reason, we strike another path by using a Green's function which automatically satisfies the impedance boundary condition (17.1) on the infinite plane. If such a Green's function is known, it can be incorporated into the HIE (17.2), and consequently, there is no need to model the infinite plane leading to an enormous reduction of the size of the boundary element model and corresponding computer time for solving the system of equations.

17.2.3 Definition of the Half–space Green's Function

For simplicity, we consider in the following a point source in the frequency domain at the source location $\boldsymbol{y} = (x_s, y_s, z_s) = (0, 0, h)$ in the three–dimensional space, given by

$$g(\boldsymbol{x}, \boldsymbol{y}) = \frac{\mathrm{e}^{\mathrm{i}kr}}{(4\pi r)}, \qquad r = \sqrt{x^2 + y^2 + (z-h)^2}, \tag{17.5}$$

where $\boldsymbol{x} = (x, y, z)$ is the receiver position. Clearly, $g(\boldsymbol{x}, \boldsymbol{y})$ solves the inhomogeneous Helmholtz equation

$$\Delta g + k^2 g = -\delta(x, y, z - h) \tag{17.6}$$

where Δ is the Laplacian, and δ is the Dirac delta function. Now, we are trying to modify Equation (17.5) in such a way, that the resulting Green's function will also satisfy the impedance boundary condition (17.1) at the infinite plane, which can be written as

$$\frac{\partial p}{\partial n} - \gamma p = 0 \quad \text{on } S_p, \tag{17.7}$$

since the gradient of the pressure is given by

$$\frac{\partial p}{\partial n} = \mathrm{i}\omega \rho_0 v_n \quad \text{on } S_p. \tag{17.8}$$

Here, ρ_0 is the fluid density, $\partial/\partial n$ is the derivative in the direction of the normal $\boldsymbol{n}$, and

$$\gamma = \frac{\mathrm{i}k}{Z_0} \tag{17.9}$$

with the normalized acoustical impedance $Z_0 = Z/(\rho_0 c)$. Now, we are looking for a solution $G(\boldsymbol{x}, \boldsymbol{y})$ of the Helmholtz equation (17.6) satisfying the boundary condition (17.7) in the half–space V_+ and the radiation condition at infinity, which is also called a half–space Green's function.

17.2.4 The Half–space Green's Function for a Rigid or a Soft Plane

For two special cases the half–space Green's function is well–known: First, for an infinite normal impedance $Z = \infty$, i.e. for $\gamma = 0$, cf. Equation (17.9), the plane S_p becomes perfectly rigid, and by a superposition of the original source (17.5) and the corresponding mirror source at the at the source location $\boldsymbol{y} = (x_s, y_s, z_s) = (0, 0, -h)$, we obtain the half–space Green's function

$$G(\boldsymbol{x}, \boldsymbol{y}) = \frac{\mathrm{e}^{\mathrm{i}kr(h)}}{4\pi r(h)} + \frac{\mathrm{e}^{\mathrm{i}kr(-h)}}{4\pi r(-h)}, \tag{17.10}$$

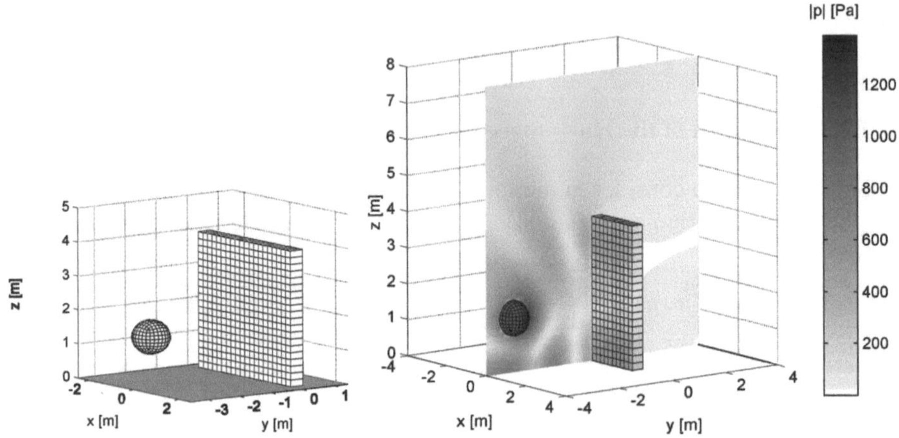

Fig. 17.2 Left: Model geometry of a pulsating sphere in front of a acoustic barrier above an infinite, rigid plane. , Right: Sound pressure distribution on a visualization plane at 136 Hz.

where we have used the abbreviation

$$r(h) = \sqrt{x^2 + y^2 + (z-h)^2}\ . \tag{17.11}$$

Secondly, the perfectly soft condition $p = 0$ on the plane S_p corresponding to $Z = 0$ can be treated by subtracting the mirror source at $\boldsymbol{y} = (x_s, y_s, z_s) = (0, 0, -h)$ leading to the exact half–space Green's function

$$G(\boldsymbol{x}, \boldsymbol{y}) = \frac{\mathrm{e}^{\mathrm{i}kr(h)}}{4\pi r(h)} - \frac{\mathrm{e}^{\mathrm{i}kr(-h)}}{4\pi r(-h)}\ , \tag{17.12}$$

for the pressure–released case.

Seybert and Soenarko have applied these Green's functions to radiation and scattering problems [39]. Both cases are often implemented in commercial boundary element codes, e.g. in SYSNOISE or VIRTUAL LAB.

Figure 17.2 shows an example configuration, where the sound field of a pulsating sphere of radius $a = 0.5$ m is reflected by a rigid plane and scattered by a rigid acoustic barrier. On the pulsating sphere a uniform surface velocity of $v_0 = 1.4 + \mathrm{i}1.9$ m/s is prescribed. The resulting sound field at 136 Hz is imaged on the visualization plane in $x = 0$. The surface pressure as well as the field point pressure are calculated by our own BEM–code BEMLAB.

17.2.5 The Half–space Green's Function for a Masslike Impedance Plane

The Mirror Source Method of Sommerfeld

Arnold Sommerfeld treated half–space problems of heat conduction by writing the total thermal field as a superposition of an original heat source, a mirror source, and

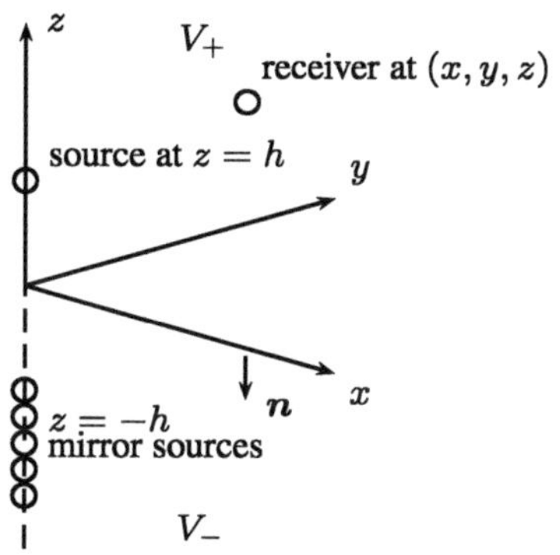

Fig. 17.3 Geometry and source distribution for the half–space Green's function $G(\boldsymbol{x}, \boldsymbol{y})$.

a line integral combined of single thermal sources placed at the z–axis below the mirror source as shown in Figure 17.3, cf. [43, Ch 3, § 13.9].

The same approach for the corresponding acoustical problem was used in [32] by attempting to find a solution of Equations (17.6) and (17.7) in the form

$$G(\boldsymbol{x}, \boldsymbol{y}) = \hat{g}(h) + A\,\hat{g}(-h) + \int_{-\infty}^{-h} a(\eta)\,\hat{g}(\eta)\,\mathrm{d}\eta \tag{17.13}$$

with the unknown coefficient A and the unknown amplitude function $a(\eta)$ of the mirror sources along the z–axis. For abbreviation, the free–space Green's function (17.5) is written as

$$\hat{g}(h) = g\left[\begin{pmatrix} x \\ y \\ z \end{pmatrix}, \begin{pmatrix} 0 \\ 0 \\ h \end{pmatrix}\right] = \frac{1}{4\pi} \frac{\mathrm{e}^{\mathrm{i}k\sqrt{x^2+y^2+(z-h)^2}}}{\sqrt{x^2+y^2+(z-h)^2}} \,. \tag{17.14}$$

Thus, $\hat{g}(-h)$ is the mirror source of $\hat{g}(h)$ with respect to the plane S_p. The second and third term on the right hand side of Equation (17.13) represent the field reflected from the plane S_p. Clearly, $G(\boldsymbol{x}, \boldsymbol{y})$ is a solution of the Helmholtz equation (17.6) with only one singularity at the source point $(0, 0, h)$. All other singularities at points $(0, 0, z)$ with $z \leq -h$ are located in the non–physical lower half–space V_-. $G(\boldsymbol{x}, \boldsymbol{y})$ can be interpreted as a representation of the half–space Green's function by a superposition of equivalent point sources.

By applying the boundary condition (17.7) on the plane S_p to formula (17.13), the coefficients A and $a(\eta)$ are obtained after some algebra

$$A = 1 \qquad \text{and} \qquad a(\eta) = 2\gamma \mathrm{e}^{-\gamma(\eta+h)} . \tag{17.15}$$

The reader interested in the details can find all steps of the calculation in [32], where it was also shown that the derivative of (17.15) is only valid, if

$$\left[\frac{a(\eta)}{r_\eta}\,\mathrm{e}^{\mathrm{i}kr_\eta}\right] \xrightarrow[\eta\to-\infty]{} 0 \tag{17.16}$$

with

$$r_\eta^2 = x^2 + y^2 + \eta^2 \; . \tag{17.17}$$

Hence, by substituting Equation (17.15) into Equation (17.13), the result for the half–space Green's function can be expressed as

$$G(\boldsymbol{x}, \boldsymbol{y}) = \hat{g}(h) + \hat{g}(-h) + 2\gamma \mathrm{e}^{-\gamma h} \int\limits_{-\infty}^{-h} \mathrm{e}^{-\gamma\eta}\, \hat{g}(\eta) \mathrm{d}\eta \tag{17.18}$$

or in a more explicit form

$$G(\boldsymbol{x}, \boldsymbol{y}) = \frac{\mathrm{e}^{\mathrm{i}kr(h)}}{4\pi r(h)} + \frac{\mathrm{e}^{\mathrm{i}kr(-h)}}{4\pi r(-h)} + 2\gamma \mathrm{e}^{-\gamma h} \int\limits_{-\infty}^{-h} \mathrm{e}^{-\gamma\eta}\, \frac{\mathrm{e}^{\mathrm{i}kr(\eta)}}{4\pi r(\eta)} \mathrm{d}\eta \tag{17.19}$$

with $r(h) = \sqrt{x^2 + x^2 + (z-h)^2}$.

If the time dependence $\mathrm{e}^{\mathrm{i}\omega t}$ is used, the half–space Green's function is simply given by the complex conjugate $G^*(\boldsymbol{x}, \boldsymbol{y})$ of Equation (17.18) or (17.19), where it should be noted that

$$\gamma^* \;=\; -\frac{\mathrm{i}k}{Z_0^*} \; . \tag{17.20}$$

Clearly, the complex conjugate impedance Z_0^* corresponds to the normalized impedance of the problem formulated with respect to the time dependence $\mathrm{e}^{\mathrm{i}\omega t}$.

For $\gamma = 0$ (see Equation (17.7)), the plane S_p becomes perfectly rigid, and Equation (17.19) gives the expected solution Equation (17.10). However, the perfectly soft condition on S_p must be treated separately by reformulating the boundary condition (17.7) as a general Robin boundary condition

$$\alpha \frac{\partial p}{\partial n} + \beta p = 0 \quad \text{on } S_p \tag{17.21}$$

and letting $\alpha \to 0$. Then, by substituting ansatz Equation (17.13) into this perfectly soft condition, one obtains

$$G(\boldsymbol{x}, \boldsymbol{y})|_{z=0} = (A+1)\frac{\mathrm{e}^{\mathrm{i}kr_h}}{4\pi r_h} + \int\limits_{-\infty}^{-h} a(\eta)\, \frac{\mathrm{e}^{\mathrm{i}kr_\eta}}{4\pi r_\eta} \mathrm{d}\eta = 0, \tag{17.22}$$

which is now satisfied for $A = -1$ and $a(\eta) = 0$ and leads to the exact half–space Green's function (17.12) for the pressure–released case.

Convergence Behavior of the Half–space Green's Function

Taking into account the definition of $\gamma = \mathrm{i}k\rho_0 c/Z$, the line integral in Equation (17.18) converges to a finite value for a real wave number k only, if

$$\Re\{\gamma\} < 0 \Leftrightarrow \Im\{Z\} < 0 \tag{17.23}$$

where $\Re\{\ \}$ or $\Im\{\ \}$ denote the real or imaginary part of the quantity in brackets, respectively. Therefore, the normal impedance of the reflecting plane is enforced to be masslike. Also, by substituting Equation (17.15) into Equation (17.16), the necessary condition (17.16) becomes

$$\left[2\gamma\,\mathrm{e}^{-\gamma(\eta+h)}\,\frac{\mathrm{e}^{\mathrm{i}kr_\eta}}{r_\eta}\right] \xrightarrow[\eta\to-\infty]{} 0 \tag{17.24}$$

which is satisfied only for masslike impedances. The physical reason for the divergence of the integral when considering surfaces with springlike impedances seems to be the occurrence of surface waves. This topic was extensively discussed in the literature [3, 10, 27, 37] starting with the work of Sommerfeld [42–44]. The approach (17.13) seems not to be adequate to model such surface waves. Unfortunately, the imaginary part of the ground impedance (the reactance) is positive for widely used models, cf. [3, p 74] or [10, p 38] for example, which motivates the extension of formula (17.19) to the case of springlike surface impedances with positive reactance.

Comparison with Three Known Solutions

From the huge number of publications about half–space Green' function, we consider three examples:

a) Sommerfeld analyzed the propagation of radio waves above the earth by using the Hankel transform technique [6]. By transferring the electromagnetic to the corresponding acoustical impedance case, the solution of Sommerfeld, cf. [44, p 228, Eq (9)], can be written (please note that Sommerfeld's constant μ_E/n^2 is equal to $-\gamma$ in our notation)

$$G(\boldsymbol{x},\boldsymbol{y}) = \frac{\mathrm{e}^{\mathrm{i}kr(h)}}{4\pi r(h)} + \frac{\mathrm{e}^{\mathrm{i}kr(-h)}}{4\pi r(-h)} + \frac{2\gamma}{4\pi}\int_0^\infty J_0(\lambda\rho)\mathrm{e}^{-\mu(z+h)}\,\frac{\lambda\mathrm{d}\lambda}{\mu(\mu-\gamma)}, \tag{17.25}$$

where $\rho^2 = x^2 + y^2$, $\mu = \sqrt{\lambda^2 - k^2}$, J_0 is the Bessel function of order zero, and $\Re\{\mu\} > 0$ must be chosen. In [32], it was shown that Equation (17.25) is in agreement with solution (17.19), if the so–called Sommerfeld identity, cf. [23], originally derived by Sommerfeld, cf. [44, p 222, p 229],

$$\frac{\mathrm{e}^{\mathrm{i}k\sqrt{\rho^2+(z+\eta)^2}}}{\sqrt{\rho^2+(z+\eta)^2}} = \int_0^\infty J_0(\lambda\rho)\mathrm{e}^{-\mu(z+\eta)}\,\frac{\lambda\mathrm{d}\lambda}{\mu} \tag{17.26}$$

was rephrased and then applied to Equation (17.25) in a certain manner. Details can be found in [32].

b) Similar to Sommerfeld, also Habault and Filippi [17] used the Hankel transform pair for investigating sound propagation over a homogeneous ground. With the ground characterized by a locally reacting surface, the authors obtained expression Equation (36) in Reference [17]. By taking into account that sign $\hat{\zeta} = -1$, i.e. $\hat{\zeta} := \Im\{Z_0\}$ in Reference [17], for a masslike surface impedance and that the right hand side of the Helmholtz equation (17.6) is not a negative but a positive Dirac delta function in Reference [17], the representation of Habault and Filippi is identical with the formula (17.19).

c) Li et al. [22] used a totally different technique, in order to derive the half–space Green's function by expanding it in a power series of the parameter γ. Again, for masslike impedances, their result is in full agreement with formula (17.19). This was shown in [32] by considering Equation (24) of Reference [22]. An interesting fact is that, Li et al. incorporated the half–space Green's function into a boundary element formulation. By using Gauss–Laguerre quadrature for evaluating the improper integral numerically, their BEM results agreed very well with analytical solutions. We have implemented $G(\boldsymbol{x}, \boldsymbol{y})$ (17.19) in BEMLAB, a self–developed BEM software code written in MATLAB [4]. Numerical results for sound radiation over masslike impedance planes can be found in Section 17.3.2. However, for springlike impedances the derivation of Li et al. [22] was not fully successful, since only an approximate Green's function could be determined, cf. [22, p 178, Eq (32)].

17.2.6 The Half–space Green's Function for a Springlike Impedance Plane by Means of the Hankel Transform

For a springlike impedance and hence for a positive reactance, $\Re\{\gamma\} > 0$, condition (17.24) is violated. Nevertheless, the approach (17.13) can be used, too, by splitting the line integral of virtual sources as follows

$$\hat{I} = \int_{-\infty}^{-h} a(\eta) g(\hat{\eta}) \mathrm{d}\eta = \int_{-\infty}^{\infty} a(\eta) g(\hat{\eta}) \mathrm{d}\eta - \int_{-h}^{\infty} a(\eta) g(\hat{\eta}) \mathrm{d}\eta =: I_1 - I_2 . \quad (17.27)$$

Proceeding along the same lines as before, the boundary condition (17.7) enforces that terms containing the integral I_2 must vanish, which again leads to expression (17.15)

$$a(\eta) = 2\gamma \mathrm{e}^{-\gamma(\eta+h)} . \quad (17.28)$$

Condition (17.16) changes to

$$\left[\frac{a(\eta)}{r_\eta} \mathrm{e}^{ikr_\eta} \right] \xrightarrow[\eta \to +\infty]{} 0 , \quad (17.29)$$

since now the contribution from the upper limit of the integral I_2 must vanish. Clearly, condition (17.29) is fulfilled for a positive reactance, cf. Equation (17.23). Consequently, the integral I_1 takes the form

$$I_1 = 2\gamma \int_{-\infty}^{\infty} \mathrm{e}^{-\gamma(\eta+h)} \frac{\mathrm{e}^{\mathrm{i}k\sqrt{x^2+y^2+(z-\eta)^2}}}{\sqrt{x^2+y^2+(z-\eta)^2}} \,\mathrm{d}\eta \,. \tag{17.30}$$

If the improper integral I_1 exists, it satisfies the boundary condition (17.7), as it was shown in [32]. Strictly speaking, the integral converges only if condition (17.29) for $\eta \to \pm\infty$ is satisfied. At a first view, that seems to be impossible. On the other hand, the integral can be solved analytically by using the generalized formula of Heine [26]. The derivation can be found in [32] leading to the solution

$$\begin{aligned} G(\boldsymbol{x},\boldsymbol{y}) &= \frac{\mathrm{e}^{\mathrm{i}kr(h)}}{4\pi r(h)} + \frac{\mathrm{e}^{\mathrm{i}kr(-h)}}{4\pi r(-h)} + \frac{1}{2}\mathrm{i}\gamma\,\mathrm{e}^{-\gamma(h+z)} H_0^{(1)}\left(k\rho\sqrt{1-\frac{1}{Z_0^2}}\right) \\ &\quad - 2\gamma\,\mathrm{e}^{-\gamma h} \int_{-h}^{\infty} \mathrm{e}^{-\gamma\eta} \frac{\mathrm{e}^{\mathrm{i}kr(\eta)}}{4\pi r(\eta)} \,\mathrm{d}\eta \,, \end{aligned} \tag{17.31}$$

where $H_0^{(1)}$ is the Hankel function of the first kind of order zero and the assumption

$$\Im\left\{k\rho\left(1 \pm \frac{1}{Z_0}\right)\right\} > 0 \tag{17.32}$$

has to be made. Unfortunately, for a real wave number k condition (17.32) is always violated.

Solution (17.31) is is in full agreement with results found in the literature: Sommerfeld, cf. Reference [42] or Reference [44, Ch 6, p 233], obtained the Hankel function by performing a contour integral and applying the residual theorem. He called the third term in Equation (17.31) the surface wave contribution and the fourth term the volume wave cf. Reference [42, Ch 5,6]. He stated, cf. Reference [42, p 695–696], that such a representation of the scattered field is only valid in the region $\rho^2 = x^2 + y^2 > 0$. In the vicinity of the z–axis, the third and the fourth term contained singularities. These singularities must cancel each other out for $\rho \to 0$, but they destroy the usefulness of formula (17.31) for a numerical evaluation. Especially, Equation (17.31) cannot be incorporated into a boundary element method due to this restricted validity, which seems to be the main drawback of representation (17.31). Looking again at Equation (36) of Habault and Filippi in [17] and choosing a positive reactance, so that sign $\hat{\zeta} = +1$ in Reference [17] follows, Equation (17.31) is obtained. Surprisingly, in their derivation no problems of convergence are mentioned and no additional conditions, such as (17.32) emerge.

17.2.7 The Half–space Green's Function for an Arbitrary Impedance Plane by Means of Sources with Complex Source Points

Representations (17.19) and (17.31) of the half–space Green's function have the advantage that their physical meaning can easily be interpreted as a superposition of acoustical sources. However, for the more important springlike impedance, the equivalent sources in the last term of Equation (17.31) are located in the lower half–space V_- $(-h \le z < 0)$ as well as in the acoustical domain V_+ $(0 \le z < \infty)$. The latter leads to singularities in the vicinity of the z–axis for $z > 0$, too. In order to find an expression better suited for numerical purposes, a completely different approach was used in [32].

Superposition of Point Sources with Complex Source Points for Constructing the Line Integral

The use of equivalent acoustic sources for modeling radiation and scattering problems is widespread, see for example [30, 31] and [28, Ch O.4], or [12, Ch 7]. In various papers, see for example [13, 14, 23]), the idea appeared to use also a discrete sum of simple acoustic sources, but to equip these sources with complex source points. This idea was originally introduced by Deschamps [7] and then enhanced by Felsen and other authors, see for example [15, 16, 18, 46]. They pointed out that spherical waves with complex source locations could be used to model the propagation of Gaussian beams. For example, if the complex location $(x, y, z) = (0, 0, ib)$ with real b is used as source point of a monopole, then the corresponding field approximately behaves like a Gaussian beam in the paraxial region near the z–axis (details can be found in [15]). In the next section, it will be shown that these complex sources are exact solutions of the Helmholtz equation.

The keystone of the approach as presented in [32] is to use the source function

$$\hat{g}(-h + \mathrm{i}\varsigma) = \frac{1}{4\pi} \frac{\mathrm{e}^{\mathrm{i}k\sqrt{x^2+y^2+(z+h-\mathrm{i}\varsigma)^2}}}{\sqrt{x^2+y^2+(z+h-\mathrm{i}\varsigma)^2}} \qquad \text{for } -\infty < \varsigma \le 0 \quad (\varsigma \text{ real}) \tag{17.33}$$

instead of

$$\hat{g}(-h + \eta) = \frac{1}{4\pi} \frac{\mathrm{e}^{\mathrm{i}k\sqrt{x^2+y^2+(z+h-\eta)^2}}}{\sqrt{x^2+y^2+(z+h-\eta)^2}} \qquad \text{for } -\infty < \eta \le 0 \tag{17.34}$$

in the line integral of Equation (17.13). Therefore, instead of using a superposition of monopole sources along the negative z–axis (starting at the mirror point $(x, y, z) = (0, 0, -h)$), we start again at the same mirror point, but now source functions are added along an "imaginary z–axis" at the points $z = -h + \mathrm{i}\varsigma$. Thus, the approach for the half–space Green's function is now, cf. Equation (17.13),

$$G(\boldsymbol{x}, \boldsymbol{y}) = \hat{g}(h) + A\,\hat{g}(-h) + \int_{-\infty}^{0} b(\varsigma)\,\hat{g}(-h + \mathrm{i}\varsigma)\,\mathrm{d}\varsigma. \tag{17.35}$$

By substituting expression (17.35) into the impedance boundary condition and proceeding just in the same way as described in section 17.2.5, we get

$$A = 1 \qquad \text{and} \qquad b(\varsigma) = 2\mathrm{i}\gamma \mathrm{e}^{-\mathrm{i}\gamma\varsigma}. \tag{17.36}$$

It follows that the entire half–space Green's function can be written as

$$G(\boldsymbol{x},\boldsymbol{y}) \;=\; \hat{g}(h) \;+\; \hat{g}(-h) \;+\; \frac{\mathrm{i}\gamma}{2\pi} \int\limits_{-\infty}^{0} \frac{\mathrm{e}^{\mathrm{i}k\sqrt{x^2+y^2+(z+h-\mathrm{i}\varsigma)^2}}}{\sqrt{x^2+y^2+(z+h-\mathrm{i}\varsigma)^2}}\,\mathrm{e}^{-\mathrm{i}\gamma\varsigma}\mathrm{d}\varsigma. \tag{17.37}$$

The change from real to complex source points has the enormous advantage that the line integral is now convergent for springlike surface impedances, too. This can be proven by considering the kernel

$$K(x,y,z;\varsigma) = \frac{b(\varsigma)}{\sqrt{x^2+y^2+(z+h-\mathrm{i}\varsigma)^2}}\mathrm{e}^{\mathrm{i}k\sqrt{x^2+y^2+(z+h-\mathrm{i}\varsigma)^2}} \tag{17.38}$$

of the integral. The first factor

$$b(\varsigma) = 2\mathrm{i}\gamma\mathrm{e}^{-\mathrm{i}\gamma\varsigma} = 2\mathrm{i}\gamma\mathrm{e}^{k\varsigma/Z_0} \tag{17.39}$$

decays exponentially for real wave numbers k and $\varsigma \to -\infty$, if

$$\Re\left\{\frac{1}{Z_0}\right\} > 0 \iff \Re\{Z_0\} > 0\,. \tag{17.40}$$

Such an assumption is true, if the surface impedance possesses a small real part, which is a reasonable assumption, since acoustical propagation always involves dissipative processes. However, Equation (17.40) is not a necessary condition, because the second factor of the kernel (17.38) also decays exponentially, thereby allowing pure imaginary surface impedances, too. This fact can be verified by observing that the square root symbol $\sqrt{\cdots}$ in Equation (17.38) must represent the root with a positive real part, in order to fulfill the radiation condition for the time dependency $\mathrm{e}^{-\mathrm{i}\omega t}$. If we now choose receiver points at the z–axis ($x = y = 0$) for simplicity, we obtain for the exponential function in Equation (17.38) ($z > 0$, $h > 0$)

$$\mathrm{e}^{\mathrm{i}k\sqrt{(z+h-\mathrm{i}\varsigma)^2}} = \mathrm{e}^{+\mathrm{i}k(z+h)}\mathrm{e}^{k\varsigma} \xrightarrow[\varsigma\to-\infty]{} 0. \tag{17.41}$$

Consequently, the improper integral in Equation (17.37) exists for all kinds of surface impedances. It converges much faster then the one in representation (17.19), since the corresponding source function (17.34) is a rapidly oscillating function for $\eta \to -\infty$, whereas the source function (17.33) decays exponentially.

Similar formulas can be found in the related literature. It is shown in [32] that the line integral of Equation (7) derived in [29] exactly agrees with the representation of the half–space Green's function (17.37) and, thus, can be interpreted as a superposition of sources with complex source points, too. Also, in the electromagnetic literature, line integrals over source points with complex source locations were considered: In [25], Lindell and Alanen obtained a line integral containing Bessel functions. Convergence of the Bessel functions can only be achieved, if the image current is located in complex space, too.

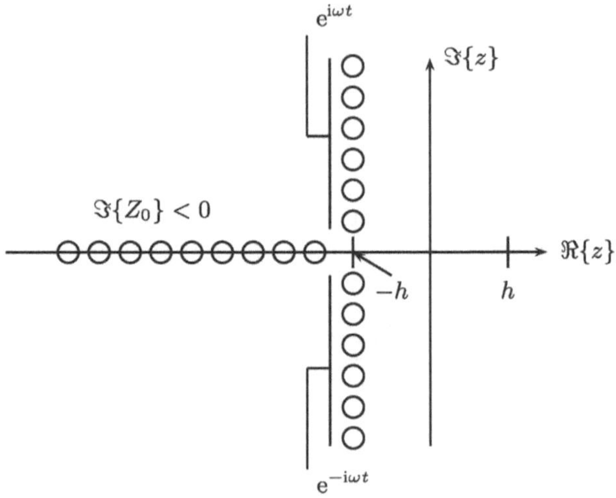

Fig. 17.4 Location of equivalent sources in the complex z–plane depending on surface impedance and time dependency.

Some care is needed, if the time dependence $e^{i\omega t}$ is used. In this case, the sources in the line integral must be located along the z–axis at points with

$$z = -h - i\varsigma \quad \text{for} \quad -\infty < \varsigma \leq 0 \tag{17.42}$$

to ensure that the kernel of the integral decays exponentially for all kinds of surface impedances. Again, the root must be taken which yields a positive real part, in order to satisfy the radiation condition at infinity. In other words, we simply have to use the complex conjugate $[\hat{g}(-h+i\varsigma)]^*$ of Equation (17.33). A graphical illustration of how to position the source points in the complex z–plane for different time dependences, is given in Figure 17.4.

The complex equivalent sources behave in a similar manner as surface waves, which explains the close relationship between formulas (17.31) and (17.37). These features will be considered in the next chapter.

Sources with Complex Source Points

Since sources with complex source points are seldom found in the acoustic literature, we describe some properties of these sources, cf. [19,33,35]. Consider, for example, the complex source

$$\varphi(x, y, z - ia) = \frac{e^{ik\sqrt{x^2+y^2+(z-ia)^2}}}{\sqrt{x^2+y^2+(z-ia)^2}} = \frac{e^{ik\tilde{r}}}{\tilde{r}} \quad \text{for a real constant } a. \tag{17.43}$$

All properties are also valid for the more general case, where the complex distance is given by

$$\tilde{r} = \sqrt{(x-ia)^2+(y-ib)^2+(z-ic)^2} \quad (a, b \text{ and } c \text{ real constants}). \tag{17.44}$$

The complex distance becomes zero on the circle

$$C = \{(x, y, z)|x^2 + y^2 + z^2 = a^2\} \tag{17.45}$$

around the z–axis at $z = 0$ with radius a. Furthermore, the complex root must be made single–valued. It can be shown, cf. [19]), that a suitable branch cut, requiring that $\Re\{\tilde{r}\} \geq 0$, is given by the disc

$$D = \{(x, y, z)|x^2 + y^2 + z^2 \leq a^2\}\,. \tag{17.46}$$

Hence, the complex source (17.43) becomes singular on the source disc D. Outside of D, the complex source is an exact solution of the Helmholtz equation with respect to the spatial coordinates (x, y, z) for a fixed complex source point. This can be shown as follows: Let ∇ and Δ be the gradient and the Laplacian with respect to (x, y, z) for a fixed. Then, we obtain

$$\begin{aligned} \nabla\varphi &= \nabla\left(\frac{\mathrm{e}^{\mathrm{i}k\tilde{r}}}{\tilde{r}}\right) = \mathrm{e}^{\mathrm{i}k\tilde{r}}\,\nabla\frac{1}{\tilde{r}} + \frac{1}{\tilde{r}}\,\nabla\,\mathrm{e}^{\mathrm{i}k\tilde{r}} \\ \Rightarrow\ \Delta\varphi &= \mathrm{e}^{\mathrm{i}k\tilde{r}}\Delta\frac{1}{\tilde{r}} + 2\nabla\frac{1}{\tilde{r}}\cdot\nabla\mathrm{e}^{\mathrm{i}k\tilde{r}} + \frac{1}{\tilde{r}}\,\Delta\,\mathrm{e}^{\mathrm{i}k\tilde{r}}, \end{aligned} \tag{17.47}$$

where the dot $\cdot$ denotes the scalar product. The first term is zero, since the holomorphic potential $\vartheta = 1/\tilde{r}$ is harmonic with respect to (x, y, z) outside the disc D, see Proposition 1 of [19]. The second term gives

$$2\,\nabla\frac{1}{\tilde{r}}\cdot\nabla\mathrm{e}^{\mathrm{i}k\tilde{r}} = -2\mathrm{i}\,k\,\mathrm{e}^{\mathrm{i}k\tilde{r}}\,\frac{1}{\tilde{r}^2} \tag{17.48}$$

since $\nabla\tilde{r}\cdot\nabla\tilde{r} = 1$. The third term in (17.47) can be written as

$$\frac{1}{\tilde{r}}\,\Delta\,\mathrm{e}^{\mathrm{i}k\tilde{r}} = \frac{\mathrm{e}^{\mathrm{i}k\tilde{r}}}{\tilde{r}}\left(-k^2 + \frac{2\mathrm{i}k}{\tilde{r}}\right)\,. \tag{17.49}$$

By combining (17.47) – (17.49) we immediately get

$$\Delta\varphi + k^2\varphi = 0 \quad \text{for } (x, y, z) \notin D. \tag{17.50}$$

Thus, the Green's function (17.37) as a superposition of real and complex sources is a solution of the Helmholtz equation, too.

Complex equivalent sources seem to behave in a similar manner as surface waves. As mentioned above, the square root in (17.38) becomes zero if and only if

$$z = -h, \quad x^2 + y^2 = \varsigma^2. \tag{17.51}$$

Whereas the ordinary monopole source becomes singular at a single point (point source), the complex point source shows a different behaviour: It becomes singular on discs enclosed by circles around the z–axis at $z = -h$ with radius ς (circle or ring sources), and for $|\varsigma| \to \infty$ the radius of these circles increases more and more like a growing circular surface wave. Figure 17.5 shows the development of a complex point source with decreasing values of ς. However, if ς changes its sign, the complex point source looks like a Gaussian beam with growing strength for increasing values of ς. This is shown in Figure 17.6. The close relationship of complex equivalent sources with Gaussian beams was described by Felsen [15], for example.

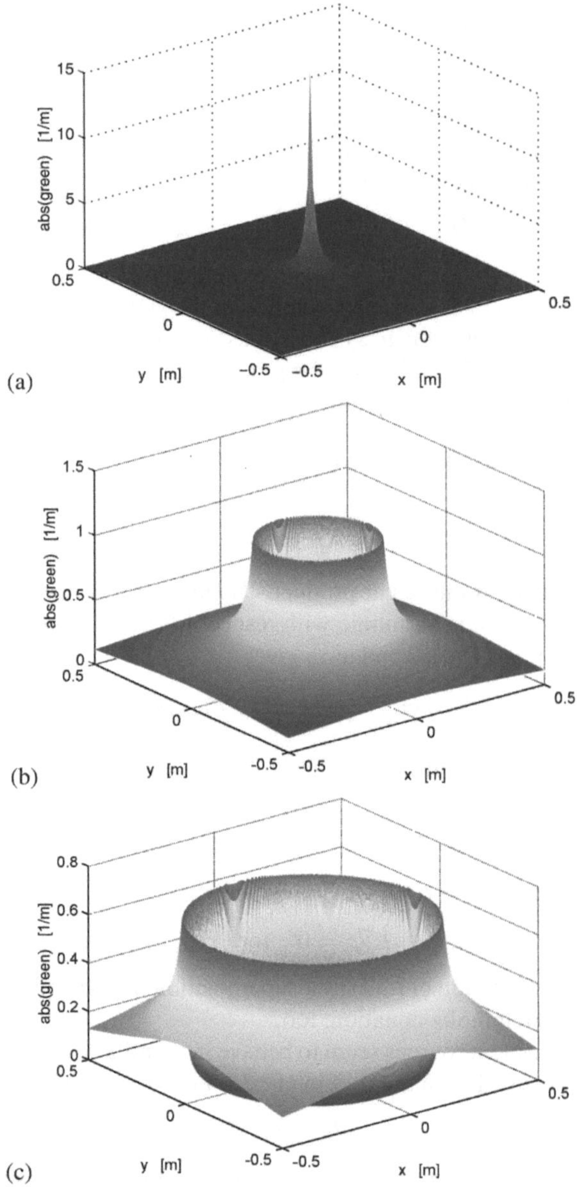

Fig. 17.5 Complex modulus $|\hat{g}|$ = abs(green) of the Green's function(17.33) for decreasing values of imaginary z–location ς over the xy–plane; $z + h = 0.005$ m and $f = 1000$ Hz; (a) $\varsigma = 0$ m,(b) $\varsigma = -0.2$ m, (c) $\varsigma = -0.4$ m.

A Closed Solution for Perpendicular Incidence

As mentioned in Section 17.2.6, expression (17.31) for the Green's function is not valid in the vicinity of the z–axis $\rho^2 = x^2 + y^2 \approx 0$, where the receiver is situated di-

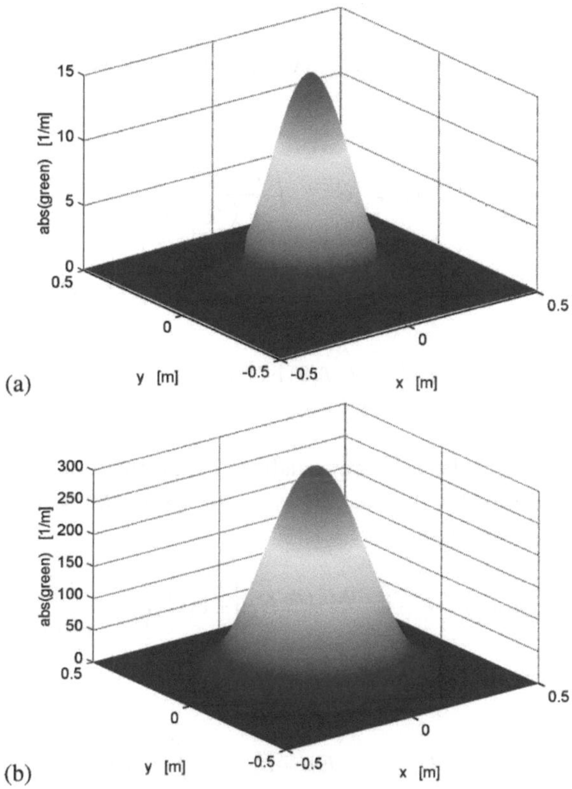

Fig. 17.6 Complex modulus $|\hat{g}| = \text{abs(green)}$ of the Green's function(17.33) for increasing values of imaginary z–location ς over the xy–plane; $z + h = 0.005$ m and $f = 1000$ Hz; (a) $\varsigma = 0.2$ m,(b) $\varsigma = 0.4$ m.

rectly below or above the source due to the appearance of singularities. Fortunately, just in this case, representation (17.37) leads to an exact formula containing an exponential integral. For $x = y = 0$, the third term in Equation (17.37) gives

$$\Gamma(\boldsymbol{x}, \boldsymbol{y}) \;=\; \frac{\mathrm{i}\gamma}{2\pi}\, I_0 \tag{17.52}$$

with

$$I_0 \;=\; \int_0^\infty \frac{\mathrm{e}^{\mathrm{i}k\sqrt{(z_h+\mathrm{i}\varsigma)^2}}}{\sqrt{(z_h+\mathrm{i}\varsigma)^2}}\,\mathrm{e}^{\mathrm{i}\gamma\varsigma}\,\mathrm{d}\varsigma \qquad \text{and} \quad z_h = z + h. \tag{17.53}$$

For satisfying the radiation condition at infinity, we have to choose the square root with a positive real part $\sqrt{(z_h + \mathrm{i}\varsigma)^2} = +(z_h + \mathrm{i}\varsigma)$ since $z_h > 0$ and $\varsigma > 0$. It should be mentioned, that $\sqrt{(z_h + \mathrm{i}\varsigma)^2}$ is not the distance between the points z_h and $-\mathrm{i}\varsigma$ in the complex plane, since this is defined by

$$\sqrt{|z_h + \mathrm{i}\varsigma|^2} = \sqrt{(z_h + \mathrm{i}\varsigma)(z_h + \mathrm{i}\varsigma)^*} \tag{17.54}$$

and not by

$$\sqrt{(z_h + \mathrm{i}\varsigma)^2} = \sqrt{(z_h + \mathrm{i}\varsigma)(z_h + \mathrm{i}\varsigma)} \,. \tag{17.55}$$

Hence, we obtain

$$I_0 = \int_0^\infty \frac{\mathrm{e}^{\mathrm{i}k(z_h+\mathrm{i}\varsigma)}}{(z_h + \mathrm{i}\varsigma)} \mathrm{e}^{\mathrm{i}\gamma\varsigma}\,\mathrm{d}\varsigma = \mathrm{e}^{\mathrm{i}kz_h} \int_0^\infty \frac{\mathrm{e}^{-k(1+1/Z_0)\varsigma}}{(z_h + \mathrm{i}\varsigma)}\,\mathrm{d}\varsigma \qquad \text{since} \quad \gamma = \frac{\mathrm{i}k}{Z_0}. \tag{17.56}$$

By setting $s := k\,(1 + 1/Z_0)$ and $f(\varsigma) = 1/(\varsigma - \mathrm{i}z_h)$, we get

$$I_0 = I_0(z) = -\mathrm{i}\,\mathrm{e}^{\mathrm{i}kz_h} \int_0^\infty f(\varsigma)\,\mathrm{e}^{-s\varsigma}\mathrm{d}\varsigma = -\mathrm{i}\,\mathrm{e}^{\mathrm{i}kz_h} L[f(\varsigma)], \tag{17.57}$$

where $L[f(\varsigma)]$ denotes the Laplace transform of the function $f(\varsigma)$. Clearly, the integral $L[f(\varsigma)]$ is convergent, since $\Re\{x\} > 0$. Using the tables of Erdélyi et al., cf. [11, p 134, Eq (6)], we obtain

$$I_0(z) = -\mathrm{i}\mathrm{e}^{-\mathrm{i}kz_h/Z_0} E_1 \left[-\mathrm{i}z_h k \left(1 + \frac{1}{Z_0} \right) \right] \tag{17.58}$$

with the the exponential integral

$$E_1 = -Ei(-z) = \int_{-\infty}^{z} \frac{\mathrm{e}^t}{t} \mathrm{d}t. \tag{17.59}$$

Hence, we get for perpendicular incidence by inserting Equations (17.52) and (17.58) into expression (17.37) the Green's function

$$G(\boldsymbol{x}, \boldsymbol{y}) = \frac{\mathrm{e}^{\mathrm{i}k(z-h)}}{z-h} + \frac{\mathrm{e}^{\mathrm{i}k(z+h)}}{z+h} + \frac{\mathrm{i}k}{2\pi Z_0} \mathrm{e}^{-\mathrm{i}kz_h/Z_0} E_1 \left[-\mathrm{i}z_h k \left(1 + \frac{1}{Z_0} \right) \right]. \tag{17.60}$$

This formula agrees with Equation (35b) of [29] and with Equation (34) of [24]. A similar solution, but based on Equation (17.19) for masslike impedances, was found by F. P. Mechel and communicated to the first author. Representation (17.60) will be used in Section 17.3.3 for calculating the pressure field radiated from a pulsating sphere directly above or below of the sphere with respect to the impedance plane.

17.3 Numerics

17.3.1 BEMLAB – a Matlab–based BEM–code

The theoretical fundamentals of the BEM were implemented as a 3D–BEM program BEMLAB, which was developed as a Matlab–toolbox. The program bases upon the time convention $e^{i\omega t}$. Therefore in this numerical section this alternative time convention is used. In combination with a renaming of γ and Z_0 as used in Section 17.2, cf. Equation (17.9), into $\tilde{\gamma}$ and $\tilde{Z}_0$ we define in accordance with the recent time convention $\gamma = -(\tilde{\gamma})^*$ and $Z_0 = (\tilde{Z}_0)^*$. The asterisk * indicates the complex conjugate.

For simplicity, it is assumed that the impedance plane S_p is always located in $z = 0$ (xy–plane).

In the following some characteristics of the BEM program will be presented. The surface of the structure S_b is represented by a set of N constant triangular or quadrilateral elements S_k ($k = 1 \ldots N$), i.e. the boundary function on each surface element is approximated by a constant value. The collocation method transforms the surface HIE (17.2) into the following system of $N \times N$ linear equations

$$DP + MV = \frac{P}{2} \tag{17.61}$$

where the matrix $D = (d_{ik})$ consists of the dipole terms

$$d_{ik} = \iint_{S_k} \frac{\partial g(\boldsymbol{x}_i, \boldsymbol{y})}{\partial n_{\boldsymbol{y}}} \, \mathrm{d}s(\boldsymbol{y}) \, , \tag{17.62}$$

and the matrix $M = (m_{ik})$ consists of the monopole terms

$$m_{ik} = \mathrm{i}\omega\rho_0 \iint_{S_k} g(\boldsymbol{x}_i, \boldsymbol{y}) \, \mathrm{d}s(\boldsymbol{y}) \, . \tag{17.63}$$

Here, P is the N–dimensional vector of pressure values in the centroid of the N elements, and V is the corresponding normal velocity vector, whose elements $v_n(\boldsymbol{y})$ are obtained from

$$v_n(\boldsymbol{y}) = -\frac{1}{\mathrm{i}\omega\rho_0} \frac{\partial p(\boldsymbol{y})}{\partial n_{\boldsymbol{y}}}. \tag{17.64}$$

The diagonal elements of the resulting matrixes, where $\boldsymbol{x}_i = \boldsymbol{y}$, become singular during the collocation process. The singularity can be resolved as described in [28, Ch O.5]. The derivative $\partial g(\boldsymbol{x}, \boldsymbol{y})/\partial n_{\boldsymbol{y}}$ for d_{ii} in (17.62) can be set to zero, which was also shown in [20]. The integral (17.63) for m_{ii} is solved using polar coordinates

$$m_{ii} = \mathrm{i}\omega\rho_0 \int_0^{2\pi} \int_0^{\tilde{R}_i} \frac{\mathrm{e}^{-\mathrm{i}kr}}{4\pi r} \, r \, \mathrm{d}r \, \mathrm{d}\varphi = -\omega\rho_0 \, \frac{\mathrm{e}^{-\mathrm{i}k\tilde{R}_i} - 1}{2k} \, , \tag{17.65}$$

where $\tilde{R}_i = \sqrt{S_i/\pi}$ is the radius of an equivalent circle area of the plane quadrilateral or triangular element S_i. The Green's function $g(\boldsymbol{x}_i, \boldsymbol{y})$ was defined in (17.4). Please note the changed time convention.

As discussed in Chapter 15 of the present book, the Helmholtz integral equation for an exterior Neumann problem does not have an unique solution at the characteristic eigenfrequencies of the associated interior Dirichlet problem and vice versa. In BEMLAB the CHIEF method (Combined integral equation formulation) was chosen to remedy the non–uniqueness at these eigenfrequencies. Additional collocation points are located in the interior domain of the object B_i, where the interior

Helmholtz integral formulation must be satisfied. With this overdetermination the surface HIE and interior HIE are solved simultaneously to enforce the finding of the unique surface solution. The CHIEF method was developed by Schenck [38], a brief description can also be found in [45, Ch 7], or [28, Ch O.5]. Several tests showed, that the CHIEF method provides satisfactory results only at the first characteristic eigenfrequencies, but so far this was sufficient for our purpose. Additionally, it is also planned to implement the Burton & Miller method as described in [36] into the BEMLAB program.

17.3.2 The Boundary Element Method for a Masslike Impedance Plane

We have implemented the Green's function for a masslike impedance plane, Equation (17.19), in our BEM–code BEMLAB and followed the suggestion from Li et al. [22] to apply the Gauss–Laguerre quadrature, cf. [4]. After a variable transformation with $\mu = -(\eta + z_s)\Re\{\gamma\}$, G reads as

$$\begin{aligned} G(\boldsymbol{x}, \boldsymbol{y}) &= \frac{\mathrm{e}^{-ikr(z_s)}}{4\pi r(z_s)} + \frac{\mathrm{e}^{-ikr(z_s)}}{4\pi r(-z_s)} \\ &\quad - 2(1 + \mathrm{i}\tan\varphi) \int\limits_0^\infty \frac{\mathrm{e}^{-ikr(-\mu/\Re\{\gamma\}-z_s)}}{4\pi r(-\mu/\Re\{\gamma\} - z_s)} \mathrm{e}^{-\mathrm{i}\mu\tan\varphi} \mathrm{e}^{-\mu} \mathrm{d}\mu \end{aligned} \tag{17.66}$$

with

$$r(z_s) = \sqrt{(x - x_s)^2 + (y - y_s)^2 + (z - z_s)^2}$$

and $\tan\varphi = \Im\{\gamma\}/\Re\{\gamma\}$. Hence, the integral in (17.66) can be solved by a Gauss–Laguerre quadrature, cf. [1, eq 25.4.45], which is defined as follows

$$\int\limits_0^\infty f(x)\mathrm{e}^{-x}\mathrm{d}x \approx \sum_{k=1}^{n} A_k^{(n)} f(x_k^{(n)}). \tag{17.67}$$

$x_k^{(n)}$ are the zeros of the Laguerre polynomials of order n and $A_k^{(n)}$ the corresponding weight factors. Tables of zeros and weight factors of Laguerre polynomials up to $n = 15$ can be found in [1, Table 25.9], up to $n = 32$ in [21, Appendix C]. To find zeros and weight factors for higher orders we used an evaluation technique by means of continued fraction as described in [41].

To verify the implementation we draw on a test configuration, introduced by Li et al. [22]. A uniformly pulsating sphere with a diameter of $a = 1$ m is situated above the impedance plane in $z = 0$ and its center is in (0, 0, 3m). A frequency–dependent impedance $Z_0 = (0.4 + \mathrm{i}0.8)ka$ is prescribed on the plane. In this test effectively a full BEM simulation is carried out for the pressure distribution on the sphere and the field point evaluation. With the BEM solution for the pressure and normal velocity at a field point on the plane its impedance Equation (17.1) can be recalculated. Figure 17.7 shows the excellent agreement between prescribed and recalculated impedance at point (0, 0, 0), directly below the sphere. In Figure 17.7

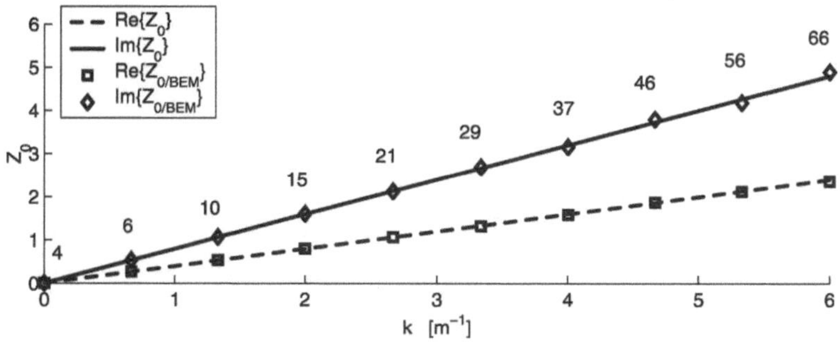

Fig. 17.7 Comparison of the prescribed impedance of the plane Z_0 and the recalculated impedance $Z_{0/\mathrm{BEM}}$. The numbers indicate the used number of zeros for the Gauss–Laguerre quadrature.

additionally the number of used integration points of the Gauss–Laguerre quadrature are annotated for each analyzed frequency step. Within a BEM calculation, $G(\boldsymbol{x}, \boldsymbol{y})$ and the number of necessary integration points vary for the different $\boldsymbol{x}$ and $\boldsymbol{y}$ combinations. Here the highest necessary number is presented. As it can be seen, the number of integration points increase dramatically with increasing wave number. This can be ascribed to the increasing fluctuation of the integrand in (17.66). But the convergence of the integral depends on other parameters as z_s as well as $\tan\varphi$, $\Re\{\gamma\}$, too.

The contour plots in Figure 17.8 should illustrate these dependencies with respect to some of the parameters. The basic configuration is $\boldsymbol{x} = (0, 0, 0), \boldsymbol{y} = (0, 0, h), \gamma = (0.5 + \mathrm{i}0.5)\,\mathrm{m}^{-1}, \ \tan\varphi = 1, k = 1\mathrm{m}^{-1}$. To study the influence of these parameters two parameters are varied, two are fixed in the basic configuration, $\boldsymbol{x}$ is always retained unchanged. The contour plots show the number of integration points n, which are necessary to solve the integral for a parameter variation of $\Re\{\gamma\}$ over h and k over h with respect to the convergence criterion $|I(n-1) - I(n)| < 5 \cdot 10^{-4}$, where

$$I(n) = \sum_{k=1}^{n} A_k^{(n)} \frac{\mathrm{e}^{-ikr(-\mu_k^{(n)}/\Re\{\gamma\}-h)}}{4\pi r(-\mu_k^{(n)}/\Re\{\gamma\} - h)} \mathrm{e}^{-\mathrm{i}\mu_k^{(n)}\tan\varphi}. \tag{17.68}$$

The study shows, that a slowly converging configuration is characterized by a high frequency, source and receiver are close to the plane, a high ratio $\tan\varphi$ and a medium value of the admittance γ. The left plot of Figure 17.8 shows the results for a variation of $\Re\{\gamma\}, \Im\{\gamma\}$ and h (please recall, that $\Re\{\gamma\} = \Im\{\gamma\}$ corresponding to $\tan\varphi = 1$). As it can be seen, the integral converges very fast in direction of the two extreme values of $\Re\{\gamma\}, \Im\{\gamma\} \to 0$ and $\Re\{\gamma\}, \Im\{\gamma\} \to \infty$, which represent the marginal cases of perfectly soft and rigid plane, respectively. The medium range around $\Re\{\gamma\}, \Im\{\gamma\} = 10^{-1}\ \mathrm{m}^{-1}$ appears to be more critical, whereas the source height h does not influence considerably the convergence of the integral. The right plot shows that the wave number definitely determines the number of necessary integration points and in combination with a very small height h the integral is hardly

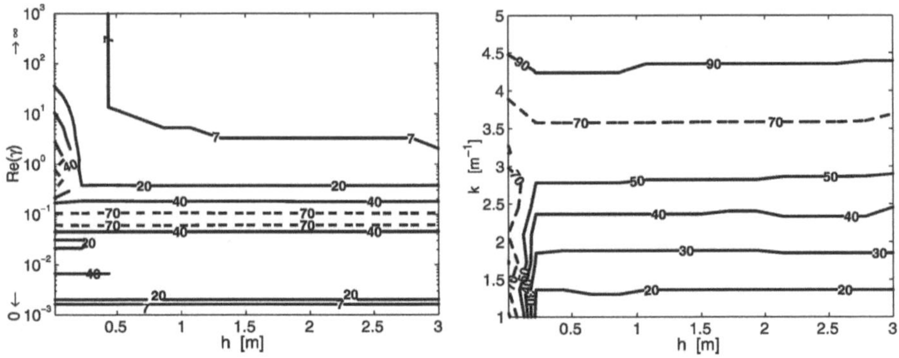

Fig. 17.8 Necessary number of integration points with respect to variations of $\Re\{\gamma\}$ over source height h (left) and of wavenumber k over source height h (right).

solvable with the presented quadrature formula due to its slow convergence. But for the noncritical configurations, the Gauss–Laguerre quadrature is found to be a stable and fast algorithm for the evaluation of the Green's function G for a masslike impedance plane within a BEM application.

17.3.3 The Boundary Element Method with Complex Sources (CBEM)

As mentioned before, the half–space Green's function (17.37) was implemented in a BEM application which bases upon the time convention $\mathrm{e}^{\mathrm{i}\omega t}$. In this case (17.37) reads

$$G(\boldsymbol{x},\boldsymbol{y})^* = \left(\hat{g}(z_s) + \hat{g}(-z_s) + \frac{\mathrm{i}\tilde{\gamma}}{2\pi}\int\limits_{-\infty}^{0} \frac{\mathrm{e}^{\mathrm{i}k\sqrt{\rho^2+(z+z_s-\mathrm{i}\varsigma)^2}}}{\sqrt{\rho^2+(z+z_s-\mathrm{i}\varsigma)^2}}\mathrm{e}^{-\mathrm{i}\tilde{\gamma}\varsigma}d\varsigma\right)^*$$

$$= G_0(\boldsymbol{x},\boldsymbol{y}) + \frac{\mathrm{i}\gamma}{2\pi}\int\limits_{-\infty}^{0} \underbrace{\frac{\mathrm{e}^{-\mathrm{i}k\sqrt{\rho^2+(z+z_s+\mathrm{i}\varsigma)^2}}}{\sqrt{\rho^2+(z+z_s+\mathrm{i}\varsigma)^2}}\mathrm{e}^{-\mathrm{i}\gamma\varsigma}}_{\Psi(\varsigma)}\,\mathrm{d}\varsigma\,. \tag{17.69}$$

Again, the vector $\boldsymbol{y}$ points at the source location $\boldsymbol{y} = (x_s,\, y_s,\, z_s)$, $\boldsymbol{x}$ points at the receiver location $\boldsymbol{x} = (x,\, y,\, z)$. ρ denotes the horizontal distance of $\boldsymbol{x}$ and $\boldsymbol{y}$, $\rho = \sqrt{(x-x_s)^2+(y-y_s)^2}$. $G_0(\boldsymbol{x},\boldsymbol{y})$ is defined as

$$G_0(\boldsymbol{x},\boldsymbol{y}) = \frac{\mathrm{e}^{-\mathrm{i}kr(z_s)}}{4\pi r(z_s)} + \frac{\mathrm{e}^{-\mathrm{i}kr(-z_s)}}{4\pi r(-z_s)} \tag{17.70}$$

with $r(z_s) = \sqrt{(x-x_s)^2+(y-y_s)^2+(z-z_s)^2}$.

The root in $\Psi(\varsigma)$, $r = \sqrt{\rho^2 + (z + z_s + \mathrm{i}\varsigma)^2} = |r|\mathrm{e}^{\mathrm{i}\phi}$, represents the complex distance with a positive real part. The imaginary part is always negative since $\phi \in (0, -\pi/2)$,

$$\Re(r) = |r|\cos(\phi) > 0, \quad \Im(r) = |r|\sin(\phi) < 0. \tag{17.71}$$

The integrand $\Psi(\varsigma)$ can be separated into a decaying envelope $\Psi_E(\varsigma)$ and an oscillating term $\Psi_O(\varsigma)$

$$\Psi(\varsigma) = \frac{\mathrm{e}^{-ikr}}{r}\mathrm{e}^{-\mathrm{i}\gamma\varsigma} = \underbrace{\frac{1}{|r|}\mathrm{e}^{[k\Im(r)+\Im(\gamma)\varsigma]}}_{\Psi_E(\varsigma)} \underbrace{\mathrm{e}^{-\mathrm{i}[k\Re(r)+\phi+\Re(\gamma)\varsigma]}}_{\Psi_O(\varsigma)}. \tag{17.72}$$

Since $\Im(\gamma)$ represents the damping of the ground, $\Im(\gamma)$ must be equal or greater than zero and the exponent in $\Psi_E(\varsigma)$ is negative for all $\varsigma \in (-\infty, 0]$. With this decaying envelope, the convergence of the integral in (17.69) is given, apart from the configuration of grazing incidence, where both source and receiver are located on the plane. In this case $z + z_s = 0$ and a singularity of the integrand occurs at $\rho = -\varsigma$ for $|r|$ becomes zero.

Hence, when source and receiver are located near the plane, $|r|$ becomes very small at $\rho = -\varsigma$ and the envelope $\Psi_E(\varsigma)$ shows a distinct peak at these points. Figure 17.9 illustrates the curve of $\Psi(\varsigma)$ for $z + z_s = 0.06$ m and the components of the envelope $\Psi_E(\varsigma)$. As it can be seen, the peak of $\Psi(\varsigma)$ results from the maximum peak of $1/|r|$, whereas the steep descent for values $\varsigma < -\rho$ is caused by the radical drop of $\mathrm{e}^{k\Im(r)}$.

With increasing distance of source and/or receiver from the plane, the discussed effects diminish and the curve of $\Psi(\varsigma)$ softens as Figure 17.10 shows. Here the combined height of source and receiver $z+z_s$ is ten times higher as in the aforementioned example, but still $z + z_s < 1$.

On the other hand, with increasing imaginary part of admittance of the plane the term, $\mathrm{e}^{\Im\{\gamma\}\varsigma}$, becomes the limiting factor of the envelope $\Psi_E(\varsigma)$. In the example shown in Figure 17.11, the steep descent of this term almost completely constrains the formation of the characteristic curve with the peak at $\varsigma = -\rho$ as seen in Figure 17.9.

The characteristics of the integrand $\Psi(\varsigma)$ depend strongly on the different configuration parameters as admittance of the plane, horizontal distance and height of source and receiver. But as discussed, with the decaying envelope $\Psi_E(\varsigma)$ a convergence of the integral is given for every possible configuration, excluding the case of grazing incidence. On the basis of the decaying characteristic it is possible to determine a lower limit ς_0, where the integration can be truncated without reasonable error. This is numerically much more comfortable than to consider the unbounded domain of the original integral. The limit ς_0 depends on the steepness of the descent of $\Psi_E(\varsigma)$.

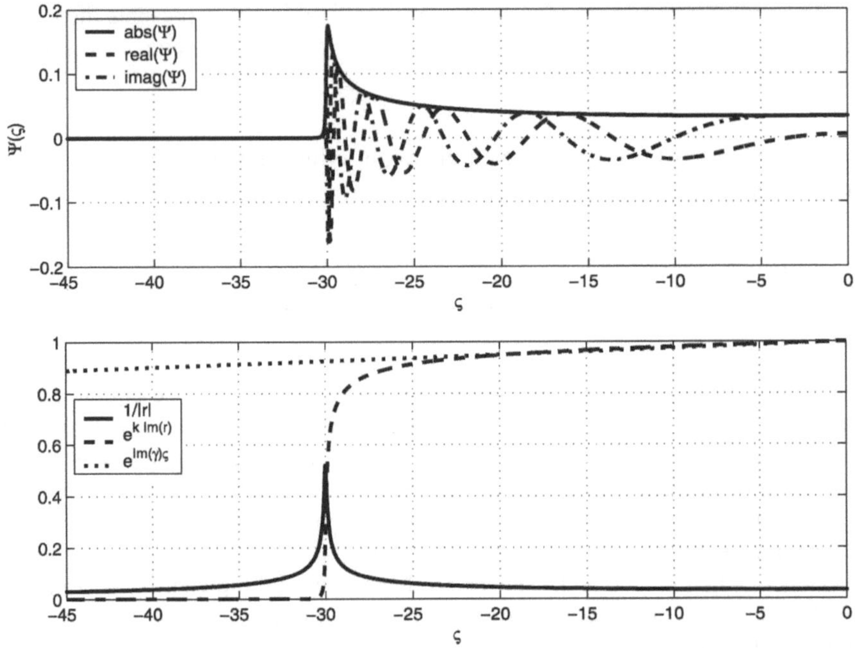

Fig. 17.9 *Upper panel*: Curve of $\Psi(\varsigma)$ in case $z + z_s << 1$ for $\rho = 30$ m, $\gamma = 0.0026 + 0.0026\text{i m}^{-1}$, $z + z_s = 0.06$ m, $k = 1\text{ m}^{-1}$. *Lower panel*: Components of the envelope $\Psi_E(\varsigma)$.

Normal Derivatives of the Green's Function

For a BEM application several normal derivatives of equation (17.69) are needed. The HIE (17.2) requires the normal derivative $\partial G(\boldsymbol{x},\boldsymbol{y})/\partial n_{\boldsymbol{y}}$. For a calculation of the normal velocity at the receiver location $v_n(\boldsymbol{x})$ the derivatives $\partial G(\boldsymbol{x},\boldsymbol{y})/\partial n_{\boldsymbol{x}}$ and $\partial^2 G(\boldsymbol{x},\boldsymbol{y})/(\partial n_{\boldsymbol{y}}\partial n_{\boldsymbol{x}})$ are necessary. For $\boldsymbol{x} \neq \boldsymbol{y}$ holds

$$\begin{aligned} v_n(\boldsymbol{x}) &= C(\boldsymbol{x})\frac{\mathrm{i}}{\omega\rho_0}\frac{\partial p(\boldsymbol{x})}{\partial n_{\boldsymbol{x}}} \\ &= C(\boldsymbol{x})\frac{\mathrm{i}}{\omega\rho_0}\int\limits_S\left[p(\boldsymbol{y})\frac{\partial^2 G(\boldsymbol{x},\boldsymbol{y})}{\partial n_{\boldsymbol{y}}\partial n_{\boldsymbol{x}}} + \mathrm{i}\omega\rho_0 v_n(\boldsymbol{y})\frac{\partial G(\boldsymbol{x},\boldsymbol{y})}{\partial n_{\boldsymbol{x}}}\right]\mathrm{d}S. \end{aligned} \tag{17.73}$$

Differentiation with Respect to $n_{\boldsymbol{y}}$ and $n_{\boldsymbol{x}}$

The normal derivative of the Green's function is given by

$$\frac{\partial G(\boldsymbol{x},\boldsymbol{y})}{\partial n_{\boldsymbol{y}}} = \frac{G_0(\boldsymbol{x},\boldsymbol{y})}{\partial n_{\boldsymbol{y}}} + \frac{\mathrm{i}\gamma}{2\pi}\int\limits_{-\infty}^{0}\frac{\partial}{\partial n_{\boldsymbol{y}}}\left(\frac{\mathrm{e}^{-\mathrm{i}kr}}{r}\right)\mathrm{e}^{-\mathrm{i}\gamma\varsigma}\mathrm{d}\varsigma. \tag{17.74}$$

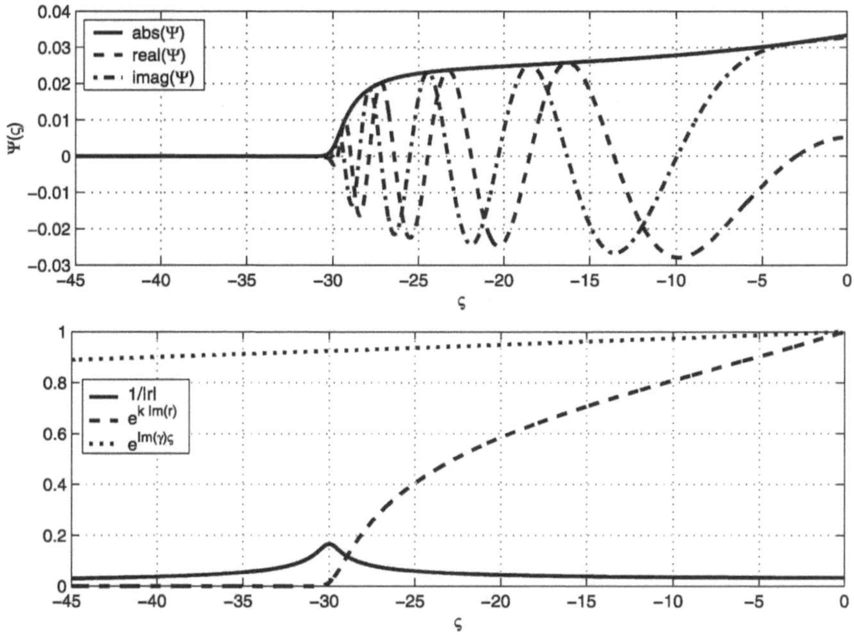

Fig. 17.10 Influence of the increased height $z + z_s$ on $\Psi(\varsigma)$. *Upper panel*: Curve of $\Psi(\varsigma)$ in case $z + z_s < 1$ for $\rho = 30$ m, $\gamma = 0.0026 + 0.0026\text{i}\ \text{m}^{-1}$, $z + z_s = 0.6$ m, $k = 1\ \text{m}^{-1}$. *Lower panel*: Components of the envelope $\Psi_E(\varsigma)$.

where $r = \sqrt{\rho^2 + (z + z_s + \text{i}\varsigma)^2}$. Differentiation of the first term is standard in half–space modeling [39]. The derivative of the kernel function $f(r) = \text{e}^{-\text{i}kr}/r$ can be obtained from

$$\frac{\partial f(r)}{\partial n_{\boldsymbol{y}}} = \frac{\partial f(r)}{\partial r}\frac{\partial r}{\partial n_{\boldsymbol{y}}} = -\frac{\text{e}^{-\text{i}kr}}{r^2}(\text{i}kr + 1)(\nabla_{\boldsymbol{y}} r \cdot \boldsymbol{n}_{\boldsymbol{y}}). \tag{17.75}$$

The dot $\cdot$ denotes the scalar product. Defining $\boldsymbol{y}_c$ as complex equivalent source point $\boldsymbol{y}_c = (x_s, y_s, -z_s - \text{i}\varsigma)$, r is represented also by $r = |\boldsymbol{x} - \boldsymbol{y}_c|$ and the derivative $\partial r/\partial n_{\boldsymbol{y}}$ is given by

$$\frac{\partial r}{\partial n_{\boldsymbol{y}}} = \nabla_{\boldsymbol{y}} r \cdot \boldsymbol{n}_{\boldsymbol{y}} = \frac{1}{r}(\boldsymbol{x} - \boldsymbol{y}_c)\boldsymbol{M} \cdot \boldsymbol{n}_{\boldsymbol{y}} \tag{17.76}$$

with the transformation matrix

$$\boldsymbol{M} = \begin{bmatrix} -1 & 0 & 0 \\ 0 & -1 & 0 \\ 0 & 0 & +1 \end{bmatrix}.$$

Please note that the normal vectors $\boldsymbol{n}_{\boldsymbol{x}}$ and $\boldsymbol{n}_{\boldsymbol{y}}$ as well as the position vectors $\boldsymbol{x}$ and $\boldsymbol{y}$ are formulated as row vectors. The derivative of the Green's functions with respect to $\boldsymbol{n}_{\boldsymbol{y}}$ is finally given by

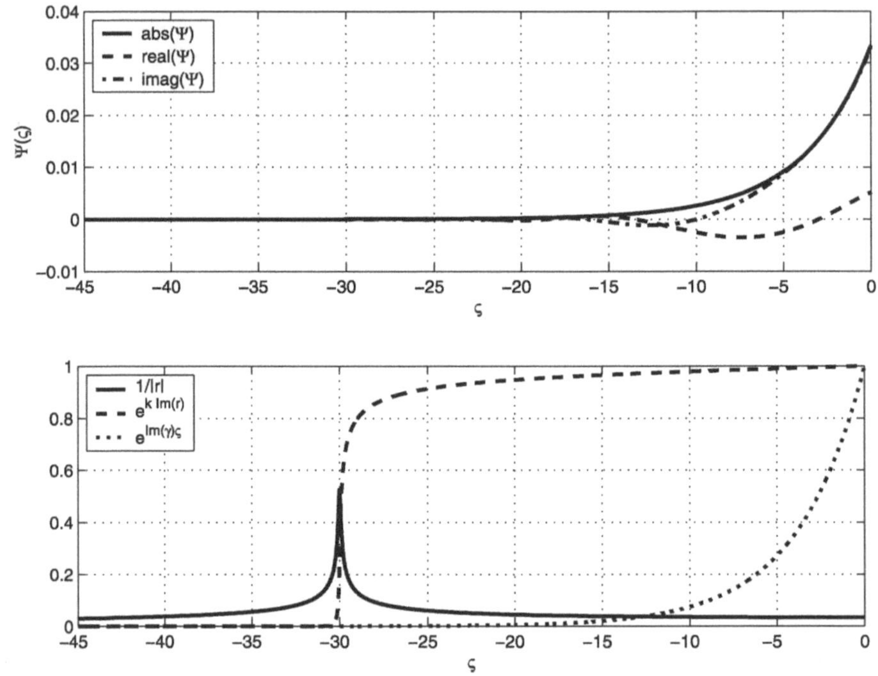

Fig. 17.11 Influence of an increased imaginary part of the plane admittance γ on $\Psi(\varsigma)$. *Upper panel*: Curve of $\Psi(\varsigma)$ for $\rho = 30$ m, $\gamma = 0.0026 + 0.26\mathrm{i}\ \mathrm{m}^{-1}$, $z + z_s = 0.06$ m, $k = 1\ \mathrm{m}^{-1}$. *Lower panel*: Components of the envelope $\Psi_E(\varsigma)$.

$$\frac{\partial G(\boldsymbol{x},\boldsymbol{y})}{\partial n_{\boldsymbol{y}}} = \frac{G_0(\boldsymbol{x},\boldsymbol{y})}{\partial n_{\boldsymbol{y}}} - \frac{\mathrm{i}\gamma}{2\pi}\int\limits_{-\infty}^{0} \frac{\mathrm{e}^{-\mathrm{i}kr}}{r}\,\mathrm{e}^{-\mathrm{i}\gamma\varsigma}\,\frac{1+\mathrm{i}kr}{r}\,\nabla_{\boldsymbol{y}} r \cdot \boldsymbol{n}_{\boldsymbol{y}}\,\mathrm{d}\varsigma. \tag{17.77}$$

The derivative with respect to $\boldsymbol{n}_{\boldsymbol{x}}$ is obtained analogously with

$$\frac{\partial r}{\partial n_{\boldsymbol{x}}} = \nabla_{\boldsymbol{x}} r \cdot \boldsymbol{n}_{\boldsymbol{x}} = \frac{1}{r}(\boldsymbol{x} - \boldsymbol{y}_c)\cdot \boldsymbol{n}_{\boldsymbol{x}}. \tag{17.78}$$

Derivative $\partial^2 G(\boldsymbol{x},\boldsymbol{y})/(\partial n_{\boldsymbol{y}}\partial n_{\boldsymbol{x}})$

The derivative of a function $f(r)$ with respect to $\partial n_{\boldsymbol{y}}\partial n_{\boldsymbol{x}}$ is generally given by

$$\frac{\partial^2 f(r)}{\partial n_{\boldsymbol{y}}\partial n_{\boldsymbol{x}}} = \frac{\partial^2 f(r)}{\partial r^2}\,\frac{\partial r}{\partial n_{\boldsymbol{y}}}\,\frac{\partial r}{\partial n_{\boldsymbol{x}}} + \frac{\partial f(r)}{\partial r}\,\frac{\partial^2 r}{\partial n_{\boldsymbol{y}}\partial n_{\boldsymbol{x}}}, \tag{17.79}$$

cf. [2, Ch 3.1.4]. For $f(r) = \mathrm{e}^{-\mathrm{i}kr}/r$ and $r = |\boldsymbol{x} - \boldsymbol{y}_c|$ it follows

$$\frac{\partial^2 f(r)}{\partial r^2} = \frac{\mathrm{e}^{-\mathrm{i}kr}}{r^3}\left[-(kr)^2 + 2\mathrm{i}kr + 2\right] \tag{17.80}$$

and

$$\frac{\partial^2 r}{\partial \boldsymbol{n_y} \partial \boldsymbol{n_x}} = -\frac{1}{r}\left(\frac{\partial r}{\partial \boldsymbol{n_y}}\frac{\partial r}{\partial \boldsymbol{n_x}} - \boldsymbol{n_y M} \cdot \boldsymbol{n_x}\right) \tag{17.81}$$

The other terms of (17.79) are found as before and the double derivative is obtained as

$$\begin{aligned}\frac{\partial^2 G(\boldsymbol{x},\boldsymbol{y})}{\partial \boldsymbol{n_y} \partial \boldsymbol{n_x}} &= \frac{\partial^2 G_0(\boldsymbol{x},\boldsymbol{y})}{\partial \boldsymbol{n_y} \partial \boldsymbol{n_x}} \\ &+ \frac{\mathrm{i}\gamma}{2\pi}\int_{-\infty}^{0} \frac{\mathrm{e}^{-\mathrm{i}kr}}{r}\,\mathrm{e}^{-\mathrm{i}\gamma\varsigma}\left[\frac{-(kr)^2+2\mathrm{i}kr+2}{r^2}\frac{\partial r}{\partial \boldsymbol{n_y}}\frac{\partial r}{\partial \boldsymbol{n_x}}\right. \\ &\left. + \frac{1+\mathrm{i}kr}{r}\frac{1}{r}\left(\frac{\partial r}{\partial \boldsymbol{n_y}}\frac{\partial r}{\partial \boldsymbol{n_x}} - \boldsymbol{n_y M}\cdot\boldsymbol{n_x}\right)\right]\mathrm{d}\varsigma .\end{aligned} \tag{17.82}$$

Convergence of the Derivatives of the Green's Function

The convergence of the integral contained in (17.77), see (17.72),

$$\int_{-\infty}^{0} \frac{\partial \Psi(\varsigma)}{\partial \boldsymbol{n_y}}\,\mathrm{d}\varsigma = \int_{-\infty}^{0} -\Psi(\varsigma)\,\frac{1+\mathrm{i}kr}{r}\,\nabla_{\boldsymbol{y}} r\cdot \boldsymbol{n_y}\,\mathrm{d}\varsigma\,, \tag{17.83}$$

with $r = \sqrt{\rho^2 + (z+z_s+\mathrm{i}\varsigma)^2}$ can be discussed concisely. The decay of $\Psi(\varsigma)$ was discussed in detail in Section 17.2.7. For $\varsigma \to -\infty$ the following limits are obtained

$$\lim_{\varsigma\to-\infty} r = \mathrm{i}\varsigma = -\mathrm{i}\infty \tag{17.84}$$

$$\Rightarrow \lim_{\varsigma\to-\infty} \frac{1+\mathrm{i}kr}{r} = \mathrm{i}k. \tag{17.85}$$

The product $\nabla_{\boldsymbol{y}} r \cdot \boldsymbol{n_y}$ can be separated into

$$\nabla_{\boldsymbol{y}} r\cdot\boldsymbol{n_y} = \frac{1}{r}\left[\Re\left\{(\boldsymbol{x}-\boldsymbol{y}_c)\,\boldsymbol{M}\cdot\boldsymbol{n_y}\right\} + \mathrm{i}\Im\left\{(\boldsymbol{x}-\boldsymbol{y}_c)\,\boldsymbol{M}\cdot\boldsymbol{n_y}\right\}\right]. \tag{17.86}$$

Using (17.84) and $\Im\{(\boldsymbol{x}-\boldsymbol{y}_c)\boldsymbol{M}\cdot\boldsymbol{n_y}\} = \varsigma \boldsymbol{n_y}_z$, where $\boldsymbol{n_y}_z$ is the z–component of $\boldsymbol{n_y}$, we have

$$\lim_{\varsigma\to-\infty} \nabla_{\boldsymbol{y}} r\cdot\boldsymbol{n_y} = \boldsymbol{n_y}_z, \tag{17.87}$$

since $\Re\{(\boldsymbol{x}-\boldsymbol{y}_c)\boldsymbol{M}\cdot\boldsymbol{n_y}\}$ is independent of ς. Thus, it is shown that the additional factors in (17.83), $(1+\mathrm{i}kr)/r$ and $\nabla_{\boldsymbol{y}} r\cdot\boldsymbol{n_y}$, have finite limits for $\varsigma\to-\infty$ and do not interfere with the exponential decay of the kernel $\Psi(\varsigma)$. Therefore the integral of the derivative of the Green's function converges, too. The same can be shown for $\partial\Psi(\varsigma)/\partial\boldsymbol{n_x}$ with

$$\lim_{\varsigma\to-\infty} \nabla_{\boldsymbol{x}} r\cdot\boldsymbol{n_x} = \boldsymbol{n_x}_z. \tag{17.88}$$

Most terms of the double derivative (17.82) are already discussed by the recent considerations. With the finite limit

$$\lim_{\varsigma \to -\infty} \frac{-(kr)^2 + 2\mathrm{i}kr + 2}{r^2} = -k^2. \tag{17.89}$$

also the integral in (17.82) converges for $\varsigma \to -\infty$. Hence, an evaluation of $\partial G / \partial n_{\boldsymbol{y}}$, $\partial G / \partial n_{\boldsymbol{x}}$ and $\partial^2 G / \partial n_{\boldsymbol{y}} \partial n_{\boldsymbol{x}}$ is achievable.

Quadrature

Since the kernel $\Psi(\varsigma)$ of the Green's functions integral is fluctuating and can have a spike pulse–shaped peak at $\varsigma = -\rho$, cf. Figure 17.9, the integration can become very tedious. Two ways of fast and accurate quadrature has been investigated: the application of a Gauss–Laguerre quadrature and an adaptive multigrid–quadrature, which will be discussed in the following for the evaluation of the integral of the Green's function $G(\boldsymbol{x}, \boldsymbol{y})$.

Gauss–Laguerre Quadrature

The Gauss–Laguerre quadrature, see (17.67) on page 478, which was already applied in Section 17.3.2, seems applicable also for the evaluation of the present integral in (17.69). By means of a variable transformation with $\eta = -\varsigma \Im\{\gamma\}$ the integral can be converted into the appropriate form

$$I = \frac{\mathrm{i} \cot \varphi - 1}{2\pi} \int_0^\infty \underbrace{\frac{\mathrm{e}^{-\mathrm{i}k\tilde{r}}}{\tilde{r}} \mathrm{e}^{\mathrm{i}\eta \cot \varphi}}_{f(\eta)} \mathrm{e}^{-\eta} \mathrm{d}\eta \tag{17.90}$$

with $\tilde{r} = \sqrt{\rho^2 + (z + z_s - \mathrm{i}\eta / \Im\{\gamma\})^2}$ and $\cot \varphi = \Re\{\gamma\} / \Im\{\gamma\}$. Contrary to (17.66) the expression for $f(\eta)$ contains also an exponentially decaying term, namely $\mathrm{e}^{k\Im\{\tilde{r}\}}$. In case the decay of the term $\mathrm{e}^{k\Im\{\tilde{r}\}}$ is considerably more steep as the decay of $\mathrm{e}^{-\eta}$, the integral can not be solved by the Gauss–Laguerre quadrature. The resistance $\Im\{\gamma\}$ has influences strongly the decay of $\mathrm{e}^{\Im(\tilde{r}\}}$ and the Gauss–Laguerre quadrature can only be applied in case $\Im\{\gamma\} >> 1$. In this case this quadrature formula is observed to solve the integral correctly and very fast, but the use is limited to the presence of an impedance plane with low resistance.

Adaptive Multigrid–Quadrature

By defining a finite lower integration limit ς_0 instead of $-\infty$ as discussed before, it is possible to apply quadrature formulae for bounded integrals to the integrals in (17.69), (17.77) and (17.82). We will discuss the procedure with respect to the integral of the Green's function $G(\boldsymbol{x}, \boldsymbol{y})$ in (17.69). In order to find the lower limit ς_0 it is necessary to evaluate the decaying envelope $\Psi_E(\varsigma)$, cf. Equation (17.72). We truncate the integral at a limit ς_0 where $\Psi_E(\varsigma_0) \leq 10^{-5}$. To solve the integral within the limits $\varsigma_0 \leq \varsigma \leq 0$ we applied an adaptive multigrid quadrature, which is described in detail in [8]. Based on a coarse subdivision of the basic interval,

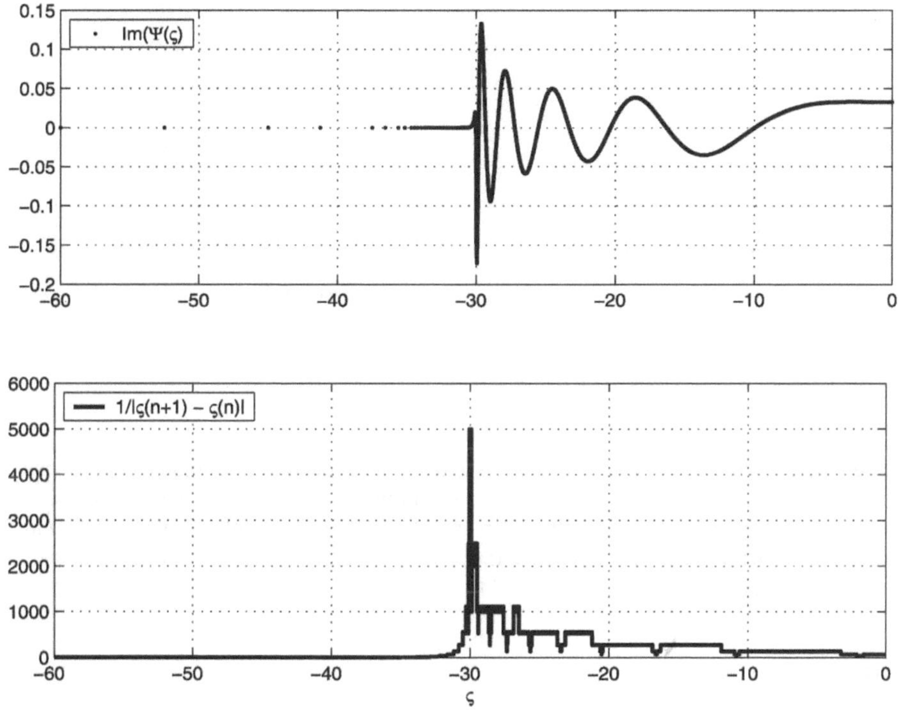

Fig. 17.12 *Upper panel*: $\Im(\Psi(\varsigma))$ at the adapted grid in case $\rho = 30$ m, $\gamma = 0.0026 + 0.0026\mathrm{i}$ m^{-1}, $z + z_s = 0.06$ m, $k = 1$ m^{-1} (example from Figure 17.9). *Lower panel*: Reciprocal value of the spacing of the adapted grid nodes.

the algorithm refines successively the grid where it is necessary for the required accuracy. In this way the adaptive multigrid quadrature is able to cope with difficult integrands as $\Psi(\varsigma)$. The grid points accumulate automatically at the critical points as the spike pulse–shaped peak at $\varsigma = -\rho$ or in strongly fluctuating zones. For the implementation of the quadrature algorithm we used the program package "mlquad" from "CodeLib", a collection of algorithms of the Konrad–Zuse–Zentrum für Informationstechnik Berlin [9], and adapt the code to the present integral.

In Figure 17.12 the functioning of the algorithm can be observed. The upper graph shows $\Im\{\Psi(\varsigma)\}$ at the adapted grid for the example configuration from Figure 17.9, which results in a quite unfavorable curve of Ψ. The refinement of the grid in the vicinity of $\varsigma = \rho$ an be clearly seen in the lower graph of Figure 17.12. For the evaluation of $\Im\{\Psi(\varsigma)\}$ in this example 11417 necessary grid points were found by the algorithm. The used adaptive multigrid quadrature is a very stable and powerful quadrature method. It is not as fast as the Gauss–Laguerre quadrature for the same configuration, but it is not limited to any impedance condition.

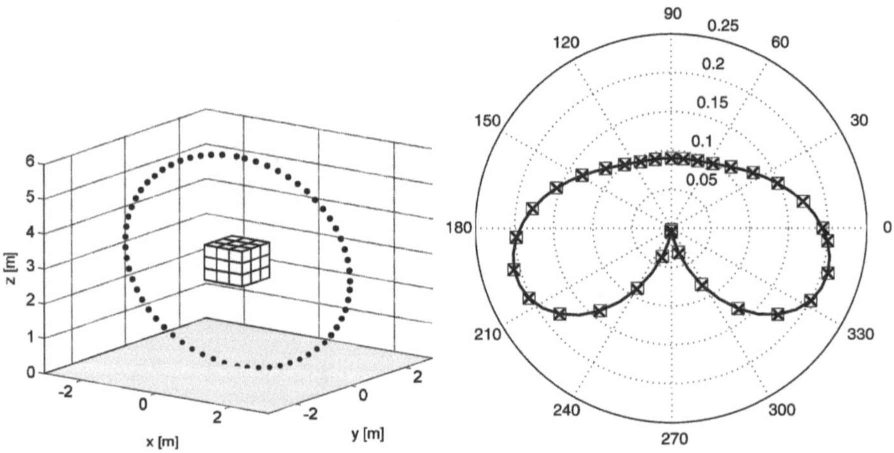

Fig. 17.13 *Left panel*: Test geometry: Field points around the radiating cube above an impedance plane. *Right panel*: Normalized sound pressure in the field points: — calculated with IMBEM in case of a perfectly soft plane ($R = -1$); □ calculated with the CBEM for a soft plane with masslike impedance ($\gamma = 10^3 + \mathrm{i}10^3\ \mathrm{m}^{-1}$) and × for a soft plane with springlike impedance ($\gamma = -10^3 + \mathrm{i}10^3\ \mathrm{m}^{-1}$).

Test Cases

The following test cases have been investigated:

- Sound radiation of a pulsating sphere above a perfectly rigid and perfectly soft plane, respectively.
- Comparison between the given and the back–calculated plane impedance.
- Sound field in case of perpendicular incidence.

Rigid and Soft Plane

The sound radiation above a rigid or soft plane can be described by an image–source Green's function, see Section 17.2.4. With the time convention $\mathrm{e}^{\mathrm{i}\omega t}$ this Green's function reads

$$G(\boldsymbol{x}, \boldsymbol{y}) = \frac{\mathrm{e}^{-ikr(z_s)}}{4\pi r(z_s)} + R\frac{\mathrm{e}^{-ikr(-z_s)}}{4\pi r(-z_s)} \qquad (17.91)$$

with $r(z_s) = \sqrt{(x - x_s)^2 + (y - y_s)^2 + (z - z_s)^2}$. The reflection coefficient R is $R = 1$ for a rigid plane and $R = -1$ for a soft plane.

In the test we compare the BEM results for the sound radiation above a rigid and soft plane by using the image–source Green's function (17.91) and the complex Green's function (17.69), respectively. The test configuration is given by a small cube with edge length of 1 m, whose surface is driven by a virtual monopole source in its center [34, 36]. The surface mesh consists of 54 four–noded quadrilateral elements. The center of the cube is located 3 m above the infinite impedance plane in, which is located in $z = 0$, see Figure 17.13. The perfectly soft and rigid plane impedance

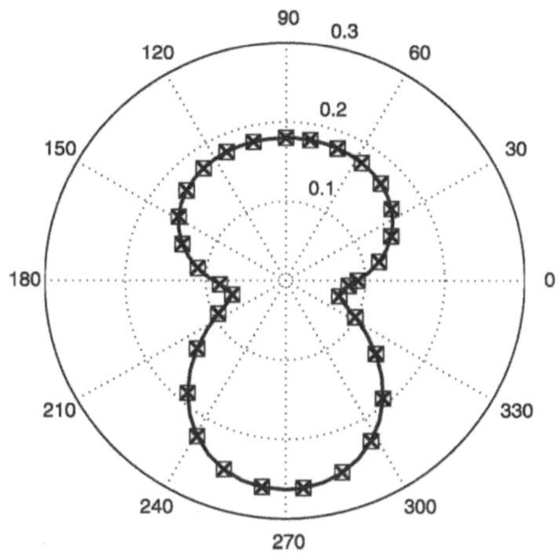

Fig. 17.14 Normalized sound pressure in the field points: — calculated with IMBEM in case of a perfectly rigid plane ($R = 1$); □ calculated with the CBEM for a rigid plane with masslike impedance ($\gamma = 10^{-3} + \mathrm{i}10^{-3}\ \mathrm{m}^{-1}$) and × for a rigid plane with springlike impedance ($\gamma = -10^{-3} + \mathrm{i}10^{-3}\ \mathrm{m}^{-1}$).

is approximated by a very high or small admittance of $|\gamma| = \sqrt{2} \cdot 10^{-3}$ and $|\gamma| = \sqrt{2} \cdot 10^{3}$, respectively. Both a masslike ($\Re(\gamma) > 0$) and a springlike ($\Re(\gamma) < 0$) plane impedance are investigated in each case. Figure 17.13 and 17.14 show the BEM results for the sound pressure in field points in a distance of 3 m around the center of the cube, see Figure 17.13, for the image–source Green's function BEM (IMBEM) and the complex Green's function BEM (CBEM). The sound pressure was normalized to $\rho_0 c|\bar{v_n}|$, with $|\bar{v_n}|$ as absolute value of the mean surface velocity of the pulsating object. For the illustrated benchmark configurations the results of the CBEM formulation match very well the results of the IMBEM.

Back–Calculated Plane Impedance

This test case was introduced by Li et al. [22] and in Section 17.3.2. It is well suited to verify a BEM formulation. At present, we will use this test to show the ability of the CBEM formulation to handle an impedance condition at the plane with a negative reactance, $\Im\{Z\} < 0$ (for time convention $\mathrm{e}^{\mathrm{i}\omega t}$). In the aforementioned configuration a frequency–dependent impedance $Z = (0.4 - \mathrm{i}0.8)\rho_0 cka$, $a = 1$ m, is prescribed to the plane. The ratio of pressure and normal velocity at field points on the plane, both calculated by the BEM application, should regain the given plane impedance. Figure 17.15 shows the given and back–calculated plane impedance at the field point $\boldsymbol{x} = (0, 0, 0)$. The agreement is superior.

Solution for Perpendicular Incidence

As discussed in Section 17.2.7, there exists an exact solution for the integral in (17.69) for perpendicular incidence, where the receiver is located directly above or

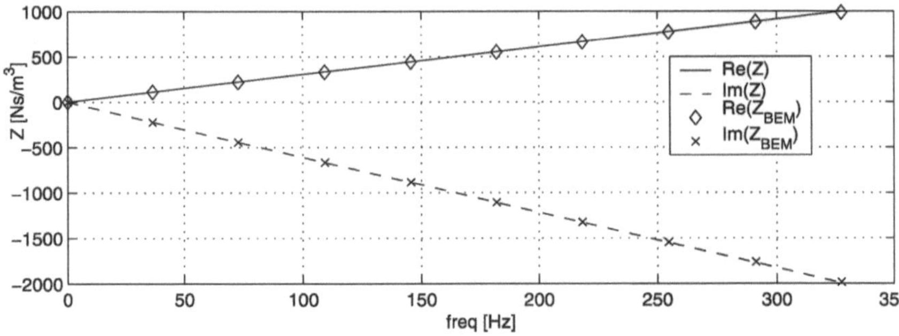

Fig. 17.15 Frequency response of the prescribed impedance Z and back–calculated impedance Z_{BEM}.

below the source ($\rho = 0$), by means of the exponential integral in Equation (17.60). For the time convention $\mathrm{e}^{\mathrm{i}\omega t}$ this solution reads

$$G_{E1}(\boldsymbol{x},\boldsymbol{y}) = G_0(\boldsymbol{x},\boldsymbol{y}) - \frac{\mathrm{i}k}{2\pi Z_0}\mathrm{e}^{\mathrm{i}k\,(z+z_s)/Z_0}\; E_1\left[\mathrm{i}k\;(z+z_s)\left(1+\frac{1}{Z_0}\right)\right]. \tag{17.92}$$

As shown in [29], for field points far above the reflecting plane, i.e. where $k(z+z_s)$ is large, (17.92) can be simplified and the total Green's function reduces to a "plane wave" solution:

$$G_{\mathrm{plane}}(\boldsymbol{x},\boldsymbol{y}) = \frac{\mathrm{e}^{-\mathrm{i}kr(z_s)}}{4\pi r(z_s)} + \frac{Z_0-1}{Z_0+1}\frac{\mathrm{e}^{-\mathrm{i}kr(z_s)}}{4\pi r(z_s)}. \tag{17.93}$$

In this test problem we consider a pulsating sphere with radius $a = 0.25$ m and a center at $(0,\ 0,\ 3\mathrm{m})$. The field points are located on a line directly below and above the sphere at $(0,\ 0,\ z)$ with $0 \le z \le 10$ m. Other parameters are $k = 1\ \mathrm{m}^{-1}$, $Z = (0.4 + \mathrm{i}0.8)\rho_0 c$. The sphere consists of 296 four–noded quadrilateral elements and its surface has a uniform normal velocity v_a.

Figure 17.16 shows the pressure values at the field points for the BEM calculation, p_{BEM} and the "plane wave" solution p_{plane} with respect to the pressure values calculated with the presented exponential integral solution p_{E1} in (17.92). Field points, which lie in the interior of the radiating sphere are not evaluated. From that results the gap in the curves. Please note, that the "plane wave" solution and the exponential integral solution are given by

$$p_{\mathrm{plane}}(\boldsymbol{x}) = A_p\, G_{\mathrm{plane}}(\boldsymbol{x},\boldsymbol{y}), \tag{17.94}$$

$$p_{\mathrm{E1}}(\boldsymbol{x}) = A_p\, G_{\mathrm{E1}}(\boldsymbol{x},\boldsymbol{y}). \tag{17.95}$$

The pressure amplitude A_p for a sphere with radius a vibrating with uniform normal surface velocity v_a reads

$$A_p = \frac{4\pi\mathrm{i}\,k\,c_0\,\rho_0\,a^2}{1+\mathrm{i}ka^2} v_a\,\mathrm{e}^{\mathrm{i}ka}. \tag{17.96}$$

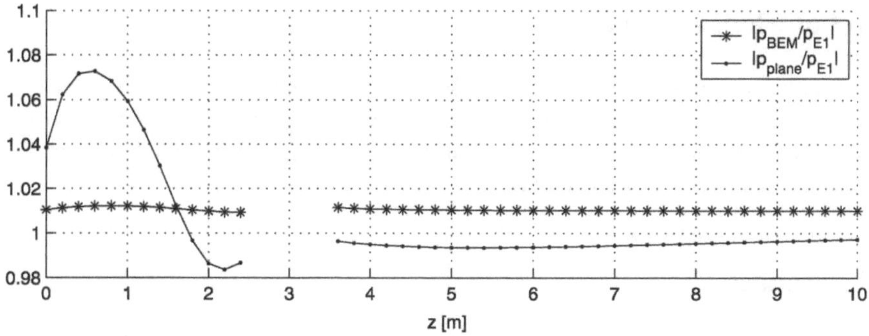

Fig. 17.16 Comparison of the field point pressure, calculated with CBEM, the exponential integral solution (exact solution) and the "plane wave" solution: –∗– ratio of the BEM calculated pressure p_{BEM} to the exact solution p_{E1}; –·– ratio of the "plane wave" solution p_{plane} to the exact solution p_{E1}.

Though a full agreement of p_{BEM} and p_{E1} can only be expected for a small sphere in high height above the sphere, where the sphere can approximated as monopole source, the BEM formulation reproduces the exact solution also for the chosen configuration very well. As expected, the "plane wave" solution fails in close vicinity to the plane, but with increasing height of the field points over the reflecting plane it also matches the exact solution. This test case additionally validates the presented CBEM formulation as well as the exact solution for perpendicular incidence.

17.4 Conclusions

It was shown that sound propagation over impedance planes can be calculated very efficiently by using a BEM formulation equipped with a Green's function built up by point sources with partially complex source locations. The half–space Green's function together with all necessary normal derivatives is convergent for every prescribed surface impedance. Hence, there is no need to model the infinite plane which may save an enormous amount of computing time. The integration of the infinite line integral appearing in the Green's function can be performed effectively and with the accuracy desired by applying an adaptive multigrid quadrature. The numerical results obtained by considering radiation from simple structures show a promising agreement with reference solutions. Therefore, future work will be devoted to the further development of our software tool BEMLAB together with the numerical investigation of sound radiation and scattering from complex structures occurring in practical industrial applications. Especially, combustion noise from flames over a flat surface will be studied.

Acknowledgements

The authors would like to thank F. P. Mechel for numerous stimulating discussions concerning particularly numerical aspects of evaluating the half–space Green's function. Also, the authors gratefully acknowledge the partial financial support by the German Research Council (DFG).

References

1. Abramowitz M, Stegun IA, editors. (1972) Handbook of mathematical functions. Dover Publications, New York
2. Amini S, Harris PJ, Wilton DT (eds) (1992) Coupled boundary and finite element methods for the solution of the dynamic fluid–structure interaction problem. Springer–Verlag, Berlin–Heidelberg–New York
3. Attenborough K (2002) Sound propagation close to the ground. Annual Review of Fluid Mechanics 34:51–82
4. Brick H, Ochmann M (2003) Eine BEM–Toolbox zur Berechnung der Schallabstrahlung schwingender Strukturen vor Hindernissen und über absorbierendem Boden (A BEM toolbox for calculation of sound radiation from vibrating structures in front of barriers and above absorbing ground). In: Deutsche Gesellschaft für Akustik (ed) Fortschritte der Akustik – DAGA'2003, Aachen, 542–543
5. Colton D, Kress R (1983) Integral equations in scattering theory. Wiley–Interscience Publication, New York
6. Davies B (2002) Integral transforms and their applications. Springer–Verlag, Berlin–Heidelberg–New York, 3rd edition
7. Deschamps GA (1971) Gaussian beams as a bundle of complex rays. Electronics Letters 7:684–685
8. Deuflhard P, Hohmann A (1993) Numerische Mathematik I – Eine algorithmisch orientierte Einführung. Chapter 9.7. 343–351, de Gruyter, Berlin–New York
9. Deuflhard P, Hohmann A (Okt. 2004) Numerical analysis and modelling–CodeLib. `http://www.zib.de/Numerik/numsoft/CodeLib/misc.en.html`
10. Embleton TFW (1996) Tutorial on sound propagation outdoors. Journal of the Acoustical Society of America 100:31–48
11. Erdélyi A (ed) (1954) Tables of integral transforms, Vol I. McGraw–Hill Book Company, New York
12. Estorff O von (ed) (2000) Boundary elements in acoustics: Advances and applications. WIT Press, Southampton–Boston
13. Fawcett JA (2000) Complex–image approximations to the half–space acousto–elastic Green's function. Journal of the Acoustical Society of America 108:2791–2795
14. Fawcett JA (2003) A method of images for a penetrable acoustic waveguide. Journal of the Acoustical Society of America 113:194–204
15. Felsen LB (1976) Complex–source–point–solutions of the field equations and their relation to the propagation and scattering of Gaussian beams. In: Instituto Nazionale di Alta Matematica (ed) Symposia Matematica. Academic Press, New York, 18:40–56
16. Felsen LB (1976) Evanescent waves. Journal of the Optical Society of America A 66:751–760
17. Habault D, Filippi PJT (1987) Ground effect analysis: Surface wave and layer potential representations. Journal of Sound and Vibration 79:529–550

18. Heyman E, Felsen LB (1989) Complex–source pulsed–beam fields. Journal of the Optical Society of America A 6:806–817
19. Kaiser G (2000) Complex–distance potential theory and hyperbolic equations. In: Ryan J, Sprößig W. (ed) Clifford Algebras and Their Applications in Mathematical Physics, Vol. 2: Clifford Analysis. Birkhäuser, Boston
20. Koopmann GH, Benner H (1982) Method for computing the sound power of machines based on the Helmholtz integral. Journal of the Acoustical Society of America 71:78–89
21. Krylov VI (1962) Approximate calculation of integrals. The Macmillan Company, New York
22. Li WL, Wu TW, Seybert AF (1994) A half–space boundary element method for acoustic problems with a reflecting plane of arbitrary impedance. Journal of Sound and Vibration 171:173–184
23. Li YL, White MJ (1996) Near–field computation for sound propagation above ground – using complex image theory. Journal of the Acoustical Society of America 99:755–760
24. Li YL, White MJ, Hwang MH (1994) Green's functions for wave propagation above an impedance ground. Journal of the Acoustical Society of America 96:2485–2790
25. Lindell IV, Alanen E (1984) Exact image theory for the Sommerfeld half–space problem. IEEE Transactions on Antennas and Propagation. Part I: Vertical magnetic dipole 32:126–133; Part II: Vertical electric dipole 32:841–847; Part III: General formulation 32:1027–1032
26. Magnus W, Oberhettinger F, Soni RP (1966) Formulas and theorems for the special functions of Mathematical Physics. Chapter 3.6.2. Springer–Verlag, Berlin–Heidelberg–New York
27. Mechel FP (1989) Schallabsorber. Vol 1, Chapter 13, Hirzel–Verlag, Stuttgard
28. Mechel FP (ed) (2002) Formulas of acoustics. Springer–Verlag, Berlin–Heidelberg–New York
29. Nobile MA, Hayek SI (1985) Acoustic propagation over an impedance plane. Journal of the Acoustical Society of America 78:1325–1336
30. Ochmann M (1995) The source simulation technique for acoustic radiation problems. Acustica 81:512–527
31. Ochmann M (1999) The full–field equations for acoustic radiation and scattering. Journal of the Acoustical Society of America 105:2574–2584
32. Ochmann M (2004) The complex equivalent source method for sound propagation over an impedance plane. Journal of the Acoustical Society of America 116:3304–3311
33. Ochmann M (2005) Complex source point method for computing the Green's function over an arbitrary impedance plane. In: Proceedings of the NOVEM 2005 (Noise and Vibration Emerging Methods), St. Raphael (USB)
34. Ochmann M, Osetrov AV (2002) Construction of analytical solutions for the error estimation of acoustical boundary element solvers. In: Proceedings of the Forum Acusticum, Sevilla, Spain, 1000–1004
35. Ochmann M, Piscoya R (2006) The source simulation technique with complex source points for computing acoustic radiation problems. In: Proceedings of the 13th International Congress on Sound and Vibration (ICSV13), Wien
36. Osetrov AV, Ochmann M (2005) A fast and stable numerical solution for acoustic boundary element method equations combined with the Burton and Miller method for models consisting of constant elements. Journal of Computational Acoustics 13:1–20
37. Raspet R, Baird GE (1989) The acoustic surface wave above a complex impedance ground. Journal of the Acoustical Society of America 85:638–640
38. Schenck HA (1968) Improved integral formulation for acoustic radiation problems. Journal of the Acoustical Society of America 44:41–58

39. Seybert AF, Soenarko B (1988) Radiation and scattering of acoustic waves from bodies of arbitrary shape in a three–dimensional half space. Transaction of the ASME, Journal of Vibration, Stress and Reliability in Design 110:112–117
40. Seybert AF, Soenarko B, Rizzo FJ, Shippy DJ (1985) An advanced computational method for radiation and scattering of acoustic waves in three dimensions. Journal of the Acoustical Society of America 77:362–368
41. Shao TS, Chena TC, Frank RM (1964) Tables of zeros and Gaussian weights of certain associated Laguerre polynomals and the related generalized hermite polynomals. Mathematics of Computation 18:598–616
42. Sommerfeld A (1909) Über die Ausbreitung der Wellen in der drahtlosen Telegraphie (The propagation of waves in wireless telegraphy). Annalen der Physik 28:665–736
43. Sommerfeld A (1926) Über die Ausbreitung der Wellen in der drahtlosen Telegraphie (The propagation of waves in wireless telegraphy). Annalen der Physik 81:1135–1153
44. Sommerfeld A (1978) Partielle Differentialgleichungen der Physik (Partial differential equations in physics). Harri Deutsch, Thun–Frankfurt/Main
45. Wu TW (ed) (2000) Boundary element acoustics: Fundamentals and computer codes. WIT Press, Southampton–Boston
46. Ziolkowski RW (1985) Exact solutions of the wave equation with complex source locations. Journal of Mathematical Physics 26:861–863

18

Time Domain Boundary Element Method

Sabine Langer[1] and Martin Schanz[2]

[1] Institute of Applied Mechanics, Technical University Clausthal
Adolph–Roemer-Str. 2a, 38678 Clausthal–Zellerfeld, Germany
s.langer@tu-clausthal.de

[2] Institute of Applied Mechanics, Graz University of Technology
Technikerstraße 4, 8010 Graz, Austria
m.schanz@tugraz.at

Summary. A wide class of acoustical problems related to wave propagation phenomena should be treated directly in time domain. Among others, the response due to a transient signal is a good example. Clearly, a solution in frequency domain with a subsequent inverse transformation works well. However, a direct evaluation in time domain has to be preferred, especially, for cases where the boundary condition can not be transformed. Hence, an initial boundary value problem has to be solved by an appropriate time domain Boundary Element formulation.

First, the classical time stepping procedure based on the analytical solution of the convolution integral is recalled. As prerequisite of this formulation the time dependent fundamental solution has to be known. A different approach is based on the Convolution Quadrature Method which utilises the Laplace domain fundamental solution and solves the convolution in time numerically. This approach has the advantages that also a viscous fluid can easily tackled and that the numerical behaviour is improved.

Finally, one test example to show the behaviour with respect to the time discretisation is presented. Further, a more realistic example of an amphitheatre is tackled and the sound pressure level as well as the reverberation time is calculated.

18.1 Introduction

For the analysis of acoustical problems, the steady–state play an important role. We observe a frequency–dependent acoustical behaviour of materials, components, buildings, or machinery. E.g., a sufficient sound insulation or acceptable noise emission has to be proven by comparing averaged acoustical characteristics over a relevant frequency–band (see *weighted sound reduction index* or *averaged sound pressure level*).

On the other hand, there are also countless problems where we have a deep interest in following the sound wave travelling from source to receiver. We mention exemplary outdoor sound propagation problems, room acoustics, or the simulation of non–destructive testing. However, the motivation for using a time domain calculation for such problems is very individual. For outdoor sound propagation moving sound sources and/or moving surfaces are very important. They can be adequately implemented only in a time domain procedure. Further, to characterise the acoustical quality of a room various parameters are used, which can be directly calculated from the time dependent response; e.g., the *reverberation time*, the *clearness*, etc. For non–destructive testing we have to observe the time signal of the sound reflected at cracks and boundaries.

The practical relevance of acoustical calculations in time domain is the motivation for this paper dealing with the time domain Boundary Element Method (BEM). To get informations about a time–dependent acoustical behaviour two totally different strategies can be chosen. First, the inverse transformation from frequency (or Laplace) domain to the time domain [9, 10]. Second, the direct time domain BEM Formulation for wave propagation [20]. The direct evaluation in time domain seems often to be preferable, not only because it is more natural to work in time domain but also because the numerical inversion formulae depend on a proper choice of their parameters; see, e.g., [24].

When performing such a direct evaluation in time domain we have to keep in mind that an adequate choice of the time step is required and that an improper chosen time step size leads to instabilities or numerical damping; see, e.g., [4,27]. Hence, procedures to improve the classical dynamic time–stepping BE formulation are in the focus of former and current research. Antes and Jäger [4] and Yu and Mansur [34] developed modified numerical time marching procedures and in [35] a modified fundamental solution is presented to improve stability of the classical dynamic time–stepping BE formulation.

Beside the problem with numerical instabilities, which is inherent in most effective numerical methods, the BEM needs a fundamental solution. This involves two main difficulties: singularities appear and the availability of a fundamental solution is not given for every problem. The first problem is solved for a lot of physical relevant linear problems and is here not discussed further. To overcome the second problem one approach is the *Dual Reciprocity Technique* [25, 26]. In this methodology the differential operator under consideration is divided into a well known linear operator (often the Laplace operator or the elastostatic operator) and a residual part. This residual part may contain the non–linearities or even the acceleration terms, etc. Bringing the residual operator applied on the field variables on the right–hand side it can be treated as a generalised source term. When the differential equation is transformed into a boundary integral equation the residual operator causes a domain integral. This domain integral can be calculated directly by integration in volume cells, using higher–order fundamental solutions [12], Fourier expansion, the Multiple Reciprocity Method (MRM) or the *Dual Reciprocity Technique*. The latter is preferable because is can be used for all types of the domain integrals.

Another possibility to overcome the problem that in any time–dependent boundary element method formulation a time–dependent fundamental solution is required, is using the *Convolution Quadrature Method* (CQM) proposed by Lubich [17–19]. It makes it possible to establish a boundary element time–stepping procedure based only on the respective Laplace domain fundamental solution. Hence, the fundamental solution not for the hyperbolic but for the corresponding (transformed) elliptic operator has to be known. This is especially advantageous for visco– and poroelastic continua [28, 29]. Additionally, this quadrature formula improves the stability behaviour of the resulting time–stepping procedure [30].

Essentially, this method is a quadrature rule for the convolution integrals appearing in time–dependent integral equations. There, the fundamental solutions and the boundary data are convoluted with respect to the time variable. The advantage of the convolution quadrature method against other numerical methods to evaluate those convolution integrals is the property that the integration weights are determined using the Laplace transformed fundamental solutions. Therefore, a boundary element time–stepping procedure can be established without the knowledge of the time domain fundamental solutions. Some details of the convolution quadrature method can be found in the appendix A.

The chapter is organised in the following way. After a short view on the fundamental equations for wave propagation in ideal and viscous fluids, Section 18.2; the Kirchhoff integral equation is recalled briefly for stationary sound sources, moved point sources, and moved surfaces in Section 18.3. In Section 18.4 the two proposed Boundary Element formulations are derived. First, the classical formulation as suggested by Mansur [20] and then the CQM based methodology as suggested by Lubich [19]. The section is closed by a comparison of both techniques. Finally, the application of the method on acoustics of an amphitheatre is given in Section 18.5.

18.2 Sound Wave Propagation

Sound is always caused by mechanical vibrations. From the physical point of view a vibration is a current transformation from kinetic to potential energy and vice versa. This transformation would never stop if no damping mechanism are considered, which is an unrealistic assumption. The sound wave propagates without losses only under the assumption of an ideal fluid. The corresponding wave equation is given in Section 18.2.1.

We only observe a compression wave in such fluids and it propagates with a constant speed of sound c that is a real–valued measure. But, modelling realistic situations requires an adequate consideration of damping mechanism. We have to distinguish between the following mechanism: *damping at boundaries*, *geometrical damping*, and *internal losses*.

To define a well posed boundary value problem Dirichlet and Neumann boundary conditions must be introduced. A third type can also be used to model more realistic boundary conditions – the so–called *Robin* boundary condition. This is a linear combination of Dirichlet and Neumann type conditions. Whereas a sound wave is reflected on both Dirichlet and Neumann boundaries without losses, the Robin boundary condition can be formulated to model a dissipative boundary.

The so–called *geometric absorption* or *geometrical damping* is not a damping in the usual sense because it does not produce a dissipation of energy. The sound amplitude decreases only because the energy is distributed over a larger volume, e.g., a semi–infinite half space.

The *material damping* is the third important damping mechanism influencing the sound propagation in non–ideal media. The simplest model of a non–ideal fluid is that of a *viscous* fluid. This introduces a time dependency of the constitutive law which results in time dependent material constants. Consequently, the wave velocity is time dependent and decreases with time. In frequency domain this can be easily modelled by introducing a complex material parameter (wave velocity). The simplest choice is a frequency independent but complex value. In more realistic models the dependency on frequency can be highly sophisticated. Such models may be denoted visco–elastic [7].

When studying acoustical problems another group of materials with high internal loss are very important, the porous and poroelastic media. Porous media consist of a rigid skeleton and fluid filled pores, e.g., fibrous materials, see, e.g. [1]. The fluid flows through the pores and due to the interstitial friction energy is dissipated. A medium with an elastic skeleton and interconnected fluid–filled pores is called a poroelastic medium [6].

18.2.1 Ideal, Compressible Fluids

The governing equation for the wave propagation in an ideal fluid can be derived by combining the hydrodynamic equations of motion, the local mass balance, and the continuity equation. With certain restrictions and after linearisation it leads to the *scalar wave equation*

$$\nabla^2 p(\mathbf{x},t) - \frac{1}{c^2}\frac{\partial^2 p}{\partial t^2}(\mathbf{x},t) = \operatorname{div}\mathbf{b} - \rho\frac{\partial a}{\partial t}\,, \tag{18.1}$$

where p denotes the sound pressure and c the speed of sound. The non–homogeneity on the right hand side depends on the divergence of the body forces $\mathbf{b}$, on the source density a, and on the density ρ of the fluid in the original state. With ∇^2 the Laplace operator is denoted.

The domain under consideration $\Omega \in \mathbb{R}^3$ is bounded by a sufficiently smooth boundary $\Gamma \subset \mathbb{R}^2$ where the boundary conditions are prescribed. Without a significant loss of generality, here, a further simplification is introduced: the initial conditions $p(\mathbf{x}, t = 0) = 0$ and $\partial p/\partial t(\mathbf{x}, t = 0) = 0$ are homogeneous. On the Dirichlet part Γ_D of the boundary the pressure p and on the Neumann part Γ_N the sound flux $q = \partial p/\partial n$ is prescribed. Further, $\Gamma = \Gamma_D \cup \Gamma_N$ and $\Gamma_D \cap \Gamma_N = 0$ holds. Clearly, the last conditions have to be changed if also a Robin type boundary condition is considered.

18.2.2 Viscous Fluids

The sound propagation in a viscous fluid can be described with a modified sound field equation [23]. A factor that is proportional to the velocity has to be added to the scalar wave equation for an ideal, compressible fluid.

The dissipative scalar wave equation for the pressure distribution is

$$\nabla^2 p(\mathbf{x},t) - \frac{1}{c^2}\frac{\partial^2 p}{\partial t^2}(\mathbf{x},t) - \frac{R}{\rho c^2}\frac{\partial p}{\partial t}(\mathbf{x},t) = \operatorname{div}\mathbf{b} - \rho\frac{\partial a}{\partial t}, \tag{18.2}$$

where the flow resistance is denoted with R.

In general a decay coefficient μ is introduced to represent the amount of damping. The definition of this coefficient can be derived transforming Equation (18.2) into the frequency domain. This yields

$$\nabla^2 \check{p} + \left[\left(\frac{\omega}{c}\right)^2 + \mathrm{i}\omega\frac{R}{\rho c^2}\right]\check{p} = 0, \tag{18.3}$$

with the Fourier transformed pressure $\check{p}$. For simplicity the inhomogeneity is set to zero for the following considerations. The factor in the square brackets in Equation (18.3) is complex. Its real part corresponds to the wave number $k = \omega/c$ of an ideal fluid. The complex wave number of a viscous fluid can be defined by a series expansion of the square root

$$\overline{k} = \sqrt{\left(\frac{\omega}{c}\right)^2 + \mathrm{i}\omega\frac{R}{\rho c^2}} \approx \frac{\omega}{c}\left(1 + \mathrm{i}\frac{R}{2\rho\omega}\right) = k + \mathrm{i}\mu\,. \tag{18.4}$$

In this approximated representation, the absorption–coefficient or decay–coefficient $\mu := R/(2\rho c)$ can be identified.

The physical effects for producing such a decay coefficient are the so–called "classic absorption" and the molecular absorption, e.g., see [22]. Beside these damping effects based on the material or also called internal losses the geometric damping in semi– or infinite domains have to be considered. In [16] these mechanism are described in detail and BEM–studies in the frequency domain concerning the influence of these mechanics on outdoor sound propagation and on the uniqueness of the boundary integral representation has been presented.

In the following derivation of the time domain BE formulation which takes these damping effects into account the Laplace transformed version of (18.2) has to be used

$$\nabla^2 \hat{p} - \left[\left(\frac{s}{c}\right)^2 + s\frac{R}{\rho c^2}\right]\hat{p} = \operatorname{div}\mathbf{b} - \rho s\hat{a}\,. \tag{18.5}$$

In Equation (18.5), the Laplace variable is denoted by s and the Laplace transformed function with $\hat{()}$. The complex valued wave number in Laplace domain can be easily identified as

$$\lambda = \frac{s}{c}\sqrt{1 + \frac{R}{\rho s}}\,. \tag{18.6}$$

Clearly, the flow resistance R and the decay coefficient μ have to be the same as given above. If a frequency dependent damping mechanism like the molecular absorption is considered this frequency dependency must be carefully changed to a dependency on s.

18.3 Boundary Integral Equations

The Boundary Element Method is a numerical solution of the boundary integral equation of the physical problem. This integral equation can be obtained from the governing partial differential equation either by formulating a weighted residual statement, using Green's second identity, or a reciprocity formulation, etc. In all possibilities an essential ingredient is the knowledge of the so–called fundamental solution. Such a solution is defined as the solution of the underlying partial differential equation with a Dirac distribution as inhomogeneity.

In the following, first the fundamental solution in time domain as well as in the transformed domain is given. We restrict ourselves to the 3–d solutions for the sake of brevity.

18.3.1 Fundamental solutions

As mentioned above the fundamental solution is a special solution of the underlying partial differential equation. Mathematically spoken we seek a solution $G(\mathbf{x}, t|\mathbf{y}, \tau)$ with the following property

$$\begin{aligned} \int_0^t \int_\Omega \mathcal{L}\left(G(\mathbf{x}, t|\mathbf{y}, \tau)\right) p\left(\mathbf{y}, \tau\right) \mathrm{d}\Omega_y \mathrm{d}\tau &= \\ = \int_0^t \int_\Omega \delta\left(\mathbf{x} - \mathbf{y}\right) \delta\left(t - \tau\right) p\left(\mathbf{y}, \tau\right) \mathrm{d}\Omega_y \mathrm{d}\tau &= p\left(\mathbf{x}, t\right) . \end{aligned} \tag{18.7}$$

In (18.7), $\mathcal{L}\,()$ symbolise the differential operator of the underlying partial differential equation and $\delta\,()$ denotes the Dirac distribution. Clearly, in the Laplace or Fourier domain the convolution integral with respect to time reduces to a simple multiplication. In such cases, the solution depends only on frequency ω or the Laplace variable s instead of the time t. This property of the fundamental solution is often reformulated such that the following equation has to be fulfilled

$$\nabla^2 G(\mathbf{x}, t|\mathbf{y}, \tau) - \frac{1}{c^2} \frac{\partial^2 p}{\partial t^2} G(\mathbf{x}, t|\mathbf{y}, \tau) = -\delta(\mathbf{x} - \mathbf{y})\delta(t - \tau) . \tag{18.8}$$

In the Laplace domain the same equation reads

$$\nabla^2 \hat{G}(\mathbf{x}, \mathbf{y}) - \lambda^2 \hat{G}(\mathbf{x}, \mathbf{y}) = -\delta(\mathbf{x} - \mathbf{y}) , \tag{18.9}$$

with the non–dissipative version of λ. The extension to the viscous case is obvious.

Fundamental solutions fulfil some fundamental properties [2]:

- Causality: $G(\mathbf{x}, t|\mathbf{y}, \tau) = 0$ if $|\mathbf{x} - \mathbf{y}| > c(t - \tau)$
- Reciprocity: $G(\mathbf{x}, t|\mathbf{y}, \tau) = G(\mathbf{y}, -\tau|\mathbf{x}, -t)$
- Translation: $G(\mathbf{x}, t|\mathbf{y}, \tau) = G(\mathbf{x}, t + t_1|\mathbf{y}, \tau + t_1)$

Causality means that the sound pressure vanishes, if the distance of source and field point is larger than the distance, a wave can travel in the time $t - \tau$. The fact that field and source point can be exchanged if the sign of the time parameter is changed to ensure the chronological order is called reciprocity. Translation means that the solution depends only on the time difference between the source and the field point and not on the absolute time.

In the following, the fundamental solutions for Laplace/frequency and the time domain for both ideal and viscous fluids are recalled. As usual, in the following, the distance between the two points $\mathbf{x}$ and $\mathbf{y}$ will be denoted by $r = |\mathbf{x} - \mathbf{y}|$.

Laplace/Frequency Domain Fundamental Solutions

From a physical point of view the above given mathematical definition can be interpreted such that a solution of the Helmholtz equation with a point source as inhomogeneity has to be found. In Laplace domain the hyperbolic differential equation (18.2) is reduced to the elliptic modified Helmholtz equation (18.5). The respective fundamental solution is

$$\hat{G}(\mathbf{x}, \mathbf{y}) = \frac{1}{4\pi r} \mathrm{e}^{\lambda r} \tag{18.10}$$

with $\lambda = \frac{s}{c}$ or in case of the viscous fluid with λ from (18.6). In Fourier domain the solution of (18.3) with a respective right hand side is

$$\check{G}(\mathbf{x}, \mathbf{y}) = \frac{1}{4\pi r} \mathrm{e}^{-ikr} \ . \tag{18.11}$$

Also in (18.11), the viscous case is found by replacing the wave number k by its complex counterpart (18.4).

Both solutions consists of two factors: a static part $\frac{1}{r}$ which describes the decay to infinity as well the singularity and a second part which impresses the wave character of the solution. Obviously, in the transformed domain the inclusion of damping is simple.

Time Domain Fundamental Solutions

The *fundamental solution* of the wave equation in time domain represents the wave propagation due to an impulse at emission time τ in point $\mathbf{y}$. The 3–d *fundamental solution* is given by

$$G(\mathbf{x}, t|\mathbf{y}, \tau) = \frac{1}{4\pi r} \delta\left(t - \tau - \frac{r}{c}\right) \ . \tag{18.12}$$

In (18.12), it is assumed that the point source is not moving. This solution may be obtained by an inverse transformation of (18.10) or (18.11).

The need of a fundamental solution when applying a space–time integral equation method can be a serious restriction, because explicitly given and sufficiently simple fundamentals are known for far less parabolic and hyperbolic equations than for the corresponding elliptic equations. In principle, one can pass from the the frequency domain to the time domain by an inverse Laplace transform. But this representation can be rather complicated as can be shown for an example with high practical relevance: The wave propagation in fluids with internal losses. As shown in Section 18.2.2 the governing equation in the Laplace domain for a viscous fluid is the same as for an ideal fluid when replacing the wave number by its complex counterpart.

Performing an inverse Laplace transformation of the fundamental solution (18.10) for a dissipative fluid, i.e., using for λ the expression (18.6), leads to the time domain fundamental solution [8]

$$\begin{aligned} G(\mathbf{x}, t|\mathbf{y}, \tau) &= \frac{\mathrm{e}^{\alpha\frac{t-\tau}{2}}}{4\pi r} \Bigg[\delta\left(t - \tau - \frac{r}{c}\right) \\ &\quad + \frac{\alpha\, r}{2\sqrt{(t-\tau)^2 - \frac{r^2}{c^2}}}\, I_1\left(\frac{\alpha}{2}\sqrt{(t-\tau)^2 - \frac{r^2}{c^2}}\right) H\left(t - \tau - \frac{r}{c}\right)\Bigg] \end{aligned} \tag{18.13}$$

with the Heaviside function $H(t)$ and the modified Bessel function of first order I_1.

In many typical acoustical situations the geometry of the problem under consideration is bounded by a plane surface (e.g., outdoor sound propagation over ground). The sound is reflected depending on the boundary conditions on such a plane surface. On a rigid plane the reflection is symmetric and on a soft boundary the reflection is antimetric. Those situations can be handled very efficiently, using the so called *half–space fundamental solution*. Such a solution can be found using a simple idea. The half space surface is substituted by a mirror plane. Hence, the fundamental solution is composed by a part describing the primary source at point $\mathbf{x}$ and the other part is the imaginary image source at the mirrored point $\mathbf{x}'$. How both parts are combined reveals the influence of bounding surface. For a perfectly reflecting boundary with zero sound flux $q = 0$ the half space fundamental solution is

$$\bar{G}(\mathbf{x}, t|\mathbf{y}, \tau) = G(\mathbf{x}, t|\mathbf{y}, \tau) + G(\mathbf{x}', t|\mathbf{y}, \tau)\ , \tag{18.14}$$

and for a soft boundary with zero pressure $p = 0$ it is

$$\bar{G}(\mathbf{x}, t|\mathbf{y}, \tau) = G(\mathbf{x}, t|\mathbf{y}, \tau) - G(\mathbf{x}', t|\mathbf{y}, \tau)\ . \tag{18.15}$$

This combination of the full space fundamental solutions to a half space solution works as well in the transformed domain.

In practise sound sources are seldom stationary. Hence, an appropriate numerical treatment of *moving sound sources* is necessary. We assume a point source located in point $\mathbf{y}$ producing noise depending on a time–depending intensity function $Q(t)$. Such a point source can be modelled as a very small pulsating sphere with changing volume V_{sphere}. We consider the rate of this change in the following manner

$$a = \frac{\partial V_{sphere}}{\partial t} \delta(\mathbf{x} - \mathbf{y}) = Q(t)\delta(\mathbf{x} - \mathbf{y}) \; . \tag{18.16}$$

The source moves with constant velocity $\mathbf{v}$. That means that the location of the sound source is also time–dependent $\mathbf{y}(t)$. Using the pressure formulation leads to a complicated right hand side of the differential equation due to the necessary time derivation of the source term. Therefore, the velocity potential formulation is used in the following. In this formulation, no derivation of the source term with respect to time has to be performed.

The potential formulation for the current problem reads

$$\nabla^2 \phi - \frac{1}{c^2} \frac{\partial^2 \phi}{\partial t^2} = -Q(t)\delta(\mathbf{x} - \mathbf{y}(\tau)) \; . \tag{18.17}$$

Using the Lorentz–Transformation the velocity potential is for a subsonic sound (Mach–number $M < 1$) referring to [23]

$$\phi(\mathbf{x}, t) = \frac{Q(t - r/c)}{4\pi r[1 - M_R]} \; . \tag{18.18}$$

with the abbreviation $M_R = \mathbf{v} \cdot \mathbf{r}/(rc)$. We get *the fundamental solution for moving sound sources* using a moved coordinate system and the result from Equation (18.18) with $Q(t) = \delta(t - \tau)$

$$G_m(\mathbf{x}, t; \mathbf{y}, \tau) = \frac{\delta(t - \tau - r/c)}{4\pi[1 - M_R]} \; . \tag{18.19}$$

In the special case $\mathbf{v} = 0$, G_m simplifies to the fundamental solution for a stationary sound source. The fundamental solution (18.19) describes the velocity potential due to a point source that moves with a subsonic velocity. The corresponding sound pressure distribution contains different terms: a term of order $1/R^2$ that can be neglected in the far field and a term of order $1/R$ that differs from the fundamental solution by the factor $(1 - M_R)^{-1}$ and by the derivation of the source function [5].

18.3.2 Boundary Integral Representation

As before, not only the time domain but also the Laplace domain integral equation is shown to present the differences. For simpler notation, in the following, it is assumed that the body force term in (18.1) vanishes.

Laplace/Frequency Domain Boundary Integral Representation

Formulating a weighted residual statement on the Helmholtz equation with the fundamental solution $\hat{G}$ as weighting function and performing two partial integrations with respect to the spatial variable the integral equation in the Laplace domain is obtained

$$d(\mathbf{x})\hat{p}(\mathbf{x})+\int_\Gamma \hat{p}(\mathbf{y})\frac{\partial \hat{G}(\mathbf{x},\mathbf{y})}{\partial n}\mathrm{d}\Gamma_y = \int_\Gamma \hat{q}(\mathbf{y})\hat{G}(\mathbf{x},\mathbf{y})\,\mathrm{d}\Gamma_y - \int_\Omega \hat{a}(\mathbf{y})\hat{G}(\mathbf{x},\mathbf{y})\mathrm{d}\Omega_y\ , \tag{18.20}$$

where the normal derivative of the pressure is replaced by the flux, i.e., $q = \partial p/\partial n$. The Laplace transformed fundamental solution $\hat{G}(\mathbf{x},\mathbf{y})$ is given in Equation (18.10). In (18.20), the limit to the boundary has been performed taking care of the weakly and strongly singular kernel functions. The latter has caused the integral free term $d(\mathbf{x})$, (e.g. $d(\mathbf{x}) = 1$ for $\mathbf{x} \in \Omega \setminus \Gamma$ or $d(\mathbf{x}) = 0$ for $\mathbf{x} \notin \Omega$, and $d(\mathbf{x}) = 1/2$ for a smooth boundary). We refer to the previous chapters of the book for a detailed derivation. A simple exchange of the fundamental solution with the respective Fourier transformed one (18.11) yields the boundary integral equation in Fourier domain.

Time Domain Boundary Integral Representation

Applying weighted residuals on the wave equations (18.1) using the fundamental solution (18.12) as weighting function leads to the time domain integral representation of the problem. Performing a partial integration with respect to the spatial variable yields

$$\begin{aligned} p(\mathbf{x},t) = &\int_0^t \int_\Gamma \left[G(\mathbf{x},t|\mathbf{y},\tau)q(\mathbf{y},\tau) - \frac{\partial G}{\partial n}(\mathbf{x},t|\mathbf{y},\tau)p(\mathbf{y},\tau)\right] \mathrm{d}\Gamma_y \mathrm{d}\tau \\ &- \int_0^t \int_\Omega a(\mathbf{y},\tau)G(\mathbf{x},t|\mathbf{y},\tau)\mathrm{d}\Omega_y\mathrm{d}\tau\ . \end{aligned} \tag{18.21}$$

Next, the explicit expressions of the time domain fundamental solution (18.12) is inserted. Using the filter property of the Dirac distribution in time and taking the time derivation of it, the 3–d *Kirchhoff's Integral Representation* for $\mathbf{x} \in \Omega \setminus \Gamma$ is obtained

$$\begin{aligned} 4\pi p(\mathbf{x},t) = &\int_\Gamma q\left(\mathbf{y},t-\frac{r}{c}\right)\frac{1}{r}\mathrm{d}\Gamma_y + \int_\Gamma p\left(\mathbf{y},t-\frac{r}{c}\right)\frac{1}{r^2}\frac{\partial r}{\partial n}\mathrm{d}\Gamma_y \\ &+ \int_\Gamma \frac{\partial p}{\partial t}\left(\mathbf{y},t-\frac{r}{c}\right)\frac{1}{rc}\frac{\partial r}{\partial n}\mathrm{d}\Gamma_y - \int_\Omega a\left(\mathbf{y},t-\frac{r}{c}\right)\frac{1}{r}\mathrm{d}\Omega_y\ . \end{aligned} \tag{18.22}$$

This 3–d formulation is advantageous in comparison with the formulation in 2–d because the time integration is due to the properties of the fundamental solution not explicitly necessary. In 2–d, this time integration has to be performed but, fortunately, it can be done analytically.

Considering a number q_{num} of discrete point sources at the location ξ_j of intensity $A_j(t)$ we can write the function of the source term a in the equation above as follows

$$a(\mathbf{y},t) = \sum_{j=1}^{q_{num}} A_j(t)\delta(\xi_j - \mathbf{y})\ . \tag{18.23}$$

Hence, the last summand in equation (18.22) is

$$\int_\Omega a\left(\mathbf{y}, t - \frac{r}{c}\right)\frac{1}{r}\mathrm{d}\Omega_y = \sum_{j=1}^{q_{num}} \int_\Omega A_j(t)\delta(\xi_j - \mathbf{y})\frac{1}{r}\mathrm{d}\Omega_y$$
$$= \sum_{j=1}^{q_{num}} -A_j\left(t - \frac{|\mathbf{x} - \xi_j|}{c}\right)\frac{1}{|\mathbf{x} - \xi_j|}, \quad (18.24)$$

which can be evaluated directly.

Shifting $\mathbf{x}$ on the boundary ($\mathbf{x} \to \Gamma$) leads to the Boundary Integral Equation for wave propagation in an ideal fluid considering moving sound sources

$$4\pi d(\mathbf{x})p(\mathbf{x}, t) = \int_\Gamma q\left(\mathbf{y}, t - \frac{r}{c}\right)\frac{1}{r}\mathrm{d}\Gamma_y + \int_\Gamma p\left(\mathbf{y}, t - \frac{r}{c}\right)\frac{1}{r^2}\frac{\partial r}{\partial n}\mathrm{d}\Gamma_y$$
$$+ \int_\Gamma \dot{p}\left(\mathbf{y}, t - \frac{r}{c}\right)\frac{1}{rc}\frac{\partial r}{\partial n}\mathrm{d}\Gamma_y + \sum_{j=1}^{q_{num}} A_j\left(t - \frac{|\mathbf{x} - \xi_j|}{c}\right)\frac{1}{|\mathbf{x} - \xi_j|}$$
$$(18.25)$$

with the same integral–free term $d(\mathbf{x})$ as in the Laplace or Fourier domain. This is due to the fact that the singular behaviour of the time domain fundamental solutions are the same as in the transformed domain. A detailed derivation can be found in Jäger [15].

Assuming moving point sources is often not sufficient when studying realistic problems. Kirchhoff's integral equation has then to be expanded to moving surfaces to consider the convective effect of the moving boundaries. Two different ways can be used: You can start from the convective wave equation and using a convective fundamental solution. Spatial coordinates are used in this approach developed by Morino/Genneratti [13] which has been extended by Baaran [3,5]. Another approach uses material coordinates as presented by Fassarat and Myers [11]. They started from the pressure formulation of the homogeneous wave equation and used the theory of distributions to derive the *general Kirchhoff–equation for a moved body*. In the following only stationary sound sources are considered.

18.4 Boundary Element formulation

The Boundary Element formulation is derived from the Boundary integral equation by approximating the field variables sound pressure p and sound flux q in space and time. The spatial discretisation described in the following Subsection 18.4.1 do not differ for the Laplace/frequency and the time domain BEM. In Subsection 18.4.2 two methods are presented to solve numerically the initial–boundary value problem: The classical approach proposed by Mansur [20] and an emerging one introduced by Lubich [17] and further extended by Schanz [30].

18.4.1 Spatial Discretisation

Beside the approximation of the two–dimensional boundary of the 3–d domain by polynomial shape functions also the known and unknown data have to be approximated. The following ansatz is used

$$p(\mathbf{x},t) = \sum_{j=1}^{n} p_j(t)\varphi_j(\mathbf{x}) \qquad \text{and} \qquad q(\mathbf{x},t) = \sum_{j=1}^{n} q_j(t)\varphi_j(\mathbf{x}) \tag{18.26}$$

where $\varphi_j(\mathbf{x})$ are polynomial spatial shape functions and $p_j(t), q_j(t)$ represent the respective time dependent coefficients. In the examples, triangles with a linear approximation assuming a piecewise plane surface are used.

18.4.2 Time Discretisation

As mentioned above, there are two possible approaches to make the time discretisation.

Classical Approach

Discretisation of the time interval $[0, T]$ is done with equidistant time–steps Δt, i.e., the time axis is $0 = t_0, t_1, ..., t_n = n\Delta t = T$. For the time discretisation – in contrast to the space discretisation – one dimensional piecewise continuous shape functions $\theta_k(t)$ are used

$$p_j(t) = \sum_{k=0}^{n} p_j^k \theta_k(t) \qquad \text{and} \qquad q_j(t) = \sum_{k=0}^{n} q_j^k \theta_k(t) \; . \tag{18.27}$$

The nodal values at all times are p_j^k and q_j^k. Investigations of Meise [21] have shown that equally spaced sub–intervals with linear approximation for the time dependence are well suited to represent the pressure p. This leads to the approximation of $\dot{p}$ by constant functions. Hence, the time derivation of the sound pressure $\dot{p}$ is approximated with a backward difference quotient

$$\dot{p}(\mathbf{x},t) = \frac{p(\mathbf{x},t) - p(\mathbf{x},t-\Delta t)}{\Delta t} \; . \tag{18.28}$$

In Jäger [4, 15] a modification of this approach is found to improve stability of the numerical scheme.

Inserting these approximations in the boundary integral equation and performing a collocation for the time parameter at the time steps $t_m = m\Delta t, m = 1,\ldots,n$ deliver for an constant or linear θ_k

$$\begin{aligned} 4\pi d(\mathbf{x})\, p((\mathbf{x},t_m) =& \sum_{k=0}^{m}\sum_{j=1}^{n} \int_\Gamma \theta_k\left(t_m - \frac{r}{c}\right) \varphi_j(\mathbf{y})\mathrm{d}\Gamma_y\, q_j^k \\ &+ \sum_{k=0}^{m}\sum_{j=1}^{n} \int_\Gamma \left(\frac{1}{r^2} + \frac{1}{cr\Delta t}\right) \theta_k\left(t_m - \frac{r}{c}\right) \frac{\partial r}{\partial n}\varphi_j(\mathbf{y})\mathrm{d}\Gamma_y\, p_j^k \\ &- \sum_{k=0}^{m}\sum_{j=1}^{n} \int_\Gamma \frac{1}{cr\Delta t}\theta_k\left(t_{m-1} - \frac{r}{c}\right) \frac{\partial r}{\partial n}\varphi_j(\mathbf{y})\mathrm{d}\Gamma_y\, p_j^k \; . \end{aligned} \tag{18.29}$$

There are two possibilities to transform this equation in a system of discrete number of equations for the spatial variable, the collocation and the Galerkin method. Here, we will use in the examples the collocation method. For a detailed description of these techniques the reader is referred to the previous chapters.

The integrals over the boundary Γ are performed with standard Gaussian quadrature. The singular integrals are treated as in the Fourier domain. For the respective techniques we refer to the special chapters in this book. They can all be transferred to the time domain formulation.

Finally, the complete system of equation can be written in a matrix form. In (18.29), the dependence on the relative time $t - \tau$ is hidden in the temporal shape function. But, the equation must have this property due to the same property of the fundamental solutions. According to $t - \tau = (m - k)\,\Delta t$, the matrices are only dependent on the difference $m - k$ and, hence, a recursion formula for $m = 1, 2, \ldots, n$

$$\mathbf{C}_0 \mathbf{d}^m = \mathbf{D}_0 \bar{\mathbf{d}}^m + \sum_{k=1}^{m} \left(\mathbf{P}_k \mathbf{q}^{m-k} - \mathbf{Q}_k \mathbf{p}^{m-k} \right) \tag{18.30}$$

can be established. In (18.30), the matrices $\mathbf{P}_k$ and $\mathbf{Q}_k$ contain the integrated fundamental solutions for the pressure and the flux, respectively, at time step k. The vectors of unknowns of the passed time steps are $\mathbf{q}^{m-k}$ and $\mathbf{p}^{m-k}$. In $\mathbf{d}^m$ and $\bar{\mathbf{d}}^m$ the known and unknown boundary data, respectively, at the actual time step m are sorted. Hence, in the matrices $\mathbf{C}_0$ and $\mathbf{D}_0$ the integrated fundamental solutions of the first time step are given sorted with respect to the known and unknown boundary data. Finally, a direct equation solver is applied.

Convolution Quadrature Method

In the second approach to perform the time integration a quadrature rule is taken. This special quadrature named *Convolution Quadrature Method* (CQM) has been developed by Lubich [17, 18]. Their largest advantage is that only the Laplace transformed fundamental solutions have to be known and not the corresponding time dependent one, see, e.g. [30]).

The time–dependent fundamental solution for wave equation of an ideal acoustical fluid is given in the previous section. Thus, there is no need to chose the CQM to solve the time dependent integral Equation (18.21). But, to show the principle procedure of a CQM based formulation and to demonstrate its positive influence on the stability of the numerical scheme such a formulation is presented. For a shorter notation, in the following the term with the moving source is neglected.

As in the previous paragraph, an equidistant time time step size Δt is assumed so that $0 = t_0, t_1, ..., t_n = n\Delta t = T$ also holds. Applying the quadrature formula (18.34) to the convolution integrals in (18.21) and inserting the spatial shape functions (18.26) a time stepping procedure for $p\,(\mathbf{x}, m\Delta t)$ and $m = 0, 1, \ldots, n$ is achieved

$$d(\mathbf{x})p\left(\mathbf{x},m\Delta t\right) = \sum_{k=0}^{m}\sum_{j=1}^{n}\left[\omega_{m-k}\left(\hat{G},\mathbf{x},\Delta t\right)q_j\left(k\Delta t\right)\right.$$
$$\left. -\,\omega_{m-k}\left(\frac{\partial\hat{G}}{\partial n_y},\mathbf{x},\Delta t\right)p_j\left(k\Delta t\right)\right] . \tag{18.31}$$

The integration weights ω_{m-k} depend on the Laplace transformed fundamental solutions (18.10), the time step size, and on the underlying multistep method; for details see Appendix A or [30]. They are calculated following Equation (18.34) using for $\hat{G}$ the appropriate fundamental solution as indicated in the argument list, e.g.,

$$\omega_n\left(\hat{G},\mathbf{x},\Delta t\right) = \frac{\mathcal{R}^{-n}}{L}\sum_{\ell=0}^{L-1}\int_\Gamma \hat{G}\left(\mathbf{x},\mathbf{y},\frac{\gamma\left(\mathcal{R}\mathrm{e}^{\mathrm{i}\ell\frac{2\pi}{L}}\right)}{\Delta t}\right)\varphi_j(\mathbf{y})\mathrm{d}\Gamma\,\mathrm{e}^{-\mathrm{i}n\ell\frac{2\pi}{L}} . \tag{18.32}$$

In this formulation it is obvious that the spatial integration can be performed as in the Laplace/Fourier domain. Also the singular integration is the same.

As in the classical formulation a recursion formula can be established. For $m = 1,2,\ldots,n$ holds

$$\omega_0\left(\mathbf{C}\right)\mathbf{d}^m = \omega_0\left(\mathbf{D}\right)\bar{\mathbf{d}}^m + \sum_{k=1}^{m}\left[\omega_k\left(\mathbf{P}\right)\mathbf{q}^{m-k} - \omega_k\left(\mathbf{Q}\right)\mathbf{p}^{m-k}\right] \tag{18.33}$$

with the time dependent integration weights ω_m containing the Laplace transformed fundamental solutions. The other matrices have the similar meaning as in (18.30). Finally, a direct equation solver is applied. It must be mentioned that the time stepping procedure in (18.33) is formally the same as in (18.30).

18.4.3 Numerical example

To compare both time domain BEM formulations of the integral equation (18.22) the pressure response in a 3–d duct is determined. The duct is loaded by $q = 1H(t)\mathrm{N/m}^2$ at one end and at the other end a zero pressure is described. The side faces have zero flux boundary conditions. Hence, it is essentially a 1–d problem for which an analytical solution exists, see, e.g., [14].

The duct is discretised with 448 linear triangles with linear shape functions for pressure and flux, see Figure 18.1. For comparison the time step size is varied using a dimensionless variable $\beta = c\Delta t/r$ which relates the time step size to the spatial discretisation. The respective length r denotes the mean element length. In a 3–d calculation this mean value is somehow arbitrary because it is not clear which length of a triangle should be used. Here, the catetus is used whereas in other publications also the hypotenuse is taken. Hence, in the following results $r = 0.25$m is inserted to calculate β.

In Figure 18.2 the pressure response in the middle of the loaded end (see the white point in Figure 18.1) is plotted versus time for different time step sizes. In

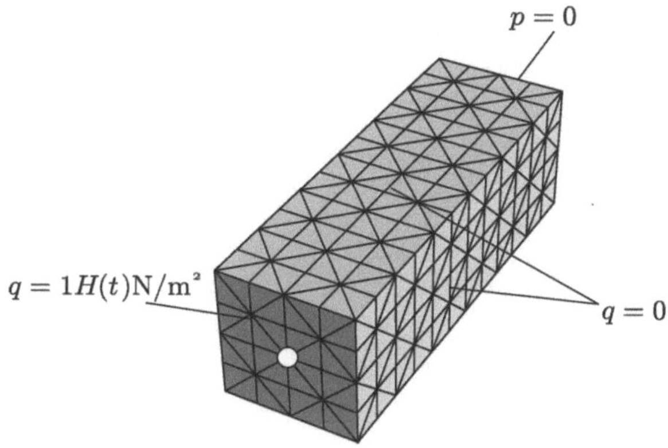

Fig. 18.1 Geometry and discretisation of the 3–d duct

Figure 18.2(a) the results are obtained with the CQM based technique whereas in Figure 18.2(b) he analytical time integration within each time step as proposed by Mansur [20] is plotted. The latter technique shows an instability at $\beta = 0.7$. Such an unstable behavior occurs also in the CQM based formulation but at a much smaller value, i.e., $\beta < 0.1$. For the classical formulation this value is not improved by a finer spatial discretisation whereas for the CQM a fine spatial mesh decreases this β–value. Further, increasing the time step size results in both formulations in a numerical damping. A physical upper restriction is $\beta = 1$ which corresponds to the time the wave travels over one element. Hence, higher values does not resolve the wave propagation correctly.

Concluding, the classical technique and the CQM have comparable result. However, the range of applicable time step sizes is for the classical formulation much narrower and this formulation is much more sensitive on the chosen time step. This insensitivity of the CQM is especially advantageous if (as in all practical applications) meshes with strongly varying element sizes are used. Further, if a coarser spatial mesh would be used the classical formulation does not yield satisfactory results at all contrary to the CQM where even meshes with a doubled size of the triangles give acceptable results. Tröndle presented comparable results in [33].

18.5 Application

The time domain BEM procedure based on the convolution quadrature method is here applied to study the acoustics of an amphitheatre to show that the method is suitable to study realistic problems. Its geometry and dimensions are depicted in Figure 18.3. It is discretised with 3474 linear triangles on 1761 nodes and a time step size of $\Delta t = 0.003125s$ is used. Sound reflecting boundary conditions are considered and the sound propagates in the air with a speed of $346\frac{m}{s}$. The sound field is excited by a prescribed sound flux of $q = 1N/m^2$ at the white point as depicted in Figure 18.3. The excitation stops after 50 time steps, i.e., after $t = 0.15625s$.

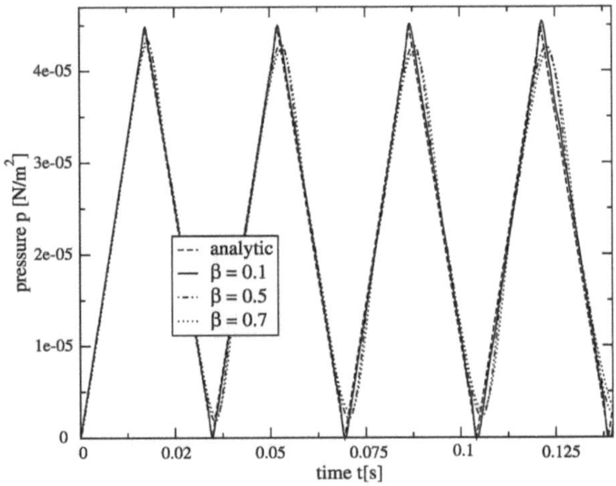

(a) CQM

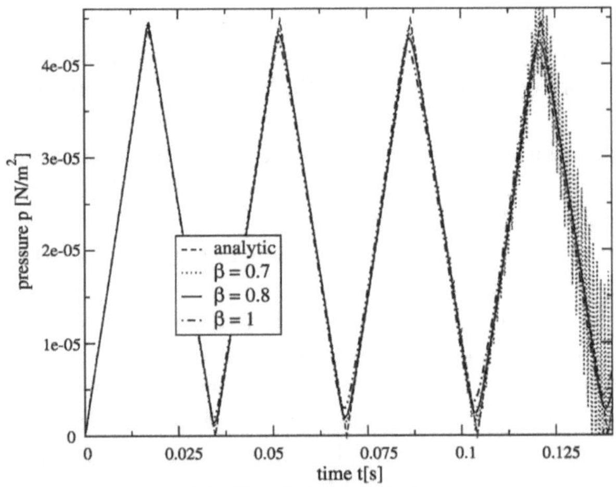

(b) Classical formulation

Fig. 18.2 Pressure response versus time at the loaded end: Variation of time step size

Contour plots of the sound pressure level distribution on the boundary of the amphitheatre are depicted in Figures 18.4 (3–D view) and 18.5 (top view) at the times $t = 0.00625s$, $t = 0.046875s$, and $t = 0.078125s$. Looking on the lowest sound pressure level it is observed that there are high negative values. Recalling the definition of the sound pressure level those values denote sound below the auditory threshold. All areas which are not reached by the sound wave at the observation time such low values are obvious. But also, after some time nothing more can be heard.

It is obvious that the sound screen behind the stage serve its purpose reflecting the sound energy to the audience. In addition we observe for the later time points

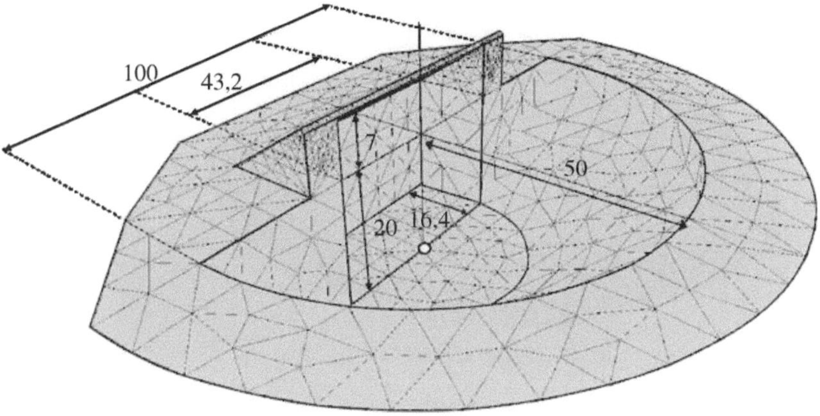

Fig. 18.3 Geometry and dimensions of amphitheatre [m] (○: location of excitation)

of observation a relevant diffraction at the edges of the sound screen so that the sound energy reaches the geometric shadow zone of the amphitheatre. This reflects one main advantage of the proposed method in comparison to standard ray–tracing calculations, They neglect the wave character of sound and cannot represent such effects as, e.g., diffraction.

In Figure 18.6, the decay of the sound pressure level over time is depicted. We can extract from this diagram the reverberation time T_{60} – one relevant measure of the quality of the configuration. It is defined as time in that after the end of excitation the sound pressure level decay is 60 dB. Approximately, we grip T_{30} – the time that is needed for a 30 dB decay – and double it to get the reverberation time T_{60} of 3 sec.

18.6 Conclusion

After recalling the basics of the governing equations for wave propagation in ideal as well as in non–ideal fluids, the respective fundamental solutions have been recalled. Based on these equations the boundary integral equation has been given.

The numerical approximation with boundary elements in a time stepping procedure can be done in two ways. Either using the time dependent fundamental solutions and perform the convolution integral analytically within each time step. Or, the other possibility is using the Convolution Quadrature Method which treats the convolution numerically utilizing the Laplace domain fundamental solution. Hence, this method can be applied on problems where time domain fundamental solutions are not known or where they are rather complicated as in the case of a viscous fluid. But, it also make sense for problems with known fundamental solution because – as shown in the numerical examples – the numerical scheme has a positive influence on the stability of the procedure. The method is here applied on acoustical studies of an amphitheatre.

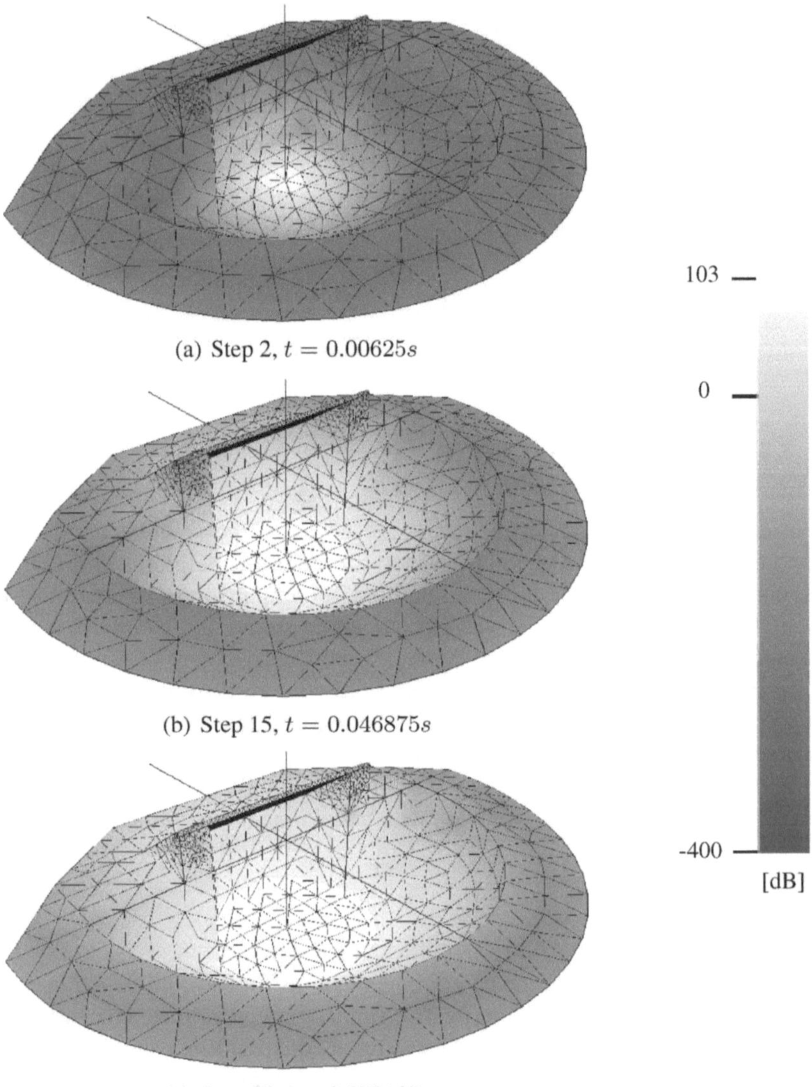

(a) Step 2, $t = 0.00625s$

(b) Step 15, $t = 0.046875s$

(c) Step 25, $t = 0.078125s$

Fig. 18.4 Sound pressure Level [dB] at the boundary of the amphitheatre

On a first view, it seems to be meaningless to apply the BEM to such studies like the amphitheatre because there are powerful alternative numerical schemes like ray–tracing available. But, these techniques have their restrictions, e.g., diffraction. The presented procedure based on the convolution quadrature methods lay an important basis for studies, e.g., in room acoustics, because it enables easy time domain BEM analysis of viscoelastic and poroelastic materials. In practice, e.g., absorbers are

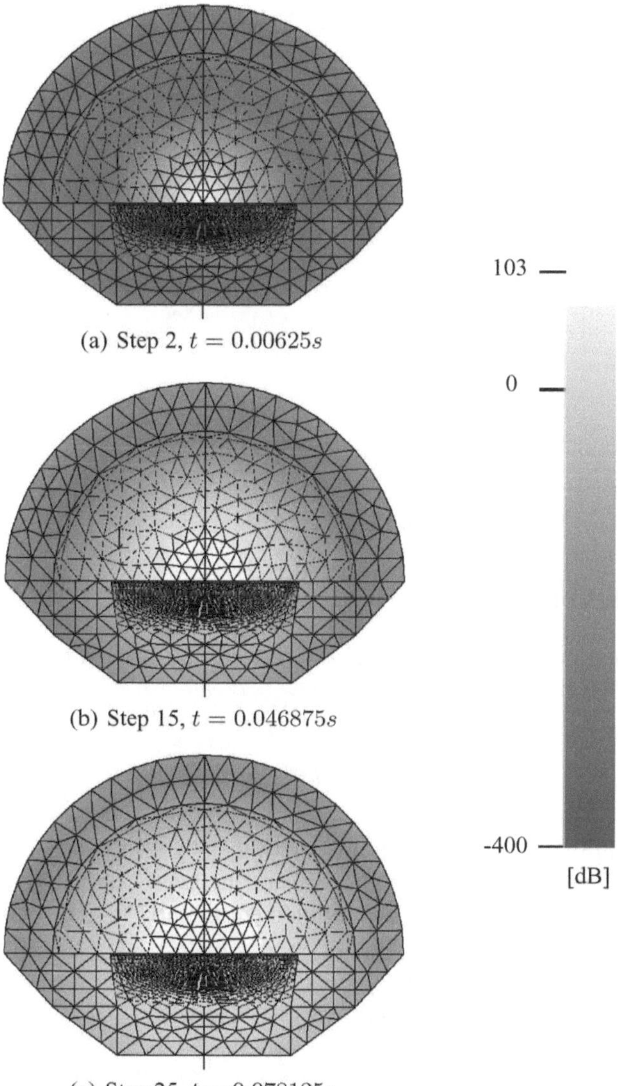

(a) Step 2, $t = 0.00625s$

(b) Step 15, $t = 0.046875s$

(c) Step 25, $t = 0.078125s$

Fig. 18.5 Sound Pressure Level [dB] on top view of the amphitheatre

used to influence the room acoustics. Combining the presented formulations for ideal and viscous fluids with those for viscoelastic and poroelastic media enables, e.g., to study complex room acoustic situations with a time domain Boundary Element formulation considering in detail the damping of sound. Generally spoken, the very important sound propagation in media with losses for practical use can be treated by the presented numerical scheme.

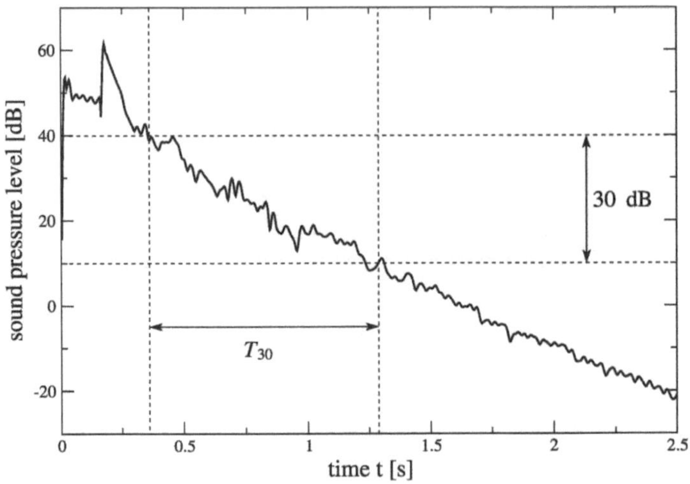

Fig. 18.6 Reverberation time

A Convolution Quadrature Method

The "Convolution Quadrature Method" developed by Lubich numerically approximates a convolution integral for $n = 0, 1, \ldots, N$

$$y(t) = \int_0^t f(t-\tau)\, g(\tau)\, \mathrm{d}\tau \quad \rightarrow \quad y(n\Delta t) = \sum_{k=0}^{n} \omega_{n-k}(\Delta t)\, g(k\Delta t)\,, \quad (18.34)$$

by a quadrature rule whose weights are determined by the Laplace transformed function $\hat{f}$ and a linear multi–step method. This method was originally published in [17] and [18]. Application to the boundary element method may be found in [30, 32]. Here, a brief overview of the method is given.

In formula (18.34), the time t is divided in N equal steps Δt. The weights $\omega_n(\Delta t)$ are the coefficients of the power series

$$\hat{f}\left(\frac{\gamma(z)}{\Delta t}\right) = \sum_{n=0}^{\infty} \omega_n(\Delta t)\, z^n \quad (18.35)$$

with the complex variable z. The coefficients of a power series are usually calculated with Cauchy's integral formula. After a polar coordinate transformation, this integral is approximated by a trapezoidal rule with L equal steps $\frac{2\pi}{L}$. This leads to

$$\begin{aligned} \omega_n(\Delta t) &= \frac{1}{2\pi i} \int_{|z|=\mathscr{R}} \hat{f}\left(\frac{\gamma(z)}{\Delta t}\right) z^{-n-1} \mathrm{d}z \\ &\approx \frac{\mathscr{R}^{-n}}{L} \sum_{\ell=0}^{L-1} \hat{f}\left(\frac{\gamma\left(\mathscr{R}\mathrm{e}^{i\ell\frac{2\pi}{L}}\right)}{\Delta t}\right) \mathrm{e}^{-in\ell\frac{2\pi}{L}}\,, \end{aligned} \quad (18.36)$$

where $\mathscr{R}$ is the radius of a circle in the domain of analyticity of $\hat{f}(z)$.

The function $\gamma(z)$ is the quotient of the characteristic polynomials of the underlying multi–step method, e.g., for a BDF 2, $\gamma(z) = 3/2 - 2z + 1/2z^2$. The used linear multi–step method must be $A(\alpha)$–stable and stable at infinity [18]. Experience shows that the BDF 2 is the best choice [28]. Therefore, it is used in all calculations in this paper.

If one assumes that the values of $\hat{f}(z)$ in (18.36) are computed with an error bounded by ϵ, then the choice $L = N$ and $\mathscr{R}^N = \sqrt{\epsilon}$ yields an error in ω_n of size $\mathcal{O}(\sqrt{\epsilon})$ [17]. Several tests conducted by the second author lead to the conclusion that the parameter $\epsilon = 10^{-10}$ is the best choice for the kind of functions dealt with in this paper [31]. The assumption $L = N$ leads to a order of complexity $\mathcal{O}(N^2)$ for calculating the N coefficients $\omega_n(\Delta t)$. Due to the exponential function at the end of formula (18.36) this can be reduced to $\mathcal{O}(N \log N)$ using the technique of the Fast Fourier Transformation (FFT).

References

1. Allard JF (1993) Propagation of sound in porous media. Elsevier Applied Science, London–New York
2. Antes H (1988) Anwendungen der Methode der Randelemente in der Elastodynamik und der Fluiddynamik. B.G. Teubner, Stuttgart
3. Antes H, Baaran J (2001) Noise radiation from moving surfaces. Engineering Analysis with Boundary Elements 25:725–740
4. Antes H, Jäger M (1995) On stability and efficiency of 3D acoustic BE procedures for moving noise sources. Computational Mechanics, Theory and Applications 2:3056–3061
5. Baaran J (1999) Schallfeldanalyse bei sich bewegenden schallerzeugenden Körpern. Braunschweiger Schriften zur Mechanik, Vol 38
6. Biot MA (1941) General theory of three–dimensional consolidation. Journal of Applied Physics, 12:155–164
7. Christensen RM (1971) Theory of viscoelasticity. Academic Press, New York
8. Costabel M (2005) Time–dependent problems with the boundary integral equation method. In: Stein E, de Borst R, Hughes TJR (eds) Encyclopedia of Computational Mechanics. Vol 1: Fundamentals. Chapter 25, John Wiley & Sons, New York–Chichester–Weinheim
9. Cruse TA (1968) A direct formulation and numerical solution of the general transient elastodynamic problem II. Journal of Mathematical Analysis and Applications 22:341–355
10. Cruse TA, Rizzo FJ (1968) A direct formulation and numerical solution of the general transient elastodynamic problem I. Journal of Mathematical Analysis and Applications 22:244–259
11. Fasserat F, Myers MK (1988) Extension of Kirchhoff's formula to radiation from moving surfaces. Journal of Sound and Vibration 123:451–460
12. Gaul L, Kögl M, Wagner M (2003) Boundary element methods for engineers and scientists. Springer–Verlag, Berlin–Heidelberg–New York-1
13. Gennaretti M, Morino L (1992) A boundary element method for the potential, compressible aerodynamics of bodies in arbitrary motion. Aeronautical Journal 96:15–19
14. Graff KF (1975) Wave motion in elastic solids. Oxford University Press, London

15. Jäger M (1994) Entwicklung eines effizienten Randelementverfahrens für bewegte Schallquellen. Braunschweiger Schriften zur Mechanik, Vol 17
16. Langer S (2004) BEM–studies of sound propagation in viscous fluids. In: Neittaanmäki P, Rossi T, Majava K, Pironneau O (eds) European Congress on Computational Methods in Applied Sciences and Engineering (ECCOMAS) 2004, Vol 1(1119), Jyväskyä
17. Lubich C (1988) Convolution quadrature and discretized operational calculus I. Numerische Mathematik 52:129–145
18. Lubich C (1988) Convolution quadrature and discretized operational calculus II. Numerische Mathematik 52:413–425
19. Lubich C (1994) On the multistep time discretization of linear initial–boundary value problems and their boundary integral equations. Numerische Mathematik 67:365–389
20. Mansur WJ (1983) A time–stepping technique to solve wave propagation problems using the boundary element method. Phd Thesis, University of Southampton
21. Meise T (1992) Randelementverfahren zur Berechnung der Ausbreitung skalarer Wellen im 3D Zeit– und Frequenzbereich. PhD Thesis, Ruhr–Universität Bochum
22. Meyer E, Neumann E–G (1979) Physikalische und technische Akustik. Vieweg, Braunschweig
23. Morse PM, Ingard KU (1986) Theoretical acoustics. Princeton University Press, Princeton
24. Narayanan GV, Beskos DE (1982) Numerical operational methods for time–dependent linear problems. International Journal for Numerical Methods in Engineering 18:1829–1854
25. Nardini D, Brebbia CA (1982) A new approach to free vibration analysis using boundary elements. In: Brebbia CA (ed) Proceedings of the 4th International Conference on BEM, Springer–Verlag, Berlin 313–326
26. Partridge PW, Brebbia CA, Wrobel LC (1992) The dual reciprocity boundary element method. Computational Mechanics Publication, Southampton
27. Peirce A, Siebrits E (1997) Stability analysis and design of time–stepping schemes for general elastodynamic boundary element models. International Journal for Numerical Methods in Engineering 40:319–342
28. Schanz M (1999) A boundary element formulation in time domain for viscoelastic solids. Communications in Numerical Methods in Engineering 15:799–809
29. Schanz M (2001) Application of 3–d boundary element formulation to wave propagation in poroelastic solids. Engineering Analysis with Boundary Elements 25:363–376
30. Schanz M (2001) Wave propagation in viscoelastic and poroelastic continua – a boundary element approach. Lecture Notes in Applied Mechanics, Vol 2, Springer–Verlag, Berlin–Heidelberg–New York
31. Schanz M, Antes H (1997) Application of "operational quadrature methods" in time domain boundary element methods. Meccanica 32:179–186
32. Schanz M, Antes H (1997) A new visco– and elastodynamic time domain boundary element formulation. Computational Mechanics 20:452–459
33. Tröndle G (1998) A comparison of two boundary element procedures in time domain. In: Fastl H (ed) Proceedings of Euronoise '98, München 129–134
34. Yu G, Mansur WJ, Carrer JAM, Gong L (1998) A linear Θ method applied to 2D time domain BEM. Journal of Mathematical Analysis and Applications 14:1171–1179
35. Yu G, Mansur WJ, Carrer JAM, Gong L (2000) Stability of Galerkin and collocation time domain boundary element methods as applied to the scalar wave equation. Computers and Structures 74:495–506

Part VI

BEM: Related Problems

19

Coupling a Fast Boundary Element Method with a Finite Element Formulation for Fluid–Structure Interaction

Lothar Gaul, Dominik Brunner, and Michael Junge

Institute of Applied and Experimental Mechanics, Pfaffenwaldring 9
70569 Stuttgart, Germany
gaul@iam.uni-stuttgart.de

Summary. Fluid–structure interaction plays a crucial role for simulating vibro–acoustic, multi–field problems. The paper starts with the fundamentals of the boundary element method for acoustics, including the Burton–Miller formulation for the solution of exterior problems. The fast multilevel–multipole algorithm is applied together with a preconditioned iterative solver yielding a quasi–linear numerical complexity. The structural part is modelled by the finite element method. Various coupling schemes are discussed depending on the influence of the feedback of the fluid onto the structure. In a first example, an expansion chamber, which is exposed to structural vibrations caused by interior pressure fluctuations, is presented. A weak coupling scheme is used for simulating the surface radiated sound. In a second example, the vibro–acoustic behavior of a totally submerged submarine–like structure is investigated. Two strong coupling formulations are compared for their numerical efficiency. The examples show, that the presented coupling algorithms are well–suited for simulating large–scale industrial applications.

19.1 Introduction

The numerical simulation of engineering applications often requires dealing with multi–field problems. Fluid–structure interaction incorporates the mutual influence of an acoustic and a structural domain. This interaction occurs at the coupling interface between the two adjacent domains [15,20,51]. The coupling is characterized by the strength of the influence between the two domains. Depending on the underlying physical problem, the correct choice of the coupling scheme is crucial for an efficient and robust numerical solution. For a high acoustic impedance mismatch between the domains, the structural and the acoustic problem can be solved subsequently one after the other. Thus, the so–called weak coupling is defined by neglecting the feedback of the acoustic pressure onto the structure. The simulation of the sound radiation of a stiff structure which is surrounded by air is a typical example where weak coupling is applied. In contrast to this, for the simulation of the sound radiation of a structure totally submerged in a dense fluid a strong coupling scheme has to be applied, since the feedback of the pressure can no longer be neglected. Here, the acoustic impedance and the structural impedance are comparable [35]. For geometrically simple structures analytical solutions are available [36,53]. More complex structures are investigated by making use of discretization methods. For coupled interior acoustic problems the finite element method (FEM) is widely used [2,33,64,66]. The boundary element method (BEM) shows its strength for the solution of the exterior acoustic problem, since the Sommerfeld radiation condition is automatically fulfilled. A common implementation is the collocation method, where the residual is set to zero at a discrete number of collocation points [18,38,65]. A second possibility are Galerkin formulations, where shape functions are used as weighting and test functions.

Classical BEM formulations suffer from fully populated matrices leading to a bottleneck both in memory consumption and computing time. One possibility to overcome this problem is the use of the fast multipole algorithm [27, 28, 55]. A review on the multipole boundary integral method for various applications is presented in [49]. Other fast methods are discussed in [54]. The important application of acoustics is considered in [21, 58, 61]. For the solution of large–scale problems in acoustics iterative solvers are well–suited [42, 50, 56]. These solvers no longer require an assembled system matrix but just the result of a matrix–vector product. In combination with the multipole algorithm this leads to a high efficiency both in memory consumption and in computing time.

In order to benefit from the advantages of both the FE and the BE method many researchers aim to combine them. There are numerous publications in the field of FE/BE coupling. To mention them all is far beyond the scope of this introduction. A symmetric Galerkin–type coupling scheme is proposed for instance in [43] for 2D and in [11,32,44] for 3D. Non–symmetric formulations are covered by [19,31,60]. As opposed to the use of conforming meshes in the aforementioned papers, non–conforming coupling schemes allow an independent mesh size adaption for each domain [3,5,22]. In this way, a high flexibility in mesh generation is combined with an efficient coupling implementation [25]. A comparison between the different approaches is found in [8]. A coupling formulation using the hybrid boundary element method for the fluid domain is proposed in [26,63]. In all previously mentioned publications the acoustic fluid is assumed to be at rest. An extension to problems with mean flow is given in [30,52].

For large structural models with a small fluid–structure interface the treatment of the FE part is expensive and may be a dominant part in the solution process, especially for densely sampled frequency sweep computations. Model reduction techniques are known to accelerate the treatment of the structural problem. The interested reader is referred to [37,41,45].

This contribution is organized as follows: First, an introduction into the boundary element method in acoustics and the efficient implementation using the fast multipole algorithm is given. In the next section, a finite element formulation for the structural domain is presented, which will provide a basis for the subsequent sections. There, different strategies for fluid–structure coupling are discussed. In the first part, a strongly coupled interior formulation is combined with a weakly coupled exterior scheme. This approach is then used to simulate the vibro–acoustic behav-

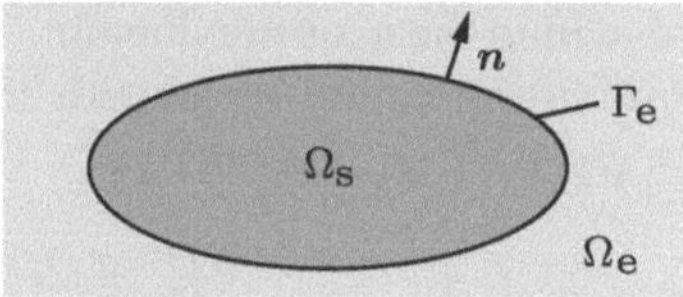

Fig. 19.1 Domains of the exterior acoustic problem. The acoustic domain Ω_e is in contact with the structure Ω_s on the interface Γ_e.

ior of a large–scale engineering application. In the second part, two strong coupling schemes for the solution of the exterior problem are investigated and compared to each other with respect to their efficiency. Finally, the contribution concludes with a brief summary.

19.2 Boundary Element Formulation for the Fluid Domain

Below, the governing equations for the fluid are derived in the frequency domain. Throughout this contribution the time–harmonic behavior $e^{-i\omega t}$ will be applied, where i is the imaginary unit, $\omega = 2\pi f$ denotes the angular frequency and f is the excitation frequency.

In the following, the exterior acoustic problem is considered as depicted in Figure 19.1. The normal on the boundary is always assumed to point into the fluid. On the interface Γ_e, the acoustic domain Ω_e is in contact with the vibrating structural domain Ω_s. The starting point of the fluid formulation is the linear, lossless wave equation [39,46]

$$\nabla^2 p(\boldsymbol{x},t) - \frac{1}{c^2}\frac{\partial^2 p(\boldsymbol{x},t)}{\partial t^2} = 0\,, \tag{19.1}$$

where c is the sound velocity of the acoustic fluid and p denotes the acoustic pressure. In the time-harmonic case (19.1) yields the Helmholtz equation

$$\nabla^2 p(\boldsymbol{x}) + \kappa^2\, p(\boldsymbol{x}) = 0, \tag{19.2}$$

which is valid for the pressure p at an arbitrary point $\boldsymbol{x}$ within the exterior acoustic domain Ω_e. The circular wavenumber is denoted by $\kappa = \omega/c$. A weak form of the Helmholtz equation is obtained by weighting with the fundamental solution

$$P^*(\boldsymbol{x},\boldsymbol{y}) = \frac{e^{i\kappa r}}{4\pi r}, \tag{19.3}$$

where $r = |\boldsymbol{x} - \boldsymbol{y}|$ denotes the distance between the load and the field point. Applying Green's second theorem yields the representation formula

$$p(\boldsymbol{x}) = -\int_{\Gamma_e} P^*(\boldsymbol{x},\boldsymbol{y})\,\frac{\partial p(\boldsymbol{y})}{\partial \boldsymbol{n}_y}\mathrm{d}s_y + \int_{\Gamma_e} \frac{\partial P^*(\boldsymbol{x},\boldsymbol{y})}{\partial \boldsymbol{n}_y}\,p(\boldsymbol{y})\,\mathrm{d}s_y, \tag{19.4}$$

which is valid for a point $\boldsymbol{x}$ within the acoustic domain. The singular boundary integral equation is obtained by shifting $\boldsymbol{x}$ onto the smooth boundary. With the definition of the acoustic flux $q(\boldsymbol{y}) = \partial p(\boldsymbol{y})/\partial \boldsymbol{n}_y$ one obtains

$$\frac{1}{2}\,p(\boldsymbol{x}) = -\underbrace{\int_{\Gamma_e} P^*(\boldsymbol{x},\boldsymbol{y})\,q(\boldsymbol{y})\,\mathrm{d}s_y}_{(Vq)(\boldsymbol{x})} + \underbrace{\int_{\Gamma_e} \frac{\partial P^*(\boldsymbol{x},\boldsymbol{y})}{\partial \boldsymbol{n}_y}\,p(\boldsymbol{y})\,\mathrm{d}s_y}_{(Kp)(\boldsymbol{x})}\,, \qquad \boldsymbol{x} \in \Gamma_e\,. \tag{19.5}$$

The single layer potential is denoted by V and the double layer potential by K, respectively. Analogously, the hypersingular boundary integral equation is derived by an additional derivative with respect to the normal $\boldsymbol{n}_x$

$$\frac{1}{2}q(\boldsymbol{x}) = -\underbrace{\int_{\Gamma_e} \frac{\partial P^*(\boldsymbol{x},\boldsymbol{y})}{\partial \boldsymbol{n}_x}\,q(\boldsymbol{y})\,\mathrm{d}s_y}_{(K'q)(\boldsymbol{x})} + \underbrace{\int_{\Gamma_e} \frac{\partial^2 P^*(\boldsymbol{x},\boldsymbol{y})}{\partial \boldsymbol{n}_x \partial \boldsymbol{n}_y}\,p(\boldsymbol{y})\,\mathrm{d}s_y}_{-(Dp)(\boldsymbol{x})}\,, \qquad \boldsymbol{x} \in \Gamma_e\,, \tag{19.6}$$

where K' denotes the adjoint double layer potential and D is the hypersingular operator.

19.2.1 Burton–Miller Formulation

For exterior acoustic problems, neither the singular boundary integral equation (19.5) nor the hypersingular boundary integral equation (19.6) have a unique solution for all frequencies. The boundary integral equation fails for frequencies representing the eigenfrequencies of the corresponding interior Dirichlet problem. On the other hand, the hypersingular integral equation breaks down at eigenfrequencies of the associated interior Neumann problem. They are of pure mathematical nature and do not represent physical eigenfrequencies of the exterior problem. Different possibilities to overcome this problem are known, including approaches of Brakhage and Werner [6], Schenk [57] and Burton and Miller [9]. In the following the latter one is applied, which uses a linear combination of both integral equations (19.5) and (19.6). Typically the factor i/κ is used, yielding

$$\left(-\frac{1}{2}I + K\right)p(\boldsymbol{x}) + \frac{\mathrm{i}}{\kappa}(Dp)(\boldsymbol{x}) = (V\,q)(\boldsymbol{x}) - \frac{\mathrm{i}}{\kappa}\left(\frac{1}{2}I + K'\right)q(\boldsymbol{x}). \tag{19.7}$$

There are various methods to derive an algebraic system of equations. Besides the well known collocation method, see e.g. [38, 58, 65], the Galerkin method is used, for which a rigorous error analysis exists. A Galerkin formulation is obtained by testing (19.7) on the interface Γ_e with linear functions ν, yielding

$$\begin{aligned}\int_{\Gamma_e} \nu(\boldsymbol{x})\left(-\frac{1}{2}I + K\right)p(\boldsymbol{x})\,\mathrm{d}\Gamma_x + \frac{\mathrm{i}}{\kappa}\int_{\Gamma_e} \nu(\boldsymbol{x})(Dp)(\boldsymbol{x})\,\mathrm{d}\Gamma_x = \\ = \int_{\Gamma_e} \nu(\boldsymbol{x})(V\,q)(\boldsymbol{x})\,\mathrm{d}\Gamma_x - \frac{\mathrm{i}}{\kappa}\int_{\Gamma_e} \nu(\boldsymbol{x})\left(\frac{1}{2}I + K'\right)q(\boldsymbol{x})\,\mathrm{d}\Gamma_x.\end{aligned} \tag{19.8}$$

The next step is to introduce admittance boundary conditions Y as they are used for many engineering applications

$$p(\boldsymbol{y})\,Y(\boldsymbol{y}) = v_{\mathrm{f}}(\boldsymbol{y}) - v_{\mathrm{s}}(\boldsymbol{y})\,. \tag{19.9}$$

Here, the particle velocity in the normal direction is denoted by v_{f} and the normal surface velocity of a vibrating structure is denoted by v_{s}. The relation between flux and particle velocity is obtained from Euler's equation $q(\boldsymbol{y}) = \mathrm{i}\omega\varrho_{\mathrm{f}}\, v_{\mathrm{f}}(\boldsymbol{y})$, where ϱ_{f} denotes the density of the fluid. Using this equation, the unknown particle velocity in (19.9) is replaced by the flux and substituted into (19.8). At the next stage, a discretization of Γ_{e} with triangular elements is introduced. Linear shape functions for p and constant shape functions for Y and v_{s} are applied to interpolate the data. This finally leads to the system

$$\underbrace{\left(-\frac{1}{2}\boldsymbol{I} + \boldsymbol{K} + \frac{\mathrm{i}}{\kappa}\boldsymbol{D} - \tilde{\boldsymbol{V}} + \frac{\mathrm{i}}{2\kappa}\tilde{\boldsymbol{I}}' + \frac{\mathrm{i}}{\kappa}\tilde{\boldsymbol{K}}'\right)}_{\boldsymbol{K}_{\mathrm{BE}}}\boldsymbol{p} = \underbrace{\left(\boldsymbol{V} - \frac{\mathrm{i}}{2\kappa}\boldsymbol{I}' - \frac{\mathrm{i}}{\kappa}\boldsymbol{K}'\right)\bar{\boldsymbol{v}}_{\mathrm{s}}}_{\boldsymbol{b}}\,. \tag{19.10}$$

Operators with tilde $(\tilde{\cdot})$ incorporate the prescribed admittance boundary conditions $\bar{\boldsymbol{Y}}$. For a pure Neumann problem with a rigid surface, these operators will disappear and only the structural velocity $\bar{\boldsymbol{v}}_{\mathrm{s}}$ remains as boundary data.

Numerical Complexity

A serious drawback of classical BE methods is, that setting up and storing the fully populated matrix $\boldsymbol{K}_{\mathrm{BE}}$ has a complexity of order $\mathcal{O}(N^2)$, where N denotes the number of degrees of freedom (DOF). If additionally a direct solver is used in combination with this fully populated matrix, the numerical expense is of order $\mathcal{O}(N^3)$. Obviously, standard BE methods are not practical for large scale problems having more than a thousand nodes. To overcome this bottleneck, a two-stage strategy is applied in the following. Instead of a direct solver it is more advantageous to apply preconditioned Krylov based methods like GMRES and BICGSTAB [21,56,58,62]. They also have the benefit, that only a matrix–vector product is needed during the solution procedure. In case of the BE method, such a matrix–vector product can be efficiently computed by the fast multilevel multipole algorithm with a numerical complexity of order $\mathcal{O}(N\log^2 N)$. With this strategy, even large scale problems with more than 500,000 DOFs can easily be simulated.

Multilevel Multipole Algorithm

For the introduced operators, see Section 19.2, one typically has to evaluate potentials of the type

$$\Phi(\boldsymbol{x}_b) = \sum_{a=1}^{A} \frac{\mathrm{e}^{\mathrm{i}\kappa|\boldsymbol{x}_b - \boldsymbol{y}_a|}}{|\boldsymbol{x}_b - \boldsymbol{y}_a|}\, q_a\,, \tag{19.11}$$

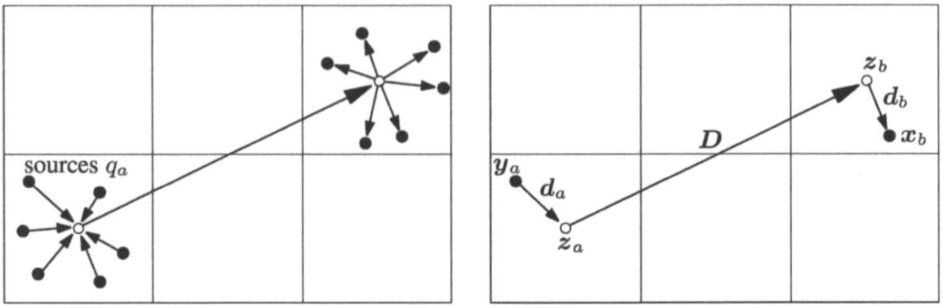

Fig. 19.2 Clustering of the sources q_a (left) and splitting up of the vector between load point and field point into three parts (right).

where q_a denotes the source strength of A sources and $|\boldsymbol{x}_b - \boldsymbol{y}_a|$ is the distance between the load and field point. Standard BE methods typically consider the interaction of every combination of a load point with a field point. In contrast to this, the multipole algorithm sets up a clustering and sums up the contribution of all sources q_a in the center $\boldsymbol{z}_a$ of a cluster, see Figure 19.2. At the next step, this so–called far–field signature is translated to the center $\boldsymbol{z}_b$ of the other clusters and from there finally distributed to $\boldsymbol{x}_b$.

From a mathematical point of view, the separation of the distance $|\boldsymbol{x}_b - \boldsymbol{y}_a|$ succeeds in the fundamental solution by using the diagonal form of the multipole expansion [55]

$$\frac{\mathrm{e}^{\mathrm{i}\kappa|\boldsymbol{x}_b-\boldsymbol{y}_a|}}{|\boldsymbol{x}_b-\boldsymbol{y}_a|} = \frac{\mathrm{i}\kappa}{4\pi}\sum_{l=0}^{\infty}(2l+1)\,\mathrm{i}^l h_l^{(1)}\left(\kappa|\boldsymbol{D}|\right)\int_{\mathbb{S}^2}\mathrm{e}^{\mathrm{i}\kappa(\boldsymbol{d}_a+\boldsymbol{d}_b)\cdot\boldsymbol{s}}P_l\left(\boldsymbol{s}\cdot\hat{\boldsymbol{D}}\right)\,\mathrm{d}s\,, \tag{19.12}$$

with the Hankel functions h_ℓ and the Legendre polynomials P_ℓ. The vectors which are local to the clusters are denoted by $\boldsymbol{d}_a$ and $\boldsymbol{d}_b$, see Figure 19.2, whereas $\boldsymbol{D}$ is defined by the centers of two interacting clusters, respectively. Since one can not compute an infinite sum, the series has to be truncated. In this case the integration over the unit sphere $\mathbb{S}^2$ and the summation can be interchanged. Introducing the translation operator

$$M_L(\boldsymbol{s},\boldsymbol{D}) = \sum_{\ell=0}^{L}(2\ell+1)\mathrm{i}^\ell h_\ell^{(1)}(\kappa|\boldsymbol{D}|)P_\ell(\boldsymbol{s}\cdot\hat{\boldsymbol{D}})\;, \tag{19.13}$$

with $\hat{\boldsymbol{D}} = \frac{\boldsymbol{D}}{|\boldsymbol{D}|}$, the original potential (19.11) can now be expressed in the form

$$\Phi(\boldsymbol{x}_b) = \frac{\mathrm{i}\kappa}{4\pi}\int_{\mathbb{S}^2}\mathrm{e}^{\mathrm{i}\kappa\,\boldsymbol{d}_b\cdot\boldsymbol{s}}M_L(\boldsymbol{s},\boldsymbol{D})\underbrace{\sum_{a=1}^{A}\mathrm{e}^{\mathrm{i}\kappa\,\boldsymbol{d}_a\cdot\boldsymbol{s}}q_a}_{F(\boldsymbol{s})}\,\mathrm{d}s. \tag{19.14}$$

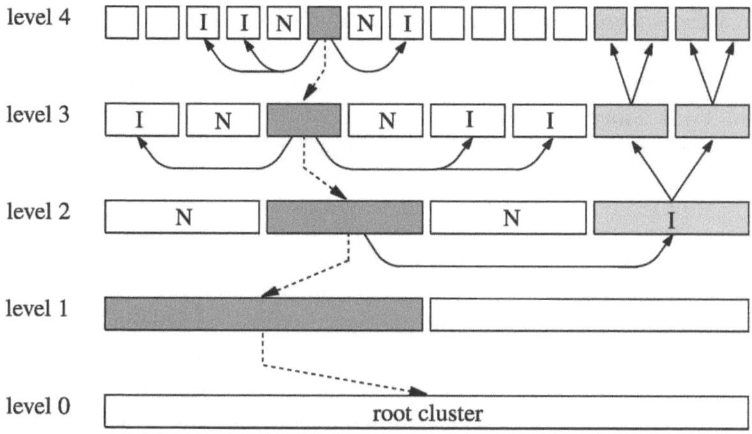

Fig. 19.3 Simplified model of a multilevel clustertree, which is created by bisectioning of the root cluster. Interaction clusters are denoted by I whereas near–field clusters are represented by N.

The choice of L, which is called the *expansion length*, has a significant influence on the accuracy and the performance of the multipole algorithm. Proper choice helps to circumvent divergence of the series and will be discussed later in this section. The sum on the right hand side of (19.14) is called the *far–field signature*. It is local to the cluster with the sources q_a, since only the vector $\boldsymbol{d}_a$ appears. In contrast to this, the translation operator M_L only depends on the vector $\boldsymbol{D}$ between two clusters' centers. Thus, if a regular cluster grid is used, the translation operators can be reused. Translating the far–field signature to another cluster using a translation operator forms the so called *near–field signature*. The solution is finally recovered by an exponential function of $\boldsymbol{d}_b$ and an integration over the unit sphere.

Since the multipole expansion is only valid for well separated load and field points, one has to split up the clusters into a near–field and far–field. All clusters which fulfill the condition

$$|\boldsymbol{D}| < c_d \frac{d}{2} \tag{19.15}$$

form the near–field. Here, d denotes the cluster diameter and c_d is a constant. The arising near–field is represented by a sparse matrix. It has to be evaluated by classical BEM. All other clusters are in the far–field and form the so called *interaction list.*

Multilevel algorithm

To obtain an optimal efficiency, a hierarchic multilevel cluster tree is used, see Figure 19.3. It is set up by consecutive bisectioning such that a mother cluster is divided into two son clusters on the next level. The procedure starts with the *root cluster*, which is the smallest parallelepiped containing all elements of the model. The division is stopped if a specified number of elements per cluster is reached. These final clusters, which do not have any sons, are called *leaf clusters*. The interaction list of every cluster is formed by those clusters, which are in the near-field of the *mother cluster* but not in its own near–field, see Figure 19.3.

Obviously, the far–field signature has to be translated to the interaction lists on different levels. Since the cluster diameters are different on every level, the expansion length L has to be adapted to every level, too. Typically the well–established semi–empirical rule [14]

$$L\left(\kappa\, d_\ell\right) = \kappa d_\ell + c_\mathrm{e} \log\left(\kappa\, d_\ell + \pi\right) \tag{19.16}$$

is used to estimate the number of series terms on level ℓ of the cluster tree. The parameter c_e has to be chosen by the user and determines the desired accuracy. In order to maintain the accuracy of the multipole expansion when the cluster diameter increases on the next level, an interpolation and filtering strategy has to be applied. It is advantageous to use a fast Fourier transform for this purpose. This is because new far–field directions have to be added, which is only possible for the original form of the multipole expansion. A more detailed description can be found in [21, 29, 58].

The resulting multilevel *fast multipole method*(FMM) has a quasi linear complexity of order $\mathcal{O}(N \log^2 N)$. For a detailed investigation of the numerical complexity, the reader is referred to [25].

The evaluation of the matrix–vector product with the FMM algorithm is similar for all operators introduced at the beginning of this section. Only slight modifications are necessary to take into account the different test and shape functions. The general procedure can be summarized with the following steps:

1. Compute the near–field part by a sparse matrix–vector multiplication.
2. Evaluate the far–field signature $F(\boldsymbol{s})$ for every leaf cluster.
3. Translate the far–field signature to all interaction cluster by means of the translation operators (Equation 19.13) and sum it up as the near–field signature $N(\boldsymbol{s})$ there.
4. Shift the far–field signature to the mother cluster and repeat step 3 until the interaction list is empty.
5. Go the opposite direction and shift the near-field signature $N(\boldsymbol{s})$ to the son clusters until the leaf clusters are reached.
6. Recover the solution by integration over the unit sphere.

19.3 Finite Element Formulation for the Structural Domain

As shown in Section 19.2 the FMM–BEM is well–suited for the solution of Neumann problems, where the normal velocity v_s is prescribed on the boundary of Γ_e. In practical applications the boundary Γ_e is represented by the surface of the considered objects. The boundary conditions may either stem from experimental measurements of the surface velocity or from simulation results. In the latter case a well–suited method for the simulation of complex structures is the finite element method. This section will provide a brief derivation of the formulas needed for the finite element formulation of a structure as it can be found in [1, 66]. These formulas will later be used for the formulation of the coupled FE/BE–simulation. Starting with the formulation of the continuous system, d'Alembert's principle is applied to the structural domain Ω_s using the dynamic equation

$$\delta\Pi_s = \int_{\Omega_s} \varrho_s \delta\boldsymbol{u} \cdot \ddot{\boldsymbol{u}}\, d\Omega + \int_{\Omega_s} \delta\epsilon : \sigma\, d\Omega - \int_{\Gamma_e} \delta\boldsymbol{u} \cdot \boldsymbol{t}\, d\Gamma = 0\,, \tag{19.17}$$

where $\delta\epsilon$ and σ are the virtual strain tensor and stress tensor respectively and $\boldsymbol{u}$ represents the structural displacements. The structure is assumed to obey linear elastic material properties. The surface integral over the boundary of the solid domain Γ_s takes into account loads on the structure expressed in terms of the surface stress vector $\boldsymbol{t}$. Γ_e represents the acoustic–structure interface as depicted in Figure 19.1. On Γ_e the law of reciprocal action must hold both in the structure and in the fluid domain, yielding

$$\boldsymbol{t} = -p\boldsymbol{n}\,. \tag{19.18}$$

The acoustic pressure p, the field variable of the acoustic domain Ω_e, is thus coupled via the surface normal vector $\boldsymbol{n}$ to the surface stress vector and influences the behavior of the structural domain. In order to obtain an approximate solution the continuous system is discretized. The structural displacements $\boldsymbol{u}$, their corresponding virtual quantities as well as the acoustic pressure p are approximated by shape functions, $\boldsymbol{N}_u$ and $\boldsymbol{N}_p$, and nodal values. Using the Galerkin method the shape functions for both the interpolating field variables as well as the virtual quantities are the same

$$\boldsymbol{u} = \boldsymbol{N}_u \boldsymbol{u}^{(e)}\,, \quad \delta\mathbf{u} = \boldsymbol{N}_u \delta\boldsymbol{u}^{(e)}\,, \quad \boldsymbol{p} = \boldsymbol{N}_p \boldsymbol{p}^{(e)}\,, \quad \delta\boldsymbol{p} = \boldsymbol{N}_p \delta\boldsymbol{p}^{(e)}. \tag{19.19}$$

The index $(\ )^{(e)}$ stands for the concentrated nodal values of one element. Instead of the strain tensor ϵ the strain vector $\boldsymbol{\varepsilon}$ is used which is obtained by differentiating the deformation field using the differential operator matrix $\mathcal{D}$

$$\boldsymbol{\varepsilon} = \mathcal{D}\boldsymbol{N}_u \boldsymbol{u}^{(e)}\,, \quad \delta\boldsymbol{\varepsilon} = \mathcal{D}\boldsymbol{N}_u \delta\boldsymbol{u}^{(e)}. \tag{19.20}$$

Equation (19.17) must hold for an arbitrary non–zero virtual displacement $\delta\boldsymbol{u}$. This leads to

$$\underbrace{\sum_e \int_{\Omega_s^{(e)}} \varrho_s \boldsymbol{N}_u^T \boldsymbol{N}_u\, d\Omega}_{\boldsymbol{M}_s}\ \ddot{\boldsymbol{u}} + \underbrace{\sum_e \int_{\Omega_s^{(e)}} (\mathcal{D}\boldsymbol{N}_u)^T \mathcal{E} \mathcal{D}\boldsymbol{N}_u\, d\Omega}_{\boldsymbol{K}_s}\ \boldsymbol{u} = -\boldsymbol{C}_{FE}\,\boldsymbol{p}\,, \tag{19.21}$$

where $\boldsymbol{u}$ denotes the nodal displacement vector of the assembled system now. The integrals on the left hand side form the element mass and stiffness matrices, $\boldsymbol{M}_s$ and $\boldsymbol{K}_s$, of the solid partition. The matrix $\boldsymbol{C}_{FE}$ is given by

$$\boldsymbol{C}_{FE} = -\int_{\Gamma_e^{(e)}} \boldsymbol{N}_u^T \boldsymbol{N}_p \boldsymbol{n}\, d\Gamma\,. \tag{19.22}$$

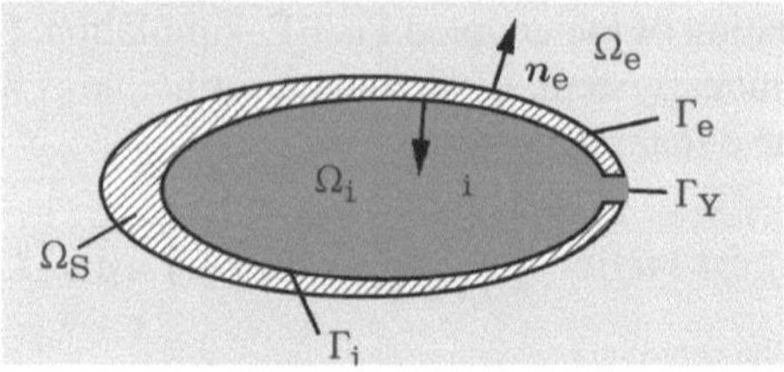

Fig. 19.4 The structural domain Ω_s encloses the volume of the fluid Ω_i defining the fluid domain for the interior problem. The structure is surrounded by the exterior fluid Ω_e. Fluid–structure interaction may occur on the interfaces Γ_i and Γ_e .

Depending on the problem formulation, $\boldsymbol{C}_{\mathrm{FE}}\,\boldsymbol{p}$ in (19.21) is either a boundary condition or a coupling condition. Its handling will be discussed in the corresponding subsections. According to an incidence table the local element matrices are assembled into global sparse matrices.

In the time–harmonic case with a given angular frequency ω, the mass and stiffness matrix are combined to the so–called dynamic stiffness matrix $\boldsymbol{K}_{\mathrm{dyn}} = -\omega^2\boldsymbol{M}_{\mathrm{s}} + \boldsymbol{K}_{\mathrm{s}}$. The matrix $\boldsymbol{K}_{\mathrm{dyn}}$ is singular for the case of eigenfrequencies of the system. Damping may be incorporated in the model for a more realistic system behavior. A simple approach is Rayleigh damping, where the damping matrix is proportional to the mass and stiffness matrix

$$\boldsymbol{D}_{\mathrm{s}} = \alpha\,\boldsymbol{M}_{\mathrm{s}} + \beta\,\boldsymbol{K}_{\mathrm{s}}\,. \tag{19.23}$$

Finally, this leads to the FE formulation

$$\underbrace{\left(-\omega^2\boldsymbol{M}_{\mathrm{s}} - \mathrm{i}\omega\boldsymbol{D}_{\mathrm{s}} + \boldsymbol{K}_{\mathrm{s}}\right)}_{\boldsymbol{K}_{\mathrm{FE}}}\boldsymbol{u} + \boldsymbol{C}_{\mathrm{FE}}\,\boldsymbol{p} = \boldsymbol{f}_{\mathrm{s}}\,. \tag{19.24}$$

In (19.24), $\boldsymbol{u}$ and $\boldsymbol{p}$ represent the nodal amplitude vectors expressed in global coordinates. In contrast to the boundary element method the mass and stiffness matrices need only be computed once and can be combined into the dynamic stiffness matrix at low numerical cost for all considered frequencies.

19.4 Combined Interior and Exterior Acoustic Problem

The following sections present a FE formulation for the fluid domain, which will then be combined with the FE formulation of the structural domain to form a fluid–structure coupled approach for the interior problem. Figure 19.4 demonstrates the different domains. The structural domain Ω_s encloses the volume of the fluid Ω_i defining the fluid domain for the interior problem. The structure is surrounded by the exterior fluid Ω_e. Fluid–structure interaction occurs on the interfaces Γ_i and Γ_e. As necessary, the normal vector on Γ_i and Γ_e will be distinguished by $\boldsymbol{n}_i$ and $\boldsymbol{n}_e$. If the structure does not totally enclose the interior fluid volume, Ω_i and Ω_e share a common boundary Γ_{Y}. On this boundary an admittance condition is applied enabling a separate computation of the interior and the exterior domain, provided that the assumed admittance yields a sufficient approximation.

19.4.1 Strong Coupling Scheme for the Interior Problem

Analog to the derivations of the finite element formulation for the structure in Section 19.3 the wave equation (19.1) is used in the weak form

$$\delta\Pi_{\mathrm{i}} = \int_{\Omega_{\mathrm{i}}} \left(\nabla^2 p - \frac{1}{c^2}\frac{\partial^2 p}{\partial t^2}\right) \delta p \, \mathrm{d}\Omega = 0, \tag{19.25}$$

with the weighting function δp. Applying Green's first theorem, (19.25) is rewritten as [66]

$$\delta\Pi_{\mathrm{i}} = \int_{\Omega_{\mathrm{i}}} \left(\nabla \delta p \cdot \nabla p + \frac{1}{c^2}\ddot{p}\, \delta p\right) \mathrm{d}\Omega - \int_{\Gamma_{\mathrm{i}} \cup \Gamma_{\mathrm{Y}}} (\, \delta p \, \nabla p \cdot \boldsymbol{n}_{\mathrm{i}} \,) \, \mathrm{d}\Gamma = 0\,. \tag{19.26}$$

On the boundary Γ_{i} both the reaction forces axiom (19.18) and the continuity of particle velocity in the normal direction $\boldsymbol{n}_{\mathrm{i}}$ must be satisfied

$$\nabla p \cdot \boldsymbol{n}_{\mathrm{i}} = -\varrho_{\mathrm{f}}\, \boldsymbol{n}_{\mathrm{i}} \cdot \ddot{\boldsymbol{u}}\,. \tag{19.27}$$

The pressure gradient in (19.26) is obtained by using the differential operator $\mathcal{D}_{\mathrm{f}}$

$$\nabla p = \mathcal{D}_{\mathrm{f}} p\,. \tag{19.28}$$

In a first step, the interface Γ_{Y} is neglected. With the shape functions (19.19), Equation (19.26) yields

$$\underbrace{\sum_{\mathrm{e}} \int_{\Omega_{\mathrm{i}}^{(\mathrm{e})}} \frac{1}{c^2} \boldsymbol{N}_{\mathrm{p}}^{\mathrm{T}} \boldsymbol{N}_{\mathrm{p}} \, \mathrm{d}\Omega}_{\boldsymbol{M}_{\mathrm{f}}} \, \ddot{\boldsymbol{p}} + \underbrace{\sum_{\mathrm{e}} \int_{\Omega_{\mathrm{i}}^{(\mathrm{e})}} (\mathcal{D}_{\mathrm{f}} \boldsymbol{N}_{\mathrm{p}})^{\mathrm{T}} \mathcal{D}_{\mathrm{f}} \boldsymbol{N}_{\mathrm{p}} \, \mathrm{d}\Omega}_{\boldsymbol{K}_{\mathrm{f}}} \, \boldsymbol{p} = \varrho_{\mathrm{f}} \boldsymbol{C}_{\mathrm{FE}}^{\mathrm{T}} \ddot{\boldsymbol{u}}\,. \tag{19.29}$$

It is worth noting that the influence of the structural domain on the fluid domain via the coupling term in (19.29) is directly linked to the coupling matrix $\boldsymbol{C}_{\mathrm{FE}}$ which quantifies the influence of the fluid on the structure in (19.21). The interface Γ_{Y}, where the exterior and interior acoustic domain adjoin to each other, necessitates special treatment. By differentiation of the specific acoustic admittance (19.9) with respect to time and by making use of the balance of momentum for a stationary fluid [39] one obtains

$$\dot{p} Y = \dot{v}_{\mathrm{f}} - \dot{v}_{\mathrm{s}} = \left(-\frac{1}{\varrho_{\mathrm{f}}} \nabla p\right) \cdot \boldsymbol{n}_{\mathrm{i}} - \dot{v}_{\mathrm{s}}\,. \tag{19.30}$$

Equation (19.30) is rewritten as

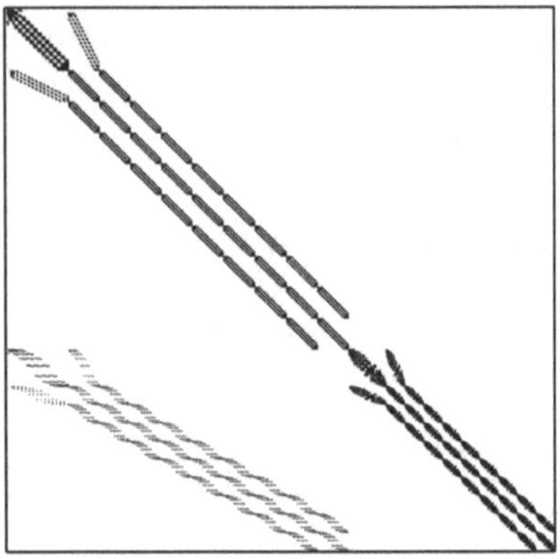

(a) mass matrix $\boldsymbol{M}$

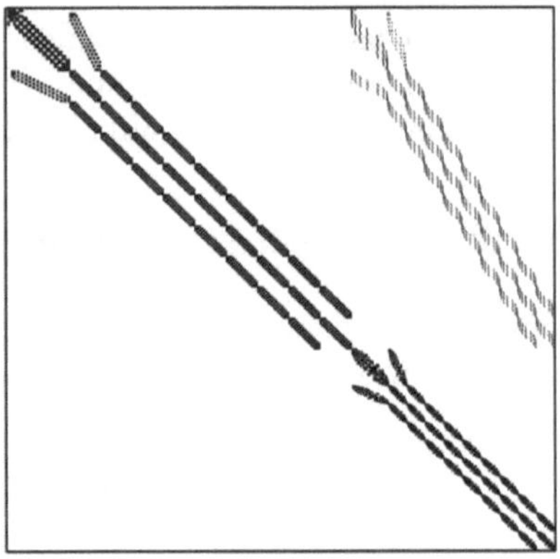

(b) stiffness matrix $\boldsymbol{K}$

Fig. 19.5 Sparsity pattern of the global mass and stiffness matrix according to (19.33) for a strongly coupled interior problem. Both matrices are non–symmetric due to the fluid–structure coupling terms.

$$\nabla p \cdot \boldsymbol{n} = -\varrho_{\mathrm{f}} \left(Y \dot{p} + \dot{v}_{\mathrm{s}} \right) \ . \tag{19.31}$$

Substituting (19.31) into (19.26) leads to an additional admittance matrix

$$\boldsymbol{D}_{\mathrm{f}} = \sum_{\mathrm{e}} \int_{\Gamma_{\mathrm{Y}}^{(\mathrm{e})}} \varrho_{\mathrm{f}} Y \, \boldsymbol{N}_{\mathrm{p}}^{\mathrm{T}} \boldsymbol{N}_{\mathrm{p}} \, \mathrm{d}\Gamma \,, \tag{19.32}$$

and a load vector $\boldsymbol{f}^{\mathrm{f}}$ which denotes an additional boundary condition, such as the influence of an acoustic source. Combining (19.21), (19.29) and (19.32) yields a coupled system of differential equations

$$\begin{bmatrix} \boldsymbol{M}_{\mathrm{s}} & 0 \\ -\varrho_{\mathrm{f}} \boldsymbol{C}_{\mathrm{FE}}^{\mathrm{T}} & \boldsymbol{M}_{\mathrm{f}} \end{bmatrix} \begin{bmatrix} \ddot{\boldsymbol{u}} \\ \ddot{\boldsymbol{p}} \end{bmatrix} + \begin{bmatrix} \boldsymbol{D}_{\mathrm{s}} & 0 \\ 0 & \boldsymbol{D}_{\mathrm{f}} \end{bmatrix} \begin{bmatrix} \dot{\boldsymbol{u}} \\ \dot{\boldsymbol{p}} \end{bmatrix} + \begin{bmatrix} \boldsymbol{K}_{\mathrm{s}} & \boldsymbol{C}_{\mathrm{FE}} \\ 0 & \boldsymbol{K}_{\mathrm{f}} \end{bmatrix} \begin{bmatrix} \boldsymbol{u} \\ \boldsymbol{p} \end{bmatrix} = \begin{bmatrix} 0 \\ \boldsymbol{f}^{\mathrm{f}} \end{bmatrix} . \tag{19.33}$$

In the equation above the damping matrix $\boldsymbol{D}_{\mathrm{s}}$ (19.23) is taken into account. It is worth noting that the global mass and stiffness matrices are non–symmetric due to the coupling matrices. With some manipulation it is possible to achieve a symmetric formulation in terms of the acoustic velocity potential [19, 45]. However, since the basic physical quantity is the acoustic pressure p, it seems more practical to use the representation in (19.33), where the acoustic pressure p is directly computed without the need of post-processing the acoustic velocity potential. In the time–harmonic case, the mass, damping and stiffness matrices in (19.33) are combined to form the non–symmetric dynamic stiffness matrix $\boldsymbol{K}_{\mathrm{dyn,i}}$ of the strongly coupled interior formulation. Figure 19.5 shows the sparsity pattern of the global mass and stiffness matrix according to (19.33) of a small piping segment. The coupling terms in the lower left block of the mass matrix and the upper right block of the stiffness matrix leading to the non–symmetric matrices can be seen.

19.4.2 Weak Coupling Scheme for the Exterior Problem

In the weak coupling case, the FE and the BE formulation is solved in two subsequent steps. For time harmonic problems (19.33) yields a dynamic stiffness matrix, which can be either solved by a direct solver or an iterative solver. For well–conditioned systems iterative solvers in many cases can result in a better performance in terms of computation time and memory usage. On the other hand, iterative solvers are less robust. Direct solvers that make use of the fact that the finite element matrices are normally sparsely populated can also be applied without further knowledge of the properties of the matrix. Since the dynamic stiffness matrix $\boldsymbol{K}_{\mathrm{dyn,i}}$ is ill–conditioned, a LU–factorization is used [17]. Having solved the interior problem according to (19.33) for a given angular frequency, the acoustic pressure amplitudes and the structural displacement amplitudes and thus the surface normal velocity are known, leading to the Neumann boundary data for the subsequent solution of the exterior problem. The Burton–Miller formulation is applied making use of the multi-level multipole expansion and the Krylov based iterative GMRES solver as discussed in Section 19.2.

Preconditioning

The condition number of the system matrix has a strong influence on the convergence of iterative solvers. The convergence behavior is significantly improved by introducing a preconditioning matrix $\boldsymbol{P}$ to reduce the condition number. Therefore, the original system (19.10) is replaced by

$$\boldsymbol{K}_{\mathrm{BE}}\,\boldsymbol{P}\boldsymbol{p}^{\mathrm{p}} = \boldsymbol{b} \qquad \text{with} \quad \boldsymbol{p} = \boldsymbol{P}\boldsymbol{p}^{\mathrm{p}}. \tag{19.34}$$

Compared to the original matrix $\boldsymbol{K}_{\mathrm{BE}}$, the modified matrix $\boldsymbol{K}_{\mathrm{BE}}\,\boldsymbol{P}$ has a better condition number, which accelerates the solution procedure. The preconditioning matrix $\boldsymbol{P}$ should be cheap to compute but it should still be a good approximation to the inverse of $\boldsymbol{K}_{\mathrm{BE}}$. For a survey on preconditioners the reader is referred to [4]. In the following, two preconditioners are investigated: First, an incomplete LU factorization (ILU) is applied [58]. The basic idea is a factorization $\boldsymbol{K}_{\mathrm{BE}} = \boldsymbol{L}\boldsymbol{U} + \boldsymbol{R}$ with a left lower triangular matrix $\boldsymbol{L}$ and a right upper triangular matrix $\boldsymbol{U}$. Since the total memory consumption of $\boldsymbol{L}$ and $\boldsymbol{U}$ should be equal to that of $\boldsymbol{K}_{\mathrm{BE}}$, the remaining matrix $\boldsymbol{R}$ is neglected. Now the inverse can be computed at low cost and serves as an approximation for $\boldsymbol{K}_{\mathrm{BE}}^{-1}$. In practice, only the sparse near–field matrix is used to set up the preconditioner. The second preconditioner is an approximate inverse preconditioner (AIP). Here, the basic idea is operator splitting. The system matrix is split up into a bounded part $\boldsymbol{K}_{\mathrm{BE,0}}$ and the remaining part $\hat{\boldsymbol{K}}_{\mathrm{BE}}$. A two element layer can for instance be used to define the bounded region. Although $\boldsymbol{K}_{\mathrm{BE,0}}$ is sparse, it would be too expensive to compute the inverse directly. An approximate inverse can be obtained by solving a least square problem. Further details can be found in [21] and [10].

The two preconditioners are compared for an L–shape test model. It consists of three cubes, each with a side length of 1 m. An artificial velocity distribution as

Table 19.1 L–shape test case: Comparison of GMRES iterations for different preconditioners

frequency [Hz]	quad elements	elements/λ	ILU	AIP	no preconditioner
419	2,300	$\approx$11	11	18	38
1,074	15,000	$\approx$11	13	20	37
2,727	96,000	$\approx$11	14	20	35

Fig. 19.6 Production series main muffler. The inner structural parts of the main muffler are removed

boundary condition is generated by three monopole sources within the L–shape. The number of iterations for the different preconditioners in combination with a GMRES are summarized in Table 19.1. A residual of 10^{-6} is used as convergence criterion. Obviously, the ILU preconditioner shows the lowest number of iteration steps. However, setting up an ILU preconditioner is typically more expensive in terms of memory and computation time compared with an AIP.

19.4.3 Application: Expansion Chamber

In this section the previously developed formulations are combined and applied to a problem in the automotive industry. The vibro–acoustic behavior of an exhaust system is investigated. Exhaust systems are exposed to large pressure pulsations due to the periodically blown out exhaust gas. These large pressure pulsations may lead to structural vibrations of the exhaust system by the transfer of acoustic energy from the fluid to the structural parts of the exhaust system. The vibrating structural parts of the exhaust system may significantly contribute to the sound radiation of the system. This phenomenon of the so called surface radiated sound is reported in [7, 24, 33]. For some frequencies the surface sound radiation may be even dominant as compared to the sound radiated at the orifice. The phenomenon of surface sound radiation is experimentally investigated in [34] on a production-model main muffler whose inner structural parts were removed. The muffler's structure is depicted in Figure 19.6.

In order to be able to predict, evaluate and optimize the sound radiation of an exhaust system the sound radiation field is simulated by the approaches presented in the previous sections 19.4.1 and 19.4.2. Physically, the interior and exterior acoustic

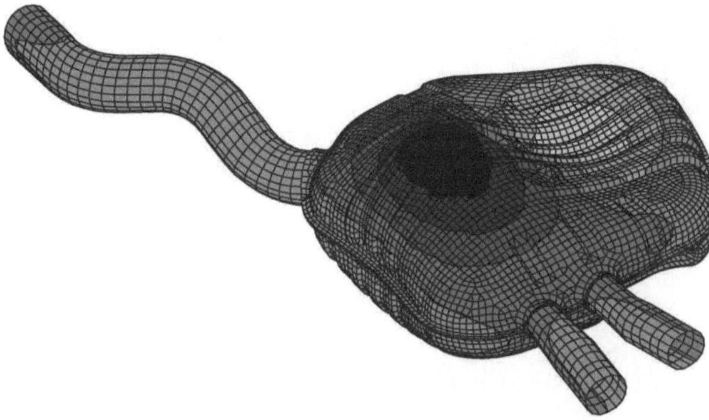

Fig. 19.7 Operational deflection shape of the main muffler at $f = 478$ Hz due to the acoustic excitation at the inlet. The gray–scaled colors represent the real part of the acoustic pressure obtained by the Fast–BE simulation.

domain are coupled at the tail pipes of the exhaust system. However, from an engineering point of view it is justified to consider the interior and exterior domains separately, while approximating the acoustical behavior at the boundary Γ_Y by an admittance boundary condition as discussed in the previous section. The value of the specific acoustic admittance at the boundary Γ_Y depends on the geometry and on the considered frequency. An analytical solution and experimental validation for the reflection coefficient of an unflanged circular pipe with and without flow can be found in [16,40,48]. For this simulation the influence of mean flow is neglected. The specific acoustic admittance is backed out directly from the reflection coefficient [47] and is applied as a boundary condition on Γ_Y. The FE–model includes only the main muffler. The internal components are not discretized, but may be approximated again by an admittance boundary condition. For the correct physical representation the input admittance at the inlet of the main muffler needs to be determined experimentally, e.g. by the two microphone method [13,23,59]. To reproduce the excitation of the system via the acoustic path, a unit specific acoustic flux $\bar{q} = \partial p / \partial \boldsymbol{n}$ is applied at the inlet of the exhaust system. The structural parts of the main muffler are modeled by approximately 10,000 shell elements with six degrees of freedom (DOF) per node. The fluid domain contains approximately 75,000 elements. This yields a total number of 88,000 DOFs. Rayleigh damping is assumed with $\alpha = 50\,\frac{1}{\mathrm{s}}$ and $\beta = 10^{-6}\,\mathrm{s}$. Figure 19.7 shows the operational deflection shape of the main muffler at $f = 478$ Hz. It is found from an experimental modal analysis (EMA) that at $f = 479$ Hz the main muffler possesses a mode shape where the deflection of the upper shell looks similar to a (2,1)-mode of a rigidly fixed plate, though it is more complex due to the given geometry [34]. As it can be seen in Figure 19.7, the operational deflection shape is dominated by this mode. Though the system is only excited in the acoustic domain at the inlet, the structural domain is also excited via the fluid–structure interface. Energy is transferred from the acoustic domain to the structural

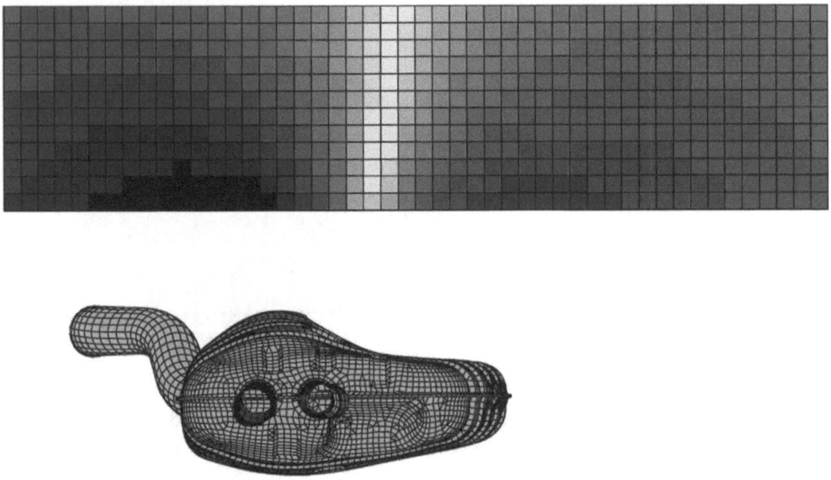

Fig. 19.8 Sound radiation of the main chamber evaluated on the depicted field point mesh. Colors in the contour plot represent the acoustic sound pressure in decibels. The shown operational deflection shape causes a dipole–like sound radiation behavior.

domain. The acoustic radiation of the main muffler is computed in a subsequent step. The gray–scaled colors in Figure 19.7 represent the magnitude of the acoustic pressure in dB on the surface of the main muffler. The radiation into the acoustic field is evaluated at several field points. The field points are located on a vertical plane arranged on an equidistant 49x13 grid with a mesh size of 0.025 m as depicted in Figure 19.8. The closest vertical distance to the main expansion chamber is 0.12 m. It can be seen that the contour lines of regions with the same sound pressure level (SPL) are approximately of a half–elliptical shape. There are two regions with high values, which are colored black. Looking at the operational deflection shape in Figure 19.8 it becomes clear that the maxima are caused by the large deflections on the upper shell of the main muffler. Since the two regions with maximum displacement are phase–shifted by π the sound field is expected to be similar to the sound field of a dipole. This also implies, that in–between the two maximum regions the SPL will be low due to destructive interference. As expected, the light–colored region in the middle of the field point plane indicates low acoustical pressures.

19.5 Strong Fluid–Structure Coupling

In contrast to Section 19.4, where a weak coupling approach for the exterior problem is presented, the feedback of the acoustic pressure on the structure is now taken into account. This is necessary if a fluid with a high density, e.g. water, and thin structures are considered. The thin structure is modeled by shell elements. Here, the FE and BE part cannot be solved consecutively any more. The investigated problem

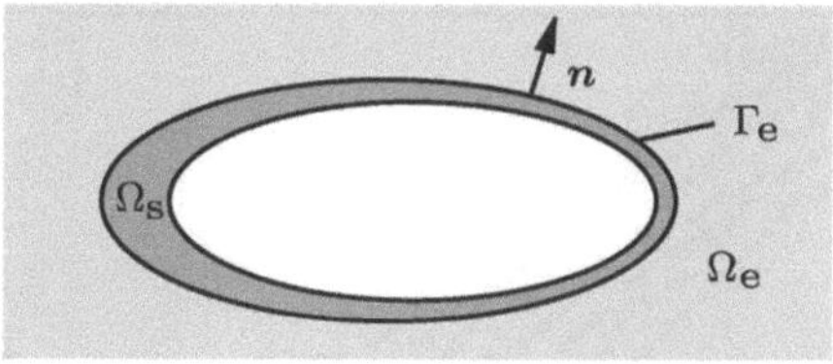

Fig. 19.9 The fluid-structure interaction occurs on the interface Γ_e. The structural domain Ω_s is assumed to be thin. Only the exterior boundary is in contact with the fluid domain Ω_e.

is shown in Figure 19.9. The fluid is assumed to have only one boundary which represents the coupling interface Γ_e. This is equivalent to the physical assumption of a totally submerged structure in an infinite fluid domain. Admittance boundary conditions could easily be incorporated but are not addressed here. In the following, two different coupling schemes are presented and compared.

19.5.1 Direct Coupling Scheme

The Burton–Miller formulation (19.8) is taken as the starting point for a strong coupling scheme. Now, the velocities v_s on the coupling interface Γ_e are not known in advance. The first part of the coupling scheme reflects the physical situation that the particle velocity on the coupling interface coincides with the local velocity of the vibrating structure. There are two alternatives to formulate the interaction:

One can first compute an averaged normal particle velocity on the elements. The boundary element formulation (19.10) can analogically be applied to the strongly coupled problem. Unlike the uncoupled problem, where a right hand side vector $\boldsymbol{b}$ of the resulting system is directly set up, a coupling matrix $\boldsymbol{C}_{\mathrm{BE}}$ with the single and the adjoint double layer potential has to be assembled.

A second possibility is to apply linear shape functions and directly use the structural displacements of the FE system. Since the integration is done elementwise, the normal which is constant on the element and the factor $\rho\omega^2$ can be pulled out of the integral. The coupling matrix $\boldsymbol{C}_{\mathrm{BE}}$ is composed of the modified operators $\hat{\boldsymbol{V}}$, $\hat{\boldsymbol{I}}'$ and $\hat{\boldsymbol{K}}'$, yielding

$$\underbrace{\left(-\frac{1}{2}\boldsymbol{I}+\boldsymbol{K}+\frac{\mathrm{i}}{\kappa}\boldsymbol{D}\right)}_{\boldsymbol{K}_{\mathrm{BE}}}\boldsymbol{p}-\underbrace{\rho\omega^2\left(\hat{\boldsymbol{V}}-\frac{\mathrm{i}}{2\kappa}\hat{\boldsymbol{I}}'-\frac{\mathrm{i}}{\kappa}\hat{\boldsymbol{K}}'\right)}_{\boldsymbol{C}_{\mathrm{BE}}}\boldsymbol{u}=0. \tag{19.35}$$

The latter approach is used in the following.

A strong coupling scheme must also include the interaction in the opposite direction. The acoustic pressure p generates surface loads on the structural part as expressed by (19.21) and (19.22). It is equivalent to every other structural pressure acting on the surface. Again, a coupling matrix $\boldsymbol{C}_{\mathrm{FE}}$ has to be assembled, since the acoustic pressure is not known in advance.

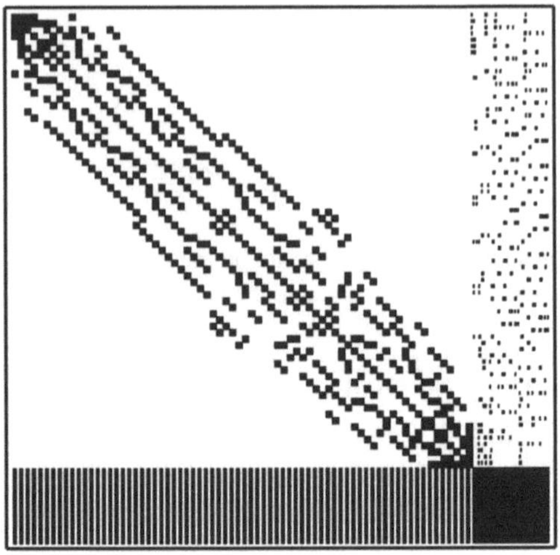

Fig. 19.10 Sparsity pattern of the overall system matrix $\boldsymbol{K}$ for a fully coupled FE/BE problem using standard BEM.

As stated by the coupling conditions, there is only an interaction between the two coupling domains in the normal direction. In the tangential direction, no forces are transmitted. From a physical point of view, this is equivalent to the assumption of an inviscid fluid. By combining (19.24) and (19.35), the coupled system is formally written as

$$\underbrace{\begin{pmatrix} \boldsymbol{K}_{\mathrm{FE}} & \boldsymbol{C}_{\mathrm{FE}} \\ \boldsymbol{C}_{\mathrm{BE}} & \boldsymbol{K}_{\mathrm{BE}} \end{pmatrix}}_{\boldsymbol{K}} \begin{pmatrix} \boldsymbol{u} \\ \boldsymbol{p} \end{pmatrix} = \begin{pmatrix} \boldsymbol{f}_{\mathrm{s}} \\ 0 \end{pmatrix}. \tag{19.36}$$

The sparsity structure of $\boldsymbol{K}$ with the non–zero elements is visualized in Figure 19.10. The upper left block illustrates the dynamic stiffness matrix $\boldsymbol{K}_{\mathrm{FE}}$. This matrix is symmetric with respect to its main diagonal and extremely sparse. A bandwidth optimized structure is obtained by rearranging the order of the degrees of freedom. It is favorable to assemble and store the mass matrix and stiffness matrix in (19.24) separately, since they are frequency independent. Thus, the dynamic stiffness matrix is built up for many different frequencies at low numerical cost. The coupling matrix $\boldsymbol{C}_{\mathrm{FE}}$ is also extremely sparse and thus the memory consumption is small. For engineering applications, structural moments as a consequence of the acoustic pressure are negligible. The matrix $\boldsymbol{K}_{\mathrm{BE}}$ on the lower right is typically fully populated for the standard BEM. Using the fast multipole method, this matrix has to be replaced by the appropriate sparse near–field matrix. The coupling matrix $\boldsymbol{C}_{\mathrm{BE}}$ incorporates the structural displacements into the BEM system. Since rotations are neglected here, this matrix has a zebra crossing like pattern. Figure 19.10 shows the situation for standard BEM. This densly populated matrix is replaced by the sparse near–field in the case of the multipole BEM.

19.5.2 Preconditioned GMRES Solver

As mentioned above, for large scale problems the solution of (19.36) by a direct solver is inefficient due to the enormous time and memory consumption. A more promising approach is the application of Krylov based iterative solvers in combination with the fast multipole method for the matrix–vector product. In the following, two schemes are presented.

The first solver applies a single GMRES to the fully coupled system (19.36) ("full–solver"). Since $\boldsymbol{p}$ and $\boldsymbol{u}$ are of totally different magnitudes when using SI units, a nondimensionalization is performed [41]. Even so, the condition number still remains poor and is typically of the magnitude 10^5. For further reduction of the condition number, a block diagonal preconditioner is proposed by Hughes et al. [31]. The sparse approximate inverse (SPAI) technique or incomplete lower upper factorization (ILU) can be applied to approximate the main diagonal blocks. The resulting preconditioning matrix takes the form

$$\boldsymbol{P} = \begin{bmatrix} \mathrm{SPAI}(\boldsymbol{K}_{\mathrm{FE}}) & 0 \\ 0 & \mathrm{SPAI}(\boldsymbol{K}_{\mathrm{BE}}) \end{bmatrix} . \tag{19.37}$$

This preconditioner is known to work for problems when the coupling has only a moderate influence on the structure. For thin shell structures and higher frequencies, where the coupling has a stronger influence, the number of GMRES iterations increases rapidly [8].

The off–diagonal block terms can be included by using a block–partitioned form of $\boldsymbol{K}$ [12]. Performing a block LU factorization, the exact solution for a prescribed right hand side vector $(\boldsymbol{f}_{\mathrm{s}}, \boldsymbol{q})^{\mathrm{T}}$ is given in the form

$$\boldsymbol{p} = \boldsymbol{S}^{-1}\left(\mathrm{q} - \boldsymbol{C}_{\mathrm{BE}}\boldsymbol{K}_{\mathrm{FE}}^{-1}\boldsymbol{f}_{\mathrm{s}}\right) \tag{19.38}$$

$$\boldsymbol{u} = \boldsymbol{K}_{\mathrm{FE}}^{-1}\left(\boldsymbol{f}_{\mathrm{s}} - \boldsymbol{C}_{\mathrm{FE}}\boldsymbol{p}\right), \tag{19.39}$$

with the Schur complement $\boldsymbol{S} = \boldsymbol{K}_{\mathrm{BE}} - \boldsymbol{C}_{\mathrm{BE}}\boldsymbol{K}_{\mathrm{FE}}^{-1}\boldsymbol{C}_{\mathrm{FE}}$. Equations (19.38) and (19.39) are employed for preconditioning. The inverse $\mathbf{S}^{-1}$ of the Schur complement may be approximated by $\mathbf{K}_{\mathrm{BE}}^{-1}$. A SPAI or ILU on the near–field can be applied for $\mathbf{K}_{\mathrm{BE}}^{-1}$. The approximate inverse of $\mathbf{K}_{\mathrm{FE}}$ is provided in the same manner. A great advantage of the mentioned solver schemes is, that one does not need an exact inverse of a single block matrix, neither for the preconditioner nor for the matrix–vector product.

A second possibility is to work directly on the reduced system defined by (19.38) and (19.39). First, one solves for the unknown pressures $\boldsymbol{p}$ with (19.38) by applying a GMRES on the Schur complement ("reduced–solver"). In this case, a nondimensionalization is implicitly included in the formulation and does not have to be performed manually. In contrast to the previous solver, the exact inverse $\boldsymbol{K}_{\mathrm{FE}}^{-1}$ within the Schur complement is needed at every iteration step. Since $\boldsymbol{K}_{\mathrm{FE}}$ is extremely sparse, it can efficiently be computed by a factorization based solver, e.g. a LU solver. This factorization has to be computed only once and is then applied at every iteration step. A preconditioner can be constructed by approximating the inverse of the Schur complement by $\boldsymbol{K}_{\mathrm{BE}}^{-1}$. Neglecting the second part of the Schur complement implies preconditioning with the weakly coupled system, see Section 19.4. The feedback effect of the pressure onto the structure is disregarded here. Again, one can use a SPAI or ILU. The structural displacements $\boldsymbol{u}$ are computed in a postprocessing step with (19.39). The memory consumption for the Krylov bases is reduced, since only the pressures are unknowns in the first solution step. If a factorization of $\boldsymbol{K}_{\mathrm{FE}}$ is affordable, the reduced solver turns out to have the best convergence [8].

19.5.3 Mortar Coupling Scheme

In this section an alternative coupling formulation is discussed. An additional variable — the Lagrange multiplier λ — is introduced to set up a coupling scheme. The pressure on the coupling interface Γ_{e} is chosen as a Lagrange multiplier, i.e. $\lambda = p$ [21]. In contrast to the formulation above, the boundary integral equation (19.5) is tested with constant test functions ν^q and the additional term

$$\int_{\Gamma_{\mathrm{e}}} \nu^q(\boldsymbol{x}) \left[p(\boldsymbol{x}) - \lambda(\boldsymbol{x})\right] \mathrm{d}\Gamma_{\boldsymbol{x}} \tag{19.40}$$

is introduced to enforce the equilibrium condition. The continuity condition between structural displacement and flux is formulated in the weak sense by

$$\int_{\Gamma_{\mathrm{e}}} \nu^\lambda(\boldsymbol{x}) \left[-\varrho_{\mathrm{f}}\omega^2 u_n(\boldsymbol{x}) + q(\boldsymbol{x})\right] \mathrm{d}\Gamma_{\boldsymbol{x}} = 0. \tag{19.41}$$

The pressure p is interpolated by linear shape functions and constant shape functions are used for the flux q. The structural domain is again modeled by the FE formulation in (19.24). This leads to the algebraic system of equations

$$\begin{pmatrix} \varrho_{\mathrm{f}}\,\omega^2 \boldsymbol{K}_{\mathrm{FE}} & \boldsymbol{0} & \boldsymbol{0} & \boldsymbol{C}_{\mathrm{FE}} \\ \boldsymbol{0} & -\tilde{\boldsymbol{V}} & \frac{1}{2}\tilde{\boldsymbol{I}} + \tilde{\boldsymbol{K}} & -\tilde{\boldsymbol{C}}_{\mathrm{BE}} \\ \boldsymbol{0} & -\frac{1}{2}\tilde{\boldsymbol{I}}^{\mathrm{T}} - \tilde{\boldsymbol{K}}^{\mathrm{T}} & -\tilde{\boldsymbol{D}} & \boldsymbol{0} \\ -\boldsymbol{C}_{\mathrm{FE}}^{\mathrm{T}} & \tilde{\boldsymbol{C}}_{\mathrm{BE}}^{\mathrm{T}} & \boldsymbol{0} & \boldsymbol{0} \end{pmatrix} \begin{pmatrix} \boldsymbol{u} \\ \boldsymbol{q} \\ \boldsymbol{p} \\ \boldsymbol{\lambda} \end{pmatrix} = \begin{pmatrix} \varrho_{\mathrm{f}}\,\omega^2 \boldsymbol{f}_{\mathrm{s}} \\ \boldsymbol{0} \\ \boldsymbol{0} \\ \boldsymbol{0} \end{pmatrix}. \tag{19.42}$$

The sparsity pattern of the resulting saddle point system is shown in Figure 19.11. Again the upper left block is the FE system matrix and the upper right block contains $\boldsymbol{C}_{\mathrm{FE}}$. Both matrices are identical compared with the direct coupling approach discussed above. In contrast to the previous formulation, the coupling matrix $\tilde{\boldsymbol{C}}_{\mathrm{BE}}$, which arises from the continuity condition (19.41), is sparse now. The matrices in the last row are simply the transpose of $\tilde{\boldsymbol{C}}_{\mathrm{FE}}$ and $\tilde{\boldsymbol{C}}_{\mathrm{BE}}$. Thus, they do not have to be stored separately. The BE part in the middle which shall be denoted by $\tilde{\boldsymbol{K}}_{\mathrm{BE}}$ is block skew symmetric. In case of the multipole BEM this fully populated part has to be replaced by the corresponding near–field.

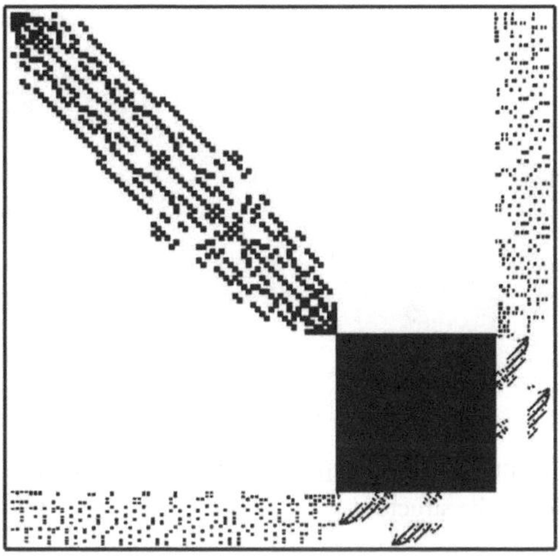

Fig. 19.11 Sparsity pattern of the system matrix for the Lagrange multiplier formulation using standard BEM.

Uzawa Algorithm

The use of an iterative solver on the system (19.42) shows poor convergence. Therefore, an Uzawa algorithm is used to introduce a reduced system which is solved for the Lagrange multiplier λ ("Lagrange–solver") [56]. Solving the first three rows for $\boldsymbol{u}$ and $(\boldsymbol{q},\boldsymbol{p})^{\mathrm{T}}$ and substituting these relations into the last row yields

$$\left[\boldsymbol{C}_{\mathrm{FE}}^{\mathrm{T}}\left(\varrho_{\mathrm{f}}\omega^2\boldsymbol{K}_{\mathrm{FE}}\right)^{-1}\boldsymbol{C}_{\mathrm{FE}}+\tilde{\boldsymbol{C}}_{\mathrm{BE}}^{\mathrm{T}}\tilde{\boldsymbol{K}}_{\mathrm{BE}}^{-1}\tilde{\boldsymbol{C}}_{\mathrm{BE}}\right]\boldsymbol{\lambda}=\boldsymbol{C}_{\mathrm{FE}}^{\mathrm{T}}\left(\varrho_{\mathrm{f}}\omega^2\boldsymbol{K}_{\mathrm{FE}}\right)^{-1}\varrho_{\mathrm{f}}\omega^2\boldsymbol{f}_{\mathrm{s}}. \tag{19.43}$$

The matrix in the square brackets is not evaluated explicitly. Instead, a GMRES is applied to compute $\tilde{\boldsymbol{K}}_{\mathrm{BE}}^{-1}$, where the matrix–vector product is provided by the FMM algorithm. The inner inverse $\left(\rho_{\mathrm{f}}\omega^2\mathbf{K}_{\mathrm{FE}}\right)^{-1}$ of (19.43) is computed by a direct LU solver. The GMRES can be preconditioned by a simple diagonal scaling or better by an ILU or AIP. The outer system is also solved using a GMRES, since standard Newton methods show a poor convergence. Instead of evaluating the exact matrix–vector product at each iteration step, an approximation can be used. For a discussion of the necessary precision of the matrix–vector the reader is referred to [21]. The structural displacements can finally be computed in a postprocessing step.

A main advantage of this formulation lies in the preconditioning of the subsystems. As the FE and the BE subsystem are separated in the inner loop, standard preconditioners can be applied. Again, the FE system can alternatively be solved by an iterative solver. Another advantage is the possibility of handling non–conforming discretizations of the BE and FE parts. The interested reader is referred to [5] and [3].

A disadvantage of this scheme arises from the nested outer–inner iterations. So, the total number of matrix–vector products for the BE system is the product of the inner times the outer iterations.

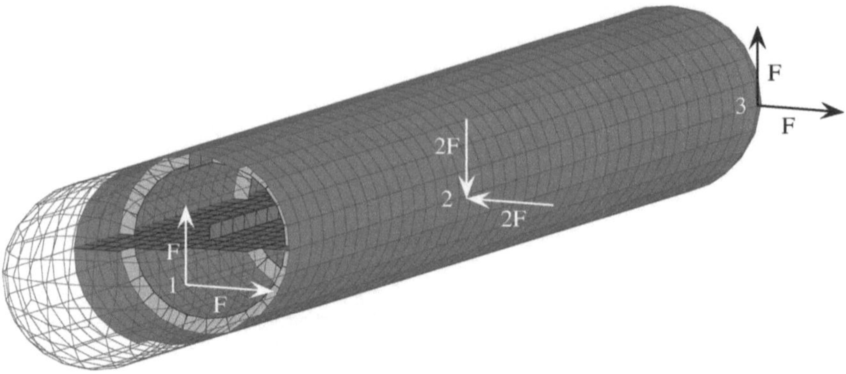

Fig. 19.12 Submarine–like structure with internal stiffeners. Six concentrated force components excite harmonically the structure.

19.5.4 Application

As an engineering application, a submarine–like structure as depicted in Figure 19.12 is presented. It consists of a cylindrical structure with a diameter of 2 m. Spherical caps are attached to both ends. The total length of the test object is 22 m. An intermediate floor is built in at half of the height with a welded on rib in longitudinal direction. Additional circular stiffeners are equidistantly positioned along the exterior cylinder. The shell thickness is 2 cm. The material data of steel are applied. Water is assumed for the acoustic fluid. The structure is discretized with quadrilateral linear elements. These elements are split into two triangular elements for the BE computations. Every FE node has six DOFs, whereas a BE node has only a single pressure DOF. For the following simulations, the total number of DOFs is 67,200. It is worth noting, that not all structural nodes are in contact with the fluid.

Six concentrated force components of magnitude F and $2F$ at nodes 1-3 on the right hand side excite the structure. The forces are chosen such that the static equilibrium condition is fulfilled. The force magnitude is $F = 1\,\mathrm{N}$.

Memory consumption

The memory consumption of the two conforming coupling schemes — the "Direct–solver" and the "Lagrange–solver" — is mainly due to the factorization of the LU solver and the near–field of the BE system. Since all formulations use a LU solver for the identical FE matrix, only the memory consumption of the BE near–fields is compared. In the case of the "Direct solver" using the averaged normal velocity, the memory consumption for $\boldsymbol{K}_{\mathrm{BE}}$ and $\boldsymbol{C}_{\mathrm{BE}}$ is 95 MB. If linear shape functions are used in combination with the FE displacements, the memory consumption increases to 127 MB. The system matrix $\boldsymbol{K}_{\mathrm{BE}}$ with a memory consumption of 35 MB is identical in both cases. If n_q denotes the number of coupling elements and n_n is the number of coupling nodes, $\boldsymbol{C}_{\mathrm{BE}}$ has the size $n_q \times n_n$ in the first case and $3\,n_n \times n_n$ in the second one. If one further takes into account that a triangular mesh with k

nodes has approximately $2k$ elements, the memory consumption for $\boldsymbol{C}_{\mathrm{BE}}$ of the first formulation is with 60 MB approximately two thirds of the allocated memory for the second one (92 MB).

The memory consumption of the Lagrange near–field is the highest with 193 MB. The $\tilde{\boldsymbol{D}}$ block of (19.42) has the same pattern as $\boldsymbol{K}_{\mathrm{BE}}$ (35 MB). The upper right block $\frac{1}{2}\tilde{\boldsymbol{I}} + \tilde{\boldsymbol{K}}$ has the same memory consumption as $\boldsymbol{C}_{\mathrm{BE}}$ with the averaged normal velocity in the first case (60 MB). But additionally, one has to allocate the $\tilde{\boldsymbol{V}}$ block with 98 MB. Obviously, this latter formulation needs more than twice as much memory for the near–field as the best direct coupling scheme.

Number of iterations

Preconditioning has a strong influence on the convergence of iterative solvers. Both the "Direct–solver" and the "Lagrange–solver" are accelerated by an ILU applied to the near–fields. If a LU solver is used for the FE part, only a single GMRES is required in case of the "Direct–solver". The most time consuming portion of an iteration step is the evaluation of the BE matrix–vector product by the multipole algorithm. Here, the number of matrix–vector products is identical to the number of GMRES iterations. In case of 65 Hz, 43 iterations are required to achieve convergence with a residual of 10^{-6}. This is in contrast to the "Lagrange–solver", where two GMRES are applied in a nested inner–outer iteration scheme. In each inner iteration a matrix–vector product has to be evaluated. The expense of this matrix–vector product is comparable to the one of the "Direct–solver". Thus, the total number of matrix–vector products is the product of the inner iterations times the outer iterations. For the same frequency the number of iterations is 127×82, which is much worse than before. An improvement could be achieved by additional preconditioning of the outer loop. Nevertheless, the performance of the nested solution scheme suffers from the multiplication of the iterations.

Frequency sweep

A comparison of the time–harmonic response for the weak and strong coupled cases is presented in Figure 19.13. The absolute value of the displacement of node 2 in vertical direction is plotted as a function of the frequency. These frequency sweeps are computed with the "Direct–solver", since it is superior to the "Lagrange–solver" in terms of efficiency. The weakly coupled solution is obtained by setting $\boldsymbol{C}_{\mathrm{FE}}$ in (19.36) to zero. One observes a strong influence of the surrounding water on the resonance frequencies. The first eigenmode is shifted from approximately 19 Hz to 13 Hz. This is due to the hydromass effect. The comparison of the weak and strongly coupled solution clearly shows the necessity of a strong coupled solution scheme.

The computed spatial pressure distribution at a frequency of 65 Hz is visualized in Figure 19.14 for the strongly coupled case. The amplitude of the acoustic pressure along concentric semi–circles of different radii in the vertical plane is plotted. The centers coincide with the center of mass of the submarine–like structure. Only the

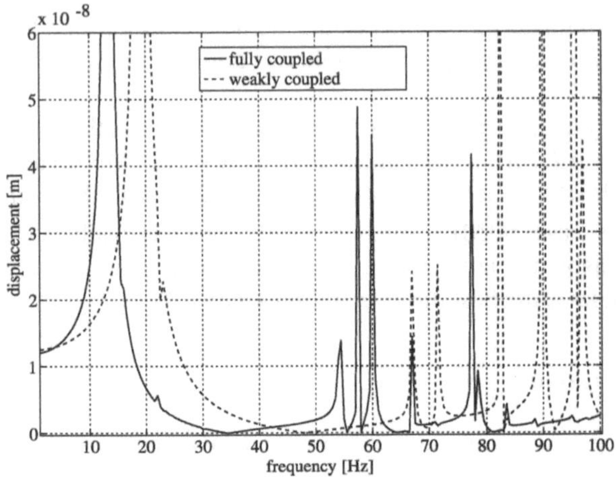

Fig. 19.13 Frequency sweep of the submarine–like structure for the weakly and strongly coupled cases.

half where the driving forces are located, see Figure 19.12, is displayed. The directivity pattern is clearly visible. One can observe a change from two lobes to three lobes with increasing distance.

19.6 Conclusion

The investigations show that fluid–structure coupled formulations are well–suited to model multi–field problems. The multipole method significantly accelerates calculations associated with the boundary element method in acoustics. The first application demonstrates that the combination of a strongly coupled interior with a weakly coupled exterior problem can efficiently model engineering applications. For an exterior problem with a dense fluid two strong coupling schemes are compared and applied to a large–scale test structure. It turns out that the correct choice of the coupling formulation and the applied solver is crucial for the overall performance and efficiency.

Acknowledgement

Results presented in the paper are partially funded by the German Research Foundation (DFG) through the transfer project SFB404/T3 and by the Friedrich–und–Elisabeth–Boysen–foundation. The authors acknowledge the valuable contributions of our co–operation partners — the Germanischer Lloyd, Hamburg and the Friedrich–Boysen GmbH & Co KG.

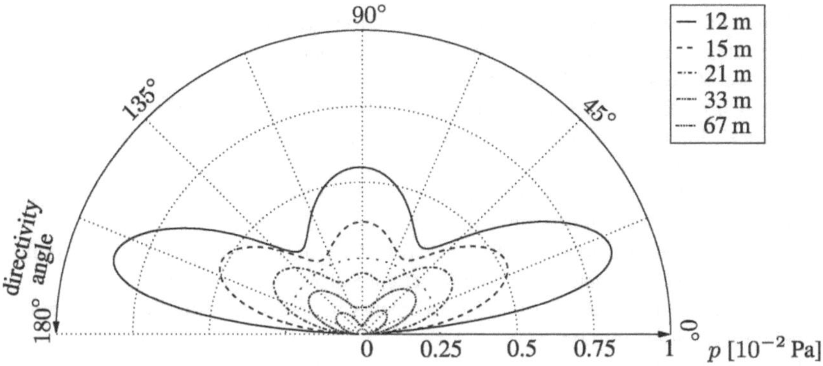

Fig. 19.14 Pressure distribution for 65 Hz along semi–circles of different radii in the vertical plane.

References

1. Bathe HJ (1995) Finite element procedures. Prentice Hall, 2nd edition
2. Bathe KJ, Natikitpaiboon C, Wang X (1995) A mixed displacement–based finite element formulation for acoustic fluid–structure interaction. Computers & Structures 56:225–237
3. Belgacem FB (1999) The mortar finite element method with Lagrange multipliers. Numerische Mathematik 84:173–197
4. Benzi M (2002) Preconditioning techniques for large linear systems: A survey. Journal of Computational Physics 182:418–477
5. Bernardi C, Maday Y, Patera AT (1994) A new nonconforming approach to domain decomposition: the mortar element method. In: Brezis H, Lions J (eds) Nonlinear Partial Differential Equations and their Applications. Collège de France Seminar, Vol XI, Pitman Research Notes in Mathematics, Longman Scientific & Technical, Harlow 299:13–51
6. Brakhage H, Werner P (1965) Über das Dirichlet'sche Außenraumproblem für die Helmholtz'sche Schwingungsgleichung. Archiv der Mathematik 16:325–329
7. Brand JF, Wiemeler D (2004) Surface radiated noise of exhaust systems – structural transmission loss test rig, part 1. In: Proceeding of the annual meeting of DAGA (CFA/DAGA), Strasbourg, France, 1003–1004
8. Brunner D, Junge M, Gaul L (2007) Strong coupling of the fast multilevel multipole boundary element method with the finite element method for vibro–acoustic problems. In: Proceedings of the 14th International Congress on Sound and Vibration (ICSV14)
9. Burton AJ, Miller GF (1971) The application of integral equation methods to the numerical solution of some exterior boundary–value problems. Proceedings of the Royal Society of London 323:201–220
10. Chen LH, Harris P (2001) Efficient preconditioners for iterative solution of the boundary element equations for the three–dimensional Helmholtz equation. Applied Numerical Mathematics 36:475–489
11. Chen ZS, Hofstetter G, Mang HA (1998) A Galerkin–type BE–FE formulation for elasto–acoustic coupling. Computer Methods in Applied Mechanics and Engineering 152:147–155
12. Chow E, Saad Y (1997) Approximate inverse techniques for block–partitioned matrices. SIAM Journal on Scientific Computing 18:1657–1675

13. Chung JY, Blaser DA (1980) Transfer function method of measuring in–duct acoustic properties. I: Theory. Journal of the Acoustical Society of America 68:907–913
14. Coifman R, Rokhlin V, Wandzura S (1993) The fast multipole method for the wave equation: A pedestrian description. IEEE Antennas and Propagation Magazine 35:7–12
15. Conca C, Planchard J, Vanninathan M (1995) Fluids and periodic structures. John Wiley & Sons, Chichester
16. Davies POAL, Coelho JLB, Bhattachary M (1980) Reflection coefficient for an unflanged pipe with flow. Journal of Sound and Vibration 72:543–546
17. Davis TA (2004) Algorithm 832: UMFPACK, an unsymmetric–pattern multifrontal method. ACM Transactions on Mathematical Software, 30:196–199
18. Estorff O von (ed) (2000) Boundary elements in acoustics: Advances and applications. WIT Press, Southampton
19. Everstine GC, Henderson FM (1990) Coupled finite element/boundary element approach for fluid–structure interaction. Journal of Sound and Vibration 87:1938–1947
20. Fahy F, Gardonio P (2007) Sound and structural vibration. Elsevier/Academic Press, 2nd edition, Amsterdam
21. Fischer M (2004) The fast multipole boundary element method and its application to structure–acoustic field interaction. PhD Thesis, University of Stuttgart
22. Flemisch B, Kaltenbacher M, Wohlmuth BI (2006) Elasto–acoustic and acoustic–acoustic coupling on nonmatching grids. International Journal for Numerical Methods in Engineering 67:1791–1810
23. European Committee for Standarization (2001) Determination of sound absorption coefficient and impedance in impedance tubes – part 2: Transfer function method (ISO 10534-2:1998), Technical report
24. Garcia P, Wiemeler D, Brand JF (2006) Oberflächenschallabstrahlung von Abgasanlagen – CAE–Methode und Entwicklungsprozess. Motortechnische Zeitung (MTZ) 76:852–859
25. Gaul L, Fischer M (2006) Large–scale simulations of acoustic–structure interaction using the fast multipole BEM. Zeitschrift für Angewandte Mathematik und Mechanik 86:4–17
26. Gaul L, Wenzel W (2002) A coupled symmetric BE–FE–method for acoustic fluid–structure interaction. Engineering Analysis with Boundary Elements 26:629–636
27. Giebermann K (1997) Schnelle Summationsverfahren zur numerischen Lösung von Integralgleichungen für Streuprobleme im $\mathbb{R}^3$. PhD Thesis, Universität Karlsruhe
28. Greengard L, Rokhlin V (1987) A fast algorithm for particle simulations. Journal of Computational Physics 73:325–348
29. Gyure MF, Stalzer MA (1998) A prescription for the multilevel Helmholtz FMM. IEEE Computational Science & Engineering 5:39–47
30. Howe MS (2004) Acoustics of fluid–structure interactions. Cambridge Monographs on Mechanics
31. Hughes MD, Chen K (2004) An efficient preconditioned iterative solver for solving a coupled fluid structure interaction problem. International Journal of Computer Mathematics 81:583–594
32. Jeans RA, Mathews JC (1990) Solution of fluid–structure interaction problems using a coupled finite element and variational boundary element technique. Jorunal of the Acoustical Society of America 88:2459–2466
33. Junge M, Fischer M, Maess M, Gaul L (July 11–14 2005) Acoustic simulation of an idealized exhaust system by coupled FEM and fast multipole BEM. In: Proceedings of the Twelfth International Congress on Sound and Vibration (ICSV12), Lisboa
34. Junge M, Schube F, Gaul L (2007) Sound radiation of an expansion chamber due to pressure induced structural vibrations. In: Proceedings of the Deutsche Jahrestagung für Akustik 07, Stuttgart

35. Junger MC (1997) Acoustic fluid–elastic structure interaction: Basic concepts. Computers & Structures 65:297–293
36. Junger MC, Feit D (1986) Sound, structures, and their interaction. The MIT Press, Cambridge, 2nd edition
37. Kergourlay G, Balmès E, Clouteau D (2000) Model reduction for efficient FEM/BEM coupling. In: Proceedings of ISMA 25, International Conference on Noise and Vibration Engineering, Leuven, 1167–1174
38. Gaul L, Kögl M, Wagner M (2003) Boundary element methods for engineers and scientists. Springer–Verlag, Berlin–Heidelberg
39. Kinsler LE, Frey AR, Coppens A, Sanders J (1999) Fundamentals of acoustics. John Wiley and Sons, Inc., New York
40. Levine H, Schwinger J (1948) On the radiation of sound from an unflanged circular pipe. Physical Review 73:383–406
41. Maess M (2006) Methods for efficient acoustic–stucture simulation of piping systems. PhD Thesis, University of Stuttgart
42. Marburg S, Schneider S (2003) Performance of iterative solvers for acoustic problems. Part I. Solvers and effect of diagonal preconditioning. Engineering Analysis with Boundary Elements, 27:727–750
43. Márquez A, Meddahi S, Selgas V (2004) A new BEM–FEM coupling strategy for two–dimensional fluid–solid interaction problems. Journal of Computational Physics 199:205–220
44. Moosrainer M (2000) Fluid–Struktur–Kopplung. PhD Thesis, Universität der Bundeswehr München
45. Morand HJP, Ohayon R (1995) Fluid structure interaction: Applied numerical methods. Wiley, New York
46. Morse PM (1948) Vibration and sound. McGraw–Hill, New York
47. Munjal ML (1987) Acoustics of ducts and mufflers. John Wiley & Sons, New York
48. Munt RM (1990) Acoustic transmission properties of a jet pipe with subsonic jet flow: I. the cold jet reflection coefficient. Journal of Sound and Vibration 142:413–436
49. Nishimura N (2002) Fast multipole accelerated boundary integral equation methods. Applied Mechanics Reviews 55:299–324
50. Ochmann M, Homm A, Makarov S, Semenov S (2003) An iterative GMRES–based boundary element solver for acoustic scattering. Engineering Analysis with Boundary Elements 27:714–725
51. Ohayon R, Soize C (1998) Structural acoustics and vibration. Academic Press, New York
52. Païdoussis MP (1998) Fluid–structure interactions: Slender structures and axial flow. Vol 1, Elsevier Science & Technology Books
53. Pretlove AJ (1966) Forced vibration of a rectangular panel backed by a closed rectangular cavity. Journal of Sound and Vibration 3:252–261
54. Rjasanow S, Steinbach O (2007) The fast solution of boundary integral equations. Springer–Verlag, New York
55. Rokhlin V (1993) Diagonal forms of translation operators for the Helmholtz equation in three dimensions. Applied and Computational Harmonic Analysis, 1:82–93
56. Saad Y (1996) Iterative methods for sparse linear systems. PWS Publishing Company, Boston
57. Schenk HA (1968) Improved integral formulation for acoustic radiation problems. Journal of the Acoustical Society of America 44:41–58
58. Schneider S (2003) Efficient usage of the boundary element method for solving the time harmonic Helmholtz equation in three dimensions. PhD Thesis, Technische Universität Dresden

59. Seybert AF, Ross DF (1977) Experimental determination of acoustic properties using a two–microphone random-excitation technique. Journal of the Acoustical Society of America 61:1362–1370
60. Seybert LH (1987) Application of the boundary element method to acoustic cavity response and muffler analysis. ASME Journal of Vibration and Acoustics 109:15–21
61. Stolper M (2004) Schnelle Randelementmethoden für die Helmholtzgleichung. PhD Thesis, Universtität Saarbrücken
62. van der Vorst HA (2003) Iterative Krylov methods for large linear systems. Cambridge University Press, Cambridge
63. Wagner N, Gaul L, Dumont NA (2002) The hybrid boundary element method in structural acoustics. Zeitschrift für Angewandte Mathematik und Mechanik 84:780–796
64. Wang X, Bathe KJ (1997) Displacement/pressure based mixed finite element formulations for acoustic fluid–structure interaction problems. International Journal for Numerical Methods in Engineering 40:2001–2017
65. Wu TW (2000) Boundary element acoustics: Fundamentals and computer codes. WIT Press, Southampton
66. Zienkiewicz OC, Taylor RL (2000) The finite element method, Vol 1. Butterworth–Heinemann, Oxford, 5th edition

20

Inverse Boundary Element Techniques for the Holographic Identification of Vibro–Acoustic Source Parameters

Jeong–Guon Ih

Centre for Noise and Vibration Control (NoViC), Department of Mechanical Engineering
Korea Advanced Institute of Science and Technology (KAIST), Science Town, Daejeon
305–701, Korea
`J.G.Ih@kaist.ac.kr`

Summary. There are various techniques for the identification sound sources and the near–field acoustical holography (NAH) is one of the highlighted methods in recent days. NAH is basically an indirect, inverse method, which exploits the field pressure data from the array sensors as input. Then the acoustic properties on the source plane can be inversely reconstructed, using the measured field pressure over the measurement or hologram plane. One can model the sound radiation, diffraction and transmission between the vibrating source and the measurement field by the vibro–acoustic transfer matrix using the boundary element method (BEM). Consequently, the distribution of the surface velocities of the arbitrary shaped source, not on the near-field "source" plane, can be reconstructed by multiplying the inverse of the calculated vibro–acoustic transfer matrix and the measured field pressure vector at any shape of near–field plane, including the conformal one. Because this type of conformal NAH technique has the ability in dealing with the complex shaped sources that cannot be described by separable coordinates, it can be applied to a lot practical problems. In this chapter, after describing the mathematical fundamentals in the NAH, the basic nature of the involved problems is explained and a procedure for realizing the inverse identification of the machine noise source is demonstrated in several practical application examples. Finally, the state–of–the–art in the study of NAH is discussed briefly and the future perspective of the development and application of the method is mentioned.

20.1 Introduction

Any noise control problem can be thought of as being comprised of three components: source, path, receiver. Although there exist many techniques to detect the position, magnitude, and other characteristics of sources, transfer paths, and receivers, the countermeasures or design changes which are applied to the primary sources are most important of all. Information on the source parameters of a vibro–acoustic system is very useful in the early stage of effective noise control, either by passive or active means. Due to this reason, the precise identification of basic source parameters, such as surface velocity, pressure, intensity, acoustic power, and field intensity vector, and so forth, can be regarded as the first step toward the effective noise control. There is no doubt about this fact in noise control area.

However, acoustic engineers meet difficulties in dealing with the extended vibro–acoustic radiators. The actual extended vibro–acoustic sources are continuous in geometry; irregular in shape with some inaccessibly small or concave spaces sometimes; wide in radiated sound spectra; complicated in modal behaviour due to varying structural thickness, internal excitations, complicated internal transfer paths, and severe interactions; harsh in operating conditions like temperature and toxic gases; sometimes too thin in thickness; sometimes combined with apertures that radiates air–borne sound; and usually located in a noisy environment during operation. Such harsh conditions of practical vibro–acoustic sources cause tremendous difficulties in source identification when the direct measurement techniques are employed using contact sensors like accelerometer, strain gage, potentiometer, etc. or sometimes non–contacting sensors like LDV, gap sensor, etc. Consequently, one needs the inverse or indirect techniques [7, 13, 19, 22] for solving various acoustic problems including the vibro–acoustic radiation case.

For the vibro–acoustic source identification, many theoretical and experimental techniques have been developed either using direct method or indirect one: selective operation, cocooning, window method, duct connection, nearfield pressure or motion measurement, direct measurement of surface vibration, directional microphones, sound intensimetry, surface intensity technique, beamforming, optical and acoustical holography, vectorial analysis or transfer path analysis, signal processing techniques, and so forth. Among these direct and indirect techniques, the most popular indirect method in the field has been the sound intensity technique. However, it should be reminded that the source activity identified by sound intensimetry is not an actual source property, but a field data. Furthermore, the excessive reactivity in the measurement field will cause a severe problem in detecting the real source. Another popular and seemingly precise technique for the source identification is the laser holography. Use of scanning laser holography (SLDV) and dual–pulsed laser Doppler interferometry are facilitated in a very controlled environment and steady state source operation condition; whereas the electronic speckle pattern interferometry can be used in 2–D painted source field without too severe restriction on the measuring environment, but, at the moment, it is implemented with a high cost.

Nearfield acoustical holography (NAH) is an indirect method using the array microphones for the inverse identification of vibro–acoustic properties of sound sources. In this technique, the acoustic properties on the source plane can be reconstructed, using the field pressure, which is measured on the measurement or hologram plane. NAH is one of the techniques for solving the inverse acoustic problems, which are to indirectly determine either the input (vibro–acoustic data) or the transfer system (radiation, reflection, scattering, diffraction) that give rise to the measurements of the output (vibro–acoustic data) which is usually contaminated by noise (instrumental, environmental, or unknown input). One should solve the inverse problems if the measured field data are available for a known acoustic system with unknown input data to be determined or for an unknown acoustical system to be determined under the restriction of known or calibrated input data. For example, on the inverse scattering problem, one can solve either for the inverse obstacle problem or the inverse medium problem. In the inverse obstacle problem, one looks for an obstacle from the far field pattern, whereas in the inverse medium problem, one looks for the continuously varying parameters of an inhomogeneous medium from the far scattered field. In many cases, the inverse restoration of target data from the indirect output measurement is inevitable because the direct measurement is not possible or not easy to practice. The target of the inverse problem is the identification of either the acoustic source properties, which are causes of radiated or propagated sound, or the physical or geometrical system parameters, which are closely related with sound propagation and transfer characteristics. It should be mentioned that, before the formulation of the vibro–acoustic NAH problem, there have been many active studies for the inverse holographic reconstruction in optics, geophysics, ultrasonics, structural acoustics, and digital image reconstruction area [7, 13, 19, 22].

Of the two major techniques that have been a great stimulus to the development of NAH, one is based on the spatial Fourier transform and the other on the acoustic boundary element method (BEM). In the former method, the field pressures, which are measured on a hologram plane, are decomposed into space and wave number domains by spatial Fourier transform. The pressure decay in propagation can be compensated and the pressure on a target plane is then reconstructed by an inverse spatial Fourier transform [48, 49, 72]. This technique was commercially implemented earlier [17]. The shape of the source surface should be regular, and, if this is not the case, then a hypothetical regular plane near the actual source surface needs to be assumed for the reconstruction. In the latter approach, the geometric and vibro–acoustic relation between the sound source and the hologram or measurement plane is modelled as the vibro–acoustic transfer matrix by numerically implementing the integral equations for acoustics [16, 35, 62, 65]. The sound radiation and transmission between the vibrating source and the measurement field (or hologram plane) can be modelled by the vibro–acoustic transfer matrix by using boundary integral equation or its discretized form of the BEM. Consequently, the distribution of the surface velocities of the source can be reconstructed by multiplying the inverse of the calculated vibro–acoustic transfer matrix and the measured field pressure vector at any shape of near–field plane, including the conformal one. This type of conformal NAH is capable of dealing with the complex shaped sources that cannot be described by separable coordinates. A similar concept named "phonoscopy" was also suggested [66].

The other technique worth mentioning is the method using a series expansion of spherical radiation functions for describing the sound field, in particular, in the hologram plane. The coefficient of each spherical function, centrally located at an origin within the source volume, is determined by the numerical inverse process [6, 73]. In a similar concept, the equivalent source method (ESM) [39] that can be also used for the reconstruction of sound field [14, 28]. The Helmholtz equation least–squares method [67], which is a special case of ESM by using spherical waves, emanating from a single point, has been suggested for reconstructing the sound field or source field in the spherical coordinates. It was shown [28, 74, 75] that this method can be combined with the BEM–based NAH method, with relatively fewer measurements, for the reconstruction of source parameters of an arbitrarily shaped object. The main topic of this paper will be confined only to the BEM–based NAH (or NAH based on the inverse BEM). Another indirect identification technique with great potential in practice is the inverse frequency response function (FRF) method [9]. This method is based on the transfer path information between inputs on a surface and outputs measured on the microphones. Once the pseudo–inverse of FRF is determined by using, in general, over–determined least–squared solution approach and singular value decomposition, the unknown inputs of the system (usually the volume velocity of a source segment) can be identified from the measured pressure. Application to an automobile interior to optimize the sound insulation material was reported [32]. An overview on the inverse reconstruction by using the NAH can be found in Reference [68, Chapters 3, 5, 7, and 8].

The basic concept of the BEM-based NAH was first explicitly studied by Gardner and Bernhard [16], who introduced the source identification method in the highly reactive field by utilizing direct BEM. Veronesi and Maynard [65] utilized the singular value decomposition of the discretized direct boundary integral equation, in order to decompose the field and source properties into the wave–vector domain. They demonstrated that the suppression of the amplification effect of the measurement noise via the rejection of higher wave–vector mode components led to an improved resolution. Bai [3] formulated the generalized holography equation based on direct BEM. He represented all possible combinations of the transfer matrix, in order to correlate the particle velocity and field pressure on the source surface and the field plane. Kim and Ih [34] described a resolution enhancement technique using the optimal selection of measurement points and regularization of the transfer matrix for the interior problem. By utilizing a trade–off relationship between variance and bias errors, the optimal rank, which produces the minimum mean–square error could be determined. Zhang, et al. [76] and Schuhmacher, et al. [59] employed the indirect formulation for the BEM–based NAH. Valdivia and Williams [63] suggested the iterative technique, which can avoid using the singular value decomposition technique that may be beneficial in dealing with a large transfer matrix. A consideration on the mathematical background of the BEM–based NAH in relation to the boundary condition can be found in Reference [54] and there are many other works on this method, for example, [11]. It should be mentioned that the finite element method can be also implemented in the holography technique [2, 12], but this will not be treated here.

As aforementioned, the BEM–based NAH technique thus provides a good opportunity for restoring the vibro–acoustic field image of many practical arbitrarily shaped sources. Its optimal feature is that only the measured field pressure is required for determining pressure, particle velocity, surface admittance, intensity, and power flow of the source and the domain of interest as well. This type of conformal NAH technique has the following advantages compared with conventional NAH based on the spatial Fourier transform. One can deal with the complex shaped sources that cannot be described by separable coordinates; the pressure need not be measured in separable coordinates, thus a reduced number of measurements with uneven spacing is possible; reflections from all directions can be considered; concave regions of the source can be reconstructed; and wrap-around error due to the finite aperture size is not involved. However, this method has some inconvenient aspects as well. The acoustic and geometric relation of the source surface and the hologram plane should be modelled via the use of the BEM and this causes problems. A considerable number of boundary elements and nodes are ultimately required for modelling the actual source surfaces involved in a practical noise problem and the amount of field pressure data increases in parallel with that of the surface nodes. The applicable frequency range can be limited by the characteristic length of the typical element. In addition, care should be taken with respect to the inversion of the fully populated vibro–acoustic transfer matrix that has a high singularity. However, if such difficulties are somehow overcome with some technical labour, the drawback of additional BEM modelling may not be a big loss after all because one needs the BEM model of the irregular "source" for the eventual forward prediction of the sound field after obtaining the source parameters, anyway. In this paper, the basic nature of the involved problems is explained and a procedure for realizing the inverse identification of the machine noise source is demonstrated in several practical applications to illustrate the actual procedure of the technique.

20.2 Summary of Theoretical Background

20.2.1 Sound Radiation from an Extended Source and its Boundary Element Modelling

Assume that the domain, V, enclosed by the boundary, S_o, is filled with an isothermal, homogeneous, inviscid, compressible, and stationary fluid medium, which is disturbed by a time harmonic ($\mathrm{e}^{\mathrm{i}\omega t}$) acoustic field. The acoustic pressure p at a position $\boldsymbol{r}$ produced by the harmonic vibration of the surface S and the distributed internal source f in a volume V is given by the following Kirchhoff–Helmholtz integral equation, which is a form of Fredholm integral equation of the first kind, known as inherently ill–posed

$$c(\boldsymbol{r})p(\boldsymbol{r}) = -\int_{S_o}\left[p\left(\boldsymbol{r}_o\right)\left(\frac{1}{R}+\mathrm{i}k\right)\frac{\mathrm{e}^{-\mathrm{i}kR}}{R}\cos\theta + \mathrm{i}\omega\rho v(\boldsymbol{r}_o)\frac{\mathrm{e}^{-\mathrm{i}kR}}{R}\right]dS + \int_V f(\boldsymbol{r}_o)G(\boldsymbol{r},\boldsymbol{r}_o)dV. \tag{20.1}$$

Here, k is the wave number, $\boldsymbol{r}_o$ the position vector of a surface point, R the distance between two points, G the Green's function, ρ the density of the fluid medium, and v the surface velocity. The first term in the integrand of Equation (20.1) has the directivity of $\cos\theta$ that implies the dipole effect by the surface pressure $p(\boldsymbol{r}_o)$. The second term in the integrand shows the monopole effect by the surface normal velocity $v(\boldsymbol{r}_o)$. The last term in the integrand is the distributed internal source effect in the domain of interest.

Equation (20.1) cannot be solved analytically in the whole domains of input or system described by nonseparable coordinates. This is due to the fact that the target input or vibro–acoustic source system is a continuous domain and, more importantly, the measured output field data are not given as continuous functions but only a finite number of data is usually available. Consequently, source parameters on the surface will be given in discrete forms and one should treat the inverse of linear system of equations that need to be solved numerically. If discrete field points and surface points are considered, due to practical reason, and $\boldsymbol{p}_s$, $\boldsymbol{v}_s$ denote pressure and normal velocity vectors on the surface nodes, respectively, the forgoing Kirchhoff–Helmholtz integral equation can be approximated by the matrix equation form as

$$\boldsymbol{D}_s\boldsymbol{p}_s = \boldsymbol{M}_s\boldsymbol{v}_s \qquad \text{on the boundary,} \qquad (20.2)$$

$$\boldsymbol{p}_f = \boldsymbol{D}_f\boldsymbol{p}_s + \boldsymbol{M}_f\boldsymbol{v}_s \qquad \text{in the domain.} \qquad (20.3)$$

Here, $\boldsymbol{D}_s$ and $\boldsymbol{M}_s$ indicate the dipole (including the solid angle information) and monopole matrices on the surface, and $\boldsymbol{D}_f$ and $\boldsymbol{M}_f$ are those corresponding to field pressures, respectively. If the Neumann, Dirichlet, and Robin (mixed) boundary conditions coexist, from the given boundary conditions, $\{\boldsymbol{p}_1, \boldsymbol{v}_2\}_s$ and z_3, Equation (20.2) can be rewritten to calculate $\{\boldsymbol{v}_1, \boldsymbol{p}_2, \boldsymbol{p}_2\}_s$ as

$$\begin{bmatrix} -\boldsymbol{M}_{11} & \boldsymbol{D}_{12} & \boldsymbol{D}_{13} - \boldsymbol{M}_{13}/z_3 \\ -\boldsymbol{M}_{21} & \boldsymbol{D}_{22} & \boldsymbol{D}_{23} - \boldsymbol{M}_{23}/z_3 \\ -\boldsymbol{M}_{31} & \boldsymbol{D}_{32} & \boldsymbol{D}_{33} - \boldsymbol{M}_{33}/z_3 \end{bmatrix}_s \begin{Bmatrix} \boldsymbol{v}_1 \\ \boldsymbol{p}_2 \\ \boldsymbol{p}_3 \end{Bmatrix}_s = \begin{bmatrix} -\boldsymbol{D}_{11} & \boldsymbol{M}_{12} \\ -\boldsymbol{D}_{21} & \boldsymbol{M}_{22} \\ -\boldsymbol{D}_{31} & \boldsymbol{M}_{32} \end{bmatrix}_s \begin{Bmatrix} \boldsymbol{p}_1 \\ \boldsymbol{v}_2 \end{Bmatrix}_s, \qquad (20.4)$$

where z_3 represents the specific acoustic impedance. If the Neumann boundary conditions are considered only, e.g., as a special case for the vibrating sources, on the condition that $\boldsymbol{D}_s^{-1}$ exists, the field pressure can be expressed as

$$\boldsymbol{p}_f = (\boldsymbol{M}_f + \boldsymbol{D}_f\boldsymbol{D}_s^{-1}\boldsymbol{M}_s)\,\boldsymbol{v}_s \equiv \boldsymbol{G}\boldsymbol{v}_s. \qquad (20.5)$$

where $\boldsymbol{G}$ is the vibro–acoustic transfer matrix that correlates the surface normal velocity with the field pressure and contains geometric information concerning the system as well. One should note that the vibro–acoustic transfer matrix $\boldsymbol{G}$ is a function of involved geometric data only.

20.2.2 Inverse Operation and Relevant Physical Meanings

Equation (20.5) should be inverted to obtain the source data; however, $\boldsymbol{G}$ is in general non–square and complex. If the field pressure is known at m points, the surface velocity at n (preferably smaller than m for a good result) nodes can be uniquely determined by utilizing a least–squared solutions approach and singular value decomposition (SVD). The SVD of $\boldsymbol{G}$ provides the acoustical modal expansion between the hologram and source field [55]. The transfer matrix $\boldsymbol{G}$ can be decomposed by

$$\boldsymbol{G} = \boldsymbol{U}\boldsymbol{\Lambda}\boldsymbol{W}^H, \qquad (20.6)$$

where

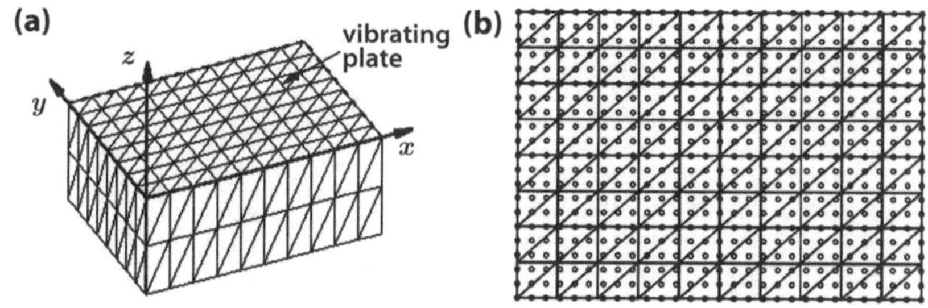

Fig. 20.1 (a) Boundary element model of a parallelepiped box (700L 520W 320H mm, 234 nodes, 63 velocity–unknown nodes, 464 linear elements), (b) 609 field points for the measurement.

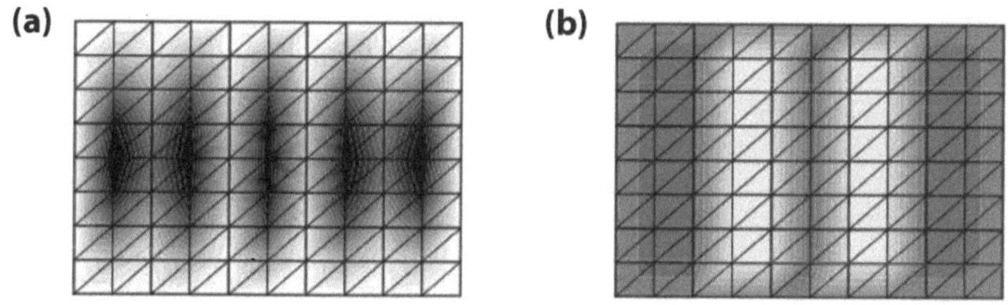

Fig. 20.2 Source velocity on the top of a box for (5,1) mode (163 Hz). (a) amplitude, (b) phase.

$$\begin{aligned} \boldsymbol{\Lambda} &= \mathbf{diag}\,(\lambda_1, \lambda_2, \ldots\ldots, \lambda_n)\ , \quad \lambda_1 \ge \lambda_2 \ge \ldots \ge \lambda_n \ge 0, \\ \boldsymbol{u}_i^H \boldsymbol{u}_j &= \delta_{ij}\ , \qquad\qquad \boldsymbol{w}_i^H \boldsymbol{w}_j = \delta_{ij}\ . \end{aligned} \tag{20.7}$$

Here, δ_{ij} is the Kronecker delta, the superscript "^{H}" signifies the Hermitian operator, and the subscript "$_n$" is the rank of $\boldsymbol{G}$, the elements of diagonal matrix $\boldsymbol{\Lambda}$ are singular values λ_i, and $\boldsymbol{U}$ and $\boldsymbol{W}$ indicate the left and right singular vectors, each of which has orthonormal columns. By virtue of the SVD, the inverse of Equation (20.5) can be expressed as

$$\boldsymbol{v}_s = \boldsymbol{G}^{+}\boldsymbol{p}_f = (\boldsymbol{G}^H\boldsymbol{G}^{-1})\boldsymbol{G}^H\boldsymbol{p}_f = \boldsymbol{W}\boldsymbol{\Lambda}^{-1}\boldsymbol{U}^H\boldsymbol{p}_f. \tag{20.8}$$

Here, $\boldsymbol{G}^{+}$ denotes the $n \times m$ pseudoinverse matrix. Equation (20.8) enables the reconstruction of the velocity field on the source surface, in principle, if the field pressures are measured and the transfer matrix is generated by the BEM. Figure 20.1 shows a parallelepiped radiator model and Figure 20.2 is the given (true) velocity of the vibrating plate on the top. Figures 20.3–20.5 illustrate typical shapes of $\boldsymbol{w}_i$, λ_i, and $\boldsymbol{u}_i$, respectively.

Physically, $\boldsymbol{w}_i$ and $\boldsymbol{u}_i$ indicate the wave–vectors which decompose the distribution of field pressure and surface velocity on the hologram and source planes for a selected frequency. Mutually orthogonal wave–vectors constitute the eigenspace of the measurement and source fields. The physical meaning of λ_i is the weighting factor for converting these field vectors from the source surface into the hologram plane.

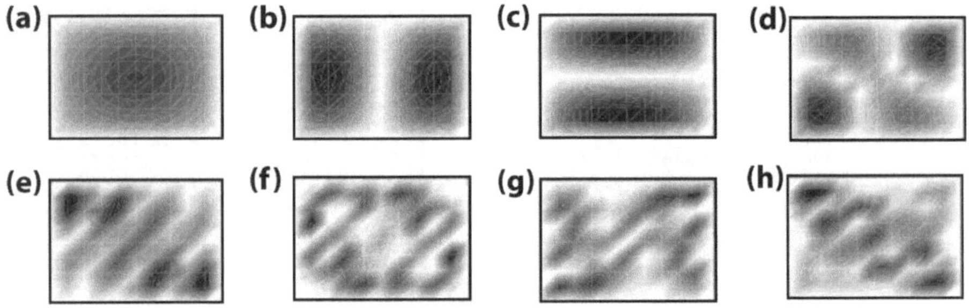

Fig. 20.3 Magnitude shapes of right singular vectors: (a) $\boldsymbol{w}_1$, (b) $\boldsymbol{w}_2$, (c) $\boldsymbol{w}_3$, (d) $\boldsymbol{w}_4$, (e) $\boldsymbol{w}_{10}$, (f) $\boldsymbol{w}_{20}$, (g) $\boldsymbol{w}_{30}$, (h) $\boldsymbol{w}_{40}$.

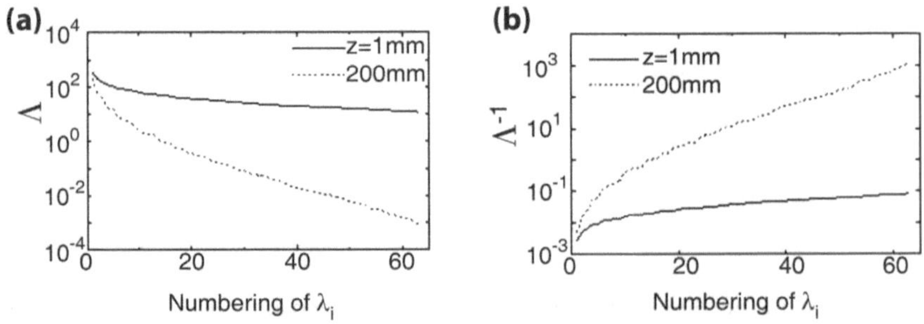

Fig. 20.4 Magnitudes of singular values of the transfer matrix. z is the distance of the hologram plane.

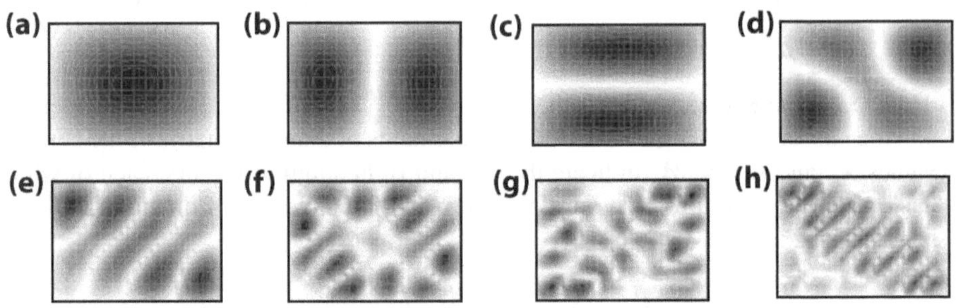

Fig. 20.5 Magnitude shapes of left singular vectors: (a) $\boldsymbol{u}_1$, (b) $\boldsymbol{u}_2$, (c) $\boldsymbol{u}_3$, (d) $\boldsymbol{u}_4$, (e) $\boldsymbol{u}_{10}$, (f) $\boldsymbol{u}_{20}$, (g) $\boldsymbol{u}_{30}$, (h) $\boldsymbol{u}_{40}$.

Conversely, each singular value represents the contribution of an acoustic mode on the source field to that of the measurement field [24]. High order modes corresponding to the components with small singular values are non–propagating wave components that decay out fast in the nearfield. Figure 20.6 shows an example sound field in a parallelepiped enclosure having a vibrating end plate. The first cross mode in the section is at 344 Hz. As can be seen, the field pressures near the source plate are not plane, which mean that evanescent wave components are rich in the near field. How-

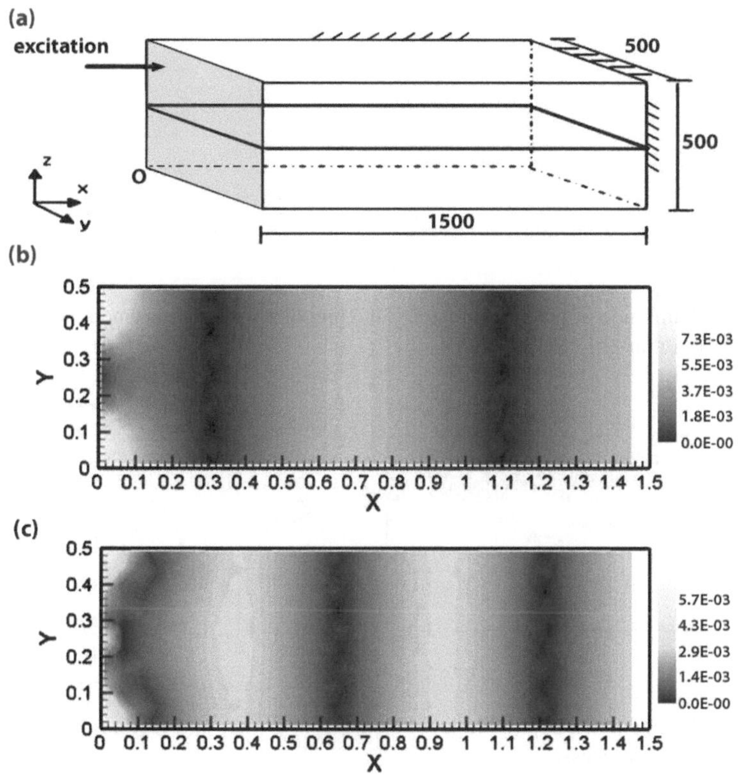

Fig. 20.6 Calculated pressure distribution in a parallelepiped with a vibrating end plate. (a) Chamber geometry, (b) pressure field by the excitation of (3,3) mode (216 Hz, $z = 0.05$ m), (c) pressure field by the excitation of (5,1) mode (304 Hz, $z = 0.25$ m).

ever, the influence of the nearfield effect does not extend to the Rayleigh distance or one wavelength. In this figure, the influence of the nonpropagating wave vectors are interfering with radiating planar component within a short distance to the source surface, say $x < \lambda/4$.

20.2.3 Ill–conditioned Nature and Regularization for Image Enhancement

Generally, the source reconstruction problem, as a linear discrete system, involves the inverse process of an ill–conditioned matrix. Numerical treatment of such equations is characterized as either rank–deficient problems or ill–posed problems [19]. This rank deficiency is due to small singular values that result in the condition number being much larger than unity. The main cause is the error generated from the measurement noise that is inevitably included in the observed field pressure and the numerical noise, which occurs in the computation. In a very ill–conditioned problem, the condition number of the matrix is very large and the problems are effectively underdetermined. For general ill–conditioned problems, the computed solution is unsta-

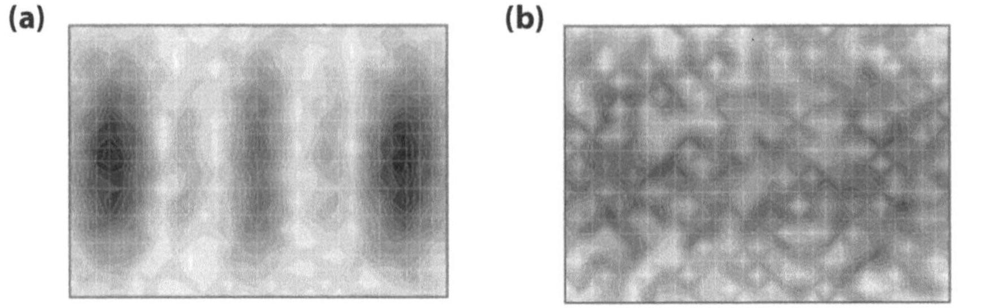

Fig. 20.7 Amplitude of given (estimated) field pressure with measurement noise (S/N ratio = 30): (a) $z = 1$ mm, (b) $z = 200$ mm.

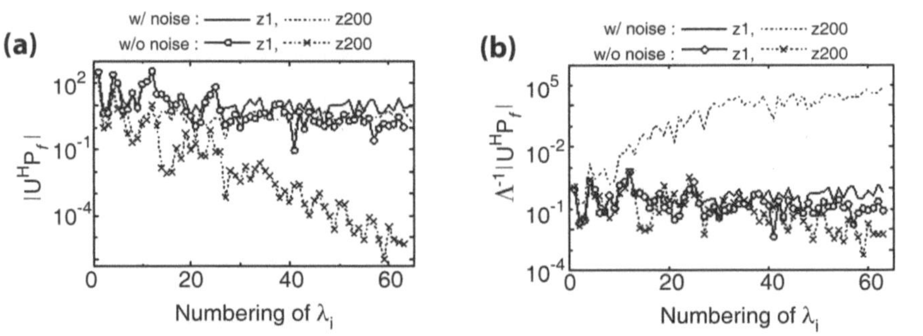

Fig. 20.8 A comparison of amplitudes of elements of $\boldsymbol{U}^H\boldsymbol{p}_f$ and vectors $\boldsymbol{\Lambda}^{-1}\boldsymbol{U}^H\boldsymbol{p}_f$ appearing in the inverse process.

ble and most likely dominated by rounding errors. However, for rank–deficient and discrete ill–posed problems, they can be solved by numerical regularization methods in which the solution is stabilized by additional information. The kernel can be represented by the singular value decomposition (SVD) such that the kernel of Fredholm integral equation can be expressed by the SVD using the orthonormal functions and singular values [19]. In addition, the usual approach for providing the initial field data set is to obtain as many independent data as possible making the problem as overdetermined one. In this way, the restored image can be far enhanced after smoothing by the regularization of the calculation result. As aforementioned, the error can be generated from the inevitable measurement noise is usually included in the field pressure and the numerical noise in the computation. In the inverse calculation, the high–order small singular values will amplify the non–propagating wave components. The contaminating noise in the measured field pressure will also be amplified during this process and this will yield highly distorted reconstructed data. Figure 20.7 depicts noise contaminated (S/N ratio=30) field pressure data at two hologram planes. Figures 20.4(b) and 20.8 show the inverse of $\boldsymbol{\Lambda}$ and its effect to the inversely projected vectors. One can find the blurred reconstruction image of the source in Figure 20.9.

To overcome the divergence phenomenon due to non–propagating wave components, proper filtering is needed during the inverse operation [55]. Because the

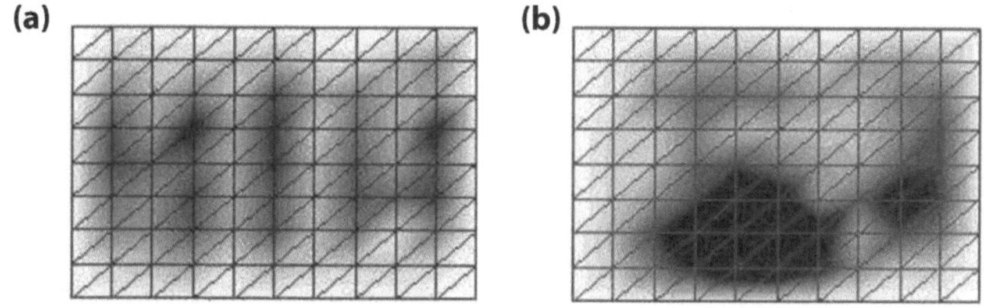

Fig. 20.9 Reconstructed velocity amplitude on the source surface before regularization: (a) $z = 1\,\text{mm}$, (b) $z = 200\,\text{mm}$.

higher–order modes in acoustical holography can be considered as non–propagating wave components, low–pass filtering such as the Wiener filtering [1, 15] can be taken in the optimal sense, and spatial regularization by using the singular value decomposition (SVD) is adopted [32, 43] for taking account of the measurement noise. Another possible approach is regularization by using an iterative inverse solution [5, 8]. This method in the image restoration area has been known that it does not require an inverse matrix and the precision of solutions can be enhanced by incorporating bounding conditions during iteration steps. In this method, the choice of iteration number is most important due to the trade-off between the variance error arising from measurement error and the bias error resulting from regularization. A method for determining the optimal iteration number was suggested by Kim and Ih [33]. Most popularly used regularization techniques are Tikhonov regularization (TR) [61] and Landweber iteration method (LITE) [38]. It is essential to find an optimal wave–vector filter shape by an optimal Tikhonov parameter or iteration number for proper regularization. Currently, there are various searching tools for an optimal parameter such as the mean–square error estimation (MSE) using a noise variance σ^2 [33, 34], sometimes referred as Morozov discrepancy principle (MDP) [69], the generalized cross–validation technique (GCV) [53], and the L–curve criterion [19], etc. To determine an optimal parameter, the golden section method (GSM) or even genetic algorithms can be adopted [4, 10, 19, 50]. A review of the regularization technique can be found in References [33, 69]. For example, a modified TR method uses the minimized Tikhonov functional Γ_2 as given by [61]

$$\Gamma_2 = \|\widehat{\boldsymbol{p}}_f - \boldsymbol{G}\boldsymbol{Q}\|^2 + \alpha\|(\boldsymbol{I} - \boldsymbol{F}_1^{\alpha})\boldsymbol{W}^H\boldsymbol{Q}\|^2, \tag{20.9}$$

where α is a regularization parameter. Using this, one can obtain the solution of Equation (20.7) as

$$\hat{\boldsymbol{Q}}_2^{\alpha} = \boldsymbol{W}\mathbf{diag}\left[\cdots, \frac{\lambda_j\left(\alpha + \lambda_j^2\right)^2}{\alpha^3 + \lambda_j^2\left(\alpha + \lambda_j^2\right)^2}, \cdots\right]\boldsymbol{U}^H\tilde{\boldsymbol{p}}_f = \boldsymbol{W}\boldsymbol{F}_2^{\alpha}\boldsymbol{\Lambda}^{-1}\boldsymbol{U}^H\tilde{\boldsymbol{p}}_f. \tag{20.10}$$

Due to the extra filtering of high–order wave–vector components, the foregoing modified TR (MTR) seems to oversmooth the solution compared with the classical

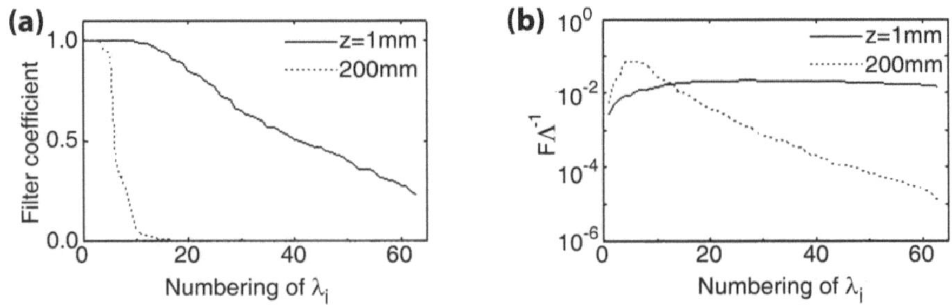

Fig. 20.10 (a) Coefficient of optimally designed wave vector filter, (b) amplitude of filtered and inverted singular values.

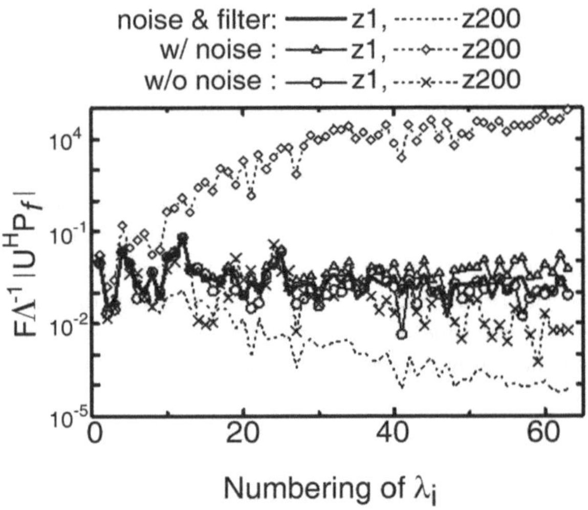

Fig. 20.11 A comparison of amplitudes of elements of $F\boldsymbol{\Lambda}^{-1}\boldsymbol{U}^H\boldsymbol{p}_f$ vectors (regularized).

TR [69]. In the LITE method, the solution can be calculated iteratively as [33, 38]

$$\hat{\boldsymbol{Q}}_3^{\alpha} = \boldsymbol{W}\,\mathbf{diag}\left[\cdots, \frac{1-\left(1-\beta\lambda_j^2\right)^{\alpha-1}}{\lambda_j}, \cdots\right]\boldsymbol{U}^H\tilde{\boldsymbol{p}}_f = \boldsymbol{W}\boldsymbol{F}_3^{\alpha}\boldsymbol{\Lambda}^{-1}\boldsymbol{U}^H\tilde{\boldsymbol{p}}_f, \tag{20.11}$$

where α denotes the iteration step and β is the convergence parameter.

Figure 20.10 shows the optimal wave vector filter shapes and filtered values of inverse singular values by using the iteration technique. In Figure 20.11, we can observe that the inverted and filtered (regularized) transfer vectors for two measurement distances have decreasing trends with the increase of the order number of singular values. Figure 20.12 shows the final restored image of the source velocity after regularization. One can find that the reconstructed source image from the nearfield data is very similar to the given velocity amplitude distribution; On the other hand, the

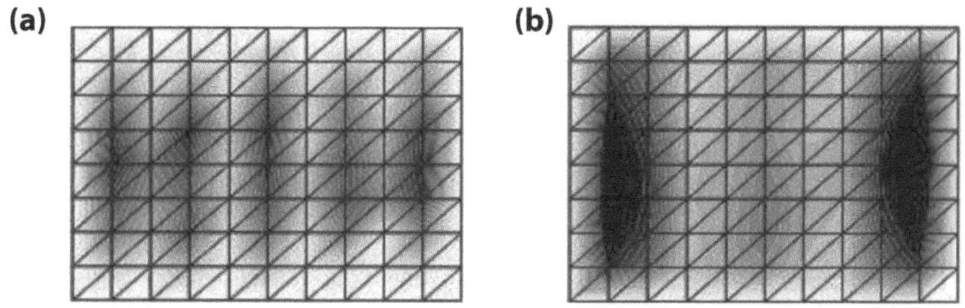

Fig. 20.12 Reconstructed velocity amplitude on the source surface after regularization: (a) $z = 1\,\text{mm}$, (b) $z = 200\,\text{mm}$.

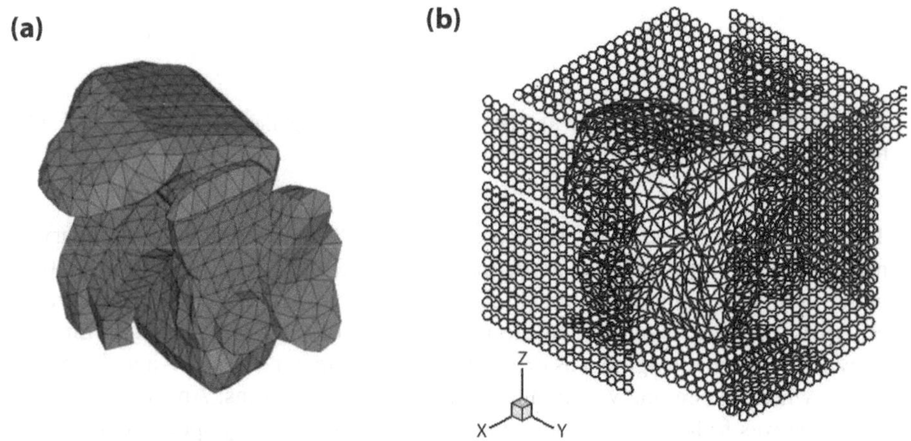

Fig. 20.13 (a) Boundary element mesh of the engine (1076 nodes), (b) sensor positions for the measurement of field pressure (1440 points) Microphone spacing was 50 mm.

source image recovered from the intermediate field data shows only the active contributing parts of the source to the field.

20.3 Practical Application Examples

Various sound sources have been tackled by the BEM–based NAH for the identification of vibro–acoustic sources. Examples include aircraft interior [71], refrigerator compressor [25], vacuum cleaner [28], internal combustion engine [28, 60], vehicle interior [55], tire exterior [1], copying machines and printers [58], etc. Because the method permits the source identification of a complex–shaped practical machinery, it is expected that the number of application cases will be increased very quickly.

Figure 20.13 shows the boundary element model, which was reduced from a very fine finite element model, of an automotive engine for the reconstruction of the surface source parameters as a demonstration example of the BEM-based NAH [28].

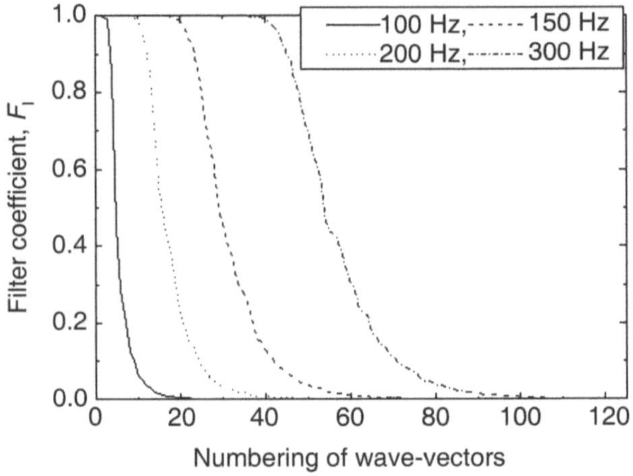

Fig. 20.14 Optimal filter shape for the regularization using the iterative technique (3000 RPM).

The engine was modeled by 1047 nodes and 2148 linear triangular boundary elements. The maximum characteristic length of this BEM model was $L_C = 98.0$ mm, which limits the applicable high frequency to about 580 Hz considering the $\lambda/4$-criterion for linear elements. In the measurement of field pressure data, this 2.5L, 6–cylinder gasoline engine was driven by a dynamometer in a semi–anechoic chamber, in which the engine was mounted on three rigid columns. An over–determined condition was followed to assure a precise result; 1440 evenly spaced field pressure data were taken as input data. Exterior noise radiated from the transmission was virtually eliminated by lead–wrapping. In the boundary element model, engine surface connected to the transmission was assigned with zero velocity (rigid boundary), that the remaining number of nodes with unknown boundary data was 860. Because the engine has 6 cylinders, the engine noise is dominant at the first firing frequency (E3) and its harmonics. The results at 150 Hz corresponds to the firing frequency (E3) at 3000 RPM. The BE model of the engine for the NAH can be later employed again for the post–processing in the prediction of sound intensity (surface and field) in magnitude and flow vector, surface admittance, sound power, radiation efficiency, directivity, field pressure map, etc. [23]. The active $\boldsymbol{I}$ and reactive $\boldsymbol{Q}$ intensity can be obtained as follows

$$\boldsymbol{I} = \frac{1}{2}\Re\left\{\boldsymbol{p}_s \boldsymbol{v}_s^*\right\}, \qquad \boldsymbol{Q} = \frac{1}{2}\Im\left\{\boldsymbol{p}_s \boldsymbol{v}_s^*\right\}. \tag{20.12}$$

Figure 20.14 illustrates an example of the shape of the regularization filter (TR) that was applied to the wave vectors in each frequency at 3000 RPM. Figure 20.15 compares measured and predicted field pressure spectra at 3000 RPM. Predicted field pressure data were based on the reconstructed surface pressure and velocity data. Four positions 1

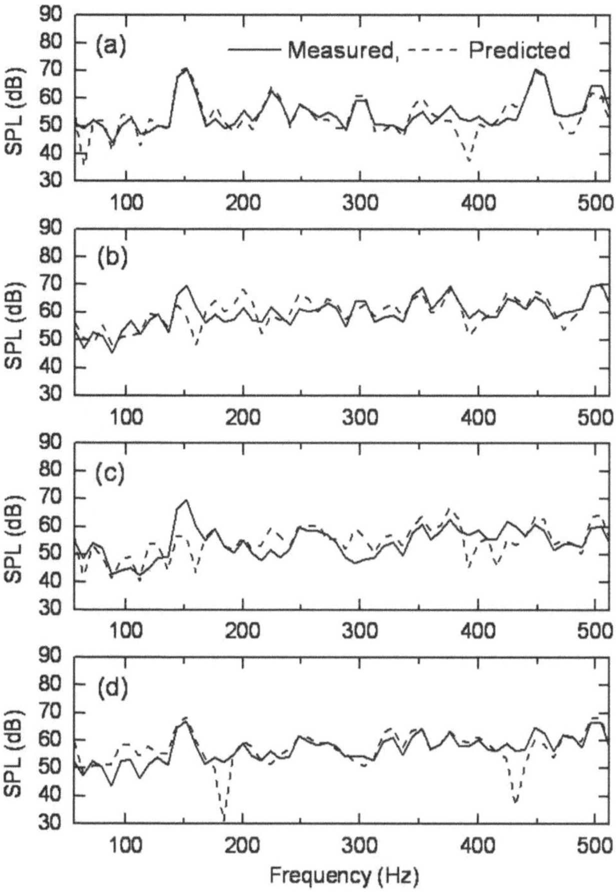

Fig. 20.15 Comparison of measured and predicted field pressure spectra (3000 RPM). Predicted field pressure data were based on the reconstructed surface pressure and velocity data. Four positions 1 m from each direction were considered: (a) intake, (b) exhaust, (c) front, (d) top directions.

shows the restored surface velocity distribution of the engine. One can easily find the acoustically hot parts in sound radiation, which would be beneficial to the noise control.

In Figure 20.17, the directivity pattern of sound radiation can be observed using the active intensity vector map. Figures 20.18 and 20.19 show the results for the identification of responsible engine parts for the spectral peak of the sound power. With the aid of such analysis, the engineer can decide which part of the engine should be modified to reduce the noise and the outcome result can be also quantified. Radiation efficiency of each part of an engine can be calculated based on the reconstructed source data. Figure 20.20 shows the result. One can find that the radiation efficiency of the exhaust manifold is far higher than the other source parts at low frequencies

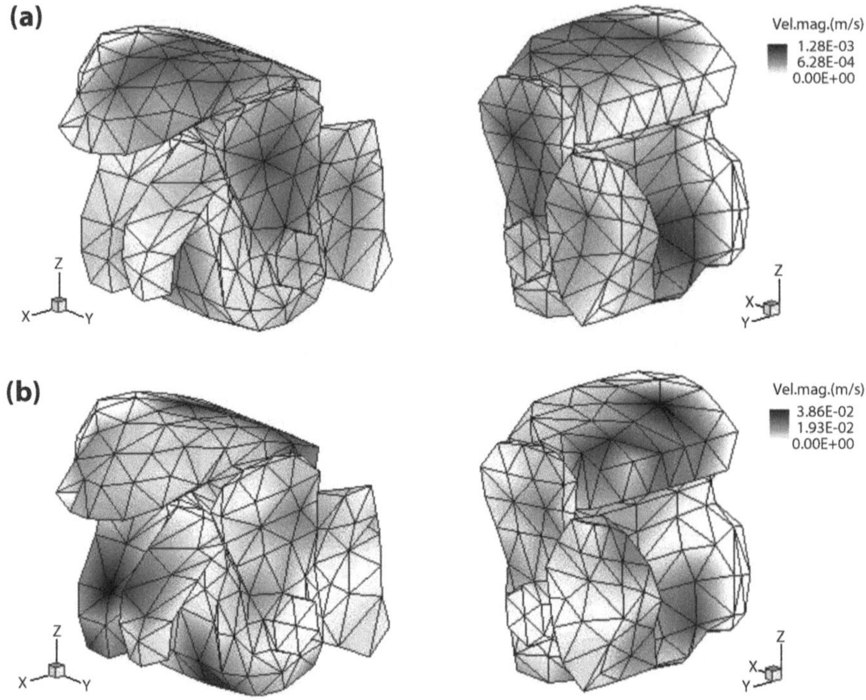

Fig. 20.16 Reconstructed normal surface velocities of an engine surface by the BEM–based NAH (3000 RPM). (a) 100 Hz, (b) 150 Hz (E3 component).

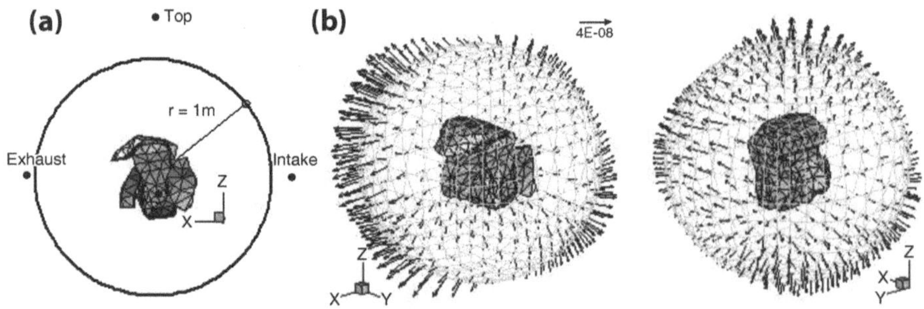

Fig. 20.17 Post processing of reconstructed source data of an engine in Figure 20.12 (300 Hz in 3000 RPM). (a) Plane for directivity definition, (b) directivity expressed by the active intensity vector.

at least by about 10 dB. It is noted that, beyond about 450 Hz, radiation efficiency becomes 1 in this case.

Another application example of the BEM–based NAH is illustrated in Figure 20.21: an irregular–shaped, three–dimensional machine, a canister–type vacuum cleaner. The vacuum cleaner was modelled by 139 nodes and 274 linear triangular elements. Number of measurement points were 352. In Figure 20.21, measurement

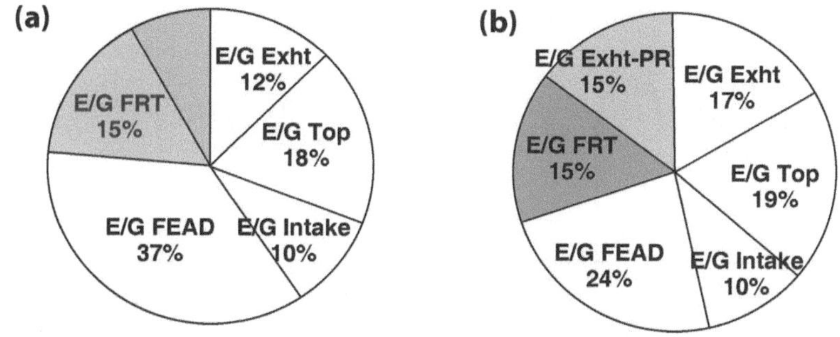

Fig. 20.18 Post processing of reconstructed source data of an engine in Figure 20.12 (3000 RPM). Contribution analysis showing: (a) SPL at 1 m position from the engine front, (b) overall radiated sound power.

points and BEM meshes can be seen, and the measured field data and reconstructed surface normal velocity distribution at 120 Hz are also shown. The parallelepiped hologram plane was separated from each nearest side of the vacuum cleaner by 60 mm. Target frequencies were 120 and 240 Hz, which were around the harmonic frequencies of the rotor installed inside the vacuum cleaner. Most intense noise was radiated at 120 Hz. Figure 20.22 maps the active and reactive intensity vector flows as a post processing of reconstructed source data of vacuum cleaner at 120 Hz. The results in Figure 20.22 suggest that some other mechanism of sound radiation also exists than the dominant exhaust discharge noise in the wheel area. As explained through application examples, once the distributions of source parameters, viz., sound pressure and particle velocity, are identified, one can easily employ the BEM model of the source again for the post–processing.

20.4 Perspective on the Further Works

Still, there are some unsolved problems for the BEM–based NAH technique to be applied conveniently to the actual problems. It is felt that we should conduct further studies to refine the technique for unrestricted and easy application to practical problems. The followings are part of features that we should consider about the enhanced command of the technique with precision or efficient practice.

20.4.1 Small Number, but Efficient Measurement Sensors

When a number of measurement sensors are used, there may be the redundancy in their signal. In this regard, the linear independence of vibro–acoustic transfer matrix should be first assured by the proper allocation of measurement points on the hologram plane, irrespective of its shape. This can be accomplished by selecting the required number of points with the aid of an effective independence method (EfI) [29]. The EfI value for each set of measurement positions can be calculated by using the singular vector in the SVD process as follows:

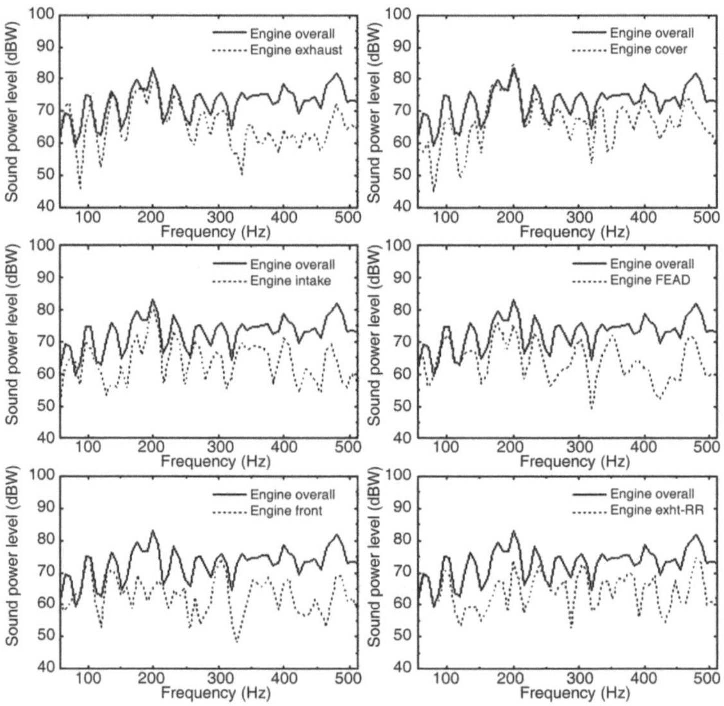

Fig. 20.19 Spectral contribution analysis using the reconstructed source data of an engine in Figure 20.12 (4000 RPM). Solid line is the overall sound power spectrum and the dashed line denotes the predicted sound power contributed from each part of the engine. One can find that 200 Hz peak is due to the engine cover and intake manifold and 300 Hz peak is mainly radiated from the front surface. It can be seen that the sound radiated from the exhaust manifold contributes to the overall sound power largely below about 250 Hz.

$$E_f = \mathbf{diag}\left[\boldsymbol{G}(\boldsymbol{G}^H\boldsymbol{G})^{-1}\boldsymbol{G}^H\right] = \mathbf{diag}\left[\boldsymbol{U}_n\boldsymbol{U}_n^H\right]. \qquad (20.13)$$

The contribution of sensor position to the linear independence of a transfer matrix can be evaluated for a frequency range of interest and a point or set of points having the smallest value is discarded from the candidate sensor positions. Then, by repeating this process, one can define the measurement positions that are the least inter–dependent for a given number of measurement locations [28, 40, 55]. Figure 20.23 explains the sorting procedure using the EfI technique.

20.4.2 Precise Reconstruction from Precise Measurement

In order to assure the measurement accuracy of the field data, sources of random and bias errors due to the mismatch of sensor positions and spacing, phase difference, small amplitude calibration, scattering from sensor fixtures, etc. should be suppressed and the influence of such measurement parameters should be monitored

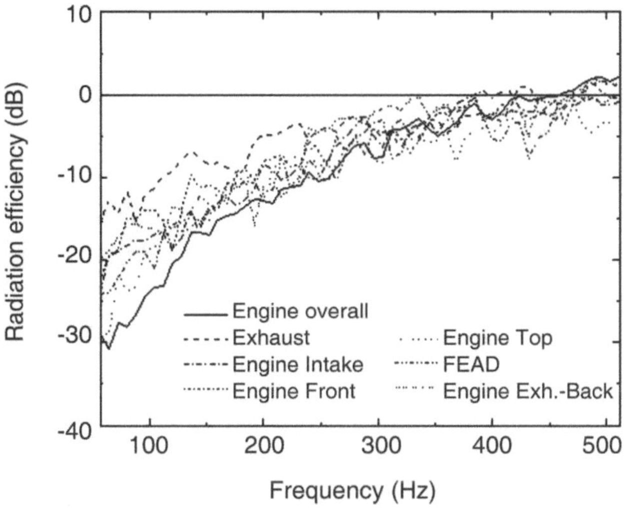

Fig. 20.20 Calculated radiation efficiency of each part of an engine in Figure 20.12, based on the reconstructed source data. One can find that the radiation efficiency of the exhaust manifold is far higher than the other source parts at low frequencies at least by about 10 dB. It is noted that, beyond about 450 Hz, radiation efficiency becomes 1.

carefully. Error analysis for the planar spatial holography [51, 57] can be partially applicable to the BEM–based NAH. Sensor proximity error to the source surface can be overcome when the non–singular or weakly singular BEM is employed in the modelling [31]. Errors associated with boundary element modelling should be taken into account [45] as well as the truncation error. If the information on a part of the source is known a priori, as is often the actual case, resolution enhancement of the restored result may be possible [30]. The effect of varying distance of the hologram plane to the source surface should be analysed and a compensation method or error analysis should be given. Most importantly, development of a new regularization technique should have a break though. One should remind that the time and spectral characteristics and magnitude of background noise are very much influential to the final reconstruction result [42]. All those aspects should be studied further in a statistical sense to obtain a fine restored result. Additional remark should be given to the use of intensity sensors [26] in the field measurement. Using such sensors, a horizon can be open to reconstruct the source field that results a three–dimensional sound field, which would be a far precise reconstruction of the source field.

20.4.3 Speedy or Convenient Measurement for Engineering Application

In the practical measurement of the source, especially for large–scaled sources or source with many ancillaries attached to the source surface or sources with inaccessible part, partial measurement of the field data is inevitable. By an analogous implementation to the BEM–based NAH, data extrapolation technique [57] or so-

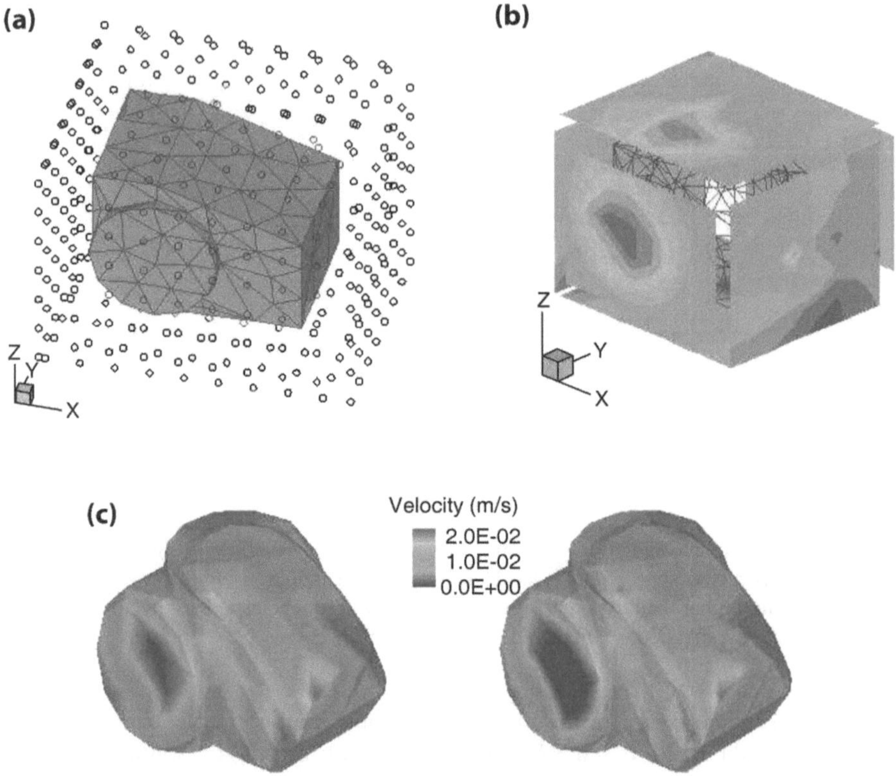

Fig. 20.21 Reconstruction source data on the vacuum cleaner by the BEM–based NAH (139 nodes, 274 linear elements). (a) 352 measurement points and BEM meshes, (b) Measured field data at 120 Hz, (c) reconstructed surface normal velocity distribution at 120 Hz.

called patched holography technique in spatial holography [18,41,70] are promising in solving the problem although one should accept certain level of errors, which may be acceptable in engineering sense. Measurement efforts and cost are too much in using the BEM–based NAH. Methods to countermeasure the measurement effort and time should be devised seriously. It was found that the use of reflecting bodies in the field, thus reducing the number of sensors, is useful [36,64] and the employment of ESM technique [28] combined with the BEM–based NAH can give a chance to virtually reduce the number of sensors and measurements, although some degradation of accuracy should be accepted in engineering sense. Measurement of pressure and particle velocity in the hologram plane and use of interpolation technique can reduce the measurement effort [21,27] although a further study is needed in the interpolation technique and sensor calibration. Quick measurement and analysis would be beneficial in many cases in engineering development or refinement of noise and vibration.

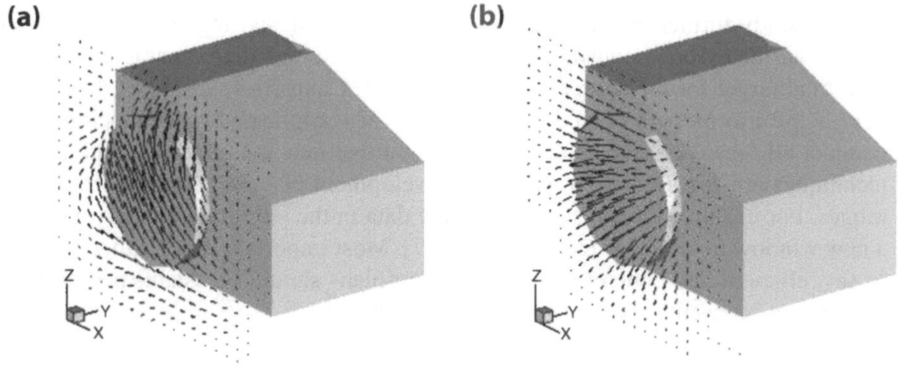

Fig. 20.22 Post processing of reconstructed source data of vacuum cleaner in Figure 20.21 (120 Hz). (a) Active intensity vector map, (b) reactive intensity vector map.

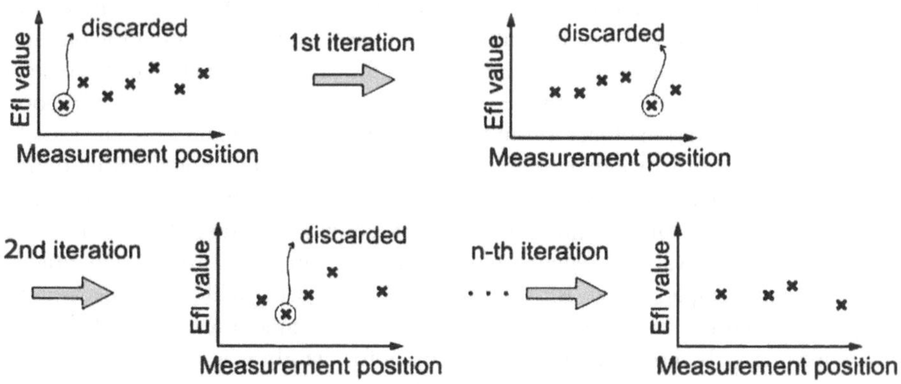

Fig. 20.23 Iteration procedure in the EfI technique for sorting out the redundant data.

20.4.4 Extended Application to Interesting Topical Area

There remain many challenging application targets for the NAH. Possible topics include: Source visualization of rotating machinery, sound transmission analysis and imaging of insulation materials, identification in transient or impulsive operating condition, characterization of electro–magnetic sound generation field separated from the structural excitation, correlated analysis of acoustic and structural wave fields, etc. Because the surface pressure and particle or normal velocity distribution are obtained by the reconstruction process, the surface impedance or admittance can be obviously estimated. Many works had been conducted on this matter, because an exact definition of the surface acoustic property will be important in the later countermeasure plan [2, 20, 37, 46, 52]. Recently, imaging of flow noise sources is one of hot research targets [47]. The structures of jet noise, mixing noise and impinging noise source mechanisms related to the flow would be very challenging topics. Identification of the surface structural wave motion or energy flow [56] is an important topic to study further on. Double stage inverse problem starting from the field data via the surface vibro–acoustic parameters to the internal source parameters would be a final target for the source identification [44, Chapters 4, 7, and 8], that would very helpful to prepare countermeasure plans of complicated noisy machinery. To conduct all these research topics, it is thought that new measurement and analysis techniques are required along with the development of new instrumentation techniques. For example, use of sound intensity data in the hologram plane would open a new window to the NAH technique [26, 27]. Most important thing is that, needless to say, efficient and concrete countermeasure plans should be established after the reconstruction result of the source image is obtained.

20.5 Conclusions

The nearfield acoustical holography based on the inverse BEM was briefly reviewed notwithstanding the fact that the author might have a limited information on the recent progress in this area. This method enables the precise estimation of the distribution of acoustical parameters on the source surface, i.e., surface pressure and normal velocity on a vibrating object. From this inverse identification process, the radiation characteristics of the irregular source can be easily predicted. Sound radiation from a top plate of a parallelepiped box was adopted as a demonstration example to show the actual stage in implementing the NAH based on the inverse BEM. An automotive engine and a vacuum cleaner, having irregular shapes, were taken as practical application examples of the BEM–based NAH. On–going and future research topics were listed for the further refinement of the technique. Further development of new ideas may be borrowed from the theories for the ultrasonic imaging, tomography, digital image enhancement, and other fields using similar techniques, but under different disciplines. There is no doubt that one can extend the concept and method of BEM–based NAH to solve many challenging vibro–acoustic problems in practical machinery. A huge opportunity is open to the acousticians in this area by virtue of the recent brilliant development of computer and measuring instrument in data processing speed and memory size.

Acknowledgements

This work was partially supported by the BK21 Project. This paper contains many works collaborated with the former and present students of Acoustics Laboratory as parts of their Ph.D. theses at KAIST: B.–K. Kim at KIMM, S.–C. Kang at Doosan Infracore Co., I.–Y. Jeon at Samsung SDI Co., J.–H. Jeong at Hyundai Motor Co., S.–I. Kim at Samsung Electronics Co., J.–Y. Kim at Georgia Inst. Tech., H.–W. Jang, and A. Oey at KAIST. Efforts in changing the manuscript in LaTex file by C.–H. Jeong, H.–W. Jang, and T.–K. Kim are very much appreciated. Contributions by Dr. S. Marburg and Dr. B. Nolte in providing some of relevant reference materials and list are acknowledged.

References

1. Andrews HC, Hunt BR (1977) Digital image restoration, Prentice–Hall, Englewood Cliffs
2. Anderssohn R, Marburg S, Großmann C (2006) FEM–based reconstruction of sound pressure field damped by partially absorbing boundary conditions. Mechanics Research Communication 33:851–859, 2006
3. Bai MR (1992) Application of BEM–based acoustic holography to radiation analysis of sound sources with arbitrarily shaped geometries. Journal of the Acoustical Society of America 92:533–549
4. Bazaraa MS, Sheriali HD, Shetty CM (1993) Nonlinear programming: Theory and algorithms. John Wiley & Sons, New York
5. Biemond J, Lagendijk RL, Mersereau RM (1990) Iterative methods for image deblurring. In: Proceedings of the IEEE 78:856–883
6. Chao YC (1987) An implicit least–square method for the inverse problem of acoustic radiation. Journal of the Acoustical Society of America 81:1288–1292
7. Colton D, Kress R (1998) Inverse acoustic and electromagnetic scattering theory. Springer–Verlag, Berlin, 2nd edition
8. Demoment G (1989) Image reconstruction and restoration: overview of common estimation structures and problems. IEEE Transaction on Acoustics, Speech & Signal Processing ASSP–37 2024–2036
9. Dumbacher S, Brown D (1997) Practical consideration of the IFRF method technique as applied to noise path analysis and acoustic imaging. In: Proceedings of International Modal Analysis Conference, Orlando 1677–1685
10. DeLillo T, Hrycak T (2006) A stopping rule for the conjugate gradient regularization method applied to inverse problems in acoustics. Journal of Computational Acoustics 14:397–414
11. DeLillo T, Hrycak T, Isakov V (2005) Theory and boundary element methods for near–field acoustic holography. Journal of Computational Acoustics 13:163–185
12. Dreckhan J, Sachau D (2005) Influence of smoothing to the inverse finite element method for acoustic hot–spot identification. Proceedings of Inter–Noise 2005, Rio de Janeiro (CD ROM)
13. Engl HW, Groetsch CW (1987) Inverse and ill–posed problems. Academic Press, Orlando
14. Fahnline JB, Koopmann GH (1991) A numerical solution for the general radiation problem based on the combined methods of superposition and singular–value decomposition. Journal of the Acoustical Society of America 90:2808–2819
15. Fleischer H, Axelrad V (1986) Restoring an acoustic source from pressure data using Wiener filtering. Acustica 60:172–175
16. Gardner K, Bernhard RJ (1988) A noise source identification technique using an inverse Helmholtz integral equation method. Transaction of the ASME: Journal of Vibration Acoustics Stress and Reliability in Design 110:84–90
17. Hald J (2003) Patch near–field acoustical holography using a new statistically optimal method. In: Proceedings of the Inter–Noise 2003, Seogwipo (CD ROM)
18. Hald J (1989) STSF–a unique technique for scan–based near–field acoustic holography without restrictions on coherence. B & K Technical Review, No. 1
19. Hansen PC (1998) Rank–deficient and discrete ill–posed problems. SIAM, Philadelphia
20. Hardtke HJ, Marburg S (1998) A boundary element method based procedure to calculate boundary admittance from measured sound pressures. Engineering Analysis with Boundary Elements 21:185–190

21. Harris MC, Blotter JD, Sommerfeldt SD (2006) Obtaining the complex pressure field at the hologram surface for use in near–field acoustical holography when pressure and in–plane velocities are mearsured. Journal of the Acoustical Society of America 119:808–816
22. Ih JG (2005) Inverse problems in the vibro–acoustics. In: Proceedings of Noise and Vibration; Emerging Methods (NOVEM), St. Raphael (USB)
23. Ih JG, Jeon IY, Kim SI (2006) Holographic reconstruction of the vibro–acoustic field of an engine using the inverse BEM and equivalent sources. In: Proceedings of International Congress on Sound and Vibration (ICSV 13), Vienna (CD ROM)
24. Ih JG, Kang SC (2000) Meanings of SVD and wave–vector filtering in the near–field acoustical holography using the inverse BEM. Journal of the Acoustical Society of America 108:2528
25. Ih JG, Kang SC, Kim SJ, Kang KS (1998) Reconstruction of the vibro–acoustic field on the surface of the refrigerator compressor by using the BEM–based acoustic holography. In: Proceedings of International Compressor Engineering Conference, Purdue 525–529
26. Jacobsen F, Jaud F (2007) Statistically optimized near field acoustic holography using an array of pressure–velocity probes. Journal of the Acoustical Society of America 121:1550–1558 (2007).
27. Jacobsen F, Liu Y (2005) Near field acoustic holography with particle velocity transducers. Journal of the Acoustical Society of America 118:3139–3144
28. Jeon IY, Ih JG (2005) On the holographic reconstruction of vibro–acoustic fields using equivalent sources and inverse boundary element method. Journal of the Acoustical Society of America 118:3473–3482
29. Kammer DC (1992) Effect of modal error on sensor placement for on–orbit modal identification of large space structures. Journal of Guidance, Control, and Dynamics 15:334–341
30. Kang SC, Ih JG (2000) The use of partially measured source data in near–field acoustical holography based on the BEM. Journal of the Acoustical Society of America 107:2472–2479
31. Kang SC, Ih JG (2001) Use of nonsingular boundary integral formulation for reducing errors due to near field measurements in the BEM–based NAH. Journal of the Acoustical Society of America 109:1320–1328
32. Kim BH, Park BH, Jeon IY, Ih JG (2003) Noise source ranking in automotive vehicle using the inverse FRF method. In: Proceedings of Inter–Noise 2003, Seogwipo (CD-ROM)
33. Kim BK, Ih JG (2000) Design of an optimal wave–vector filter for enhancing the resolution of reconstructed source field by near–field acoustical holography (NAH). Journal of the Acoustical Society of America 107:3289–3297
34. Kim BK, Ih JG (1996) On the reconstruction of the vibro–acoustic field over the surface enclosing an interior space using the boundary element method. Journal of the Acoustical Society of America 100:3003–3016
35. Kim GT, Lee BH (1990) 3–D sound source reconstruction and field reprediction using the Helmholtz integral equation. Journal of Sound and Vibration 136:245–261
36. Kim SI, Ih JG, Jeong JH (2007) Use of rigid reflectors for the virtual increase of field data in the near–field acoustical holography. Journal of Computational Acoustic 15:49-61
37. Kim Y–K, Kim Y–H (1999) Holographic reconstruction of active sources and surface admittance in an enclosure. Journal of the Acoustical Society of America 105:2377–2383
38. Kirsch A (1996) An introduction to the mathematical theory of inverse problems. Springer–Verlag, New York

39. Koopmann GH, Song L, Fahnline JB (1989) A method for computing acoustic fields based on the principle of wave superposition. Journal of the Acoustical Society of America 86:2433–2438
40. la Grange MR, Berhaar BT, Roozen NB (2004) Efficient BEM based acoustic imaging using the EfI method. In: Proceedings of the International Congress on Acoustics (ICA2004), Kyoto (CD ROM)
41. Lee M, Bolton JS (2005) Patch near–field acoustical holography in cylindrical geometry. Journal of the Acoustical Society of America 118:3721–3732
42. Lee M, Bolton JS (2006) Scan–based near–field acoustical holography and partial field decomposition in the presence of noise and source level variation. Journal of the Acoustical Society of America 119:382–393
43. Lee H, Sullian DP (1988) Fundamental limitation of resolution enhancement by wave–field extrapolation. Journal of the Acoustical Society of America 84:611–617
44. Lyon RH, (1987) Machinery noise and diagnostics. Butterworths, Stoneham
45. Marburg S (2002) Six boundary elements per wavelength: Is that enough? Journal of Computational Acoustics 10:25–51
46. Marburg S, Hardtke H–J (1999) A study on the acoustic boundary admittance. Determination, results and consequences. Engineering Analysis with Boundary Elements 23:737–744
47. Martinus F, Herrin DW, Tao Z, Seybert AF (2005) Identification of aeroacoustic noise sources using inverse boundary element method. SAE paper 2005-01-2497
48. Maynard JD (1988) Acoustic holography for wideband, odd–shaped noise sources. In: Proceedings of the Inter–Noise 88, Avignon 223–231
49. Maynard JD, Williams EG, Lee Y (1985) Nearfield acoustic holography: I. Theory of generalized holography and the development of NAH. Journal of the Acoustical Society of America 78:1395–1413
50. Michalewicz Z (1996) Genetic algorithms+data structures=evolution programs, Springer Verlag, New York
51. Nam KU, Kim YH (1999) Errors due to sensor and position mismatch in planar acoustic holography. Journal of the Acoustical Society of America 106:1655–1665
52. Nava GP, Sato Y, Yasuda Y, Sakamoto S (2006) An inverse method for in–situ estimation of acoustic surface impedance targeting inverse sound rendering in rooms. In: Proceedings of the Inter–Noise 2006, Honolulu (CD ROM)
53. Nelson PA, Yoon SH (2000) Estimation of acoustic source strength by inverse methods: Part I, conditioning of the inverse problem. Journal of Sound and Vibration 233:643–668
54. Nolte B (2005) Reconstruction of sound sources by means of and inverse boundary element formulation. Journal of Computational Acoustic 13:187–201
55. Photiadis DM (1990) The relationship of singular value decomposition to wave–vector filtering in sound radiation problems. Journal of the Acoustical Society of America 88:1152–1159
56. Saijyou K (2007) Measurement of structural intensity using boundary element method–based nearfield acoustical holography. Journal of the Acoustical Society of America 121:3493–3500
57. Saijyou K, Yoshikawa S (2001) Reduction methods of the reconstruction error for large–scale implementation of near–field acoustical holography. Journal of the Acoustical Society of America 110:2007–2023
58. Sasaki Y, Mori A, Takanashi A (2004) Noise source identification using I–BEM method. In: Proceedings of the International Congress on Acoustics (ICA2004), Kyoto (CD ROM)

59. Schumacher A, Hald J, Rasmussen KB, Hansen PC (2003) Sound source reconstruction using inverse boundary element calculations. Journal of the Acoustical Society of America 113:114–127
60. Seybert AF (2002) Forward and inverse numerical acoustics for NVH applications. In: Proceedings of International Congress on Sound and Vibration (ICSV 9), Orlando (CD ROM)
61. Tikhonov AN, Arsenin VY (1977) Solutions of ill–posed problems. Halsted Press, New York
62. Valdivia NP, Williams EG (2004) Implicit methods of solution to integral formulation in boundary element method based nearfield acoustic holography. Journal of the Acoustical Society of America 116:1559–1572
63. Valdivia NP, Williams EG (2005) Krylov subspace iterative methods for boundary element method based near–field acoustic holography. Journal of the Acoustical Society of America 117:711–724
64. Valdivia NP, Williams EG (2006) Study of the comparison of the methods of equivalent sources and boundary element methods for near–field acoustic holography. Journal of the Acoustical Society of America 120:3694-3705
65. Veronesi WA, Maynard JD (1989) Digital holographic reconstruction of source with arbitrarily shaped surfaces. Journal of the Acoustical Society of America 85:588–598
66. Villot M, Chaveriat G, Roland J (1992) Phonoscopy: an acoustical holography technique for plane structures radiating in enclosed spaces. Journal of the Acoustical Society of America 91:187–195
67. Wang Z, Wu SF (1997) Helmholtz equation–least–squares method for reconstructing the acoustic pressure field. Journal of the Acoustical Society of America 102:2020–2032
68. Williams EG (1999) Fourier acoustics. Sound radiation and nearfield acoustical holography. Academic Press, San Diego–London
69. Williams EG (2001) Regularization methods for near–field acoustical holography. Journal of the Acoustical Society of America 110:1976–1988
70. Williams EG, Houston BH, Herdic PC (2003) Fast Fourier transform and singular value decomposition formulations for patch nearfield acoustical holography. Journal of the Acoustical Society of America 114:1322–1333
71. Williams EG, Houston BH, Herdic PC, Raveendra ST, Gardner B (2000) Interior near–field acoustical holography in flight. Journal of the Acoustical Society of America 108:1451–1463
72. Williams EG, Maynard JD (1980) Holographic imaging without the wavelength resolution limit. Physical Review Letters 45:554–557
73. Williams W, Parke NG, Moran DA, Sherman CH (1964) Acoustic radiation from a finite cylinder. Journal of the Acoustical Society of America 36:2316–2322
74. Wu SF (2004) Hybrid near–field acoustic holography. Journal of the Acoustical Society of America 115:207–217
75. Wu SF, Zhao X (2002) Combined Helmholtz equation least–squares method for reconstructing acoustic radiation from arbitrary shaped objects. Journal of the Acoustical Society of America 112:179–188
76. Zhang Z, Vlahopoulos N, Raveendra ST, Allen T, Zhang KY (2000) A computational acoustic field reconstruction process based on an indirect boundary element formulation. Journal of the Acoustical Society of America 108:2167–2178

Index

MIX
Papier aus verantwortungsvollen Quellen
Paper from responsible sources
FSC® C105338

If you have any concerns about our products,
you can contact us on
ProductSafety@springernature.com

In case Publisher is established outside the EU,
the EU authorized representative is:
Springer Nature Customer Service Center GmbH
Europaplatz 3, 69115 Heidelberg, Germany

Printed by Libri Plureos GmbH
in Hamburg, Germany